CLYMER®

MERCURY/MARINER

OUTBOARD SHOP MANUAL

2.5-60 HP TWO-STROKE • 1998-2002

The world's finest publisher of mechanical how-to manuals

P.O. Box 12901, Overland Park, KS 66282-2901

FIRST EDITION
First Printing May, 2002
Second Printing July, 2008

Printed in U.S.A.

CLYMER and colophon are registered trademarks of Penton Business Media, Inc.

ISBN-10: 0-89287-785-5

ISBN-13: 978-0-89287-785-0

Library of Congress: 2002106386

AUTHOR AND TECHNICAL PHOTOGRAPHY: Mark Rolling.

TECHNICAL ILLUSTRATIONS: Bob Meyer.

WIRING DIAGRAMS: Robert Caldwell.

EDITORS: Dustin Uthe and Fred Hunter.

PRODUCTION: Shara Pierceall.

COVER: Photo courtesy of Godfrey Marine, Elkhart, Indiana.

General Information 1

Tools and Techniques 2

Troubleshooting 3

Lubrication, Maintenance and Tune-Up 4

Synchronization and Linkage Adjustments 5

Fuel System 6

Ignition and Electrical Systems 7

Power Head 8

Gearcase 9

Jet Drive 10

Trim and Tilt Systems 11

Oil Injection Systems 12

Manual Rewind Starters 13

Remote Control 14

Index 15

Wiring Diagrams 16

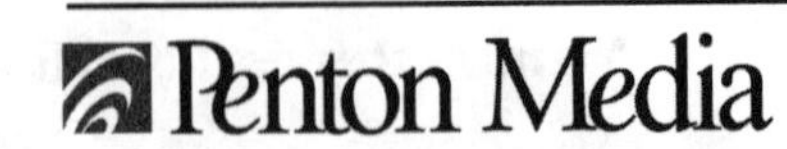

P.O. Box 12901, Overland Park, KS 66282-2901 • 800-262-1954 • 913-967-1719

More information available at *clymer.com*

Contents

QUICK REFERENCE DATA. IX

CHAPTER ONE
GENERAL INFORMATION 1

Manual organization 1
Warnings, cautions and notes 1
Engine operation 2
Torque specifications 2
Fasteners 2
Lubricants 6
Gasket sealant 7
Galvanic corrosion 8
Propellers 10

CHAPTER TWO
TOOLS AND TECHNIQUES 17

Safety 18
Basic hand tools 19
Test equipment 22
Service hints 25
Special tips 27
Mechanics techniques 27

CHAPTER THREE
TROUBLESHOOTING 30

Safety precautions 31
Terminology and test equipment 32
Operating requirements 35
Starting system 36
Lighting coil 46
Battery charging system 47
Electrical accessories 51
Ignition system 51
Capacitor discharge ignition
(2.5, 3.3, 4 and 5 hp) 53
Alternator-driven ignition
(6-25 hp except 1998 20 jet) 58
Alternator-driven ignition (1998 20 jet) 63
Capacitor discharge module (CDM) ignition
(30 and 40 hp two-cylinder models) 67
Capacitor discharge module (CDM) ignition
(40-60 hp three-cylinder models) 72
Fuel system 76
Fuel primer and enrichment systems 78
Engine temperature and overheating 80
Engine 83
Boat wiring harness standard wire colors 87
Battery cable recommendations 87
Starting system troubleshooting 87
Charging system troubleshooting 88
Ignition system troubleshooting 88
Fuel system troubleshooting 88
Electric starter specifications 89
Charging system specifications 89
Stator resistance specifications
(battery charging/lighting circuit) 90
CDI ignition system specifications 90
Ignition stator resistance specifications 91
Ignition system identification 92

CHAPTER FOUR
LUBRICATION, MAINTENANCE AND TUNE-UP 93

Hour meter 93
Fuels and lubrication 94
Off-season storage 104
Anticorrosion maintenance 107
Engine submersion 108
Cooling system flushing 109
Tune-up 110
Torque specifications 118
Maintenance schedule 118
Spark plug recommendations 119
Lubricant capacity 119

CHAPTER FIVE
SYNCHRONIZATION AND LINKAGE ADJUSTMENTS 120

Safety precautions 120
General information 121
Synchronization and linkage procedures 123
General specifications 138

CHAPTER SIX
FUEL SYSTEM 142

Fuel pump 142
Carburetors 146
Fuel primer solenoid (6-25 hp and 20 jet remote control models) 162
Fuel primer valve (30-60 hp remote control models) 163
Engine mounted primer bulb (30-55 hp manual start models) 163
Antisiphon devices 163
Fuel filters 164
Fuel tanks 164
Fuel hose and primer bulb 164
Reed valve service 167
Fuel bleed (recirculation) system 172
Torque specifications 174
Carburetor specifications 175
Reed valve specifications 177

CHAPTER SEVEN
IGNITION AND ELECTRICAL SYSTEMS 178

Battery 178
Charging system 185
Fuses 188
Starting systems 188
Capacitor discharge ignition (2.5-5 hp models) 196
Alternator driven capacitor discharge ignition system(6-25 hp models) 202
Capacitor discharge module (CDM) ignition (30-60 hp models) 208
Torque specifications 213
Battery capacity (hours) 214
Battery cable recommendations 214
Battery state of charge 215
Boat wiring harness standard wire colors 215

CHAPTER EIGHT
POWER HEAD 216

Service considerations 216
Model identification 217
Power head break-in 217
Service recommendations 217
Lubricants, sealants and adhesives 218
Sealing surfaces 219
Fasteners and torque 220
Power head removal/installation 221
Power head disassembly 229
Cleaning and inspection 247
Power head assembly 257
Torque specifications 288
Connecting rod service specifications 290
Crankshaft service specifications 290
Cylinder block service specifications 290
Piston service specifications 291
Model number codes 291

CHAPTER NINE
GEARCASE. 292
Gearcase operation 292
Gear ratio 293
High-altitude operation 294
Service precautions 294
Corrosion control 295
Gearcase lubrication 296
Propeller 296
Trim tab adjustment 299
Gear housing 300
Water pump 309
Gearcase disassembly/assembly 321
Torque specifications 375
Gear ratio and approximate lubricant capacity 377
Gearcase service specifications 377

CHAPTER TEN
JET DRIVE 378
Jet pump unit service 384
Torque specifications 388

CHAPTER ELEVEN
TRIM AND TILT SYSTEMS 389
Power trim and tilt systems 393
Power trim and tilt system service 403
Torque specifications 410

CHAPTER TWELVE
OIL INJECTION SYSTEMS 412
Oil injection system capacity 420
Oil pump output specification 420
Torque specifications 420

CHAPTER THIRTEEN
MANUAL REWIND STARTERS 422
Torque specifications 436

CHAPTER FOURTEEN
REMOTE CONTROL 437
Commander 3000 control torque specifications 444

INDEX 445

WIRING DIAGRAMS 448

Quick Reference Data

MAINTENANCE SCHEDULE

Maintenance Interval	Maintenance required
2.5-5 hp models	
Before each use (saltwater or freshwater use)	Check the lanyard switch operation* Inspect the fuel system for leakage Check the outboard mounting bolts Check the steering for binding or looseness Check the shift and throttle control Inspect the propeller for damage
First 10 days of operation (saltwater or freshwater use)	Check gearcase lubricant level and condition
First 25 hours of operation	Drain and refill the gearcase lubricant
Every 30 days (freshwater use)	Check gearcase lubricant level and condition
Every 30 days (saltwater use)	Lubricate throttle and shift linkages Lubricate reverse lock hooks Lubricate swivel and tilt pins Check gearcase lubricant level and condition
Every 60 days (saltwater use)	Lubricate the propeller shaft
Every 60 days (freshwater use)	Lubricate throttle and shift linkages Lubricate reverse lock hooks Lubricate swivel and tilt pins
100 hours or once a year	Drain and refill the gearcase lubricant Lubricate the propeller shaft
Before long term storage	Drain and refill the gearcase lubricant
6-50 hp models	
Before each use (saltwater or freshwater use)	Check the lanyard switch operation* Inspect the fuel system for leakage Check the outboard mounting bolts Check the steering for binding or looseness Check the shift and throttle control Inspect the propeller for damage
After each use (saltwater use)	Flush the cooling system
100 hours or once a year	Lubricate the throttle and shift linkages Lubricate the steering system Lubricate the tiller control pivots Lubricate the tilt and swivel shaft Lubricate the propeller shaft Lubricate the starter motor pinion* Lubricate the propeller shaft bearing carrier Drain and refill the gearcase lubricant Lubricate the drive shaft splines
	(continued)

MAINTENANCE SCHEDULE (continued)

Maintenance Interval	Maintenance required
6-50 hp models (cont.)	
100 hours or once a year	Inspect and test the battery*
	Service the fuel filter
	Adjust the carburetor(s)
	Adjust the ignition timing
	Adjust the synchronization and linkages
	Inspect the sacrificial anodes*
	Check the power trim fluid level*
	Check control cable adjustments*
	Remove carbon deposits from the cylinder(s)
	Check for loose or damaged fasteners
	Clean remote fuel tank filter*
300 hours or 3 years	Replace the water pump impeller

*This maintenance item does not apply to all models.

SPARK PLUG RECOMMENDATIONS

Model	Plug part No.	Gap in. (mm)
2.5 and 3.3 hp		
NGK plug	BPR6HS-10	0.040 (1.02)
Champion plug	RL87YC	0.040 (1.02)
4 and 5 hp		
NGK plug	BP7HS-10	0.040 (1.02)
Champion plug	L82YC	0.040 (1.02)
6 and 8 hp		
NGK standard plug	BP8H-N-10	0.040 (1.02)
NGK radio suppression plug	BPZ8H-N-10	0.040 (1.02)
9.9 and 15 hp		
NGK standard plug	BP8HS-15	0.060 (1.52)
NGK radio suppression plug	BPZ8H-N-10	0.060 (1.52)
20 hp, 20 jet and 25 hp		
NGK standard plug	BP8H-N-10	0.040 (1.02)
NGK radio suppression plug	BPZ8H-N-10	0.040 (1.02)
30 and 40 hp (two-cylinder)		
NGK standard plug	BP8H-N-10	0.040 (1.02)
NGK radio suppression plug	BPZ8H-N-10	0.040 (1.02)
40-60 hp (three-cylinder)		
NGK standard plug	BP8H-N-10	0.040 (1.02)
NGK radio suppression plug	BPZ8H-N-10	0.040 (1.02)

LUBRICANT CAPACITY

Outboard model	Gearcase capacity	Oil reservoir capacity
2.5 hp	3.0 oz. (89 ml)	*
3.3 hp	2.5 oz. (74 ml)	*
4 and 5 hp	6.6 oz. (195 ml)	*
6, 8, 9.9 15 hp	6.8 oz. (201 ml)	*
20 and 25 hp	7.8 oz. (231 ml)	*
30 and 40 hp (two-cylinder)	14.9 oz. (441 ml)	50.5 (1.5)
40 and 50 hp (three-cylinder)	14.9 oz. (441 ml)	96 (3.5)
55 and 60 hp		
Standard gearcase	11.5 oz. (340 ml)	96 (3.5)
Bigfoot gearcase	22.5 oz. (665 ml)	96 (3.5)

*This model is not equipped with oil injection.

STANDARD TORQUE SPECIFICATIONS—U.S. STANDARD AND METRIC FASTENERS

Screw or nut size	ft.-lb.	in.-lb.	N•m
U.S. standard fasteners			
6-32	–	9	1
8-32	–	20	2.3
10-24	–	30	3.4
10-32	–	35	4.0
12-24	–	45	5.1
1/4-20	6	72	8.1
1/4-28	7	84	9.5
5/16-18	13	156	17.6
5/16-24	14	168	19
3/8-16	23	270	31.2
3/8-24	25	300	33.9
7/16-14	36	–	48.8
7/16-20	40	–	54
1/2-13	50	–	67.8
1/2-20	60	–	81.3
Metric fasteners			
M5	–	36	4.1
M6	6	72	8.1
M8	13	156	17.6
M10	26	312	35.3
M12	35	–	47.5
M14	60	–	81.3

Chapter One

General Information

This detailed, comprehensive manual contains complete information on maintenance and overhaul. Hundreds of photos and drawings illustrate every step-by-step procedure.

Troubleshooting, tune-up, maintenance and repair procedures are not difficult with the proper tools, equipment and instructions. Anyone with some mechanical ability can perform most of the procedures in this manual. See Chapter Two for more information on tools and techniques.

A shop manual is a reference. Clymer books are designed to make finding information quick and easy. All chapters are thumb tabbed and important topics are indexed at the end of the manual. All procedures, tables, photos and instructions in this manual assume the reader may be working on the machine or using the manual for the first time.

Store the manual with other tools in the workshop or boat. It provides a better understanding of how the boat runs and lowers repair and maintenance costs.

MANUAL ORGANIZATION

Chapter One provides general information useful to boat owners and mechanics.

Chapter Two discusses the tools and techniques for preventative maintenance, troubleshooting and repair.

Chapter Three provides troubleshooting procedures for all engine systems and individual components.

Chapter Four provides maintenance, lubrication and tune-up instructions.

Additional chapters cover storage, adjustment and specific repair instructions. All disassembly, inspection and assembly instructions are in step-by-step form. Specifications are included at the end of the appropriate chapters.

WARNINGS, CAUTIONS AND NOTES

The terms WARNING, CAUTION and NOTE have specific meanings in this manual.

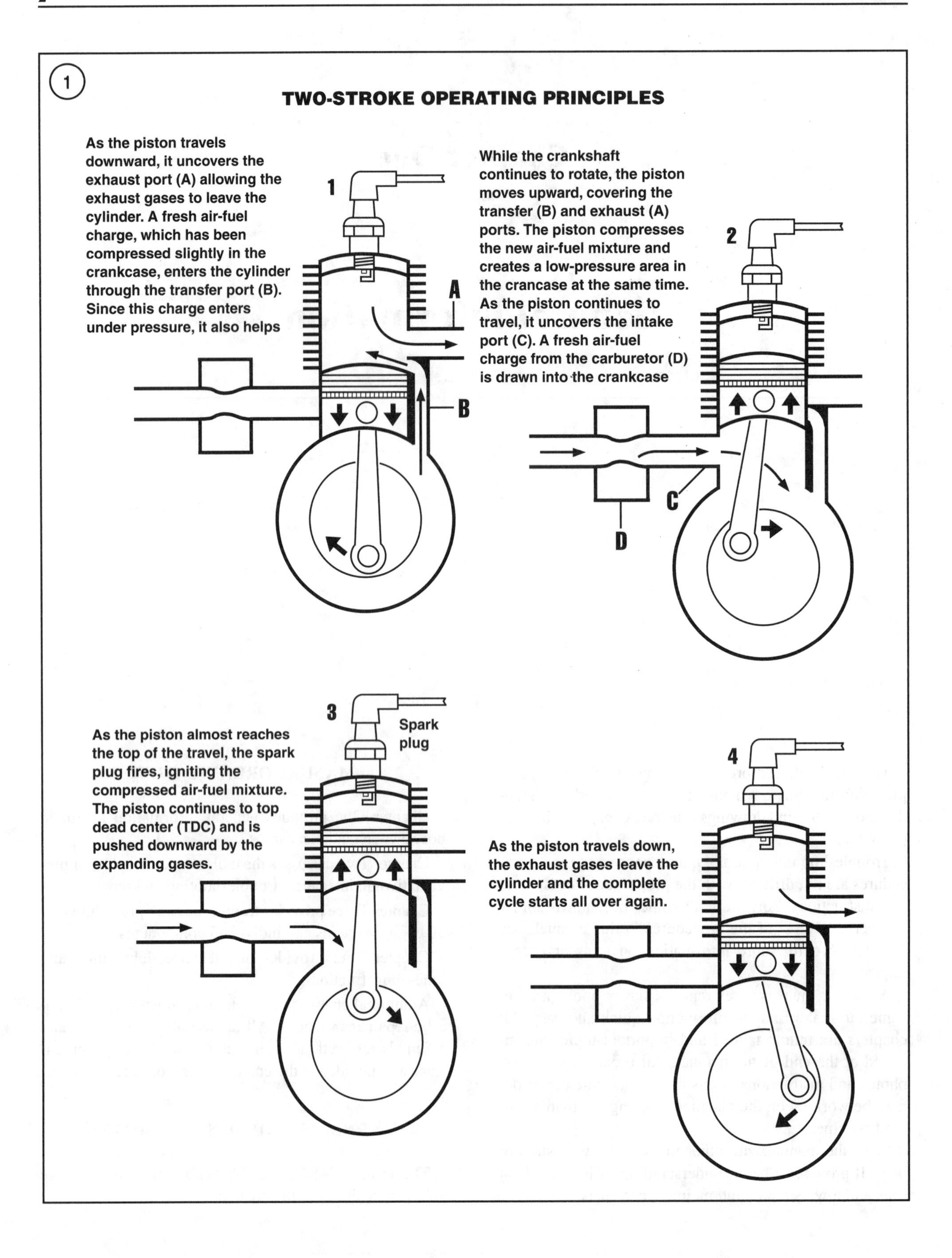
1
TWO-STROKE OPERATING PRINCIPLES
As the piston travels downward, it uncovers the exhaust port (A) allowing the exhaust gases to leave the cylinder. A fresh air-fuel charge, which has been compressed slightly in the crankcase, enters the cylinder through the transfer port (B). Since this charge enters under pressure, it also helps
1
A
B
While the crankshaft continues to rotate, the piston moves upward, covering the transfer (B) and exhaust (A) ports. The piston compresses the new air-fuel mixture and creates a low-pressure area in the crancase at the same time. As the piston continues to travel, it uncovers the intake port (C). A fresh air-fuel charge from the carburetor (D) is drawn into the crankcase
2
C
D
As the piston almost reaches the top of the travel, the spark plug fires, igniting the compressed air-fuel mixture. The piston continues to top dead center (TDC) and is pushed downward by the expanding gases.
3
Spark plug
As the piston travels down, the exhaust gases leave the cylinder and the complete cycle starts all over again.
4

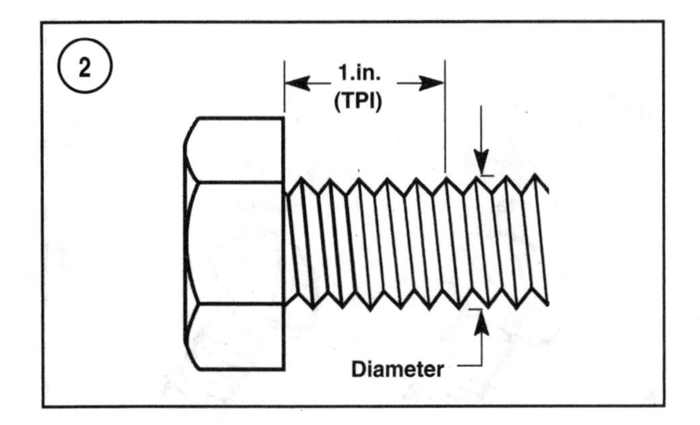

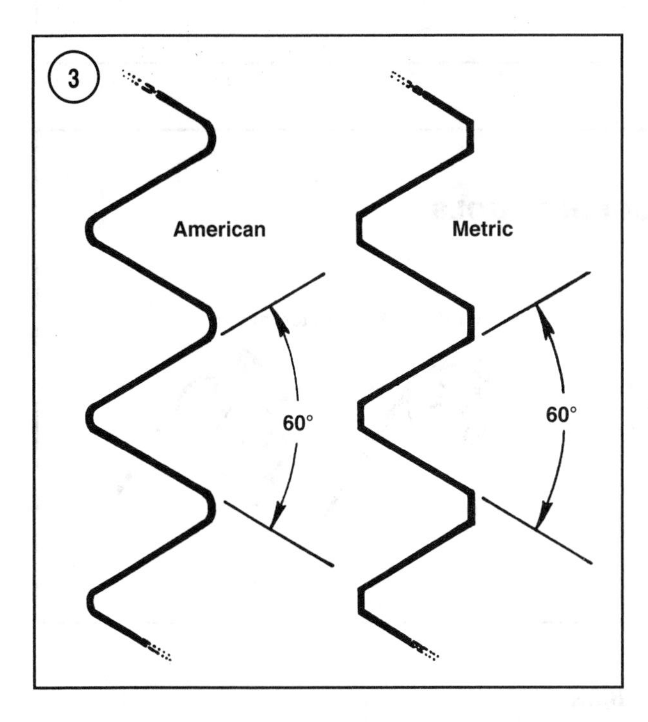

A WARNING emphasizes areas where injury or even death could result from negligence. Mechanical damage may also occur. WARNINGS *are to be taken seriously.*

A CAUTION emphasizes areas where equipment damage could result. Disregarding a CAUTION could cause permanent mechanical damage, though injury is unlikely.

A NOTE provides additional information to make a step or procedure easier or clearer. Disregarding a NOTE could cause inconvenience, but would not cause equipment damage or injury.

ENGINE OPERATION

All marine engines, whether two or four-stroke, gasoline or diesel, operate on the Otto cycle of intake, compression, power and exhaust phases. A two-stroke engine requires one crankshaft revolution (two strokes of the piston) to complete the Otto cycle. All Mercury and Mariner engines covered in this manual are of the two-stroke design. **Figure 1** shows gasoline two-stroke engine operation.

TORQUE SPECIFICATIONS

The materials used in the manufacture of the engine may be subjected to uneven stresses if the fasteners of the various subassemblies are not installed and tightened correctly. Fasteners that are improperly installed can work loose and cause extensive damage. It is essential to use an accurate torque wrench, described in Chapter Two, with the torque specifications in this manual.

Specifications for torque are provided in Newton-meters (N•m), foot-pounds (ft.-lb.) and inch-pounds (in.-lb.). Torque specifications for specific components are at the end of the appropriate chapters.

FASTENERS

The materials and design of the various fasteners used on marine equipment are specifically chosen for performance and safety. Fastener design determines the type of tool required to work with the fastener. Fastener material is carefully selected to decrease the possibility of physical failure or corrosion. See *Galvanic Corrosion* in this chapter for information on marine materials.

Threaded Fasteners

Nuts, bolts and screws are manufactured in a wide range of thread patterns. To join a nut and bolt, the diameter of the bolt and the diameter of the hole in the nut must be the same. It is just as important that the threads are compatible.

The easiest way to determine if the threads on the fasteners are compatible is to turn the nut on the bolt (or bolt into its respective opening) with fingers only. Make sure both pieces are clean. If much force is required, check the thread condition on each fastener. If the thread condition is good but the fasteners jam, the threads are not compatible.

Four important specifications describe the thread:

1. Diameter.
2. Threads per inch.
3. Thread pattern.
4. Thread direction.

Figure 2 shows the first two specifications. Thread pattern is more subtle. Italian and British standards exist, but the most commonly used by marine equipment manufactures are American and metric. The root and top of the thread are cut differently as shown in **Figure 3**.

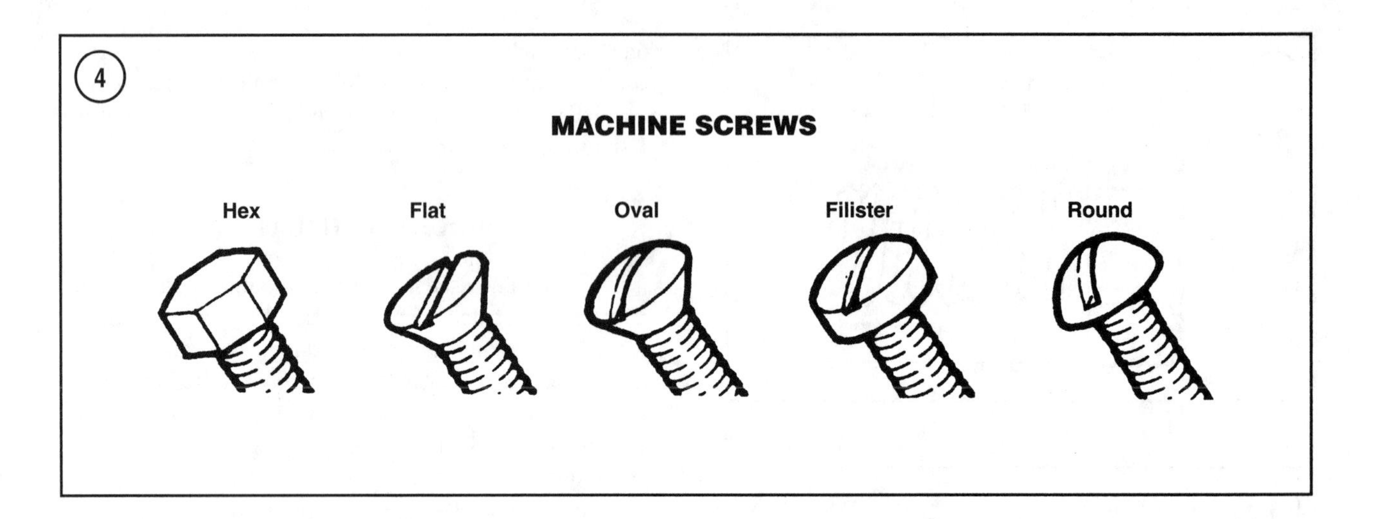

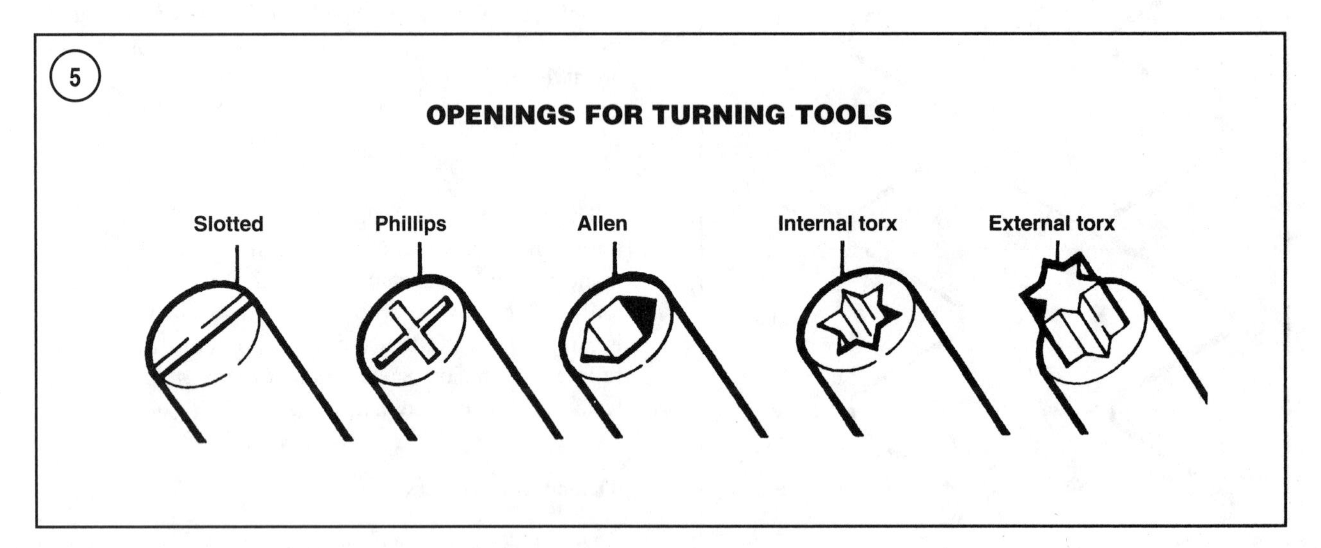

Most threads are cut so the fastener must be turned clockwise to tighten it. These are called right-hand threads. Some fasteners have left-hand threads; they must be turned counterclockwise for tightening. Left-hand threads are used in locations where normal rotation of the equipment would loosen a right-hand threaded fastener. *Assume all fasteners use right-hand threads unless the instructions specify otherwise.*

Machine Screws

There are many different types of machine screws (**Figure 4**). Most are designed to protrude above the secured surface (rounded head) or be slightly recessed below the surface (flat head). In some applications the screw head is recessed well below the fastened surface. **Figure 5** shows a number of screw heads requiring different types of turning tools. See Chapter Two for detailed information.

Bolts

Commonly called bolts, the technical name for these fasteners is cap screw. They are normally described by diameter, threads per inch and length. For example, 1/4-20 × 1 indicates a bolt 1/4 in. in diameter with 20 threads per inch and 1 in. long. The measurement across two flats on the head of the bolt indicates the proper wrench size to use.

Nuts

Nuts are manufactured in a variety of types and sizes. Most are hexagonal (six-sided) and fit on bolts, screws and studs with the same diameter and threads per inch.

Figure 6 shows several types of nuts. The common nut is usually used with some type of lockwasher. Self-locking nuts have a nylon insert that helps prevent the nut from

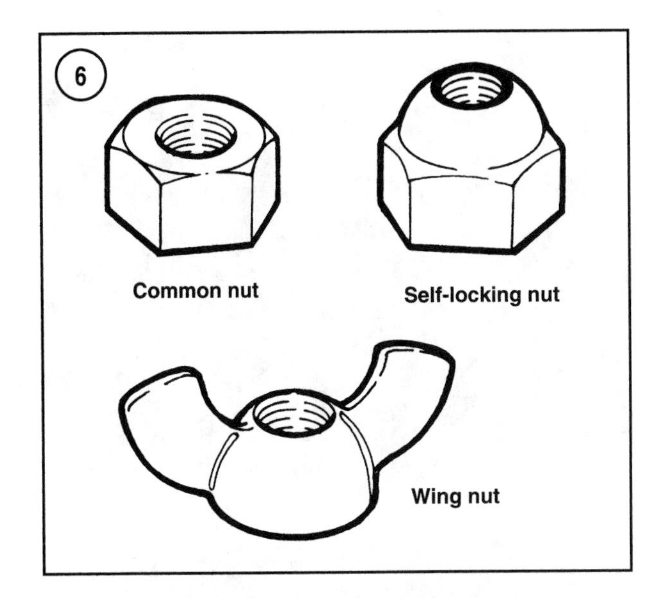

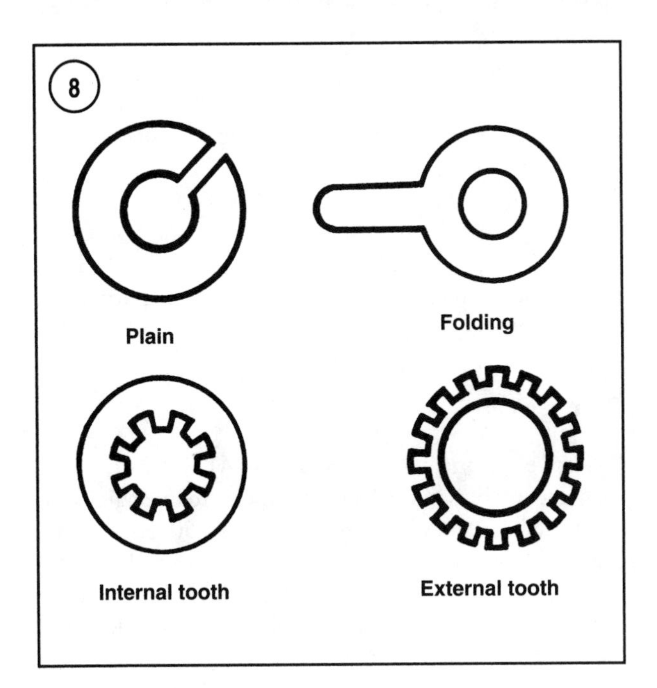

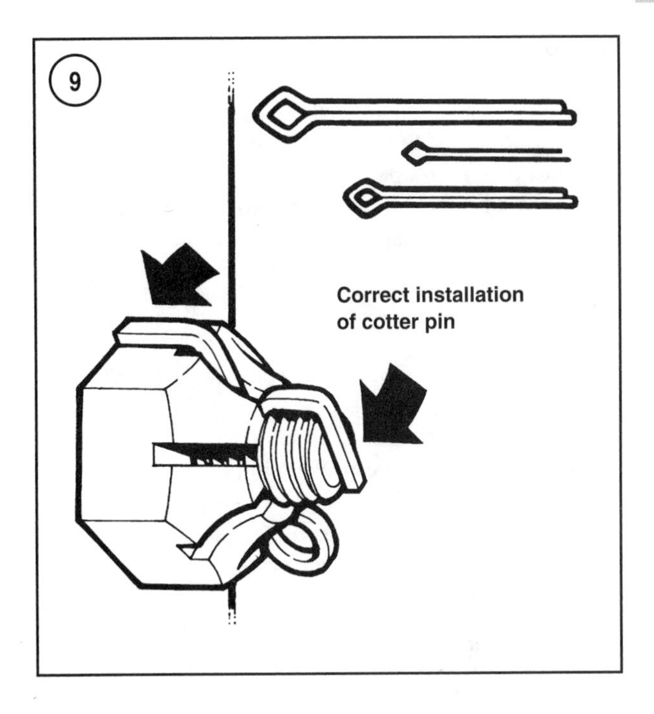

loosening; no lockwasher is required. Wing nuts are designed for fast removal by hand. Wing nuts are used for convenience in non-critical locations.

To indicate the size of a nut, manufacturers specify the diameter of the opening and the threads per inch. This is similar to a bolt specification, but without the length dimension. The measurement across two flats on the nut indicates the wrench size to use.

Washers

There are two basic types of washers: flat washers and lockwashers. A flat washer (**Figure 7**) is a simple disc with a hole that fits the screw or bolt. Lockwashers are designed to prevent a fastener from working loose due to vibration, expansion and contraction. **Figure 8** shows several types of lockwashers. Flat washers are often used between a lockwasher and a fastener to provide a smooth bearing surface. This allows the fastener to be turned easily with a tool.

Cotter Pins

Cotter pins (**Figure 9**) are used to secure special kinds of fasteners. The threaded stud, bolt or shaft has a hole for the cotter pin; the nut or nut lock piece has projections for the cotter pin. This type of nut is called a *castellated nut*. Always replace the cotter pin if it is removed.

Snap Rings

Snap rings can be an internal or external (**Figure 10**) design. They are used to retain components on shafts (external type) or within openings (internal type). Snap rings can be reused if they are not distorted during removal. In some applications, snap rings of varying thickness can be selected to position or control end play of parts assemblies.

LUBRICANTS

Periodic lubrication ensures long service life for any type of equipment. It is especially important with marine equipment because of exposure to salt, brackish or polluted water and other harsh environments. The *type* of lubricant used is just as important as the lubrication service itself; although in an emergency, the wrong type of lubricant is better than none at all. The following paragraphs describe the types of lubricants most often used on marine equipment. Be sure to follow the equipment manufacturer's recommendations for the lubricant types.

Generally, all liquid lubricants are called *oil*. They may be mineral-based (including petroleum bases), natural-based (vegetable and animal bases), synthetic-based or emulsions (mixtures). *Grease* is an oil thickened with additives, and maybe enhanced with anticorrosion, antioxidant and extreme pressure (EP) additives. Grease is often classified by the type of thickener added; lithium and calcium soap are the most commonly used.

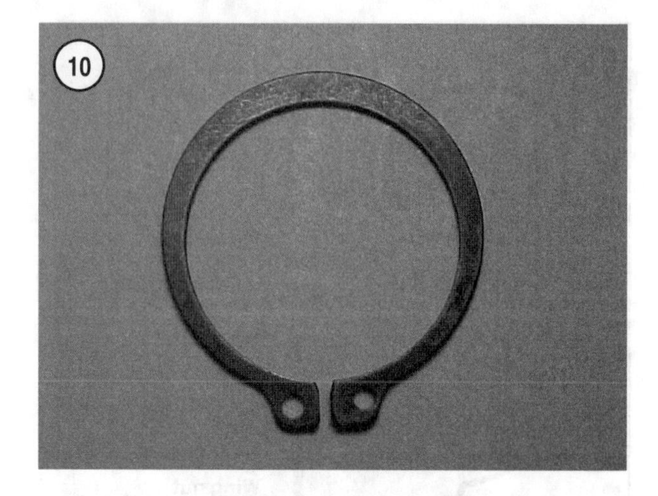
10

11

Two-stroke Engine Oil

Lubrication for a two-stroke engine is provided by oil mixed into the incoming air-fuel mixture. Some of the oil mist settles out in the crankcase, lubricating the crankshaft, bearings and lower end of the connecting rod. The rest of the oil enters the combustion chamber to lubricate the piston, rings and the cylinder wall. This oil is burned with the air-fuel mixture during the combustion process.

Engine oil must have several special qualities to work well in a two-stroke engine. It must mix easily and stay in suspension with gasoline. When burned, it cannot leave behind excessive deposits. It must also withstand the high operating temperatures associated with two-stroke engines.

The National Marine Manufacturer's Association (NMMA) has set standards for oil used in two-stroke, water-cooled engines. This is the NMMA TC-W (two-cycle, water-cooled) grade (**Figure 11**). It indicates the oil's performance in the following areas:

1. Lubrication (prevention of wear and scuffing).
2. Spark plug fouling.
3. Piston ring sticking.
4. Preignition.
5. Piston varnish.
6. General engine condition (including deposits).
7. Exhaust port blockage.
8. Rust prevention.
9. Mixing ability with gasoline.

In addition to oil grade, manufacturers specify the ratio of gasoline to oil required during break-in and normal engine operation.

Gearcase Oil

Gearcase lubricants are assigned SAE viscosity numbers under the same system as four-stroke engine oil. Gearcase lubricant falls into the SAE 72-250 range. Some gearcase lubricants, such as SAE 85-90, are multigrade.

Three types of marine gearcase lubricant are generally available: SAE 90 hypoid gearcase lubricant is designed for older manual-shift units; Type C gearcase lubricant

contains additives designed for the electric shift mechanisms; high viscosity gearcase lubricant is a heavier oil designed to withstand the shock loading of high performance engines or units subjected to severe duty use. Always use a gearcase lubricant of the type specified by the gearcase manufacturer.

Grease

Greases are graded by the National Lubricating Grease Institute (NLGI). Greases are graded by number according to the consistency of the grease. These ratings range from No. 000 to No. 6, with No. 6 being the most solid. A typical multipurpose grease is NLGI No. 2. For specific applications, equipment manufactures may require grease with an additive such as molybdenum disulfide (MOS^2).

GASKET SEALANT

Gasket sealant is used instead of pre-formed gaskets on some applications, or as a gasket dressing on others. Three types of gasket sealant are commonly used: gasket sealing compound, room temperature vulcanizing (RTV) and anaerobic. Because these materials have different sealing properties, they cannot be used interchangeably.

Gasket Sealing Compound

Gasket sealing compound is a non-hardening liquid used primarily as a gasket dressing. Gasket sealing compound is available in tubes or brush top containers. When exposed to air or heat it forms a rubber-like coating. The coating fills in small imperfections in gasket and sealing surfaces. Do not use gasket sealing compound that is old, has began to solidify or has darkened in color.

Applying Gasket Sealing Compound

Carefully scrape residual gasket material, corrosion deposits or paint from the mating surfaces. Use a blunt tip scraper and work carefully to avoid damaging the mating surfaces. Use quick drying solvent and a clean shop towel to wipe residual oil or other contaminants from the surfaces. Wipe or blow loose material or contaminants from the gasket. Brush a light coat on the mating surfaces and both sides of the gasket. Do not apply more compound than needed. Excess compound is squeezed out as the surfaces mate and may contaminate other components. Do not allow compound into bolt or aligning pin openings. A *hydraulic lock* can occur as the bolt or pin compresses the compound, resulting in incorrect bolt torque.

RTV Sealant

RTV sealant is a silicone gel supplied in tubes. Moisture in the air causes RTV to cure. Always place the cap on the tube as soon as possible after using RTV. RTV has a shelf life of approximately one year and will not cure properly if the shelf life has expired. Check the expiration date on the tube and keep partially used tubes tightly sealed. RTV can generally fill gaps up to 1/4 in. (6.3 mm) and works well on slightly flexible surfaces.

Applying RTV Sealant

Carefully scrape all residual sealant and paint from the mating surfaces. Use a blunt tip scraper and work carefully to avoid damaging the mating surfaces. Remove residual sealant from bolt or aligning pin openings. Use a quick drying solvent and a clean shop towel to wipe all oil residue or other contaminants from the surfaces.

Apply RTV sealant in a continuous bead 0.08-0.12 in. (2-3 mm) thick. Circle all mounting bolt or aligning pin holes unless otherwise specified. Do not allow RTV sealant into bolt or aligning pin openings. A *hydraulic lock* can occur as the bolt or pin compresses the sealant, resulting in incorrect bolt torque. Tighten the mounting fasteners within 10 minutes after application.

Anaerobic Sealant

Anaerobic sealant is a gel supplied in tubes (**Figure 12**). It cures only in the absence of air, as when squeezed tightly between two machined mating surfaces. For this reason, it will not spoil if the cap is left off the tube. Do not use anaerobic sealant if one of the surfaces is flexible. Anaerobic sealant can fill gaps up to 0.030 in. (0.8 mm) and

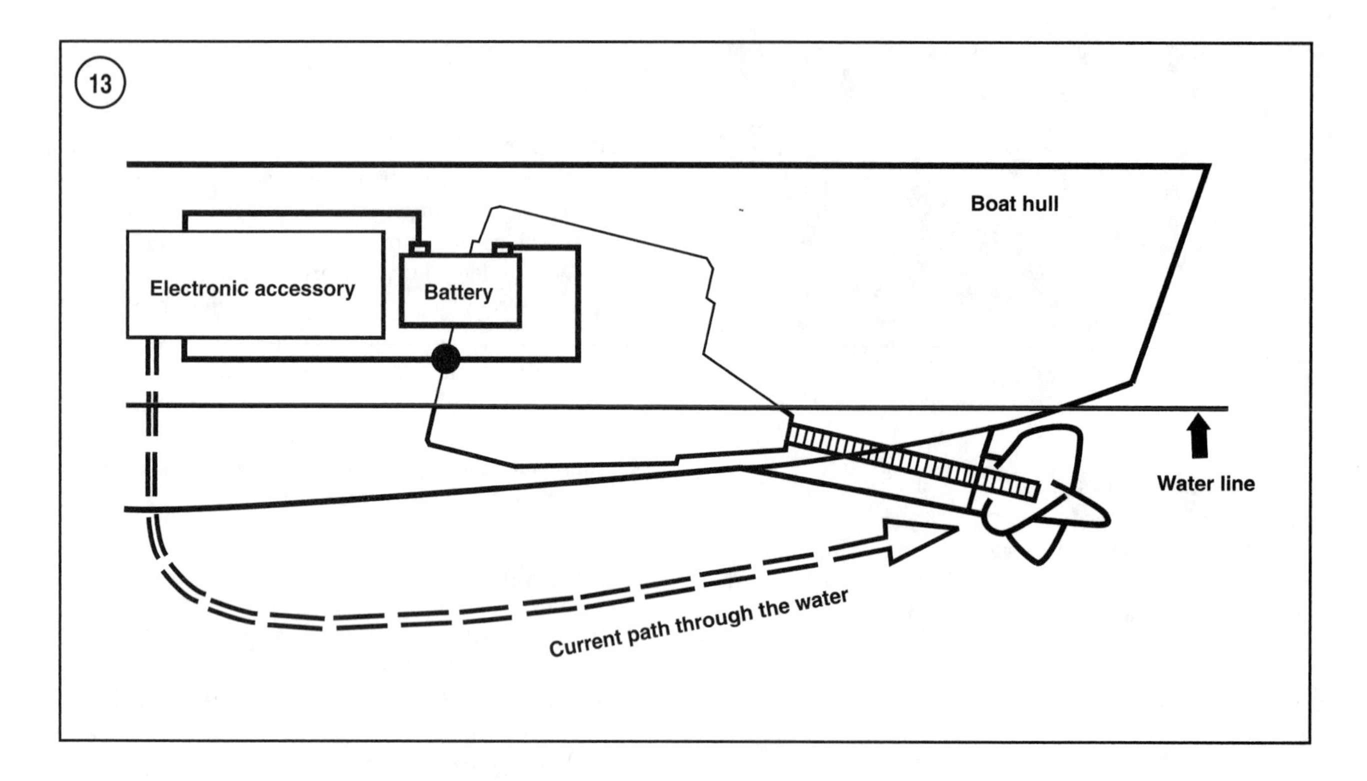

generally works best on rigid, machined flanges or surfaces.

Applying Anaerobic Sealant

Carefully scrape all residual sealant from the mating surfaces. Use a blunt tip scraper and work carefully to avoid damaging the mating surfaces. Clean sealant from bolt or aligning pin openings. Use a quick drying solvent and shop towel to wipe oil residue or other contaminants from the surfaces. Apply anaerobic sealant in a 0.04 in. (1 mm) thick continuous bead onto one of the surfaces. Circle bolt and aligning pin openings. Do not get sealant into bolt or aligning pin openings. A *hydraulic lock* can occur as the bolt or pin compresses the sealant, resulting in incorrect bolt torque. Tighten the mounting fasteners within 10 minutes after application.

GALVANIC CORROSION

A chemical reaction occurs whenever two different types of metal are joined by an electrical conductor and immersed in electrolyte. Electrons transfer from one metal to the other through the electrolyte and return through the conductor.

The hardware on a boat is made of many different types of metal. The boat hull acts as a conductor between the metals. Even if the hull is wooden or fiberglass, the slightest film of water (electrolyte) within the hull provides conductivity. This combination creates a good environment for electron flow (**Figure 13**). Unfortunately, this electron flow results in galvanic corrosion, causing one of the metals to corrode. The amount of electron flow (corrosion) depends on the following factors:

1. The types of metal involved.
2. The efficiency of the conductor.
3. The strength of the electrolyte.

Metals

The chemical composition of the metals used in marine equipment has a significant effect on the amount and speed of galvanic corrosion. Certain metals are more resistant to corrosion than others. These electrically negative metals are commonly called *noble*; they act as the cathode in any reaction. Metals that are more subject to corrosion are electrically positive; they act as the anode in a reaction. The more *noble* metals include titanium, 18-8 stainless steel and nickel. Less *noble* metals include zinc, aluminum and magnesium. Galvanic corrosion becomes more severe as the difference in electrical potential between the two metals increases.

In some cases, galvanic corrosion can occur within a single piece of metal. Common brass is a mixture of zinc and copper. When immersed in electrolyte, the zinc portion of the mixture corrodes away as a reaction occurs between the zinc and copper particles.

Conductors and Insulators

The hull of the boat often acts as the conductor between different types of metal. Marine equipment, such as the drive unit can act as the conductor. Large masses of metal, firmly connected together, are more efficient conductors than water. Rubber mountings and vinyl-based paint can act as insulators between pieces of metal.

Electrolyte

The water in which a boat operates acts as the electrolyte for the corrosion process.

Cold, clean freshwater is the poorest electrolyte. Pollutants increase conductivity; brackish or saltwater is an efficient electrolyte. The better a conductor, the more severe and rapid the corrosion. This is one of the reasons that most manufacturers recommend a freshwater flush after operating in polluted, brackish or saltwater.

Slowing Corrosion

Because of the environment in which marine equipment operates, preventing galvanic corrosion is practically impossible. There are several ways to slow the corrosion process. These are *not* substitutes for the corrosion protection methods discussed under *Sacrificial Anodes* and *Impressed Current Systems* in this chapter, but they can help these methods reduce corrosion.

Use fasteners of a metal more noble than the parts they secure. If corrosion occurs, the parts they secure may suffer but the fasteners will be protected. The larger secured parts are more able to withstand the loss of material. Major problems could arise if the fasteners corrode to the point of failure.

Keep all painted surfaces in good condition. If paint is scraped off and bare metal exposed, corrosion rapidly increases. Use a vinyl- or plastic-based paint, which acts as an electrical insulator.

Do not apply metal-based antifouling paint to metal parts of the boat or the drive unit. If applied onto metal surfaces, it will react with the metal and result in corrosion between the metal and the layer of paint. Maintain a minimum 1 in. (25 mm) border between the painted surface and metal parts. Organic-based paints are available for use on metal surfaces.

Corrosion protection devices must be immersed in the electrolyte with the boat to provide protection. If the gearcase is raised out of the water when the boat is docked, anodes on the gearcase may be removed from the corrosion process and rendered ineffective. Never paint or apply any coating to anodes or other protection devices. Paint or other coatings insulate them from the corrosion process.

Changes in the boat's equipment, such as the installation of a new stainless steel propeller, changes the electrical potential and may cause increased corrosion. Keep this in mind when adding equipment or changing exposed materials. Add additional anodes or other protection as required. The expense to repair corrosion damage usually far exceeds that of additional corrosion protection.

Sacrificial Anodes

Sacrificial anodes are specially designed to corrode. They act as the anode in *any* galvanic reaction that occurs; the other metal in the reaction acts as the cathode and is not damaged.

Anodes are usually made of zinc. Later model Quicksilver anodes are an aluminum and indium alloy. This alloy is less noble than the aluminum alloy in drive system components, providing the desired sacrificial properties. The aluminum and indium alloy is more resistant to oxide coating than zinc anodes. Oxide coating occurs as the anode material reacts with oxygen in the water. An oxide coating acts as an insulator, dramatically reducing corrosion protection.

Proper anode selection and placement is critical to providing adequate protection. First determine how much surface area requires protection and use the Military Specification MIL-A-818001 as a general rule. The specification states that 1 sq. in. (6.4 square cm) of new anode protects either:

1. 800 sq. in. (5161 sq. cm.) of freshly painted steel.
2. 250 sq. in. (1613 sq. cm.) of bare steel or bare aluminum alloy.
3. 100 sq. in. (645 sq. cm.) of copper or copper alloy.

This applies to a boat at rest. When operating, additional anode area is required to protect the same surface area.

The anode must have a good electrical connection with the metal that it protects. If possible, attach an anode to all metal surfaces requiring protection.

Good quality anodes have inserts made of a more noble material around the anode fastener holes. Otherwise, the anode could erode away around the fastener hole, allowing the anode to loosen and possibly fall off, loosing needed protection.

Impressed Current System

An impressed current system can be added to any boat. The components of the commonly used Quicksilver Mercathode System (**Figure 14**) include the anode, controller and reference electrode. The anode in this system is coated with a very noble metal, such as platinum. It is almost corrosion-free and can last indefinitely. The reference electrode, under the boat's waterline, allows the control module to monitor the potential for corrosion. If electrical current flow reaches the point indicating galvanic corrosion, the control module applies positive battery voltage to the anode. Current then flows from the anode to all other metal components, regardless of how noble or non-noble. The electrical current from the battery counteracts the galvanic reaction to dramatically reduce corrosion.

Only a small amount of current is needed to counteract corrosion. Using input from the sensor, the control module provides only the amount of current needed to suppress galvanic corrosion. Most systems consume a maximum of 0.2 Ah at full demand. Under normal conditions, these systems can provide protection for 8-12 weeks without recharging the battery. Keep in mind that this system must have constant connection to the battery. Frequently, the battery supply to the system is connected to a battery switching device causing the operator to inadvertently shut off the system while docked.

An impressed current system is more expensive to install than sacrificial anodes, but has low maintenance requirements and provides superior protection.

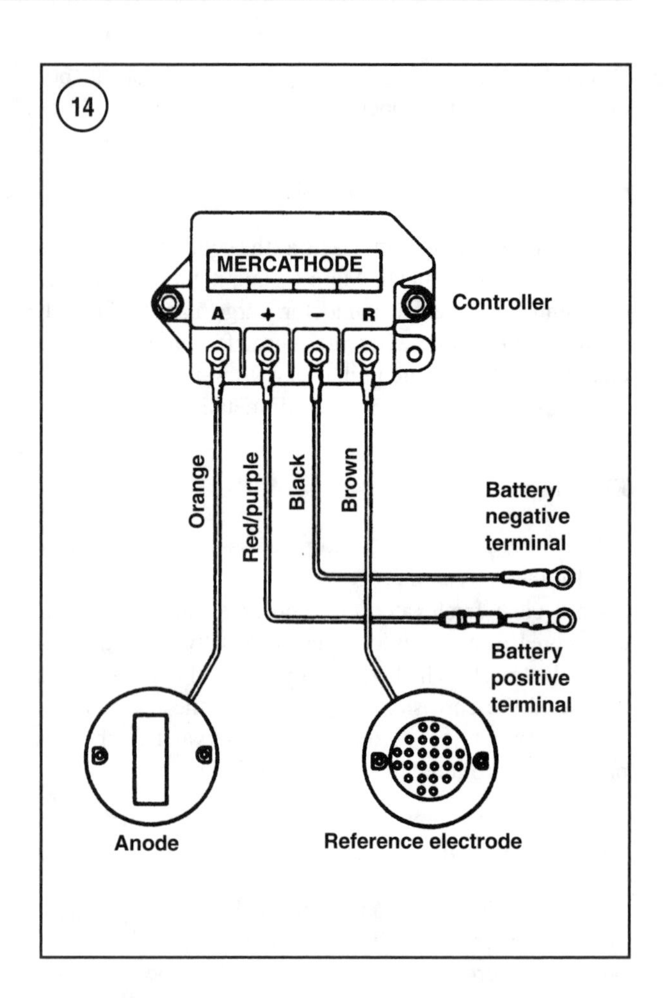

PROPELLERS

The propeller is the final link between the boat's drive system and the water. A perfectly maintained engine and hull are useless if the propeller is the wrong type or has deteriorated. Although propeller selection for a specific application is beyond the scope of this manual, the following provides the basic information needed to make an informed decision. A marine dealership is the best source for a propeller recommendation.

Propeller Operation

As the curved blades of a propeller rotate through the water, a high-pressure area forms on one side of the blade and a low-pressure area forms on the other side of the blade (**Figure 15**). The propeller moves toward the low-pressure area, carrying the boat with it.

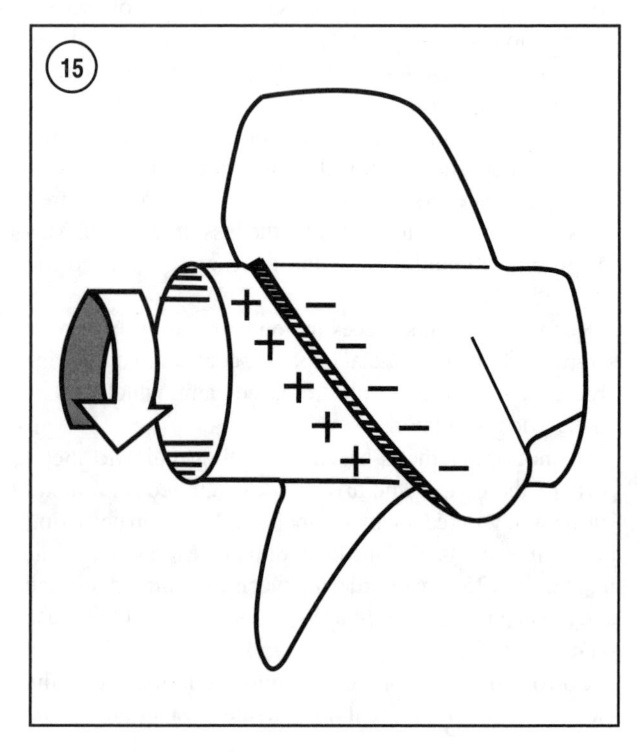

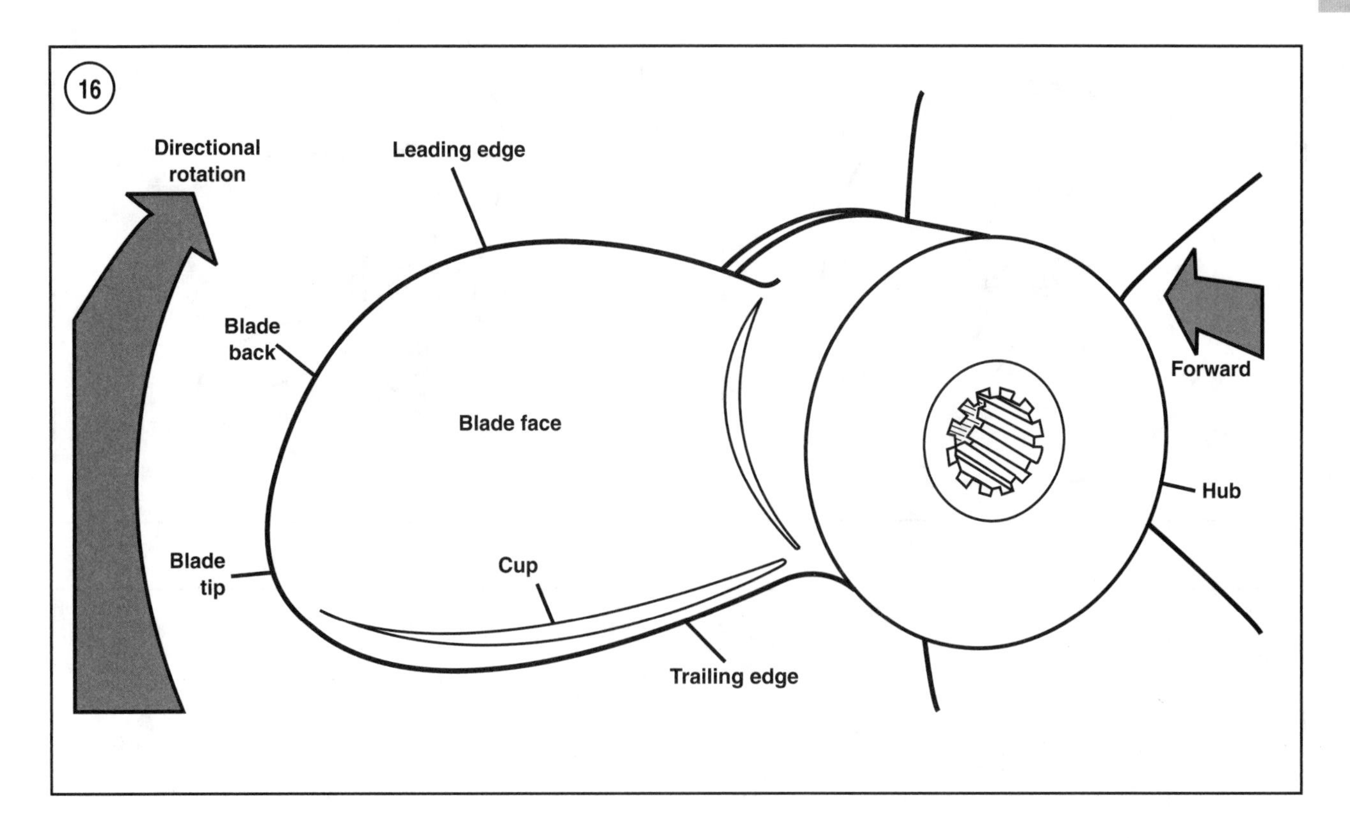

Propeller Parts

Although a propeller is usually a one-piece unit, it contains several different parts (**Figure 16**). Variations in the design of these parts make different propellers suitable for different applications.

The blade tip is the point on the blade furthest from the center of the propeller hub or propeller shaft bore. The blade tip separates the leading edge from the trailing edge.

The leading edge is the edge of the blade nearest the boat. During forward gear operation, this is the area of the blade that first cuts through the water.

The trailing edge is the surface of the blade furthest from the boat. During reverse gear operation, this is the area of the blade that first cuts through the water.

The blade face is the surface of the blade that faces away from the boat. During forward gear operation, high-pressure forms on this side of the blade.

The blade back is the surface of the blade that faces toward the boat. During forward gear operation, low-pressure forms on this side of the blade.

The cup is a small curve or lip on the trailing edge of the blade. Cupped propeller blades generally perform better than noncupped propeller blades.

The hub is the central portion of the propeller. It connects the blades to the propeller shaft. On most drive systems, engine exhaust is routed through the hub. On these systems, the hub is made up of an outer and inner portion, connected by ribs.

A diffuser ring is used on hub exhaust models to prevent exhaust gasses from entering the blade area.

Propeller Design

Changes in length, angle, thickness and material of propeller parts make different propellers suitable for different applications.

Diameter

Propeller diameter is the distance from the center of the hub to the blade tip, multiplied by two. It is the diameter of the circle formed by the blade tips during propeller rotation (**Figure 17**).

Pitch and rake

Propeller pitch and rake describe the placement of the blades in relation to the hub (**Figure 18**).

Pitch describes the theoretical distance the propeller would travel in one revolution. In A, **Figure 19**, the propeller would travel 10 inches in one revolution. In B, **Figure 19**, the propeller would travel 20 inches in one

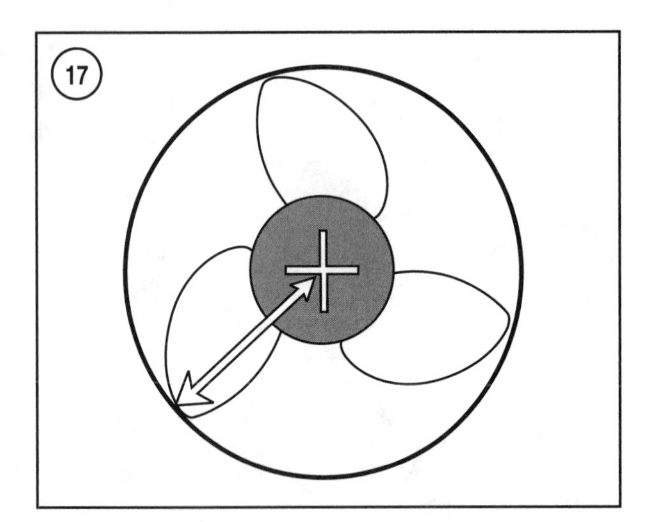

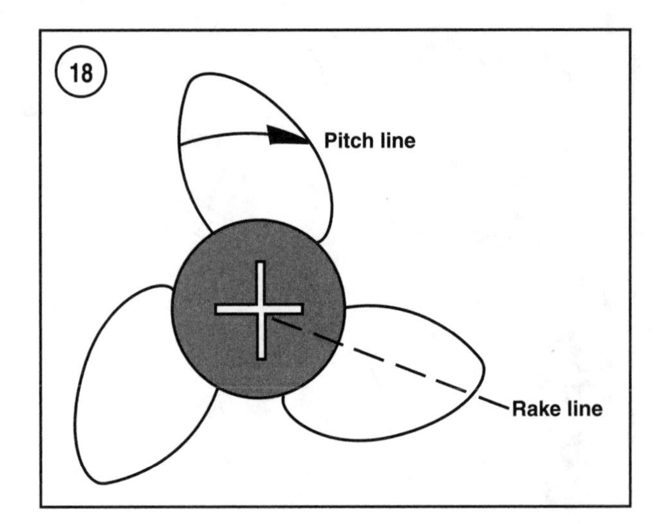

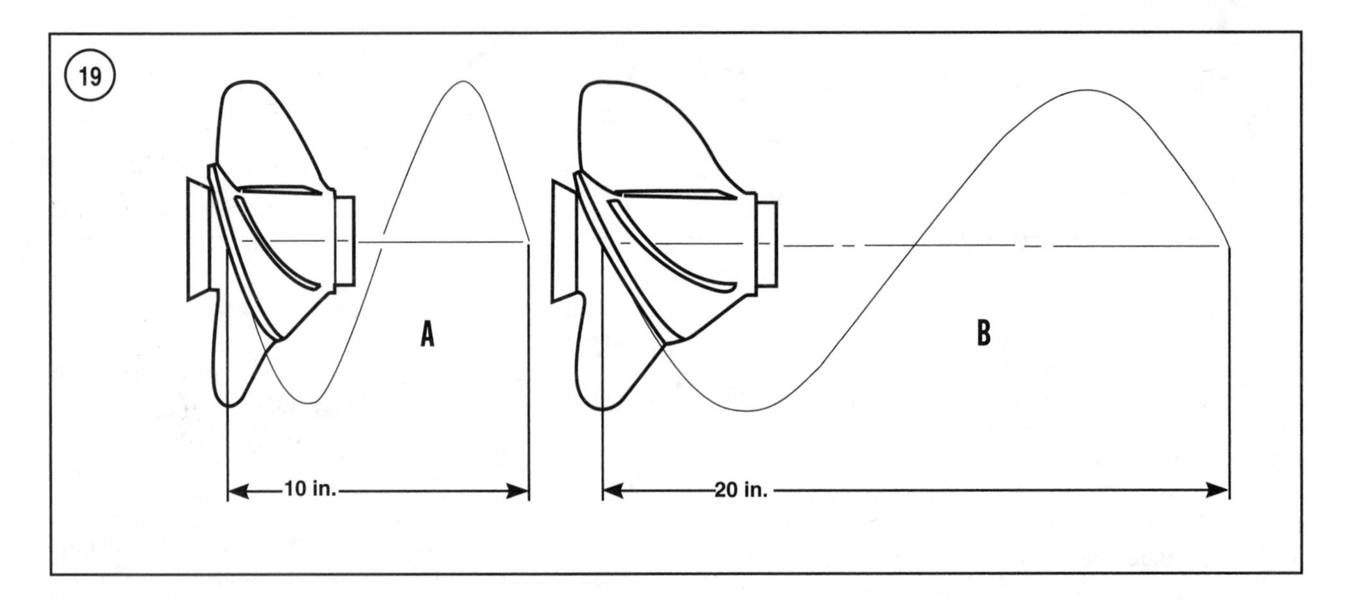

revolution. This distance is only theoretical; during operation, the propeller achieves only 75-85% of its pitch. Slip rate describes the difference in actual travel relative to the pitch. Lighter, faster boats typically achieve a lower slip rate than heavier, slower boats.

Propeller blades can be constructed with constant pitch (**Figure 20**) or progressive pitch (**Figure 21**). Progressive pitch starts low at the leading edge and increases toward the trailing edge. The propeller pitch specification is the average of the pitch across the entire blade. Propellers with progressive pitch usually provide better overall performance than constant pitch propellers.

Blade rake is specified in degrees and is measured along a line from the center of the hub to the blade tip. A blade that is perpendicular to the hub (**Figure 22**) has 0° rake. A blade that is angled (**Figure 22**) has a rake expressed by its difference from perpendicular. Most propellers have rakes ranging from 0-20°. Lighter, faster boats generally perform better with propellers that have a greater amount of rake. Heavier, slower boats generally perform better using a propeller with less rake.

Blade thickness

Blade thickness is not uniform at all points along the blade. For efficiency, blades are as thin as possible at all points while retaining enough strength to move the boat. Blades are thicker where they meet the hub and thinner at the blade tips (**Figure 23**) to support the heavier loads at the hub section of the blade. Overall blade thickness is dependent on the strength and the material used.

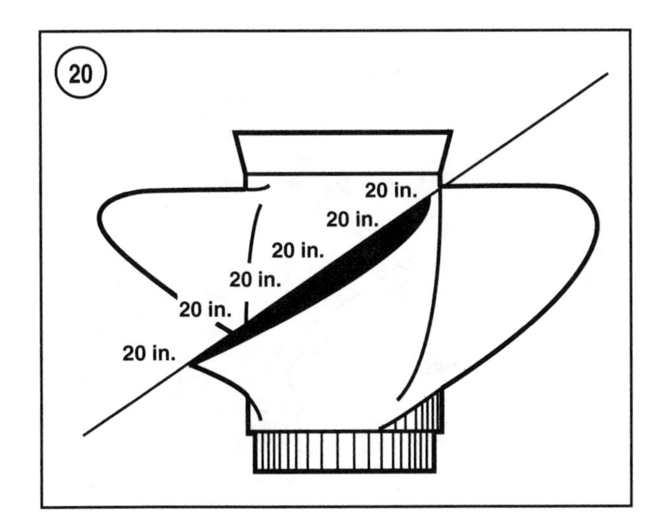

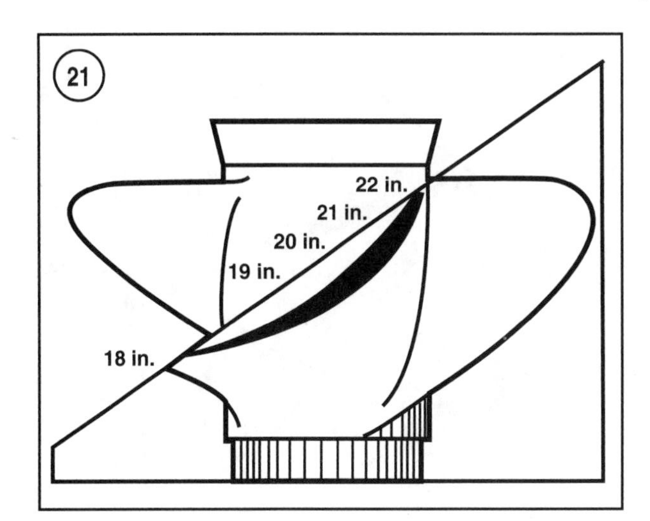

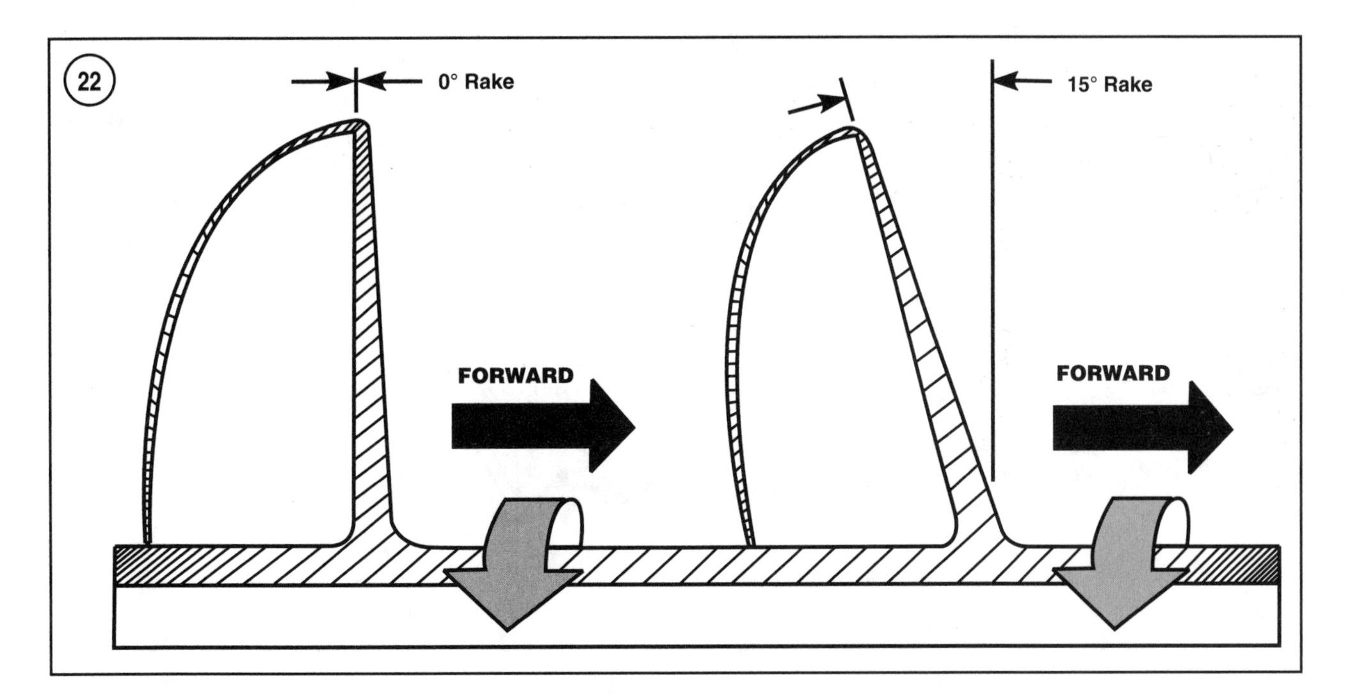

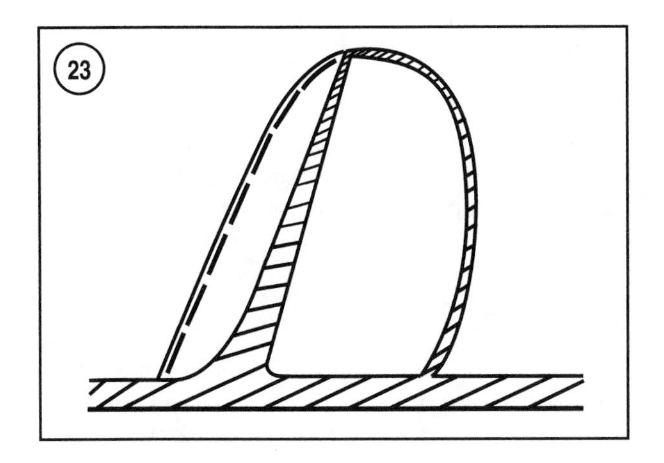

When cut from the leading edge to the trailing edge in the central portion of the blade (**Figure 24**), the propeller blade resembles an airplane wing. The blade face, where high-pressure exists during forward gear rotation, is almost flat. The blade back, where low-pressure exists during forward gear rotation, is curved with the thinnest portions at the edges and the thickest portion at the center.

Propellers that run only partially submerged, as in racing applications, may have a wedge shaped cross-section (**Figure 25**). The leading edge is very thin; the blade thickness increases toward the trailing edge, where it is thickest. This type of propeller is very inefficient when completely submerged.

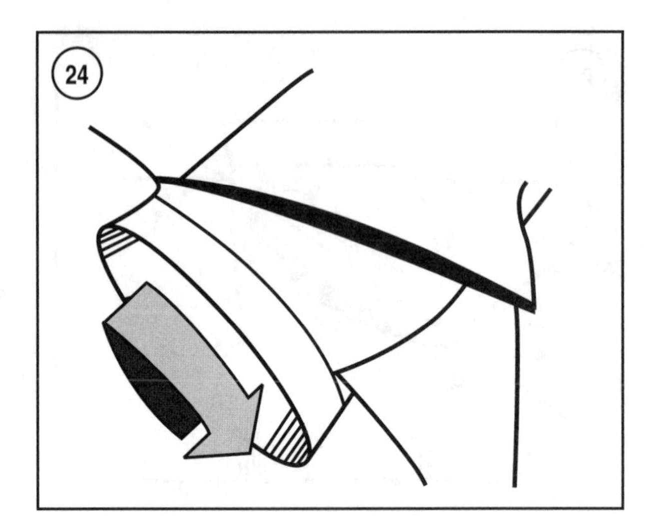

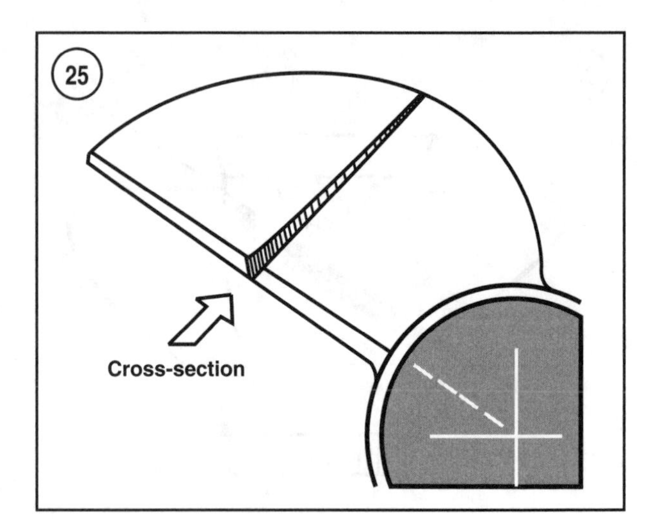

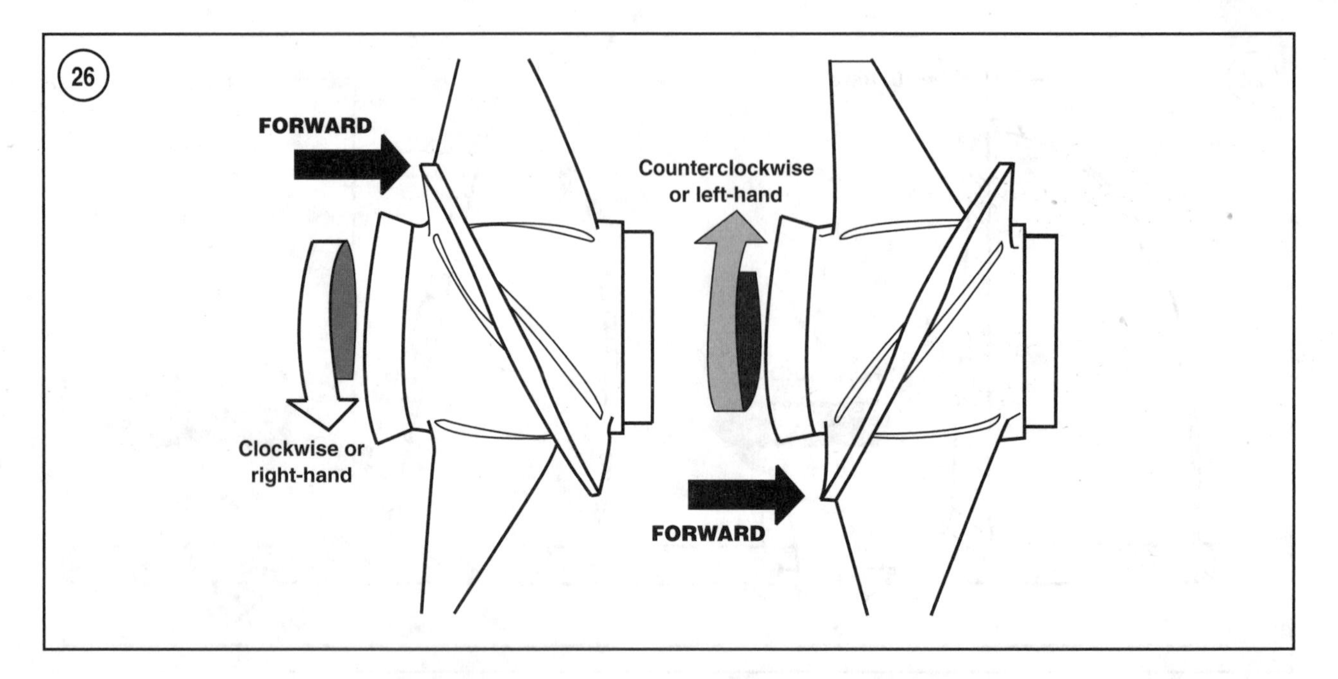

Number of blades

The number of blades used on a propeller is a compromise between efficiency and vibration. A one-bladed propeller would be the most efficient, but it would create and unacceptable amount of vibration. As blades are added, efficiency decreases, but so does vibration. Most propellers have three or four blades, representing the most practical trade-off between efficiency and vibration.

Material

Propeller materials are chosen for strength, corrosion resistance and economy. Stainless steel, aluminum, plastic and bronze are the most commonly used materials. Bronze is quite strong but rather expensive. Stainless steel is more common than bronze because of its combination of strength and lower cost. Aluminum alloy and plastic materials are the least expensive but usually lack the strength of stainless steel. Plastic propellers are more suited for low horsepower applications.

Direction of rotation

Propellers are made for both right-hand and left-hand rotations. Right-hand is the most commonly used. As viewed from the rear of the boat while in forward gear, a

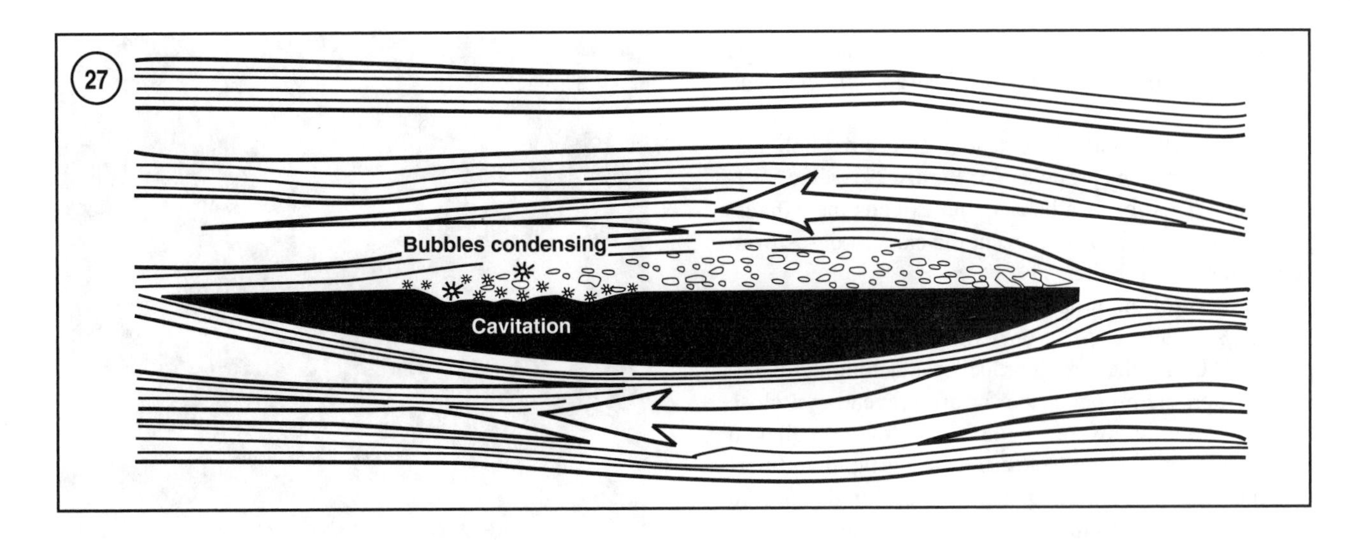

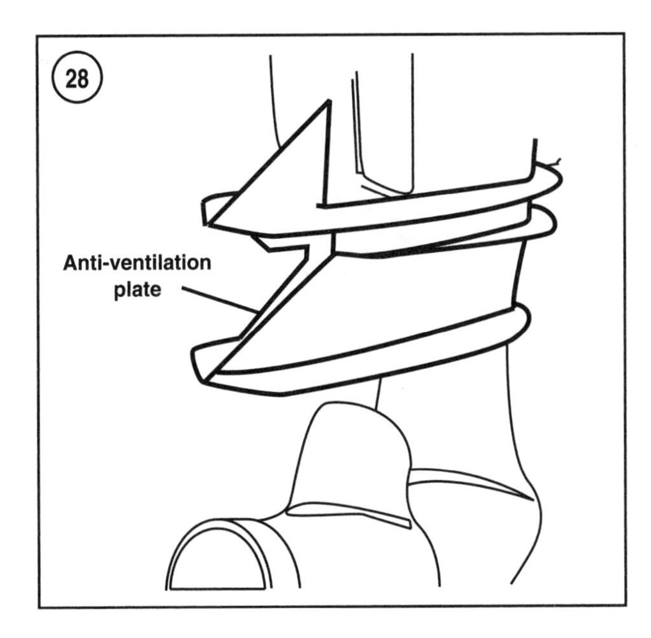

right-hand propeller turns clockwise and a left-hand propeller turns counterclockwise. When detached from the boat, the propeller's direction of rotation is determined by the angle of the blades (**Figure 26**). A right-hand propeller's blade slants from the upper left to the lower right; a left-hand propeller's blades are the opposite.

Cavitation and Ventilation

Cavitation and ventilation are *not* interchangeable terms. They refer to two distinct problems encountered during propeller operation.

To understand cavitation, consider the relationship between pressure and the boiling point of water. At sea level, water boils at 212° F (100° C). As pressure increases, such as within an engine cooling system, the boiling point of the water increases. The opposite is also true. As pressure decreases, water boils at a temperature lower than 212° F (100° C). If the pressure drops low enough, water boils at normal room temperature.

During normal propeller operation, low pressure forms on the blade back. Normally the pressure does not drop low enough for boiling to occur. However, poor propeller design, damaged blades or the wrong propeller can cause unusually low pressure on the blade surface (**Figure 27**). If the pressure drops low enough, boiling occurs and bubbles form on the blade surfaces. As the boiling water moves to a higher pressure area on the blade, the boiling ceases and bubbles collapse. The collapsing bubbles release energy that erodes the surface of the propeller blade.

Corroded surfaces, physical damage or even marine growth combined with high speed operation can cause low pressure and cavitation on gearcase surfaces. In such cases, low pressure forms as water flows over a protrusion or rough surface. Boiling water causes bubbles to form and collapse as they move to a higher pressure area toward the rear of the surface imperfection.

This entire process of pressure drop, boiling and bubble collapse is called *cavitation*. The ensuing damage is called *cavitation burn*. Cavitation is caused by a decrease in pressure, *not* an increase in temperature.

Ventilation is not as complex as cavitation. Ventilation refers to air entering the blade area, either from above the surface of the water or from a through-hub exhaust system. As the blades meet the air, the propeller momentarily looses it contact to the water, losing most of its thrust. Then the propeller and engine over rev, causing very low pressure on the blade back and massive cavitation.

All Mercury and Mariner engines have a plate (**Figure 28**) above the propeller area to prevent air from entering the blade area. This plate is called an *antiventilation plate*, although it is often incorrectly called an *anticavitation plate*.

Most propellers have an extended and flared hub section at the rear of the propeller (A, **Figure 29**) called a diffuser ring. This forms a barrier, and extends the exhaust passage far enough aft to prevent the exhaust gases from ventilating the propeller.

A close fit of the propeller to the gearcase (B, **Figure 29**) keeps exhaust gasses from exiting and ventilating the propeller. Using the wrong propeller attaching hardware can position the propeller too far aft, preventing a close fit. The wrong hardware can also allow the propeller to rub heavily against the gearcase, causing rapid and permanent wear to both components. Wear or damage to these surfaces allows the propeller to ventilate.

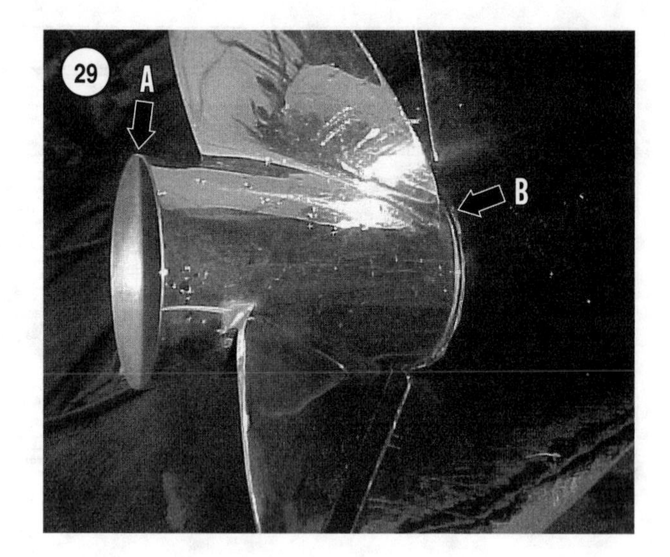

Chapter Two

Tools and Techniques

This chapter describes the common tools required for Mercury and Mariner engine repairs and troubleshooting. It also describes techniques that make the work easier and more effective. Some procedures in this book require special skills or expertise; in some cases it is better to entrust the job to a Mercury/Mariner dealership or qualified specialist.

SAFETY

Professional mechanics can work for years and never sustain a serious injury or mishap. Follow these guidelines and practice common sense to safely service the vehicle.

1. Do not operate the engine in an enclosed area. The exhaust gasses contain carbon monoxide, an odorless, colorless, and tasteless poisonous gas. Carbon monoxide levels build quickly in enclosed areas and can cause unconsciousness and death in a short time. Make sure the work area is properly ventilated or operate the engine outside.
2. *Never* use gasoline or any extremely flammable liquid to clean parts. Refer to *Cleaning Parts* and *Handling Gasoline Safely* in this chapter.
3. *Never* smoke or use a torch in the vicinity of flammable liquids, such as gasoline or cleaning solvent.
4. If welding or brazing, make sure that any fuel source is at least 50 ft. (15 m) away.
5. Use the correct type and size of tools to avoid damaging fasteners.
6. Keep tools clean and in good condition. Replace or repair worn or damaged equipment.
7. When loosening a tight fastener, be guided by what would happen if the tool slips.
8. When replacing fasteners, make sure the new fasteners are of the same size and strength as the original ones.
9. Keep the work area clean and organized.
10. Wear eye protection *anytime* the safety of your eyes is in question. This includes procedures involving drilling, grinding, hammering, compressed air and chemicals.

11. Wear the correct clothing for the job. Tie up or cover long hair so it cannot get caught in moving equipment.
12. Do not carry sharp tools in clothing pockets.
13. Always have a Coast Guard approved fire extinguisher available. Make sure it is rated for gasoline (Class B) and electrical (Class C) fires.
14. Do not use compressed air to clean clothes, the engine or the work area. Debris may be blown into your eyes or skin. *Never* direct compressed air at yourself or someone else. Do not allow children to use or play with any compressed air equipment.
15. When using compressed air to dry rotating parts, hold the part so it cannot rotate. Do not allow the force of the air to spin the part. The air jet is capable of rotating parts at extreme speed. The part may be damaged or disintegrate, causing serious injury.
16. Do not inhale the dust or particles created when removing gaskets. In most cases these particles contain asbestos. Inhaling asbestos particles is hazardous to your health.
17. Never work on the machine while someone is working under it.
18. If placing the machine on a stand or securing it with a lift, make sure it is secure before walking away.

Handling Gasoline Safely

Gasoline is a volatile flammable liquid and is one of the most dangerous items in the shop. However, because gasoline is used so often, many people forget that it is hazardous. Only use gasoline as fuel for gasoline internal combustion engines. Keep in mind, when working on the engine, gasoline is always present in the fuel tank, fuel line and carburetor. To avoid a disastrous accident when working around the fuel system, carefully observe the following precautions:
1. *Never* use gasoline to clean parts. See *Cleaning Parts* in this chapter.
2. When working on the fuel system, work outside or in a well-ventilated area.
3. Do not add fuel to the fuel tank or service the fuel system while the machine is near open flames, sparks or where someone is smoking. Gasoline vapor is heavier than air; it collects in low areas and is more easily ignited than liquid gasoline. Always turn the engine off before refueling.
4. Allow the engine to cool completely before working on any fuel system component.
5. When draining the carburetor, catch the fuel in a plastic container and then pour it into an approved gasoline storage device.
6. Do not store gasoline in glass containers. If the glass breaks, a serious explosion or fire may occur.

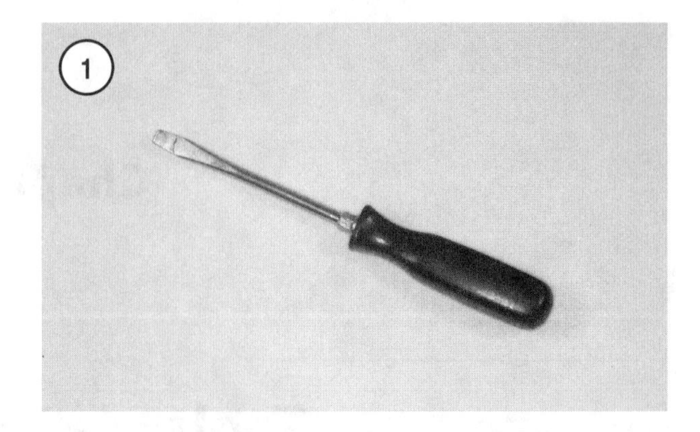

7. Immediately wipe up spilled gasoline with rags. Store the rags in a metal container with a lid until they can be properly disposed of, or place them outside in a safe place for the fuel to evaporate.
8. Do not pour water onto a gasoline fire. Water spreads the fire and makes it more difficult to put out. Use a class B, BC or ABC fire extinguisher to extinguish the fire.

Cleaning Parts

Cleaning parts is one of the more tedious and difficult service jobs performed in the home garage. There are many types of chemical cleaners and solvents available for shop use. Most are poisonous and extremely flammable. To prevent chemical exposure, vapor buildup, fire and serious injury, observe each product warning label and note the following:
1. Read and observe the entire product label before using any chemical. Always know what type of chemical is being used and whether it is poisonous and/or flammable.
2. Do not use more than one type of cleaning solvent at a time. If mixing chemicals is called for, measure the proper amounts according to the manufacturer.
3. Work in a well-ventilated area.
4. Wear chemical-resistant gloves.

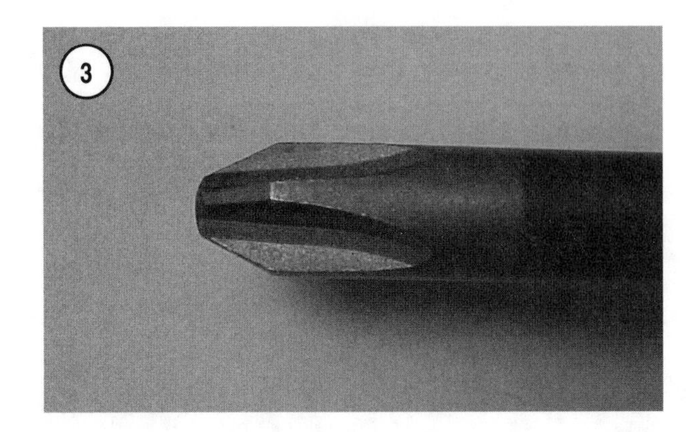

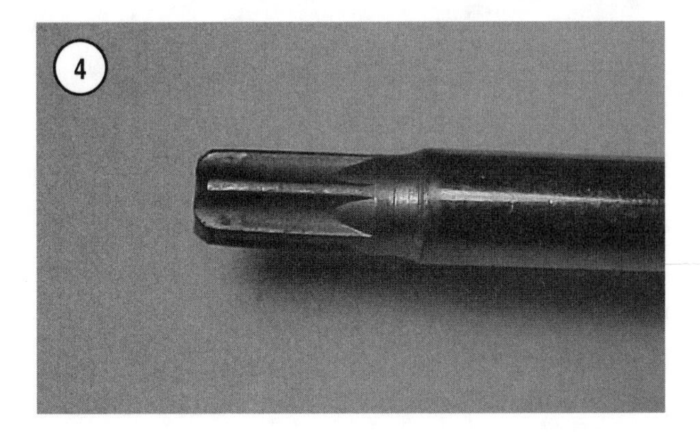

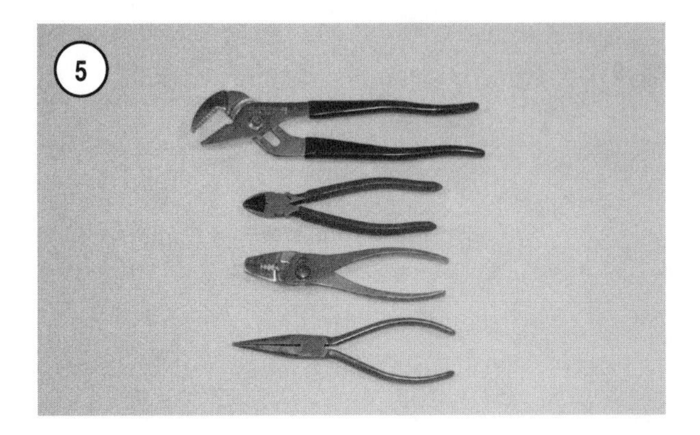

5. Wear safety glasses.
6. Wear a vapor respirator if the instructions call for it.
7. Wash hands and arms thoroughly after cleaning parts.
8. Keep chemical products away from children and pets.
9. Thoroughly clean all oil, grease and cleaner residue from any part that must be heated.
10. Use a nylon brush when cleaning parts. Metal brushes may cause a spark.
11. When using a parts washer, only use the solvent recommended by the manufacturer. Make sure the parts washer is equipped with a metal lid that will lower in case of fire.

BASIC HAND TOOLS

A number of tools are required to maintain and repair a Mercury or Mariner engine. Most of these tools are also used for home and automobile repairs. Some tools are made especially for working on Mercury and Mariner engines; these tools can be purchased from a Mercury or Mariner dealership. Having the required tools always makes the job easier and more effective.

Keep the tools clean and in a suitable box. Keep them organized and stored with related tools. After using a tool, wipe off dirt and grease with a shop towel.

The following tools are required to perform virtually any repair job. Each tool is described and the recommended size is given for starting a tool collection. Additional tools and some duplication may be added as you become more familiar with the equipment. You may need standard U.S. size tools, metric size tools or a mixture of both.

Screwdrivers

A screwdriver (**Figure 1**) is a very basic tool, but if used improperly, it can do more damage than good. The slot on a screw has a definite dimension and shape. Always use a screwdriver that comforms to the shape of the screw. Use a small screwdriver for small screws and a large one for large screws, or the screw head will be damaged.

Three types of screwdrivers are commonly required: a slotted (flat-blade) screwdriver (**Figure 2**), Phillips screwdriver (**Figure 3**) and Torx screwdriver (**Figure 4**).

Use screwdrivers only for driving screws. Never use a screwdriver for prying or chiseling. Do not attempt to remove a Phillips, Torx or Allen head screw with a slotted screwdriver; the screw head can be damaged to the point that the proper tool is unable to remove it.

Keep screwdrivers in the proper condition so they will last longer and perform better. Always keep the tip of a common screwdriver in good condition. Carefully regrind the tip to the proper size and taper if it becomes worn or damaged. The sides of the blade must be parallel and the blade tip must be flat. Replace a Phillips or Torx screwdriver if its tip is worn or damaged.

Pliers

Pliers come in a wide range of types and sizes. Pliers are useful for cutting, gripping, bending and crimping. Never use pliers to cut hardened objects, or to turn bolts or nuts. **Figure 5** shows several types of pliers.

Each type of pliers has a specialized function. General purpose pliers are mainly used for gripping and bending.

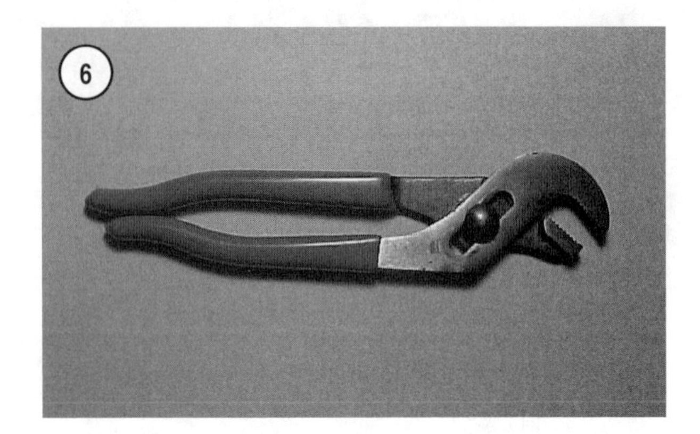

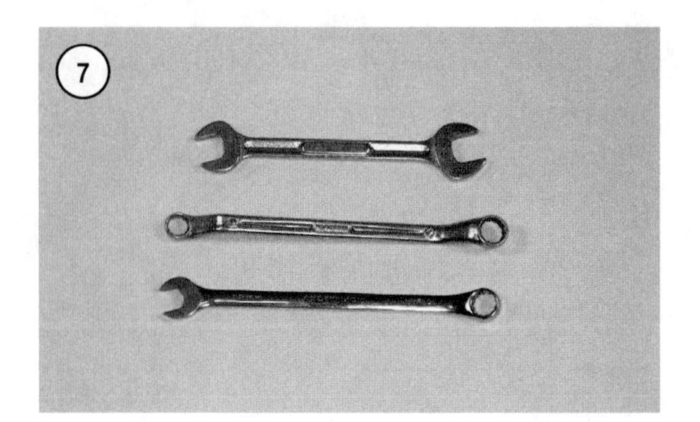

Locking pliers are used for gripping objects very tightly, like a vise. Needlenose pliers are used to grip or bend small objects. Adjustable or slip-joint pliers (**Figure 6**) can be adjusted to grip various sized objects; the jaws remain parallel for gripping objects such as pipe or tubing. There are many more types of pliers. The ones described here are the most common.

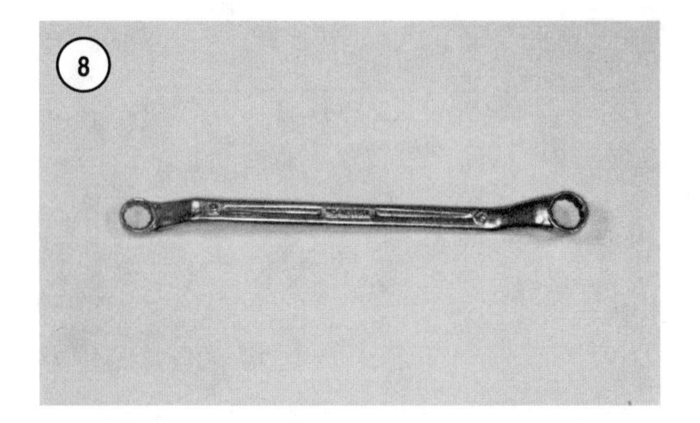

Box- and Open-end Wrenches

Box- and open-end wrenches (**Figure 7**) are available in sets in a variety of sizes. The number stamped near the end of the wrench refers to the distance between two parallel flats on the hex head bolt or nut.

Box-end wrenches (**Figure 8**) provide a better grip to the nut and are stronger than open-end wrenches. An open-end wrench (**Figure 9**) grips the nut on only two flats. Unless it fits well, it may slip and round off the points on the nut. A box-end wrench grips all six flats. Box-end wrenches are available with 6-point or 12-point openings. The 6-point opening provides superior holding power; the 12-point allows a shorter swing for working in tight quarters.

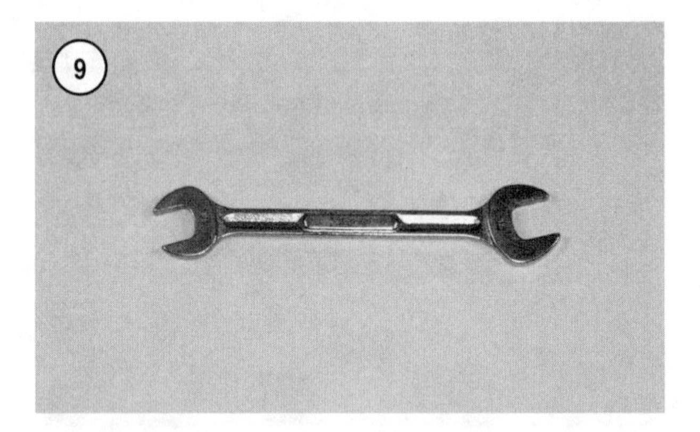

Use an open-end wrench if a box-end wrench cannot be positioned over the nut or bolt. To prevent damage to the fastener, avoid using an open-end wrench if a large amount of tightening or loosening torque is required.

A combination wrench (**Figure 10**) has both a box-end and open-end. Both ends are the same size.

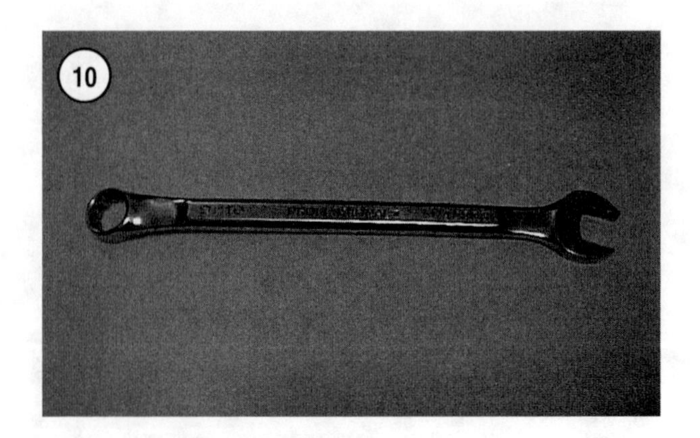

Adjustable Wrenches

An adjustable wrench (**Figure 11**) can be adjusted to fit virtually any nut or bolt head. However, it can loosen and slip from the nut or bolt, causing damage to the nut and possible physical injury. Use an adjustable wrench only if a proper size open- or box-end wrench is not available.

11

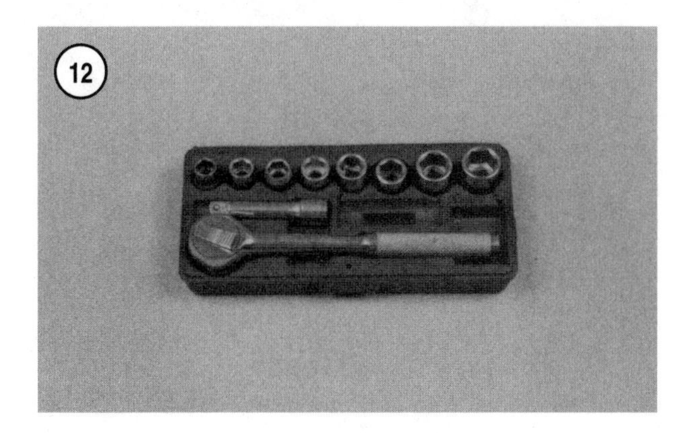

12

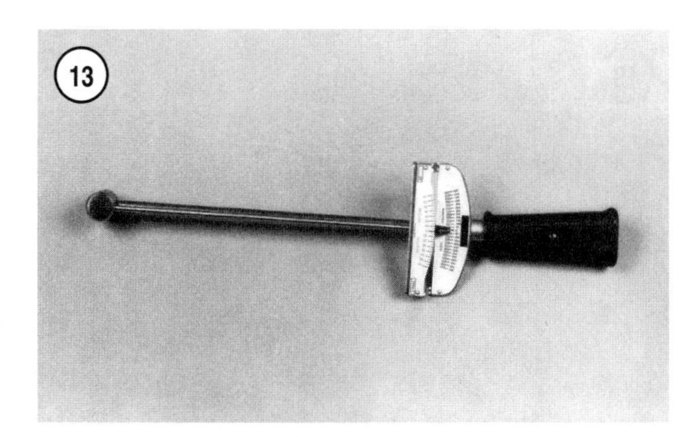

13

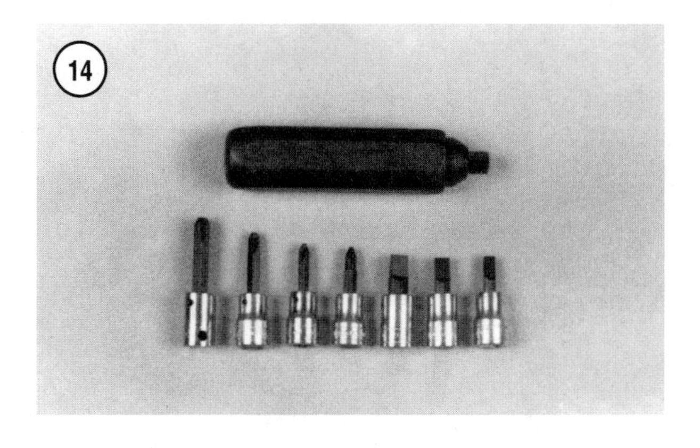

14

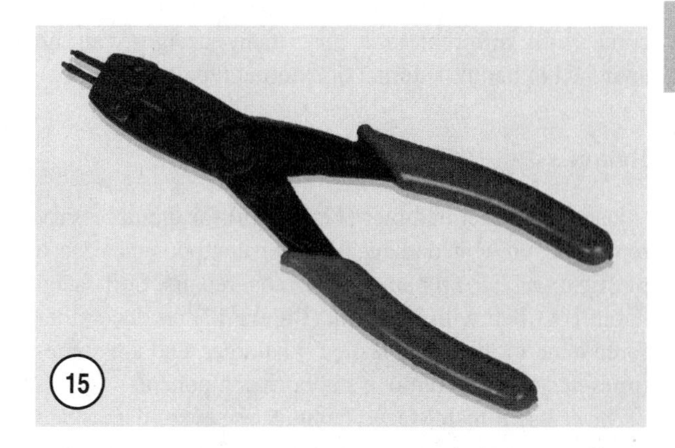

15

Avoid using an adjustable wrench if a large amount of tightening or loosening torque is required.

Adjustable wrenches come in sizes ranging from 4 to 18 in. overall length. A 6 or 8 in. size is recommended as an all purpose wrench.

Socket Wrenches

A socket wrench (**Figure 12**) is generally faster, safer and more convenient to use than a common wrench. Sockets, which attach to a suitable handle, are available with 6-point or 12-point openings, and use 1/4, 3/8, and 1/2 in. drive sizes. The drive size corresponds with the square hole that mates with the ratchet or flex handle.

Torque Wrench

A torque wrench (**Figure 13**) is used with a socket to measure how tight a nut or bolt is installed. They come in 1/4, 3/8, and 1/2 in. drive sizes. The drive size corresponds with the square hole that mates with the socket.

A typical 1/4 in. drive torque wrench measures torque in in.-lb. increments, and has a range of 20-150 in.-lb. (2.2-17 N•m). A typical 3/8 or 1/2 in. torque wrench measures torque in ft.-lb. increments, and has a range of 10-150 ft.-lb. (14-203 N•m).

Impact Driver

An impact driver (**Figure 14**) makes removal of tight fasteners easy, and reduces damage to bolts and screws. Interchangeable bits allow use on a variety of fasteners.

Snap Ring Pliers

Snap ring pliers, also called circlip pliers, are necessary to remove snap rings. Snap ring pliers (**Figure 15**) usually come with different size tips; many designs can be switched to handle internal or external type snap rings.

Hammers

Use the correct hammer (**Figure 16**) for the necessary repairs to prevent damage to other components. Use a plastic or rubber tip hammer for most repairs. Soft-faced hammers filled with buckshot (**Figure 17**) produce more force than a rubber or plastic tip hammer, and are sometimes necessary to remove stubborn components.

Never use a metal-faced hammer as severe damage to engine components or tools will occur. The same amount of force can be produced with a soft-faced hammer.

Feeler Gauges

A feeler gauge has either flat or wire measuring gauge (**Figure 18**). Wire gauges are used to measure spark plug gap; flat gauges are used for other measurements. A non-magnet (brass) gauge may be specified when working around magnetized components.

Other Special Tools

Many of the maintenance and repair procedures require special tools (**Figure 19**). Most of the tools are available from marine dealerships. The remainder is available from tool suppliers. Instruction on their use and the manufacture's part numbers are included in the appropriate chapter.

Some special tools can be made locally by a qualified machinist, often at a lower price. Many marine dealerships and rental outlets will rent some of the required tools. Using makeshift tools may result in damaged parts that cost far more than the recommended tool.

TEST EQUIPMENT

This section describes equipment for testing, adjustments and measurements on Mercury and Mariner engines. Most of these tools are available from a local marine dealership or automotive parts store.

Multimeter

A multimeter is invaluable for electrical troubleshooting and service. It combines a voltmeter, ohmmeter and an ammeter in one unit. It is often called a VOM.

Two types of mutimeter are available: analog and digital. On analog meters (**Figure 20**), a moving needle and

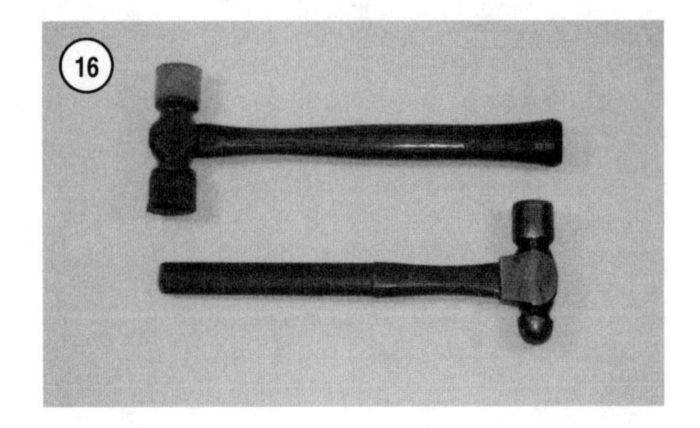
16

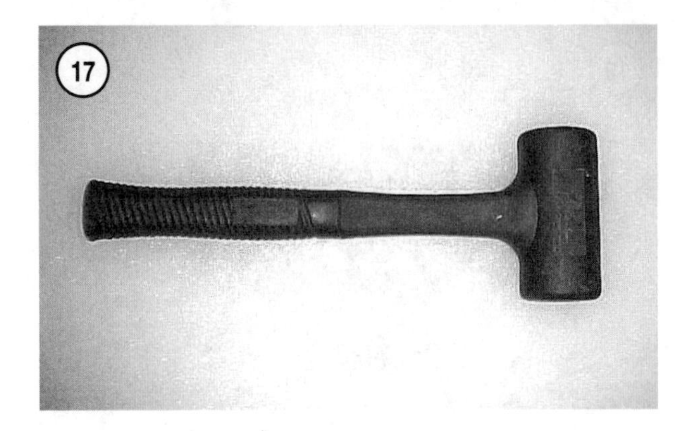
17

18

19

20

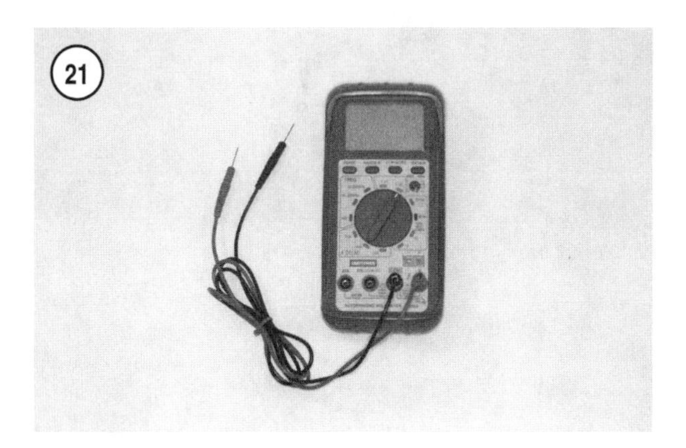

21

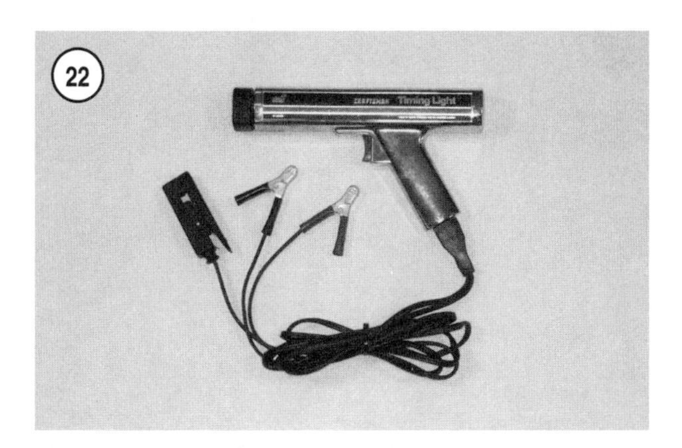

22

marked bands on the meter face indicate the volt, ohm and amperage scales. An analog meter must be calibrated each time the battery or scale is changed.

A digital meter (**Figure 21**) is ideal for electrical troubleshooting because it is easy to read and more accurate than an analog meter. Most models are autoranging, have automatic polarity compensation and have internal overload protection circuits.

Either type of meter is suitable for most electrical testing described in this manual. An analog meter is better suited for testing pulsing voltage signals such as those produced by the ignition system. A digital meter is better suited for testing involving a very low resistance or voltage reading (less than 1 volt or 1 ohm). The test instructions indicate if a specific type of meter is required.

The ignition system produces electrical pulses that are too short in duration for accurate measurement with a common multimeter. Use a meter with peak volt reading capability for ignition system testing. This type of meter utilizes special circuits that capture and display the peak voltage reached during the pulse. Unless specified otherwise, use the appropriate scale in the DVA function when testing the ignition system.

Scale selection, meter specifications and test lead connection points vary by the manufacturer and model of the meter. Read the instructions supplied with the meter before performing any test. The meter and certain electrical components on the engine can be damaged if tested incorrectly. Have the test performed by a qualified professional if you are unfamiliar with the testing or general meter usage. The expense to replace damaged equipment can far exceed the cost of having the test performed by a professional.

Timing Light

A timing light is necessary to set ignition timing while the engine is running. By flashing a light at the precise instant the spark plug fires, the position of the timing mark can be seen. The flashing light makes a moving mark appear to stand still next to a stationary timing mark.

Suitable timing lights (**Figure 22**) range from inexpensive models to expensive models with a built-in tachometer and timing advance compensator. A built-in tachometer is very useful as most ignition timing specifications refer to a specific engine speed.

A timing advance compensator can delay the strobe enough to bring the timing mark to a certain place on the scale. Although useful for troubleshooting, this feature should not be used to check or adjust the base ignition timing.

Tachometer/Dwell Meter

A portable tachometer (**Figure 23**) is needed for tuning and testing of Mercury and Mariner engines. Ignition timing and carburetor adjustments must be performed at a specified engine speed. Tachometers are available with either an analog or digital display.

Carburetor adjustments are performed at idle speed. If using an analog tachometer, choose one with a low range of 0-1000 rpm or 0-2000 rpm range and a high range of 0-6000 rpm. The high range setting is necessary for testing but lacks the accuracy needed at lower speeds. At lower speeds the meter must be capable of detecting changes of 25 rpm or less.

Digital tachometers (**Figure 24**) are generally easier to use than most analog tachometers. Their measurements are accurate at all speeds without the need to change the range or scale. Many of these use an inductive pickup to receive the signal from the ignition system.

A dwell meter is often incorporated into the tachometer to allow testing and/or adjustments to engines with a breaker point ignition system. The engines covered in this manual are all equipped with a transistorized ignition system and a dwell meter is not needed.

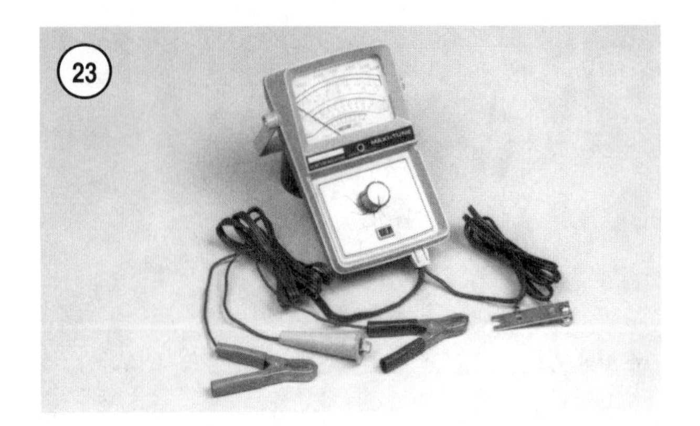

Vacuum Gauge

The vacuum gauge measures the intake manifold vacuum created during the engine intake stroke. Manifold air leakage, improper carburetor or timing adjustment and valve problems can be identified by interpreting the readings. If combined with compression gauge readings, other engine mechanical problems can be diagnosed.

Compression Gauge

A compression gauge (**Figure 25**) measures the amount of pressure created in the combustion chamber during the compression stroke. Compression indicates the general engine condition making it one of the most useful troubleshooting tools.

The easiest gauge to use has screw-in adapters that fit the spark plug holes. Press-in rubber-tipped gauges are also available. This gauge must be held firmly into the spark plug opening to prevent air leakage and inaccurate compression measurements. Only use a good quality gauge and check its accuracy by measuring the compression on a well performing engine. False measurement can lead to needless and expensive power head repair.

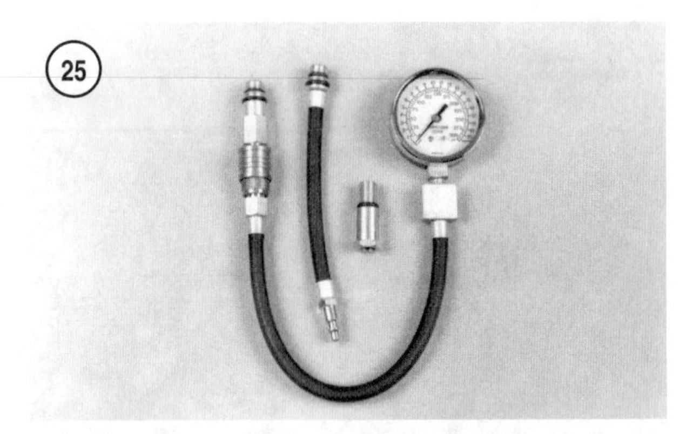

Hydrometer

A hydrometer measures the specific gravity of the electrolyte in the battery. The specific gravity indicates the battery's state of charge by measuring the density of the electrolyte as compared to pure water. Choose a hydrometer (**Figure 26**) with automatic temperature compensation; otherwise, the electrolyte temperature must be measured to determine the actual specific gravity.

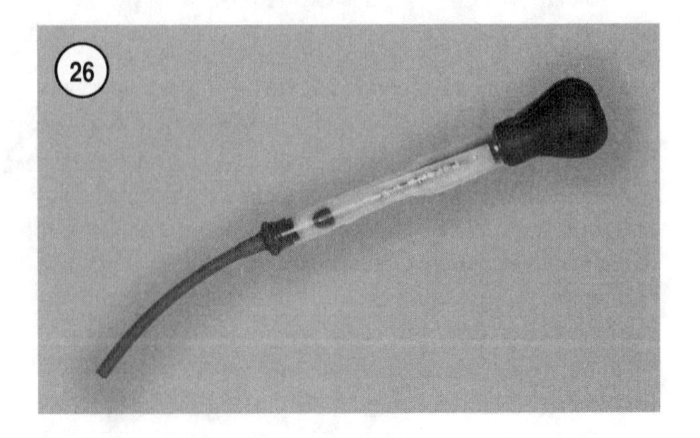

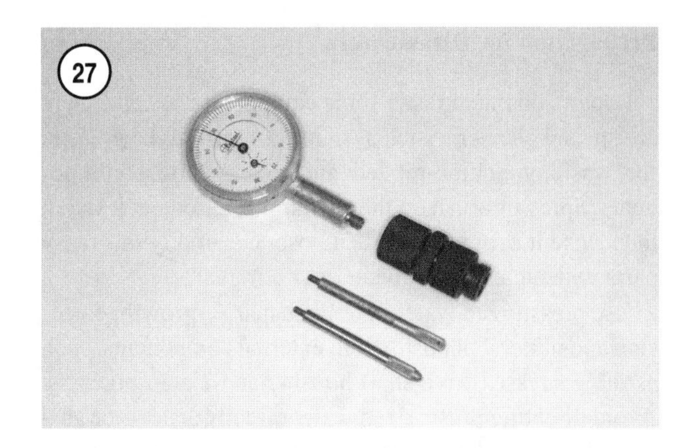

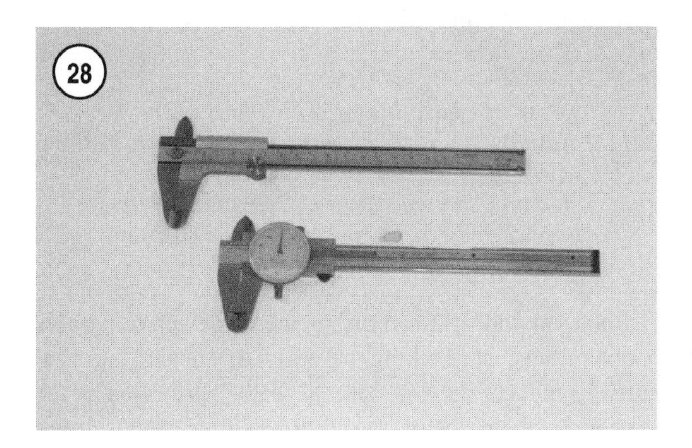

Precision Measuring Tools

Various tools are required to make precision measurements. A dial indicator (**Figure 27**), for example, determines piston position in the cylinder, runout and end-play of parts assemblies. It also measures free movement between the gear teeth (backlash) in the drive unit.

Vernier calipers (**Figure 28**), micrometers (**Figure 29**) and other precision tools are used to measure the size of parts, such as the piston.

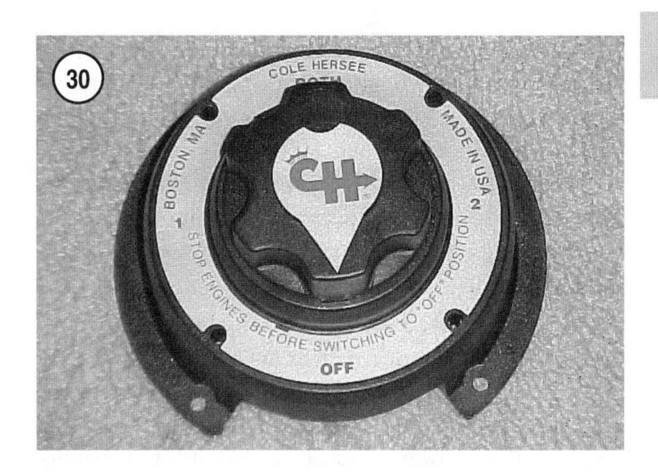

Precision measuring equipment must be stored, handled and used carefully or it will not remain accurate.

SERVICE HINTS

Most of the service procedures are straightforward and can be performed by anyone handy with tools. Consider the necessary skill, tools and equipment before attempting repairs involving major disassembly of the engine or drive unit.

Some operations, for example, require the use of a press. Other operations require precision measurement. Have the procedure(s) or measurement(s) performed by a professional if the correct equipment or experience is lacking.

Working With Electrical Components

All models covered in this manual use a transistorized ignition system. The components of the system are able to withstand a rigorous marine environment. However they can be damaged under certain circumstances, such as operating the engine with the spark plug lead(s) disconnected or improperly disconnecting the battery or wire harness connection. Always ground disconnected spark plug leads before operating the starter.

Battery Precautions

Disconnecting or connecting the battery can create a surge of current throughout the electrical system. This surge can damage certain components of the charging system. Always make sure the ignition switch is in the *off* position before connecting or disconnecting the battery or changing the selection on a battery switch (**Figure 30**).

Always disconnect both battery cables and remove the battery from the boat for charging. If the battery cables are connected, the charger may induce a damaging surge of current into the electrical system. During charging, batteries produce explosive and corrosive gasses. These gasses can cause corrosion within the battery compartment and create an extremely hazardous condition.

The cables must be disconnected from the battery prior to testing, adjusting or repairing many of the systems or components on the engine. This prevents damage to test equipment, ensures accurate testing or adjustment, and ensures safety. Always disconnect or connect the battery cables in the appropriate order.

1. To disconnect the battery cables, disconnect the negative then the positive cable.
2. To connect the battery cables, connect the positive then negative cable.

Electrostatic Discharge Damage

Sliding across the vinyl-covered upholstery used in most boats can generate as much as 25,000 volts of static electricity in the human body. Electrostatic discharge occurs if an electrostatic charged individual (or object) contacts a non-charged surface. It takes a minimum of 4000 volts for the average person to even feel an electrostatic discharge. The engine control unit and many of the sensors can be damaged if subjected to an electrostatic discharge of as little as 100 volts.

Automotive technicians often use a special grounding strap attached to the wrist to prevent the buildup of an electrostatic charge. Grounding straps are available from automotive parts stores and most tool suppliers. If a ground strap is not available, first verify that no flammable gas or liquid is present in the work area then momentarily touch a known engine ground. Touch the ground prior to disconnecting, testing or connecting any wire harness or electrical component. Electrostatic discharge usually occurs only in very dry conditions. Electrostatic discharge seldom occurs under humid conditions.

WARNING
Arcing produced by an electrostatic discharge can ignite flammable gas or liquids resulting in a fire or explosion. Never allow an electrostatic discharge near fuel or flammable material.

Preparation for Disassembly

Repairs go much faster if the equipment is clean. There are special cleaners, such as Gunk or Bel-Ray Degreaser, for washing engine-related and non-electrical components. Spray or brush on the cleaning solution, let it stand, then rinse it off with a garden hose. Clean oily or greasy parts with a cleaning solvent after removal.

Use pressurized water to remove marine growth, corrosion and mineral deposits from external components, such as the gearcase, drive shaft housing and clamp brackets. Avoid directing pressurized water directly at seals or gaskets. Pressurized water can flow past seal and gasket surfaces and contaminate lubricating fluids.

WARNING
Never use gasoline as a cleaning agent. It presents an extreme fire hazard. Always work in a well-ventilated area if using cleaning solvent. Keep a Coast Guard approved fire extinguisher, rated for gasoline fires, in the work area.

Removal and disassembly to access defective parts is much of the cost of taking the vessel to a dealership. Frequently, most of the disassembly can be performed by the owner and the defective part or assembly can be taken to the dealership for repair.

Before beginning repair work, read the appropriate procedure in this manual. Study the illustrations and text to fully understand what is involved to complete the repair. Make arrangements to purchase or rent all required special tools and equipment before starting the repair.

Disassembly Precautions

During disassembly, keep a few general precautions in mind. Force is rarely needed to separate parts. If parts fit tightly, such as a bearing on a shaft, there is usually a tool designed to separate them. Never use a screwdriver to separate parts with machined mating surfaces, such as cylinder heads and manifolds. The surfaces will be damaged and could cause a leak.

Make diagrams or take photographs wherever similar-appearing parts are found. For example, cylinder head bolts are often different lengths. Disassembled parts can be left for several days or longer before work is resumed. Carefully arranged parts may get disturbed.

Cover all openings after removing parts to keep dirt, insects or other parts from entering.

Tag all similar internal parts for location and mounting direction. Reinstall all internal components in the same locations and mounting directions as removed. Record the thickness and locations of shims as removed. Place small bolts and parts in plastic sandwich bags. Seal and label the bags with masking tape.

Tag all wires, hoses and connection points, and make a sketch of the routing. Never rely on memory alone; it may be several days or longer before work is resumed.

Protect all painted surfaces from physical damage. Never allow gasoline or cleaning solvent on these surfaces.

Assembly Precautions

No parts, except those assembled with a press fit, require unusual force during assembly. If a part is hard to remove or install, find out why before proceeding.

When assembling parts, start all fasteners, then tighten them evenly in an alternating or crossing pattern unless a specific tightening sequence or procedure is given.

When assembling parts, install all shims, spacers and washers in the same position and location as removed.

Whenever a rotating part butts against a stationary part, look for a shim spacer or washer. Use new gaskets, seals and O-rings if there is any doubt about the condition of the used ones. Unless otherwise specified, a thin coat of oil on gaskets may help them seal more effectively. Use heavy grease to hold small parts in place if they tend to fall out during assembly.

Use emery cloth and oil to remove high spots from piston surfaces. Use a dull scraper to remove carbon deposits from the cylinder head, ports and piston crown. *Do not* scratch or gouge these surfaces. Wipe the surfaces clean with a *clean* shop towel when finished.

If the carburetor must be repaired, completely disassemble it and soak all metal parts in a commercial carburetor cleaner. Never soak gaskets and rubber or plastic parts in these cleaners. Clean rubber or plastic parts in warm soapy water. Never use a wire to clean out jets and small passages; they are easily damaged. Use compressed air to blow debris from all passages within the carburetor body.

Take the time to do the job right. The break-in procedure for a newly rebuilt engine or drive is the same as a new engine. Use the recommended break-in oil and follow the instructions provided in the appropriate chapter.

SPECIAL TIPS

Because of the extreme demands placed on marine equipment, keep several points in mind when performing service and repairs. Following these general suggestions can improve the overall life of the machine and help avoid costly problems.

1. Unless otherwise specified, apply a locking compound, such as Locktite Threadlocker, to all bolts and nuts even if they are secured with a lockwasher. Use only the specified grade of threadlocking compound. A screw or bolt lost from an engine cover or bearing retainer could easily cause serious and expensive damage before the loss is noticed. When applying threadlocking compound, use only enough to lightly coat the threads. If too much is used, it can work its way down the threads and affect parts not meant to be stuck together.
2. Be careful when using air tools to remove stainless steel nuts or bolts. The threads of stainless steel fasteners are easily damaged by the heat generated if spun rapidly. To prevent thread damage, apply a penetrating oil as a cooling agent and loosen or tighten them slowly.
3. When straightening the tab of a fold-over lockwasher, use a wide-blade chisel, such as an old and dull wood chisel. This chisel provides a better contact surface than a screwdriver or pry bar, making straightening easier. During installation, use a new fold-over lockwasher. If a new washer is not available, fold over a tab on the washer that was not previously used. Reusing the same tab may cause the washer to break, resulting in a loss of locking ability and a loose piece of metal in the engine. When folding the tab into position, carefully pry it toward the flat on the bolt or nut. Use pliers to bend the tab against the fastener. Do not use a punch and hammer to drive the tab into position. The resulting fold may be too sharp, weakening the washer and increasing its chance of failure.
4. Use only authorized replacement parts when replacing missing or damaged bolts, screws or nuts. Many fasteners are specially hardened for the application. The wrong bolt can easily cause serious and expensive damage.
5. Install only authorized gaskets. Unless specified otherwise, install them without sealant. Many gaskets are made with a material that swells when it contacts oil. Gasket sealer prevents them from swelling as intended and can result in oil leakage. Authorized gaskets are cut from material of a precise thickness. Installation of a too thick or too thin gasket in a critical area could cause expensive damage.

MECHANICS TECHNIQUES

Marine engines are subjected to conditions very different from most engines. They are repeatedly subjected to a corrosive environment followed by long periods of non-use. This increases corrosion damage to fasteners, causing difficulty or breakage during removal. This sec-

tion provides information for removing stuck or broken fasteners and repairing damaged threads.

Removing Stuck Fasteners

When a nut or bolt corrodes and cannot be removed, several methods may be used to loosen it. First, apply penetrating oil, such as Liquid Wrench or WD-40 available at hardware or automotive supply stores. Apply it liberally to the threads and allow it to penetrate for 10-15 minutes. Tap the fastener several times with a small hammer; do not hit it hard enough to cause damage. Reapply the penetrating oil if necessary.

For stuck screws, apply penetrating oil as described, then insert a screwdriver in the slot. Tap the top of the screwdriver with a hammer. This loosens the corrosion in the threads. If the screw head is too damaged to use a screwdriver, grip the head with locking pliers to loosen it.

A Phillips, Allen or Torx screwdriver may start to slip in the screw during removal. If this occurs, stop immediately and apply a dab of valve lapping compound to the tip of the screwdriver. Valve lapping compound or a special screw removal compound (**Figure 31**) is available from most hardware and automotive parts stores. Insert the driver into the screw and apply downward pressure while turning. The gritty material in the compound improves the grip to the screw, allowing more force to be applied before it will slip. Keep the compound away from other engine components. It is very abrasive and can cause rapid wear if applied to moving or sliding surfaces.

Avoid applying heat unless specifically instructed. Heat can melt, warp or remove the temper from parts.

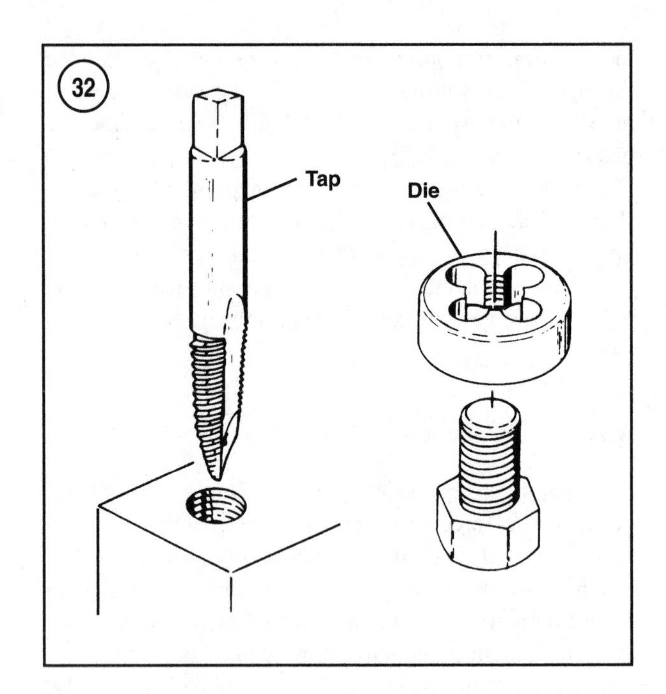

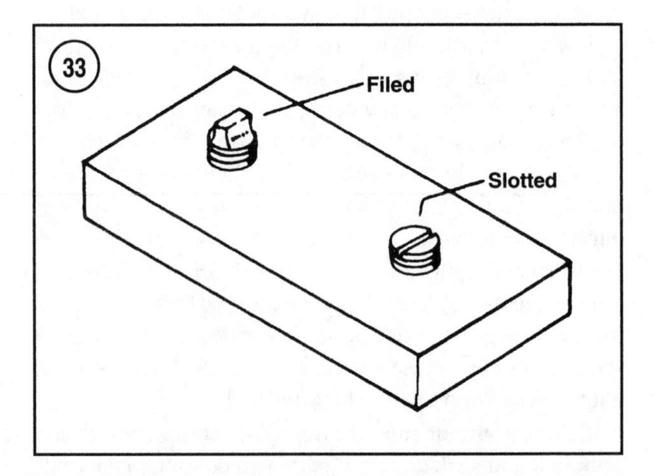

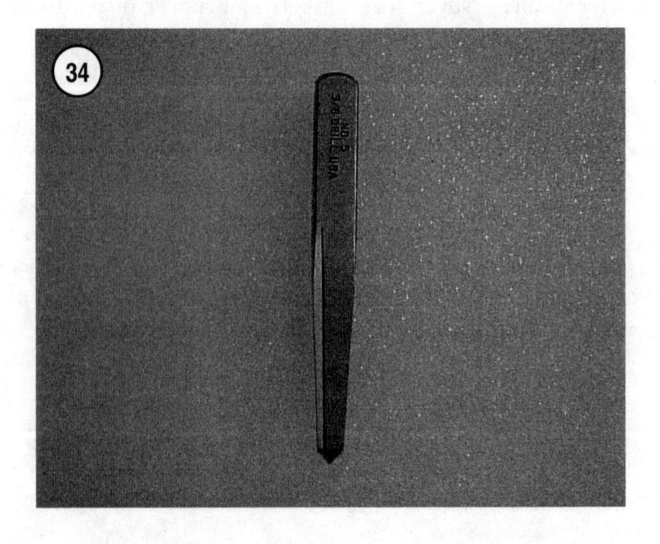

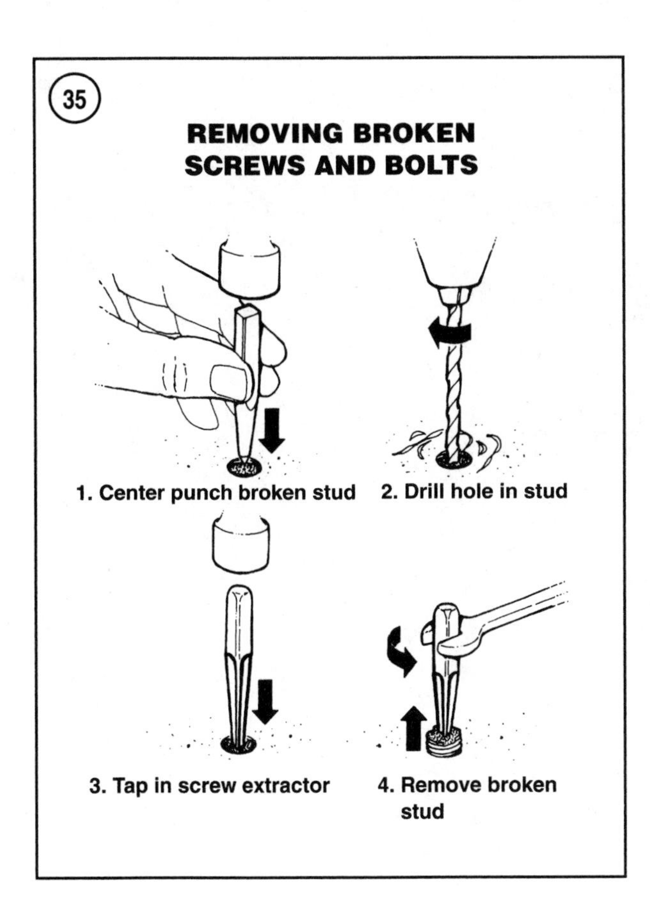

Stripped Threads Repair

Occasionally, threads are stripped through carelessness or impact damage. Often the threads can be repaired by using a tap (for internal threads) or die (for external threads) (**Figure 32**).

Damaged threads in a housing or component can often be repaired by installing a threaded insert.

Removing Broken Bolts or Screws

The head of a bolt or screw may unexpectedly twist off during removal. Several methods are available for removing the remaining portion of the bolt or screw.

If a large portion of the bolt or screw projects out, try gripping it with locking pliers. If the projecting portion is too small, file it to fit a wrench or cut a slot in it to fit a screwdriver (**Figure 33**). If the head breaks off flush or cannot be turned with a screwdriver or wrench, use a screw extractor (**Figure 34**). To do this, center punch the remaining portion of the screw or bolt (**Figure 35**). Select the proper size extractor for the fastener. Using the drill size specified on the extractor, drill a hole into the fastener. Do not drill deeper than the remaining fastener. Carefully tap the extractor into the hole (**Figure 35**). Back the remnant out with a wrench on the extractor.

Chapter Three

Troubleshooting

This chapter contains troubleshooting procedures. Once the defective component is identified, refer to the appropriate chapter for component removal and installation procedures.

Table 1 lists standard wire color codes. **Table 2** list battery cable recommendations. **Tables 3-6** cover typical symptoms and solutions for the starting, charging, ignition and fuel systems. **Tables 7-13** list specifications for the starting, charging and ignition systems. **Table 14** covers ignition system identification. **Tables 1-14** are located at the end of this chapter.

Troubleshooting is testing individual systems to quickly isolate good systems from the defective or inoperative system(s). When a system is identified as defective, troubleshooting continues with testing of the individual components in that system. Perform one test procedure at a time to determine the condition of each component. Occasionally a component in a system cannot be isolated for testing. In this case, other components are tested and eliminated until the suspect component is identified as defective by the process of elimination.

When troubleshooting, always test systems before components and be methodical. Haphazardly jumping from one system or component to another may eventually solve the problem, but will waste time and effort, and cause good parts to be replaced needlessly.

Use the various system diagrams provided in this manual to identify all components in a system. Test each component in a rational order to determine which component has caused the system to fail.

After noticing a symptom, such as a noticeable decrease in performance or unsatisfactory operating characteristic, consider the following questions:

1. Did the problem occur all at once (suddenly) or was its onset gradual?
2. Is there a specific rpm or load at which the problem occurs?
3. Does the weather (extreme cold, heat or dampness) affect the symptom?
4. Has any recent service work been performed by anyone else?
5. Has the engine recently come out of storage?
6. Is the engine using a different brand or grade of fuel?
7. Is the manufacturer's recommended oil being used?
8. Have any accessories been added to the boat or motor recently?

Some problems can be caused by simple mistakes, such as failing to prime the fuel system or attach the safety lanyard, and incorrect starting procedures.

Before beginning a troubleshooting procedure, perform a thorough visual inspection of the unit. Check the condition of the battery cable connections at both ends, all electrical harness connectors and terminals, fuel quantity, quality and supply, indications of engine overheating, evidence of leaks (fuel, oil and water), and mechanical integrity (loose fasteners, cracked or broken castings).

Be realistic about the level of skill required to complete a procedure. Service departments tend to charge heavily to reassemble an engine which was dissassembled by someone else; some refuse to take on such a job.

Performing lubrication, maintenance and engine tune-up procedures as described in Chapter Four reduces the need for troubleshooting. However, because of the harsh and demanding environment in which the outboard motor operates, troubleshooting at some point in the motors serviceable life is inevitable.

SAFETY PRECAUTIONS

Wear approved eye protection at all times (**Figure 1**), especially when machinery is in operation and hammers are being used. Wear approved ear protection during all running tests. Keep loose clothing tucked in and long hair tied back. Refer to *Safety First* in Chapter Two for additional safety guidelines.

When making or breaking an electrical connection, always disconnect the negative battery cable first. When performing tests that require cranking the engine without starting, disconnect and ground all spark plug leads to prevent accidental starts and sparks.

Securely cap or plug all disconnected fuel lines to prevent fuel discharge when the motor is cranked or the primer bulb is squeezed.

Thoroughly read all manufacturer's instructions and safety sheets for test equipment and special tools being used.

Do not substitute parts unless they meet or exceed the original manufacturer's specifications.

Never run an outboard motor without an adequate water supply. Never run an outboard motor at wide-open throttle without an adequate load. Do not exceed 3000 rpm in *neutral* (with no load).

Safely performing on-water tests requires two people. One person to operate the boat, the other to monitor the gauges or test instruments. All personnel must remain seated inside the boat at all times. Do not lean over the transom while the boat is under way. Use test wire extensions to allow all gauges and meters to be located in the normal seating area.

A test propeller is an economical alternative to the dynometer. A test propeller is also a convenient alternative to on-water testing. Test propellers are made by turning down the diameter of a standard low pitch aluminum propeller until the recommended wide-open throttle engine speed can be obtained with the motor in a test tank or backed into the water on the trailer. Be careful when tying the boat to a dock as considerable thrust is developed by the test propeller. Some docks may not be able to withstand the load.

Propeller repair stations can provide this modification service. Normally, approximately 1/3 to 1/2 of the blades are removed. However, it is better to remove too little than too much. It may take several tries to achieve the correct full throttle speed, but once it is achieved, no further modifications is required. Many propeller repair stations have experience with this type of modification and may be able to recommend a starting point.

Test propellers also allow simple tracking of engine performance. The full throttle test speed of an engine fitted with a correctly modified test propeller can be tracked from season to season. It is not unusual for a new or rebuilt engine to show a slight increase in test propeller engine speed as complete break-in is achieved. The engine generally holds this speed over the normal service life of the engine.

Starting Difficulties

Occasionally an outboard motor is plagued by hard starting and generally poor performance (especially at low speeds) for which there seems to be no identifiable cause. Fuel and ignition systems test satisfactorily and a compression test indicates that the combustion chamber (piston, rings, cylinder walls and head gasket) is in good condition.

The next step is to test the crankcase sealing. A two-cycle engine cannot function unless the crankcase is adequately sealed. As the piston travels downward, the crankcase must pressurize and push the air/fuel mixture into the combustion chamber as the intake ports are uncovered. As the piston travels upward, the crankcase must create a vacuum to pull the air/fuel mixture into the crankcase from the carburetor in preparation for the next cycle. Refer to Chapter Two for operational diagrams of a typical two-cycle engine.

Leaks in the crankcase cause the air/fuel charge to leak into the atmosphere under crankcase compression. During the intake cycle, crankcase leaks will cause air from the atmosphere to be drawn into the crankcase, diluting the air/fuel charge. The result is inadequate fuel in the combustion chamber. On multiple cylinder engines, each crankcase must be sealed from all other crankcases. Internal leaks allow the air/fuel charge to leak to a different cylinder's crankcase, rather than travel to the correct combustion chamber.

The function of the lower piston ring on most two-cycle engines is crankcase compression. It is difficult to test this ring. Compression tests typically test the upper compression ring, not the lower ring. A classic symptom of lower ring failure is the inability to idle at the recommended idle speed. The engine generally performs fine at higher speeds, but slowly dies out at idle speed.

External crankcase leakage can be identified with a visual inspection for fuel residue leaking from the crankcase parting lines, upper and lower crankshaft seals, reed valves or intake manifolds. Pressure leaking out of the crankcase can be identified by applying a soap and water solution to the surfaces. Air leaking into the crankcase can be found by applying oil to the suspect sealing area. The oil is drawn into the crankcase at the point of the leak.

Internal leakage is difficult to identify. If there are fittings on each crankcase for fuel pumps, primers or recirculation systems, a fuel pressure/vacuum gauge can be attached. As the engine is cranked over, a repeating pressure/vacuum cycle should be observed on the gauge. The pressure reading should be substantially higher than the vacuum reading. All cylinders should read basically the same. If this test is not possible, test the fuel and ignition systems. As a final resort, disassemble and inspect the power head.

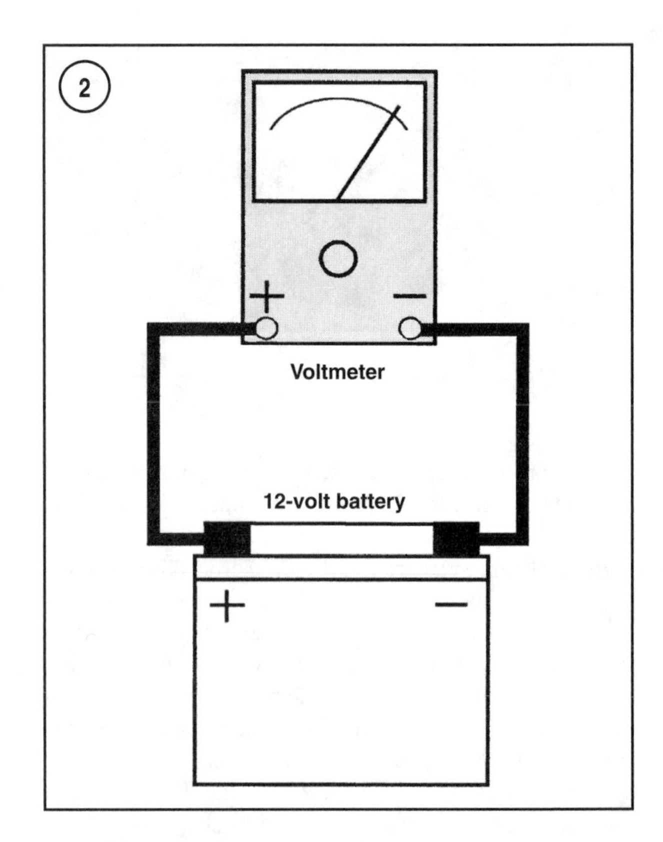

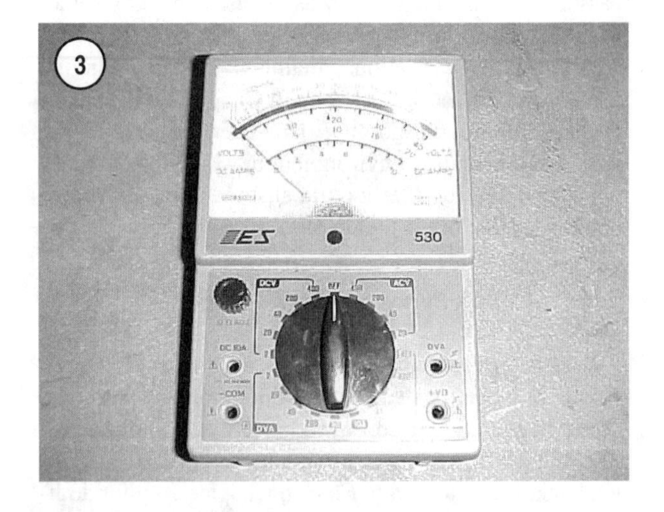

TERMINOLOGY AND TEST EQUIPMENT

Voltage

Voltage is the pressure in an electrical circuit. The more pressure, the more work that can be done. Voltage can be visualized as water pressure in a garden hose. The more pressure, the further the water can be sprayed. You can have water present in the hose, but without pressure, you cannot accomplish anything. If the water pressure is too high, the hose will burst. When voltage is excessive it will leak past insulation and arc to ground. Voltage is always measured with a voltmeter in a simple parallel connection. The connection of a voltmeter directly to the negative and positive terminals of a battery is an example of a parallel connection (**Figure 2**). Nothing has to be disconnected to make a parallel connection. A voltmeter is an electrical pressure gauge that taps into the electrical circuit.

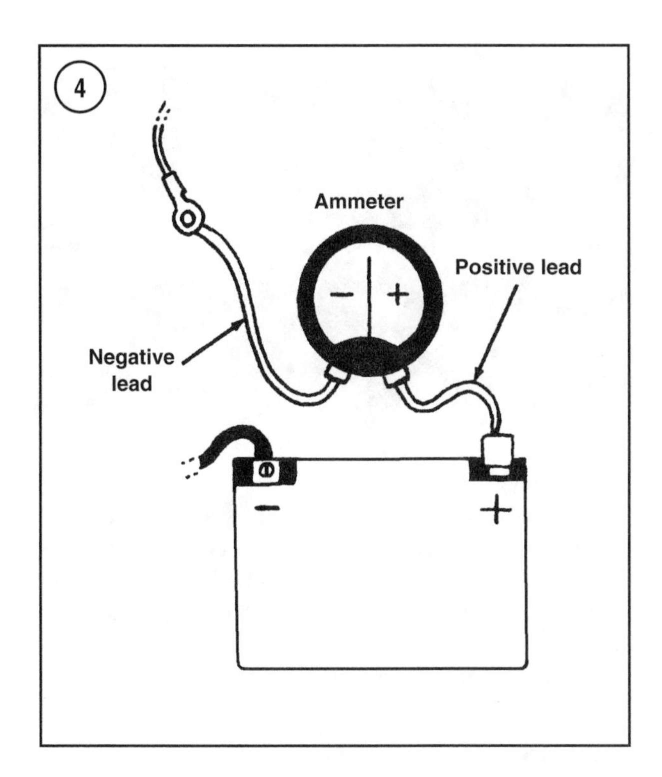

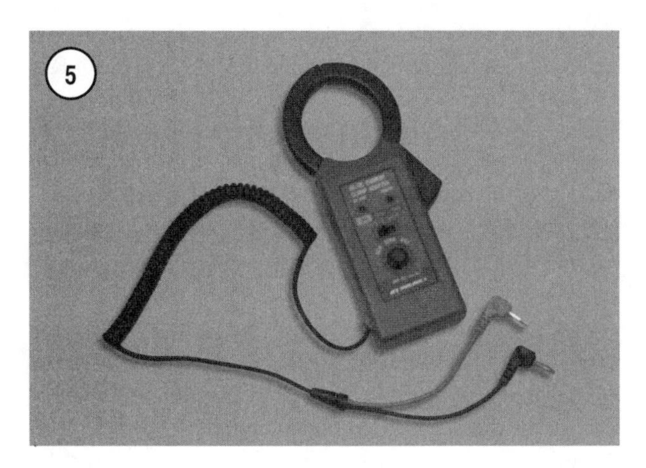

DC voltage

DC voltage is direct current voltage. The electricity always flows in one direction. All circuits associated with the battery are DC circuits.

AC voltage

AC voltage is alternating current. The current flows in one direction momentarily, then switches to the opposite direction. The frequency with which AC voltage changes direction is hertz or cycles per second. Household wiring is 110 volts AC and typically 60 hertz (the average value of electrical pressure is 110 volts and the electricity changes direction 60 times per second). In typical outboard motors, the charging system's stator output is AC voltage. In larger, inboard motors, AC voltage is typically created by a dedicated AC generator (genset) that powers high load devices such as air-conditioning and appliances. Shore power is also AC voltage. Standard AC voltmeters take an average reading of the fluctuating voltage signal. RMS (root mean square) AC voltmeters use a different mathematical formula to come up with a value of the voltage signal. RMS meters should only be used where specified, since the difference in readings between a standard AC meter and a RMS AC meter is significant.

DVA voltage

DVA stands for direct voltage adapter. This measures the AC voltage at the absolute peak or highest value of the fluctuating AC voltage signal. Peak readings are substantially higher than standard or RMS AC values and are typically used when testing marine CD (capacitor discharge) ignition systems. Failure to use a meter with a DVA scale can cause good ignition components to be incorrectly diagnosed as faulty. See **Figure 3** for a typical multimeter with a DVA scale.

Amperes

Amperes (amps) are current. Current is the actual flow of electricity in a circuit. Current can be visualized as water flowing from a garden hose. There can be pressure in the hose, but if we do not let it flow, no work can be done. The higher the flow of current, the more work that can be done. However, when too much current flows through a wire, the wire will overheat and melt. Melted wires are caused by excessive current, not excessive volts. Amps are measured with an ammeter in a simple series connection. A circuit must be disconnected and the ammeter spliced into the circuit. An ammeter must have all of the current flow through it. Always use an ammeter that can read higher than the anticipated current flow. Always connect the positive lead of the ammeter to where the electricity is coming from (electrical source) and the negative lead of the ammeter to where the electricity is going (electrical load). See **Figure 4**.

Many digital multimeters can use inductive or clamp-on ammeter probes (**Figure 5**). These probes read the magnetic field strength created by current flowing through a wire. No electrical connection is required, simply slip the probe over the lead.

A simple ammeter is the direct reading inductive ammeter (**Figure 6**). These meters directly read the magnetic field strength created from current flowing through a wire. No electrical connection is required. Slip the meter over the lead so the lead is in the channel or groove on the rear of the meter. See **Figure 6**.

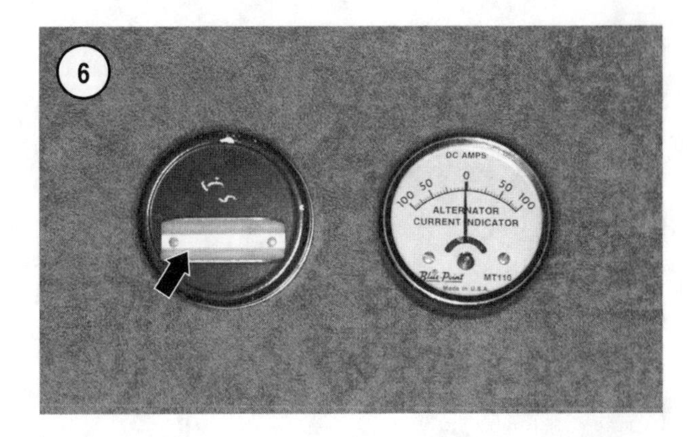

Watts

Watts (W) is the measurement unit for power in an electrical circuit. Watts rate the ability to do electrical work. The easiest formula for calculating watts is to multiply the system voltage times the amps flowing. A 12-volt system multiplied by a 10 amp alternator equals a 120 watt maximum load. Amp load can be reversed by dividing watts by voltage. A 12 watt radio divided by a 12 volt system uses 1 amp of current. When calculating load on a charging system, remember the system cannot carry more load than it is rated for or the battery will constantly discharge.

Ohms

An ohm is the measurement unit for resistance in an electrical circuit. Resistance causes a reduction in current flow and a reduction in voltage. Visualized as a kink in a garden hose, which would cause less water (current) to flow, it would also cause less pressure (volts) to be available downstream from the kink. Ohms are measured with self-powered ohmmeters. Ohmmeters send a small amount of electricity into a circuit and measure how hard they have to push to return the electricity to the meter. An ohmmeter must only be used on a circuit or component that is disconnected from any other circuit or component and has no voltage present. Ohmmeters are technically connected in series.

Voltage drop test

Since resistance causes voltage to drop, resistance can be measured on an active circuit with a voltmeter. A voltage drop test measures the difference in voltage from the beginning of the tested circuit to the end of the tested circuit while the circuit is operated. If the circuit has no resistance, there is no voltage drop and the meter will read zero volts. The more resistance the circuit has, the higher the voltmeter reading. Generally, voltage drop readings of one or more volts are considered unsatisfactory. The chief advantage to the voltage drop test over an ohmmeter resistance test is that the circuit is tested while under operation. A zero reading on a voltage drop test is good, while a battery voltage reading would signify an open circuit.

The voltage drop test provides an excellent means of testing solenoids relays, battery cables and high current positive and negative electrical leads. As with the ammeter, always connect the positive lead of the voltmeter to where the electricity is coming from (electrical source) and the negative lead of the voltmeter to where the electricity is going (electrical load).

Multipliers

When using an analog multimeter to measure ohms, the scale choices will typically be labeled R × 1, R × 10, R × 100 and R × 1k. These are resistance scale multipliers. R × 100 means to multiply the meter reading by 100. If the needle indicated a reading of 75 ohms while set to the R × 100 scale, the actual resistance is 75 × 100 or 7500 ohms. Note the scale multiplier when using an analog ohmmeter.

Diodes

Diodes are one-way electrical check valves. A series of diodes used to change AC current to DC current is a rectifier. Single diodes used to prevent reverse flow of electricity are typically called blocking diodes. Diodes can be tested with an analog meter set to any ohmmeter scale other than low or with a digital multimeter set to the diode test scale. A diode tested with an analog ohmmeter indicates a relatively low reading in one polarity and a relatively high reading in the opposite polarity. A diode tested with a digital multimeter reads a voltage drop of approximately 0.4-0.9 volts in one polarity and an open circuit (OL or OUCH) in the opposite polarity.

Analog Multimeter

When using an analog meter to read ohms, recalibrate (zero) the meter each time the scale or range is changed. Normally the ohmmeter leads are connected for calibra-

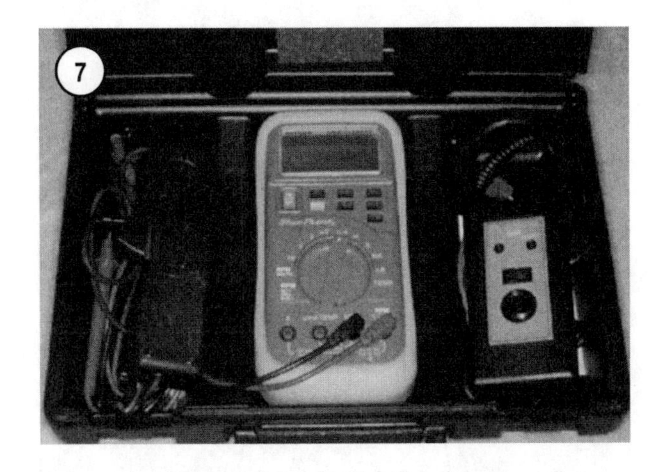

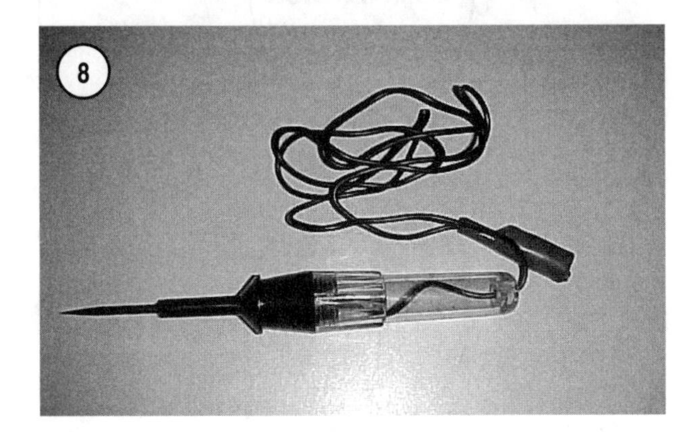

tion; however, some meters require the leads to be separated for calibration when using the low ohms scale. Always follow the manufacturer's instructions for calibration. When checking for shorts to ground, calibrate on the highest scale available. When checking diodes, calibrate on the R × 10 scale or higher. Never use the *low* scale if so equipped to test a diode or short to ground. When checking for a specific ohm value, calibrate the ohmmeter on a scale that allows reading the specification as near the middle of the meter movement as possible. Analog meters allow easy visual identification of erratic or fluctuating readings.

Digital Multimeter

Digital multimeter displays are easy to read. Most digital meters are autoranging. They automatically shift to the scale most appropriate for displaying the value being read. Be careful to read the scale correctly. Fluctuating readings can be frustrating to read as the display changes several times a second. Quality digital multimeters typically have a bar graph located below the digital number display. The bar graph allows easy interpretation of fluctuating readings. The scale range and multiplier (if applicable) is displayed alongside the actual reading. Most quality digital meters have a special diode test scale that measures the voltage drop of the diode, instead of the resistance. Do not attempt to use the digital multimeter's ohms scale to test diodes, as the readings are inconsistent. The digital multimeter is protected by internal fuses.

Adapters are available for temperature readings, inductive ammeter readings and many other functions. **Figure 7** shows a digital multimeter in a protective case with several adapters.

Test Light

The test light (**Figure 8**) is a useful tool for troubleshooting, such as starter circuits. A test light should not be used on the ignition system. The current draw of the test light can damage delicate electronic circuits. Do not use a test light in tests where specific voltage values are needed. Before beginning troubleshooting with a test light, connect the test light directly to the battery and note the bulb brightness. Reference other readings against this test. If the bulb is not as bright as when attached to the battery, there is a problem.

Connect the test light lead directly to the positive battery terminal to check ground circuits. When the test light probe is connected to any ground circuit, the light should glow brightly.

Electrical Repairs

Check all electrical connections for corrosion, mechanical damage, heat damage and loose connections. Clean and repair all connections as necessary. All wire splices or connector repairs must be made with waterproof marine grade connectors and heat shrink tubing. A Quicksilver electrical hardware repair kit (part No. 86-813937A 1) and crimping plier (part No. 91-808696) are available to repair the serviceable connectors and make wire splices on the engine. Heat shrink connectors and heat shrink tubing for making other waterproof connections and repairs are also available from Quicksilver. Marine and industrial suppliers are also good sources for electrical repair equipment.

OPERATING REQUIREMENTS

All two-stroke engines require three basic conditions to run properly: correct air and fuel mixture from the carburetor, crankcase and combustion chamber compression, and adequate spark delivered to the spark plug at the correct time. When troubleshooting, it is helpful to remember: fuel, compression and spark (**Figure 9**). If any of these are lacking, the motor will not run. First, verify the mechanical integrity of the engine by performing a compression test (Chapter Four). Once compression has been verified, test the ignition system with an air gap spark tester and finally check on the fuel system. Troubleshooting in this order provides the quickest results.

If the motor has been sitting for any length of time and refuses to start, check the condition of the battery to make sure it is adequately charged, then inspect the battery cable connections at the battery and the engine. Examine the fuel delivery system. This includes the fuel tank, fuel pump, fuel lines, fuel filters and carburetor(s). Rust or corrosion may have formed in the tank, restricting fuel flow. Gasoline deposits may have gummed up carburetor jets and air passages. Gasoline may have lost its potency after standing for long periods. Condensation may have contaminated the fuel with water. Connect a portable tank containing fresh fuel mix to help isolate the problem. Do not drain the old gasoline unless you are sure it is at fault. Always dispose of old gasoline in accordance with EPA regulations.

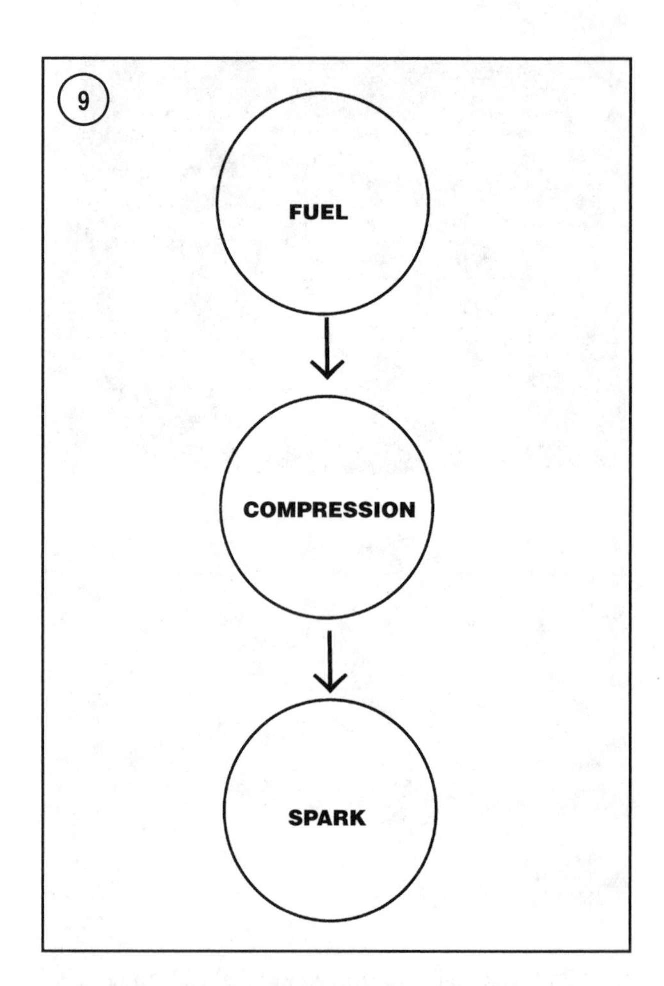

STARTING SYSTEM

Mercury/Mariner 6 hp and larger models may be equipped with electric start systems. The starter motor is mounted vertically on the engine. When battery current is supplied to the starter motor, its pinion gear (**Figure 10**, typical) is thrust upward to engage the teeth on the engine flywheel. Once the engine starts, the pinion gear disengages from the flywheel.

The starting system requires a fully charged battery to provide the large amount of electrical current necessary to operate the starter motor. Electric start models are equipped with an alternator to charge the battery during operation.

The electric starting system consists of the battery, starter switch, neutral safety switch, starter solenoid, starter motor, starter drive and related wiring. See **Figure 11**. The neutral safety switch allows starter engagement only when the gearshift is in NEUTRAL. Tiller handle models have the neutral safety switch mounted on the engine. Remote control models have the neutral safety switch mounted in the remote control box. Remote control models incorporate a 20 amp fuse to protect the remote control key switch circuits. The fuse is located on the engine between the starter solenoid and the boat main harness connector.

The neutral safety switch is on the positive side of the solenoid on remote control models and the negative side of the solenoid on tiller handle models. Engaging the starter switch allows current to flow to the starter solenoid coil windings. When the current path is complete, the solenoid contacts close, allowing current to flow from the battery through the solenoid to the starter motor.

Solenoid design varies, but all solenoids use two large terminal studs (battery positive and starter cables) and two small terminal studs (black and yellow/red primary wires).

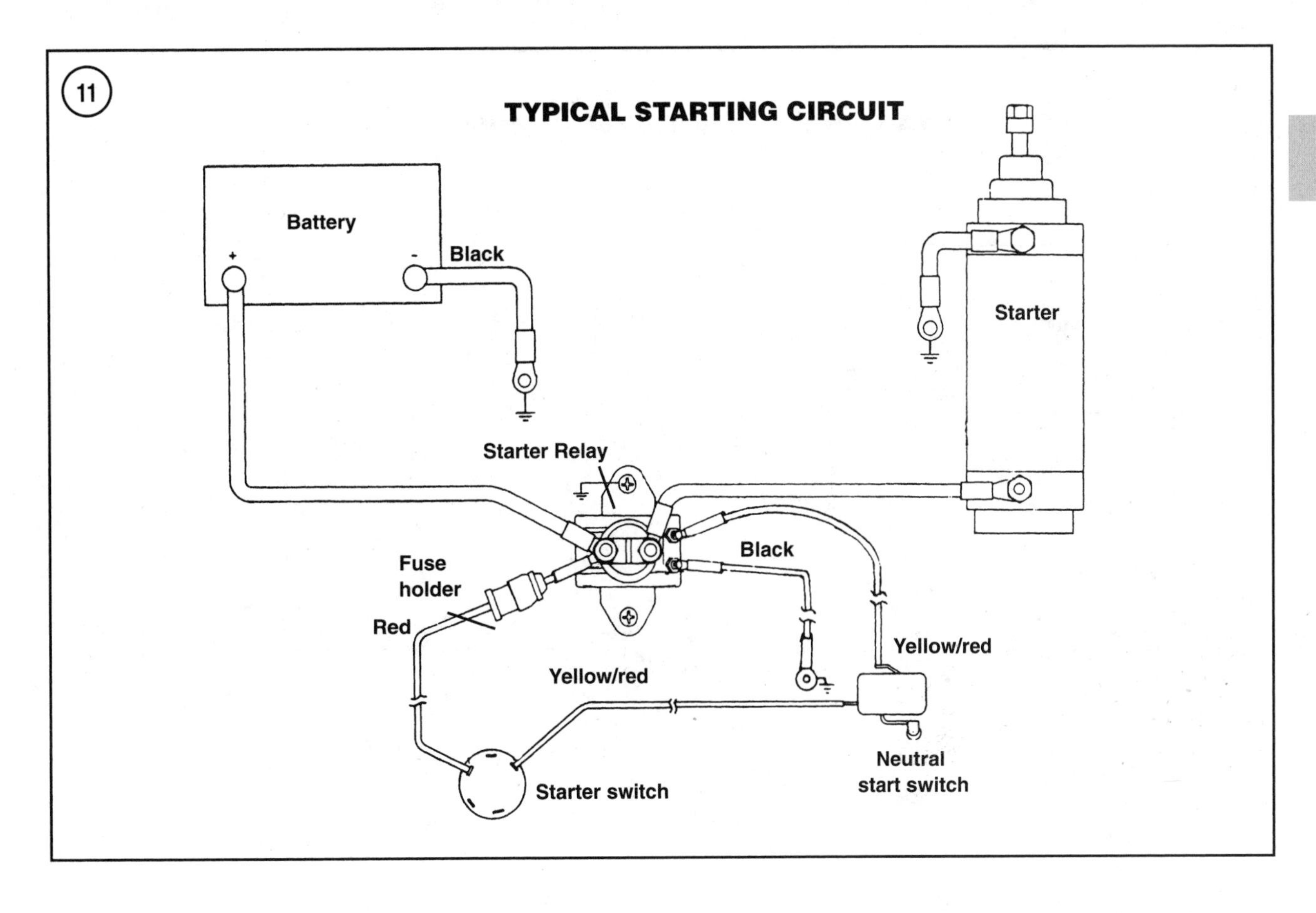

If the following procedures do not locate the problem, refer to **Table 3** for additional information. Before troubleshooting the starting circuit, check for the following:

1. The battery is fully charged.
2. The shift control lever is in NEUTRAL.
3. All electrical connections are clean and tight.
4. The wiring harness is in good condition with no worn or frayed insulation.
5. The fuse protecting the starter switch is not blown on all remote control models and 30 hp-up tiller models.
6. The power head and gearcase are not faulty.

CAUTION

To prevent starter damage, never operate the starter motor continuously for more than ten seconds. Allow the starter motor to cool for at least two minutes between attempts to start the engine.

NOTE

Unless otherwise noted, perform all tests with the terminals connected.

NOTE

The cable connecting the battery to the starter solenoid is red or black with a red sleeve. The cable connecting the starter solenoid to the starter motor is yellow, black with yellow sleeved ends or black with red sleeved ends.

Starter Motor Turns Slowly

1. Make sure the battery is in acceptable condition and fully charged.
2. Inspect all electrical connections for looseness or corrosion. Clean and tighten them as necessary.
3. Check for the proper size and length of battery cables. Refer to **Table 2** for recommended minimum cable gauge sizes and lengths. Replace cables that are too short or move the battery to shorten the distance between the battery and starter solenoid.
4. Disconnect and ground the spark plug leads to the engine to prevent accidental starting. Turn the flywheel clockwise by hand and check for mechanical binding. If mechanical binding is evident, remove the lower gearcase to determine if the binding is in the power head or the lower gearcase. If no binding is evident, other than normal compression and water pump impeller drag, continue to Step 5.

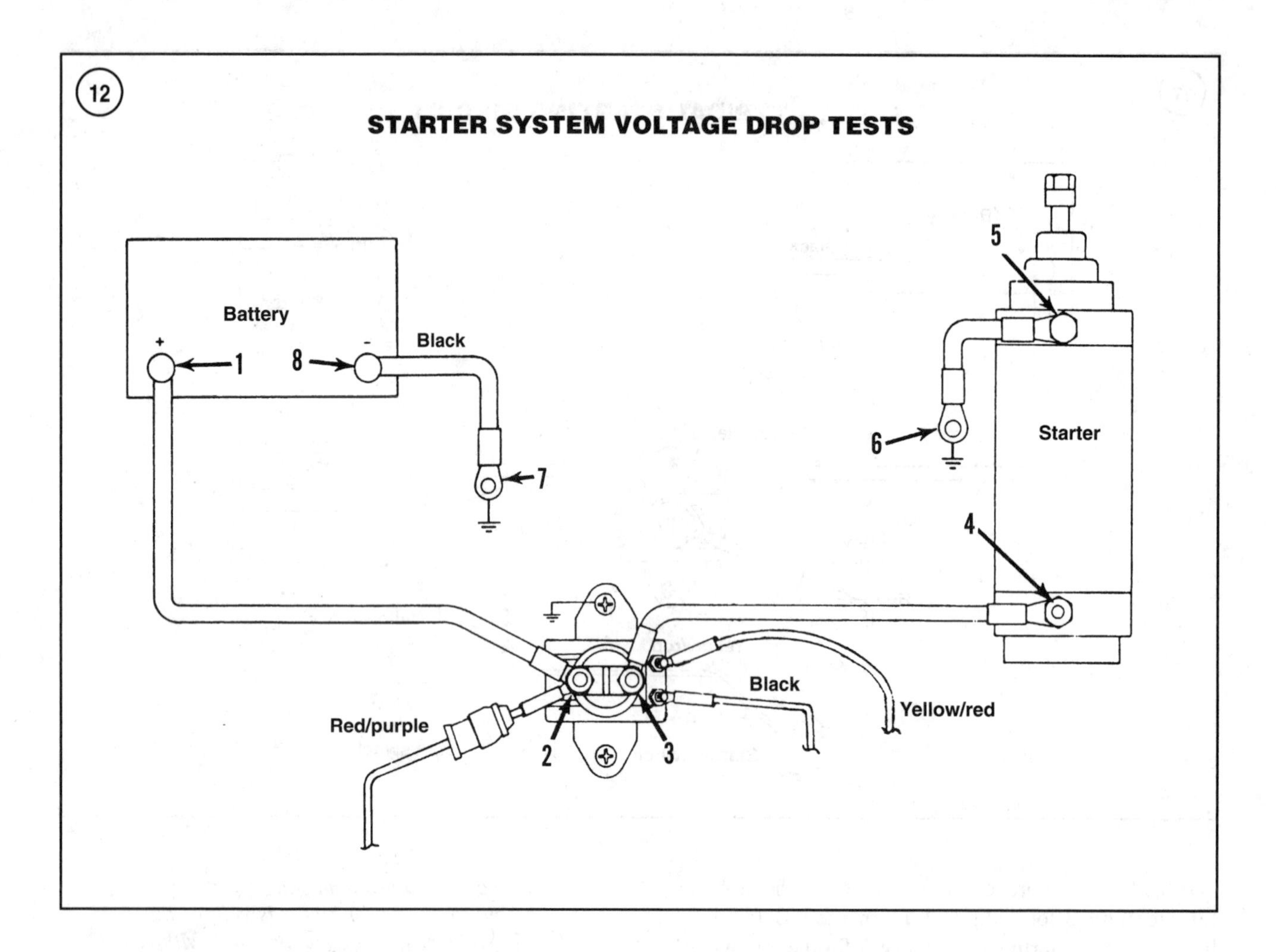

5. Perform the starting system voltage drop test as described in the next section.
6. Check the starter motor current draw as described in this chapter.

Starting system voltage drop test

As described in this chapter, resistance causes a reduction in current flow and causes voltage to drop. Excessive resistance in the battery cables, starter solenoid and starter cable can restrict the current flow to the starter, causing the starter to turn the motor slowly. Slow cranking speeds cause low ignition system output and hard starting.

Use the following procedure to determine if any of the cables or the starter solenoid is the source of a voltage drop causing slow cranking speeds. If the problem is intermittent, try gently pulling, bending and flexing the cables and connections during the test. Sudden voltmeter fluctuations indicate a poor connection has been located.

A voltage drop test measures the difference in voltage from the beginning of a circuit or component to the end of the circuit or component. If there is resistance in the circuit, the voltage at the end of the circuit is less than the voltage at the beginning. The starter must be engaged for a voltage drop reading. A voltmeter reading of 0 (zero) means that there is no resistance in the test circuit. A reading of battery voltage means that the circuit is completely open.

Refer to **Figure 12** for this procedure. Clean, tighten, repair or replace cables or solenoids with excessive voltage drop.

1. Disconnect and ground the spark plug wires to the engine to prevent accidental starting.
2. Connect the positive voltmeter wire to the positive battery terminal (1, **Figure 12**). Connect the negative voltmeter wire to the positive solenoid terminal (2, **Figure 12**).
3. Engage the electric starter and observe the meter. If the meter indicates more than 0.3 volts, resistance in the positive battery cable is excessive. Clean the connections, repair the terminal ends or replace the positive battery cable.

CAUTION
Do not connect the positive voltmeter test wire in Step 3-6 until after the engine begins cranking. The open solenoid provides battery voltage that can damage a voltmeter set to a low volts scale. In addition, disconnect the voltmeter test wires before stopping cranking.

4. Connect the negative voltmeter wire to the starter side of the solenoid (3, **Figure 12**). Engage the electric starter. While the engine is cranking, touch the positive voltmeter wire to the battery cable positive solenoid terminal (2, **Figure 12**). Note the meter reading, remove the positive voltmeter wire and stop cranking. If the meter indicated more than 0.2 volts, the starter solenoid has excessive internal resistance and must be replaced.
5. Connect the positive voltmeter wire to the starter side of the solenoid (3, **Figure 12**) and the negative voltmeter wire to the starter motor terminal (4). Engage the electric starter and observe the meter. If the meter indicates more than 0.2 volts, resistance in the starter motor cable is excessive. Clean the connections, repair the terminal ends or replace the starter motor cable. If the starter motor has a ground cable, repeat this test with the positive voltmeter lead hooked to the starter end of starter ground cable (5, **Figure 12**) and the negative voltmeter wire connected to the engine end of the ground cable (6).
6. Connect the positive voltmeter wire to the engine end of the negative battery cable (7, **Figure 12**) and the negative voltmeter wire to the negative battery terminal (8). Engage the electric starter and observe the meter. If the meter indicates more than 0.3 volts, resistance in the battery negative cable is excessive. Clean the connections, repair the terminal ends or replace the negative battery cable.

Starter Motor Does Not Turn

A test light or voltmeter are both acceptable tools for troubleshooting the starter circuit. If using a voltmeter, all test readings should be within 1 volt of battery voltage. Readings of 1 volt or more below battery voltage indicate problems (excessive resistance) with the circuit being tested. If using a test light, first connect the test light directly to the battery and note the bulb brightness. Reference other readings against this test. If the bulb is not as bright as when attached to the battery, the resistance is excessive.

CAUTION
Disconnect and ground the spark plug wire to the engine to prevent accidental starting during all test procedures.

Remote control models and 50-60 hp tiller handle models

Refer to **Figure 13** for this procedure. Refer to the wiring diagrams at the end of this manual.

1. Connect the test light wire to the positive terminal of the battery and touch the test light probe to metal anywhere on the engine block. If the test light does not illuminate or is dim, the battery ground cable connections are loose or corroded, or there is an open circuit in the battery ground cable. Check connections on both ends of the ground cable.
2. Place the shift lever in NEUTRAL and connect the test light wire to a good engine ground for Steps 3-12.
3. Connect the test light probe to the starter solenoid battery terminal (1, **Figure 13**). If the test light does not illuminate or is very dim, the positive battery cable connections are loose or corroded, or there is an open in the cable between the battery and the solenoid. Clean and tighten the connections or replace the battery cable as required.
4. Remove the 20 amp fuse and connect the test light probe to the battery side of the fuse holder (2, **Figure 13**). If the light does not illuminate, repair or replace the wire between the starter solenoid and the fuse holder.
5. Inspect the 20 amp fuse. Install a known good fuse into the fuse holder. Unplug the main 8-pin connector and connect the test light probe to pin No. 8 of the main engine harness connector. If the test light does not illuminate, repair or replace the wire between the fuse holder and the main engine harness connector.

6A. *Remote control models*—Reconnect the main harness connector and access the key switch on the dash or in the remote control box. Connect the test light probe to terminal B on the key switch (3, **Figure 13**). If the light does not illuminate, repair or replace the wire between pin No. 8 of the main boat harness connector and the key switch terminal B.

6B. *Tiller control models*—Reconnect the main harness connector. Connect the test light probe to terminal B on the key switch (3, **Figure 13**). If the light does not illuminate, repair or replace the wire between pin No. 8 of the main harness connector and the key switch terminal B.

7. Connect the test light probe to the key switch terminal S (4, **Figure 13**). Turn the key switch to the start position. If the test light does not illuminate, replace the key switch.

8A. *Remote control models*—Remove the cover from the remote control box and connect the test light probe to the key switch side of the neutral safety switch (5, **Figure 13**). Turn the key switch to the start position. If the test light does not illuminate, repair or replace the wire between the neutral safety switch and the key switch.

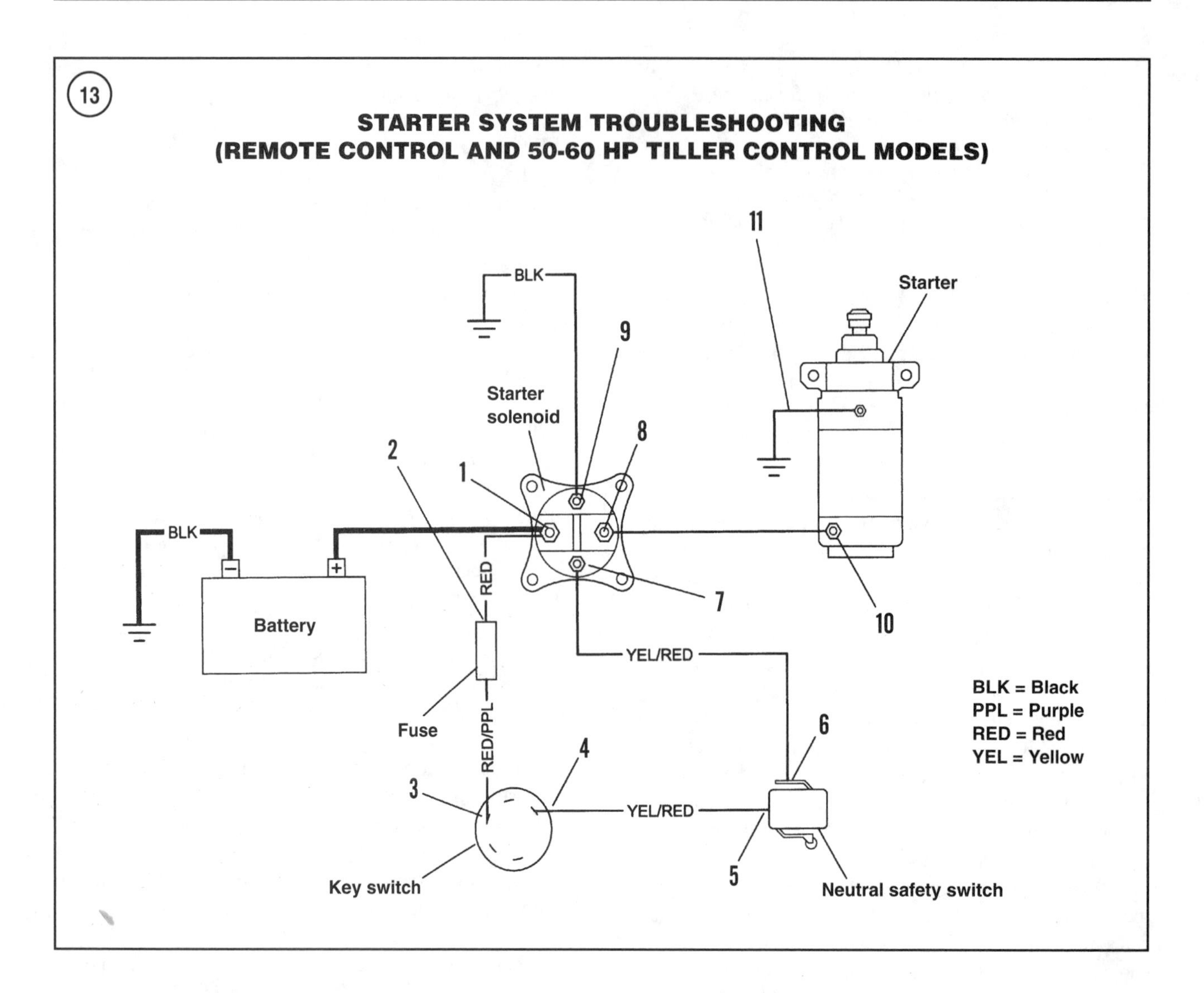

8B. *50-60 hp tiller control models*—Connect the test light probe to the key switch side of the neutral safety switch (5, **Figure 13**). Turn the key switch to the start position. If the test light does not illuminate, repair or replace the wire between the neutral safety switch and the starter switch.

9. Move the test light probe to the solenoid side of the neutral safety switch (6, **Figure 13**). Turn the key switch to the start position. If the test light does not illuminate, make sure the shift control is still in NEUTRAL and retest. Replace the neutral safety switch if the light does not illuminate.

10. Connect the test light probe to the yellow/red terminal on the starter solenoid (7, **Figure 13**). Turn the key switch to the start position. If the test light does not illuminate, repair or replace the wire between the neutral start switch and the starter solenoid. This includes the main harness connector pin No. 7.

11. Connect the test light probe to the starter solenoid terminal leading to the starter motor (8, **Figure 13**). Turn the key switch to the start position. If the test light does not illuminate, connect the test light lead to the positive battery terminal and connect the test light probe to the small black (ground) terminal of the starter solenoid (9, **Figure 13**). If the test light does not illuminate, repair or replace the ground wire between the starter solenoid and the engine block. If the test light only illuminates during the ground wire test, replace the starter solenoid.

12. Connect the test light lead to a good engine ground. Connect the test light probe to the starter motor terminal (10, **Figure 13**). Turn the key switch to the start position. If the test light does not illuminate, repair or replace the cable between the starter solenoid and the starter motor. If the test light illuminates, proceed to Step 13.

13A. *Starter equipped with a ground cable*—Inspect the ground cable (11, **Figure 13**) for loose connections, corro-

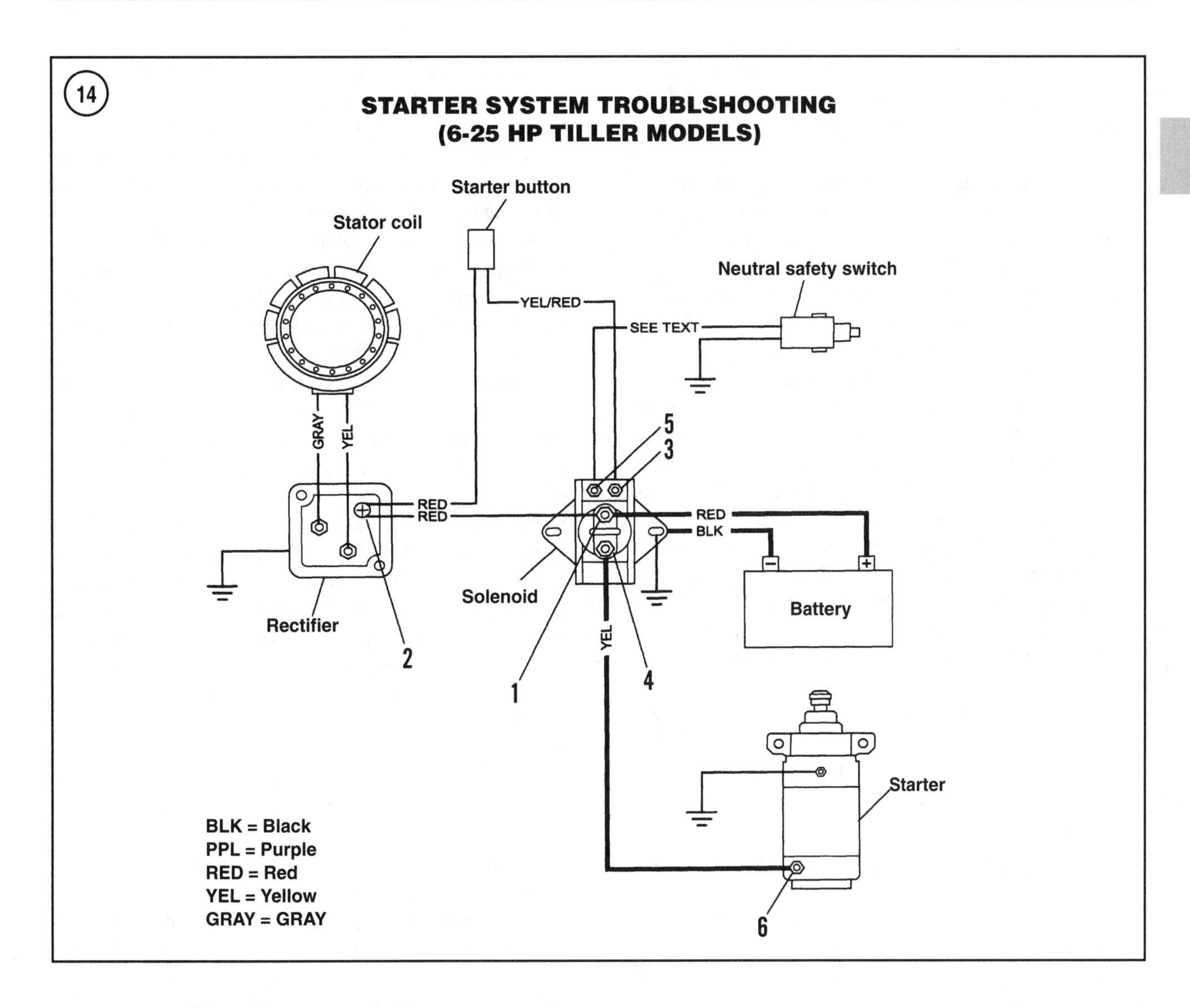

sion and damage. Clean, tighten or repair it as necessary. If the starter still will not engage, remove the starter for replacement or repair.

13B. *Starter not equipped with a ground cable*—Remove the starter and inspect it for paint or corrosion on the mounting bolts and bosses. If paint or corrosion is found, clean the mounting bolts and bosses. Reinstall the starter and test the starter engagement. If the starter still will not engage, remove the starter for replacement or repair.

Tiller handle 6-25 hp models

Refer to **Figure 14** for this procedure. Refer to the individual model wiring diagrams at the end of the book.

NOTE

*The starter solenoid ground terminal (5, **Figure 14**) has either a black wire with a yellow sleeve or a yellow/red wire connecting it to the neutral safety switch.*

1. Connect the test light lead to the positive terminal of the battery and touch the test light probe to metal anywhere on the engine block. If the test light does not illuminate or is dim, the battery ground cable connections are loose or corroded, or there is an open circuit in the battery ground cable. Check connections on both ends of the ground cable.

2. Place the shift lever into NEUTRAL and connect the test light lead to a good engine ground for Steps 3-10.

3. Connect the test light probe to the starter solenoid battery terminal (1, **Figure 14**). If the test light does not illuminate or is very dim, the battery cable connections are loose or corroded, or there is an open in the cable between the battery and the solenoid. Clean and tighten the connections or replace the battery cable.

4. Connect the test light probe to the rectifier red lead terminal (2, **Figure 14**). If the test light does not illuminate, repair or replace the red wire between the starter solenoid and the rectifier.
5. Connect the test light probe to the starter solenoid yellow/red input terminal (3, **Figure 14**). Depress the starter switch. If the test light does not illuminate, replace the starter switch assembly.
6. Connect the test light probe to the starter solenoid output terminal (4, **Figure 14**). Depress the starter switch. If the test light does not illuminate, connect a suitable jumper wire between the starter solenoid ground wire terminal (5, **Figure 14**) and a good engine ground. Depress the starter switch. If the test light now illuminates, or the engine cranks, replace the neutral start switch assembly. If the test light only illuminates during the ground wire terminal test, replace the starter solenoid ground wire.
7. Connect the test light probe to the starter motor terminal (6, **Figure 14**). Depress the starter switch. If the test light does not illuminate, repair or replace the cable between the starter solenoid and the starter motor. If the test light illuminates, remove the starter and inspect for paint or corrosion on the mounting bolts and bosses. If paint or corrosion is found, clean the mounting bolts and bosses. Reinstall the starter and test starter engagement. If the starter still does not operate, remove the starter for replacement or repair.

Tiller handle 30-40 hp models (two-cylinder)

Refer to **Figure 15** for this procedure. Refer to the individual model wiring diagrams at the end of the book.
1. Connect the test light wire to the positive terminal of the battery and touch the test light probe to metal anywhere on the engine block. If the test light does not light or is dim, the battery ground cable connections are loose or corroded, or there is an open circuit in the battery ground cable. Check the connections on both ends of the ground cable.
2. Place the engine shift lever into NEUTRAL and connect the test light wire to a good engine ground.
3. Connect the test light probe to the starter solenoid battery terminal (1, **Figure 15**). If the test light does not illuminate or is very dim, the battery cable connections are loose or corroded, or there is an open in the cable between the battery and the solenoid. Clean and tighten the connections or replace the battery cable as required.
4. Remove the 20 amp fuse and connect the test light probe to the input side (starter soleniod side) of the fuse holder (2, **Figure 15**). If the test light does not illuminate, repair or replace the wire between the starter solenoid and the fuse holder.
5. Reinstall the fuse. Connect the test light probe to the starter button red wire bullet connector (3, **Figure 15**). If the test light does not illuminate, replace the 20 amp fuse (**Figure 15**) and retest. If the light does not illuminate with a known good fuse, repair or replace the wire between the fuse terminal on the starter solenoid and the starter button switch red wire bullet connector (3, **Figure 15**).
6. Connect the test light probe to the starter button yellow/red terminal (4, **Figure 15**). Depress the starter button. If the test light does not illuminate, replace the starter button.
7. Connect the test light probe to the yellow/red terminal on the starter solenoid (5, **Figure 15**). Depress the starter button. If the test light does not illuminate, repair or replace the wire between the starter button and the starter solenoid.
8. Connect the test light probe to the starter solenoid terminal leading to the starter motor (6, **Figure 15**). Depress the starter button. If the test light does not illuminate, connect the test light wire to the positive battery terminal and connect the test light probe to the ground terminal of the starter solenoid (7, **Figure 15**). If the test light illuminates, replace the starter solenoid. If it does not illuminate, go to Step 9.
9. With the test light probe still connected to the ground terminal of the starter solenoid, disconnect the neutral start switch and connect a suitable jumper wire between the engine harness bullet connectors. If the test light now illuminates, replace the neutral start switch.
10. Connect the test light probe to the starter motor terminal (8, **Figure 15**). Depress the starter button. If the test light does not illuminate, repair or replace the cable between the starter solenoid and the starter motor. If the test light does not illuminate, remove the starter and inspect for paint or corrosion on the mounting bolts and bosses. If paint or corrosion is found, clean the mounting bolts and bosses. Reinstall the starter and test starter engagement. Make sure the ground cable is securely attached to the starter frame. If the starter still does not engage, remove the starter for replacement or repair.

Start Button Test (6-40 hp Tiller Models)

Refer to the wiring diagrams at the end of the book.
1. Disconnect the negative battery cable from the battery.
2A. *6-25 hp models*—Disconnect the start button yellow/red ring terminal from the starter solenoid (3, **Figure 14**) and the starter button red wire from the rectifier (2).
2B. *30 and 40 hp models (two-cylinder)*—Disconnect the starter button red wire (3, **Figure 15**) and yellow/red wire connector (4) from the starter button terminal.

(15)

**STARTING SYSTEM TROUBLESHOOTING
(30 AND 40 HP TWO-CYLINDER TILLER CONTROL MODELS)**

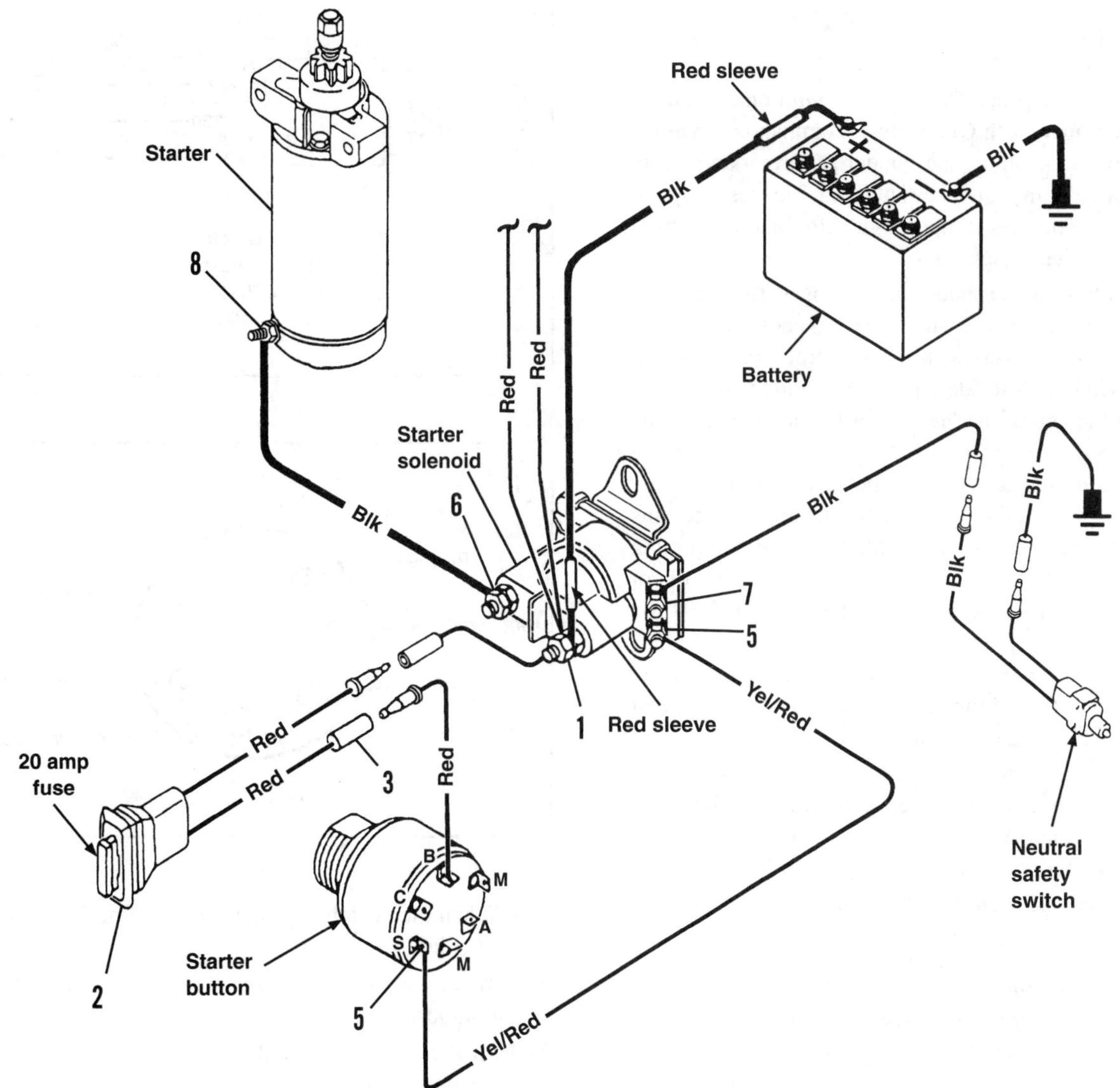

1. Starter solenoid terminal
2. Fuse input terminal
3. Starter button terminal connector
4. Starter button yellow/red terminal
5. Starter solenoid yellow/red terminal
6. Starter solenoid terminal (to starter motor)
7. Starter solenoid black terminal
8. Starter motor terminal

3. Connect an ohmmeter set on the R × 1 scale between the start button terminals. If the meter reads continuity, replace the start button.
4. Depress the start button. If the meter does not read continuity, replace the start button.

Ignition Switch Test

The following procedure tests the ignition switch on models equipped with Quicksilver Commander style remote control assemblies and standard aftermarket remote controls or dash-mounted key switches. This test may not be valid on *all* models equipped with aftermarket controls and electrical harnesses.

The ignition switch and main wiring harness can be tested at the main engine wiring harness connector, eliminating the need to disassemble the control box or remove the key switch from the dash panel. Refer to the wiring diagrams at the end of the manual for the main engine wiring harness connector pin locations and wire color code identification. When testing at the main engine harness connector, connect the ohmmeter to the appropriate pins based on the wire color codes called out in the following text. Testing at the main engine harness connector will not isolate a bad wiring harness from the key switch. If the switch and harness fails the test procedure at the main harness connector, the key switch must be disconnected and retested to verify that the main harness is not faulty. Test the key switch by itself as follows:

Use an ohmmeter calibrated on the R × 1 scale to test the key switch circuits. Refer to **Figure 16** for Mercury/Mariner factory switches or **Figure 17** for typical aftermarket switches.

1. Disconnect the negative battery cable from the battery.

NOTE
Mercury/Mariner factory switches typically use short color-coded wires with bullet connectors. Aftermarket switches typically use screw terminals labeled with abbreviations. When testing at the main harness connector, consult the wiring diagram at the end of the manual for pin locations at the main engine harness connector.

2. Access the key switch and disconnect the wires from the bullet connectors or key switch terminals. Note the color code and terminal markings of aftermarket switches.
3. Connect one lead of the ohmmeter to the switch red/purple wire (BAT, B or B+ terminal) and the other ohmmeter lead to the purple wire (A, ACC or IGN terminal). When the switch is in the off position, there should be no continuity.

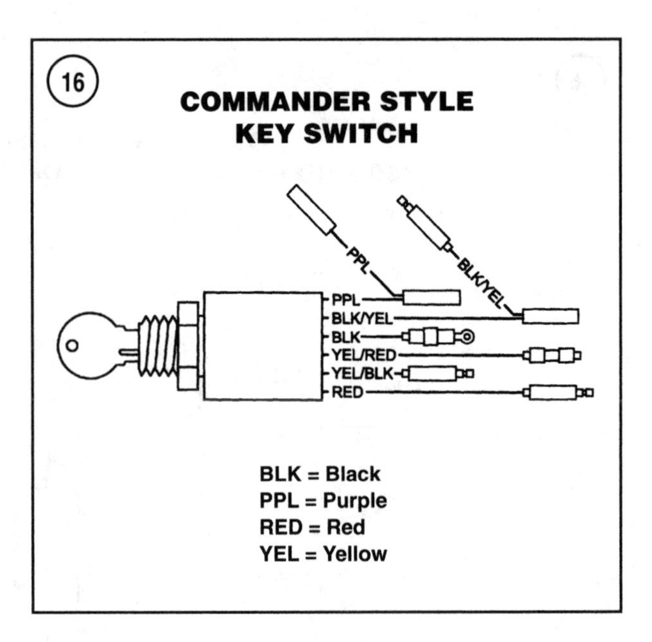

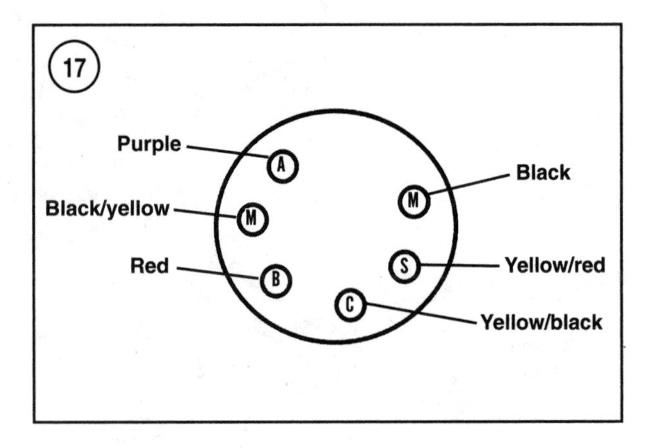

4. Turn the switch to the on position. The ohmmeter should indicate continuity.
5. Turn the switch to the start position. The ohmmeter should indicate continuity.
6. Turn the switch to the off position. Connect one ohmmeter lead to the black/yellow wire (first M terminal) and the other ohmmeter lead to the black wire (second M terminal). The ohmmeter should indicate continuity.
7. While observing the meter, turn the switch to the on and start positions. The ohmmeter should read no continuity in both positions.
8. Turn the switch to the off position. Connect one ohmmeter lead to the red/purple wire (BAT, B or B+ terminal) and the other ohmmeter lead to the yellow/black wire (C terminal). The ohmmeter should read no continuity.

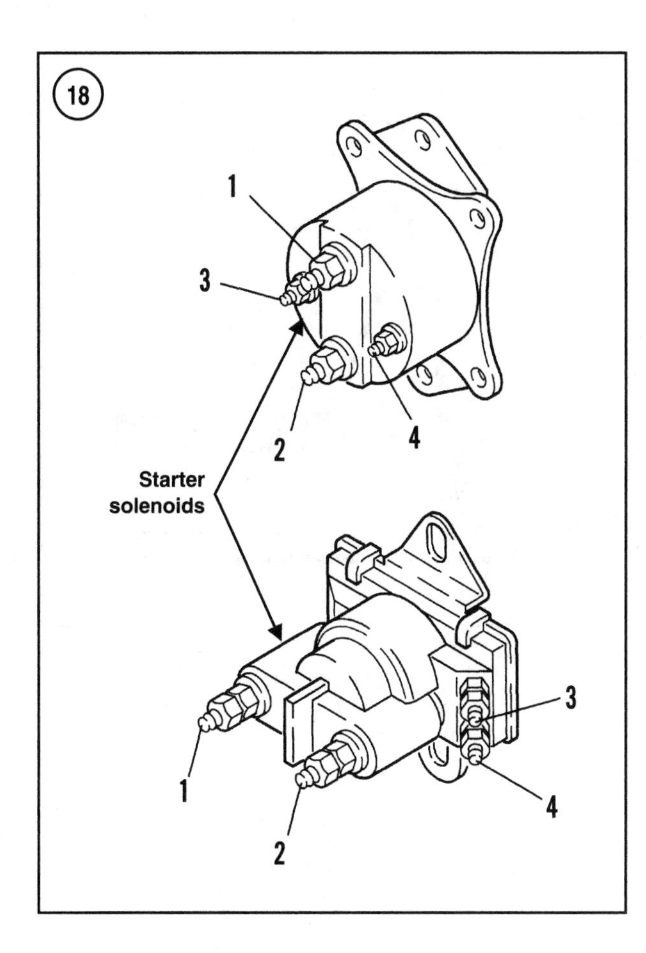

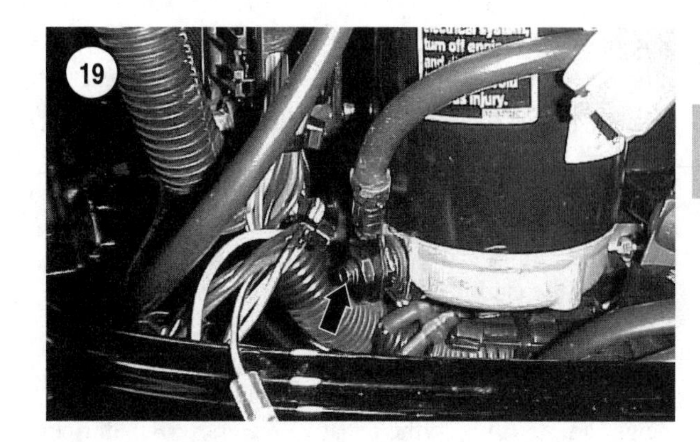

9. Turn the switch to the on position. The ohmmeter should read no continuity. Press in on the key to engage the choke or prime. The ohmmeter should read continuity.
10. Turn the switch to the start position. The ohmmeter should read no continuity. Press in on the key to engage the choke or prime. The ohmmeter should read continuity.
11. Replace the key switch if it is faulty.

Starter Solenoid Bench Test

NOTE
All engine wiring harness leads must be disconnected from the solenoid for this test.

Solenoids vary from engine to engine, but all solenoids have two large terminal studs and two small terminal studs. Refer to **Figure 18** for this procedure.
1. Disconnect the negative battery cable from the battery.
2. Disconnect all wires from the solenoid terminal studs. If necessary, remove the solenoid from the engine.
3. Connect an ohmmeter set on the R × 1 scale to each of the two large terminal studs (1 and 2, **Figure 18**). If the ohmmeter indicates continuity, replace the solenoid.
4. Attach a 12-volt battery with suitable jumper leads to the two small terminal studs (3 and 4, **Figure 18**). Polarity is not important. An audible click should be heard as the solenoid engages. If the ohmmeter does not indicate continuity between the large terminals with jumpers attached, replace the solenoid.
5. Reconnect all wires. Connect the negative battery cable last.

Starter Motor Current Draw Tests

Load test

A clip-on or inductive ammeter, if available, is simplest to use as no electrical connections are required. Make sure the ammeter can read higher than the anticipated highest amp reading (**Table 7**). The spark plugs must be installed for the load test.
1. When using a conventional ammeter, disconnect the negative battery cable from the battery.
2. Disconnect the starter motor lead from the starter motor (**Figure 19**, typical) and securely connect the ammeter positive lead to the starter motor cable. Insulate the connection with electrical tape to prevent accidental arcing.
3. Securely connect the ammeter negative lead to the starter motor terminal. Disconnect and ground the spark plug leads to the engine to prevent accidental starting.
4. Reconnect the negative battery cable.
5. Crank the engine to check the current draw. If current draw is excessive (**Table 7**), repair or replace the starter motor.

No-load test

If starter system troubleshooting indicates that additional starter motor tests are necessary, use the starter no-load current draw test as an indication of internal starter condition. A clip-on or inductive ammeter, if available, is simplest to use as no electrical connections are required. Make sure the ammeter can read higher than the anticipated highest amp reading (**Table 7**).

1. Remove the starter motor assembly from the power head. Securely fasten the starter motor in a vise or another holding fixture. Do not damage the starter motor by crushing it in the vise.
2. Obtain a fully charged battery that meets the requirements for the engine being tested.
3. Using heavy gauge jumper cable, connect a conventional ammeter in series with the positive battery cable and the starter motor terminal. Connect another heavy gauge jumper cable to the negative battery terminal. See **Figure 20**.

WARNING
Make the last battery connection to the starter frame in Step 4. Do not create any sparks near the battery or a serious explosion could occur.

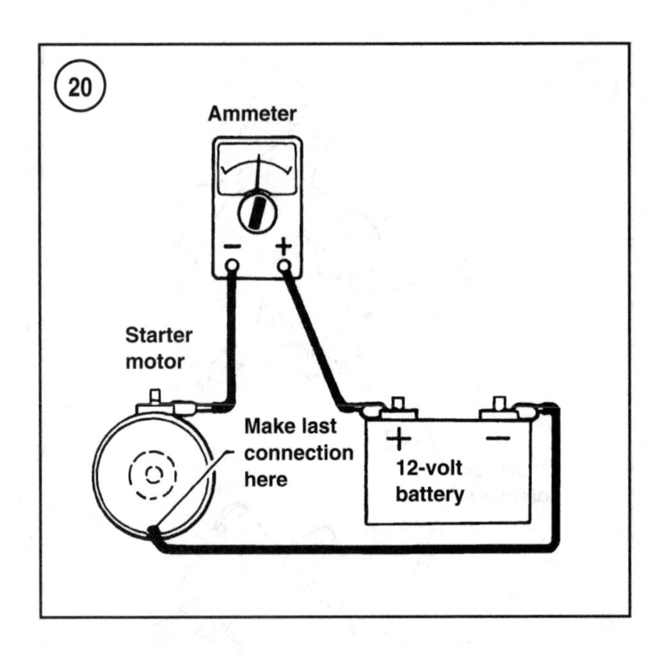

4. When ready to perform the no-load test, quickly and firmly connect the remaining connection to the starter motor frame (**Figure 20**). Observe the amperage reading, then disconnect the jumper cable from the starter motor frame.
5. If the motor does not perform to specification in **Table 7**, repair or replace the motor. See Chapter Seven. Refer to **Table 3** for additional starter motor symptoms and remedies.

LIGHTING COIL

Many manual start models are equipped with an AC lighting coil. The lighting coil is located under the flywheel and is a standard stator coil without a rectifier or rectifier/regulator. The output of the AC lighting coil is directly proportional to engine speed.

A typical application of the AC lighting coil is to power the running lights on small work boats where the weight of a battery is not desired. The running lights are wired directly to the two yellow lighting coil wires in a series circuit. The faster the motor runs, the brighter the lights. If one light fails as an open circuit, all of the lights go out. If a light fails as a short circuit, the remaining lights burn brighter and fail prematurely as they absorb the extra current. The load of the light bulbs must be matched to the output of the AC lighting coil. If the load is too great, the lights burn dimly. If the load is too small, the lights burn out at high speeds.

The output is generally rated in watts, which is the rating for electrical power. Watts (W) can be calculated by multiplying the amps times the voltage. For example, a 5 amp AC lighting coil can carry a 60 watt load (5 amps × 12 volts = 60 watts).

Troubleshooting

1. Make sure there are no open or short circuits in the wires going to the running lights or AC accessories. Inspect all of the bulbs in the lighting system. One bulb blowing open will cause all of the remaining bulbs to cease to function because of the open circuit.
2. The connections at the engine are on a terminal strip on the side of the power head. The two AC lighting coil (stator) wires are generally yellow, but one may be yellow and one gray. Disconnect all wires from the stator terminal strip and perform stator resistance and short-to-ground tests.
3. To perform the stator resistance test, set an ohmmeter on the appropriate scale to read the stator ohms specification (**Table 9**). Connect an ohmmeter wire to each of the stator yellow wires and note the reading. Compare to the specifications in **Table 9**. Replace the AC lighting coil (stator) if it is not within specifications.
4. To perform a stator short-to-ground test, set an ohmmeter on the highest scale available. Connect one ohmmeter wire to a clean engine ground. Connect the other ohmmeter wire to one of the stator yellow wires. If the meter reads continuity, inspect the yellow wires for damaged insulation and repair if possible. If the wires are not damaged, replace the AC lighting coil (stator). Repeat this step for the remaining stator yellow wires.
5. Reconnect all wires.

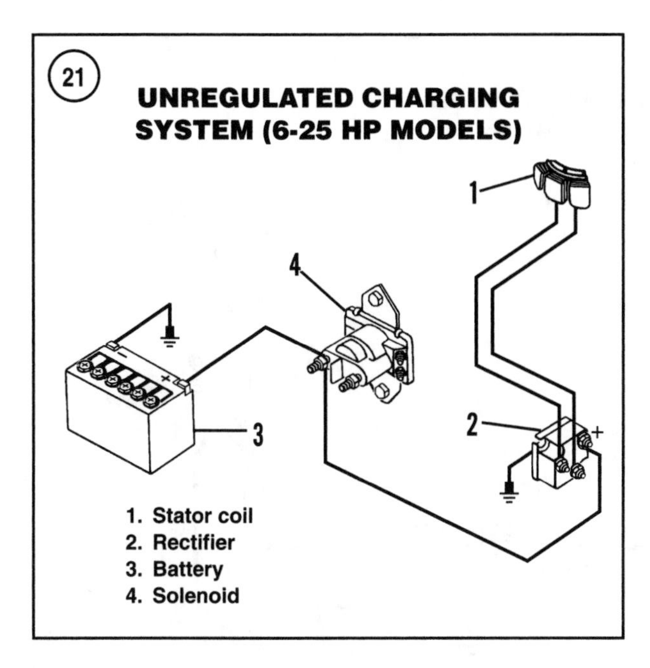

BATTERY CHARGING SYSTEM

An alternator charging system is used on all electric start models. The charging system keeps the battery fully charged and supplies current to run accessories. Charging systems can be divided into two basic designs: internal unregulated and integral regulated. Refer to **Table 8** and **Table 9** for charging system specifications and applications.

Integral systems use permanent magnets mounted in the flywheel and a stator coil winding mounted to the power head. As the flywheel rotates, the magnetic fields in the flywheel pass by the stator coil windings, inducing AC (alternating current). Unregulated systems use a rectifier (a series of four diodes) to change the AC to DC (direct current). See **Figure 21**. The output from a unregulated charging system is directly proportional to engine speed. Because an unregulated system has the potential to overcharge the battery during long periods of wide-open throttle operation, maintenance-free batteries are not recommended. Overcharging a battery causes the electrolyte level to drop, leading to premature battery failure. Vented batteries allow removal of the vent caps and refilling of the electrolyte as needed for longer service life.

Integral regulated systems use the same type flywheel magnets and stator coil windings as the unregulated system, but the rectifier is replaced with a rectifier/regulator. The rectifier changes the AC current to DC current, while the regulator monitors system voltage and adjusts the charging system output. Batteries that are maintained at 13-15 volts will stay fully charged without excessive venting. The regulator controls the output of the charging system to keep system voltage at approximately 14.5 volts. The large red wire of the rectifier/regulator is DC output. The small red wire is the sensing terminal which allows the regulator to monitor system voltage. See **Figure 22**.

Another function of the integral charging system is to provide the signal for the tachometer. The tachometer counts AC voltage pulses coming out of the stator before the AC voltage is rectified to DC. Tachometer failure on models with integral charging systems is related to the charging system, not the ignition system. The tachometer connects to one of the stator yellow wires on unregulated systems and connects to the rectifier/regulator gray wire (**Figure 22**) on regulated systems.

Refer to **Table 4** for typical charging system problems and solutions. Refer to the wiring diagrams at the end of the book.

Malfunctions in the charging system generally cause the battery to be undercharged and the tachometer to read erratically or not at all. The following conditions will cause rectifier or rectifier/regulator failure.

1. Reversing the battery cables.
2. Disconnecting the battery cables while the engine is running.
3. Loose connections in the charging system circuits, including battery connections and ground circuits.

CAUTION

If an integral unregulated or integral regulated charging system equipped outboard must be operated with the battery removed or disconnected, disconnect and insulate both stator leads (two yellow or one yellow and one gray) on both ends of the connection.

Inspection (All Models)

Before performing the troubleshooting procedure, check the following.

1. Make sure the battery is properly connected. If the battery polarity is reversed, the rectifier or voltage regulator will be damaged.
2. Check for loose or corroded connections. Clean and tighten them as necessary.
3. Check the battery condition. Recharge or replace the battery as necessary.
4. Check the wiring harness between the stator and battery for cut, chafed or deteriorated insulation, and cor-

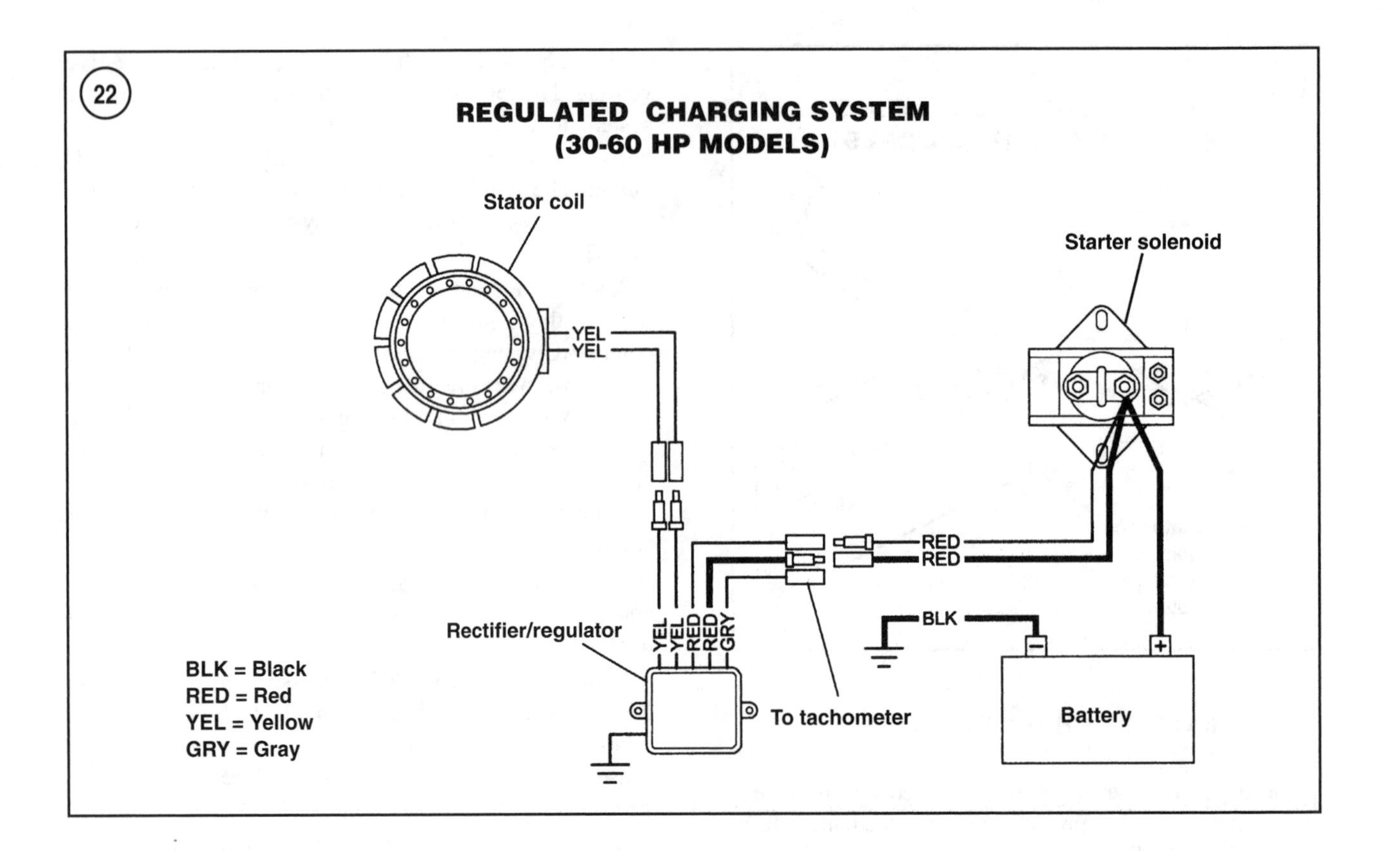

roded, loose or disconnected connections. Repair or replace the wiring harness as necessary.

5. Visually inspect the stator windings for discoloration and burned windings. Replace any stator that shows evidence of overheating.

CAUTION
Unless otherwise noted, perform all voltage tests with the wires connected, but with the terminals exposed to accommodate the test lead connection. All electrical components must be securely grounded to the power head any time the engine is cranked or started. Otherwise, electrical components may be damaged.

Current Draw Test

Use this test to determine if the total load of the engine electrical system and boat accessories exceeds the capacity of the charging system.

NOTE
If a clip-on or inductive ammeter is used, install the probe on the positive battery cable near the battery and go to Step 3. If a conventional ammeter is used, make sure the ammeter is rated for at least 50 amps.

1. Disconnect the negative battery cable from the battery.
2. Disconnect the positive battery cable from the battery. Securely connect the ammeter between the positive battery post and the positive battery cable. Reconnect the negative battery cable.
3. Turn the ignition switch to the on position and turn on all accessories. Note the ammeter reading. Turn the ignition switch to the off position and turn off all accessories. If the ammeter reading exceeds the rated capacity of the charging system, reduce the accessory load connected to the charging system.

Troubleshooting Integral Unregulated Models

Refer to **Figure 21** for this procedure. Refer to **Table 8** and **Table 9** for specifications, and the end of the manual for specific wiring diagrams.

NOTE
If a clip-on or inductive ammeter is used, install the probe on the rectifier red wire and go to Step 4.

1. Disconnect the negative battery cable from the battery.

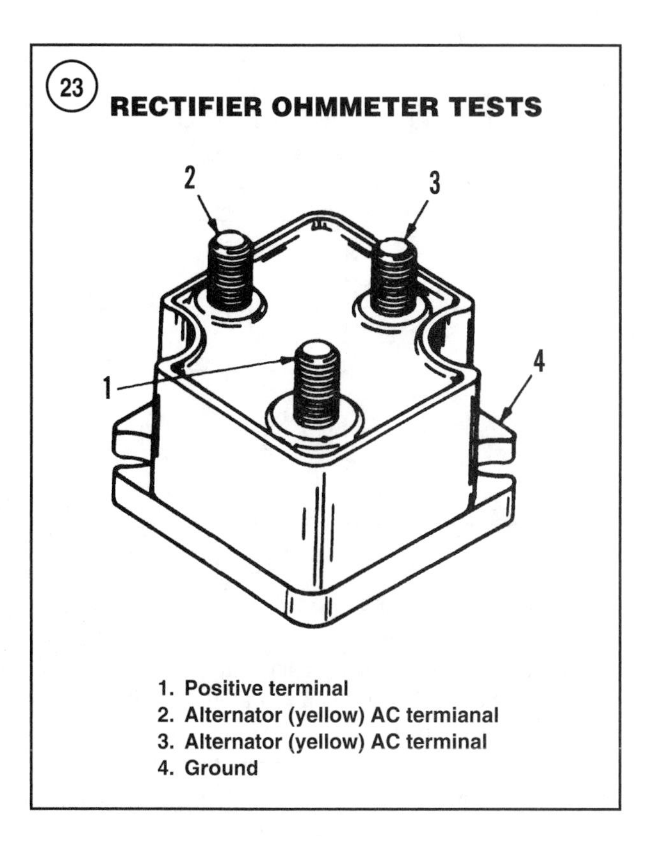

(23) RECTIFIER OHMMETER TESTS

1. Positive terminal
2. Alternator (yellow) AC termianal
3. Alternator (yellow) AC terminal
4. Ground

2. Connect an ammeter of sufficient size to measure the maximum rated output of the charging system in *series* between the positive output terminal of the rectifier and the red wire that was connected to the positive output terminal of the rectifier. Connect the positive wire of the ammeter to the rectifier terminal and the negative wire of the ammeter to the rectifier engine harness wire. Make sure the connections are secure and insulated from other wires or grounds.
3. Reconnect the negative battery cable.
4. Install a shop tachometer according to its manufacturer's instructions.

CAUTION
*Do not run the engine without an adequate water supply and do not exceed 3000 rpm without an adequate load. Refer to **Safety Precautions** at the beginning of this chapter.*

5. Start the engine and run it to the engine speed specified in **Table 8** while observing the ammeter readings. If amperage output is less than specified, continue with Step 6. If amperage output is within specification, the charging system is functioning correctly.
6. To check the resistance of the stator, disconnect the negative battery cable. Disconnect the two yellow (or one yellow and one gray) stator wires from the AC terminals of the rectifier. Set the ohmmeter on the correct scale and test the stator resistance. Connect one lead of the ohmmeter to each of the stator wires. Note the reading. Replace the stator if its resistance is out of the specified range (**Table 9**).
7. To check the stator for shorts, connect one wire of an ohmmeter to a clean engine ground. Connect the other wire alternately to each of the two stator wires. If the ohmmeter reads *continuity*, the stator is shorted and must be replaced.
8. To check the diodes in the rectifier, disconnect the remaining rectifier wires. Set the ohmmeter on the appropriate scale to test diodes. Connect one ohmmeter wire to the rectifier mounting base (4, **Figure 23**) and the other wire to one of the AC terminals (2 or 3, **Figure 23**). Note the ohmmeter reading. Reverse the ohmmeter wires and note the reading. The reading should be high in one polarity and low in the other. If the reading was high in both polarities or low in both polarities, replace the rectifier. Repeat the test for the other AC terminal (2 or 3, **Figure 23**).
9. Connect one wire of the ohmmeter, set to test diodes, to the rectifier positive terminal (1, **Figure 23**) and the other wire to one of the AC terminals (2, **Figure 23**). Note the ohmmeter reading. Reverse the ohmmeter wires and note the reading. The reading should be high in one polarity and low in the other. If the reading was high in both polarities or low in both polarities, replace the rectifier. Repeat the test for the other AC terminal (3, **Figure 23**).
10. To check the continuity of the rectifier positive wire to the battery, make sure the negative wire of the battery is disconnected. Make sure the rectifier positive wire (1, **Figure 23**) is disconnected from the rectifier. Set the ohmmeter on a high-ohms scale. Connect one wire of the ohmmeter to the battery positive terminal. Connect the other wire of the ohmmeter to the rectifier end of the lead that connects to the rectifier positive terminal. Note the ohmmeter reading. A good circuit has a zero or very low resistance reading. If the circuit is faulty, repair or replace the wire , connections and fuse (if equipped) between the rectifier and the battery.
11. Reconnect all wires.

Troubleshooting Integral Regulated Models

Refer to **Table 8** and **Table 9** at the end of this chapter for specifications and the end of the manual for specific wiring diagrams.

NOTE
A regulated charging system only puts out the current necessary to maintain 14.5 volts at the battery while the engine is running. If

the battery is fully charged, the alternator will not produce its rated output unless there is enough accessory demand.

If an inductive ammeter is used, install the probe onto the rectifier/regulator large red wire and go to Step 4.

1. Disconnect the negative battery cable from the battery.
2. Connect an ammeter with wires of sufficient size in series between the output terminal of the rectifier/regulator and the large red wire that was hooked to the output terminal of the rectifier/regulator (**Figure 22**). Hook the positive wire of the ammeter to the rectifier/regulator wire (male bullet connector) and the negative wire of the ammeter to the rectifier/regulator engine harness wire (female bullet connector). Make sure the connections are secure and insulated from other wires or grounds.
3. Reconnect the negative battery cable.
4. Install a shop tachometer according to its manufacturer's instructions.
5. Connect a voltmeter to the battery terminals.

CAUTION
*Do not run the engine without an adequate water supply and do not exceed 3000 rpm without and adequate load. Refer to **Safety Precautions** at the beginning of this chapter.*

6. Start the engine and run it at the engine speed specified in **Table 8** while noting both the ammeter and voltmeter readings. If the voltage exceeds 12.5 volts, turn on accessories or attach accessories to the battery to maintain battery voltage at 12.5 volts or less. If amperage output is less than specified, continue to Step 7. If amperage output is within specification, turn off or disconnect the accessories and run the engine at approximately 3000 rpm while observing the voltmeter. As the battery approaches full charge, the voltage should rise to approximately 14.5 volts and stabilize. If the voltage stabilizes at approximately 14.5 volts, the voltage regulator is functioning correctly. If the voltage exceeds 15 volts, go to Step 7 and check the rectifier/regulator sensing circuit voltage.
7. To test the rectifier/regulator sensing circuit, disconnect the rectifier/regulator small red wire from the engine harness. Connect the positive wire of a voltmeter to the engine harness side of the small red wire (male bullet connector) and the negative wire of the voltmeter to the negative battery terminal. The voltmeter should read battery voltage. If the voltage is more than 0.5 volt below battery voltage, clean and tighten the connections, or repair or replace the wire between the regulator small red wire and the battery. If the voltage is within 0.5 volt of battery voltage, check the stator for shorts to ground as described in Step 9. If the stator tests good, replace the rectifier/regulator.

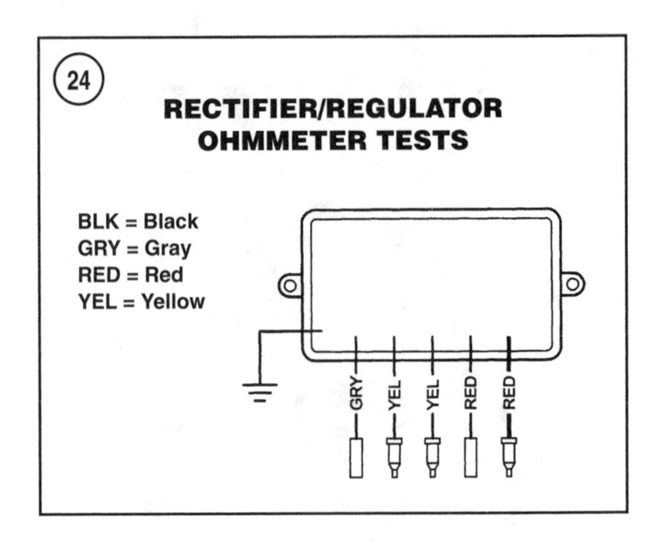

8. To check the resistance of the stator, disconnect the negative battery cable. Disconnect the two yellow stator wires from the rectifier/regulator. Set an ohmmeter on the appropriate scale to read the stator resistance specification (**Table 9**). Connect one wire of the ohmmeter to each of the stator wires. Note the reading. Replace the stator if its resistance is out of the specification range.
9. To check the stator for shorts to ground, set the ohmmeter on its highest scale. Connect one wire of the ohmmeter to a clean engine ground. Connect the other wire alternately to each of the two yellow stator wires. If the ohmmeter reads continuity, the stator is shorted to ground and must be replaced.
10. To check the diodes in the rectifier/regulator, disconnect the negative battery cable. Disconnect the two red, two yellow and one gray rectifier/regulator wires (**Figure 24**). Set the ohmmeter to read 100-400 ohms. Connect the ohmmeter positive wire to the rectifier/regulator large red wire and connect the ohmmeter negative wire to one of the rectifier/regulator yellow AC wires. Note the ohmmeter reading. The reading should be very low (100-400 ohms). If it is not, replace the rectifier/regulator. Repeat the test for the other yellow AC wire.
11. Set the ohmmeter to read 40,000 ohms or greater. Connect the ohmmeter negative lead to the rectifier regulator large red wire and connect the ohmmeter positive lead to one of the rectifier/regulator yellow AC wires (**Figure 24**). Note the meter reading. Repeat the test for the other yellow AC wires. One yellow wire should read *no continuity*, while the other wire should read 40,000 ohms or more. If it does not, replace the rectifier/regulator.
12. To check the rectifier/regulator silicon controlled rectifier (SCR) in each yellow wire, set the ohmmeter on the

highest ohms scale. Connect the ohmmeter negative wire to one of the rectifier/regulator yellow AC wires. Connect the ohmmeter positive wire to the rectifier/regulator metal case (**Figure 24**). If the ohmmeter does not read 10,000 ohms or more, replace the rectifier/regulator. Repeat the test for the other yellow AC wire.

13. To check the rectifier/regulator tachometer circuit, set an ohmmeter on the high-ohms scale. Connect the ohmmeter negative wire to the rectifier/regulator metal case (**Figure 24**). Connect the ohmmeter positive wire to the rectifier/regulator tachometer gray wire. If the ohmmeter does not read 10,000-50,000 ohms, replace the rectifier/regulator.

14. To check the continuity of the rectifier/regulator positive wire to the battery, make sure the negative battery cable is disconnected. Make sure the rectifier/regulator large red wire is disconnected from the rectifier/regulator. Set the ohmmeter on a high-ohms scale. Connect one wire of the ohmmeter to the battery positive terminal. Connect the other wire of the ohmmeter to the rectifier/regulator end of the wire that connects to the rectifier/regulator large red wire. Note the ohmmeter reading. If the meter does not indicate continuity or very low resistance (less than 1 ohm), repair or replace the wire, connections and fuse between the rectifier/regulator and/or the battery.

15. Reconnect all wires.

Stator Ohmmeter Tests (Regulated Charging System)

1. To check the resistance of the stator, disconnect the negative battery cable. Disconnect the two yellow stator wires connected to the terminal block or rectifier/regulator. Set the ohmmeter on the appropriate scale to read the stator ohms specification (**Table 9**). Connect one wire of the ohmmeter to each of the stator wires. Note the reading. Replace the stator if the resistance is out of specification.
2. To check the stator for shorts, set the ohmmeter on the high ohms scale. Connect one wire of the ohmmeter to a clean engine ground. Connect the other wire alternately to each of the two stator wires. If the ohmmeter reads continuity, the stator is shorted to ground and must be replaced
3. Reconnect all wires.

ELECTRICAL ACCESSORIES

The wiring harness used between the ignition switch and outboard motor is adequate to handle the electrical requirements of the outboard motor. It cannot handle the electrical requirements of accessories.

Whenever an accessory is added, run new wiring between the battery and the accessory. Install a separate fuse panel on the instrument panel.

If the ignition switch requires replacement, *never* install an automotive switch. Only use a switch approved for marine use.

IGNITION SYSTEM

This section deals with troubleshooting the various ignition systems used on Mercury and Mariner outboard motors. Once the defective component is identified, refer to Chapter Seven for component removal and replacement procedures.

Troubleshooting Notes and Precautions (All Models)

Observe the following troubleshooting precautions to avoid injuries and/or damaging the ignition system.

1. Never reverse the battery connections. Reverse battery polarity damages electronic components.
2. Never disconnect the battery cables while the engine is running.
3. Never crank or run the outboard unless all electrical components are grounded to the power head.
4. Never touch or disconnect ignition components while the outboard is running, while the ignition switch is on or while the battery cables are connected.
5. Never rotate the flywheel when performing ohmmeter tests or the meter will be damaged.
6. If a sudden unexplained timing change is noted:
   a. Check the trigger magnets in the hub of the flywheel, if so equipped, for damage or a possible shift in magnet position. If the magnets are cracked, damaged or have shifted position, replace the flywheel. See Chapter Seven.
   b. Check the flywheel key for wear or damage. See Chapter Seven.
7. The ignition system on electric start models requires the electric starter to crank the engine at normal speed in order for the ignition system to produce adequate spark. If the starter motor cranks the engine slowly or not at all, refer to *Starting System* in this chapter and correct the starting system problem before continuing.
8. The spark plug(s) must be installed during the troubleshooting process. The ignition system must produce adequate spark at normal cranking speed. Removing the spark plug(s) artificially raises the cranking speed and may prevent diagnoses of a problem in the ignition system.

9. Check the battery cable connections, on models so equipped, for secure attachments to both battery terminals and the engine. Clean corrosion from all connections. Discard wing nuts and install corrosion resistant hex-nuts at all battery cable connections. Place a corrosion-resistant locking washer between the battery terminal stud and battery cable terminal end to ensure a positive connection.

10. Check all ignition component ground wires for secure attachments to the power head. Clean and tighten all ground wires, connections and fasteners as necessary. Loose ground connections and loose component mounting hardware can cause many different symptoms.

CAUTION
If a charging system equipped outboard must be operated with the battery removed or disconnected, disconnect and insulate both yellow (or one gray and one yellow) wires at both ends of the connection.

Resistance (Ohmmeter) Tests

The resistance values specified in the following test procedures are based on tests performed at room temperature. Actual resistance readings obtained during testing are generally slightly higher for hot components. In addition, resistance readings may vary depending on the manufacturer of the ohmmeter. Be careful when failing a component that is only slightly out of specification. Many ohmmeters have difficulty reading less than 1 ohm accurately. If this is the case, specifications of less than 1 ohm generally appear as a very low continuity reading.

Direct Voltage Tests

Direct voltage tests are designed to check the voltage output of the ignition stator and the switch box (CD module). The test procedures check voltage output at normal cranking speed. If an ignition misfire or failure occurs only when the engine is running and cranking speed tests do not show any defects, perform the output tests at the engine speed at which the ignition symptom or failure occurs. **Tables 10-13** list the cranking and high speed running voltages for all applicable tests. When checking the DVA voltage output of a component, observe the meter needle for fluctuations, which indicates erratic voltage output. The voltage output of the ignition stator and switch box (CD module) changes with engine speed, but it should not be erratic.

CAUTION
Do not run the engine without an adequate water supply and do not exceed 3000 rpm without an adequate load. Refer to Safety Precautions at the beginning of this chapter.

The term *peak volts* is used interchangeably with *DVA* (Direct Volts Adapter). The Mercury Marine 91-99750 multimeter has a DVA scale that must be used whenever the specification is in peak volts or DVA. A DVA adapter (part No. 91-98045) is available to adapt any analog voltmeter capable of reading at least 400 DC volts. If the DVA or peak volts specification is listed as polarity sensitive, reverse the meter test leads and retest if the initial reading is unsatisfactory.

WARNING
High voltage is present during ignition system operation. Never touch ignition system components, wire harness leads or test wires while cranking or running the engine.

CAUTION
Unless otherwise noted, all direct voltage test must be performed with all wires connected, but with the terminals exposed to accommodate test wire connections. All electrical components must be grounded to the power head when the engine is cranked or started. Otherwise, electrical components will be damaged.

Push Button Stop Switch Test (Models So Equipped)

Refer to the end of this manual for specific wiring diagrams.

NOTE
If the push button stop switch incorporates a safety lanyard, the safety lanyard must be installed during testing.

1A. *2.5-5 hp models*—Disconnect the stop switch brown/white or brown and black wires from the engine harness and ground.
1B. *6-60 hp models*—Disconnect the stop switch black and black/yellow wires from the engine harness and ground.
2. Connect an ohmmeter set on the R × 1 scale between the disconnected stop switch terminals.
3. With the stop switch released, the meter should indicate *no continuity*.
4. With the stop switch depressed, the meter should indicate *continuity*.

5. Replace the stop switch if it fails to perform as specified.

Lanyard Safety Switch Test (Models So Equipped)

Refer to the end of the book for wiring diagrams.

1. Disconnect the lanyard safety switch black and black/yellow wires from the electrical harness and ground.
2. Connect an ohmmeter set on the R × 1 scale between the disconnected lanyard switch terminals.
3. With the lanyard switch in the *run* position, the meter should indicate *no continuity*.
4. With the lanyard switch in the *off* position, the meter should indicate *continuity*.
5. Replace the lanyard safety switch if it fails to perform as specified.

CAPACITOR DISCHARGE IGNITION (2.5, 3.3, 4 AND 5 HP)

The CDI (capacitor discharge ignition) system used on 2.5-5 hp models consists of the flywheel, charge coil, trigger coil (4 and 5 hp), CD module, ignition coil, spark plug and a combination push button stop/safety lanyard switch.

The 2.5 and 3.3 hp models feature fixed timing with no spark advance or timing adjustments. Because the charge coil and trigger coil are combined, the ignition coil sparks every 180° of crankshaft rotation.

The 4 and 5 hp models incorporate a nonadjustable electronic spark advance built into the CD module. These models feature a separate trigger coil. The ignition coil sparks once every 360° of crankshaft rotation.

Troubleshooting

Refer to **Figure 25** for 2.5 and 3.3 hp models and **Figure 26** for 4 and 5 hp models. Ignition system specifications are in **Table 10**. The recommended troubleshooting procedure is:

1. Preliminary checks.
   a. Spark test.
   b. Stop circuit isolation.
2. CD module stop circuit test.
3. CD module diode test (2.5 and 3.3 hp).
4. Charge coil output test.
5. Charge coil resistance test.
6. Trigger coil resistance test (4 and 5 hp).
7. CD module output test.
8. Ignition coil ohmmeter tests.

WARNING
High voltage is present in the ignition system. Never touch or disconnect ignition components while the engine is running.

CAUTION
Do not run the engine without an adequate water supply and do not exceed 3000 rpm without an adequate load. Refer to Safety Precautions at the beginning of this chapter.

Preliminary checks

1. Disconnect the spark plug lead from the spark plug and install an air gap spark tester, part No. 91-63998A-1 or equivalent, (**Figure 27**) to the spark plug wire. Adjust the air gap, if applicable, to 7/16 in. (11.1 mm). Then connect the alligator clip of the spark tester to a clean engine ground.
2. Make sure the safety lanyard is installed on the push button stop switch.

NOTE
If a 2.5 or 3.3 hp model does not return to idle when the throttle valve closes, replace the CD module. Failure of the bias circuit in the CD module causes over advanced ignition timing and unusually high idle speed. Also, spark advance on 4 and 5 hp models is controlled by the CD module. If spark output is good, yet the timing does not advance correctly, replace the CD module.

3. Crank the engine while observing the tester. If there is a crisp, blue spark, the ignition system is functioning correctly. If the engine will not start or does not run correctly, check the spark plug and ignition timing (4 and 5 hp). If the engine backfires or pops when attempting to start, remove the flywheel and check for a sheared flywheel key. If there is no spark, a weak spark or an erratic spark, continue with Step 4.
4. Disconnect the push button stop switch wire from the CD module bullet connector. See **Figure 25** or **Figure 26**. Crank the engine while observing the spark tester. If there is now a good spark, replace the push button stop switch.

CD module stop circuit test

WARNING
To prevent accidental starting, disconnect the spark plug wire from the spark plug and install a spark gap tester, part No. 91-63998-1 or equivalent, to the spark plug

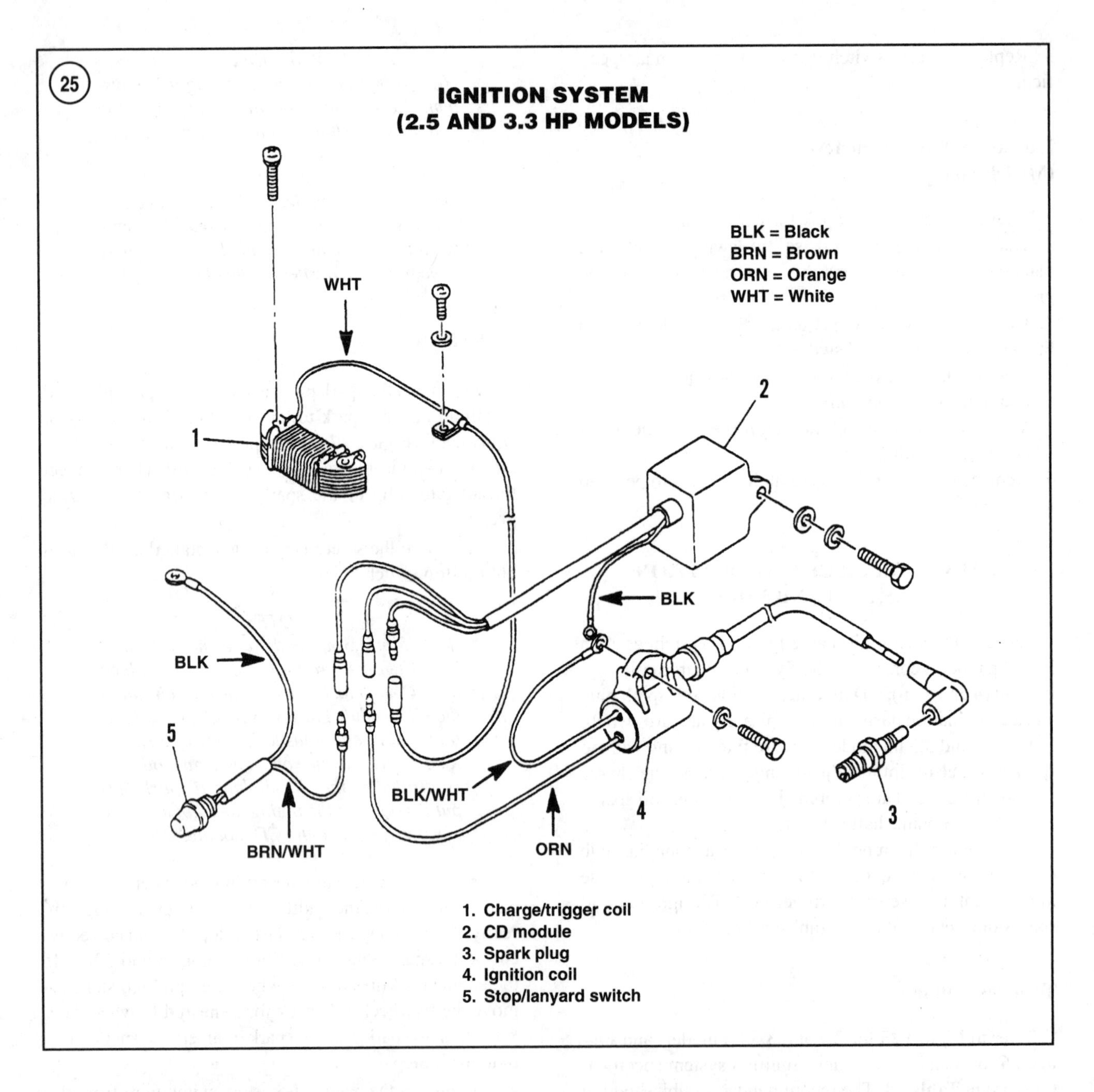

1. Charge/trigger coil
2. CD module
3. Spark plug
4. Ignition coil
5. Stop/lanyard switch

wire. Connect the alligator clip of the spark tester to a clean engine ground.

1A. *2.5 and 3.3 hp models*—Connect the meter positive lead to a good engine ground and the meter negative lead to the brown/white wire leading to the stop switch.

1B. *4 and 5 hp models*—Connect the meter positive lead to the brown wire leading to the stop switch and the meter negative lead to a good engine ground.

2. Set the meter selector switch to 400 DVA.

3. Crank the engine while noting the meter reading.

4. The stop circuit voltage must be within the specification in **Table 10**. If the voltage is correct, refer to *Charge coil output* in this chapter.

5A. *2.5 and 3.3 hp models*—If the voltage is above the specification, either the CD module or charge/trigger coil is defective. Test the charge/trigger coil output and resistance test as described in this chapter. If the charge/trigger coil tests correctly, replace the CD module.

5B. *4 and 5 hp models*—If the stop circuit voltage is above specification, either the CD module or the trigger coil is defective. Test the trigger coil resistance as de-

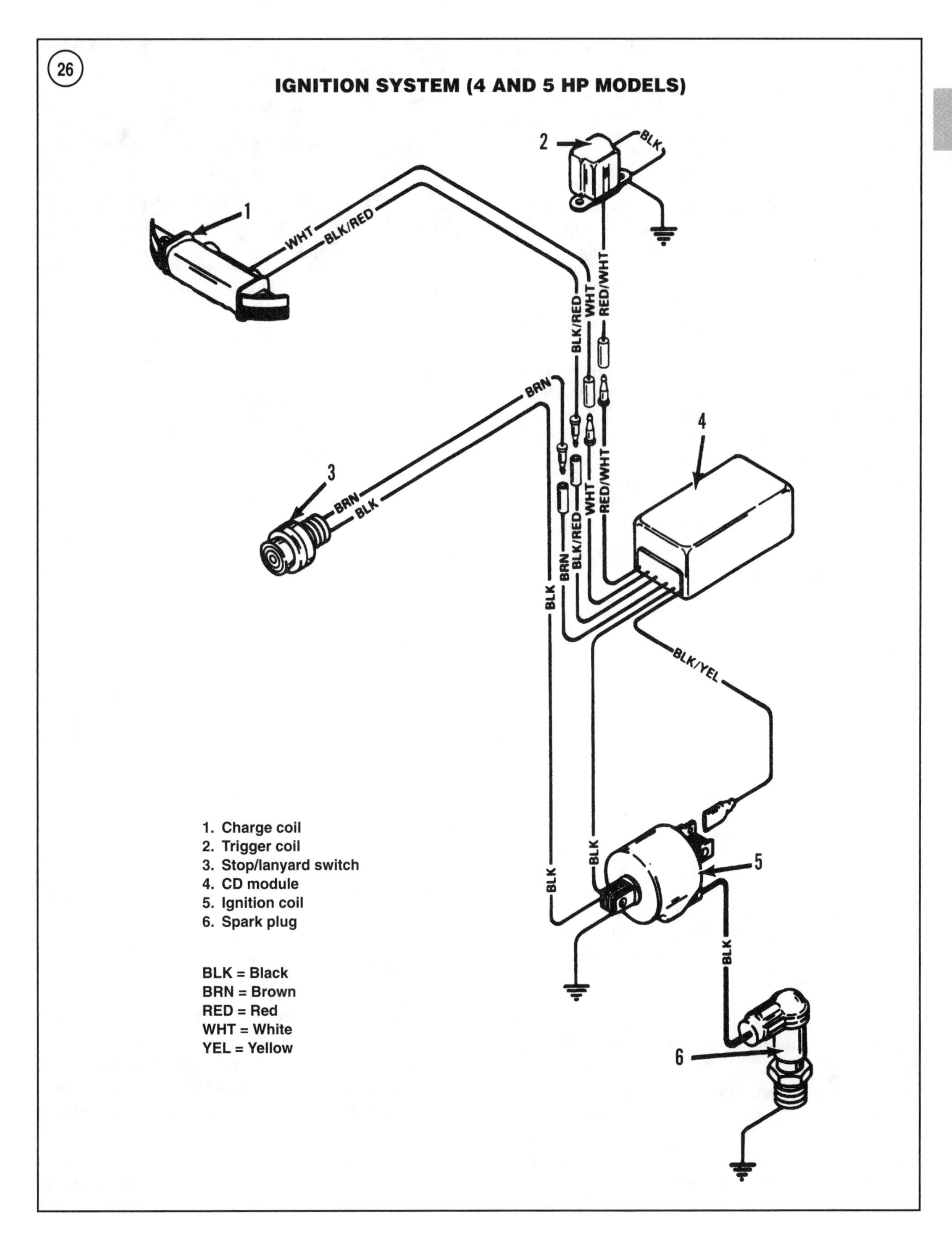
26
IGNITION SYSTEM (4 AND 5 HP MODELS)
1
2
3
4
5
6
BLK
WHT
BLK/RED
RED/WHT
BRN
BLK/YEL
1. Charge coil
2. Trigger coil
3. Stop/lanyard switch
4. CD module
5. Ignition coil
6. Spark plug
BLK = Black
BRN = Brown
RED = Red
WHT = White
YEL = Yellow

scribed in this chapter. If the trigger coil tests correctly, replace the CD module.

6. If the stop circuit voltage is below the specification, disconnect the brown (4 and 5 hp) or brown/white (2.5 and 3.3 hp) stop switch wire from the CD module at the bullet connector.

7. Repeat Step 3. If the stop circuit voltage is now within the specified range, the push button stop switch is defective and must be replaced. If the stop circuit voltage is still below specification, test the charge coil output as described in this chapter.

CD module diode test (2.5 and 3.3 hp)

1. Disconnect the white charge coil wire and the orange ignition coil wire from the CD module bullet connectors.

2. Set the ohmmeter on the appropriate scale to test a diode. Connect one meter lead to the orange CD module wire and the other meter lead to the CD module white wire. Note the ohmmeter reading. Reverse the ohmmeter leads and note the reading. The reading should be relatively high in one polarity and relatively low in the other. If the reading is high in both polarities or low in both polarities, replace the CD module.

Charge coil output test

WARNING

To prevent accidental starting, disconnect the spark plug wire from the spark plug and install a spark gap tester, part No. 91-63998-1 or equivalent, to the spark plug wire. Connect the alligator clip of the spark tester to a clean engine ground.

NOTE

The charge coil and trigger coil are combined in one assembly on 2.5 and 3.3 hp models.

1A. *2.5 and 3.3 hp models*—Connect the meter positive lead to a good engine ground and the meter negative lead to the white charge/trigger coil wire at the bullet connector. See **Figure 25**.

1B. *4 and 5 hp models*—Connect the meter positive lead to the black/red charge coil wire leading to the CD module. Connect the meter negative lead to a good engine ground.

2. Set the meter selector switch to 400 DVA.

3. Crank the engine and note the meter reading.

4A. *2.5 and 3.3 hp models*—The charge coil output voltage should be within the specification in **Table 10**. If it is not, test the charge/trigger coil resistance as described in the next section.

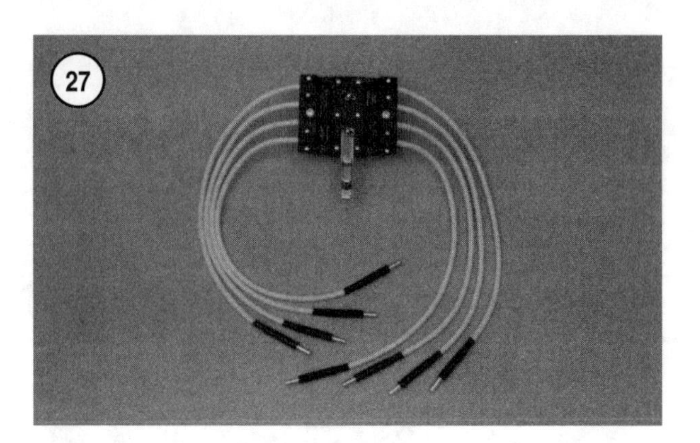

27

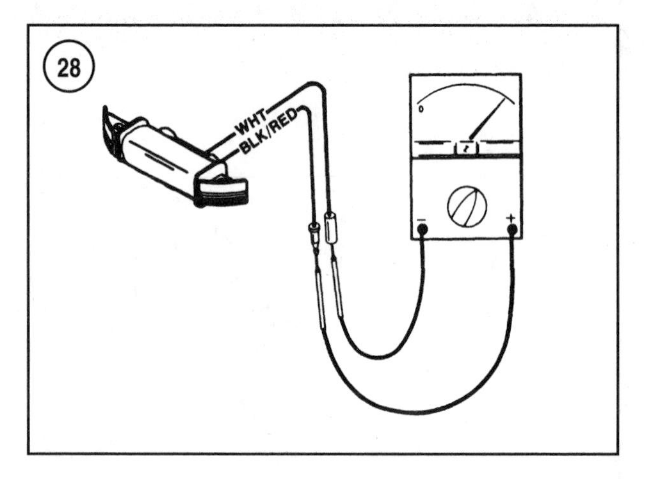

28

4B. *4 and 5 hp models*—The charge coil output voltage should be within the specification in **Table 10**. If it is not, test the charge coil resistance as described in the next section.

Charge coil resistance test

1A. *2.5 and 3.3 hp models*—Disconnect the white charge coil wire from the CD module bullet connector. See **Figure 25**.

1B. *4 and 5 hp models*—Disconnect the white and black/red charge coil wires from the CD module bullet connectors. See **Figure 26**.

2. Calibrate an ohmmeter on the scale specified in **Table 10**.

3A. *2.5 and 3.3 hp models*—Connect the ohmmeter negative lead to a good engine ground and the ohmmeter positive lead to the white charge/trigger coil wire at the bullet connector.

3B. *4 and 5 hp models*—Connect one ohmmeter lead to the black/red charge coil wire and the other ohmmeter lead to the white charge coil wire. See **Figure 28**.

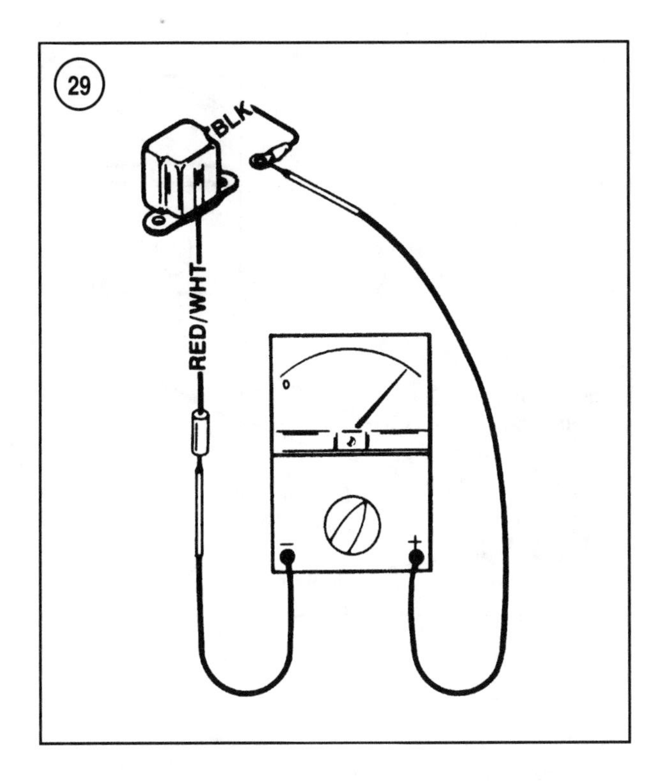

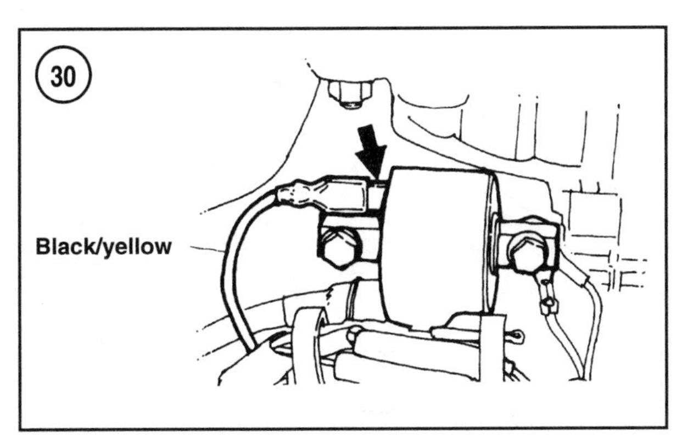

4A. *2.5 and 3.3 hp models*—The coil resistance should be within the specification in **Table 10**. If it is not, replace the charge/trigger coil.

4B. *4 and 5 hp models*—The coil resistance should be within the specification in **Table 10**. If it is not, replace the charge coil.

Trigger coil resistance test (4 and 5 hp)

1. Disconnect the red/white trigger coil wire from the CD module bullet connector. See **Figure 26**.

2. Set the ohmmeter on the scale specified in **Table 10**.

3. Connect one ohmmeter wire to the trigger coil red/white wire and the other ohmmeter wire to the trigger coil ground wire. See **Figure 29**.
4. The coil resistance should be within the specification in **Table 10**. If it is not, replace the trigger coil.

CD module output test

WARNING
To prevent accidental starting, disconnect the spark plug wire from the spark plug and install a spark gap tester, part No. 91-63998-1 or equivalent, to the spark plug wire. Connect the alligator clip of the spark tester to a clean engine ground.

1. Connect the meter positive wire to a good engine ground.
   a. *2.5 and 3.3 hp models*—Connect the meter negative wire to the orange wire at the bullet connector located between the ignition coil and CD module (**Figure 25**).
   b. *4 and 5 hp models*—Connect the meter negative wire to the black/yellow ignition coil primary terminal (**Figure 30**).
2. Set the meter selector switch to 400 DVA.
3. Crank the engine while noting the meter reading.
4. The CD module output voltage should be within the specification in **Table 10**. If it is not, proceed as follows:
   a. *Output voltage below the specification*—Perform the ignition coil resistance test as described in this section. If the ignition coil resistance tests are within specification, replace the CD module.
   b. *Voltage within the specification*—If the output voltage is within the specification and there is still no spark, a weak spark or an erratic spark, replace the ignition coil.

Ignition coil resistance tests

1. Set the ohmmeter on the primary resistance scale specified in **Table 10**.
2A. *2.5 and 3.3 hp models*—Disconnect the spark plug wire. Disconnect the orange bullet connector between the CD module and the ignition coil. See **Figure 25**.
2B. *4 and 5 hp models*—Disconnect the spark plug wire. Disconnect the CD module black/yellow wire from the ignition coil. See **Figure 30**.
3A. *2.5 and 3.3 hp models*—Connect the ohmmeter negative wire to the ignition coil ground wire terminal. Connect the ohmmeter positive wire to the ignition coil orange wire. See **Figure 25**.

3B. *4 and 5 hp models*—Connect the ohmmeter negative wire to the ignition coil laminations or a mounting bolt. Connect the ohmmeter positive wire to the ignition coil primary terminal blade. See **Figure 31**.

4. If the meter reading is within primary resistance specification in **Table 10**, go to Step 5. If the meter reading is not within specification, replace the ignition coil.

5. Set the ohmmeter on the secondary resistance scale specified in **Table 10**.

6A. *2.5 and 3.3 hp models*—Connect the ohmmeter positive wire to the spark plug terminal inside the spark plug cap. Connect the ohmmeter negative wire to the ignition coil ground wire terminal. See **Figure 25**.

6B. *4 and 5 hp models*—Connect the ohmmeter positive wire to the spark plug terminal inside the spark plug cap. Connect the ohmmeter negative wire to the ignition coil primary terminal blade. See **Figure 32**.

7. The meter reading should be within the specification in **Table 10**. If it is not, replace the ignition coil.

8. If the meter reading (Step 7) is within the specification (**Table 10**) and all other ignition components are within specification, but there is still no spark, a weak spark or an erratic spark, replace the ignition coil.

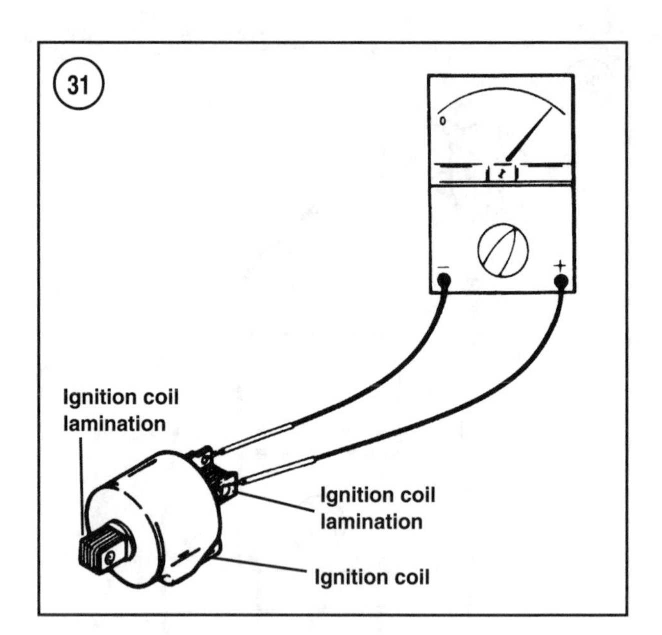

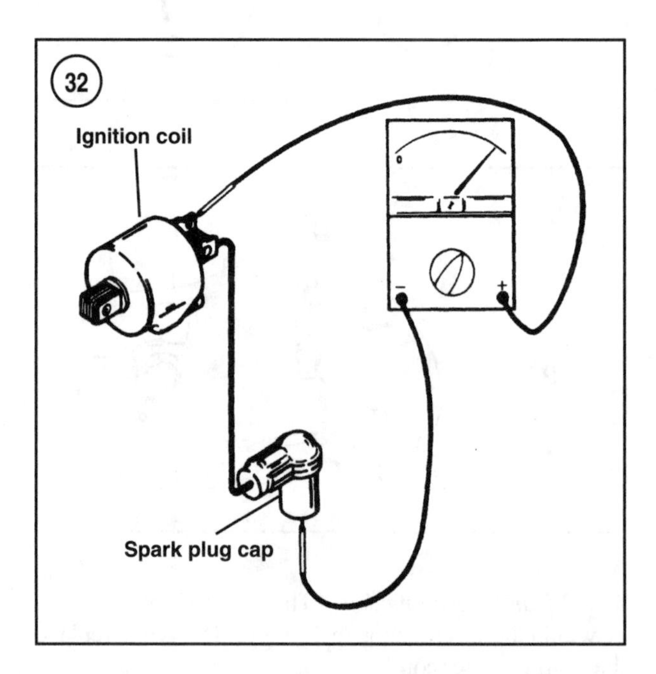

ALTERNATOR DRIVEN IGNITION MECHANICAL ADVANCE (6-25 HP EXCEPT 1998 20 JET)

This AD-CDI ignition is an alternator driven, capacitor discharge ignition system with mechanical spark advance. It is used on 2-25 hp models except for the 20 Jet model. Refer to **Figure 33** for an operational diagram of this ignition system.

NOTE
A red stator upgrade kit is available for 1998 6-25 hp models. The kit includes a red color stator, flywheel and switchbox with an rpm limiting circuit. All 1999-on 6-25 hp models are equipped with a red stator. Manual start models use a switchbox with an rpm limiting circuit. Electric start models (1999-on) use a switchbox without an rpm limiting circuit.

1. *Flywheel*—The flywheel inner magnets are for the trigger coil timing information. The outer magnets are for the ignition stator and battery charging stator, if equipped.
   a. *1998 models* use a flywheel equipped with two outer magnets. The two magnet flywheel must be used with a black stator.
   b. *1999-on models* use a flywheel equipped with four outer magnets. The four-magnet flywheel must be used with a red stator.

2. *Ignition stator*—The ignition stator provides the power the switch box needs to operate the ignition system. Stator output is always AC (alternating current) voltage.
   a. *1998 models* use a black stator equipped with low- and high-speed windings and a ground wire. The ignition stator low-speed windings provide most of the electricity for low-speed, cranking and idle, operation. High-speed windings provide most of the

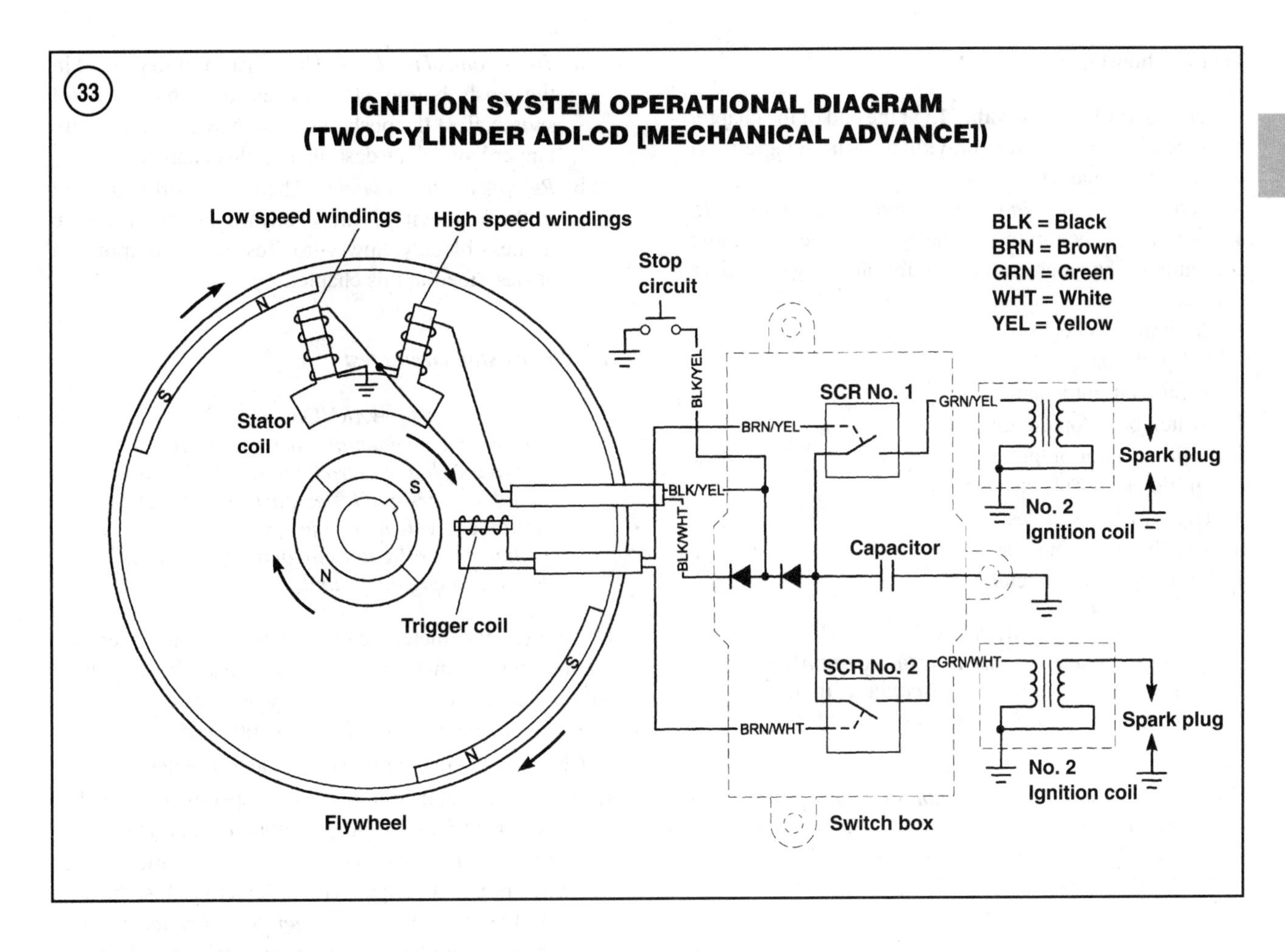

electricity for high-speed, cruising and wide-open throttle, operation.

b. *1999-on models* use a red stator equipped with a single ignition charging bobbin. This stator produces more voltage than the ignition system requires under normal operation. Special circuits in the switchbox regulate the stator voltage to approximately 300 volts. The red stator upgrade kit can be retrofitted to 1998 models.

3. *Trigger coil*—The trigger coil tells the switch box when to fire the ignition coils. The trigger coil is rotated by mechanical linkage to change the trigger's position relative to the flywheel. This advances or retards the ignition spark timing. The trigger coil has one winding with two wires (brown/white and brown/yellow). A trigger failure results in no spark on both cylinders.

4. *Switch box*—The switch box stores the electricity from the ignition stator until the trigger coil(s) tells it to send the electricity to the ignition coils. The switch box uses an internal rectifier to change the ignition stator AC voltage to DC voltage. The DC voltage is stored in a capacitor, until it is released by a SCR (silicon controlled rectifier), which is an electronic switch. There is one SCR for each cylinder. The SCR is controlled by the trigger coil.

5. *Ignition coils*—One ignition coil is used for each cylinder. The ignition coil transforms the relatively low voltage from the switch box into voltage high enough (35,000 volts) to jump the spark plug gap and ignite the air/fuel mixture.

6. *Spark plugs*—One spark plug is used for each cylinder. Use only the recommended spark plugs or serious engine damage may result. Resistor or suppressor plugs are designed to reduce RFI (radio frequency interference) emissions that can cause interference with electrical accessories. Use the recommended RFI spark plug if RFI emissions are suspected of causing interference or malfunction of electrical accessories.

7. *Stop circuit*—The stop circuit connects to one end of the capacitor in the switch box. Whenever the stop circuit is connected to ground, the capacitor is shorted to ground and cannot store electricity. There is no voltage available to send to the ignition coil and the ignition system ceases producing spark. The stop circuit must have an open circuit for the engine to run.

Troubleshooting

Refer to **Table 11** and **Table 13** at the end of the chapter for specifications and the individual wiring diagrams at the end of the manual.

Read *Troubleshooting Notes and Precautions (All Models)* at the beginning of the ignition section before continuing. The recommended troubleshooting procedure is listed below.

1. Preliminary checks.
   a. Spark test.
   b. Stop circuit isolation.
2. Switch box stop circuit test.
3. Ignition stator output test.
4. Ignition stator resistance test.
5. Trigger resistance test.
6. Switch box output test.
7. Ignition coil ohmmeter test.

WARNING
High voltage is present in the ignition system. Never touch or disconnect ignition components while the engine is running.

CAUTION
*Do not run the engine without an adequate water supply and do not exceed 3000 rpm without an adequate load. Refer to **Safety Precautions** at the beginning of this chapter.*

Preliminary checks

1. Disconnect the spark plug wires from the spark plugs and install an air gap spark tester (part No. FT-11295 or equivalent) to the spark plug wires. See **Figure 27**. Connect the alligator clip of the spark tester to a clean engine ground. Set the spark tester air gap to 7/16 in. (11.1 mm).
2. Make sure the safety lanyard is installed on the safety lanyard switch and the ignition switch (remote models) is in the *run* position.
3. Crank the engine while observing the tester. If there is a crisp, blue spark at each air gap, the ignition system is functioning correctly. If the engine will not start or does not run correctly, check the spark plugs and ignition timing. If the engine backfires or pops when attempting to start, remove the flywheel and check for a sheared flywheel key. If there is no spark, weak spark or erratic spark, continue with Step 4.
4. Disconnect the stop circuit black/yellow wire from the switch box bullet connector. Crank the engine while observing the spark tester. If there is a good spark, the stop circuit is shorted to ground.
   a. *Tiller control models*—There is a short to ground in the push button stop switch or safety lanyard switch. Test the push button stop switch and safety lanyard switch as described in this chapter.
   b. *Remote control models*—There is a short to ground in the key switch, safety lanyard switch or engine harness black/yellow wire. Test these components as described in this chapter.

Switch box stop circuit test

WARNING
To prevent accidental starting, disconnect the spark plug wire from the spark plug and install a spark gap tester, part No. FT11295 or an equivalent, to the spark plug wire. Connect the alligator clip of the spark tester to a clean engine ground.

1. Connect the meter negative wire to a good engine ground and the meter positive wire to the black/yellow wire bullet connector nearest the switch box.
2. Set the meter selector switch to 400 DVA.
3. Crank the engine while noting the test meter.
4. If the stop circuit voltage is within specification (**Table 11**), continue at *Ignition stator output test* in this section.
5. If the stop circuit voltage is above specification, either the switch box or the trigger coil is defective. Test the trigger coil(s) as described in *Trigger coil ohmmeter test* in this section. If the trigger coil(s) is within specification, replace the switch box.
6. If the stop circuit voltage is below specifications, disconnect the black/yellow stop circuit wire from the switch box bullet connector.
7. Repeat Step 3. If the stop circuit voltage is now within the specified range, the stop circuit is defective (partially shorted to ground) and must be repaired. If the stop switch voltage is still below specification, continue with the *Ignition stator output test* in this section.

Ignition stator output test

WARNING
To prevent accidental starting, disconnect the spark plug wire from the spark plug and install a spark gap tester, part No. FT11295 or equivalent, to the spark plug wire. Connect the alligator clip of the spark tester to a clean engine ground.

1A. *Black ignition stator (1998 models)*—Connect the meter positive wire to a good engine ground and the meter

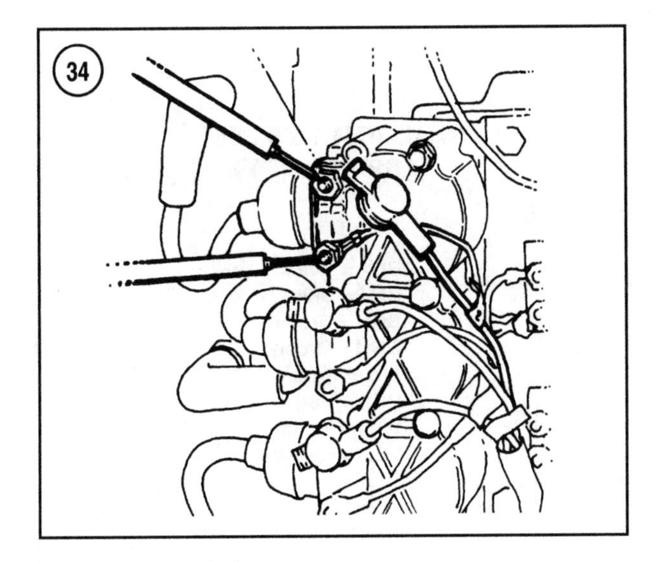

negative wire to the black/yellow ignition stator wire bullet connector.

1B. *Red ignition stator (1999-on models)*—Connect the meter negative wire to a good engine ground and the positive wire to the green/white ignition stator bullet connector.

2. Set the meter selector switch to 400 DVA.
3. Crank the engine and note the meter.

4A. *Black ignition stator (1998 models)*—If the reading is not within low speed specification (**Table 11**), test the resistance of the ignition stator windings as described in the next section.

4B. *Red ignition stator (1999-on models)*—If the reading is not within specification (**Table 11**), test the resistance of the ignition stator winding as described in the next section.

5. *Red ignition stator (1999-on models)*—Repeat the stator output test using the white/green ignition stator wire. See Steps 1-4.
6. *Black ignition stator (1998 models)*—Test the stator high speed winding output as follows:
   a. Connect the meter positive wire to a good engine ground and the meter negative wire to the black/white ignition stator wire bullet connector.
   b. Verify the meter is still set to the 400 DVA scale.
   c. Crank the engine and note the meter.
   d. The stator high speed winding output should be within the specification in **Table 11**. If it is not, test the resistance of the stator windings as described in this section.

Ignition stator resistance tests

1A. *Black ignition stator (1998 models)*—Disconnect the ignition stator black/yellow and black/white wires from the switch box bullet connectors.

1B. *Red ignition stator (1999-on models)*—Disconnect the ignition stator green/white and white/green wires from the switch box bullet connectors.

2A. *Black ignition stator (1998 models)*:
   a. Calibrate the ohmmeter on the appropriate scale to read the specifications in **Table 13** prior to measuring each circuit.
   b. Connect the positive test wire to the black/white stator wire. Connect the negative test wire to a good engine ground. Note the meter reading.
   c. Connect the positive test wire to the black/yellow stator wire. Connect the negative test wire to a good engine ground. Note the meter reading.
   d. Connect the positive test wire to the black/yellow stator wire. Connect the negative test wire to the black/white stator wire. Note the meter reading.

2B. *Red ignition stator (1999-on models)*—Connect the positive test wire to the green/white stator wire. Connect the negative test wire to the white/green stator wire. Note the meter reading.

3. All resistance readings should be within the specifications in **Table 13**. If they are not, replace the ignition stator as described in Chapter Seven.

Trigger coil resistance test

1. Disconnect the brown/white and brown/yellow trigger coil wires from the switch box bullet connectors.
2. Calibrate a digital ohmmeter on the R × 1 scale.
3. Connect one ohmmeter wire to the trigger coil brown/white wire and the other ohmmeter lead to the trigger coil brown/yellow wire .
4. The reading should be within the specification in **Table 11**. If it is not, replace the trigger coil as described in Chapter Seven.

Switch box output test

WARNING
To prevent accidental starting, disconnect the spark plug lead from the spark plug and install a spark gap tester, part No. FT11295 or an equivalent, to the spark plug lead. Connect the alligator clip of the spark tester to a clean engine ground.

1. Connect the meter positive lead to the cylinder No. 1 ignition coil positive primary terminal. Connect the meter negative lead to the cylinder No. 1 ignition coil negative primary terminal. See **Figure 34**.
2. Set the meter selector switch to 400 DVA.

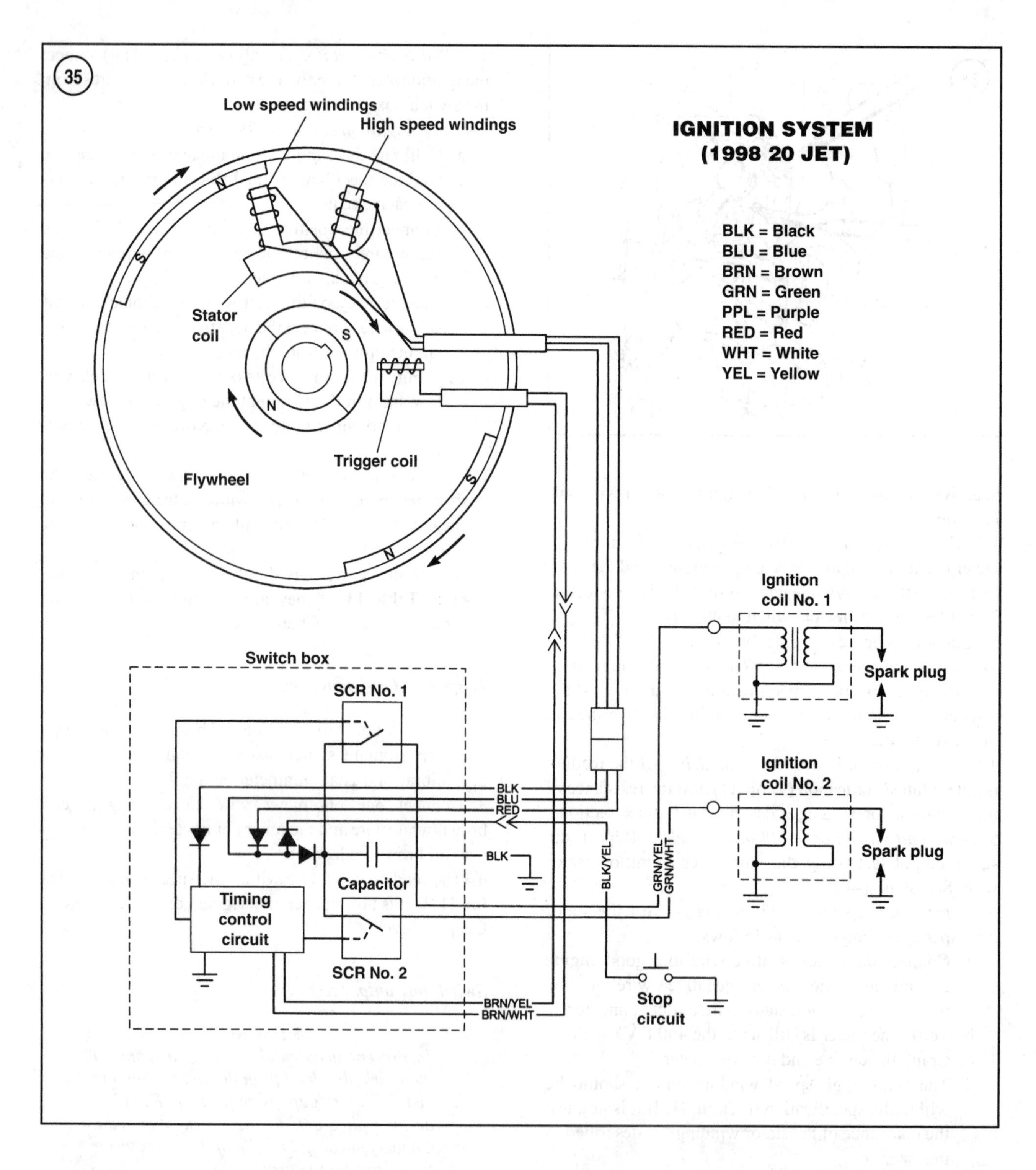

3. Crank the engine while noting the meter reading.
4. The meter reading should be within the specification in **Table 11**. If it is not, proceed as follows:
   a. *Meter reading below the specification*—Perform the ignition coil resistance test as described in this section. If the coil test correctly, replace the switchbox.
   b. *Meter reading within the specification*—If there is no spark, weak spark or erratic spark, the ignition coil or wiring is faulty. Perform the ignition coil ohmmeter test as described in this section.
5. Repeat the test procedure for the remaining ignition coil.

Ignition coil resistance test

1. Set an ohmmeter on the appropriate scale to read the primary resistance specification in **Table 11**.
2. Disconnect the spark plug wires. Carefully remove the spark plug wires from the ignition coils. Disconnect the switch box primary wire from each ignition coil positive primary terminal stud.
3. Connect the ohmmeter negative wire to the cylinder No. 1 ignition coil negative terminal. Connect the ohmmeter positive wire to the cylinder No. 1 ignition coil positive terminal stud.
4. The primary resistance reading should be within the specification in **Table 11**. If it is not, replace the ignition coil and retest spark output. Repeat the test for the remaining ignition coil(s).
5. Set the ohmmeter on the appropriate scale to read the secondary resistance specification in **Table 11**.
6. Connect the ohmmeter positive wire to the cylinder No. 1 ignition coil spark plug tower. Connect the ohmmeter negative wire to the cylinder No. 1 ignition coil primary positive terminal stud.
7. The secondary resistance reading should be within the specification in **Table 11**. If it is not, replace the ignition coil and retest for spark output.
8. If the meter reading is within specification and all other ignition components are within specification, but there is no spark, a weak spark or an erratic spark, replace the ignition coil(s) and retest spark output. The coil windings may arc internally during use, yet deliver correct resistance test readings.
9. Set an ohmmeter on the R × 1 scale. Connect one ohmmeter lead to each end of the spark plug wire. Gently twist and flex the spark plug wire while observing the meter. The meter should indicate a very low reading. If the meter indicates a high reading or fluctuates when the wire is flexed, replace the spark plug wire. Repeat the test for the remaining spark plug wires.

ALTERNATOR-DRIVEN IGNITION (1998 20 JET)

This AD-CDI ignition is an alternator driven, capacitor discharge ignition system with electronic spark advance. It is used on 1998 20 jet models.

Refer to **Figure 35** for an operational schematic of this ignition system. The major components include:

1. *Flywheel*—The flywheel inner magnets are for the trigger coil timing information. The outer magnets are for the ignition stator and battery charging stator, if equipped.
2. *Ignition stator*—The ignition stator provides the power the switch box needs to operate the ignition system. Stator output is always AC voltage. This model uses a black stator equipped with low- and high-speed windings and a ground wire. The ignition stator low-speed windings provide most of the electricity for low-speed, cranking and idle, running. High-speed windings provide most of the electricity for high-speed, cruising and wide-open throttle, running. The stator must be grounded to operate. A black wire carries the ground from the stator, through the three-pin connector and into the switch box.
3. *Trigger coil*—The trigger coil has two wires and is mounted in a fixed position. This model uses electronic spark advance. The trigger coil tells the switch box when to fire the ignition coils.
4. *Switch box*—The switch box stores the electricity from the ignition stator until the trigger coil tells it to send the electricity to the ignition coils. The switch box uses an internal rectifier to change the ignition stator AC voltage to DC voltage. The DC voltage is stored in a capacitor, until it is released by an SCR (silicon controlled rectifier), which is an electronic switch. There is one SCR for each cylinder. The SCRs are controlled by the trigger coil. As the engine accelerates, timing is advanced electronically. Maximum advance occurs at approximately 2500 rpm. Timing is retarded 10° any time engine speed exceeds 5800 rpm. Timing will advance up to 10° BTDC any time idle speed falls below 600 rpm. It is normal for timing to fluctuate 2-3° at idle speed.
5. *Ignition coils*—There is one ignition coil for each cylinder. The ignition coil transforms the relatively low voltage from the switch box into voltage high enough (35,000 volts) to jump the spark plug gap and ignite the air/fuel mixture.
6. *Spark plugs*—There is one spark plug for each cylinder. Use only the recommended spark plugs or serious engine damage may occur. Resistor or suppressor plugs are designed to reduce RFI (radio frequency interference) emissions that can cause interference with electrical accessories. Use the recommended RFI spark plug if RFI is suspected of causing interference or malfunction of electrical accessories.
7. *Stop circuit*—The stop circuit is connected to one end of the capacitor in the switch box. Whenever the stop circuit is connected to ground, the capacitor is shorted to ground and cannot store electricity. There is no voltage available to send to the ignition coil and the ignition system ceases producing spark. The stop circuit must have an open circuit for the engine to run.

Troubleshooting

Refer to **Figure 36** for this procedure. Refer to **Table 11** and **Table 13** for specifications and the individual wiring

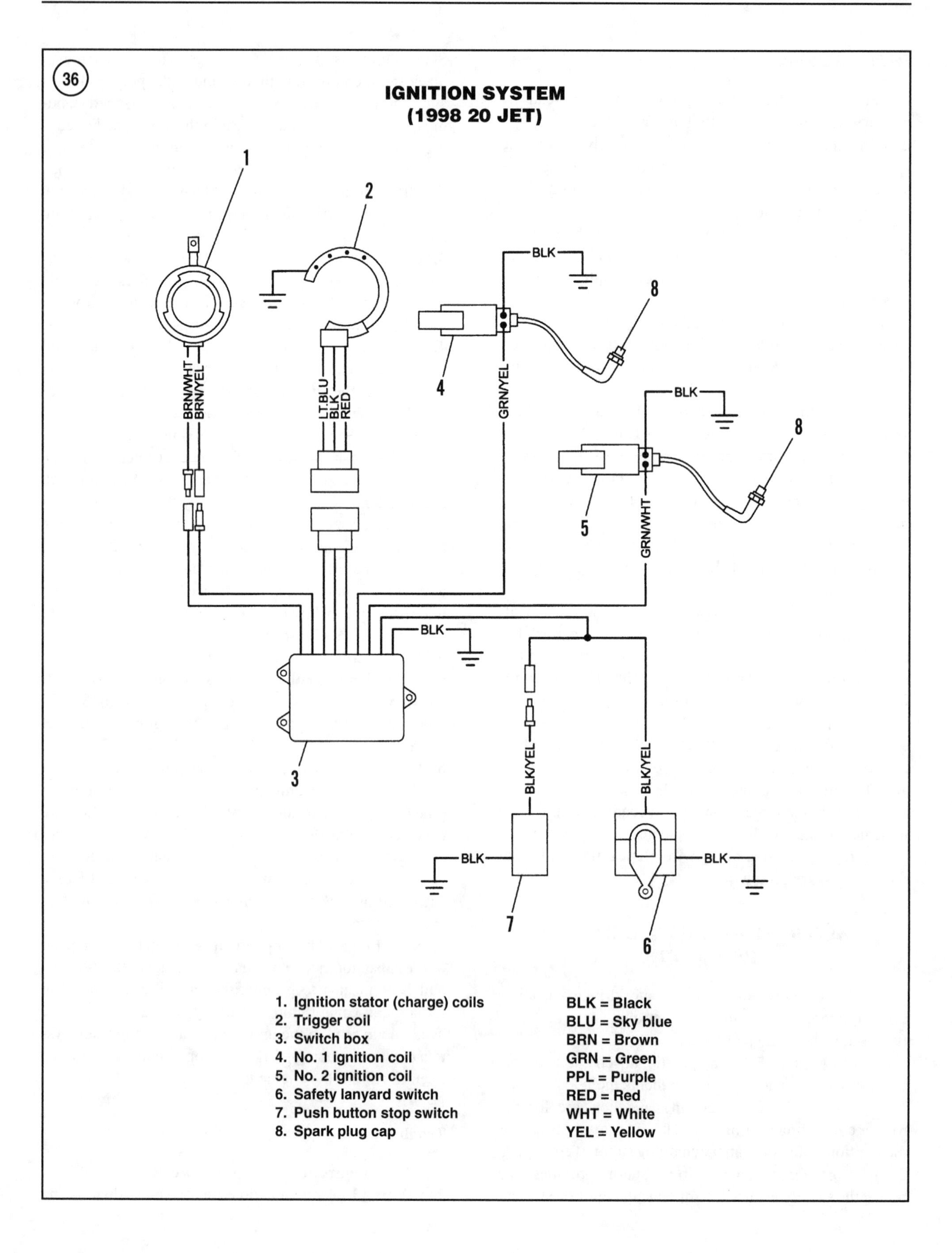
36
IGNITION SYSTEM
(1998 20 JET)
1
2
BLK
8
4
BRN/WHT
BRN/YEL
LT.BLU
BLK
RED
GRN/YEL
BLK
8
5
GRN/WHT
BLK
3
BLK/YEL
BLK/YEL
BLK
BLK
7
6
1. Ignition stator (charge) coils
2. Trigger coil
3. Switch box
4. No. 1 ignition coil
5. No. 2 ignition coil
6. Safety lanyard switch
7. Push button stop switch
8. Spark plug cap
BLK = Black
BLU = Sky blue
BRN = Brown
GRN = Green
PPL = Purple
RED = Red
WHT = White
YEL = Yellow

diagrams at the end of the manual. Read *Troubleshooting Notes and Precautions (All Models)* at the beginning of the ignition section before continuing. The recommended troubleshooting procedure is listed below.

1. Preliminary checks.
   a. Spark test.
   b. Stop circuit isolation.
2. Switch box stop circuit test.
3. Ignition stator output test.
4. Ignition stator resistance test.
5. Trigger resistance test.
6. Switch box output test.
7. Ignition coil resistance test.

WARNING
High voltage is present in the ignition system. Never touch or disconnect ignition components while the engine is running.

CAUTION
Do not run the engine without an adequate water supply and do not exceed 3000 rpm without an adequate load. Refer to Safety Precautions at the beginning of this chapter.

Preliminary checks

1. Disconnect the spark plug wires from the spark plugs and install an air gap spark tester (FT-11295 or equivalent) to the spark plug wires. Connect the alligator clip of the spark tester to a clean engine ground. Set the spark tester air gap to 7/16 in. (11.1 mm)
2. Make sure the safety lanyard is installed on the safety lanyard switch and the ignition switch (remote control models) is in the *run* position.
3. Crank the engine while observing the tester. If there is a crisp, blue spark at each air gap, the ignition system is functioning correctly. If the engine will not start or does not run correctly, check the spark plugs and ignition timing. If the engine backfires or pops when attempting to start, remove the flywheel and check for a sheared flywheel key. If there is no spark, a weak spark or an erratic spark, continue with Step 4.

NOTE
On 1998 20 jet models, it is normal for timing to fluctuate 2-3° at idle speed. Timing is retarded 10° any time engine speed exceeds 5800 rpm. Timing advances up to 10° any time idle speed falls below 600 rpm. All timing advance functions are controlled by circuits within the switchbox.

4. Disconnect the stop circuit black/yellow wire from the switch box bullet connector. Crank the engine while observing the spark tester. If there is a good spark, the stop circuit is shorted to ground.
   a. *Tiller control models*—The push button stop switch or safety lanyard is shorted to ground. Test the push button stop switch and safety lanyard switch as described in this chapter. Replace the defective switch and retest spark output.
   b. *Remote control models*—The key switch, safety lanyard switch or main engine harness black/yellow wire is shorted to ground. Test, repair or replace the circuit or component as necessary.

Switch box stop circuit test

WARNING
To prevent accidental starting, disconnect the spark plug wires from the spark plug and install a spark gap tester, part No. FT11295 or an equivalent, to the spark plug wires. Connect the alligator clip of the spark tester to a clean engine ground.

Refer to **Figure 36** for this procedure.

1. Connect the meter negative wire to a good engine ground and the meter positive wire to the black/yellow wire bullet connector nearest the switch box.
2. Set the meter selector switch to 400 DVA.
3. Crank the engine while noting the test meter.
4. If the stop circuit voltage is within specification (**Table 11**), continue at *Ignition stator output tests* in this section.
5. If the stop circuit voltage is above specification, either the switch box or the trigger coil is defective. Test the trigger coil as described in *Trigger coil resistance tests* in this section. If the trigger coil test correctly, replace the switch box.
6. If the stop circuit voltage is below specification, disconnect the black/yellow stop circuit wire from the switch box bullet connector.
7. Repeat Step 3. If the stop circuit voltage is now within the specified range, the stop circuit is defective (shorted to ground) and must be repaired. If the stop circuit voltage is still below specification, perform the *Ignition stator output tests* as described in this section.

Ignition stator output test

WARNING
To prevent accidental starting, disconnect the spark plug wires from the spark plug and install a spark gap tester, part No. FT11295 or equivalent, to the spark plug wires. Con-

3

nect the alligator clip of the spark tester to a clean engine ground.

Refer to **Figure 36** for this procedure.

1. Connect the meter negative wire to a good engine ground and the meter positive wire to the blue ignition stator wire at the switch box three-pin connector.
2. Set the meter selector switch to 400 DVA.
3. Crank the engine and note the meter.
4. The stator output voltage reading should be within the low speed winding specification in **Table 11**. If it is not, test the resistance of the ignition stator windings as described in the next section.
5. Connect the meter negative wire to a good engine ground and the meter positive wire to the red ignition stator wire at the switch box three-pin connector.
6. Make sure the meter is still set to the 400 DVA scale.
7. Crank the engine and note the meter.
8. The stator output voltage reading should be within the high speed winding specification in **Table 11**. If it is not, test the resistance of the ignition stator windings as described in the next section.

Ignition stator resistance tests

Refer to **Figure 36** for this procedure.

1. Disconnect the ignition stator three-pin connector from the switch box.
2. Set the ohmmeter on the appropriate scale to read the low speed winding specification in **Table 13**.
3. Connect the ohmmeter negative wire to the black stator wire at the three-pin connector and the ohmmeter positive wire to the blue ignition stator wire at the connector.
4. The meter reading should be within the specification in **Table 13**. If it is not, replace the stator coil assembly as described in Chapter Seven.
5. Set an ohmmeter on the appropriate scale to read the high speed winding specification in **Table 13**.
6. Connect the ohmmeter negative wire of the three-pin connector and the ohmmeter positive wire to the red stator wire at the connector.
7. The meter reading should be within the specification in **Table 13**. If it is not, replace the stator coil assembly as described in Chapter Seven.
8. Set the ohmmeter on the appropriate scale to read 2800-3400 ohms.
9. Connect the ohmmeter negative wire to the blue ignition stator wire and the ohmmeter positive wire to the red ignition stator wire at the ignition stator three-pin connector.
10. The reading should be within the specification in **Table 13**. If it is not, replace the ignition stator as described in Chapter Seven.

Trigger coil resistance test

1. Disconnect the brown/white and brown/yellow trigger coil wires from the switch box bullet connectors.
2. Set a digital ohmmeter on the R × 1 scale.
3. Connect one ohmmeter wire to the trigger coil brown/white wire and the other ohmmeter wire to the trigger coil brown/yellow wire.
4. The reading should be within the specification in **Table 11**. If it is not, replace the trigger coil as described in Chapter Seven.

Switch box output test

WARNING
To prevent accidental starting, disconnect the spark plug wires from the spark plugs and install a spark gap tester, part No. FT11295 or equivalent, to the spark plug wire. Connect the alligator clip of the spark tester to a clean engine ground.

1. Connect the meter positive wire to the cylinder No. 1 ignition coil positive primary terminal. Connect the meter negative wire to the cylinder No. 1 ignition coil negative primary terminal. See **Figure 34**.
2. Set the meter selector switch to 400 DVA.
3. Crank the engine while noting the meter reading.
4. The meter reading should be within the specification in **Table 11**. If it is not, proceed as follows:
   a. *Meter reading below the specification*—Perform the ignition coil resistance test as described in this section. If the coil test correctly, replace the switchbox.
   b. *Meter reading within the specification*—If there is no spark, a weak spark or an erratic spark, the ignition coil or wiring is faulty. Perform the ignition coil ohmmeter test as described in this section.
5. Repeat the test procedure for the remaining ignition coil.

Ignition coil ohmmeter test

1. Set an ohmmeter on the appropriate scale to read the primary resistance specification in **Table 11**.
2. Disconnect the spark plug wires from the spark plugs. Carefully remove the spark plug wires from the ignition coils. Disconnect the switch box primary

wire from each ignition coil positive primary terminal stud.

3. Connect the ohmmeter negative wire to the cylinder No. 1 ignition coil negative terminal. Connect the ohmmeter positive wire to the cylinder No. 1 ignition coil positive terminal stud.

4. The primary resistance reading should be within the specification in **Table 11**. If it is not, replace the ignition coil and retest spark output. Repeat the test for the remaining ignition coil.

5. Set the ohmmeter on the appropriate scale to read the secondary resistance specification in **Table 11**.

6. Connect the ohmmeter positive wire to the cylinder No. 1 ignition coil spark plug tower. Connect the ohmmeter negative wire to the cylinder No. 1 ignition coil primary positive terminal stud.

7. The secondary resistance reading should be within the specification in **Table 11**. If it is not, replace the ignition coil and retest for spark output.

8. If the meter reading is within specification and all other ignition components are within specification, but there is no spark, a weak spark or an erratic spark, replace the ignition coil(s) and retest spark output. The coil windings may arc internally during use, yet deliver correct resistance test readings.

9. Set the ohmmeter on the R × 1 scale. Connect one ohmmeter wire to each end of the spark plug wire. Gently twist and flex the spark plug wire while observing the meter. The meter should indicate a very low reading. If the meter indicates a high reading or fluctuates when the wire is flexed, replace the spark plug wire. Repeat the test for the remaining spark plug wires.

CAPACITOR DISCHARGE MODULE (CDM) IGNITION (30 AND 40 HP TWO-CYLINDER MODELS)

This CDM ignition is an alternator driven, capacitor discharge module system with mechanical spark advance. It is used on 30 and 40 hp two-cylinder models.

An ignition test harness, part No. 84-825207A2, is required to test the CDM system without damaging the wiring harness and connectors.

The major components include:

1. *Flywheel*—The flywheel inner magnet is for the trigger coil timing information. The outer magnets are for the ignition stator and battery charging stator.
2. *Ignition stator*—The stator (1, **Figure 37**) consists of one winding around three ignition charging bobbins. The ignition stator is not grounded to the power head. The ignition stator provides power to the CDM modules. Stator output is always AC voltage.
3. *Trigger coil*—The trigger coil (2, **Figure 37**) tells the CDM modules when to fire. The trigger coil is rotated by mechanical linkage to change the trigger's position relative to the flywheel. This advances or retards the ignition spark timing.
4. *CDM modules*—There is one CDM module (3 and 4, **Figure 37**) for each cylinder. The CDM module integrates the CD module (switch box) and ignition coil into one unit. The rectifier in each CDM transforms the ignition stator AC voltage into DC voltage so it can be stored in the CDM module capacitor. The capacitor holds the voltage until the SCR (silicon controlled rectifier), which is an electronic switch, releases the voltage to the integral ignition coil primary windings. The SCR is controlled by the trigger coil signal. The ignition coil transforms the relatively low voltage from the capacitor into voltage high enough (45,000 volts) to jump the spark plug gap and ignite the air/fuel mixture.
5. *Spark plugs*—There is one spark plug for each cylinder. Use only the recommended spark plugs or engine damage may result. Resistor or suppressor plugs are designed to reduce RFI (radio frequency interference) emissions that can cause interference with electrical accessories. Use the recommended RFI spark plug if RFI is suspected of causing interference or malfunction of electrical accessories.
6. *Stop circuit*—The stop circuit is connected by the black/yellow wire to one end of the capacitor in each CDM module. Whenever the stop circuit connects to ground, the capacitor is shorted and cannot store electricity. Since no voltage is available to send to the ignition coil windings, the ignition system ceases producing spark. The stop circuit must have an open circuit for the engine to run.

Troubleshooting

Refer to **Table 12** for ignition system specifications and the wiring diagrams at the end of the manual. Read *Troubleshooting Notes and Precautions (All Models)* at the beginning of the ignition section before continuing.

The recommended troubleshooting procedure is listed below.

1. Preliminary checks.
2. Ground circuit verification test.
3. Stop circuit isolation test.
4. Ignition stator output test
5. Ignition stator resistance test.
6. Trigger output test.

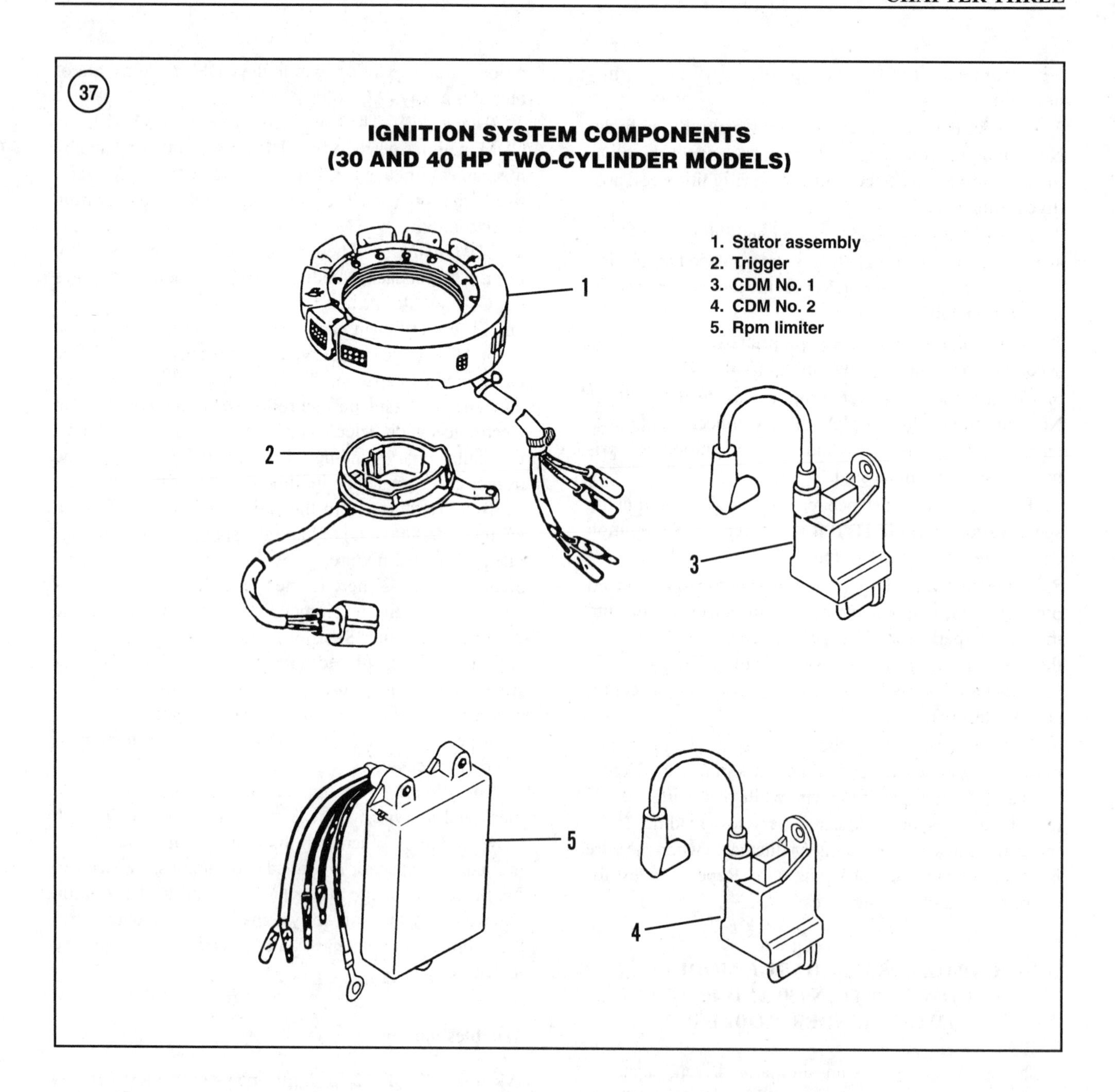

7. CDM module resistance test (optional).
8. Rpm-limit module test.

WARNING
High voltage is present in the ignition system. Do not touch or disconnect ignition components while the engine is running.

CAUTION
*Do not run the engine without an adequate water supply and do not exceed 3000 rpm without and adequate load. Refer to **Safety Precautions** at the beginning of this chapter.*

NOTE
Both CDM modules must be connected to the wiring harness during any output test.

Preliminary checks

1. Disconnect the spark plug wires from the spark plugs and install an air gap spark tester (part No. FT-11295 or equivalent) to the spark plug wires. Connect the alligator clip of the spark tester to a clean engine ground. Set the spark tester air gap to 7/16 in. (11.1 mm).

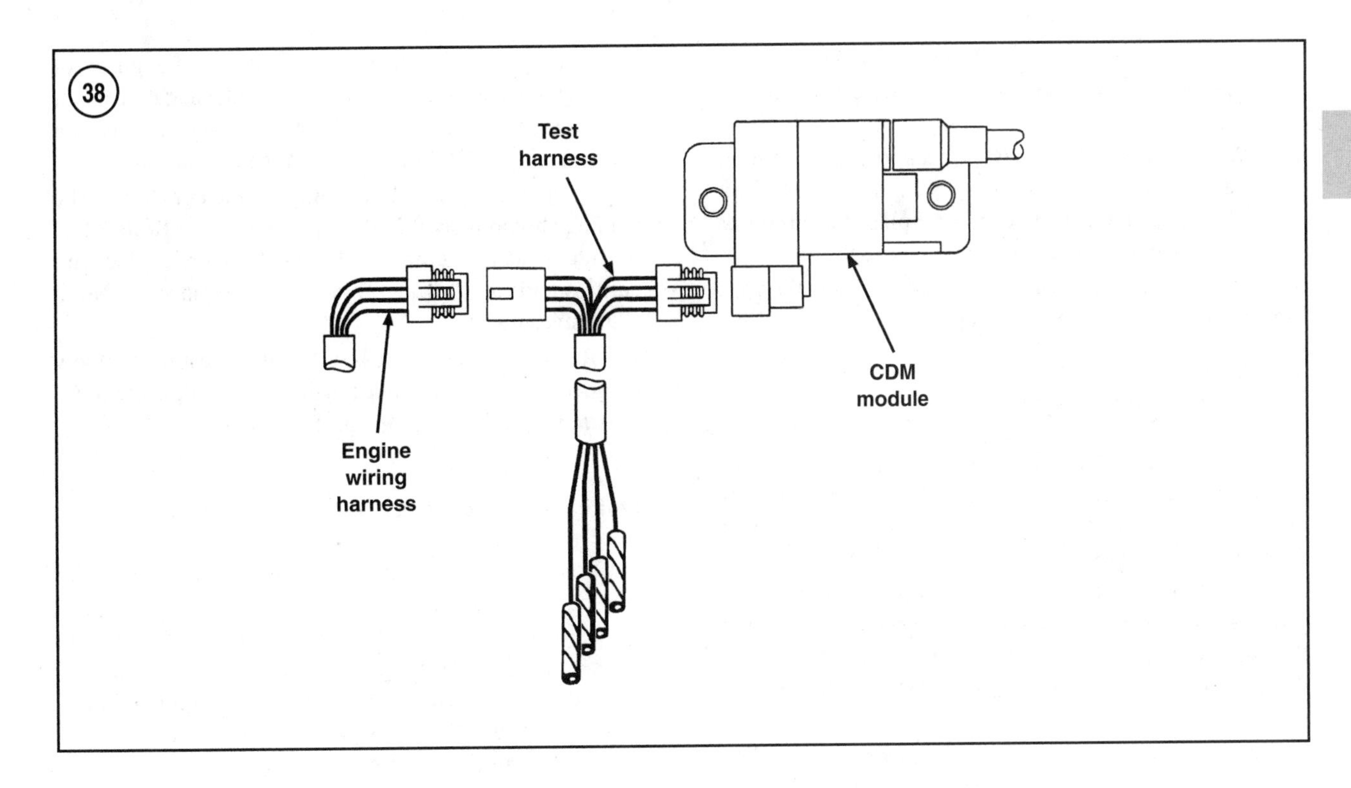

2. Make sure the safety lanyard is installed on the safety lanyard switch and the ignition switch is in the *run* position.

3. Crank the engine while observing the tester. If there is a crisp, blue spark at each air gap, the ignition system is functioning correctly. If the engine will not start or does not run correctly, check the spark plugs and ignition timing. If the engine backfires or pops when attempting to start, remove the flywheel and check for a sheared flywheel key. If there is no spark, a weak spark or an erratic spark, continue with *Ground circuit verification* in the next section.

Ground circuit verification test

1. Disconnect both CDM module plugs. Connect the test harness adapter (part No. 84-825207A2) to the wiring harness connector of the cylinder No. 1 CDM module. See **Figure 38**. Do not connect the test harness to the CDM module. Set an ohmmeter on the highest scale available. Connect one wire of the ohmmeter to a clean engine ground, connect the other ohmmeter wire to the test harness black wire. The ohmmeter should indicate continuity. If it does not, repair or replace the engine harness black wire or CDM module connector as necessary. Repeat the test for the cylinder No. 2 CDM wiring harness connector.

2. Continue to Step 3 once both CDM module ground circuits have been verified.

3. Verify the grounding of the ignition component mounting plate by connecting one wire of an ohmmeter set on the highest scale to a clean engine ground and the other wire to the ground wire terminal on the ignition component mounting plate. The ohmmeter should indicate continuity. If it does not, repair or replace the ground path wires and/or ground stud as required.

Stop circuit isolation test

WARNING

To prevent accidental starting, disconnect the spark plug wires from the spark plugs and install a spark gap tester, part No. FT11295 or equivalent, to the spark plug wire. Connect the alligator clip of the spark tester to a clean engine ground.

1. Isolate the stop circuit from the CDM ignition system by disconnecting the black/yellow bullet connector near the CDM modules. Refer to the wiring diagrams at the end of the manual for connector location. Make sure the black/yellow wire is not touching any other wire or ground.

2. Crank the motor and observe the spark tester. If there is good spark now, a short to ground is present in the stop circuit (black/yellow).

a. *Tiller control models*—Check the safety lanyard and push button stop switch as described in this chapter.

b. *Remote control models*—Check the key switch, safety lanyard switch and main engine harness for a short to ground. Test, repair or replace the circuit or component as necessary.

3. If the switches and wiring test correctly, test the rpm limit module as described in this section.

Ignition stator output test

1. Connect the test harness adapter (part No. 84-825207A2) to the wiring harness connector of the cylinder No. 1 CDM module and the engine wire harness connector. See **Figure 38**.
2. Set the multimeter to the 400 DVA scale. Connect the positive meter wire to the red test harness wire. Connect the negative meter wire to an engine ground. Do not disconnect the ignition stator wires or CDM module from the test harness.
3. Crank the engine while noting the meter reading. The reading should be within the specification in **Table 12**. If it is not, disconnect the No. 2 CDM from the engine wire harness and retest. If the voltage rises, replace the No. 2 CDM module.
4. Connect the positive meter wire to the green or white/green test harness lead. Connect the negative meter wire to an engine ground. Do not disconnect the ignition stator wires or CDM module from the test harness.
5. Crank the engine while noting the meter reading. The reading should be within the specification in **Table 12**. If it is not, disconnect the No. 2 CDM from the engine wire harness and retest. If the voltage rises, replace the No. 2 CDM module.
6. Remove the test harness from the No. 1 CDM module and engine wire harness. Reconnect the wire harness to the No. 1 CDM module.
7. Connect the test harness adapter (part No. 84-825207A2) to the wiring harness connector of the cylinder No. 2 CDM module and the engine wire harness connector. See **Figure 38**.
8. Connect the positive meter wire to the red test harness wire. Connect the negative meter wire to an engine ground. Do not disconnect the ignition stator wires or CDM module from the test harness.
9. Crank the engine while noting the meter reading. The reading should be within the specification in **Table 12**. If it is not, disconnect the No. 1 CDM from the engine wire harness and retest. If the voltage rises, replace the No. 1 CDM module.
10. Connect the positive meter wire to the green or white/green test harness wire. Connect the negative meter wire to an engine ground. Do not disconnect the ignition stator wires or CDM module from the test harness.
11. Crank the engine while noting the meter reading. The reading should be within the specification in **Table 12**. If it is not, disconnect the No. 2 CDM from the engine wire harness and retest. If the voltage rises, replace the No. 2 CDM module.
12. Remove the test harness. If the meter reading is below the specification in **Table 12** in the tests, continue with *Ignition stator resistance test* in this section.

Ignition stator resistance test

1. Disconnect the green/white and white/green stator wire bullet connectors. Set an ohmmeter on the appropriate scale to read the specification in **Table 13**. Connect an ohmmeter wire to each ignition stator wire. Note the meter reading. The meter reading should be within the specification in **Table 13**. If it is not, replace the ignition stator as described in Chapter Seven.
2. Calibrate an ohmmeter on the highest scale available. Connect one ohmmeter wire to a good engine ground and the other ohmmeter wire alternately to the ignition stator green/white and white/green wires while noting the meter. The meter should indicate no continuity. If it does not, replace the ignition stator as described in Chapter Seven.

Trigger output test

1. Connect the test harness adapter (part No. 84-825207A2) to the wiring harness connector of the cylinder No. 1 CDM module and the engine wire harness connector. See **Figure 38**.
2. Set the multimeter to the 20 DVA scale. Connect the positive meter wire to the white test harness wire. Connect the negative meter wire to the black test harness wire. Do not disconnect the ignition stator wires or CDM module from the test harness.
3. Crank the engine while noting the meter reading. The reading should be within the trigger output specification in **Table 12**. If the reading is below the specification, replace the trigger as described in Chapter Seven and retest. If a low trigger voltage persist, replace the No. 1 CDM module.
4. Remove the test harness and reconnect the engine harness to the No. 1 CDM module.
5. Connect the test harness adapter (part No. 84-825207A2) to the wiring harness connector of the cyl-

inder No. 2 CDM module and the engine wire harness connector. See **Figure 38**.

6. Set the multimeter to the 20 DVA scale. Connect the positive meter wire to the white test harness lead. Connect the negative meter wire to the black test harness wire. Do not disconnect the ignition stator wires or CDM module from the test harness.

7. Crank the engine while noting the meter reading. The reading should be within the trigger output specification in **Table 12**. If the reading is below the specification, replace the trigger as described in Chapter Seven and retest. If a low trigger voltage persist, replace the No. 2 CDM module.

CDM module resistance test

Refer to **Table 12** for specifications. Use an analog multimeter for this procedure. Refer to the letter stamped on the side of the CDM terminal housing to identify the test wire connection points.

1. Calibrate an ohmmeter to the R × 100 scale.

2. Connect the ohmmeter positive wire to the CDM *C* terminal and the ohmmeter negative wire to the CDM *A* terminal. The resistance reading should be within the specification in **Table 12**. If it is not, replace the CDM module.

NOTE
Steps 3A-4B are the diode test. The ohmmeter readings may be reversed depending on the polarity of the ohmmeter used. The test is correct if the first part of the test is opposite of the second part.

3A. Connect the ohmmeter positive wire to the CDM *B* terminal. Connect the ohmmeter negative wire to the CDM *D* terminal. The meter should read continuity.

3B. Connect the ohmmeter positive wire to the CDM *D* terminal. Connect the ohmmeter negative wire to the CDM *B* terminal. The meter should read no continuity.

4A. Connect the ohmmeter positive wire to the CDM *D* terminal. Connect the ohmmeter negative wire to the CDM *A* terminal. The meter should read continuity.

4B. Connect the ohmmeter positive wire to the CDM *A* terminal. Connect the ohmmeter negative wire to the CDM *D* terminal. The meter should read no continuity.

5. Connect the ohmmeter positive wire to the spark plug lead terminal inside spark plug boot. Connect the negative test wire to the CDM *A* terminal. The meter reading should be within the specification in **Table 12**. If it is not, replace the CDM module.

RPM Limit Module

All 1998-on 30 and 40 hp models are equipped with a rpm limit module. The rpm limit module is connected to the CDM stop circuit. If the engine speed exceeds the preprogrammed limit, the rpm limit module momentarily shorts the black/yellow wire to ground, limiting engine speed. There are four wires going to the rpm limit module. The purple wire is power for the module from the key switch on electric start models or the ignition stator on manual start models.

The brown wire connects to the brown trigger coil wire and is an rpm signal for the module. The black/yellow wire connects to the switch box stop circuit and is shorted to ground by the module to control engine speed by switching the ignition system on and off. The black wire is the ground path for the module.

Troubleshooting the rpm limit module

Refer to the end of the manual for individual wiring diagrams.

CAUTION
*Do not run the engine without an adequate water supply and do not exceed 3000 rpm without an adequate load. Refer to **Safety Precautions** at the beginning of this chapter before continuing.*

NOTE
If the rpm limit module is suspected of causing a high-speed misfire, check the engine speed with an accurate shop tachometer. The dash mounted tachometer may be faulty or inaccurate. The high-speed misfire may be caused by engine over-speeding without the operator's knowledge.

1. If the rpm limit module is suspected of causing a lack of spark, a weak spark, an erratic spark or a misfire while running, disconnect the rpm limit module black/yellow and brown engine harness bullet connectors from the rpm limit module.

2. Retest spark output or run the engine at the speed at which the misfire occurs. If the spark output is now satisfactory or the misfire is no longer present, the rpm limit module is defective and must be replaced.

3. If the rpm limit module is suspected of not functioning when needed, check the ground lead for a secure attachment to the power head. Clean and tighten the connection as necessary to assure a good ground path.

4. Disconnect the purple wire bullet connector from the rpm limit module. Connect the positive lead of a voltme-

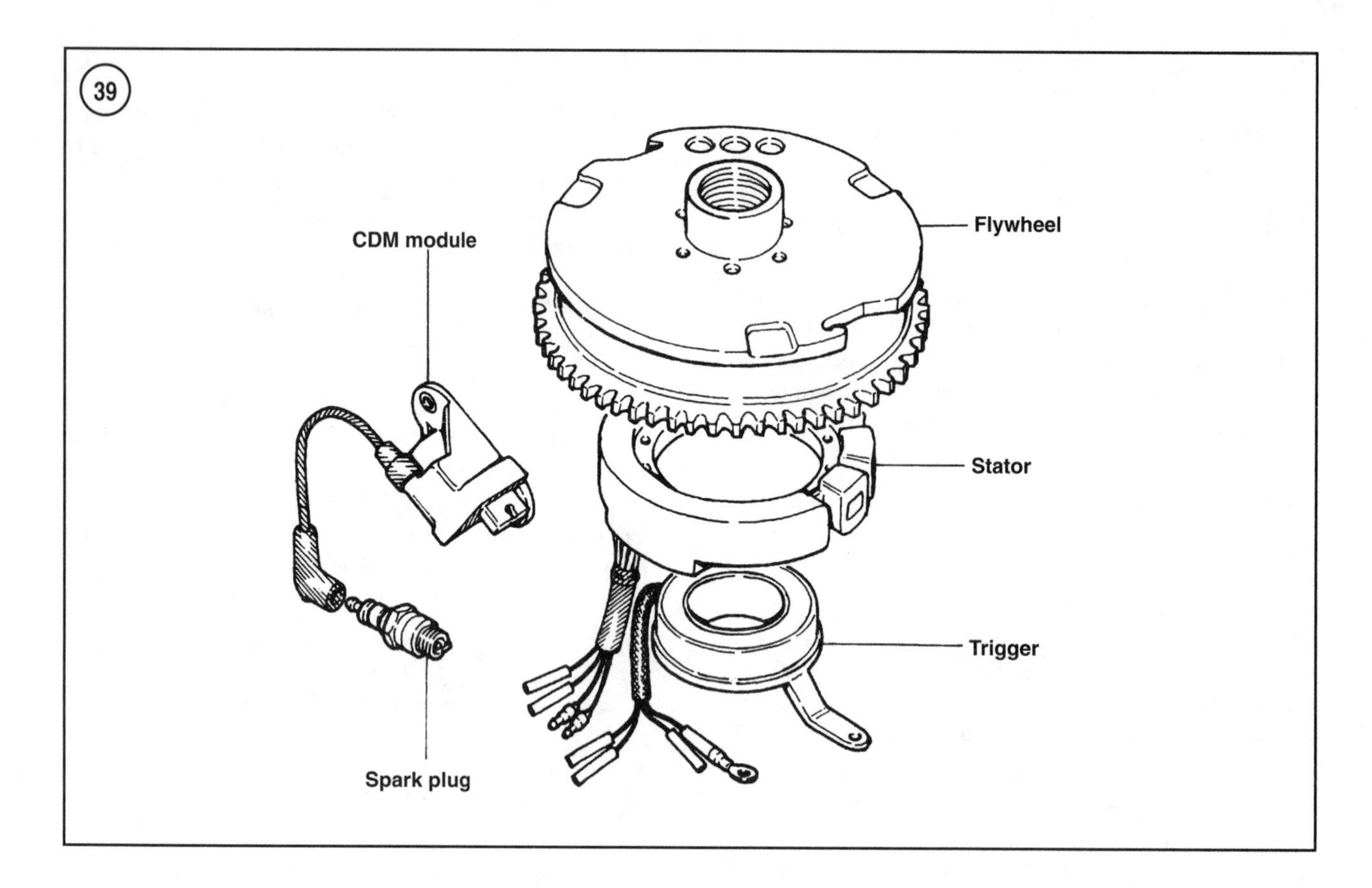

ter set to read 20 volts DC to the engine harness purple or light blue/white wire connector. Connect the negative wire of the voltmeter to a good engine ground.

a. *Electric start models*—With the ignition key *on*, the voltmeter should read within 1 volt of battery voltage. When the key is *off*, the voltmeter should read zero volts. If the meter reading is below specification, repair or replace the purple wire from the key switch to the engine harness rpm limit module bullet connector.
b. *Manual start models*—With the engine running, the voltmeter should read 10 volt or higher. If it does not, replace the ignition stator.

CAPACITOR DISCHARGE MODULE (CDM) IGNITION (40-60 HP THREE-CYLINDER MODELS)

This CDM ignition is an alternator driven, capacitor discharge module system with mechanical spark advance. It is used on 40-60 hp, three-cylinder models. An ignition test harness part No. 84-825207A2 is required to test the CDM system without damaging the wiring harness and connectors.

The major components (**Figure 39**) include:

1. *Flywheel*—The flywheel inner magnet is for the trigger coil timing information. The outer magnets are for the ignition stator and battery charging stator.

NOTE
The ignition stator circuit must be complete from the stator to a CDM module and back to the stator through a different CDM module in order for the system to function.

2. *Ignition stator*—The stator consists of one winding around three bobbins. The ignition stator is not grounded to the power head. The ignition stator provides power to the CDM modules. Stator output is always AC voltage. The voltage return path for cylinder No. 1 CDM module is either cylinder No. 2 or No. 3 CDM module. The voltage return path for cylinder No. 2 and No. 3 CDM modules is through cylinder No. 1 CDM module. See **Figure 40**.
3. *Trigger*—The trigger coil tells the CDM modules when to fire. The trigger coil is rotated by mechanical linkage to change the trigger's position relative to the flywheel. This advances or retards the ignition spark timing. The trigger coil has three windings grounded by a common black wire. The three individual trigger wires are brown, white and purple. Failure of the common black wire can cause erratic spark on all cylinders. Failure of a

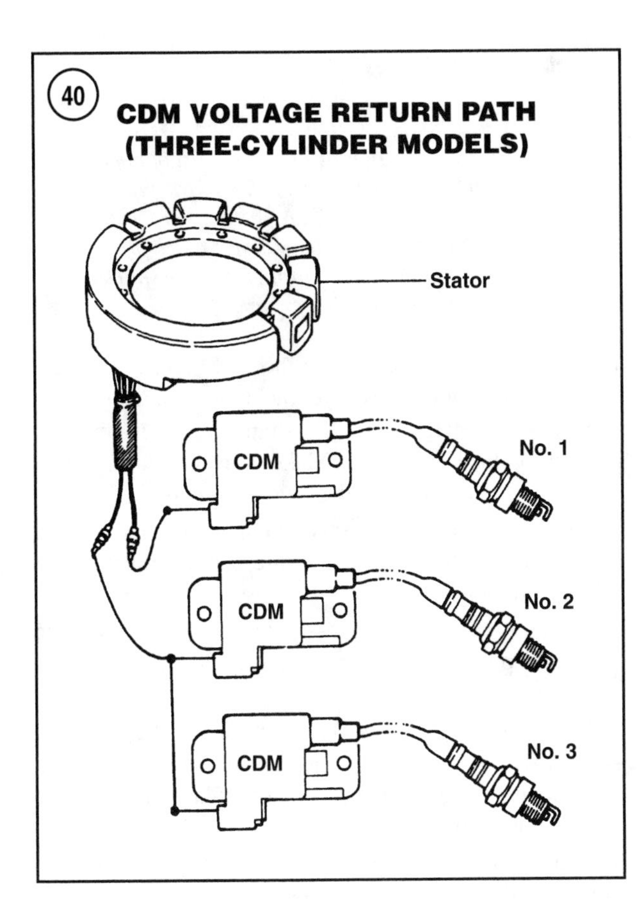

single trigger wire causes the loss of spark on one cylinder.

4. *CDM modules*—There is one CDM module for each cylinder. The CDM module integrates the CD module switch box and ignition coil into one unit. The rectifier in each CDM transforms the ignition stator AC voltage into DC voltage so it can be stored in the CDM module capacitor. The capacitor holds the voltage until the SCR (silicon controlled rectifier), which is an electronic switch, releases the voltage to the integral ignition coil primary windings. The SCR is triggered by the trigger coil signal. The ignition coil transforms the relatively low voltage from the capacitor into voltage high enough (45,000 volts) to jump the spark plug gap and ignite the air/fuel mixture.

5. *Spark plugs*—There is one spark plug for each cylinder. Use only the recommended spark plugs or engine damage may result. Resistor or suppressor plugs are designed to reduce RFI (radio frequency interference) emissions that can cause interference with electrical accessories. Use the recommended RFI spark plug if RFI is suspected of causing interference or malfunction of electrical accessories.

6. *Stop circuit*—The stop circuit is connected by the black/yellow wire to one end of the capacitor in each CDM module. Whenever the stop circuit connects to ground, the capacitor is shorted and cannot store electricity. Since no voltage is available to send to the ignition coil windings, the ignition system ceases producing spark. The stop circuit must have an open circuit for the engine to run.

NOTE

The 40-60 hp models are equipped with an rpm limit module. The rpm limit module connects to the CDM module's stop circuit (black/yellow). If the engine speed exceeds the preprogrammed limit, the rpm limit module momentarily shorts the black/yellow wire to ground, interrupting ignition and limiting engine speed.

Troubleshooting

Refer to **Table 12** for specifications and the wiring diagrams at the end of the manual. Read *Troubleshooting Notes and Precautions (All Models)* at the beginning of the ignition section before continuing.

The recommended troubleshooting procedure is listed below.

1. Preliminary checks.
2. Ground circuit verification test.
3. Stop circuit isolation test.
4. Ignition stator output test
5. Ignition stator resistance test.
6. Trigger output test.
7. CDM module resistance test (optional).
8. Rpm-limit module test.

WARNING

High voltage is present in the ignition system. Do not touch or disconnect ignition components while the engine is running.

CAUTION

*Do not run the engine without an adequate water supply and do not exceed 3000 rpm without an adequate load. Refer to **Safety Precautions** at the beginning of this chapter.*

NOTE

All CDM modules must be connected when troubleshooting the system. Disconnecting the No. 1 CDM module causes a loss of spark on all cylinders. Disconnecting either stator wire causes loss of spark on all cylinders.

Preliminary checks

1. Disconnect the spark plug wires from the spark plugs and install an air gap spark tester (part No. FT-11295 or equivalent) to the spark plug wires. Connect the alligator clip of the spark tester to a clean engine ground. Set the spark tester air gap to 7/16 in. (11.1 mm).
2. Make sure the safety lanyard is installed on the safety lanyard switch and the ignition switch is in the *run* position.
3. Crank the engine while observing the tester. If there is a crisp, blue spark at each air gap, the ignition system is functioning correctly. If the engine will not start or does not run correctly, check the spark plugs and ignition timing. If the engine backfires or pops when attempting to start, remove the flywheel and check for a sheared flywheel key. If there is no spark, a weak spark or an erratic spark, continue with *Ground circuit verification test* in the next section.

Ground circuit verification test

NOTE

To prevent damage to the connector pins, use test harness adapter (part No. 84-825207A 2) for all tests involving connection to the CDM modules and engine harness connectors. See ***Figure 38****.*

1. Disconnect both CDM module plugs. Connect the test harness adapter (part No. 84-825207A2) to the wiring harness connector of the cylinder No. 1 CDM module. See **Figure 38**. Do not connect the test harness to the CDM module. Set an ohmmeter on the highest scale available. Connect one wire of the ohmmeter to a clean engine ground. Connect the other ohmmeter wire to the test harness black wire. The ohmmeter should indicate continuity. If it does not, repair or replace the engine harness black wire or CDM module connector as necessary. Repeat the test for cylinder No. 2 CDM wiring harness connector.
2. Repeat Step 1 for each of the remaining CDM module harness connectors. Continue to Step 3 once all CDM module grounds are verified.
3. Check the grounding of the ignition component mounting plate by connecting one wire of an ohmmeter calibrated on the highest scale to a clean engine ground and the other wire to the ground wire terminal on the ignition component mounting plate. The ohmmeter should indicate continuity. If it does not, repair or replace the ground path wires and/or ground stud.

Stop circuit isolation test

WARNING

To prevent accidental starting, disconnect the spark plug wires from the spark plugs and install a spark gap tester, part No. FT11295 or an equivalent, to the spark plug wires. Connect the alligator clip of the spark tester to a clean engine ground.

1. Isolate the stop circuit from the CDM ignition system by disconnecting the black/yellow bullet connector. This connector is located in the tie strapped bundle of wires near the voltage regulator. Make sure the black/yellow wire is not touching any other wire or ground.
2. Crank the motor and observe the spark tester. If there is a good spark, a short to ground is present in the stop circuit (black/yellow wire).
   a. *Tiller control models*—Check the safety lanyard and push button stop switch as described in this chapter.
   b. *Remote control models*—Check the key switch, safety lanyard switch and main engine harness for a short to ground. Test, repair or replace the circuit or component as necessary.
3. If the switches and wiring test correctly, test the rpm limit module as described in this section.

Ignition stator output test

WARNING

To prevent accidental starting, disconnect the spark plug wires from the spark plugs and install a spark gap tester, part No. FT11295 or an equivalent, to the spark plug wires. Connect the alligator clip of the spark tester to a clean engine ground.

1. Connect the test harness adapter (part No. 84-825207A2) to the wiring harness connector of the cylinder No. 1 CDM module and the engine wire harness connector. See **Figure 38.**
2. Set the multimeter to the 400 DVA scale. Connect the positive meter wire to the white test harness wire. Connect the negative meter wire to the black test harness wire. Do not disconnect the ignition stator wires or CDM module from the test harness.
3. Crank the engine while noting the meter reading. The reading should be within the specification in **Table 12**. If it is not, disconnect the No. 2 CDM module from the engine wire harness and retest. If the voltage rises, replace the No. 2 CDM module. Disconnect the No. 3 CDM mod-

ule from the engine wire harness and retest. If the voltage rises, replace the No. 3 CDM module.

4. Remove the test harness from the No. 1 CDM module and engine wire harness. Reconnect the wire harness to the No. 1 CDM module.

5. Connect the test harness adapter (part No. 84-825207A2) to the wiring harness connector of the cylinder No. 2 CDM module and the engine wire harness connector. See **Figure 38**.

6. Connect the positive meter wire to the white test harness lead. Connect the negative meter wire to the black test harness wire .

7. Crank the engine while noting the meter reading. The reading should be within the specification in **Table 12**. If it is not, disconnect the No. 1 CDM module from the engine wire harness and retest. If the voltage rises, replace the No. 1 CDM module. Disconnect the No. 3 CDM module from the engine wire harness and retest. If the voltage rises, replace the No. 3 CDM module.

8. Remove the test harness from the No. 2 CDM module and engine wire harness. Reconnect the wire harness to the No. 2 CDM module.

9. Connect the test harness adapter (part No. 84-825207A2) to the wiring harness connector of the cylinder No. 3 CDM module and the engine wire harness connector. See **Figure 38**.

10. Connect the positive meter wire to the white test harness wire. Connect the negative meter wire to the black test harness wire.

11. Crank the engine while noting the meter reading. The reading should be within the specification in **Table 12**. If it is not, disconnect the No. 1 CDM module from the engine wire harness and retest. If the voltage rises, replace the No. 1 CDM module. Disconnect the No. 2 CDM module from the engine wire harness and retest. If the voltage rises, replace the No. 2 CDM module. Remove the test harness and reconnect the engine wire harness.

12. Remove the test harness. If the meter reading in all steps are below the specification in **Table 12**, continue with *Ignition stator resistance test* in this section.

Ignition stator resistance tests

1. Disconnect the green/white and white/green stator wire bullet connectors. Calibrate an ohmmeter on the appropriate scale to read the specification in **Table 13**. Connect an ohmmeter wire to each ignition stator wire. Note the meter reading. The meter reading should be within the specification in **Table 13**. If it is not, replace the ignition stator as described in Chapter Seven.

2. Calibrate an ohmmeter on the highest scale available. Connect one ohmmeter wire to a good engine ground and the other ohmmeter wire alternately to the ignition stator green/white and white/green wires while noting the meter. The meter should indicate no continuity. If it does not, replace the ignition stator as described in Chapter Seven.

3

Trigger output test

WARNING

To prevent accidental starting, disconnect the spark plug wires from the spark plugs and install a spark gap tester, part No. FT11295 or an equivalent, to the spark plug wires. Connect the alligator clip of the spark tester to a clean engine ground.

NOTE

Pin C of the CDM module connector is the trigger lead. The trigger wires on the ignition harness are color coded for each cylinder. Cylinder No. 1 uses a purple trigger wire. Cylinder No. 2 uses a white trigger wire and cylinder No. 3 uses a brown trigger wire.

1. Install the test harness part No. 84-825207A2 between the cylinder No. 1 CDM module and the ignition harness. See **Figure 38**.

2. Set the multimeter to the lowest DVA scale. Connect the meter positive wire to the test harness white wire and the meter negative wire to the test harness black wire. Crank the engine while noting the meter reading. The meter reading should be within the cranking speed specification in **Table 12**.

   a. *Trigger voltage below the specification*—Replace the trigger and retest.

   b. *Trigger voltage above the specification*—Replace the No. 1 CDM module. Repeat the test for each remaining CDM module.

NOTE

If a trigger voltage reading remains low after installing a new trigger, replace the CDM module to which the low reading trigger wire connects.

CDM module resistance test

Refer to **Table 12** for specifications. Use an analog multimeter for this procedure. Refer to the letter stamped on the side of the CDM terminal housing to identify the test wire connection points.

1. Set an ohmmeter to the R × 100 scale.

2. Connect the ohmmeter positive wire to the CDM *C* terminal and the ohmmeter negative wire to the CDM *A* ter-

minal. The resistance reading should be within the specification in **Table 12**. If it is not, replace the CDM module.

NOTE
Steps 3A-4B are the diode test. The ohmmeter readings may be reversed depending on the polarity of the ohmmeter used. The test is correct if the first part of the test is opposite of the second part.

3A. Connect the ohmmeter positive wire to the CDM *B* terminal. Connect the ohmmeter negative wire to the CDM *D* terminal. The meter should read continuity.

3B. Connect the ohmmeter positive wire to the CDM *D* terminal. Connect the ohmmeter negative wire to the CDM *B* terminal. The meter should read no continuity.

4A. Connect the ohmmeter positive wire to the CDM *D* terminal. Connect the ohmmeter negative wire to the CDM *A* terminal. The meter should read continuity.

4B. Connect the ohmmeter positive wire to the CDM *A* terminal. Connect the ohmmeter negative wire to the CDM *D* terminal. The meter should read no continuity.

5. Connect the ohmmeter positive wire to the spark plug lead terminal inside spark plug boot. Connect the negative test wire to the CDM *A* terminal. The meter reading should be within the specification in **Table 12**. If it is not, replace the CDM module.

RPM Limit Module

All 1998 40-60 hp models are equipped with a rpm limit module. The rpm limit module connects to the CDM stop circuit. If the engine speed exceeds the preprogrammed limit, the rpm limit module momentarily shorts the black/yellow wire to ground, limiting engine speed. There are four wires going to the rpm limit module. The purple wire is power for the module from the key switch on electric start models or the ignition stator on manual start models.

The brown wire connects to the brown trigger coil wire and is an rpm signal for the module. The black/yellow wire connects to the switch box stop circuit and is shorted to ground by the module to control engine speed by switching the ignition system on and off. The black wire is the ground path for the module.

Troubleshooting the rpm limit module

Refer to the end of the manual for individual wiring diagrams.

CAUTION
*Do not run the engine without an adequate water supply and do not exceed 3000 rpm without an adequate load. Refer to **Safety Precautions** at the beginning of this chapter before continuing.*

NOTE
If the rpm limit module is suspected of causing a high-speed misfire, check the engine speed with an accurate shop tachometer. The dash mounted tachometer may be faulty or incorrectly set. The high-speed misfire may be caused by engine over-speeding without the operator's knowledge.

1. If the rpm limit module is suspected of causing a lack of spark, a weak spark, an erratic spark or a misfire while running, disconnect the rpm limit module black/yellow and brown engine harness bullet connectors from the rpm limit module.

2. Retest spark output or run the engine at the speed at which the misfire occurs. If the spark output is now satisfactory or the misfire is no longer present, the rpm limit module is defective and must be replaced.

3. If the rpm limit module is suspected of not functioning when needed, check the ground wire for secure attachment to the power head. Clean and tighten the connection as necessary to assure a good ground path.

4. Disconnect the purple wire bullet connector from the rpm limit module. Connect the positive wire of a voltmeter set to read 20 volts DC to the engine harness purple or light blue/white wire connector. Connect the negative wire of the voltmeter to a good engine ground.

   a. *Electric start models*—With the ignition key *on*, the voltmeter should read within 1 volt of battery voltage. When the key is *off*, the voltmeter should read zero volts. If the meter reading is below specification, repair or replace the purple wire circuit from the key switch to the engine harness rpm limit module bullet connector.

   b. *Manual start models*—With the engine running, the voltmeter should read 10 volts or higher. If it does not, replace the ignition stator.

FUEL SYSTEM

Outboard owners often assume the carburetor(s) is at fault when the engine does not run properly. While fuel system problems are not uncommon, carburetor adjustment is seldom the solution. In many cases, adjusting the carburetor only compounds the problem by making the engine run worse.

Never attempt to adjust the carburetor(s) idle speed and idle mixture unless:

1. The ignition timing is correctly adjusted.
2. The engine throttle and ignition linkage is correctly synchronized and adjusted.
3. The engine is running at normal operating temperature.
4. The outboard is in the water, running in *forward* gear with the correct propeller installed.

If the engine appears to be running lean or starving for fuel, fuel system troubleshooting should be divided into determining whether the boat fuel system or the engine fuel system is causing the problem. Engines that appear to be running rich or receiving excessive fuel usually have a problem in the engine fuel system, not the boat fuel system.

The boat fuel system consists of the fuel tank, fuel vent line and vent fitting, the fuel pickup tube and antisiphon valve, the fuel distribution lines, boat-mounted water separating fuel filter (recommended), and the primer bulb.

The typical engine fuel system consists of an engine fuel filter, crankcase pulse driven fuel pump, carburetor(s), primer valve, and the necessary lines and fittings.

Troubleshooting

Make sure fresh fuel is present in the fuel tank. If the fuel is stale or sour, drain the fuel tank and dispose of the fuel in an approved manner. Clean fuel filters and flush fuel lines to remove traces of the stale or sour fuel. Inspect fuel lines for evidence of leaks or deterioration. Replace suspect components. Make sure the fuel tank vent is open and not restricted. Refer to Chapter Six for component removal, rebuilding and replacement procedures and component illustrations.

2.5 and 3 hp models

2.5 and 3 hp models use a gravity feed fuel system. Make sure fuel is in the fuel tank and the fuel tank vent is open. Disconnect the fuel line from the fuel tank to the carburetor and direct the line into a suitable container. Turn on the fuel valve and check for free fuel flow into the container. If the fuel flow is restricted, clean or replace the fuel valve and filter (Chapter Four). If the fuel flows freely and the symptom persists, rebuild the carburetor as described in Chapter Six.

4-25 hp models

NOTE

When troubleshooting 4 hp and larger engines, connect a substitute fuel tank and fuel line filled with fresh fuel to the engine. If the symptom is eliminated, the problem is in the original fuel tank and/or fuel line. If the symptom persists, the problem is in the engine.

4-25 hp models use a combination fuel pump/carburetor. The fuel pump is driven by crankcase pressure and vacuum pulses. If the cylinder that drives the fuel pump suffers a mechanical failure, the fuel pump cannot operate. Check the cranking compression before continuing. Disconnect the fuel inlet line from the fuel pump/carburetor and direct the line into a suitable container. Squeeze the primer bulb and observe the fuel flow into the container. If the fuel flow appears restricted, check the portable fuel tank vent and pickup screen, both fuel line quick-connectors, and the primer bulb for restrictions or defects. It may be easier to install a new fuel line, quick-disconnects and primer bulb assembly if the condition of the original assembly is in question. If the fuel flows freely and the symptom is still believed to be fuel related, rebuild and adjust the fuel pump/carburetor assembly.

30-60 hp models

30-60 hp models use a mechanical fuel pump(s) driven by crankcase pressure and vacuum pulses. If the cylinder(s) that drives the fuel pump(s) fails, the fuel pump(s) cannot operate. Check the cranking compression before continuing.

If the boat is equipped with a permanent fuel system, make sure the fuel tank vent line is not kinked or obstructed. If all visual checks are satisfactory, continue with the mechanical fuel pump pressure and vacuum tests.

Mechanical fuel pump pressure and vacuum tests (30-60 hp models)

CAUTION

Do not run the engine without an adequate water supply and do not exceed 3000 rpm without an adequate load. Refer to ***Safety Precautions*** *at the beginning of this chapter before continuing.*

1A. *Permanent fuel tank*—Make sure the fuel vent fitting and vent line are not obstructed.

1B. *Portable fuel tank*—Open the fuel tank vent to relieve pressure that may be present and make sure the tank is no more than 24 in. (61 cm) below the level of the fuel pump.

2. Disconnect the fuel *inlet* hose from the fuel pump. See **Figure 41**.
3. Connect a combination vacuum and fuel pressure gauge between the fuel pump and the inlet line using a T-fitting, a short piece of clear vinyl hose, and the appropriate fittings and clamps. See **Figure 42**.
4. Disconnect the fuel pump output hose that leads to the carburetor(s). See **Figure 41**.
5. Connect a fuel pressure gauge between the fuel pump and the output line using a T-fitting and appropriate fuel line and clamps. See **Figure 42**. Squeeze the primer bulb and check for leaks.
6. Start the engine and allow it to reach operating temperature. Run the engine in *forward* gear at idle speed. Observe the fuel pressure gauge, vacuum gauge and clear hose. Run the engine at wide-open throttle while observing the gauges and hoses.
7. If air bubbles are visible in the clear hose at any of the test speeds, check the fuel supply line back to the pickup tube in the fuel tank for loose fittings, loose clamps, a defective primer bulb, damaged filters or any problems that would allow air to leak into the fuel line. The fuel pump cannot develop the specified pressure if air is present. Correct any problems and retest.
8. If no air bubbles are visible in the clear hose and the vacuum gauge reading does not exceed 4 in. Hg (13.5 kPa), the fuel supply system is in satisfactory condition.
9. The fuel pressure should be 2.5-4 psi (17.2-27.6 kpa) at idle speed and 4-7 psi (27.6-48.3 kPa) at wide-open throttle. If the fuel pressure is below specification at either of the test speeds, repair or replace the fuel pump(s) as necessary.
10. If no air bubbles are visible in the clear hose, the vacuum did not exceed 4 in. Hg (13.5 kPa) and fuel pressure is within specification, the mechanical fuel pump and fuel supply system are not faulty.
11. Inspect the fuel lines and fittings from the fuel pump(s) to the carburetor(s) for leaks, deterioration, kinks and blockages. Correct any problems found. If the lines and fittings are in satisfactory condition and the problem is still believed to be fuel related, rebuild and adjust the carburetor(s). On multicarburetor engines, service all carburetors, even if only one is malfunctioning.

FUEL PRIMER AND ENRICHMENT SYSTEMS

Fuel Primer Solenoid (6-25 hp Electric Start Models)

The fuel primer solenoid (A, **Figure 43**) on 6-25 hp models is a simple spring-loaded, electromechanical,

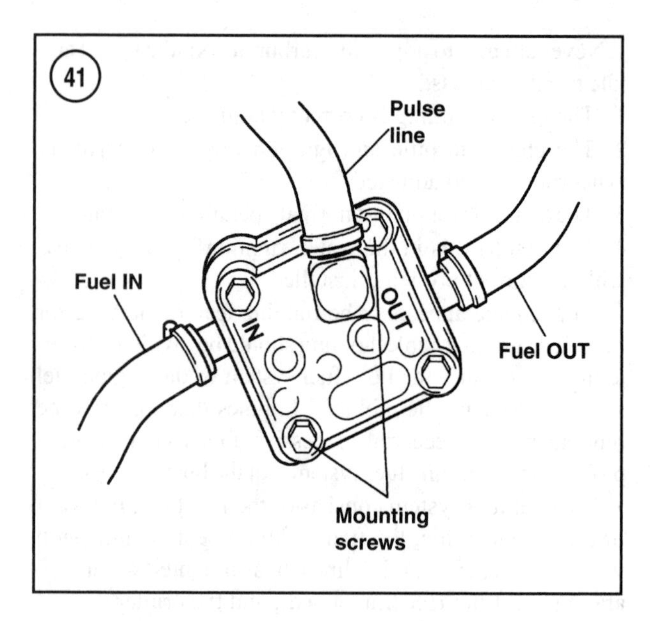

plunger type solenoid. When activated, the plunger head (B, **Figure 43**) covers an air bleed valve on the carburetor body. This activates the primer circuit in the carburetor, enriching the air/fuel mixture. The primer circuit works only when the engine is cranked and the throttle plate is fully closed. The solenoid plunger is exposed to allow for manual activation. Pressing the plunger toward the carburetor activates the primer circuit and the solenoid spring returns the plunger when released. Electricity applied to the yellow/black wire energizes the solenoid and moves the plunger toward the carburetor.

Troubleshooting

NOTE

The ignition switch must be held in the choke or prime position for all of the following tests.

1. Manually activate the solenoid plunger. Make sure the plunger moves freely and covers the carburetor air bleed valve. Make sure the spring returns the solenoid to the off position. Replace the solenoid if the plunger is frozen, binds or does not return to the off position.
2. Make sure the black wire coming from the fuel primer solenoid is connected to a good ground. Clean and tighten the ground connection as necessary.
3. Connect the test light wire to a good engine ground.
4. Access the ignition switch yellow/black wire, terminal C. Connect the test light probe to this wire. With the ignition switch in the choke or prime position, the test light should illuminate. If it does not, check the ignition switch red/purple wire, terminal B or BAT, for power. If there is

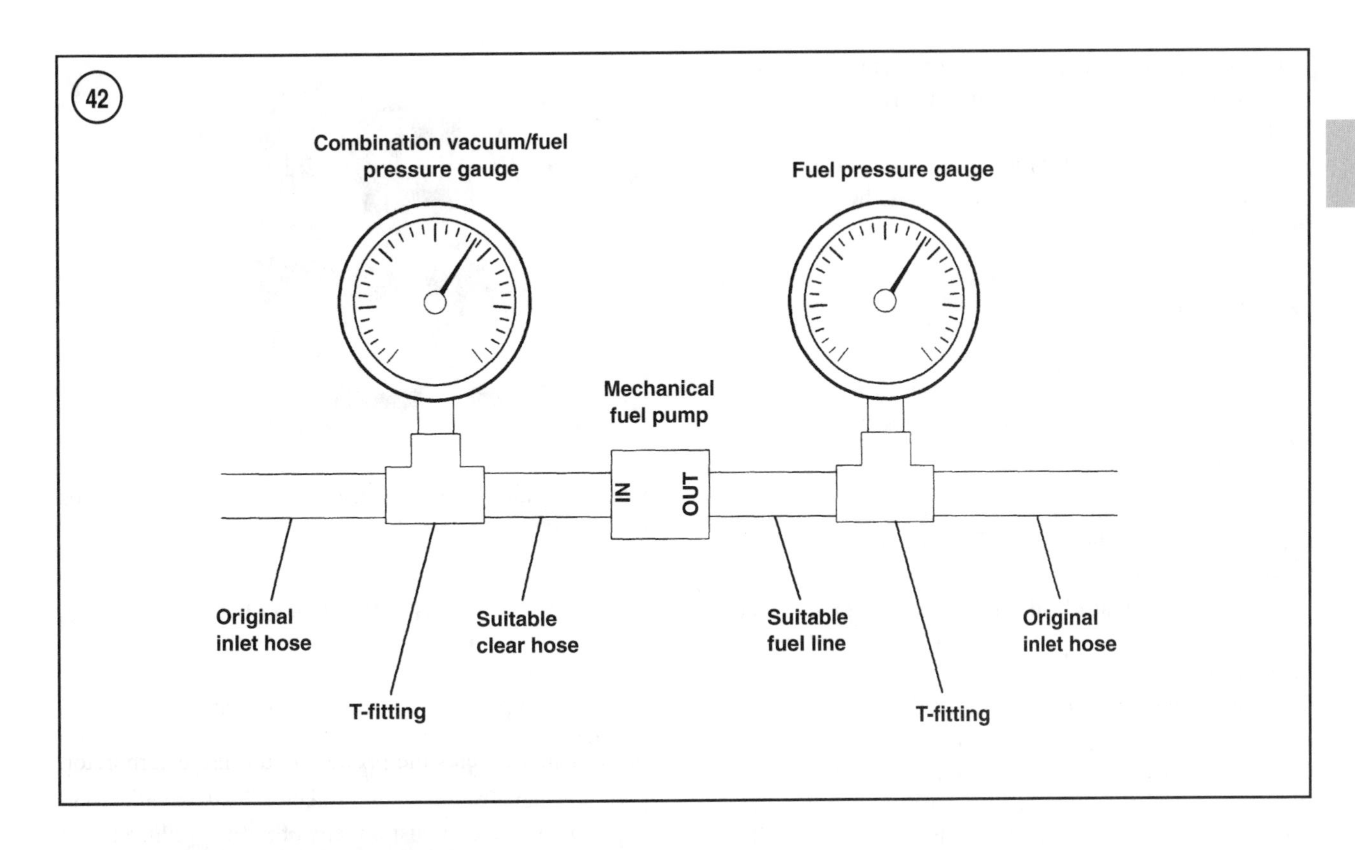

no power at terminal B, repair or replace the red or red/purple wire from terminal B of the ignition switch through the main 20 amp fuse to the starter solenoid.

5. Disconnect the yellow/black wire from the fuel primer solenoid. Connect the test light probe to the engine harness side yellow/black wire. With the ignition switch in the choke or prime position the test light should illuminate. If it does not, repair or replace the yellow/black wire from the fuel primer valve to the ignition switch. If the test light lights, replace the fuel primer solenoid and recheck solenoid function.

Fuel Primer Valve (30-60 hp Models)

Fuel primer valves (**Figure 44**) have replaced choke valves as the primary means of enriching the air/fuel mixture for cold starting. The fuel primer valve typically enriches the air/fuel mixture by flowing fuel directly into the intake manifold.

The fuel primer valve is an electrical solenoid valve that opens and closes. The fuel primer valve does not pump fuel; fuel must be supplied by the primer bulb or engine fuel pump. Applying electricity to the yellow/black wire, energizes the solenoid and the fuel valve opens. An internal spring forces the valve closed when the solenoid is not energized.

The fuel primer valve is equipped with a button to allow manual operation of the valve if the electrical circuit fails. Fuel flows as long as the button is depressed.

Refer to Chapter Six for fuel primer valve hose routing diagrams and to the wiring diagrams at the end of the manual.

Troubleshooting

NOTE
The ignition switch must be held in the choke or prime position for all of the following tests.

1. Check the fuel lines going to and from the fuel primer valve for deterioration and obstructions. Correct any problems.
2. Make sure the black lead coming out of the fuel primer valve is connected to a clean ground. Clean and tighten the connection as necessary.
3. Connect the test light lead to a good engine ground.
4. Access the ignition switch yellow/black wire, terminal C. Connect the test light probe to this wire. With the ignition switch in the choke or prime position, the test light should illuminate. If the test light illuminates, go to Step 6. If the test light does not illuminate, go to Step 5.
5. Connect the test light probe to the ignition switch red/purple wire, terminal B or BAT. The test light should illuminate regardless of the ignition switch position. If the test light illuminates, replace the ignition switch and repeat Step 4. If the test light does not illuminate, repair or replace the red or red/purple wire from terminal B of the ignition switch through the main 20 amp fuse to the starter solenoid.
6. Disconnect the yellow/black wire at the fuel primer valve. Connect the test light probe to the yellow/black wire on the engine harness side. With the ignition switch in the choke or prime position, the test light should illuminate. If the test light does not illuminate, repair or replace the yellow/black wire from the fuel primer valve to the ignition switch.
7. Disconnect the yellow/black and black wires from the fuel primer valve. Connect an ohmmeter set on the appropriate scale to read 10-12 ohms. Connect one meter lead to the fuel primer valve yellow/black wire and the other meter lead to the fuel primer valve black wire. Replace the solenoid if the resistance is not within 10-12 ohms.
8. Disconnect the fuel hoses from the fuel primer valve and connect a short length of hose to one of the valve ports. Blow into the hose while depressing the manual valve button on top of the valve. Air should flow through the valve with the button depressed, and should not flow when the button is released. Replace the valve if it does not perform as specified.

ENGINE TEMPERATURE AND OVERHEATING

Proper temperature is critical to good engine operation. Internal engine damage occurs if the engine is overheated. Engines that are overcooling will have fouled spark plugs, poor acceleration and idle quality, excessive carbon buildup in the combustion chamber and reduced fuel economy. All recreational outboard models 30 hp and larger are equipped with thermostat(s).

A thermostat (**Figure 45**) is standard on 9.9 and 15 hp models and optional on 6-8 hp and 20-25 hp models. The addition of the optional thermostat is recommended if the engine is experiencing any of the previously mentioned symptoms.

A thermostat cannot be added to the 2.5-5 hp models.

The thermostat(s) controls the flow of water leaving the cylinder block. Many engines with thermostat(s) also incorporate a poppet or pressure relief valve. These valves may be mounted on the power head or in the power head adapter plate. These valves control the pressure in the cooling system. When water pressure exceeds the spring pressure of the valve, cooling water is bypassed out the water discharge on some models. This has the same effect as bypassing the thermostat. On these models, thermostats control the water temperature at idle and low speeds. Once the engine speed has increased to the point that water pressure overcomes the poppet valve, the thermostats are no longer in primary control of engine temperature. Since an engine at high power settings is producing a lot of combustion chamber heat, this system helps keep the combustion chambers cool enough to prevent preignition and detonation.

On all thermostat equipped engines, the thermostat only controls the minimum engine temperature. The number

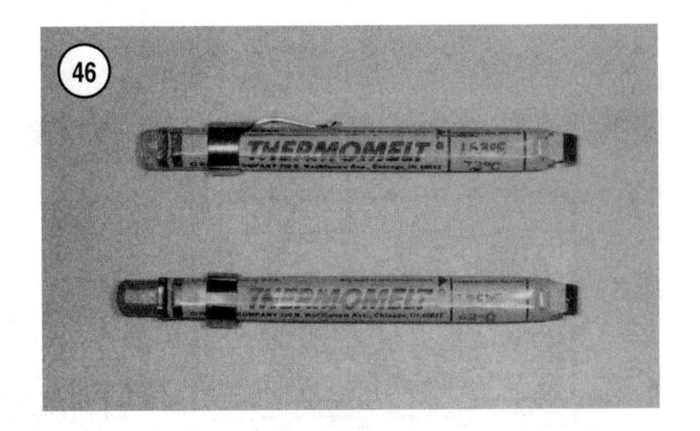

46

stamped on the thermostat is the initial opening or *cracking* temperature (Fahrenheit). The thermostat will not reach full opening (full water flow) until 15-20° F above the thermostat rated temperature. If the engine temperature drops at high engine speeds, but returns to thermostat controlled temperature at idle and low speeds, the system is working correctly. Engine temperature should not exceed the thermostat rated temperature by more than 20° F.

Troubleshooting

Engine temperature can be checked with thermomelt sticks. Thermomelt sticks (**Figure 46**) come in 100, 125, 131, 163 and 175° F ratings. The thermomelt stick looks similar to a piece of chalk. When the engine is marked with the thermomelt stick, the mark remains dull and chalky until the temperature of the engine exceeds the rating of the stick, at which point the mark becomes liquid and glossy.

The preferred and most accurate way to check engine temperature is with a pyrometer. A pyrometer is an electronic thermometer. Pyrometers come as a single instrument or as an adapter module designed to fit a standard digital multimeter. When using a pyrometer, a dab of silicone grease on the end of the probe helps create a good thermal bond between the probe and the power head.

The temperature should be measured on the cylinder block side of the thermostat housing. Measure the temperature just before the water reaches the thermostat.

Engine temperature check

To be accurate, the cooling water inlet temperature must be within 60-80° F (18-24° C). Extreme variation in water inlet temperature can affect engine operating temperature. Refer to the previously mentioned number stamping for thermostat specifications. Engine temperature checks cannot be performed on an outboard running on a flushing device.

CAUTION

Do not run the engine without an adequate water supply and do not exceed 3000 rpm without an adequate load. Refer to ***Safety Precautions*** *at the beginning of this chapter.*

1. Start the engine and run it at 2000-3000 rpm in forward gear for 5-10 minutes to allow it to reach operating temperature.
2. Reduce the speed to the specified idle speed, or a maximum of 900 rpm, for 5 minutes.

3A. *Using thermomelt sticks*—Mark the cylinder block with the two thermomelt sticks that are just above and just below the thermostat(s) rated temperature.

3B. *Using a pyrometer*—Note the temperature of the cylinder block and compare to the thermostat rated temperature. The engine should operate within 5° F of the thermostat rated temperature at idle.

4A. If the lower temperature thermomelt sticks do not melt or if the pyrometer reads more than 5° F below the thermostat rated temperature, the engine is overcooling. Check the thermostat(s) and poppet valve, if so equipped, for debris preventing the thermostat and poppet valve from closing. If the thermostat and poppet valve pass a visual inspection, test the thermostat(s) as described in the following section.

4B. If the lower temperature thermomelt sticks melt or the pyrometer reads within 5° F of the thermostat rated temperature, the engine is at least reaching operating temperature and is not overcooling.

5A. If the higher temperature thermomelt sticks do not melt or the pyrometer does not read more than 5° F above the thermostat rated temperature, the engine is not overheating.

5B. If the higher temperature thermomelt sticks melt, the engine is overheating. Test the thermostat(s) as described in the following section. Check the water inlet screens or cast inlet holes in the lower gearcase for debris or corrosion. Clean as necessary. Remove the lower gearcase and inspect the water pump assembly. Inspect the water tube from the water pump to the power head for damage or corrosion. If necessary, remove the cylinder head(s) or cylinder cover and check for debris and corrosion in the water jackets surrounding the cylinder(s).

NOTE

It is normal for the engine to run cooler than the thermostat rated temperature at higher engine speed if the engine is equipped with a poppet valve. However, regardless of whether the engine is equipped with a poppet valve or not, the engine temperature must not exceed the thermostat rated temperature by more than 20° F.

6. Repeat the temperature test at 3000 rpm and wide-open throttle. It is acceptable for the pyrometer to read up to 15-20° F above the thermostat rated temperature during the higher speed tests.

CAUTION

If the engine is run in a test tank, the 3000 rpm and wide-open throttle test may not be accurate due to aeration of the test tank water and subsequent overheating.

Thermostat test

1. Remove the screws securing the thermostat cover to the cylinder block or cylinder head(s). Carefully remove the thermostat cover(s).
2. Clean all gasket material from the thermostat cover(s) and cylinder block or cylinder head(s).
3. Check thermostat covers for cracks or corrosion damage and replace as necessary.
4. Wash the thermostat with clean water. Remove the thermostat grommet, if so equipped. Read the thermostat rated opening temperature stamped on the thermostat.
5. Manually open the thermostat and insert a thread gauge or narrow feeler gauge through the valve and seat. Let the thermostat close, pinching the thread or feeler gauge in the valve.
6. Suspend the thermostat from a string attached to the feeler gauge in a container of water that can be heated. Then suspend an accurate thermometer in the container of water. See **Figure 47**.

NOTE

The thermostat and thermometer must not touch the sides or bottom of the container.

7. Heat the water and note the temperature at which the thermostat falls from the feeler gauge. The thermostat must begin to open within ±5° F of the rated temperature.
8. Continue to heat the water until the thermostat fully opens. The thermostat must fully open within 15-20° F above the rated temperature.
9. Replace the thermostat(s) if it fails to open at the specified temperature or if it does not open completely. Install a new gasket and grommet, if so equipped.

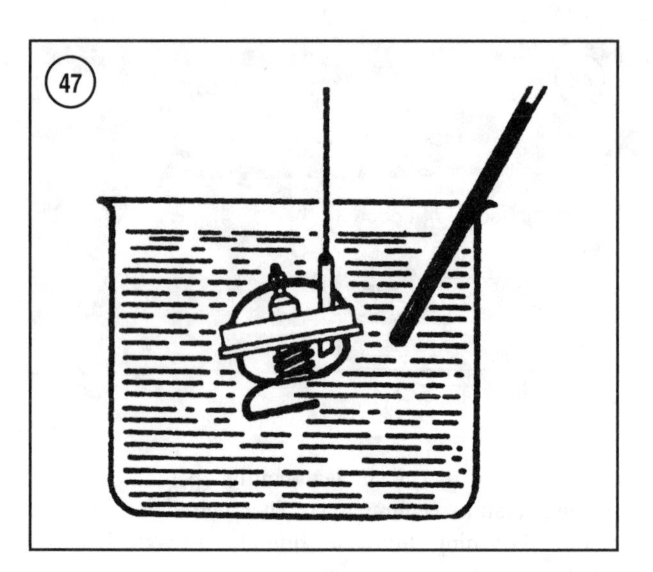

NOTE

If the engine is equipped with a poppet valve or spring in the thermostat housing, make sure the poppet valve, spring and seat are correctly oriented and not damaged before installing the thermostat cover. Replace damaged or worn parts as described in Chapter Eight.

Engine temperature switches and sensors

All 30-60 hp models are equipped with a temperature switch and warning horn to alert the operator if the engine overheats. Temperature switches are normally open and short to ground when the engine temperature exceeds the rated temperature of the switch. The switch grounds through direct contact with the cylinder head.

Temperature switch test

Refer to the wiring diagrams at the end of the manual for wire routing and connection points.

1. Disconnect the engine wiring harness tan/sky blue wire from the temperature switch terminal.
2. Set an ohmmeter on the highest ohms scale available. Connect one meter wire to a clean engine ground and the other ohmmeter wire to the temperature switch terminal. The ohmmeter should read *no continuity*. If it does not, replace the temperature switch.

CAUTION

Do not run the engine without an adequate water supply and do not exceed 3000 rpm without an adequate load. Refer to ***Safety***

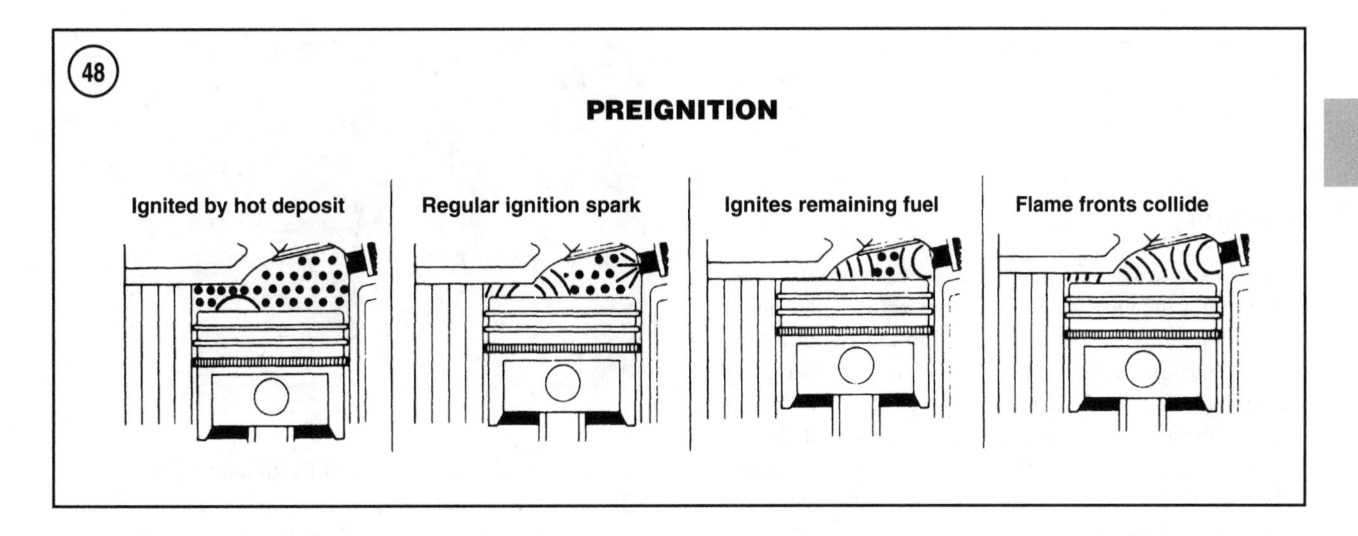

Precautions at the beginning of this chapter.

3. Warm the engine to operating temperature as described in *Engine temperature check* in this chapter. The ohmmeter should continue to indicate *no continuity* when the engine is at normal operating temperature. The meter should indicate continuity only if the engine is actually overheating. Replace the switch if the meter indicates continuity at normal engine temperature or fails to indicate continuity if the engine overheats.

ENGINE

Engine problems are generally caused by a failure in another system, such as the ignition, fuel and lubrication, or cooling systems. If a power head is properly cooled, lubricated, timed correctly and given the correct air/fuel ratio, the engine should experience no mechanical problems other than normal wear. If the power head fails, determine why the power head failed. Replacing failed components accomplishes nothing if the cause of the failure is not corrected.

Overheating and Lack of Lubrication

Overheating and lack of lubrication cause the majority of engine mechanical problems. Anytime an outboard motor is run, adequate cooling water must be supplied by immersing the gearcase water inlets in a test tank or lake or using an approved flushing device. Only run the engine at low speeds on a flushing device. Never start the engine without a water supply. Irreparable water pump damage occurs in seconds.

Carbon buildup in a two-stroke outboard motor causes premature power head failure. Carbon buildup comes from the lubricating oil and fuel. Use a premium quality outboard motor power head lubricant. The current TCW-3 specification for power head lubricants ensures that maximum lubrication is delivered with minimal carbon deposit buildup. Using gasoline from a major brand manufacturer ensures that the fuel contains the detergents necessary to minimize carbon build-up. Fuels that contain alcohol tend to build carbon up at an accelerated rate. Avoid alcohol blended fuels whenever possible. Refer to Chapter Four for additional fuel and oil recommendations.

Use Quicksilver Power Tune (part No. 92-15104) periodically to remove carbon deposits from the combustion chamber and piston rings before they can contribute to high combustion chamber temperatures.

Preignition

Preignition is the premature ignition of the air/fuel charge in the combustion chamber. Preignition is caused by hot spots in the combustion chamber. See **Figure 48**. Anything in the combustion chamber that gets hot enough to light the air/fuel charge causes preignition. Glowing carbon deposits, inadequate cooling, improperly installed thread inserts, incorrect head gaskets, sloppy machine work, previous combustion chamber damage, such as nicks and scratches, or overheated (incorrect) spark plugs can all cause preignition. Preignition is usually first noticed as power loss, but eventually results in extensive damage to the internal engine components because of excessive combustion chamber pressure and temperature. Preignition damage typically looks like an acetylene torch was used to melt away the top of the piston. Sometimes the piston will have a hole melted through the piston crown. See **Figure 49**. Remember that preignition can lead to detonation and detonation can lead to preignition. Both types of damage may be evident when the engine is disassembled.

Detonation

Commonly referred to as *spark knock or fuel knock*, detonation is the violent, spontaneous explosion of fuel in the combustion chamber as opposed to the smooth, progressive burning of the air/fuel mixture that occurs during normal combustion. See **Figure 50**. When detonation occurs, combustion chamber pressure and temperature rise dramatically, creating severe shock waves in the engine. This causes severe engine damage. It is not unusual for detonation to break a connecting rod or crankshaft.

Detonation occurs when the octane requirements of the engine exceed the octane of the fuel being used. It does not necessarily mean that the wrong fuel is being used. It means that at the time of detonation, the engine needed a higher octane fuel than is being used. All fuel spontaneously explodes when subjected to enough heat and pressure.

The fuel octane requirements of an engine are generally determined by:

1. *Compression ratio*—Higher compression ratios require higher octane fuel. Carbon buildup raises compression ratios.
2. *Combustion chamber temperature*—Higher temperature requires higher octane fuel. Water pump and thermostat malfunctions typically raise the combustion chamber temperature.
3. *Air/fuel mixture*—Leaner mixtures require higher octane fuels. Richer mixtures require lower octane fuels.
4. *Spark advance*—Spark occurring too early causes excessive combustion chamber pressures. Spark occurring extremely late causes the combustion flame front to spread along a larger surface area of the cylinder walls, exceeding the cooling system capacity to remove the heat. Both of these situations raise the octane requirements.
5. *Engine operating speed*—Propping an engine so it cannot reach the recommended operating speed range is considered *lugging or overpropping* the engine. When an engine is overpropped to the point that it cannot reach its recommended speed, combustion chamber temperature skyrockets, increasing the octane requirement.

Fuel degrades over time in storage causing the actual octane rating of the fuel to drop. Even though the fuel may have exceeded the manufacturer's recommendations when the fuel was fresh, it may have dropped below recommendations over time. Use a fuel stabilizer, such as Quicksilver gasoline stabilizer to prevent octane deterioration. The fuel stabilizer must be added to fresh fuel. It will not raise the octane of stale or sour fuel.

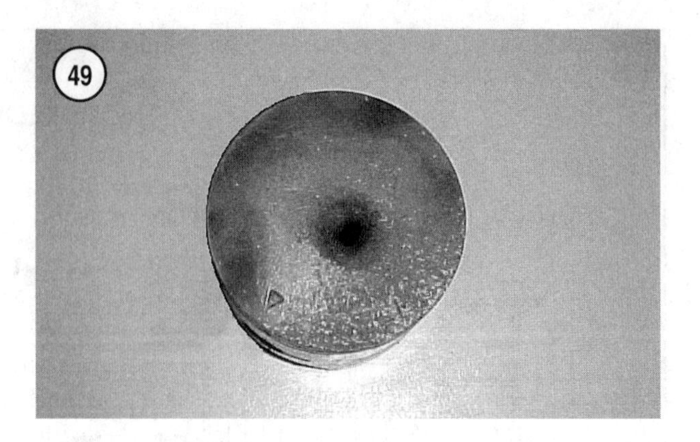
49

Properly dispose of questionable fuel and start with a fresh tank rather than risk a power head failure. Power head failure typically occurs in a few seconds or less when an engine is detonating. An operator rarely can detect detonation and reduce engine speed in time to save the power head. Detonation can lead to preignition and preignition can lead to detonation. Both types of damage may be evident when the engine is disassembled.

Poor Idle Quality

Poor idle quality can usually be attributed to one of the following conditions.

1. *Overcooling*—If the power head does not reach the recommended operating temperature, fuel tends to puddle in the crankcase, resulting in a lean air/fuel ratio in the combustion chamber. This produces a lean spit or backfire through the carburetor at idle. Overheating is usually caused by debris caught in the thermostat(s) or poppet valve, if so equipped. A few models are not equipped with thermostats. Refer to *Engine Temperature and Overheating* in this chapter for engine temperature checks.
2. *Crankcase seal failure*—A two-stroke engine cannot function unless the crankcase is adequately sealed. As the piston travels downward, the crankcase must pressurize and push the air/fuel mixture into the combustion chamber as the transfer ports are uncovered. As the piston travels upward, the crankcase must create a vacuum to pull the air/fuel mixture into the crankcase from the carburetor in preparation for the next cycle. Leaks in the crankcase cause the air/fuel charge to escape into the atmosphere under crankcase compression. During the intake cycle, crankcase leakage allows air from the atmosphere to be drawn into the crankcase, diluting the air/fuel charge. This causes inadequate fuel in the combustion chamber. On multiple cylinder engines, each crankcase must be sealed

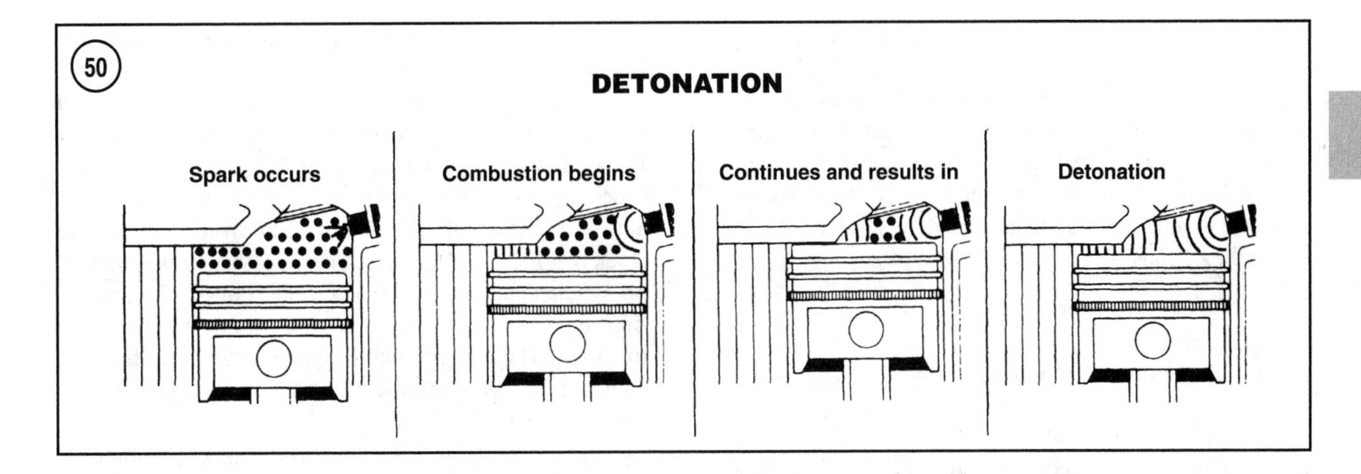

from all other crankcases. Internal leakage allows the air/fuel charge to leak to another cylinder's crankcase, rather than travel to the correct combustion chamber. Refer to *Starting Difficulties* at the beginning of this chapter for additional information.

3. *Crankcase bleed system failure*—Multiple cylinder motors are equipped with a fuel recirculation system designed to collect unburned fuel and oil from the low spots of the crankcase. Since the intake system used by two-stroke engines does not completely transfer all of the fuel sent through the crankcase to the combustion chamber, especially during low-speed operation, the recirculation system collects the fuel and oil pooled in the low spots of the crankcase and transfers it to the intake ports or intake manifold where it can be recycled. Correct recirculation system operation is important to efficient engine operation. If the system fails, excessive amounts of fuel and oil will puddle in the crankcase and not reach the combustion chamber during low-speed operation causing a lean mixture. When the engine is accelerated, the puddles of fuel and oil are drawn into the engine causing a temporary rich mixture. This results in poor low speed performance, poor acceleration, spark plug fouling, stalling or spitting at idle, and excessive smoke on acceleration. Refer to Chapter Six for bleed system service.

4. *Incorrect carburetor adjustments or carburetor malfunction*—The carburetor idle mixture screw must be correctly adjusted for the engine to idle and accelerate properly. An engine that is too lean at idle spits or backfires through the carburetor and hesitate on acceleration. Refer to Chapter Six for carburetor adjustments.

Misfiring

True misfiring is an ignition system malfunction, generally caused by weak or erratic spark or defective spark plugs. The ignition system is not able to deliver enough spark energy to fire the spark plug at the time of the misfire.

Four-stroking is a form of misfire caused by an air/fuel ratio so rich it cannot consistently ignite. The term four-stroking comes from the fact that the engine is typically firing every other revolution instead of every revolution. Four-stroking is caused by a fuel system malfunction. Check for excessive fuel pump pressure, carburetor(s) with a leaking inlet needle and seats or fuel primer systems stuck on.

Insufficient compression can cause a misfire at all speeds. It will often cause a cylinder to not fire at idle and low speed, and begin firing at mid-range and high speed. Always perform a compression test to verify the mechanical integrity of the combustion chamber.

Flat Spots and Hesitation During Acceleration

If the engine seems to hesitate or bog down when the throttle is opened, then recovers, check for a restricted main jet in the carburetor(s), water in the fuel, or an excessively lean fuel mixture. Faulty accelerator pump operation, on models so equipped, and incorrect synchronization of the spark advance to the throttle opening, on models with adjustments, can cause flat spots and hesitation on acceleration.

Water Leakage Into Cylinder(s)

Check for water leakage into a cylinder by checking the spark plugs. Water in the combustion chamber tends to clean the spark plug. If one spark plug in a multicylinder engine is clean and the others have normal deposits, a water leak is likely in the cylinder with the clean spark plug.

A compression test can also be used to check the mechanical integrity of the combustion chamber. The piston crown can be visually inspected for the absence of carbon deposits. A cylinder crown that looks steam cleaned typically indicates water leakage into that combustion chamber. If the exhaust port area can be accessed, look for evidence of hard mineral deposits and the absence of soft, wet carbon deposits.

Water Damage in Power Head Lower Cylinder(s)

While water leakage into the combustion chambers is generally caused by defective or failed head gaskets, if so equipped, water can also enter the lower cylinder(s) of a power head through the exhaust ports and carburetor(s). When a steep unloading ramp or tilted trailer bed is used to launch the boat from a trailer and the boat enters the water too quickly, water can be forced into the drive shaft housing and up through the exhaust chamber into the cylinders if the pistons are not covering the exhaust ports.

Sudden deceleration with the engine shut off can cause a wave to swamp the engine and enter the exhaust ports or enter through the lower carburetor(s). This is most common in stern heavy boats. Operating a boat with twin engines with one engine shut off is considered hazardous because there is no exhaust back pressure to keep water out of the engine that is not running. Leave the engine that is not being used for docking or low speed maneuvering running at idle speed to reduce the risk of water entry.

Water entering a cylinder can result in a bent connecting rod, a broken piston and/or piston pin, a cracked cylinder and/or cylinder head, or any combination of these conditions. Even if no immediate physical damage is done to the power head, the entry of water results in rust and corrosion of all internal surfaces such as bearings, crankshaft, cylinder walls, connecting rods and piston rings.

Power Loss

Several factors can cause a loss of power. An engine needs three things to run properly: compression, fuel and ignition. Check the mechanical integrity of the combustion chamber by performing a cranking compression test. Test the ignition system with an air gap tester and verify ignition timing at wide-open throttle. Check the fuel system for air leaks into the fuel lines and fittings, and test the fuel pump for adequate output pressure at wide open throttle. Clean or replace all fuel filters. Remove a carburetor and inspect the float chamber for water in the fuel and gum or varnish buildup in the metering passages and jets. Clean all of the carburetors if debris or buildup is found in one carburetor.

If the compression test reveals a mechanical defect in a combustion chamber, treat the engine with Quicksilver Power Tune Engine Cleaner (part No. 92-15104). Many times the piston rings are stuck to the piston and cannot adequately seal to the cylinder walls. Power Tune can free stuck piston rings and prevent unnecessary disassembly if no mechanical damage has occurred. Follow the instructions on the can and retest cranking compression after the treatment. If the compression is now within specification, consider changing lubricant and fuel to a higher quality brand. See Chapter Four.

If the compression is still not within specification after the Power Tune treatment, disassemble the motor, and locate and repair the defect. After the power head rebuild, make sure the carburetors and fuel pump are rebuilt, a new water pump and thermostat(s) are installed and all synchronization and linkage adjustments are made (Chapter Five).

Marine growth on the bottom of the hull and lower gearcase reduces the top speed and fuel economy of a boat. If the motor is in a good state of tune and has no apparent malfunction, yet fuel economy and top speed have suffered, inspect the bottom of the hull and lower gearcase for marine growth and clean as necessary.

Piston Seizure

Piston seizure can be caused by insufficient piston-to-cylinder bore clearance, improper piston ring end gap, inadequate or inferior lubrication, cooling system failure (overheating), preignition or detonation.

Excessive Vibration

Excessive vibration can be caused by an engine misfiring on one or more cylinders, loose or broken motor mounts and worn or failed bearings. Gearcase problems that can cause excessive vibration are bent propeller shafts, damaged propellers or propellers with marine growth on the blades. A propeller ventilating because of damage or defects on the leading edge of the gearcase, an improperly mounted speedometer or depth finder sending unit, or any hull deformity that disturbs the water flow to the propeller can also cause excessive vibration.

Engine Noise

Experience is necessary to diagnose engine noises accurately. Noises are difficult to differentiate and hard to describe. Even a properly assembled two-stroke power head produces much more mechanical noise than a four-stroke. A two-stroke power head produces substantial intake (induction) noise. Deep knocking noises usually mean crankshaft main or rod bearing failure. A light slapping noise generally comes from a loose piston; however, some piston noise is normal, especially during warm-up. Any knocking noise during acceleration or at high speed could be preignition or detonation and must be investigated immediately.

Table 1 BOAT WIRING HARNESS STANDARD WIRE COLORS

Circuit	Wire color
Starter engagement	Yellow/red
Tachometer	Gray
Stop circuit (ignition side)	Black/yellow
Stop circuit (ground side)	Black
Choke or primer	Yellow/black
Overheat/low oil warning	Tan/blue
Switched B+	Purple
Protected B+	Red/purple
Temperature gauge	Tan
Fuel sender circuit	Pink
Grounds	Black
Trim motor up	Blue
Trim motor down	Green
Trim switching circuit up	Blue/white
Trim switching circuit down	Green/white
Trim switching circuit B+	Red/purple
Trim trailer circuit	Purple/white
Trim sender circuit	Brown/white
Trim system grounds	Black

Table 2 BATTERY CABLE RECOMMENDATIONS

Cable length	Minimum cable gauge size (AWG)
To 3-1/2 ft.	4
3-1/2 to 6 ft.	2
6 to 7-1/2 ft.	1
7-1/2 to 9-1/2 ft.	0
9-1/2 to 12 ft.	00
12 to 15 ft.	000
15 to 19 ft.	0000

Table 3 STARTING SYSTEM TROUBLESHOOTING

Symptom	Probable cause	Remedy
Low no-load speed with high current draw	Tight or dirty bushings	Clean and lubricate bushings
	Shorted armature	Test armature (Chapter Seven)
Low no-load speed with low current draw	High resistance in the armature circuit	Check brushes and springs
		Test armature (Chapter Seven)
		Clean and inspect commutator
No rotation with high current draw	Stuck armature	Clean and lubricate bushings
	Internal short to ground	Check brush leads for shorts
Starter continues after key is released	Starter solenoid stuck	Replace solenoid
	Key switch failure	Test key switch
	Yellow or yellow/red wire malfunction	Disconnect suspect wires

(continued)

Table 3 STARTING SYSTEM TROUBLESHOOTING (continued)

Symptom	Probable cause	Remedy
Starter turns motor over too slowly	High resistance in solenoid	Measure voltage drop (Chapter Three)
	Mechanical failure of gearcase	Check for debris on drain/fill plug
	Mechanical failure of power head	Manually rotate flywheel
	Battery cables too small	Check battery cable size (Table 2)
Starter spins but drive does not engage flywheel	Corroded starter drive	Inspect starter drive (Chapter Seven)
	Drive needs lubrication	Lubricate starter drive
	Faulty battery	Check battery (Chapter Seven)
	Faulty battery cables	Check voltage drop (Chapter Three)
	Faulty starter	Disassemble and inspect starter

Table 4 CHARGING SYSTEM TROUBLESHOOTING

Symptom	Probable cause	Remedy
Battery overcharging		
Unregulated system	Extended high speed operation	Switch on accessories during extended high speed operation
Regulated system	Regulator failure	Test wiring/replace regulator
	Stator shorted to ground	Test stator
Battery looses charge		
While running	Stator failure	Test stator
	Excessive accessory load	Perform current draw test
While in storage	Faulty regulator	Test wiring/replace regulator
	Current draw from accessories	Switch off accessories
	Defective battery	Test battery

Table 5 IGNITION SYSTEM TROUBLESHOOTING

Symptom	Probable cause	Remedy
Fails to start	Fouled spark plug(s)	Clean or replace spark plugs
(spark test good)	Incorrect timing	Check for sheared flywheel key
Engine backfires	Incorrect timing	Check ignition timing (Chapter Five)
	Incorrect firing order	Check primary and secondary coil wiring
	Cracked spark plug insulator	Inspect spark plugs
High speed misfire	Insufficient spark	Check for strong blue spark
	Incorrect spark plug gap	Check spark plug gap (Chapter Four)
	Loose electrical connection	Check all wiring and terminals
Pre-ignition	Wrong type of spark plug	Check spark plugs (Chapter Four)
	High operating temperature	Check for overheating
	Incorrect ignition timing	Check ignition timing (Chapter Five)
	Lean fuel mixture	Check fuel system for blockage
Spark plug failure	Incorrect spark plug	Check spark plugs (Chapter Four)
	Loose spark plugs	Torque spark plugs
	Fuel system malfunction	Check for rich or lean condition
	High operating temperature	Check operating temperature
	Heavy carbon deposits	Clean carbon from engine
Ignition component failure	Loose electrical connection	Check wiring and terminals
	Loose mounting fasteners	Check component fasteners
	High operating temperature	Check for overheating
	Corrosion	Check for source of water

Table 6 FUEL SYSTEM TROUBLESHOOTING

Symptom	Probable cause	Remedy
Engine fails to start	No fuel to carburetor(s)	Verify fuel in tank
		Check fuel tank vent
		Check fuel tank pickup or filter
		Clean all fuel filters

(continued)

Table 6 FUEL SYSTEM TROUBLESHOOTING (continued)

Symptom	Probable cause	Remedy
Engine fails to start (continued)		Verify primer bulb operation
	Carburetor(s) failure	Rebuild carburetor(s)
Flooding at carburetor	Float or needle malfunction	Rebuild carburetor(s)
	Excessive fuel pressure	Check fuel pump operation
Loss of power	Restricted fuel supply	Clean fuel filters
		Check fuel hoses
		Check hose connections
	Blocked carburetor passages	Rebuild carburetor(s)
	Air leakage in fuel supply	Check hoses and connections
	Low fuel pressure	Check fuel pump operation
Hesitation on acceleration	Improper carburetor adjustment	Adjust carburetor
	Improper synchronization	Check carburetor synchronization
	Restricted fuel supply	Clean fuel filters
		Check fuel hoses
		Check hose connections
	Blocked carburetor passages	Rebuild carburetor(s)
	Air leakage in fuel supply	Check hoses and connections
	Low fuel pressure	Check fuel pump operation
Excessive fuel consumption	Improper carburetor adjustment	Adjust carburetor
	Carburetor float malfunction	Rebuild carburetor(s)
	Blocked carburetor passages	Rebuild carburetor(s)
	High fuel pressure	Check fuel pump operation
Spark plug fouling	Improper carburetor adjustment	Adjust carburetor
	Improper synchronization	Check carburetor synchronization
	Excessive oil in fuel	Adjust oil injection linkage
		Mix fuel at recommended ratio
Engine detonation	Low fuel octane rating	Use higher octane fuel
	Carbon deposits	Remove deposits from engine
Engine pre-ignition	Restricted fuel supply	Clean fuel filters
		Check fuel hoses
		Check hose connections
	Blocked carburetor passages	Rebuild carburetor(s)
	Air leakage in fuel supply	Check hoses and connections
	Low fuel pressure	Check fuel pump operation

Table 7 ELECTRIC STARTER SPECIFICATIONS

Model	Current draw (amperage)
6-25 hp	
Loaded*	55
No load	15
30 and 40 hp (2-cylinder)	
Loaded*	95
No load	20
40-60 hp (3-cylinder)	
Loaded*	125
No load	not available

*The spark plugs must be installed for this test. The no-load test is a bench test.

Table 8 CHARGING SYSTEM SPECIFICATIONS

Model	Maximum output (amperage @ rpm)	No. of poles
6-25 hp		
Black stator (1998)	4 @ 6000	8
Red stator (1999-On)	6 @ 6000	10
30 and 40 hp (2-cylinder)	18 @ 3000-5000	12
40-60 hp (3-cylinder)	16 @ 2000	12

Table 9 STATOR RESISTANCE SPECIFICATIONS (BATTERY CHARGING/LIGHTING CIRCUIT)

Model	Specification (ohm)
4 and 5 hp	0.31-0.47
6-25 hp	0.65
30 and 40 hp (two-cylinder)	0.16-0.18
40-60 hp (three-cylinder)	0.16-0.19

Table 10 CDI IGNITION SYSTEM SPECIFICATIONS (2.5-5 HP MODELS)

Test (output/resistance) (cranking to 2000 rpm)	Meter scale	Specification
2.5 and 3.3 hp		
CD module output	400 DVA	100-320 volts
Charge/trigger coil output	400 DVA	120-320 volts
Stop circuit output	400 DVA	120-320 volts
Charge/trigger coil resistance	R x 100	300-400 ohms
Ignition coil primary winding	R x 1	less than 1 ohms
Ignition coil secondary winding	R x 1K	3000-4000 ohms
4 and 5 hp		
CD module output	400 DVA	120-300 volts
Charge coil output	400 DVA	150-325 volts
Stop circuit output	400 DVA	175-300 volts
Charge coil resistance	R x 10	93-142 ohms
Trigger coil resistance	R x 10	80-115 ohms
Ignition coil primary winding resistance	R x 1	0.02-0.38 ohms
Ignition coil secondary winding resistance	R x 1K	3000-4400 ohms

Table 11 CDI IGNITION SYSTEM SPECIFICATIONS (6-25 HP MODELS)

Test (output/resistance)	Meter scale	Specification
Stop circuit output		
Black stator (1998 except 20 jet)		
300-1000 rpm	400 DVA	150-300 volts
1000-4000 rpm	400 DVA	250-360 volts
20 Jet (1998)		
300-3000 rpm	400 DVA	220-320 volts
3001-4000 rpm	400 DVA	300-350 volts
Red stator (1999-On)		
300-1000 rpm	400 DVA	150-330 volts
1000-4000 rpm	400 DVA	250-330 volts
Ignition stator output		
Black stator (1998 except 20 jet)		
Low speed winding		
300-1000 rpm	400 DVA	150-300 volts
1000-4000 rpm	400 DVA	250-360 volts
High speed winding		
300-1000 rpm	400 DVA	10-75 volts
1000-4000 rpm	400 DVA	50-300 volts
20 Jet (1998)		
Low speed winding		
300-3000 rpm	400 DVA	220-320 volts
3001-4000 rpm	400 DVA	300-350 volts
High speed winding		
300-3000 rpm	400 DVA	30-220 volts
3001-4000 rpm	400 DVA	200-280 volts
Red stator (1999-On)		
300-1000 rpm	400 DVA	150-330 volts
1000-4000 rpm	400 DVA	250-330 volts

(continued)

Table 11 CDI IGNITION SYSTEM SPECIFICATIONS (6-25 HP MODELS) (continued)

Test (output/resistance)	Meter scale	Specification
Switchbox output		
Black stator (1998 except 20 jet)		
300-1000 rpm	400 DVA	125-260 volts
1000-4000 rpm	400 DVA	200-360 volts
20 Jet (1998)		
300-3000 rpm	400 DVA	160-250 volts
3001-4000 rpm	400 DVA	200-280 volts
Red stator (1999-on)		
300-1000 rpm	400 DVA	125-320 volts
1000-4000 rpm	400 DVA	200-320 volts
Trigger coil resistance	R × 1	6.5-8.5 ohms
Ignition coil resistance		
Primary winding	R × 1	0.02-0.04 ohms
Secondary winding	R × 1K	8000-11,000 ohms

Table 12 CDM IGNITION SYSTEM SPECIFICATIONS (30-60 HP MODELS)

Test (output/resistance)	Meter scale	Specification
Ignition stator output		
30 and 40 hp models (2-cylinder)		
Red or green white to ground	400 DVA	190-320 volts
Green or green white to ground	40 DVA	20-40 volts
40-60 hp models (3-cylinder)		
Cranking speed	400 DVA	100-350 volts
Idle speed	400 DVA	200-350 volts
Trigger output		
30 and 40 hp models (2-cylinder)	20 DVA	2-8 volts
40-60 hp models (3-cylinder)		
Cranking speed	2 DVA	0.2-2 volts
Idle speed	20 DVA	2-8 volts
CDM resistance		
(+) to C, (-) to A	R × 100	1000-1250 ohms
(+) to B, (-) to D	R × 100	Continuity
(+) to D, (-) to B	R × 100	No continuity
(+) to D, (-) to A	R × 100	Continuity
(+) to A, (-) to D	R × 100	No continuity
(+) to Spark plug terminal, (-) to A	R × 100	900-1200 ohms

Table 13 IGNITION STATOR RESISTANCE SPECIFICATIONS

Model	Meter scale	Specification (ohms)
6-25 hp except 20 jet (1998)		
Black stator		
Black/white wire to ground(high speed winding)	R × 1	120-180
Black/yellow to ground(low speed winding)	R × 100	32-38
Black/yellow to black/white	R × 100	31-37
Red stator (1999-on)		
Green/white to white/green	R × 100	370-445
20 Jet (1998)		
Red to black (high speed winding)	R × 10	100-180
Blue to black (low speed winding)	R × 100	29-35
Red to blue	R × 100	28-34
30-60 hp		
Green/white to white/green	R × 10	660-710

Table 14 IGNITION SYSTEM IDENTIFICATION

Model	System	Timing advance	Features
2.5 and 3.3 hp	CDI	None	Integrated charge and trigger coil
4 and 5 hp	CDI	Electronic	Separate charge and trigger coils
6-15 hp (1998)	CDI	Mechanical	Black ignition stator
6-15 hp (1999-on)	ADI	Mechanical	Red ignition stator (rev limit on manual start models)
20 Jet (1998)	ADI	Electronic	Black ignition stator/ RPM limiting spark retard
20 Jet (1999-on)	ADI	Mechanical	Red ignition stator/RPM limit circuit in switchbox
20 and 25hp (1998)	ADI	Mechanical	Black ignition stator
20 and 25 hp (1999-on)	ADI	Mechanical	Red ignition stator(rev limit on manual start models)
30-60 hp	CDM	Mechanical	Separate rev limit module

Chapter Four

Lubrication, Maintenance and Tune-Up

Proper lubrication, maintenance and tune-ups are important to maintain a high level of performance, extend engine life and extract the maximum economy during operation.

The owner's operation and maintenance manual is a helpful supplement to this service manual and a valuable resource for anyone operating or maintaining the engine. If it is missing, purchase an owner's manual through a Mercury or Mariner dealership. The complete serial number of the outboard motor is required to obtain the correct owner's manual.

The following information is based on recommendations from Mercury Marine to help keep the Mercury or Mariner outboard motor operating at its peak performance level. **Table 1** provides torque specifications. **Table 2** provides the recommended preventive maintenance schedule. **Table 3** contains the recommended spark plugs. **Table 4** list gearcase lubricant and oil injection system capacity. **Tables 1-4** are located at the end of this chapter.

HOUR METER

Since a boat is not equipped with an odometer, service schedules for outboard motors are based on hours of engine operation. An engine hour meter (**Figure 1**, typical) is highly recommended to help keep track of the actual hours of engine operation. Many types of hour meters are available. The most accurate type for maintenance purposes is triggered by a spark plug lead. This makes sure that only actual running time is recorded. If an hour meter is operated by the key switch, artificial running hours are recorded if the operator forgets to turn off the key.

The Quicksilver Service Monitor (part No. 79-828010A-x1) is a spark plug wire driven hour meter that can also be set to flash an alarm at any time interval set by the operator. Quicksilver also offers many models of ignition switch operated hour meters.

A sample service log is included in the back of this manual to help with record keeping purposes.

FUELS AND LUBRICATION

Proper Fuel Selection

Two-stroke engines are lubricated by mixing oil with the fuel. The oil is either premixed in the fuel tank by the operator or automatically mixed by an oil injection system. The various components of the engine are lubricated as the fuel/oil mixture passes through the crankcase and cylinders. Since two-stroke fuel serves the dual function of producing combustion and distributing the lubrication, never use marine white gasoline or any other fuel not intended to be used in modern gasoline powered engines. Any substandard fuel and lubricating oil increases combustion chamber deposits, which leads to piston ring sticking, exhaust port blockage and abnormal combustion (preignition and detonation).

NOTE

Use the highest quality fuel without alcohol and lubricating oil available to reduce combustion chamber deposits and the resulting problems.

The recommended fuel is regular unleaded gasoline from a major supplier with a minimum pump posted octane rating of 87 with no alcohol. The minimum fuel requirements are regular unleaded gasoline with a minimum pump posted octane rating of 87 with no more than 10% ethanol. The use of methanol in any quantity is not recommended.

Recently reformulated fuels have been introduced into regions that have not achieved federally mandated reductions in emissions. Reformulated fuels are specifically blended to reduce emissions. Reformulated fuels normally contain oxygenates, such as ethanol, methanol or MTBE (methyl tertiary butyl ether). Reformulated fuels may be used as long as they do not contain methanol and normal precautions for ethanol extended fuels are taken. See *Alcohol Extended Gasoline* in this chapter.

If the engine is used for commercial service, or if detonation is suspected to be caused by a poor grade gasoline, use mid-grade gasoline of 89-91 pump posted octane with no alcohol from a major supplier.

The installation of a Quicksilver Water Separating Fuel Filter is recommended as a preventive measure on all permanently installed fuel systems. The *manufacturer* specifically recommends the installation of the Quicksilver Water Separating Fuel Filter if using alcohol blended or alcohol extended gasoline.

Sour Fuel

Fuel should not be stored for more than 60 days under ideal conditions. As gasoline ages, it forms gum and varnish deposits that restrict carburetor and fuel system passages, causing the engine to starve for fuel. The octane rating of the fuel also deteriorates over time, increasing the likelihood of preignition or detonation. Use a fuel additive such as Quicksilver Gasoline Stabilizer on a regular basis to stabilize the octane rating and prevent gum and varnish formation. Gasoline stabilizers must be added to fresh fuel. Gasoline stabilizers cannot rejuvenate old fuel. If the fuel is known to be sour or stale, drain and replace it with fresh gasoline. Dispose of the sour fuel in an approved manner. Always use fresh gasoline when mixing outboard engine fuel.

Alcohol Extended Gasoline

Although the *manufacturer* does not recommend the use of gasoline that contains alcohol, the minimum gasoline specification allows for a maximum of 10% ethanol to be used. Methanol is not recommended since the detrimental effects of methanol are more extreme than ethanol. If alcohol extended gasoline is used, consider the following:

1. Alcohol extended gasoline promotes leaner air/fuel ratios, which can:
   a. Raise combustion chamber temperatures, leading to preignition and/or detonation.
   b. Cause hesitation or stumbling during acceleration.
   c. Cause hard starting, hot and cold.
   d. Cause the engine to produce slightly less horsepower.
2. Alcohol extended gasoline attracts moisture, which can:
   a. Cause a water buildup in the fuel system.
   b. Block fuel filters.

c. Block fuel metering components.
d. Cause corrosion of metallic components in the fuel system and power head.

3. Alcohol extended gasoline deteriorates nonmetallic components, such as:
a. Rubber fuel lines.
b. Primer bulbs.
c. Fuel pump internal components.
d. Carburetor internal components.
e. Fuel recirculation components.

4. Alcohol extended gasoline promotes vapor lock and hot soak problems.

5. Alcohol extended fuel tends to build up combustion chamber deposits more quickly, which leads to:
a. Higher compression ratios, increasing the likelihood of preignition or detonation.
b. Piston ring sticking, which causes elevated piston temperatures, loss of power and ultimately preignition or detonation.
c. Exhaust port blockage or obstruction on engines with small or multiple exhaust ports.

NOTE
If the moisture content of the fuel reaches 0.5%, the water separates from the fuel and settles to the low points of the fuel system. This includes the fuel tank, fuel filters and carburetor float chambers. Alcohol extended fuels aggravate this situation.

If any or all of these symptoms are regularly occurring, consider testing the fuel for alcohol or changing to a different gasoline supplier. If the symptoms are no longer present after the change, continue using the gasoline from the new supplier.

If usage of alcohol extended fuel is unavoidable, perform regular maintenance and inspections more often than normal recommendations. Pay special attention to changing or cleaning the fuel filters, inspecting rubber fuel system components for deterioration, inspecting metallic fuel system components for corrosion and monitoring the power head for warning signs of preignition and/or detonation. It is sometimes necessary to enrich the carburetors metering circuits to compensate for the leaning effect of these gasolines.

Reformulated gasolines that contain MTBE (methyl tertiary butyl ether) in normal concentrations have no side effects other than those listed under Step 1. This does not apply to reformulated gasoline that contains ethanol or methanol.

The following procedure used for detecting alcohol in gasoline. Gasoline must be checked prior to mixing with oil. Use a small transparent bottle or tube that can be capped and mark it at approximately 1/3 full. A pencil mark on a piece of adhesive tape is sufficient.

1. Fill the container with water to the 1/3 full mark.
2. Add gasoline until the container is almost full. Leave a small air space at the top.
3. Shake the container vigorously, then allow it to sit for 3-5 minutes. If the volume of water appears to have increased, alcohol is present. If the dividing line between the water and gasoline becomes cloudy, use the center of the cloudy band as a reference.

This procedure cannot differentiate between ethanol or methanol, and it is not absolutely accurate, but it is accurate enough to determine if enough alcohol is present to require precautions.

Gasoline Additives

The following are the only recommended fuel additives and the benefits from their use.

1. *Quicksilver Fuel System Treatment*—When added to fresh fuel, this additive stabilizes the octane rating preventing fuel degradation and oxidation, prevents the formation of gum and varnish in the fuel system components, and prevents moisture buildup in the fuel tank, fuel system and carburetor(s).
2. *Quicksilver Gasoline Stabilizer*—Same benefits as Quicksilver Fuel System Treatment and Stabilizer, but much more concentrated. It is used to treat large quantities of fuel.
3. *Quicksilver Quickleen Fuel Treatment*—Quickleen is designed to help prevent combustion chamber deposits, and protect the internal fuel system and power head mechanical surfaces against corrosion. Use Quickleen if substandard or questionable fuel or oil is being used, or if combustion chamber deposits are a continual problem.

Unless the boat is consistently operated with fresh fuel in the fuel tank, using a fuel stabilizer is recommended.

CAUTION
Some marinas blend valve recession additives into their fuel to accommodate owners of older four-stroke marine engines. Valve recession additives are designed to help prevent premature valve seat wear on older four-stroke engines. The valve recession additives may react with some outboard motor oils. The reaction may cause certain two-stroke oil additives to precipitate (gel), plugging fuel and oil system passages. Avoid using fuel containing valve recession additives in an outboard motor.

Recommended Fuel/Oil Mixtures

The recommended oil for all Mercury/Mariner outboard motors is Quicksilver Premium or Premium Plus, Two-Cycle Outboard Oil. This oil meets or exceeds TCW-3 (two-cycle, water cooled, third revision) standards set by the NMMA (National Marine Manufacturers Association). If Quicksilver Premium Blend is not available, use a NMMA certified TCW-3 outboard oil from another engine manufacturer. Look for the NMMA certification insignia (**Figure 2**, typical) on the container label.

TCW-3 oils are designed to improve lubrication over previous standards (TCW and TCW-II) and reduce combustion chamber deposits caused by the lubricating oil. Do not use any oil other than a NMMA approved TCW-3 outboard motor oil.

CAUTION
Do not use automotive crankcase oil in a two-stroke outboard. Using this lubricant causes serious power head failure.

The recommended fuel/oil ratio for normal operation in all models without oil injection is 50:1 (50 parts fuel to 1 part oil). This is the standard 6 gal. (22.7 L) of gasoline to 16 fl. oz. (473 mL) of oil.

Models 25 hp (20 jet) and smaller are not equipped with oil injection and require a fuel and oil premix in the fuel tank.

Most 30-60 hp models are equipped with an engine mounted oil injection system. Oil injection models can be identified by an engine mounted oil reservoir.

Power Head Break-In Procedure

For a new outboard motor, rebuilt power head or replacement power head, follow the manufacturer's recommended break-in procedure. During the first hour of engine operation, change the engine speed frequently and avoid extended full-throttle operation. After the first hour of operation, the engine can be operated as desired within normal operating guidelines.

Mercury Marine also recommends that a new motor, rebuilt power head or replacement power head be operated on a 25:1 fuel/oil mixture for the first tank or an amount of fuel based on the engine's horsepower.

The formula is 1 gal. (3.8 L) of 25:1 fuel/oil mix for every 10 hp, rounded to the nearest gallon (liter). Small engines, such as 2.5-5 hp, with integral fuel tanks should use a 25:1 fuel/oil mixture for the first tank of fuel. It is acceptable to exceed the formula in order to accommodate the fuel tank size especially on mid-size motors, but the amount of 25:1 fuel/oil mix should not be less than the formula recommends.

2

Non-oil injected models

Use a 25:1 fuel/oil mixture during the engine break-in period. The mixture is 16 fl. oz. (473 mL) of oil for every 3 gal. (11.4 L) of fuel.

Use a 50:1 fuel/oil mixture after break-in and for all normal operation. The mixture is 8 fl. oz. (237 mL) of oil for every 3 gal. (11.4 L) of fuel. On models with an integral fuel tank, mix the fuel and oil in a separate container. On models equipped with a remote fuel tank, mix the fuel and oil in the remote fuel tank.

Oil injected models

Use a 25:1 fuel/oil mixture during the engine break-in period. The mixture is 8 fl. oz. (237 mL) of oil for every 3 gal. (11.4 L) of fuel. This provides a 50:1 fuel/oil mixture in the fuel tank that supplements the oil delivered by the engine mounted oil injection system. The final result provides a 25:1 fuel/oil mixture to the engine.

Do not mix any oil in the fuel tank after break-in and for all normal operation. The engine mounted oil injection system provides the engine with the required fuel/oil mixture. Monitor the engine mounted oil tank and the boat mounted oil tank, if so equipped, and keep the tank(s) filled with the recommended oil.

Fuel Mixing Procedure

WARNING
Gasoline is an extreme fire hazard. Never use gasoline near heat, spark or flame. Never smoke while mixing fuel.

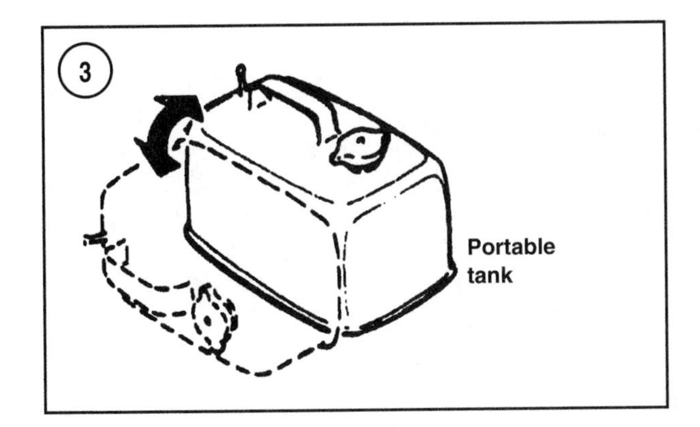

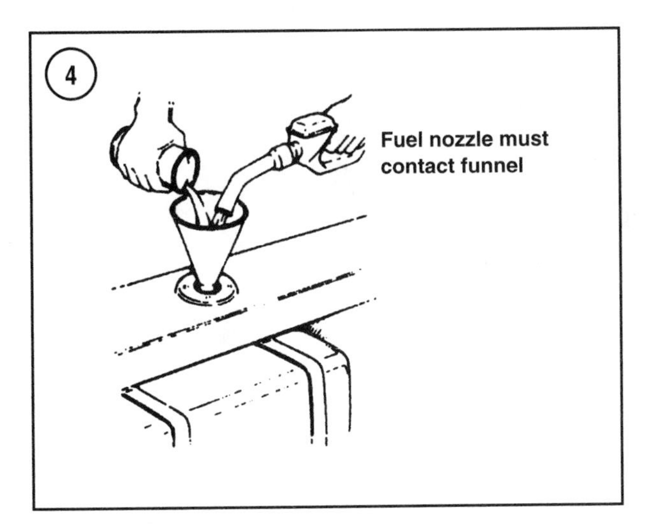

Mix the fuel and oil outside or in a well-ventilated area. Mix the fuel and oil to the recommended fuel/oil ratio. Using less than the specified amount of oil can result in insufficient lubrication and serious engine damage. Using more oil than specified causes spark plug fouling, erratic fuel metering, excessive smoke and accelerated carbon accumulation.

Cleanliness is important when mixing fuel. Small particles of dirt can restrict fuel metering passages.

Only use fresh fuel. If the fuel is sour, dispose of the fuel in an appropriate manner and start over with fresh fuel. Add fuel stabilizer to the fuel mix if it is not going to be used within two weeks.

Above 32° F (0° C)

Measure the required amount of gasoline and recommended outboard oil accurately. Pour the oil into the remote tank and add the fuel. Install the tank fill cap and mix the fuel by tipping the tank from side-to-side several times. See **Figure 3**.

If a built-in tank is used, insert a large filter-type funnel into the tank fill neck. Carefully pour the specified oil and gasoline into the funnel at the same time. See **Figure 4**.

Below 32°F (0° C)

Measure the required amount of gasoline and outboard oil. Pour approximately 1 gal. (3.8 L) of the gasoline into the tank, then add the required amount of oil. Install the tank fill cap and shake the tank vigorously to mix thoroughly the fuel and oil. Remove the cap, add the rest of the fuel and shake the tank again.

For a built-in tank, insert a large filter-type funnel into the tank fill neck. Mix the required amount of oil with one gallon of gasoline in a separate container. Carefully pour the mixture into the funnel while filling the tank with gasoline.

Consistent fuel mixture

The carburetor idle mixture adjustment is sensitive to fuel mixture variations caused by using different oils and gasoline or inaccurate measuring and mixing. This may require constant readjustment of the idle mixture screw(s). To prevent the necessity of carburetor readjustment or erratic running from one fuel batch to another, always be consistent when mixing fuel. Prepare each batch of fuel exactly the same as previous ones.

Be careful when using premixed fuel sold at marinas, since the quality and consistency of premixed fuel can vary greatly. The possibility of engine damage caused by an incorrect or substandard fuel/oil mixture often outweighs the convenience of premixed fuel. Ask the operator of the marina or fuel supply station about the specifications for the oil and fuel being used. Do not use premixed fuel if the fuel does not meet or exceed the engine fuel and oil requirements.

Variable Ratio Oil Injection (30 and 40 hp Two-Cylinder Models)

CAUTION

On a boat mounted electric fuel supply pump, fuel pressure must not exceed 6 psi (41.4 kPa) at the engine fuel line connector. If necessary, install a fuel pressure regulator between the electric fuel pump and the engine fuel line connector. Adjust the fuel pressure to a maximum of 6 psi (41.4 kPa). The electric fuel pump installation must conform to Coast Guard safety standards for permanently installed fuel systems.

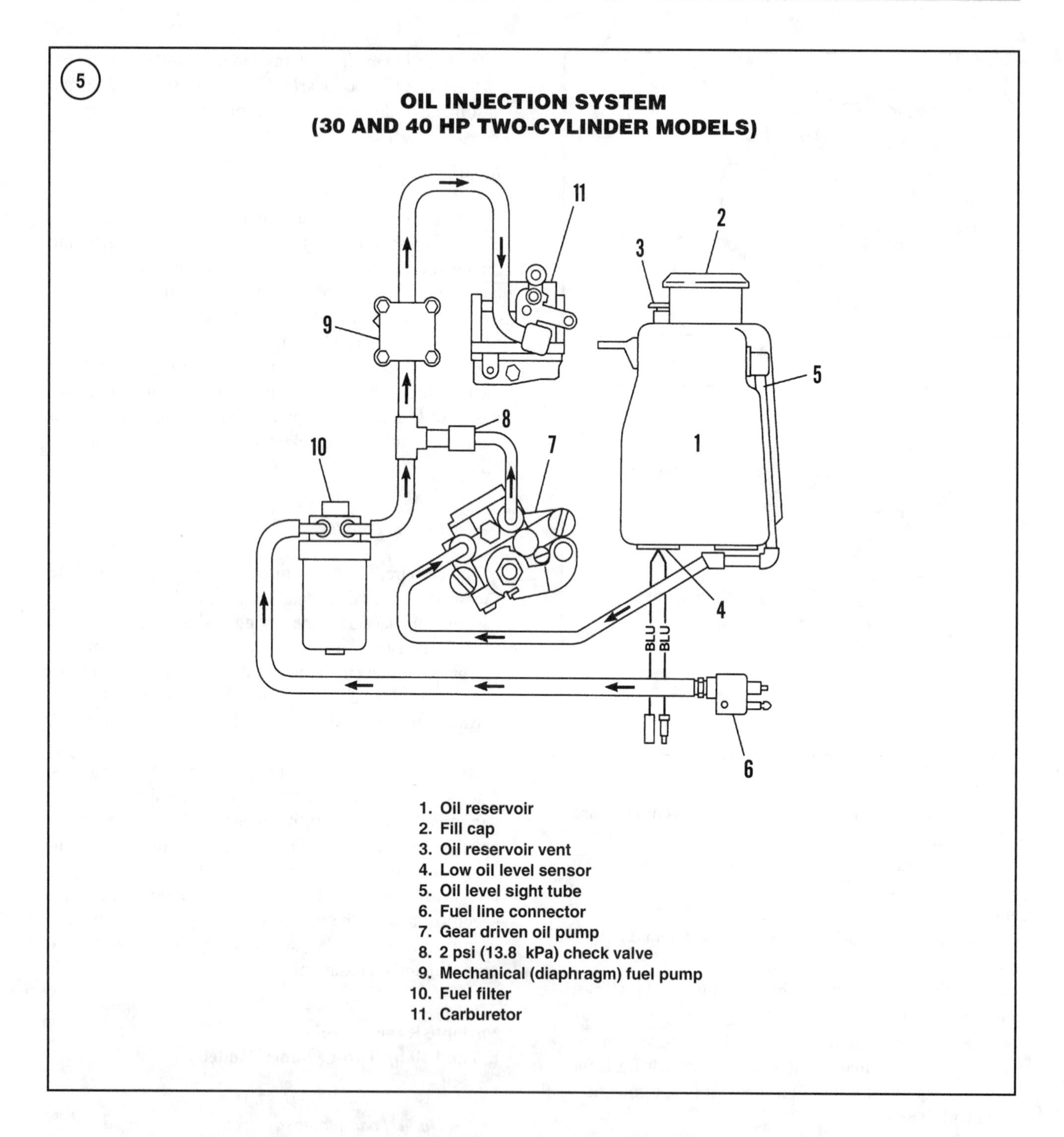

The oil pump is driven by the engine crankshaft and injects oil into the fuel stream at the fuel pump as shown in **Figure 5**. The oil pump delivers an amount of oil relative to carburetor throttle valve opening and engine speed. The fuel/oil ratio is approximately 100:1 at idle and approximately 50:1 at wide-open throttle.

The engine mounted oil tank capacity is 50.5 fl. oz. (1.5 L) and provides approximately 4.7 hours operation at wide-open throttle. The oil tank is equipped with an oil level sight gauge visible through an opening in the engine cowl. See **Figure 5**.

An oil level sensor in the oil reservoir is designed to activate the warning horn if the oil level drops to 7 fl. oz. (207 mL).

The oil tank can be filled without removing the engine cowl. Remove the fill cap and fill it with the recommended outboard oil. Reinstall the fill cap securely. Refer to Chapter Twelve for additional system information and service procedures.

Variable Ratio Oil Injection (40-60 hp Three-Cylinder Models)

CAUTION

On a boat mounted electric fuel supply pump, fuel pressure must not exceed 6 psi (41.4 kPa) at the engine fuel line connector. If necessary, install a fuel pressure regulator between the electric fuel pump and the engine fuel line connector. Adjust the fuel pressure to a maximum of 6 psi (41.4 kPa). The electric fuel pump installation must conform to Coast Guard safety standards for permanently installed fuel systems.

The oil pump is driven by the engine crankshaft and injects oil into the fuel stream at the fuel pump as shown in **Figure 6**. The oil pump delivers an amount of oil relative to carburetor throttle valve opening and engine speed. Maximum oil delivery occurs at wide-open throttle.

The engine mounted oil tank capacity is 3 qt. (2.8 L) and provides approximately 7 hours operation at wide-open throttle. The oil tank is equipped with an oil level sight gauge visible through an opening in the engine cowl. See **Figure 6**.

An oil level sensor in the oil reservoir is designed to trigger the warning horn if the oil level drops to 14.5 fl. oz. (435 mL). There is no warning system module or warning system self-test.

The oil tank can be filled without removing the engine cowl. Remove the fill cap and fill it with the recommended outboard oil. Reinstall the fill cap securely. Refer to Chapter Twelve for additional system information and service procedures.

Checking Lower Gearcase Lubricant

Check the lower gearcase lubricant level after 20 hours of operation or 10 days, then every 50 hours of operation or once a month thereafter. The recommended lubricant is Quicksilver Premium Blend Gear Lubricant. If the gearcase is subjected to severe use, consider using Quicksilver High Performance Gear Lubricant.

CAUTION

Never use regular automotive gear lubricant in the gear housing. The expansion, foam characteristics and water tolerance of automotive gear lubricant are not suitable for marine use.

1. Place the outboard motor in an upright position. Place a suitable container under the gear housing. Loosen the gearcase drain/fill plug (**Figure 7**). Allow a small amount of lubricant to drain. If water is present inside the gear housing, it will drain before the lubricant, or the lubricant will have a white or cream tint to the normal lubricant color. If the lubricant looks satisfactory, retighten the drain/fill plug securely. If water was present in the lubricant, or if the lubricant is dirty, fouled or contains substantial metal shavings, allow the remaining lubricant to drain completely from the gearcase.

CAUTION

If water is present in the gearcase and it is not possible to repair it at this time, completely drain the contaminated lubricant and refill the gearcase with fresh lubricant. Crank the engine several revolutions and spin the propeller shaft several times to distribute the fresh lubricant on the internal components. Repair the gearcase before operating the engine or placing it in the water.

NOTE

The presence of a small amount of metal filings and fine metal particles in the lubricant is normal, while an excessive amount of metal filings and larger chips indicate a problem. Remove and disassemble the gearcase to determine the source and cause of metal filings and chips. See Chapter Nine.

2. Remove the gearcase vent plug. See **Figure 7**. Replace the sealing washer on each plug. The lubricant level must be even with the vent plug hole.

CAUTION

The vent plug vents displaced air while lubricant is added to the gearcase. Never attempt to fill or add lubricant to the gearcase without first removing the vent plug.

3. If the lubricant level is low, temporarily reinstall the vent plug(s) and remove the drain/fill plug. Insert the gearcase filling tube into the drain/fill plug hole, then remove the vent plug(s) again.
4. Replace the sealing washer on the drain plug.
5. Inject the recommended lubricant into the drain/fill plug hole until excess lubrication flows from the vent plug hole (**Figure 7**). Install and tighten the vent plug. Remove the gearcase filling tube and quickly install the drain/fill

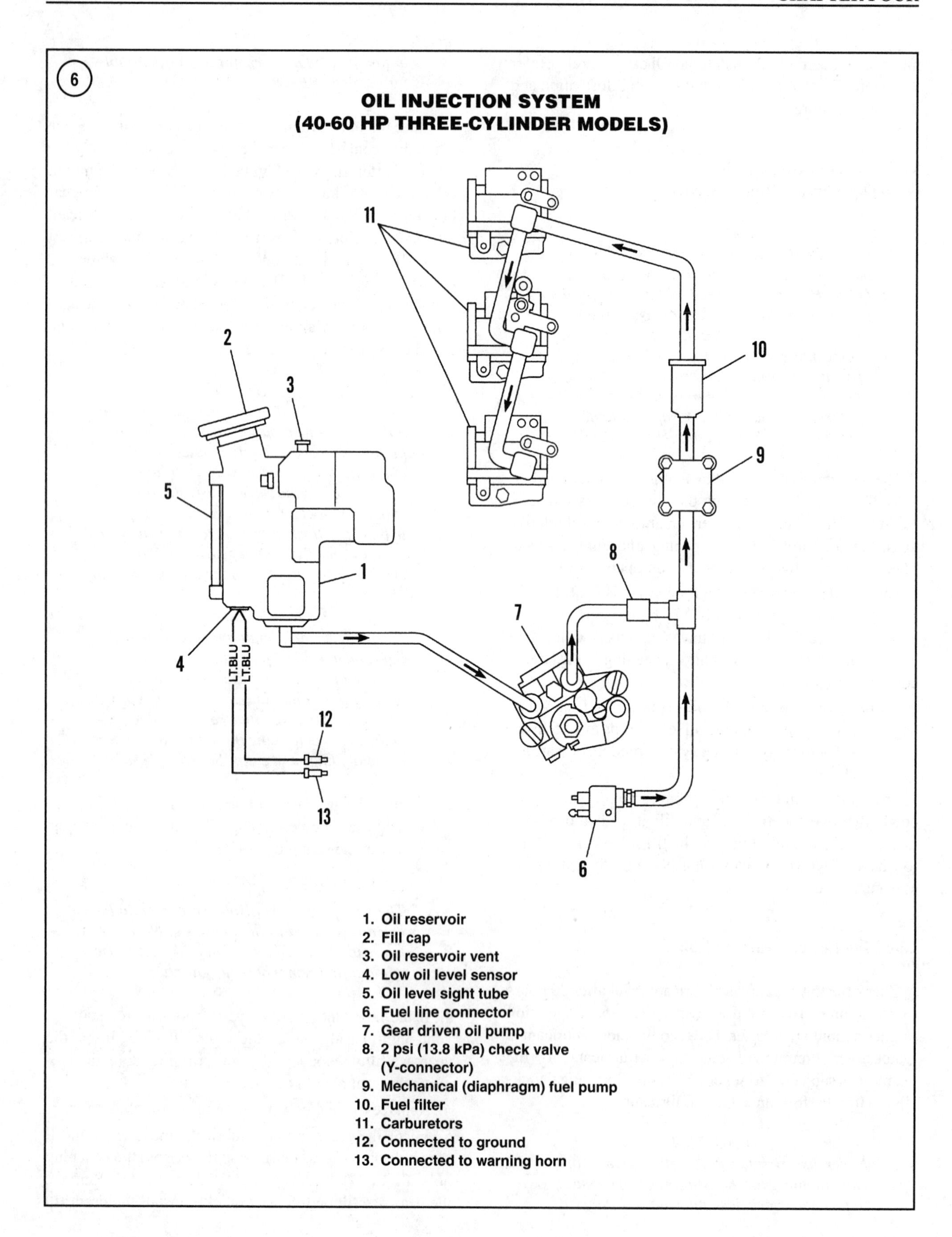
6
OIL INJECTION SYSTEM
(40-60 HP THREE-CYLINDER MODELS)
11
2
3
10
9
5
8
1
7
4
LT.BLU
LT.BLU
12
13
6
1. Oil reservoir
2. Fill cap
3. Oil reservoir vent
4. Low oil level sensor
5. Oil level sight tube
6. Fuel line connector
7. Gear driven oil pump
8. 2 psi (13.8 kPa) check valve (Y-connector)
9. Mechanical (diaphragm) fuel pump
10. Fuel filter
11. Carburetors
12. Connected to ground
13. Connected to warning horn

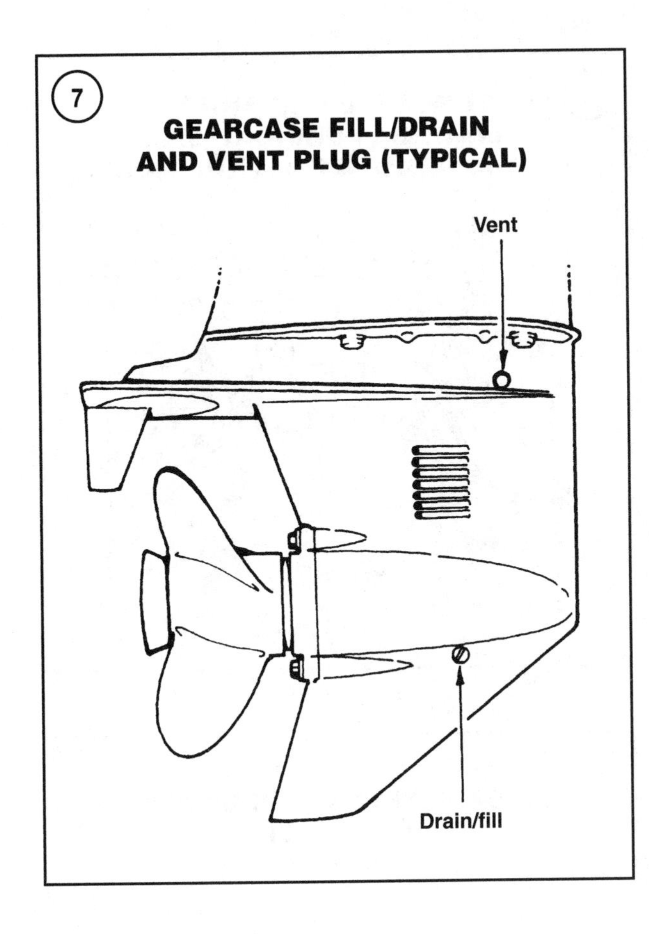

plug. Tighten the drain/fill plug and vent plug to the specification in **Table 1**.

Changing Lower Gearcase Lubricant

Change the lower gearcase lubricant after the first 20 hours of operation, and every 100 hours or every season thereafter. Refer to **Table 4** for gearcase capacities.

Refer to **Figure 7** for this procedure.

1. Remove the engine cover. To prevent accidental starting, disconnect and ground the spark plug leads to the power head.
2. Place the outboard motor in an upright position. Place a suitable container under the lower gearcase.
3. Remove the drain/fill plug, then the vent plug. Allow the lubricant to drain fully into the container.
4. Inspect the drained lubricant. Lubricant contaminated with water has a white or cream tint as compared to the normal lubricant color. The presence of a small amount of metal filings and fine metal particles in the lubricant is normal, while an excessive amount of metal filings and larger chips indicates a problem.

CAUTION
If water is present in the gearcase and it is not possible to repair it at this time, completely drain the contaminated lubricant and refill the gearcase with fresh lubricant. Crank the engine several revolutions and spin the propeller shaft several times to distribute the fresh lubricant on the internal components. Repair the gearcase before operating the engine or placing it in the water.

CAUTION
If there is an excessive amount of metal filings and larger chips in the lubricant, remove and disassemble the gearcase to determine the cause. See Chapter Nine for removal and disassembly. Operating the gearcase in this condition will increase the amount of damage to the internal components.

5. Refill the gearcase with the recommended lubricant as described under *Checking Lower Gearcase Lubricant* in this chapter. Refer to **Table 4** for lower gearcase lubricant capacity.

Jet Pump Maintenance

The jet pump unit on 20, 30 and 45 jet models requires that the drive shaft bearings be lubricated daily or every 10 hours of operation, whichever comes first. After every 30 hours of operation, pump in extra lubricant to purge the old grease and moisture that may have accumulated.

The clearance of the jet pump impeller to the impeller liner must be approximately 0.030 in., (0.8 mm). If the jet pump is operated in silt laden or sandy waters, check the clearance frequently. Shims above and below the impeller can be repositioned to set the required clearance. Refer to Chapter Ten for jet pump service procedures.

1. To lubricate the drive shaft bearings, disconnect the vent hose (1, **Figure 8**) from the grease fitting.
2. Connect a grease gun filled with Quicksilver 2-4-C with teflon or Lubriplate 630-AA lubricant to the grease fitting (2, **Figure 8**).
3. Pump in grease until grease exits the vent hose (3, **Figure 8**).
4. Reconnect the vent hose back to the grease fitting.

NOTE
Every 30 days, pump in extra grease until all old grease is purged from the bearings and fresh grease exits the vent hose. If more than slight traces of water are present in the exiting grease, replace the drive shaft seals. See Chapter Ten.

8 **JET PUMP DRIVE SHAFT SERVICE (ALL MODELS)**

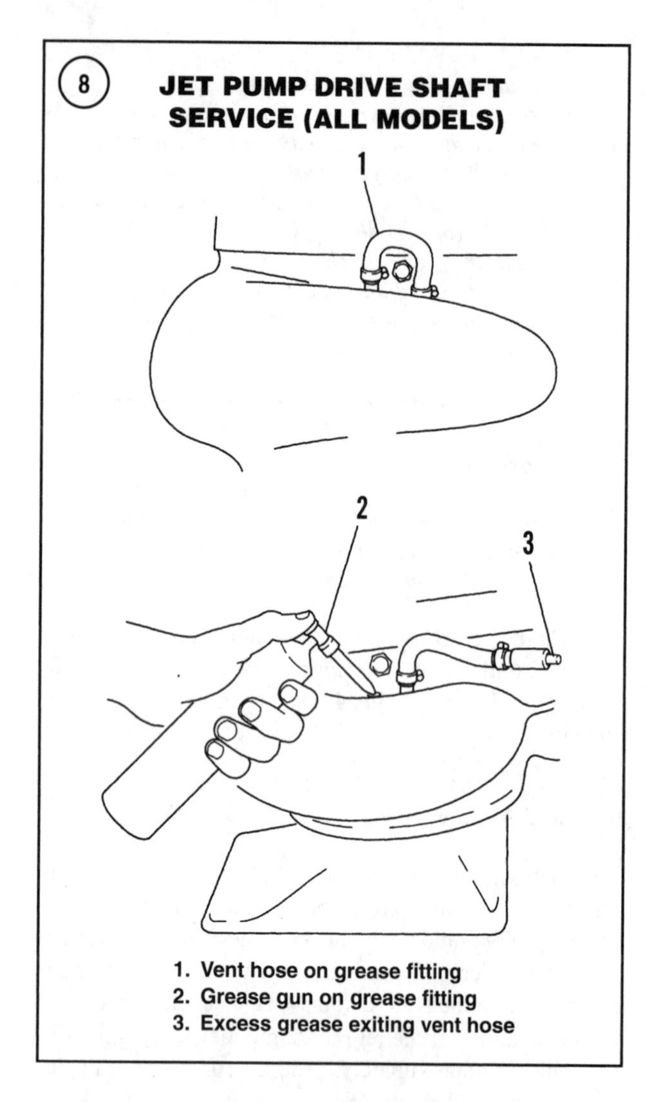

1. Vent hose on grease fitting
2. Grease gun on grease fitting
3. Excess grease exiting vent hose

9 **CLAMP SCREW LUBRICATION (PORTABLE MODELS)**

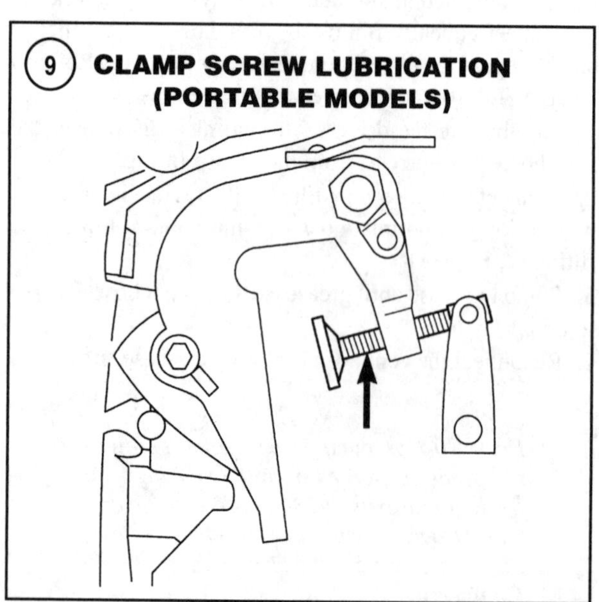

10 **THROTTLE AND SHIFT LINKAGE LUBRICATION (TYPICAL REMOTE CONTROL MODEL)**

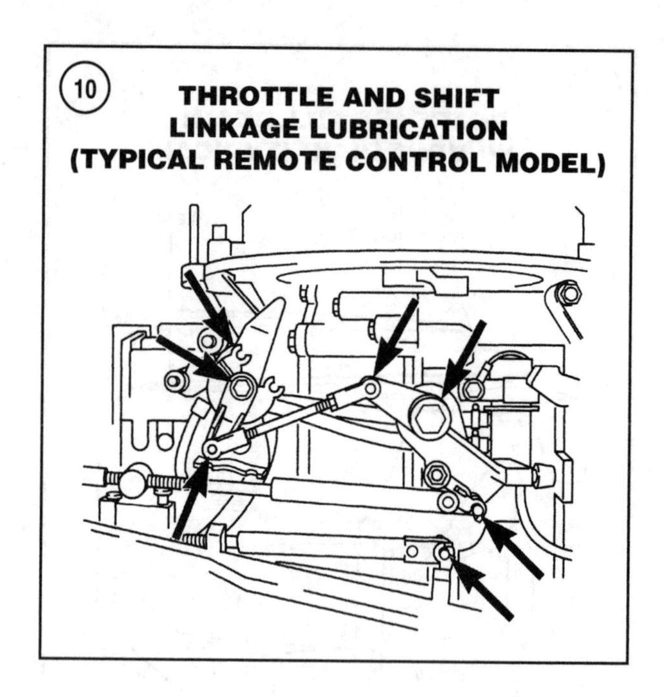

11 **MID-SECTION LUBRICATION (TYPICAL)**

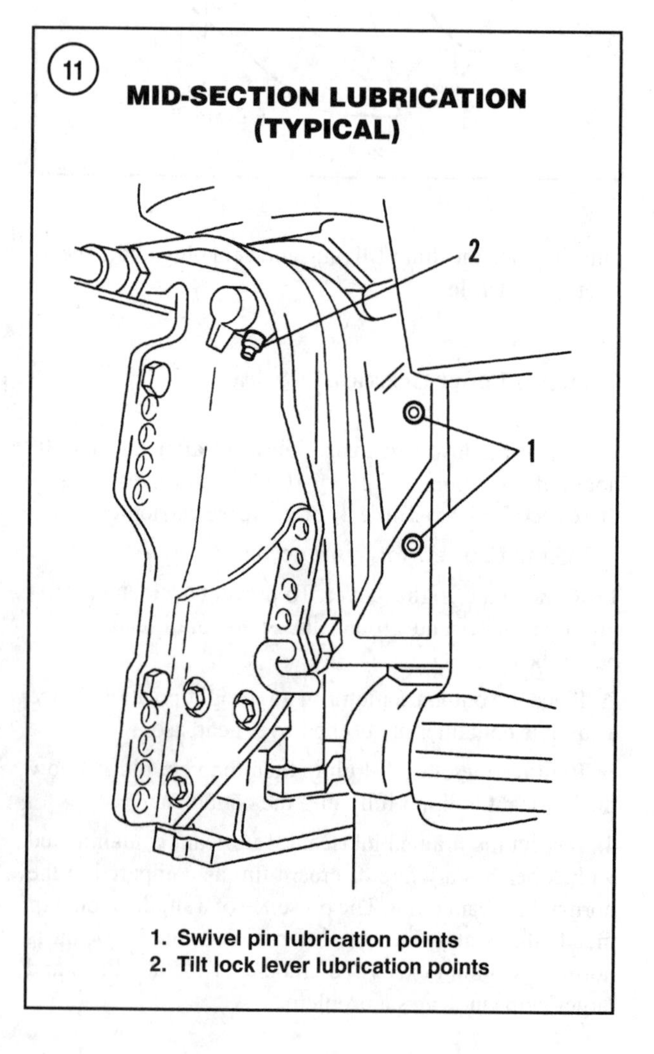

1. Swivel pin lubrication points
2. Tilt lock lever lubrication points

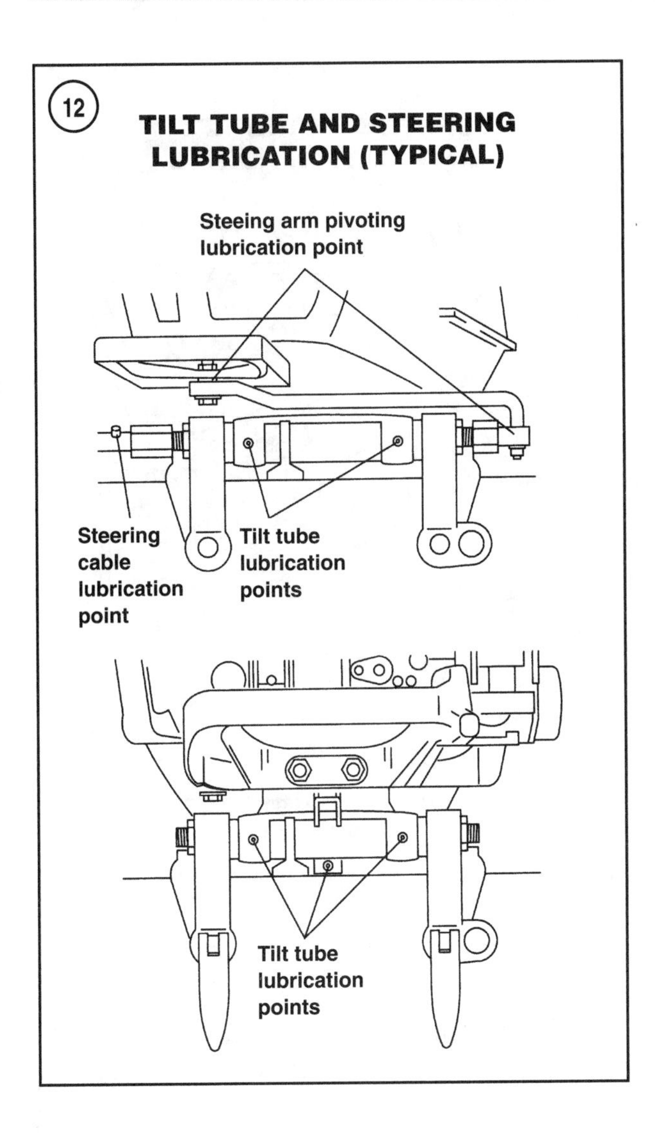

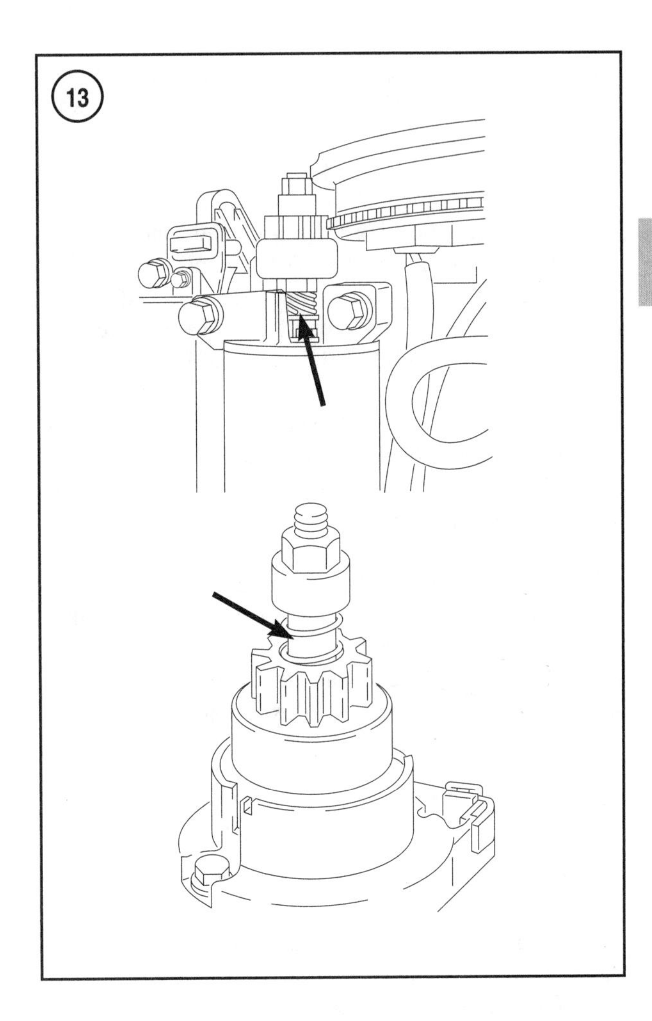

Propeller Shaft

To prevent corrosion and to ease the future removal of the propeller, lubricate the propeller shaft at least once each season during freshwater operation or every 60 days of saltwater operation. Remove the propeller (Chapter Nine) and remove corrosion or dried grease, then coat the propeller shaft splines with Quicksilver Special Lube 101 (part No. 92-1387A1), Quicksilver 2-4-C Marine Lubricant (part No. 92-825407) or a suitable waterproof anticorrosion grease.

Recommended Preventive Maintenance and Lubrication

Refer to **Table 2** for recommended preventive maintenance procedures. Typical lubrication points for the more common engines are shown in **Figures 9-15**.

Lubricate every grease fitting on the midsection with 2-4-C grease (part No. 92-825407). Lubricate all pivoting or sliding throttle, shift and ignition linkages with 2-4-C grease. Lubricate the steering arm pivot points on remote control models with SAE 30 engine oil. Lubricate the steering cable sliding surfaces with 2-4-C grease.

CAUTION

When lubricating the steering cable, make sure its core is fully retracted into the cable housing. Lubricating the cable while extended can result in hydraulic lock, hard steering or loss of steering control.

Corrosion of the Propeller Shaft Bearing Carrier

Saltwater corrosion between the propeller shaft bearing carrier and gearcase housing can eventually split the housing and destroy the lower gearcase assembly. If the outboard motor is operated in saltwater, remove the bearing carrier retaining hardware and the propeller shaft bearing carrier (**Figure 16**) at least once per season. Refer to Chapter Nine for bearing carrier removal procedures for all models.

Thoroughly clean all corrosion and dried lubricant from each end of the propeller shaft bearing carrier.

Clean the gear housing internal threads and retaining ring external threads on models so equipped. Replace the bearing carrier O-rings and propeller shaft seals if the carrier is removed. Apply a liberal coat of Quicksilver Perfect Seal (part No. 92-34227-1), Special Lube 101 (part No. 92-13872A1) or 2-4-C Marine Lubricant (part No. 92-825407) to each end of the carrier and to the gear housing and cover nut threads, if so equipped. Do not allow any Quicksilver Perfect Seal into the propeller shaft bearings. Reinstall the bearing carrier and retaining ring as described in Chapter Nine.

Make sure all available anodes are installed and securely grounded to the gearcase or midsection. Replace any anode that is deteriorated to half of its original size. Refer to Chapter One for anode information. Refer to *Anticorrision maintenance* in this chapter for additional corrosion prevention procedures.

OFF-SEASON STORAGE

Prepare an outboard motor for storage to protect it from rust, corrosion, dirt or other contamination and to protect it from physical damage. Mercury Marine recommends the following procedure.

1. Remove the air intake cover, if so equipped (**Figure 17**, typical).
2. Add a good quality fuel stabilizer to all fuel tanks. Mix according to the manufacturer's instructions for storage. Fuel stabilizer, when added to fresh fuel:
   a. Prevents gum and varnish from forming in the fuel system.
   b. Controls moisture in the fuel system.
   c. Prevents modern fuels from reacting with brass and copper fuel system components.
   d. Stabilizes the fuel to prevent octane loss and prevents the fuel from going sour.

CAUTION

Do not run the engine without an adequate water supply and do no exceed 3000 rpm without an adequate load. Refer to ***Safety Precautions*** *at the beginning of Chapter Three.*

3. Start the engine and run it at fast idle at least 15 minutes until warmed up to operating temperature. This ensures the gasoline stabilizer has had time to reach the carburetor(s).

14

UPPER SHIFT SHAFT LUBRICATION POINTS

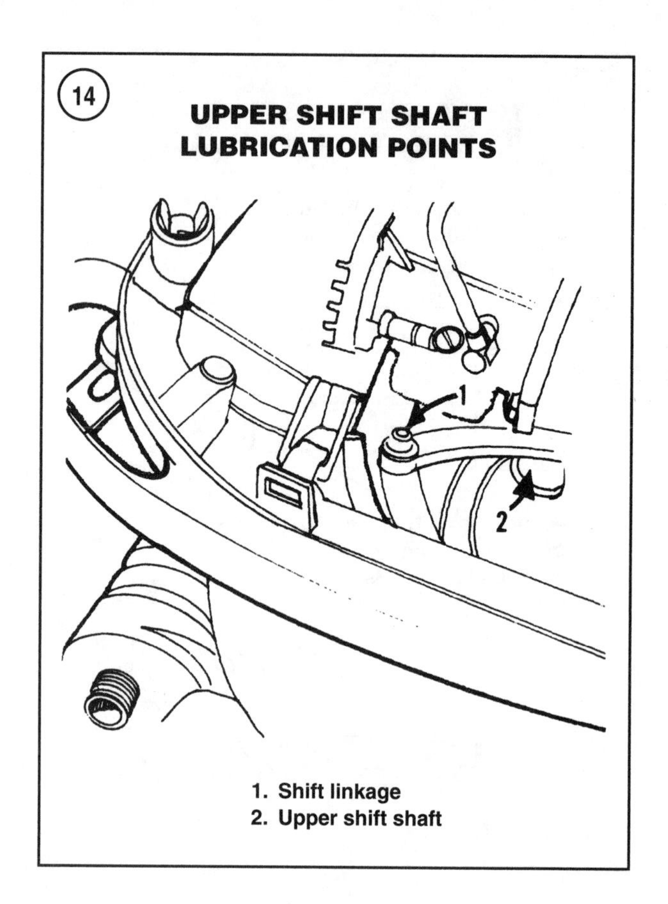

1. Shift linkage
2. Upper shift shaft

15

UPPER SHIFT SHAFT LUBRICATION

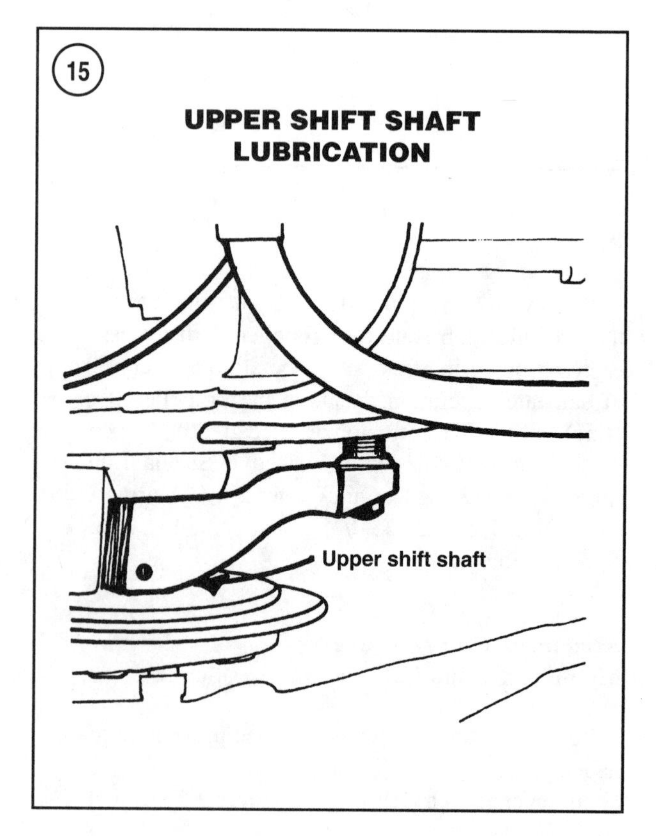

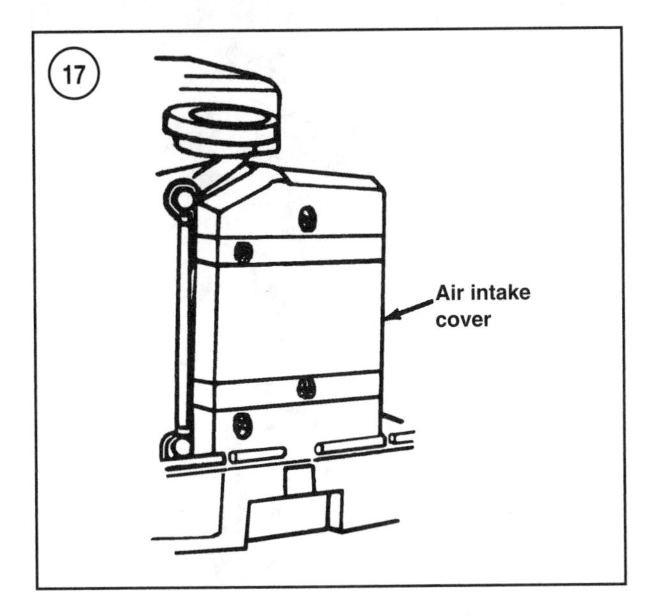

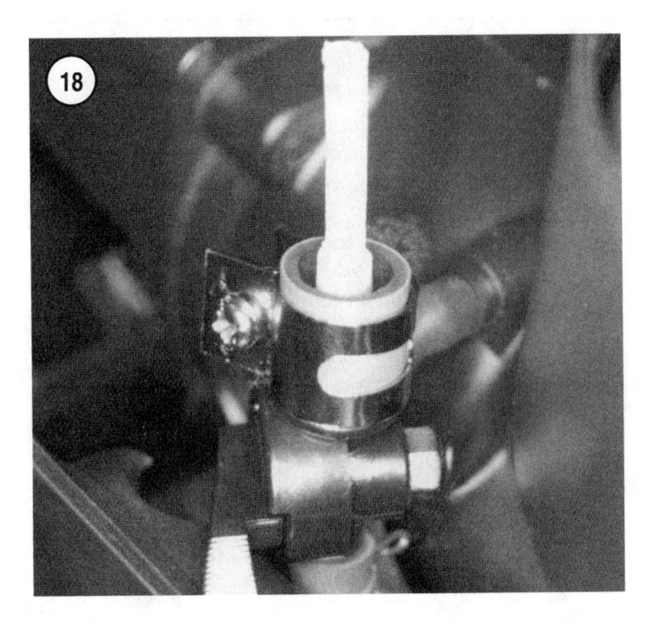

NOTE
If the fuel in the fuel tank is fresh and properly stabilized, it is not necessary to run the engine fuel system dry. However, some small portable engines may be transported or stored in a position that could cause fuel to spill. Drain the fuel tank in these instances. Engines permanently mounted to the transom and connected to a built-in fuel tank gain no benefit from running the fuel system dry.

4. With the engine running at fast idle and at operating temperature, spray the recommended quantity of Quicksilver Storage Seal (part No. 92-86145A12) or an equivalent into each carburetor throat following its manufacturer's instructions. Remove the motor from the water supply when the application of Quicksilver Storage Seal is complete.
5. Remove the spark plugs as described in this chapter. Spray about 1 oz. (30 mL) of Quicksilver Storage Seal into each spark plug hole. Crank the engine clockwise by hand several revolutions to distribute the storage seal throughout the cylinders. Reinstall the spark plugs.
6. On portable fuel tanks, service the filter by detaching the fuel hose from the tank. Unthread the pickup tube assembly or pickup tube retaining ring from the tank. Remove the pickup tube assembly. Clean or replace the fine mesh filter as necessary. Replace any gaskets, seals or O-rings. Thread the pickup assembly into the tank and tighten it or the retaining ring securely.

NOTE
All models covered in this manual use one of three types of engine mounted fuel filters. 2.5-5 hp models use a filter in the integral fuel tank outlet, and all other models use either a sight bowl filter or an inline filter.

7. *2.5-4 hp models*—Service the fuel filter as follows:
   a. Remove the fuel tank as described in Chapter Six. Drain the fuel into a suitable container. If the fuel is contaminated or sour, dispose of it in an approved manner.
   b. Loosen the clamp screw securing the fuel valve and filter assembly to the fuel tank. Remove the fuel valve and filter assembly (**Figure 18**) from the fuel tank.
   c. Clean dirt or debris from the fuel tank.
   d. Clean the filter screen in the fuel outlet valve assembly with solvent and low pressure compressed air. If the screen cannot be cleaned, replace the outlet valve assembly.

e. Reinstall the valve assembly to the fuel tank and tighten the clamp screw securely.
f. Reinstall the fuel tank to the motor and check for fuel leaks.

8. *Sight bowl fuel filter (**Figure 19**)*—Service the fuel filter as follows:

a. Unscrew the sight bowl from the filter base (**Figure 20**). Safely dispose of any fuel in the bowl.
b. Remove the filter element (**Figure 20**) from the bowl. Clean the element and fuel bowl in clean solvent, and dry them with compressed air. Replace the element if it cannot be satisfactorily cleaned.
c. Install a new bowl seal into the fuel bowl. Reinstall the filter into the bowl and reinstall the bowl assembly onto the filter base. Hand-tighten the bowl.
d. Test the installation by squeezing the primer bulb and checking for leaks.

9. *5 hp and other models with an inline fuel filter*—Service the fuel filter (**Figure 21** or **Figure 22**) as follows:

a. Carefully compress the spring clamps or cut the tie-strap clamps from each end of the filter.
b. Disconnect the fuel lines from the filter. Discard the filter and replace any fuel lines damaged in the filter removal process.
c. Connect the fuel lines to the new filter. Make sure the arrow is pointing in the direction of fuel flow toward the carburetor. Fasten the hoses to the filter securely with new tie-straps.
d. Test the installation by squeezing the primer bulb and checking for leaks.

10. Drain and refill the lower gearcase with the recommended lubricant as described in this chapter. Install new sealing washers on all drain and vent plugs.

11. *Jet models*—Service the drive shaft bearings as described in this chapter. See *Jet Pump Maintenance*.

12. Refer to **Figures 9-15** and **Table 2**, for preventive maintenance and general lubrication recommendations.

13. Clean the exterior areas of the outboard motor, including all accessible power head parts. Spray the entire power head, including all electrical connections, with Quicksilver Corrosion Guard (part No. 92-815869A12). Install the engine cowling and spray a thin film of Quicksilver Corrosion Guard on all remaining metal painted surfaces of the midsection and lower gearcase. This is especially important if the engine is operated or stored in saltwater or polluted water.

14. Remove the propeller as described in Chapter Nine. Lubricate the propeller shaft with Quicksilver Special Lube 101 (part No. 92-13872A1) or Quicksilver 2-4-C Marine Lubricant (part No. 92-825407) and reinstall the propeller.

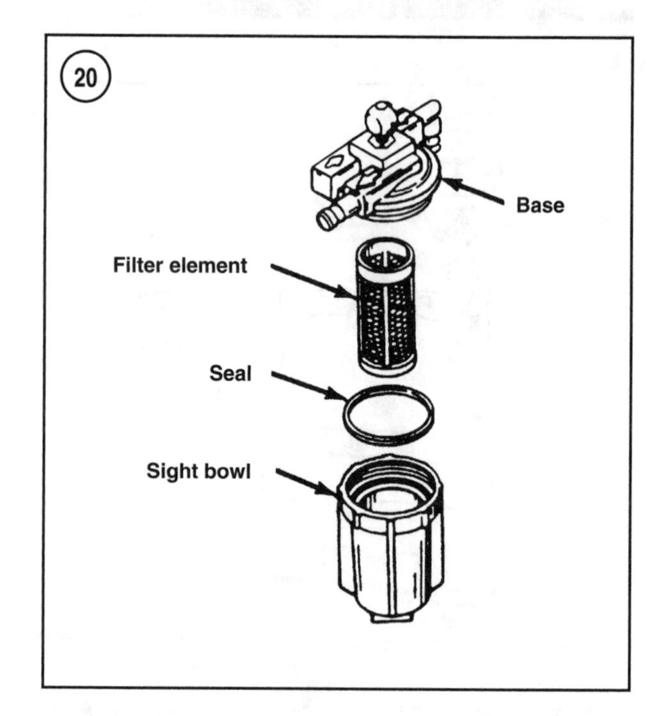

CAUTION

All water drain holes in the gear housing must be open for water draining. Water expands as it freezes and can crack the gear housing or water pump. If the boat is equipped with a speedometer, disconnect the water pickup tube and allow it to completely drain, then reconnect the tube.

15. Drain the cooling system completely to prevent freeze damage by positioning the motor in a vertical position. Check water drain holes for blockage.

16. Store the motor in a vertical position. Do not store an outboard motor with the power head below the lower gearcase. The power head must be higher than the lower gearcase to prevent water from entering the engine through the exhaust ports.

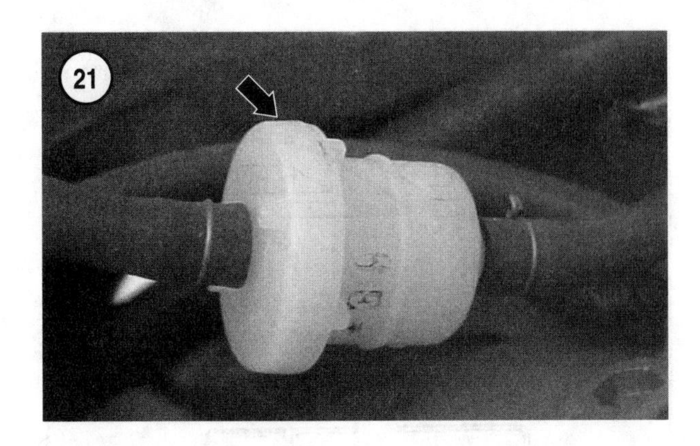

17. Prepare the battery for storage as follows:
   a. Disconnect the negative, then the positive, battery cables.
   b. Clean all grease, sulfate or other contamination from the battery case and terminals.
   c. Remove the vent caps if possible and check the electrolyte level of each cell. Add distilled water to the level recommended by the battery manufacturer. Do not overfill.
   d. Lubricate the terminals and terminal fasteners with Quicksilver Corrosion Guard (part No. 92-815869 A12) or an equivalent.

CAUTION

A discharged battery can be damaged by freezing. Consider using a battery float style charger to maintain the battery charge during storage. A float charger is an easy and inexpensive way to keep the battery at peak charge without excessive venting or gassing and subsequent water loss. The Guest Battery Pal (part No. 2602) is available from any Mercury/Mariner dealership and most marine supply stores.

   e. Make sure the battery is fully-charged with a specific gravity at 1.260-1.280. Store the battery in a *cool, dry* location where the temperature will not drop below freezing.
   f. Recharge the battery every 45 days or whenever the specific gravity drops below 1.230. Maintain the recommended electrolyte level. Add distilled water as necessary to maintain the level recommended by the battery manufacturer. For maximum battery life, avoid charge rates over 6 amps. Discontinue charging when the specific gravity reaches 1.260 at 80° F (27°C).
   g. Remove the grease on the battery terminals prior to returning the battery to service. Make sure the battery is fully-charged.

ANTICORROSION MAINTENANCE

NOTE

Magnesium anodes are available for extra corrosion protection in freshwater. Never use magnesium anodes in saltwater. The engine will be over-protected, causing the paint to blister or peel.

1. Flush the cooling system with freshwater as described in this chapter after each outing in saltwater. Wash the exterior with freshwater.
2. Dry the exterior of the outboard and apply primer over paint nicks and scratches. Use Mercury Marine recommended touch-up paint. Never use paint containing mercury or copper. Never apply paint to sacrificial anodes or anodic trim tabs.
3. Apply Quicksilver Corrosion Guard (part No. 92-815869A12) to the power head surfaces and electrical connections.
4. Inspect all of the sacrificial anodes and the trim tab, if it is anodic. Replace anodes that have deteriorated to less than half their original size.
5. Check the anodes for proper grounding as follows:
   a. Set and ohmmeter on the highest available scale.
   b. Connect one meter test lead to a power head ground. Connect the other test lead to the anode surface.
   c. The ohmmeter should indicate continuity. If it does not, remove the anode and thoroughly clean the anode, mounting bolts, bolts openings and mounting surface.
   d. Reinstall the anode and retest it as previously described. Replace the anode and check the gearcase-to-midsection and midsection-to-power

head mating surfaces and fasteners for corrosion if incorrect test results persist.

6. If the engine is operated in saltwater, polluted or brackish water, perform routine lubrication twice as often as specified in **Table 2**.

ENGINE SUBMERSION

Recover and attend to an outboard motor that has been lost overboard as quickly as possible. A delay results in rust and corrosion damage to internal components. Perform the following emergency steps immediately if the motor is submerged in freshwater.

NOTE

If the outboard falls overboard in saltwater or heavily polluted water, completely disassemble and clean the outboard before attempting to start the engine. If it is not possible to disassemble and clean the engine immediately, briefly flush and resubmerge the outboard in freshwater to minimize rust and corrosion until the engine can be serviced.

1. Wash the outside of the motor with clean water to remove weeds, mud and other debris.
2. Rinse the power head clean of weeds, mud and other debris with freshwater.
3. Remove, clean and dry the spark plug(s).
4. Drain the carburetor float bowl(s). Do not reinstall the float chamber plugs. See Chapter Six.
5. On oil injected models, drain and clean the oil tanks. See Chapter Twelve. Flush out all lines. Refill the system with the recommended oil. Bleed as much air out of the system as possible at this time.
6. Connect a clean fuel tank to the engine fuel line connector. Squeeze the primer bulb repeatedly to flush fresh fuel through the entire fuel system and purge the system of water.
7. Replace *all* fuel filters.
8. Reinstall the carburetor drain plugs.

CAUTION

If sand may have entered the power head, do not attempt to start the engine. Sand and other gritty debris can quickly damage the internal power head components. If the engine is lost overboard while running, internal engine damage is probable. Do not force the engine if it fails to turn over easily with the spark plugs removed. Binding indicates internal damage such as a bent connecting rod or broken piston.

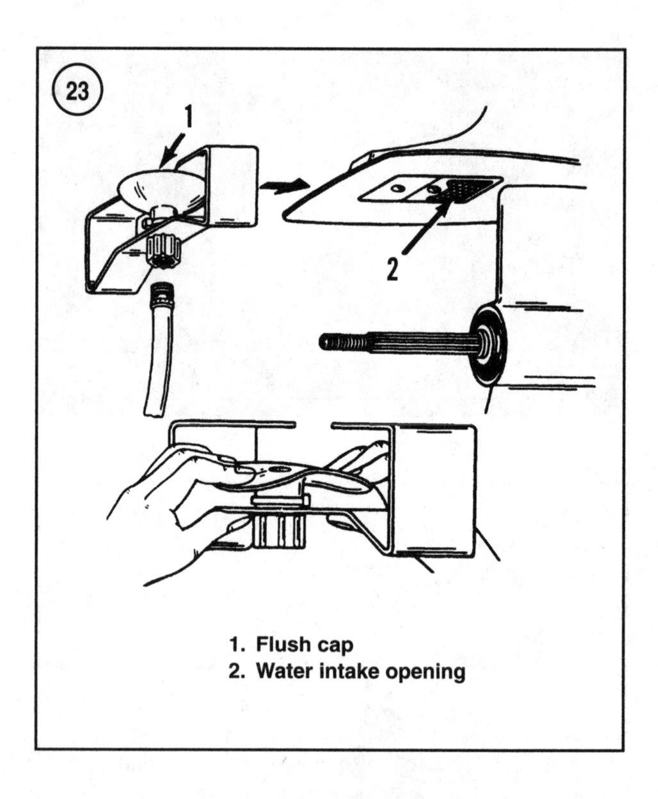

1. Flush cap
2. Water intake opening

9. Drain as much water as possible from the power head by placing the motor in a horizontal position. Position the spark plugs facing downward and manually rotate the flywheel to expel water from the cylinder(s).
10. Pour liberal amounts of isopropyl rubbing alcohol into each carburetor or throttle body throat while rotating the flywheel to help absorb the remaining water or moisture.
11. Disconnect all electrical connectors and dry them with electrical contact cleaner or isopropyl alcohol. Lubricate all electrical connectors with Quicksilver Dielectric silicone grease (part No. 92-823506-1).
12. On electric start models, remove and disassemble the electric starter as described in Chapter Seven. Dry all components with electrical contact cleaner or isopropyl alcohol. Reassemble and install the starter (Chapter Seven).
13. Pour approximately one teaspoon of engine oil into each cylinder through the spark plug hole(s). Rotate the flywheel by hand to distribute the oil.
14. Position the outboard with the induction system facing upward. Pour engine oil into each carburetor or throttle body throat while rotating the flywheel by hand to distribute the oil.
15. Reinstall the spark plug(s).
16. Attempt to start the engine using a fresh tank of 50:1 fuel/oil mixture. If the outboard motor starts, allow it to run at least one hour to evaporate the remaining water in-

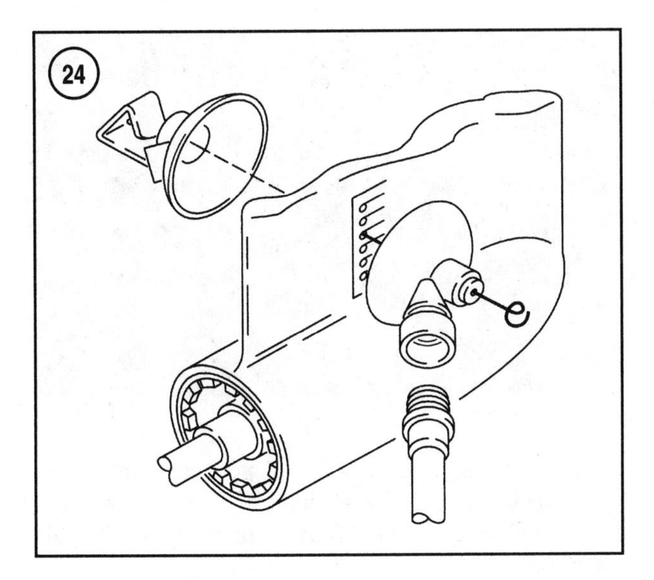

side the engine. Purge the remaining air from the oil injection system on models so equipped.

17. If the motor will not start, attempt to diagnose the cause as fuel, electrical or mechanical, and repair as necessary. If the engine cannot be started within two hours, completely disassemble, clean and oil all internal components as soon as possible.

COOLING SYSTEM FLUSHING

Periodic flushing with clean freshwater prevents salt or silt deposits from accumulating in the cooling system passageways. Perform the flushing procedure after each outing in saltwater, polluted or brackish water.

Keep the motor upright during and after flushing. This prevents water from passing into the power head through the drive shaft housing and exhaust ports during the flushing procedure. It also eliminates the possibility of residual water being trapped in the drive shaft housing or other passages.

2.5-3.3 hp Models

WARNING
The 2.5 hp engine is equipped with a direct drive gearcase. The propeller rotates any time the engine is running. Remove the propeller before beginning flushing procedures.

The 2.5 and 3.3 hp models do not have a flush kit adapter available.

1. Remove the propeller as described in Chapter Nine.
2. Place the outboard motor in a test tank or immerse the gearcase in a suitable container of freshwater, such as a 55 gal. drum. Position the outboard in the vertical position.
3. Start the engine and operate at idle to 1500 rpm.
4. Make sure cooling water is being discharged from the water pump indicator hose or fitting. If it is not, stop the motor immediately and determine the cause of the problem.
5. Flush the motor for 5-10 minutes. If the outboard was last used in saltwater, flush for 10 minutes minimum.
6. Remove the outboard motor from the water supply.
7. Keep the outboard vertical to allow the water to drain from the drive shaft housing. If water is not drained, water can enter the power head through the exhaust ports.
8. Reinstall the propeller as described in Chapter Nine.

4-15 hp Models

4-15 hp models require a special flushing adapter, Quicksilver part No. 12612A-2 (**Figure 23**).

1. Remove the propeller as described in Chapter Nine.
2. Position the outboard in the normal operating position.
3. Compress the flush cup and slide the flush adapter over the antiventilation plate. Locate the cup over the water intake opening. See **Figure 23**.
4. Connect a garden hose between a water tap and the flushing device.
5. Open the water tap to provide full water pressure.
6. Shift the outboard into neutral and start the engine. Adjust the engine speed to approximately 1000-1500 rpm.
7. Make sure water is being discharged from the water pump indicator hose or fitting. If it is not, stop the motor immediately and determine the cause of the problem.
8. Flush the motor for 5-10 minutes or until the discharged water is clear. If the outboard was last used in saltwater, flush for a minimum of 10 minutes.
9. Stop the engine, then shut off the water supply. Remove the flushing device from the outboard by compressing the cup and sliding it off the rear of the antiventilation plate.
10. Keep the outboard in the normal operating position to allow water to drain from the drive shaft housing. If the water is not drained, it can enter the power head through the exhaust ports.
11. Reinstall the propeller as described in Chapter Nine.

20-60 hp Models

The recommended flushing adapter is Quicksilver part No. 44357A-2 (**Figure 24**).

1. Remove the propeller as described in Chapter Nine.

2. Position the outboard in the normal operating position.
3. Attach the flushing device to the lower gearcase as shown in **Figure 24**.
4. Connect a garden hose (1/2 in. or larger) between a water tap and the flushing device.
5. Open the water tap partially. Adjust the water pressure until a significant amount of water escapes from around the flushing cups, but do not apply full pressure.
6. Shift the outboard into neutral and start the engine. Adjust the engine speed to approximately 1000-1500 rpm.
7. Adjust the water flow to maintain a slight loss of water around the rubber cups of the flushing device.
8. Make sure water is being discharged from the water pump indicator hose or fitting (**Figure 25**). If it is not, stop the motor immediately and determine the cause of the problem.
9. Flush the motor for 5-10 minutes or until the discharged water is clear. If the outboard was last used in saltwater, flush for a minimum of 10 minutes.
10. Stop the engine, then shut off the water supply. Remove the flushing device from the outboard.
11. Keep the outboard in the normal operating position to allow the water to drain from the drive shaft housing. If the water is not drained, it can enter the power head through the exhaust ports.
12. Reinstall the propeller as described in Chapter Nine.

20-45 Jet Models

The 20 and 30 jet models do not have a flushing port or adapter. Submerge the jet pump unit in freshwater and run the engine for at least 10 minutes at approximately 1000 rpm.

The 45 jet models have a flushing port built into the jet pump unit. See **Figure 26**. A Quicksilver flushing adapter part No. 24789A-1 is available.

1. Position the outboard in the normal operating position.
2. Remove the flushing port plug and washer (**Figure 26**), install the flushing adapter and connect a 1/2 in. or larger garden hose to it.
3. Open the water tap approximately halfway. It is not necessary to use full water pressure.
4. Shift the outboard into neutral and start the engine. Adjust the engine speed to approximately 1000 rpm.
5. Make sure water is being discharged from the water pump indicator hose or fitting (**Figure 25**). Adjust the water tap as necessary. If no water is being discharged, stop the motor immediately and determine the cause of the problem.
6. Flush the motor for 5-10 minutes or until the discharged water is clear. If the outboard was last used in saltwater, flush for a minimum of 10 minutes.

7. Stop the engine, then shut off the water supply. Remove the flushing device from the outboard and reinstall the plug and washer in the flushing port. Tighten the plug securely.
8. Thoroughly rinse the intake grate area and all outer surfaces of the pump unit with the garden hose. Direct a garden hose into the grate area and over the outer surfaces of the pump unit after flushing the cooling system.
9. Keep the outboard in the normal operating position to allow the water to drain from the drive shaft housing. If the water is not drained, it can enter the power head through the exhaust ports.

TUNE-UP

A tune-up is a series of inspections, adjustments and part replacements to compensate for normal wear and deterioration of the outboard motor components. Regular tune-ups maintain(s) performance and fuel economy. Mercury Marine recommends performing tune-up procedures at least once a season or every 100 hours of operation. Individual operating conditions may require more frequent tune-ups. Also perform a tune-up any time the outboard exhibits a substantial performance loss.

Since proper outboard motor operation depends upon a number of interrelated systems, performing only one or two of the recommended tune-up procedures will seldom provide satisfactory results. For best results, a thorough inspection, analysis and correction procedure is necessary.

Prior to performing a tune-up, flush the outboard cooling system as described in this chapter to check water pump operation.

The recommended tune-up procedure is listed below.

1. Removal of combustion chamber deposits.
2. Compression test.
3. Cylinder head bolt torque (2.5-5 hp models).
4. Electrical wiring harness inspection.

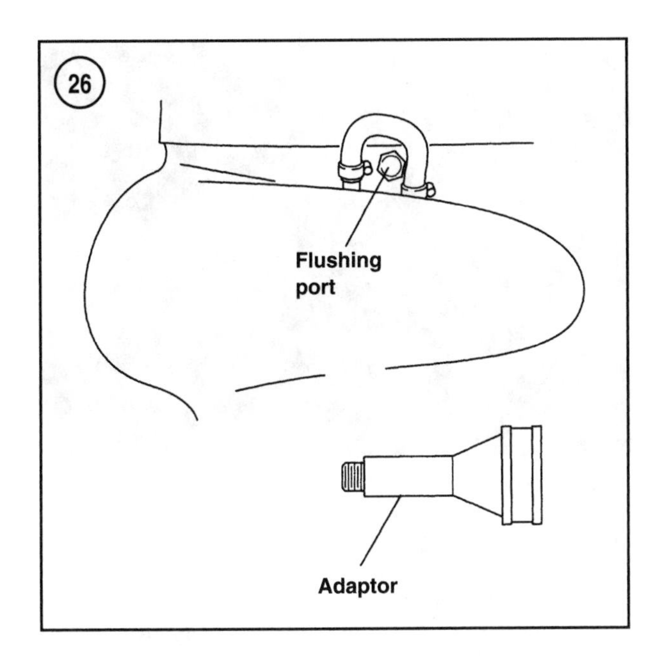

5. Spark plug service.
6. Gearcase lubricant change and propeller shaft spline lubrication.
7. General engine lubrication at all applicable lubrication points. See **Table 2**.
8. Fuel filter service.
9. Water pump service.
10. Fuel system and oil injection system, if so equipped, service.
11. Ignition system service.
12. Charging system service, if so equipped.
13. Battery and starter system service, if so equipped.
14. Synchronization and linkage adjustments. See Chapter Five.
15. On-water performance test.

When the fuel or ignition system is adjusted or defective parts replaced, check engine synchronization and linkage adjustments. These procedures are described in Chapter Five. Perform all synchronization and linkage adjustments *before* running the on-water performance test.

Removing Combustion Chamber Deposits

During operation, carbon deposits accumulate on the piston(s), rings, cylinder head(s) and exhaust ports. If the carbon builds up, the effective compression ratio increases, raising the fuel octane requirements of the power head.

If the carbon builds up in the piston ring area, the piston rings stick in the piston ring grooves causing a loss of compression and the loss of heat transfer to the cylinder walls and water passages. When the piston rings stick, performance suffers and combustion chamber temperatures increase dramatically, leading to preignition and detonation. All of these situations eventually lead to power head failure.

Quicksilver Power Tune (part No. 92-15104) is designed to remove combustion chamber deposits and free stuck piston rings, restoring engine performance and lowering the risk of engine failure.

NOTE
Use quality fuel and an NMMA approved TCW-3 outboard oil to minimize combustion chamber deposits and piston ring sticking. If the use of poor quality fuel and/or oil is unavoidable and combustion chamber deposits are a continual problem, use Quickleen Fuel Treatment on a regular basis.

For effective preventive maintenance, remove combustion chamber deposits with the Quicksilver Power Tune Engine Cleaner (part No. 92-15104) every 100 hours of operation or as required. Follow the instructions on the Quicksilver Power Tune container.

Compression Test

An accurate cylinder cranking compression check provides an indication of the mechanical condition of the combustion chamber. It is an important preliminary step in a tune-up as a motor with low or unequal compression between cylinders *cannot* be satisfactorily tuned. Correct any compression problems before continuing with the tune-up procedure. Use a thread-in compression tester for best results.

A variation of more than 15 psi (103.4 kPa) between any two cylinders indicates a problem. If the compression is unacceptable, remove the cylinder head, if applicable, and inspect the cylinder wall(s), piston(s) and head gasket(s). If the cylinder wall(s), piston(s) and head gasket show no evidence of damage or failure, the piston rings are stuck, worn or damaged and the power head must be repaired.

CAUTION
*Do not run the engine without an adequate water supply and do not exceed 3000 rpm without an adequate load. Refer to **Safety Precautions** at the beginning of Chapter Three.*

1. Run the engine until it reaches operating temperature.
2. Remove the spark plug(s) as described in this chapter.

3. Securely ground the spark plug lead(s) to the engine to disable the ignition system, prevent accidental starting and possible ignition system damage.
4. Following the manufacturer's instructions for the compression gauge, connect the gauge to the No. 1 cylinder spark plug hole (**Figure 27**, typical).
5. Manually hold the throttle plates in the wide-open throttle position. Crank the engine through at least four compression strokes and record the gauge reading.
6. Repeat Step 4 and Step 5 for all remaining cylinders. A variation of more than 15 psi (103.4 kPa) between two cylinders indicates a problem with the lower reading cylinder, such as worn or sticking piston rings and/or scored pistons or cylinder walls. Pour a tablespoon of engine oil into the suspect cylinder and repeat Step 4 and Step 5. If the compression increases by 10 psi (69 kPa) or more, the rings are worn or damaged and the power head must be disassembled and repaired.

If the compression is within specification, but the outboard motor is difficult to start or has poor idle quality, refer to *Starting Difficulties* in Chapter Three.

Cylinder Head Bolt Torque

CAUTION
The power head must be cool to the touch before performing this procedure. Excessive torque distorts the cylinder bore and cylinder head. Insufficient torque allows the cylinder head gasket to leak.

Retorque the cylinder head bolts at each tune-up interval. To retorque the cylinder head bolts, loosen each cylinder head bolt slightly and retorque, in the proper sequence (Chapter Eight), to the specification in **Table 1**.

Electrical Wiring Harness Inspection

Inspect all harnesses, leads, connectors and terminals for loose connections, corrosion, mechanical damage, damaged insulation and improper routing. Check harnesses close to moving components for chafing or rubbing damage. Reroute, retape and secure harnesses as necessary. Inspect all harnesses, wires and components on or near the cylinder head and exhaust passages for heat damage. Repair any damage found. Refer to Chapter Three for recommended tools and repair kits.

Spark Plug Replacement

Improper installation and incorrect application are common causes of poor spark plug performance in outboard motors. The gasket on the plug must be fully compressed against a clean plug seat for heat transfer to take place effectively. If heat transfer cannot take place, the spark plug will overheat and fail. This may also lead to preignition and detonation. Make sure the spark plugs are correctly torqued.

27

Incorrect application can also lead to ignition system symptoms or failure from RFI (radio frequency interference). Always use the recommended spark plugs.

If the engine does not require inductor or suppression spark plugs, yet ignition interference with onboard accessories occurs, install the recommended inductor or suppression spark plug as recommended in **Table 3**.

CAUTION
When the spark plug(s) are removed, dirt or other foreign material surrounding the spark plug openings can fall into the cylinder(s). Foreign material inside the cylinders can cause damage when the engine is started.

1. Clean the area around the spark plug(s) using compressed air or an appropriate brush.
2. Disconnect the spark plug lead(s) by twisting the boot back and forth on the spark plug insulator while pulling outward. Pulling on the wire instead of the boot can cause internal damage to the wire.
3. Remove the spark plugs using an appropriate size spark plug socket. Arrange the spark plugs in order of the cylinders from which they were removed.
4. Examine each spark plug and compare the condition to those in **Figure 28** or **Figure 29**. Spark plug condition is a good indicator of piston, rings and cylinder condition, and can warn of developing problems.
5. Check each plug for make and heat range. All spark plugs must be identical. Refer to **Table 3** and make sure the spark plugs are correct for the application.

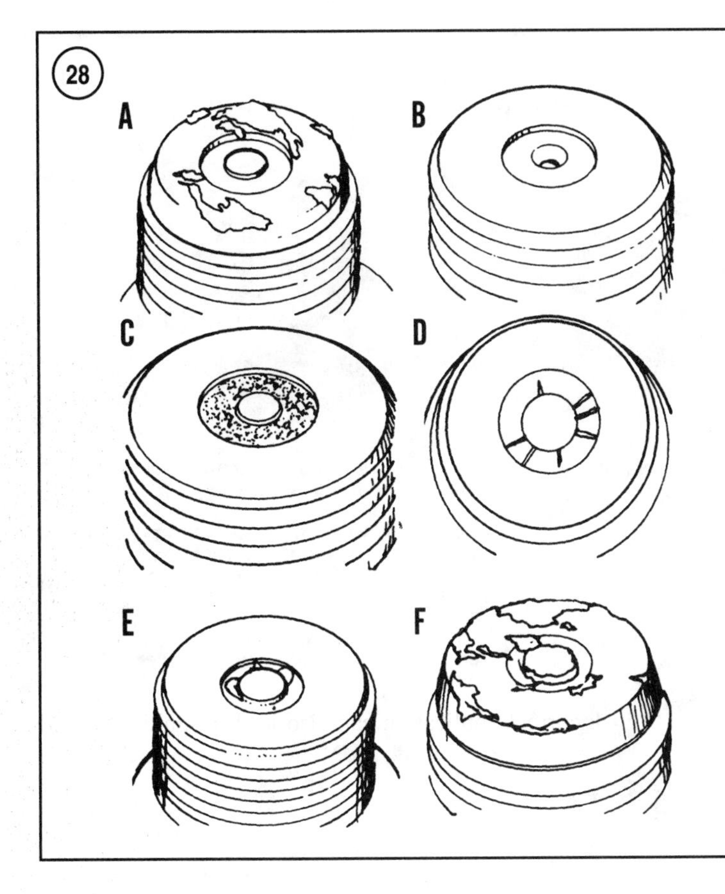

SPARK PLUG ANALYSIS (SURFACE GAP SPARK PLUGS)

A. Normal—Light tan or gray colored deposits indicate that the engine/ignition system condition is good. Electrode wear indicates normal spark rotation.
B. Worn out—Excessive electrode wear can cause hard starting or a misfire during acceleration.
C. Cold fouled—Wet oil or fuel deposits are caused by "drowning" the plug with raw fuel mix during cranking, overich carburetion or an improper fuel:oil ratio. Weak ignition will also contribute to this condition.
D. Carbon tracking—Electrically conductive deposits on the firing end provide a low-resistance path for the voltage. Carbon tracks form and can cause misfires.
E. Concentrated arc—Multi-colored appearance is normal. It is caused by electricity consistently following the same firing path. Arc path changes with deposit conductivity and gap erosion.
F. Aluminum throw off—Caused by preignition. This is not a plug problem but the result of engine damage. Check engine to determine cause and extent of damage.

6. If the spark plugs are in good condition, they may be cleaned and regapped as described in this section. Install new spark plugs if there is any question as to the condition of the spark plugs.
7. Inspect the spark plug threads in the engine and clean them with a thread chaser (**Figure 30**) if necessary. Wipe the spark plug seats clean before installing new spark plugs.
8. Install the spark plugs with new gaskets and tighten them to the specification in **Table 1**. If a torque wrench is not available, seat the plugs finger-tight, then tighten them an additional 1/4 turn with a wrench.
9. Inspect each spark plug lead before reconnecting it to its spark plug. If the insulation is damaged or deteriorated, install a new plug wire. Push the boot onto the plug terminal and make sure it is fully seated.

Spark Plug Gap Adjustment (Conventional Gap Only)

Carefully set the electrode gap on new spark plugs to ensure a reliable, consistent spark (**Table 3**). Use a special spark plug gapping tool with wire gauges. **Figure 31** shows a common type of gapping tool.

1. Make sure the gaskets are installed on the spark plugs, except taper seat spark plugs.

NOTE
*On some spark plug brands, the terminal end must be screwed on the plug before installation. See **Figure 32**.*

2. Insert the appropriate size wire gauge (**Table 3**) between the electrodes (**Figure 33**). If the gap is correct, a slight drag is felt as the wire is pulled through. To adjust the gap, bend the side electrode with the gapping tool (**Figure 34**), then remeasure the gap.

CAUTION
Never attempt to close the gap by tapping the spark plug on a solid surface. This can damage the spark plug. Always use the proper adjusting tool to open or close the gap.

Water Pump

Overheating and extensive power head damage can result from a faulty water pump. Replace the water pump impeller, seals and gaskets at the following intervals.

(29)

SPARK PLUG CONDITION

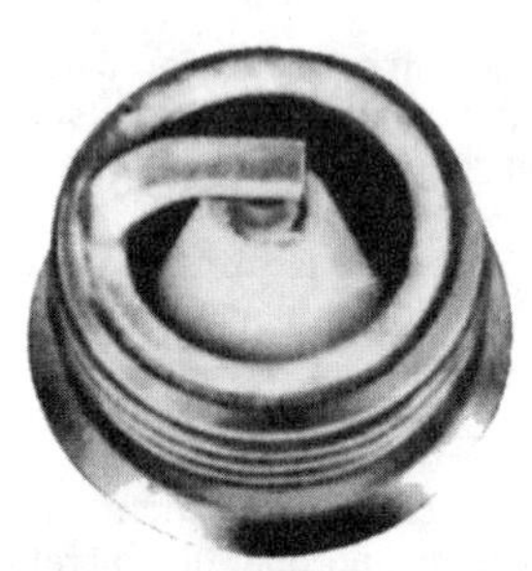

NORMAL

- Identified by light tan or gray deposits on the firing tip.
- Can be cleaned.

GAP BRIDGED

- Identified by deposit buildup closing gap between electrodes.

OIL FOULED

- Identified by wet black deposits on the insulator shell bore and electrodes.
- Caused by excessive oil entering combustion chamber through worn rings and pistons, excessive clearance between valve guides and stems or worn or loose bearings. Can be cleaned. If engine is not repaired, use a hotter plug.

CARBON FOULED

- Identified by black, dry fluffy carbon deposits on insulator tips, exposed shell surfaces and electrodes.
- Caused by too cold a plug, weak ignition, dirty air cleaner, too rich fuel mixture or excessive idling. Can be cleaned.

ADDITIVE FOULED

- Identified by dark gray, black, yellow or tan deposits or a fused glazed coating on the insulator tip.
- Caused by using gasoline additives. Can be cleaned.

WORN

- Identified by severely eroded or worn electrodes.
- Caused by normal wear. Should be replaced.

FUSED SPOT DEPOSIT

- Identified by melted or spotty deposits resembling bubbles or blisters.
- Caused by sudden acceleration. Can be cleaned.

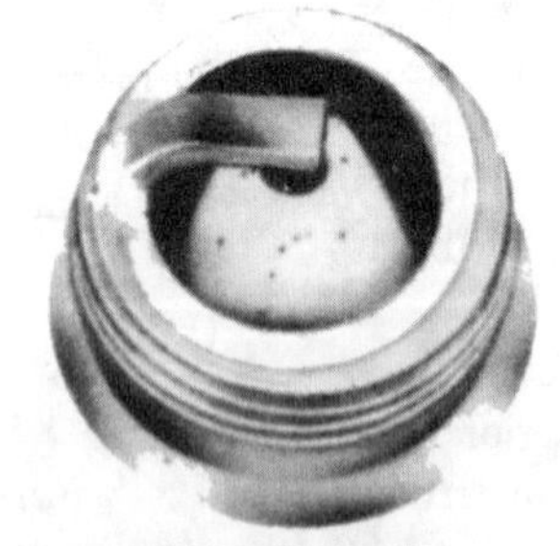

OVERHEATING

- Identified by a white or light gray insulator with small black or gray brown spots with bluish-burnt appearance of electrodes.
- Caused by engine overheating, wrong type of fuel, loose spark plugs, too hot a plug or incorrect ignition timing. Replace the plug.

PREIGNITION

- Identified by melted electrodes and possibly blistered insulator. Metallic deposits on insulator indicate engine damage.
- Caused by wrong type of fuel, incorrect ignition timing or advance, too hot a plug, burned valves or engine overheating. Replace the plug.

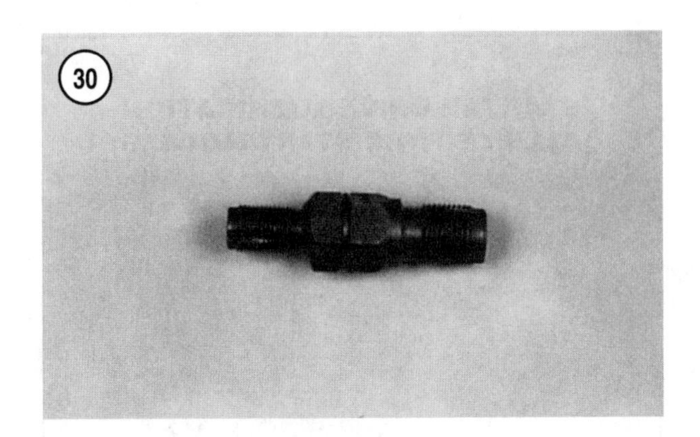
30

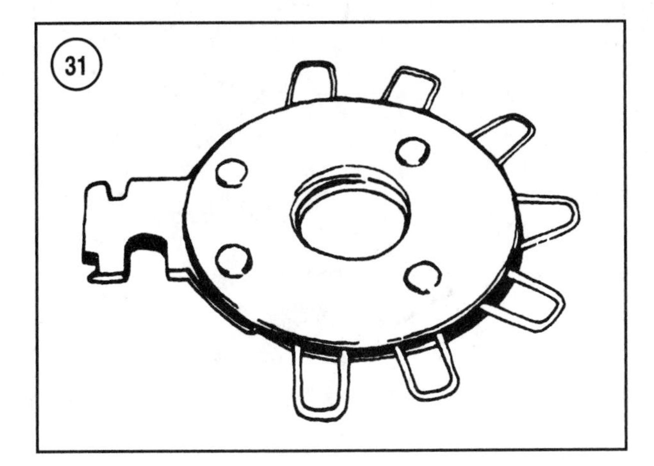
31

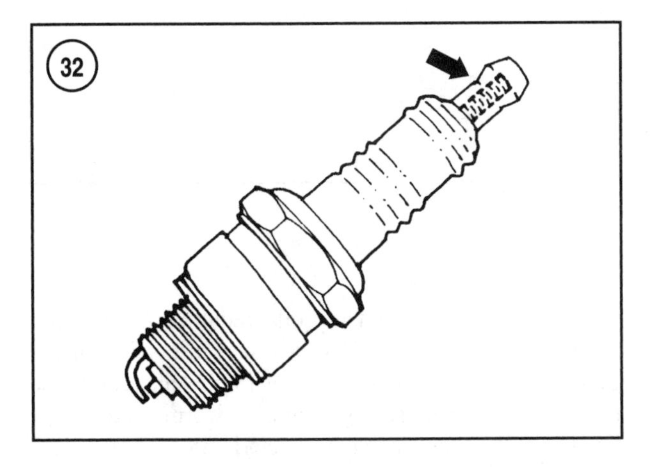
32

NOTE
Individual operating conditions may make more frequent water pump service necessary. Always service the water pump anytime the lower gearcase or jet pump assembly is removed for service.

1. *High-pressure water pump*—Replace the water pump after every 100 hours of operation or once a year. Refer to

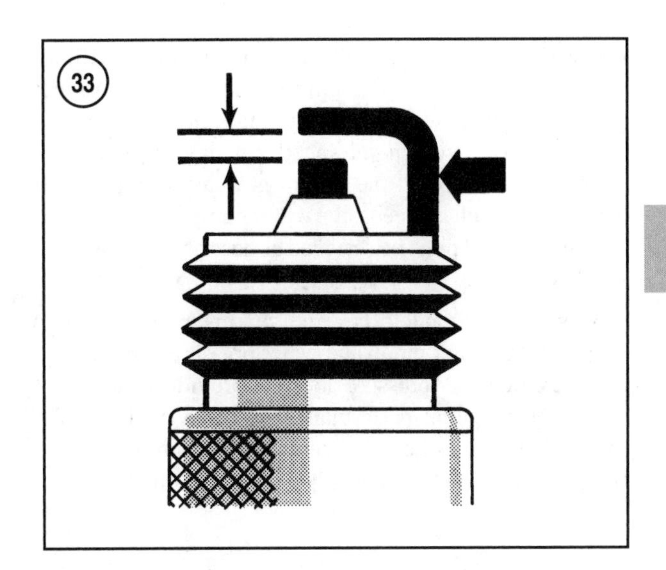
33

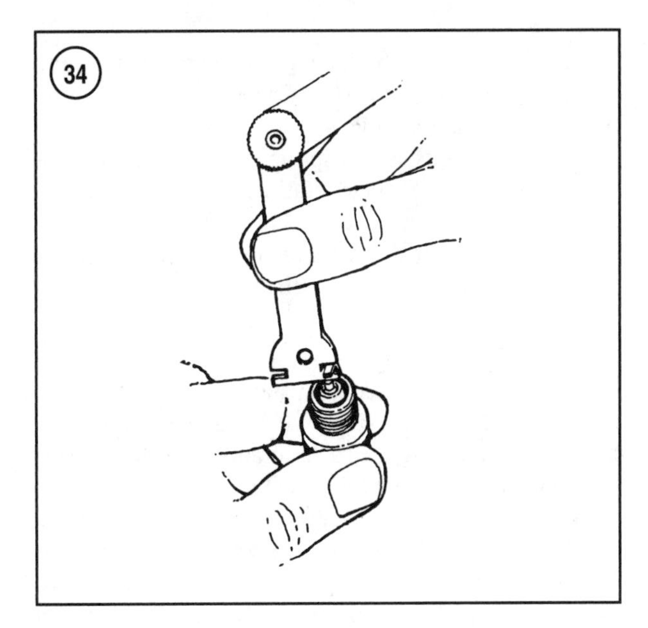
34

Chapter Nine for pump identification and service procedures.

2. *High-volume water pump*—Replace the water pump after every 300 hours of operation or every 3 years. Refer to Chapter Nine for pump identification and service procedures.

NOTE
When the lower gearcase is removed, clean and lubricate the drive shaft splines with 2-4-C (part No. 92-825407) or Special Lube 101 (part No. 92- 113872A 1). Do not apply lubricant to the end of the drive shaft as this may prevent the drive shaft from fully seating into the crankshaft.

Check the water pump indicator stream (**Figure 25**) to verify operation of water pump. The stream must be present anytime the engine is running. The water stream pressure varies with engine speed and may be somewhat weak at idle speed. If the stream is weak or missing at off-idle and higher speeds, the water pump is likely worn or damaged and must be serviced as described in Chapter Nine.

Most of the models covered in this manual use a high-volume water pump and water pressure is relatively low. Testing water pressure does not provide an accurate indication of pump performance on these models. The manufacturer does not provide water pressure specifications for the models using a high-volume pump.

Fuel and Oil Injection Systems

During a tune-up, check all synchronization and linkage adjustments as described in Chapter Five. Clean or replace all fuel and air filters. Inspect all fuel lines, fuel system components and all spring clamps, worm clamps or tie-straps for leaks, deterioration, mechanical damage and secure mounting. All replacement fuel lines must be alcohol resistant. If the fuel system is suspected of not functioning correctly, refer to Chapter Three for troubleshooting procedures.

Refer to *Fuel and Lubrication* in this chapter for basic oil injection system description and component function. Check the oil injection system, if so equipped, for leaks, loose lines and fittings and deterioration. Synchronize the oil pump linkage with the throttle linkage (Chapter Five). Inspect the warning buzzer and sensor wiring, if so equipped, as described under *Electrical Wiring Harness Inspection* in this chapter. If the oil injection system is suspected of not functioning correctly, refer to Chapter Twelve for oil injection system troubleshooting and service procedures.

Fuel filters

Service the engine mounted fuel filter as described previously in this chapter.

Service the portable fuel tank filter by detaching the fuel hose from the tank. Unthread the pickup tube assembly or pickup tube retaining ring from the tank. Remove the pickup tube assembly. Clean or replace the fine mesh filter as necessary. Replace any gaskets, seals or O-rings. Thread the pickup assembly into the tank and tighten it or the retaining ring securely.

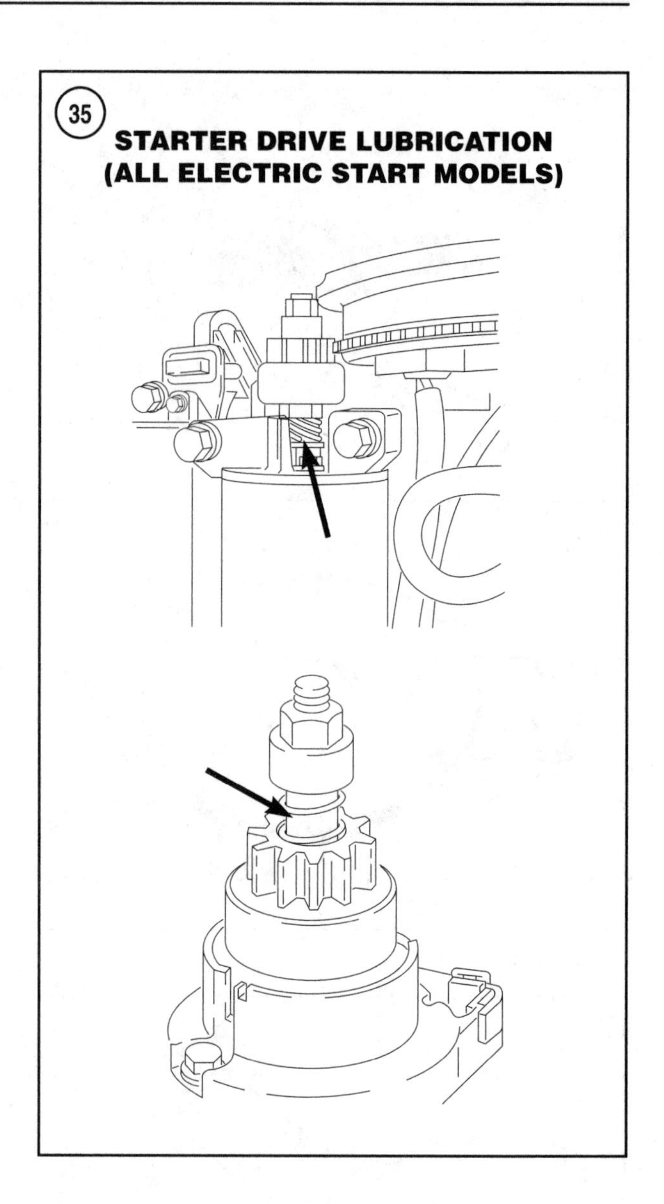

Fuel pump

The fuel pump does not generally require service during a tune-up. However, conduct a visual inspection of the fuel pump, pump mounting hardware, fuel and crankcase pulse hoses, and spring clamps, worm clamps or tie-straps. Replace damaged or deteriorated components. If the fuel pump or fuel system is suspected of not functioning correctly, refer to Chapter Three for troubleshooting procedures.

Ignition System Service

Other than inspecting or replacing the spark plugs, the ignition system is relatively maintenance free. During a tune-up, check all synchronization and linkage adjust-

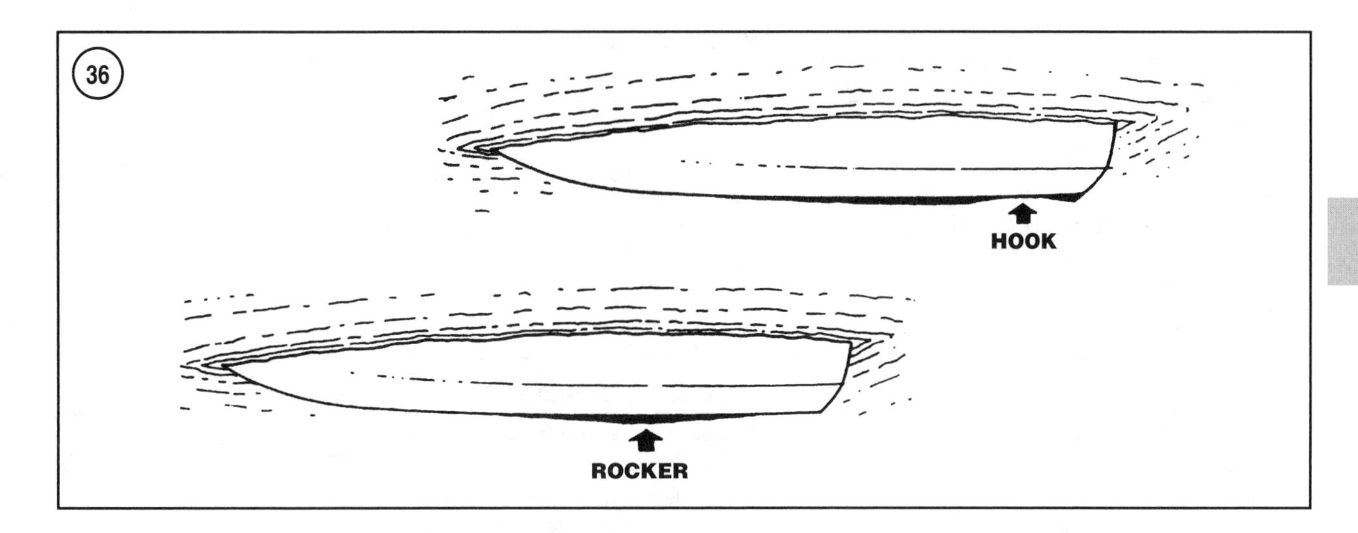

ments as described in Chapter Five. If the ignition system is suspected of not functioning correctly, refer to Chapter Three for troubleshooting procedures.

Charging System Service

If the charging system is suspected of not functioning correctly, refer to Chapter Three for troubleshooting procedures.

Battery and Electric Starter System

The electric starter system requires minimal maintenance during a tune-up. Clean and lubricate the starter motor shaft and starter drive *lightly* with SAE 30 engine oil as shown in **Figure 35**.

Inspect the battery cable connections for corrosion, loose connections or mechanical damage. If wing-nuts are present, discard them and replace them with corrosion resistant hex nuts and lockwashers. Place a lockwasher under each battery cable to ensure positive contact with the battery terminal. Tighten the battery connections securely. Check the electric starter system as follows:

1. Check the battery state of charge as described in Chapter Seven.
2. Disable the ignition system by removing the spark plug wire from each spark plug and grounding the spark plug wires to the power head.
3. Connect a multimeter set to the 20-volt DC scale to the battery positive and negative terminals.
4. Turn the ignition switch to the start position and note the meter reading while the engine is cranking for several seconds.
   a. If the voltage is 9.5 volts or higher and the cranking speed is normal, the starting system is functioning normally and the battery is of sufficient capacity for the engine.
   b. If the voltage is below 9.5 volts and/or the cranking speed is below normal, the starting system is malfunctioning. Refer to Chapter Three for troubleshooting procedures.
5. Reconnect the spark plug leads when finished.

On-Water Performance Testing

Before performance testing the outboard motor, make sure the boat bottom is free of marine growth and no hook or rocker is present in the boat bottom (**Figure 36**). These conditions substantially reduce the boat performance.

Performance test the boat with an average load on board. Tilt or trim the outboard motor at an angle that produces optimum performance and balanced steering control. If equipped with an adjustable trim tab, adjust it to allow the boat to steer in either direction with an equal amount of effort at the boat's normal cruising speed.

CAUTION
Mercury/Mariner outboard motors perform best when propped toward the upper limit of the recommended rpm range. Do not allow the engine to operate above or below the recommended speed range at wide-open throttle.

Check the engine speed at wide-open throttle. If engine speed is not within the specified range (Chapter Five) and the engine has been properly synchronized and adjusted (Chapter Five), change the propeller. Use a higher pitch propeller to reduce engine speed or a lower pitch propeller to increase engine speed.

Table 1 TORQUE SPECIFICATIONS

Component	ft.-lb.	in.-lb.	N•m
Spark plugs	20	240	27.1
Gearcase drain/fill plugs			
20 and 25 hp	–	60	6.8
30-50 hp	–	55	6.2
55 and 60 hp (standard gearcase)	–	35-80	4.0-9.0
60 hp (with Bigfoot gearcase)	–	60	6.8
Cylinder head bolt			
2.5 and 3.3 hp	–	85	9.6
4 and 5 hp	18	216	24.4

Table 2 MAINTENANCE SCHEDULE

Maintenance interval	Maintenance required
2.5-5 hp models	
Before each use (saltwater or freshwater use)	Check the lanyard switch operation*
	Inspect the fuel system for leakage
	Check the outboard mounting bolts
	Check the steering for binding or looseness
	Check the shift and throttle control
	Inspect the propeller for damage
First 10 days of operation (saltwater or freshwater use)	Check gearcase lubricant level and condition
First 25 hours of operation	Drain and refill the gearcase lubricant
Every 30 days (freshwater use)	Check gearcase lubricant level and condition
Every 30 days (saltwater use)	Lubricate throttle and shift linkages
	Lubricate reverse lock hooks
	Lubricate swivel and tilt pins
	Check gearcase lubricant level and condition
Every 60 days (saltwater use)	Lubricate the propeller shaft
Every 60 days (freshwater use)	Lubricate throttle and shift linkages
	Lubricate reverse lock hooks
	Lubricate swivel and tilt pins
100 hours or once a year	Drain and refill the gearcase lubricant
	Lubricate the propeller shaft
Before long term storage	Drain and refill the gearcase lubricant
6-60 hp models	
Before each use (saltwater or freshwater use)	Check the lanyard switch operation*
	Inspect the fuel system for leakage
	Check the outboard mounting bolts
	Check the steering for binding or looseness
	Check the shift and throttle control
	Inspect the propeller for damage
After each use (saltwater use)	Flush the cooling system
100 hours or once a year	Lubricate the throttle and shift linkages
	Lubricate the steering system
	Lubricate the tiller control pivots
	Lubricate the tilt and swivel shaft
	Lubricate the propeller shaft
	Lubricate the starter motor pinion*
	Lubricate the propeller shaft bearing carrier
	Drain and refill the gearcase lubricant
	Lubricate the drive shaft splines
	Clean and inspect the spark plug(s)
	Inspect and test the battery*

(continued)

Table 2 MAINTENANCE SCHEDULE (continued)

Maintenance interval	Maintenance required
6-60 hp models	
100 hours or once a year	Service the fuel filter
	Adjust the carburetor(s)
	Adjust the ignition timing
	Adjust the synchronization and linkages
	Inspect the sacrificial anodes*
	Check the power trim fluid level*
	Check control cable adjustments*
	Remove carbon deposits from the cylinder(s)
	Check for loose or damaged fasteners
	Clean remote fuel tank filter*
300 hours or 3 years	Replace the water pump impeller

*This maintenance item does not apply to all models.

Table 3 SPARK PLUG RECOMMENDATIONS

Model	Plug part No.	Gap in. (mm)
2.5 and 3.3 hp		
NGK plug	BPR6HS-10	0.040 (1.02)
Champion plug	RL87YC	0.040 (1.02)
4 and 5 hp		
NGK plug	BP7HS-10	0.040 (1.02)
Champion plug	L82YC	0.040 (1.02)
6 and 8 hp		
NGK standard plug	BP8H-N-10	0.040 (1.02)
NGK radio suppression plug	BPZ8H-N-10	0.040(1.02)
9.9 and 15 hp		
NGK standard plug	BP8HS-15	0.060 (1.52)
NGK radio suppression plug	BPZ8H-N-10	0.060(1.52)
20 hp, 20 Jet and 25 hp		
NGK standard plug	BP8H-N-10	0.040 (1.02)
NGK radio suppression plug	BPZ8H-N-10	0.040(1.02)
30 and 40 hp (2-cylinder)		
NGK standard plug	BP8H-N-10	0.040 (1.02)
NGK radio suppression plug	BPZ8H-N-10	0.040(1.02)
40-60 hp (3-cylinder)		
NGK standard plug	BP8H-N-10	0.040 (1.02)
NGK radio suppression plug	BPZ8H-N-10	0.040(1.02)

Table 4 LUBRICANT CAPACITY

Outboard model	Gearcase capacity	Oil reservoir capacity fl. oz. (L)
2.5 hp	3.0 oz. (89 ml)	*
3.3 hp	2.5 oz. (74 ml)	*
4 and 5 hp	6.6 oz. (195 ml)	*
6, 8, 9.9 15 hp	6.8 oz. (201 ml)	*
20 and 25 hp	7.8 oz. (231 ml)	*
30 and 40 hp (2-cylinder)	14.9 oz. (441 ml)	50.5 (1.5)
40 and 50 hp (3-cylinder)	14.9 oz. (441 ml)	96 (2.86)
55 and 60 hp		
Standard gearcase	11.5 oz. (340 ml)	96 (2.8)
Bigfoot gearcase	22.5 oz. (665 ml)	96 (2.8)

*This model is not equipped with oil injection.

Chapter Five

Synchronization and Linkage Adjustments

For an outboard motor to deliver maximum efficiency, performance and reliability, the ignition and fuel systems must be correctly adjusted.

Failure to properly synchronize and adjust an engine will cause a loss of engine performance and efficiency, and can lead to power head damage. Perform synchronization and linkage adjustments during a tune-up or whenever replacing ignition or fuel system components.

On a typical engine, a synchronization and linkage procedure involves the following:

1. Synchronizing and adjusting the ignition and fuel systems linkages.
2. Verifying that the fuel system throttle plate(s) fully open and close, and that all throttle plates open and close at exactly the same time on 40-60 hp models.
3. Synchronizing the ignition system spark advance with throttle plate(s) operation for optimum off-idle acceleration and smooth part-throttle operation.
4. Adjusting the ignition timing at idle and wide-open throttle.
5. Setting the idle speed correctly and verifying the wide-open throttle engine speed.

Synchronization and linkage procedures for Mercury/Mariner outboard motors differ according to engine model and the ignition and fuel systems used. This chapter is divided into sections by model. **Tables 1-6** provide the general specifications. All tables are located at the end of the chapter.

WARNING

Read the safety precaution and general information in the next two sections before performing the adjustments.

SAFETY PRECAUTIONS

Wear eye protection at all times, especially when machinery is in operation. Wear approved ear protection during all running tests and in the presence of noisy machinery. Keep loose clothing tucked in and long hair tied back. Refer to Chapter Two, *Safety First* for additional safety guidelines.

When making or breaking an electrical connection, always disconnect the negative battery cable. When performing tests that require cranking the engine without starting, disconnect and ground the spark plug leads to prevent accidental starts and sparks.

Securely cap or plug all disconnected fuel lines to prevent fuel discharge when the motor is cranked or the primer bulb is squeezed.

Read all manufacturer's instructions and safety sheets provided with the test equipment and special tools.

Do not substitute parts unless they meet or exceed the original manufacturer's specifications.

Never run an outboard motor without an adequate water supply. Never run an outboard motor at full throttle without an adequate load. Do not exceed 3000 rpm in neutral (no load).

Safely performing on-water tests requires two people; one person to operate the boat, the other to monitor the gauges or test instruments. All personnel must remain seated inside the boat at all times. Do not lean over the transom while the boat is under way. Use test equipment wire extensions as needed to position gauges and meters in the normal seating area.

A test propeller is an economical alternative to the dynometer and a convenient alternative to on-water testing. Test propellers are made by turning down the diameter of a standard low pitch aluminum propeller until the recommended wide-open throttle speed can be obtained with the motor in a test tank or on the trailer backed into the water. Be careful of tying the boat to a dock as considerable thrust is developed by the test propeller. Some docks may not be able to withstand the load.

Propeller repair stations can make this modification. Normally, approximately 1/3 to 1/2 of the outer blade surface is removed. It is better to remove too little than too much. It may take several tries to achieve the correct full throttle speed, but once achieved, no further modifications are required. Many propeller repair stations have experience with this type of modification and may be able to recommend a starting point.

Test propellers also allow tracking of engine performance. The full-throttle test speed of an engine fitted with a correctly modified test wheel can be tracked from season to season. It is not unusual for a new or rebuilt engine to show a slight increase in test propeller speed as complete break-in is achieved and to hold that speed over the normal service life of the engine. As the engine begins to wear out, the test propeller speed shows a gradual decrease.

GENERAL INFORMATION

For accuracy, perform synchronization and linkage adjustments with the engine running under actual operating conditions. Carburetor idle mixture and idle speed adjustments are very sensitive to engine load and exhaust system back pressure. If the adjustments are made with the engine running on a flushing device, the adjustments are incorrect when the motor is operated in the water under load.

CAUTION
Do not run the engine without an adequate water supply and do not exceed 3000 rpm without an adequate load. Refer to ***Safety Precautions*** *at the beginning of this chapter.*

Ignition Timing

All models 4 hp and larger use timing marks for checking ignition timing with a strobe timing light. On models with adjustable timing, a linkage adjustment brings the timing into specification. If the timing is not within specification on models with nonadjustable timing, either a mechanical or electrical defect is present in the system. Chapter Three covers ignition troubleshooting for all models.

The maximum timing advance specification is best checked at wide-open throttle. The outboard must be operated at full throttle in forward gear under load to verify maximum timing advance. This requires a test tank or test propeller, as timing an engine while speeding across open water is not safe and is not recommended. Refer to *Safety Precautions* in the previous section.

The maximum timing advance on some models can be set by holding the ignition linkage in the full-throttle position while the engine is being cranked. While acceptable where noted, this procedure is not considered as accurate as the wide-open throttle check. Whenever possible, check the maximum timing specification at wide-open throttle with a test propeller.

High-Speed Air/Fuel Mixture

CAUTION
Operating the engine with an incorrect fuel jet(s) can result in serious power head damage and/or increased exhaust emissions. Unless operating the engine at higher than standard elevations, never change the fuel jet sizes.

The high speed air/fuel ratio is controlled by a fixed high speed jet (main jet). Only change the high speed jet size to compensate for changes in elevation. Operating the engine with the wrong fuel jets for the elevation can result in serious power head damage and/or increased exhaust emissions. Refer to Chapter Six for fuel jet size specifications.

Full Throttle RPM Verification

All outboard motors have a specified full throttle speed range (**Tables 1-6**). When the engine is mounted on a boat and run at full throttle, the engine speed must be within the specified range. If the engine speed is above or below the specified range, the engine could be damaged.

NOTE
Use an accurate shop tachometer for checking full throttle engine rpm. Do not use the boat's tachometer.

Operating an engine with a propeller that does not allow the engine to reach its specified range is called over-propping. This causes the combustion chamber temperature to rise dramatically, leading to preignition and detonation (Chapter Three).

Operating an engine with a propeller that allows an engine to exceed its specified range is called over-revving. Over-revving an engine leads to mechanical failure of the reciprocating engine components. Some engines are equipped with an rpm limit module that shorts out the ignition system to limit engine speed. Over-revving these engines can cause ignition misfire symptoms that can cause troubleshooting difficulty.

Changing the pitch, diameter or style of propellers changes the load on the engine and the resulting full throttle engine speed. If the full throttle engine speed exceeds the specified range, install a propeller with more pitch or a larger diameter and recheck. If the full throttle engine speed is below the specified range, install a propeller with less pitch or smaller diameter and retest.

Required Equipment

Static adjustment of the ignition timing and/or verification of the timing pointer requires a suitable dial indicator to position the No. 1 piston at top dead center (TDC) accurately before making timing adjustments. DetermineTDC by removing the No. 1 spark plug and installing the dial indicator in the spark plug hole. Refer to **Figure 1**.

Ignition timing checks and adjustments require a strobe timing light (**Figure 2**) connected to the No. 1 spark plug lead. As the engine is cranked or operated, the light flashes each time the spark plug fires. When the light is pointed at the moving flywheel, the mark on the flywheel

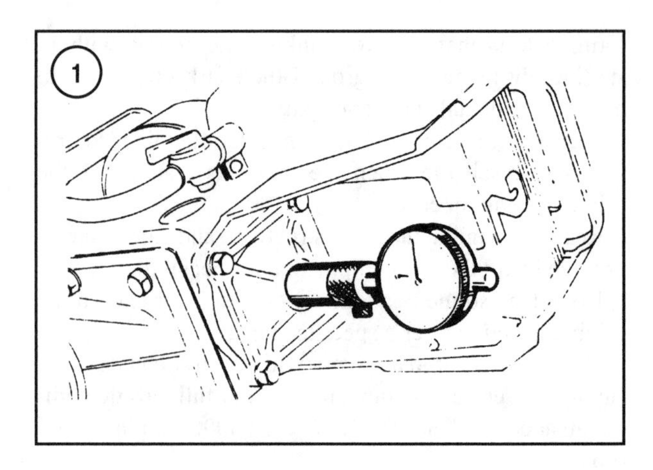

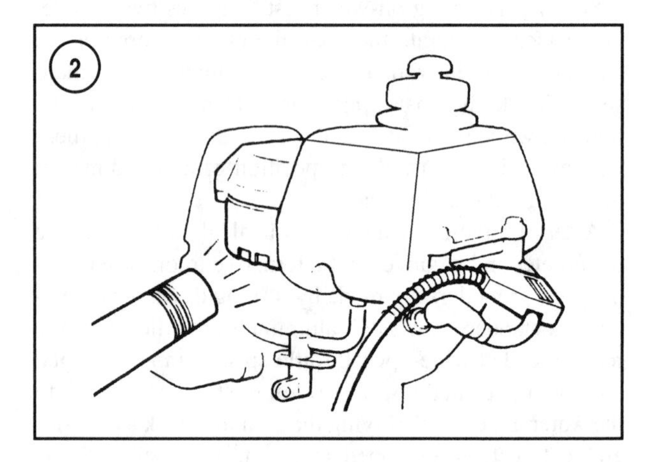

appears to stand still. The appropriate timing marks align if the timing is correct.

CAUTION
Factory timing specifications provided by Mercury Marine are in the tables at the end of the chapter. However, Mercury Marine has modified their specification during production. If the engine has a decal attached to the power head or air box, always follow the specification on the decal.

NOTE
Timing lights with built-in features, such as a timing advance function, are not recommended for outboard motors. Use a basic high-speed timing light, such as Mercury Marine timing light (part No. 91-99379), with an inductive pickup.

Use an accurate shop tachometer to determine engine speed during timing adjustment. Do not rely on the tachometer installed in the boat to provide accurate engine speed readings.

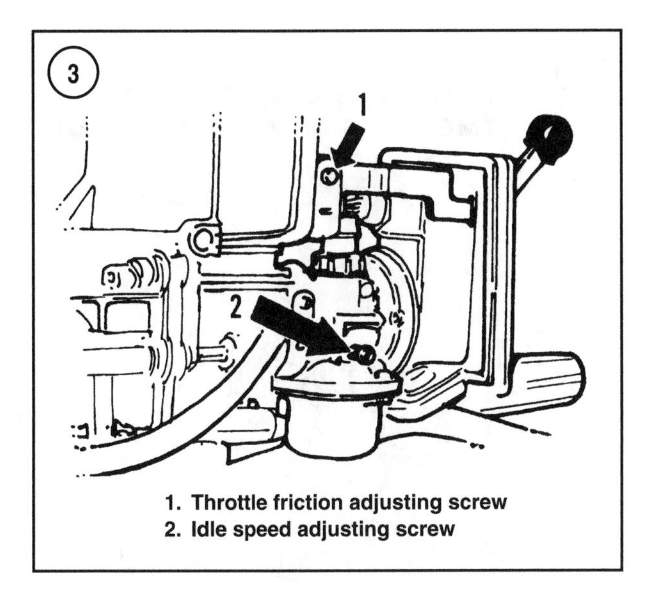

1. Throttle friction adjusting screw
2. Idle speed adjusting screw

SYNCRONIZATION AND LINKAGE PROCEDURES

2.5 and 3.3 hp Models

Ignition timing is not adjustable and no specifications are given. Timing is determined by the mechanical mounting of the ignition components and the flywheel. If timing is suspected of being incorrect, remove the flywheel and inspect the ignition components mounting hardware, the flywheel key and the flywheel magnets.

Refer to **Table 1** for specifications. Make the synchronization and linkage adjustments as follows:

1. Set the throttle lever friction.
2. Adjust the idle speed.
3. Verify full throttle engine speed.

Setting throttle lever friction

Adjust the throttle lever friction screw (1, **Figure 3**) as necessary to provide the desired throttle lever friction.

Idle speed adjustment

CAUTION
Do not run the engine without an adequate water supply and do not exceed 3000 rpm without an adequate load. Refer to ***Safety Precautions*** *at the beginning of this chapter.*

1. Start the engine and allow it to warm to normal operating temperature.
2. Connect an accurate shop tachometer to the spark plug lead.
3. Place the throttle lever in the slowest speed position.
4. On 3.3 hp models, shift the gearcase into forward gear.
5. Adjust the idle speed screw (2, **Figure 2**) to obtain the idle speed in **Table 1**.
6. Slowly advance the throttle to approximately 2000 rpm and return to idle speed. Allow the idle speed to stabilize for a few minutes and recheck the idle speed. Readjust the idle speed as necessary.

5

Full throttle engine speed verification

1. Connect an accurate shop tachometer to the spark plug lead.
2. With the engine mounted on a boat, the boat unrestrained in the water and the engine running at full throttle in forward gear, record the maximum engine speed noted on the tachometer.
3. If the maximum speed exceeds the recommended range in **Table 1**, check the propeller for damage or incorrect application. Repair or replace the propeller as necessary.
4. If the maximum speed does not reach the recommended range in **Table 1**, check the propeller for incorrect application. If the correct propeller is installed, the engine is not producing its rated horsepower. Start troubleshooting by checking the cranking compression as described in Chapter Four.

4 and 5 hp Models

The ignition timing is advanced electronically and is not adjustable. Verify the timing to ensure the ignition system is functioning correctly. Refer to **Table 2** for specifications. Make synchronization and linkage adjustments as follows:

1. Make preliminary adjustments.
2. Verify timing.
3. Adjust idle speed.
4. Adjust idle mixture.
5. Verify full throttle engine speed.

Preliminary adjustments

Refer to **Figure 4** for this procedure.

1. Move the throttle control to the idle position.
2. Loosen the throttle wire retaining screw (3, **Figure 4**).
3. Back out the idle speed screw (5, **Figure 4**) until it no longer touches the throttle lever (2).

4. Turn the idle speed screw inward until it just contacts the throttle arm (1, **Figure 4**), then turn the screw an additional two turns inward.
5. Pull up on the throttle wire (4, **Figure 4**) with a pair of needlenose pliers to remove all slack from the throttle cable. Make sure the throttle lever is against the screw contact point (1, **Figure 4**), then securely tighten the throttle wire retaining screw (3).
6. Operate the throttle control and make sure the throttle plate fully opens and returns against the idle speed screw. Readjust the throttle wire if necessary.
7. Turn the idle mixture screw (6, **Figure 4**) clockwise until it is lightly seated. Do not force the screw tightly into the carburetor or the tip of the screw and the carburetor will be damaged.
8. Back out the idle mixture screw to the specification in **Table 2**.

Timing verification

There are two cast-in timing marks on the top surface or outer diameter of the flywheel, TDC and 30° BTDC. See **Figure 5**. The timing marks align with the cylinder block parting line (1, **Figure 5**) on the port side of the engine. When the flywheel marks are viewed from the port side of the engine and the flywheel is rotated clockwise as viewed from above, the 30° BTDC mark (3, **Figure 4**) appears first followed by the TDC mark (2). Verify the engine timing as follows:

1. Remove the spark plug and install the dial indicator, part No. 91-58222A-1 or an equivalent, into the spark plug hole. See **Figure 1**.
2. Rotate the flywheel in the normal direction of rotation (clockwise) to position the piston at TDC.
3. With the piston positioned at TDC, the TDC mark on the flywheel should align with the cylinder block-to-crankcase cover parting line (1, **Figure 5**). If it does not, proceed as follows:
   a. The dial indicator is incorrectly installed, set up or misread. Make sure the dial indicator is set up correctly.
   b. The flywheel is incorrectly installed or the flywheel key is sheared. Remove the flywheel and inspect as it described in Chapter Seven.

NOTE

Do not proceed unless the timing marks align as described in Step 3.

4. Remove the dial indicator and reinstall the spark plug and lead.

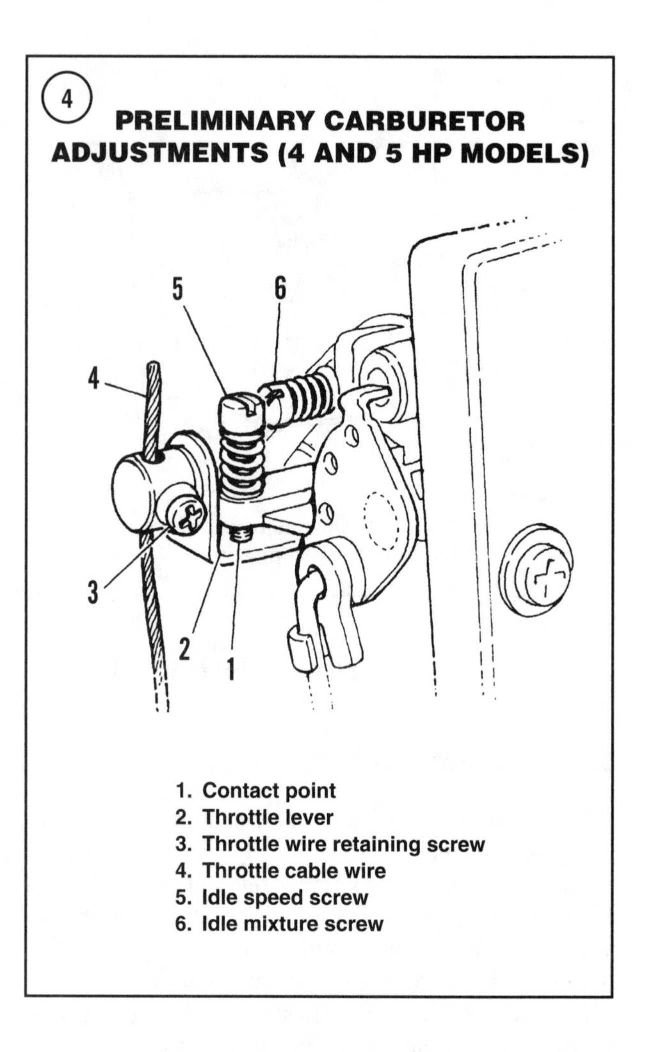

1. Contact point
2. Throttle lever
3. Throttle wire retaining screw
4. Throttle cable wire
5. Idle speed screw
6. Idle mixture screw

5. Connect a suitable timing light to the spark plug lead (**Figure 2**).

CAUTION

*Do not run the engine without an adequate water supply and do not exceed 3000 rpm without an adequate load. Refer to **Safety Precautions** at the beginning of this chapter.*

6. Start the engine and allow it to warm to normal operating temperature.
7. With the engine running at idle speed, shift into forward gear.
8. Point the timing light at the flywheel. With the engine idling at 800-900 rpm in forward gear, the TDC mark on the flywheel should align to a point approximately 1/4 in. (6.4 mm) to the right of the cylinder block-to-crankcase cover split line (5° BTDC).
9. Advance the throttle to wide-open throttle.
10. With the engine running at wide-open throttle in forward gear, the 30° BTDC mark (3, **Figure 5**) on the fly-

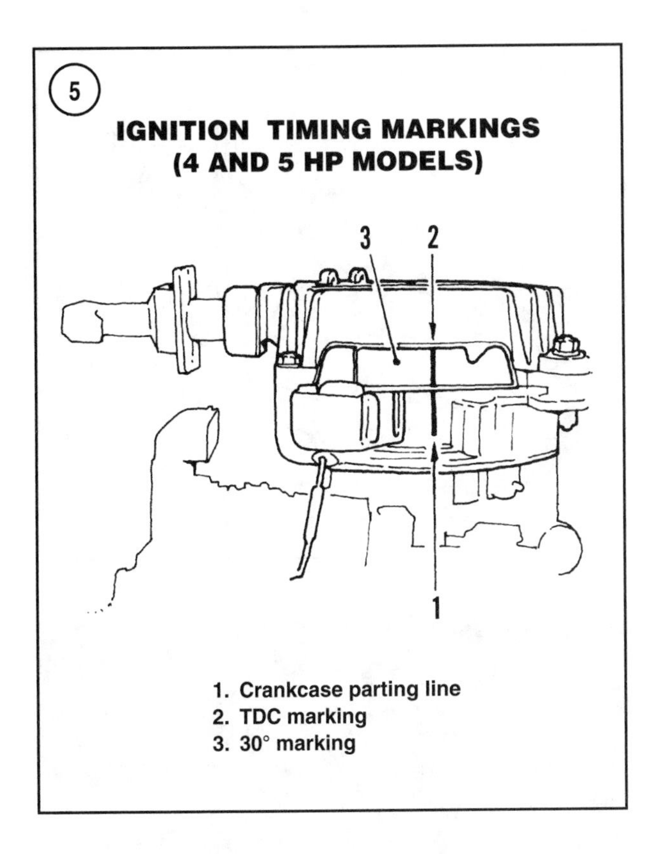

1. Crankcase parting line
2. TDC marking
3. 30° marking

wheel should align with the cylinder block-to-crankcase cover parting line (1, **Figure 5**). If it does not, test the ignition system as described in Chapter Three. Ignition system spark advance is controlled by the CD module.

Idle speed adjustment

1. Connect an accurate shop tachometer to the spark plug lead.
2. Start the engine and run at 2000 rpm until it is warmed to normal operating temperature. Allow the motor to idle 1-2 minutes to stabilize the motor.
3. Adjust the idle speed screw (5, **Figure 4**) to the idle speed specification in **Table 1**.
4. Slowly advance the throttle to approximately 2000 rpm and return to idle speed. Allow the idle speed to stabilize for a few minutes and recheck the idle speed. Readjust the idle speed as necessary.

Idle mixture screw adjustment

1. Connect an accurate shop tachometer to the spark plug lead.
2. Start the engine and run at 2000 rpm until it is warmed to normal operating temperature. Allow the motor to idle 1-2 minutes to stabilize the motor.
3. Adjust the idle speed screw as described previously in this chapter.
4. Slowly turn the idle mixture screw (6, **Figure 4**) counterclockwise in 1/8 turn increments, pausing at least 10 seconds between turns. Continue until the idle speed decreases and idle becomes rough due to an excessively rich mixture. Note the position of the mixture screw slot.
5. Slowly turn the idle mixture screw clockwise in 1/8 turn increments, pausing at least 10 seconds between turns. Continue until the engine speed begins to slow again and/or misfires due to the excessively lean mixture. Note the position of the mixture screw slot.
6. Position the mixture screw at a midpoint between the settings of Step 4 and Step 5. Do not position the screw outside of the adjustment specification in **Table 2**.
7. Quickly accelerate the engine to wide-open throttle and back to idle. The engine accelerates cleanly and without hesitation if the mixture is adjusted correctly. Readjust as necessary.
8. Adjust the idle speed screw as described previously in this section.

Full throttle engine speed

1. Connect an accurate shop tachometer to the spark plug lead.
2. With the engine mounted on a boat, the boat unrestrained in the water and the engine running at full throttle in forward gear, record the maximum engine speed noted on the tachometer.
3. If the maximum engine speed exceeds the recommended speed range listed in **Table 2**, check the propeller for damage or incorrect application. Repair or replace the propeller as necessary.
4. If the maximum engine speed does not reach the recommended speed range in **Table 2**, check the propeller for incorrect application. If the correct propeller is installed, the engine is not producing its rated horsepower. Start troubleshooting by checking the cranking compression as described in Chapter Four.

6-15 hp Models

The ignition timing is advanced mechanically and requires adjustment of both the idle and full throttle settings. The 6 hp model (1998-1999) does not have an idle speed adjustment. The carburetor is calibrated to the specified idle speed. Later 6 hp models (2000-on) are equipped with an idle speed screw like the 8-15 hp models. Refer to **Table 3** for specifications. Make the synchronization and linkage adjustments as follows:

1. Make preliminary adjustments.
2. Adjust ignition timing.
3. Adjust fast idle speed.
4. Adjust idle speed.
5. Adjust idle mixture.
6. Verify full throttle engine speed.

Preliminary adjustments

1. On tiller control models, loosen the throttle cable jam nuts. Adjust the nuts for equal full-throttle movement in forward and reverse gears, and remove slack from the cables. See **Figure 6**, typical. Securely tighten the cable jam nuts.
2. Push the primer/fast idle knob (**Figure 7**) in, then turn the knob fully counterclockwise.
3. On 6 hp models (1998-1999), loosen the throttle cam locking screw (C, **Figure 8**). Push the cam follower (D, **Figure 8**) down until it contacts the throttle cam. Tighten the cam locking screw.
4. For 6 hp (2000-on) and 8-15 hp models, turn the idle speed screw (**Figure 9**) counterclockwise until it no longer touches the cam follower. Turn the idle speed screw clockwise until it just touches the cam follower, then turn the screw an additional 1/2 turn to slightly open the throttle plate.
5. Remove the access plug from the front top of the carburetor intake cover, if so equipped. Turn the idle mixture screw clockwise until it is lightly seated. Do not force the screw tightly into the carburetor or the tip of the screw and the carburetor will be damaged. Back out the mixture screw to the middle of the turns out range in **Table 3**. **Figure 10** shows the idle mixture screw with the intake cover removed for clarity.

Ignition timing adjustments

CAUTION
*Do not run the engine without an adequate water supply and do not exceed 3000 rpm without an adequate load. Refer to **Safety Precautions** at the beginning of this chapter.*

1. Connect a timing light to the top, cylinder No. 1 spark plug lead.
2. Start the engine and allow it to warm to normal operating temperature.

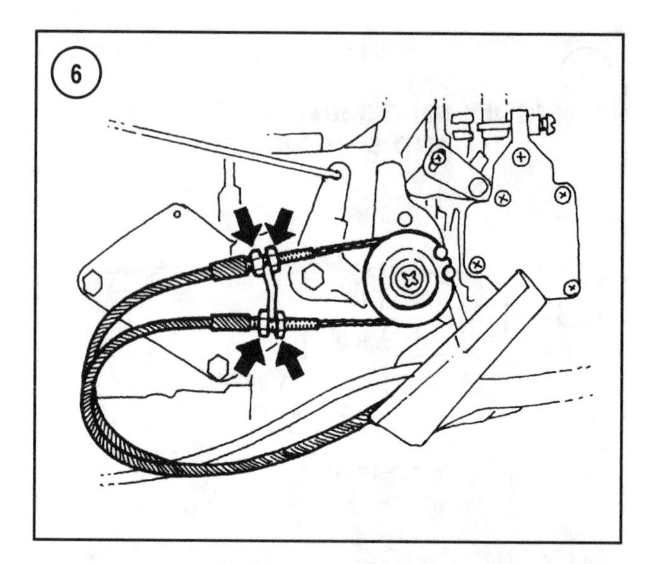

3. Reduce engine speed to idle and shift the outboard into forward gear.
4. Point the timing light at the timing pointer above and to the port side of the carburetor. Advance the throttle to wide-open and note the timing. The timing mark should align with full throttle timing mark in **Table 1**.
5. If adjustment is necessary, stop the engine and loosen the jam nut on the maximum advance screw (A, **Figure 8**). Restart the engine and advance the throttle to wide-open. Adjust the screw to align the specified timing mark and the timing pointer as described in Step 4. Stop the engine and securely tighten the jam nut.
6. Push the primer/fast idle knob (**Figure 7**) in, then turn the knob fully counterclockwise.
7. Start the engine, reduce the engine speed to idle and shift into forward gear.

NOTE
It may be necessary to adjust the idle speed screw temporarily to obtain the stable idle

8

9

10

*speed required for checking idle timing. See **Figure 9**.*

8. Adjust the idle timing screw (B, **Figure 8**) to align the timing pointer with the specified timing mark (**Table 3**) on the flywheel.

Fast idle speed adjustment

1. Push the primer/fast idle knob (**Figure 7**) in, then turn the knob fully counterclockwise.
2. Shift the outboard into neutral.
3. Adjust the screw (E, **Figure 8**) until there is no clearance between the idle wire (F) and the ignition system trigger assembly.
4. Connect an accurate shop tachometer to the spark plug lead.
5. Start the engine and run at idle speed until the engine reaches normal operating temperature.
6. Rotate the fast idle/primer knob fully clockwise. The idle speed should increase to 1500-2000 rpm. If it does not, recheck the ignition timing and fast idle speed adjustment.

Idle speed adjustment

1. Connect an accurate shop tachometer to a spark plug lead.
2. Start the engine and run at 2000 rpm until the engine reaches normal operating temperature. Set the throttle control to the idle position, shift the gearcase into forward gear. Allow the motor to idle 1-2 minutes to stabilize the motor.
3. Push the primer/fast idle knob (**Figure 7**) in, then turn the knob fully counterclockwise.

NOTE

The carburetor used on 6 hp models (1998-1999) is calibrated for proper air flow for the idle speed of 575-725. If the idle speed is not correct, make sure the idle timing is correct, the throttle plate is fully closed and the idle mixture screw is correctly set. The carburetor used on 6 hp models (2000-On) is equipped with an idle speed adjustment screw like all 8-15 hp models.

4. On 6 hp (2000-on) and 8-15 hp models, adjust the idle speed screw (**Figure 9**) to achieve the idle speed in **Table 3**.
5. Reinstall the access plug into the carburetor intake cover, if so equipped.

Idle mixture adjustment

1. Connect an accurate shop tachometer to the spark plug lead. Remove the access plug from the front top of the carburetor intake cover, if so equipped.
2. Start the engine and run at 2000 rpm until the engine is warmed to normal operating temperature. Allow the motor to idle 1-2 minutes to stabilize.
3. Adjust the idle speed screw as described previously in this chapter.
4. Slowly turn the idle mixture screw (**Figure 10**) counterclockwise in 1/8 turn increments, pausing at least 10

seconds between turns. Continue until the idle speed decreases and idle becomes rough due to an excessively rich mixture. Note the position of the mixture screw slot.
5. Slowly turn the idle mixture screw clockwise in 1/8 turn increments, pausing at least 10 seconds between turns. The idle gradually becomes smoother and speed increases. Continue until the engine speed begins to slow again and/or misfires due to the excessively lean mixture. Note the position of the mixture screw slot.
6. Position the mixture screw at a midpoint between the settings of Step 4 and Step 5. Do not position the screw outside of the adjustment specification in **Table 3**.
7. Quickly accelerate the engine to wide-open throttle and back to idle. The engine accelerates cleanly and without hesitation if the mixture is correctly adjusted. Readjust as necessary.
8. Adjust the idle speed screw as described previously in this chapter.

Full throttle engine speed verification

1. Connect an accurate shop tachometer to a spark plug lead.
2. With the engine mounted on a boat, the boat unrestrained in the water and the engine running at wide-open throttle in forward gear, record the maximum rpm noted on the tachometer.
3. If the maximum speed exceeds the recommended range in **Table 3**, check the propeller for damage. Repair or replace the propeller as necessary. If the propeller is in good condition, install a propeller with more pitch or a larger diameter and recheck the speed.
4. If the maximum speed does not reach the recommended range in **Table 3**, install a propeller with less pitch or a smaller diameter and recheck the speed.

20-25 hp and 20 Jet Models

On 1998 20 jet models, the ignition timing is advanced electronically. A *setup* timing specification is given to allow setting the timing without running the engine at full throttle. Incorrect setup timing affects the idle timing and full throttle timing.

All 20 hp, 25 hp and 1999-on 20 jet models use a mechanical timing advance that requires adjustment of both the idle and wide-open throttle positions.

Refer to **Table 4** for general specifications.

Make the synchronization and linkage adjustments as follows:
1. Make preliminary adjustments.
2. Adjust wide-open throttle stop.

11

THROTTLE STOP ADJUSTMENT (20 JET, 20 HP AND 25 HP MODELS)

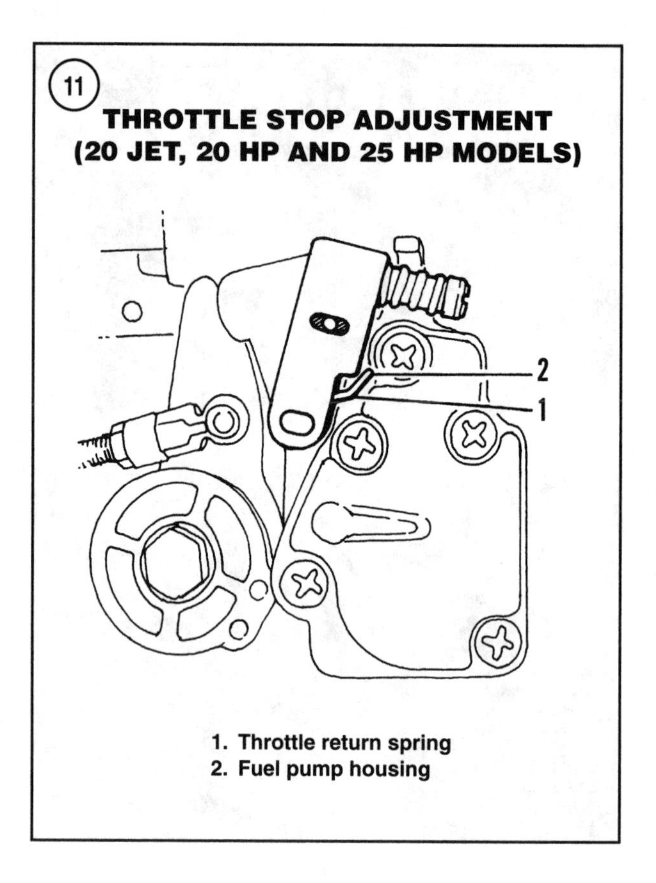

1. Throttle return spring
2. Fuel pump housing

3. Adjust maximum spark advance.
4. Adjust idle timing.
5. Adjust idle speed.
6. Adjust idle mixture.
7. Adjust fast idle speed.
8. Adjust shift and throttle cable on remote control models.
9. Verify full throttle engine speed.

Preliminary adjustments

1. For tiller control models, loosen the throttle cable jam nuts and adjust the nuts for equal full-throttle travel in forward and reverse gears, and to remove slack in the cables. See **Figure 6**, typical. Securely tighten the cable jam nuts.
2. Push the primer/fast idle knob (**Figure 7**), then turn the knob fully counterclockwise.
3. Turn the idle speed screw (**Figure 9**) counterclockwise until it no longer touches the cam follower. Turn the idle speed screw clockwise until it just touches the cam follower, then turn the screw an additional 1/2 turn to open the throttle plate.
4. Remove the access plug from the front top of the carburetor intake cover, if so equipped. Turn the idle mixture screw clockwise until it is lightly seated. Do not force the

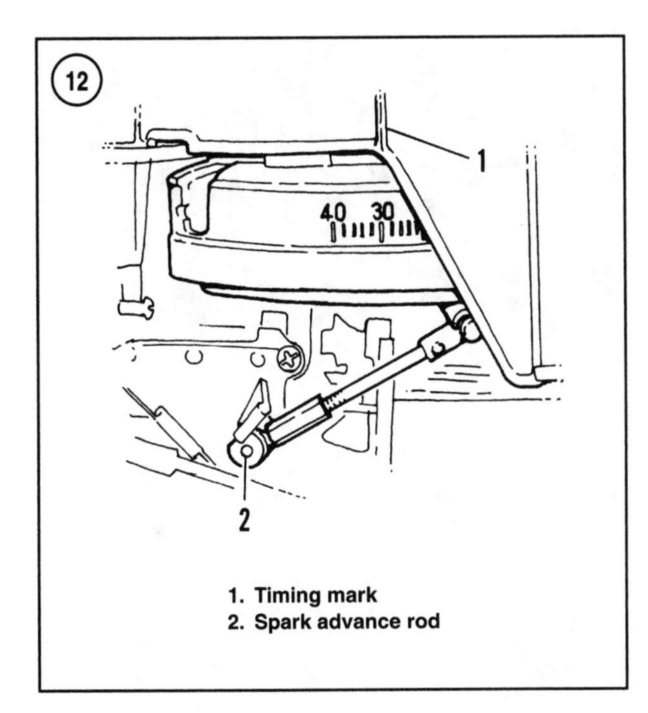

1. Timing mark
2. Spark advance rod

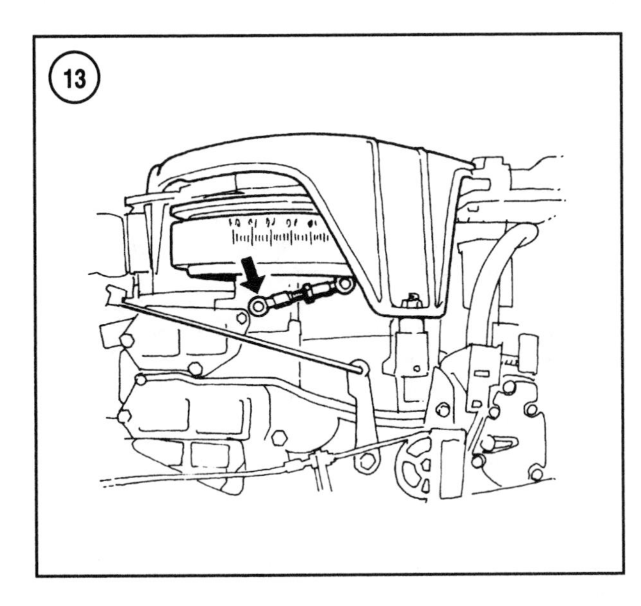

screw tightly into the carburetor or the tip of the screw and the carburetor will be damaged. Back out the mixture screw to the middle of the turns out specification in **Table 4**. **Figure 10** shows the idle mixture screw with the intake cover removed for clarity.

Full throttle stop adjustment

1. Shift the gearcase into forward gear. Move the twist grip or remote control to the full throttle position.

2. The throttle return spring (1, **Figure 11**) should just contact the fuel pump housing (2, **Figure 11**). If it does not, correct the adjustment as follows:

   a. *Tiller shift and remote control models*—Adjust the throttle cam link rod as necessary to ensure the throttle plate opens fully and the throttle return spring is not binding against the fuel pump housing.

   b. *Side shift models*—Adjust the throttle cable jam nuts (**Figure 6**, typical) as necessary to ensure the throttle plate opens fully and the throttle return spring is not binding against the fuel pump housing. Make sure there is no slack in the throttle cables when the adjustment is completed.

Maximum spark advance adjustment

WARNING
Do not attempt to adjust the timing with the engine running. Serious personal injury could occur if contact is made with the moving flywheel.

CAUTION
*Do not run the engine without an adequate water supply and do not exceed 3000 rpm without an adequate load. Refer to **Safety Precautions** at the beginning of this chapter.*

1. Connect a timing light to the top, cylinder No. 1 spark plug wire.
2. Connect an accurate shop tachometer to the engine.
3. Start the engine and allow it to warm to normal operating temperature.
4. Reduce engine speed to idle and shift the outboard into forward gear.
5. Point the timing light at the timing pointer (1, **Figure 12**) on the starboard side of the rewind starter assembly.

6A. *20 jet (1998) models*—Advance the throttle to 3000 rpm and note the timing. Timing should be 28° BTDC at 3000 rpm. The switch box automatically retards timing to 25° BTDC at 5500 rpm. If adjustment is necessary, proceed as follows:

   a. Stop the engine and remove the trigger link rod from engine mounted ball stud (**Figure 13**).
   b. Lengthen the rod to advance timing or shorten the rod to retard timing.
   c. Reinstall the link rod and recheck the timing.

6B. *20 jet (1999-on), 20 hp and 25 hp models*—Advance the throttle to the full throttle position and note the timing. Refer to **Table 4** for specifications. If adjustment is necessary, proceed as follows:

a. Stop the engine and loosen the jam nut (D, **Figure 14**) on the maximum advance screw (E).
b. Restart the engine and adjust the screw as necessary to align the specified timing mark and the timing pointer as described in Step 6B.
c. Stop the engine and securely tighten the jam nut.

Idle timing adjustment (pickup timing)

1. Connect a timing light to the top, cylinder No. 1 spark plug wire.
2. Connect an accurate shop tachometer to the engine.
3. Start the engine and allow it to warm to normal operating temperature.
4. Reduce the engine speed to idle and shift the outboard into forward gear.
5. Point the timing light at the timing pointer (1, **Figure 12**) on the starboard side of rewind starter assembly.

6A. *20 jet (1998) models*—The idle timing should be within the range in **Table 4**. If the timing is incorrect, check the maximum spark advancement as described previously in this section. If the maximum spark advance is correct but the idle timing is incorrect, replace the switch box (Chapter Seven).

6B. *20 jet (1999-on), 20 hp and 25 hp models*—The idle timing should be within the range in **Table 4**. If timing is incorrect, proceed as follows:

a. Stop the engine and loosen the idle timing jam nut (A, **Figure 14**).
b. Restart the engine and adjust the idle timing screw (B, **Figure 14**) to obtain the idle timing in **Table 4**.
c. Stop the engine and retighten the idle timing jam nut.

Idle speed adjustment

1. Connect an accurate shop tachometer to a spark plug wire.
2. Start the engine and run at 2000 rpm until the engine is warmed to normal operating temperature. Set the throttle control to the idle position, shift the gearcase into forward gear and allow the motor to idle 1-2 minutes to stabilize the motor and allow the fuel recirculation system to begin functioning.
3. Push the primer/fast idle knob (**Figure 7**) in, then turn the knob fully counterclockwise.
4. Adjust the idle speed screw (C, **Figure 14**) to obtain the idle speed in **Table 4**.

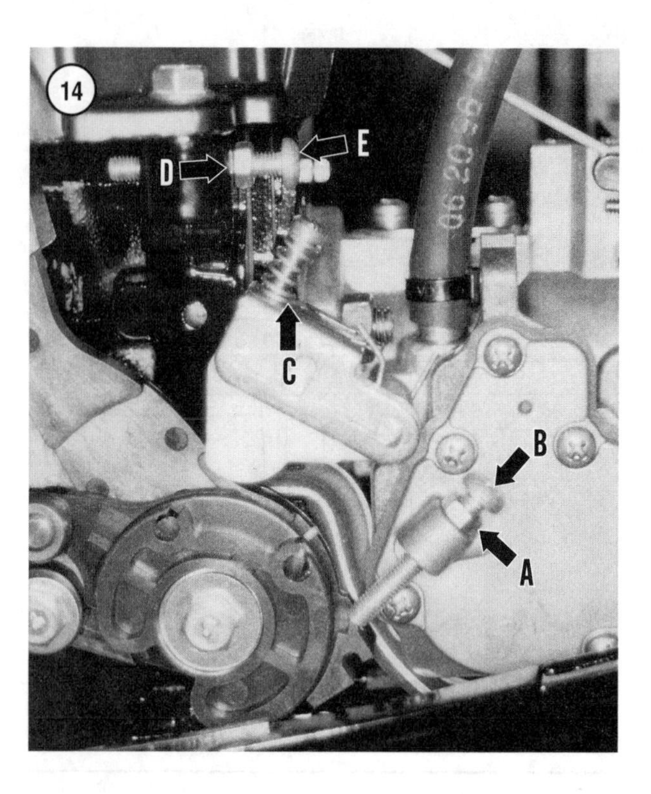

Idle mixture adjustment

NOTE

Idle mixture cannot be properly set unless the carburetor is operating on the idle circuit. Setting the idle mixture at higher speeds results in incorrect adjustment. It may be necessary to switch back and forth between the idle mixture and idle speed adjustment several times to achieve proper adjustment. Always adjust the idle speed last.

1. Connect an accurate shop tachometer to the spark plug wire. Remove the access panel from the front top of the carburetor intake cover, if so equipped.
2. Start the engine and run at 2000 rpm until the engine is warmed to normal operating temperature. Allow the motor to idle 1-2 minutes to stabilize.
3. Adjust the idle speed screw as described previously in this section.
4. Slowly turn the idle mixture screw (**Figure 10**) counterclockwise in 1/8 turn increments, pausing at least 10 seconds between turns. Continue until the idle speed decreases and idle becomes rough due to an excessively rich mixture. Note the position of the mixture screw slot.
5. Slowly turn the idle mixture screw clockwise in 1/8 turn increments, pausing at least 10 seconds between turns. Continue until the engine speed begins to slow

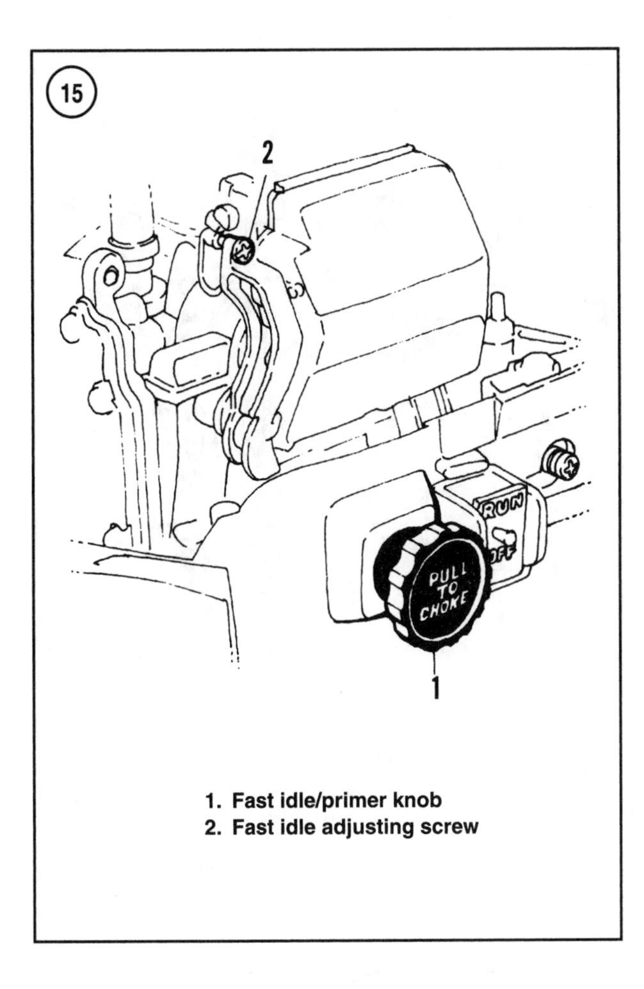

1. Fast idle/primer knob
2. Fast idle adjusting screw

again and/or misfires due to the excessively lean mixture. Note the position of the mixture screw slot.

6. Position the mixture screw at a midpoint between the settings of Step 4 and Step 5. Do not position the screw outside of the adjustment specification in **Table 4**.

7. Quickly accelerate the engine to wide-open throttle and back to idle. The engine accelerates cleanly and without hesitation if the mixture is adjusted correctly. Readjust as necessary.

8. Adjust the idle speed screw as described previously in this section.

Fast idle speed adjustment

1. Connect an accurate shop tachometer to a spark plug wire.

2. Start the engine and run at 2000 rpm until the engine is warmed to normal operating temperature. Set the throttle control to the idle position and allow the motor to idle 1-2 minutes to stabilize the motor and allow the fuel recirculation system to begin functioning.

3. Push the primer/fast idle knob (1, **Figure 15**) in, then turn the knob fully clockwise.

4. Shift the gearcase into neutral.

5. Adjust the fast idle speed screw (2, **Figure 15**) to achieve the fast idle speed in **Table 4**.

Shift and throttle cable adjustments (remote control models)

Refer to Chapter Fourteen for remote control cable adjustments.

Wide-open throttle engine speed

1. Connect an accurate shop tachometer to a spark plug wire.

2. With the engine mounted on a boat, the boat unrestrained in the water and the engine running at wide-open throttle in forward gear, record the maximum engine speed noted on the tachometer.

3. If the recorded engine speed exceeds the recommended speed range in **Table 4**, check the propeller for damage. Repair or replace the propeller as necessary. If the propeller is in good condition, install a propeller with more pitch or a larger diameter and recheck the engine speed.

4. If the maximum engine speed does not reach the recommended speed range in **Table 4**, install a propeller with less pitch or a smaller diameter and recheck the engine speed.

30 and 40 hp Models (Two-Cylinder)

The ignition system is mechanically advanced and requires the cam follower, idle timing and maximum spark advance to be set. The idle timing must be set with the engine running.

Refer to **Table 5** for general specifications.

Make the synchronization and linkage adjustment as follows:

1. Make preliminary adjustments.
2. Adjust cam follower.
3. Adjust idle speed.
4. Adjust idle timing.
5. Adjust maximum spark advance.
6. Adjust idle mixture screw.
7. Adjust oil pump linkage.
8. Adjust shift and throttle cable (remote control models).
9. Verify full throttle engine speed.

Preliminary adjustments

1. Loosen the cam follower screw (1, **Figure 16**). Turn the idle speed screw (2, **Figure 16**) counterclockwise until it no longer touches the throttle arm allowing the throttle plate to fully close.
2. Turn the idle mixture screw (3, **Figure 16**) clockwise until it is lightly seated. Do not force the screw tightly into the carburetor or the tip of the screw and the carburetor will be damaged. Back the mixture screw out to the middle of the specification in **Table 5**.

Cam follower adjustments

1. On tiller control models, position the twist grip in the idle position. Loosen the throttle cable jam nuts (4, **Figure 16**) and adjust the nuts to center the raised mark on the throttle cam (A, **Figure 17**) with the cam follower roller (B).

NOTE
Do not excessively tighten the throttle cables in Step 2. Excessive tension causes binding.

2. On tiller control models, adjust the nuts to maintain the cam position described in Step 1. The tension should be enough to maintain 1/16-1/8 in. (1.59-3.18 mm) free movement measured at the link rod ball (C, **Figure 17**). Tighten the nuts securely when all adjustments are correct.

3A. On tiller control models, rest the cam follower on the throttle cam and tighten the cam follower screw (D, **Figure 17**) as shown.

3B. On remote control models, hold the throttle lever in the full-idle position and rest the cam follower on the throttle cam. Tighten the cam follower screw (D, **Figure 17**) as shown.

4. Make sure the throttle cam is in the idle position. Turn the idle speed screw (**Figure 18**) clockwise until the throttle cam follower is 0.005-0.040 in. (0.13-1.02 mm) from the throttle cam as shown in **Figure 18**.

Idle speed adjustment

1. Connect an accurate shop tachometer to a spark plug lead.
2. Start the engine and run at 2000 rpm until the engine reaches normal operating temperature. Set the throttle control to idle, shift the gearcase into forward gear and allow the motor to idle 1-2 minutes to stabilize the motor

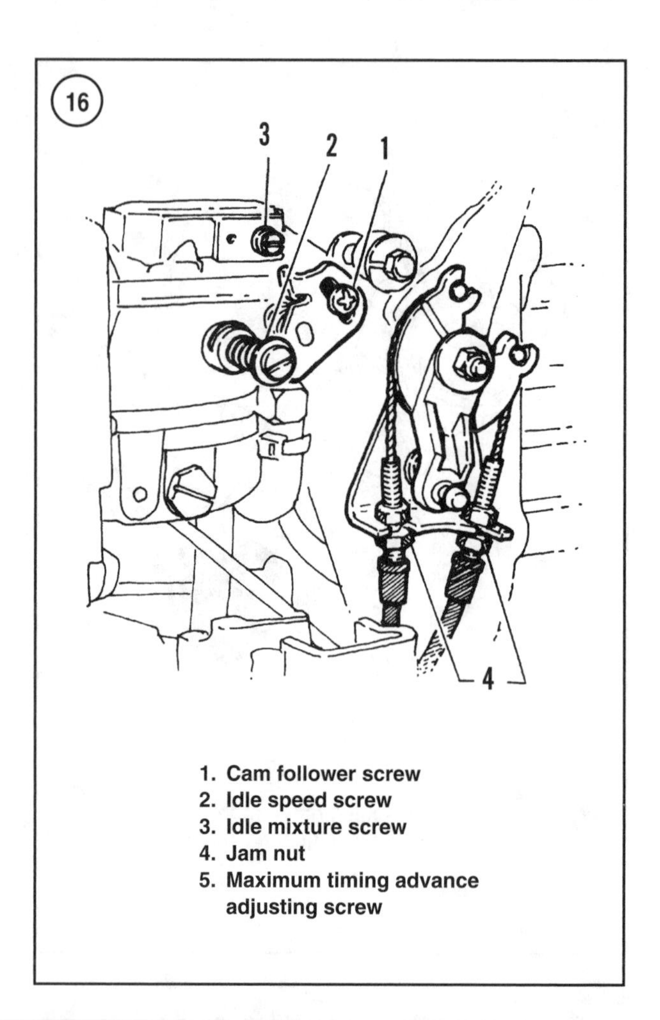

1. Cam follower screw
2. Idle speed screw
3. Idle mixture screw
4. Jam nut
5. Maximum timing advance adjusting screw

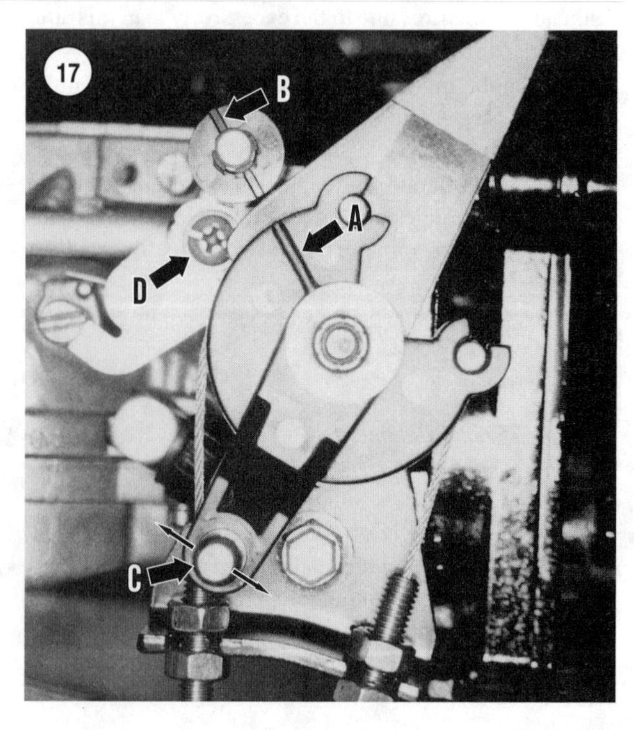

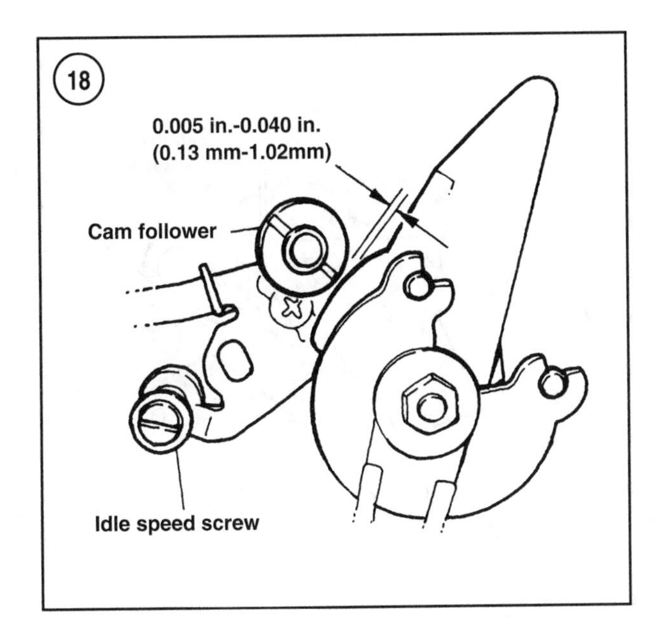

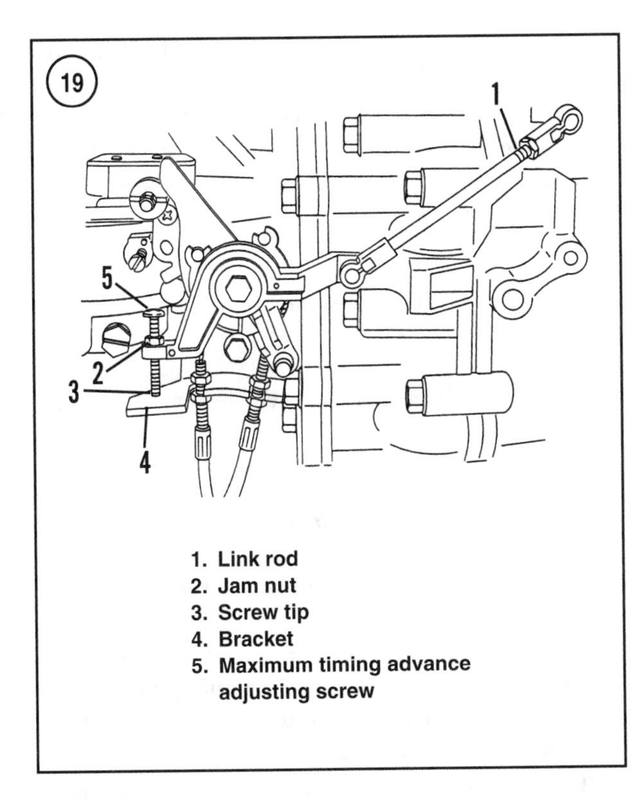

1. Link rod
2. Jam nut
3. Screw tip
4. Bracket
5. Maximum timing advance adjusting screw

and allow the fuel recirculation system to begin functioning.

3. Adjust the idle speed screw to achieve the idle speed in **Table 5**.

4. Quickly accelerate the engine to full throttle and decelerate to idle. Allow the idle speed to stabilize. Check the idle speed and readjust as necessary.

Idle timing adjustment

1. Connect a suitable timing light to the top, cylinder No. 1 spark plug lead.

CAUTION
*Do not run the engine without an adequate water supply and do not exceed 3000 rpm without an adequate load. Refer to **Safety Precautions** at the beginning of this chapter.*

2. Start the engine and allow it to warm to normal operating temperature.
3. Reduce the engine speed to idle and shift the gearcase into forward gear.
4. Point the timing light at the flywheel and timing pointer. The idle timing should be within the range in **Table 5**. If the idle timing is incorrect, proceed as follows:
   a. Stop the engine and carefully snap the throttle link rod (1, **Figure 19**) from the throttle arm.
   b. Lengthen the rod to advance timing or shorten the rod to retard timing.
   c. Reinstall the link rod and recheck timing.
   d. Repeat Steps 4a-4d until the idle timing is within the specification in **Table 5**.

Maximum spark advance adjustment

1. Switch the engine off. Place the tiller or remote control in wide-open full throttle position.
2. Loosen the jam nut (2, **Figure 19**). Rotate the maximum spark advance screw (5, **Figure 19**) counterclockwise until a gap exists between the tip of the screw (3) and the bracket (4).
3. Rotate the maximum spark advance screw clockwise until the tip just contacts the bracket (4, **Figure 19**).
4. Rotate the maximum spark advance screw exactly one turn clockwise.
5. Securely tighten the jam nut (2, **Figure 19**).
6. Connect a suitable timing light to the top, cylinder No. 1 spark plug wire.

CAUTION
*Do not run the engine without an adequate water supply and do not exceed 3000 rpm without an adequate load. Refer to **Safety Precautions** at the beginning of this chapter.*

7. Start the engine and allow it to warm to normal operating temperature.
8. Advance the throttle to the wide-open position.

9. Point the timing light at the flywheel and timing pointer. The maximum timing advance should be within the range in **Table 5**. If the idle timing is incorrect, readjust the idle timing as described previously in this section.

Idle mixture screw adjustment

NOTE
Idle mixture cannot be properly set unless the carburetor is operating on the idle circuit. Setting the idle mixture at higher speeds results in incorrect adjustment. It may be necessary to switch back and forth between the idle mixture and idle speed adjustment several times to achieve proper adjustment. Always adjust the idle speed last.

1. Connect an accurate shop tachometer to the spark plug wire. Remove the access plug from the front top of the carburetor intake cover, if so equipped.
2. Start the engine and run at 2000 rpm until the engine reaches normal operating temperature. Allow the motor to idle 1-2 minutes to stabilize.
3. Adjust the idle speed screw as described previously in this section.
4. Slowly turn the idle mixture screw (3, **Figure 16**) counterclockwise in 1/8 turn increments, pausing at least 10 seconds between turns. Continue until the idle speed decreases and idle becomes rough due to an excessively rich mixture. Note the position of the mixture screw slot.
5. Slowly turn the idle mixture screw clockwise in 1/8 turn increments, pausing at least 10 seconds between turns. Continue until the engine speed begins to slow again and/or misfires due to the excessively lean mixture. Note the position of the mixture screw slot.
6. Position the mixture screw at a midpoint between the settings of Step 4 and Step 5. Do not position the screw outside of the adjustment specification in **Table 5**.
7. Quickly accelerate the engine to wide-open throttle and decelerate to idle. The engine accelerates cleanly and without hesitation if the mixture is adjusted correctly. Readjust as necessary.
8. Adjust the idle speed screw as described previously in this section.

Oil pump linkage adjustment

Any time the throttle linkage is adjusted, the oil pump linkage must be synchronized to the throttle linkage.

1A. *Tiller control models*—Position the twist grip to the full-idle position.

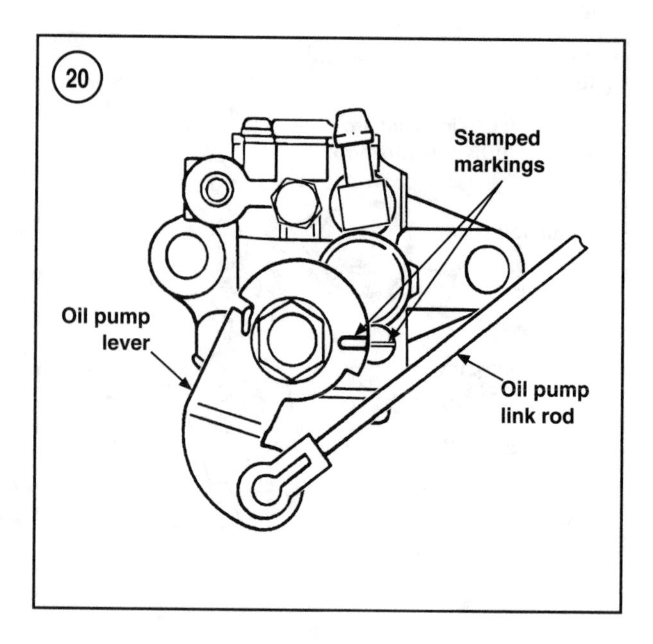

1B. *Remote control models*—Position the throttle lever to the idle stop position.
2. With the throttle linkage held in the idle position, adjust the oil pump link rod length to align the stamped mark on the oil pump lever with the stamped mark on the oil pump body. See **Figure 20**.

Shift and throttle cable adjustments (remote control models)

Refer to Chapter Fourteen for remote control cable adjustments.

Full throttle engine speed verification

1. Connect an accurate shop tachometer to a spark plug lead.
2. With the engine mounted on a boat, the boat unrestrained in the water and the engine running at wide-open throttle in forward gear, record the maximum engine speed noted on the tachometer.
3. If the maximum speed exceeds the recommended speed range in **Table 5**, check the propeller for damage. Repair or replace the propeller as necessary. If the propeller is in good condition, install a propeller with more pitch or a larger diameter and recheck the engine speed.
4. If the maximum engine speed does not reach the recommended speed range in **Table 5**, install a propeller with less pitch or a smaller diameter and recheck the engine speed.

21

LINKAGE ADJUSTMENT POINTS (40-60 HP, 30 JET AND 45 JET MODELS [THREE-CYLINDER])

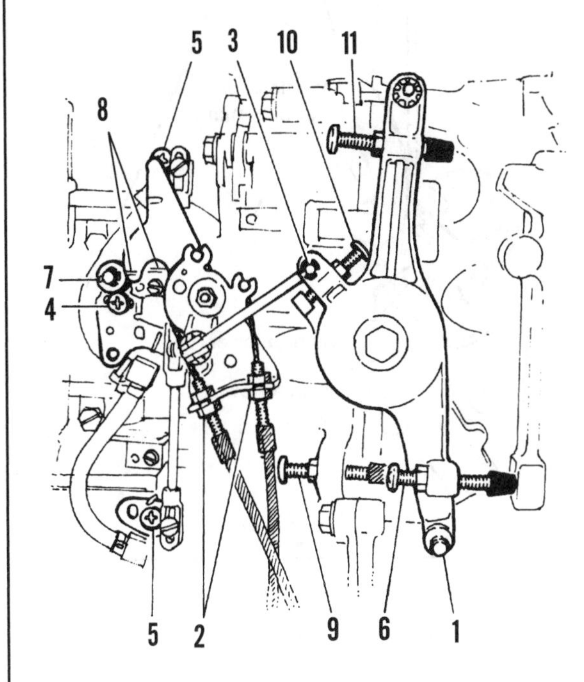

1. Remove control throttle cable attaching point
2. Throttle cable adjusting nuts (tiller control 40-50 hp models)
3. Free movement check point (tiller control 40-50 hp models)
4. Throttle cam follower adjustment screw
5. Throttle plate synchronization screws
6. Idle stop screw
7. Throttle cam follower roller
8. Throttle cam alignment marking
9. Wide-open throttle stop screw
10. Idle timing screw
11. Maximum spark advance adjusting screw

40-60 hp Models (Three-Cylinder), 30 Jet and 45 Jet

The ignition timing is mechanically advanced and requires adjustment of the idle and maximum timing. Timing can be set at cranking speed or while running. Setting the timing while running is more accurate.

Refer to **Table 6** for general specifications. Make the synchronization and linkage adjustments as follows:

22

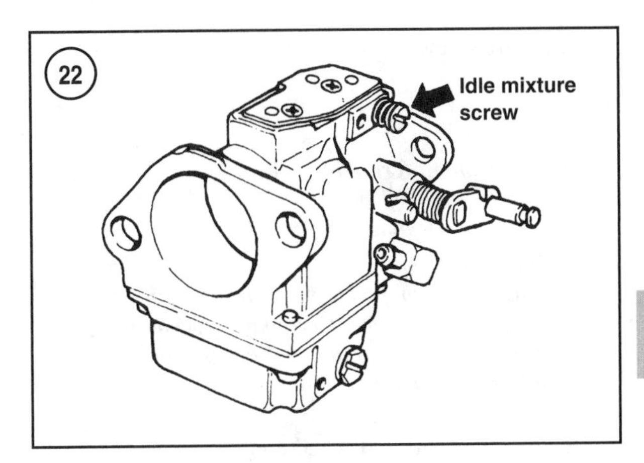

1. Make preliminary adjustments.
2. Adjust timing (cranking speed).
3. Synchronize throttle plate.
4. Adjust throttle cam.
5. Adjust idle speed.
6. Adjust idle mixture.
7. Adjust oil pump linkage.
8. Adjust timing (running).
9. Adjust shift and throttle cable (remote control models).
10. Verify full throttle engine speed.

Preliminary adjustments

1A. *Remote control and 55/60 hp tiller models*—Disconnect the throttle cable from the throttle lever arm (1, **Figure 21**).

NOTE

Do not excessively tighten the throttle cables in the next step. Excessive tension results in throttle arm binding.

1B. *40 and 50 hp tiller control models*—Adjust the throttle cable jam nuts (2, **Figure 21**) to provide full throttle arm travel and enough tension to maintain 1/16-1/8 in. (1.59-3.18 mm) free movement measured at the throttle arm (3). Tighten the nuts securely when all adjustments are correct.

2A. *40 and 50 hp models*—Remove the screw from the top of the carburetor air box cover. Remove the air box cover.

2B. *55 and 60 hp models*—Remove the four screws from the carburetor air box cover. Remove the air box cover.

3. Turn the idle mixture screw (**Figure 22**) on each carburetor clockwise until it is lightly seated. Do not force the screws tightly into the carburetors or the tips of the screws and the carburetors will be damaged. Back out

each mixture screw to the middle of the turns out range in **Table 6**.

Timing adjustment (cranking speed)

NOTE

The battery must be fully charged and the starting system functioning properly for accurate adjustment at cranking speed. Removing the spark plugs increases cranking speed and timing accuracy.

1. Remove the spark plugs and connect an air gap spark tester to the spark plug leads.
2. Connect a suitable timing light to the top, cylinder No. 1 spark plug wire.
3. Hold the idle stop screw (6, **Figure 21**) against the cylinder block stop.
4. Crank the engine with the electric starter while noting the timing.
5. Adjust the idle timing screw (10, **Figure 21**) to align the 0° TDC mark on the flywheel with the V-notch in the timing window.
6. Tighten the idle timing screw jam nut when finished.

NOTE

*Due to the electronic spark advance characteristics of this ignition system, the timing retards approximately 2° when running at wide-open throttle (5000 rpm). The maximum timing must be adjusted to the cranking specification in **Table 6** to obtain the correct ignition timing when running at 5000 rpm. Readjust the timing if necessary to achieve the correct timing while running.*

7. Hold the throttle arm so the maximum advance screw (11, **Figure 21**) is against the cylinder block stop.
8. Crank the engine while noting the timing. The V-notch in the timing window must align with the maximum timing advance specification in **Table 6**.
9. If adjustment is necessary, loosen the jam nut and adjust the maximum spark advance screw (11, **Figure 21**) to align the V-notch with the specified timing mark on the flywheel.
10. Tighten the maximum spark advance screw jam nut when finished.
11. Remove the timing light and the air gap spark tester. Reinstall the spark plugs and reconnect the spark plug leads.

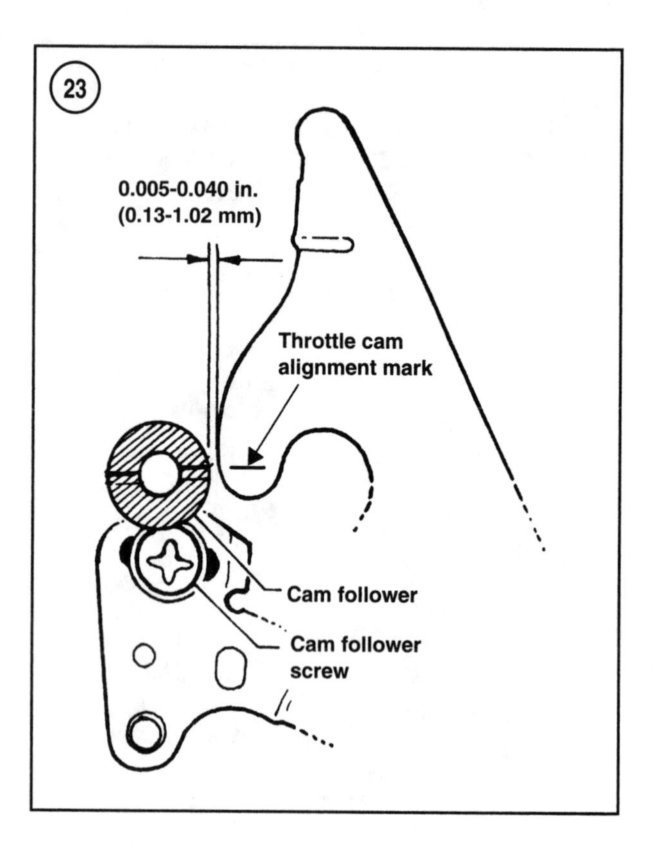

Throttle plate synchronization

1. Loosen the cam follower adjusting screw (4, **Figure 21**).
2. Loosen the two throttle plate synchronizing screws on the upper and lower carburetor throttle shafts (5, **Figure 21**).
3. Make sure the throttle valves in all three carburetors are fully closed, then retighten the screws (5, **Figure 21**). Do not tighten the cam follower screw (4, **Figure 21**) at this time.
4. Make sure all throttle plates open and close together. Readjust as necessary.

Throttle cam adjustment

1. Hold the idle stop screw (6, **Figure 21**) against its stop by pushing the throttle arm.
2. Position the cam follower roller (7, **Figure 21**) against the throttle cam. Adjust the idle stop screw (6, **Figure 21**) to align the throttle cam mark (8) with the center of the follower roller (7), then tighten the idle stop screw locknut.
3. While holding the throttle arm in the idle position, adjust the cam follower to set a 0.005-0.040 in. (0.13-1.02 mm) clearance between the cam follower roller and throt-

tle cam. See **Figure 23**. Make sure the throttle cam mark aligns with the center of the roller, then securely tighten the cam follower screw (**Figure 23**).

4. Move the throttle arm until it contacts the wide-open throttle stop screw (9, **Figure 21**). Loosen the jam nut then adjust the wide-open throttle stop screw so the carburetor throttle valves are fully open while allowing approximately 0.015 in. (0.38 mm) play in the throttle linkage. Make sure the throttle valves do not bottom out at wide-open throttle.
5. Reinstall the carburetor air box cover.

Idle mixture adjustment

NOTE
*The idle mixture must be properly set on all carburetors. Adjust the top carburetor first, the middle carburetor second and the bottom carburetor last. It may be necessary to switch back and forth between the carburetors several times to correct the mixture. If necessary, reset the mixture screws to the initial settings (**Table 6**) and try again.*

1. Connect an accurate shop tachometer to a spark plug wire.
2. Start the engine and run at 2000 rpm until the engine reaches normal operating temperature. Shift the gearcase into forward gear and allow the motor to idle 1-2 minutes to stabilize the motor and allow the fuel recirculation system to begin functioning.
3. Set the throttle lever control to idle. If necessary, adjust the idle timing screw (10, **Figure 21**) temporarily to obtain 650-700 rpm in forward gear.

NOTE
Idle mixture cannot be properly set unless the carburetors are operating on the idle circuit(s). Make sure the throttle plates are fully closed during this adjustment.

4. Slowly turn the idle mixture screw counterclockwise in 1/8 turns increments, pausing at least 10 seconds between turns. Continue until the idle speed decreases and idle becomes rough due to an excessively rich mixture. Note the position of the mixture screw slot.
5. Slowly turn the idle mixture screw clockwise in 1/8 turns increments, pausing at least 10 seconds between turns. Continue until the engine speed begins to slow again and/or misfires due to the excessively lean mixture. Note the position of the mixture screw slot.
6. Position the mixture screw at a midpoint between the settings of Step 4 and Step 5. Do not position the screw outside of the adjustment specification in **Table 6**. Repeat Steps 4-6 for the remaining carburetors.
7. Quickly accelerate the engine to wide-open throttle then throttle back to idle. The engine accelerates cleanly without hesitation if the mixture is adjusted correctly. Readjust as necessary.

Idle speed adjustment

NOTE
Adjust the idle mixture prior to adjusting the idle speed.

1. Connect an accurate shop tachometer to a spark plug wire.
2. Start the engine and run at 2000 rpm until the engine reaches normal operating temperature. Shift the gearcase into forward gear and allow the engine to idle 1-2 minutes to stabilize the motor and allow the fuel recirculation system to begin functioning.
3. Set the throttle lever control to idle.
4. On remote control and 55/60 hp tiller models, remove the throttle cable barrel from the barrel retainer on the cable anchor bracket.
5. Adjust the idle screw (10, **Figure 21**) to obtain the idle speed in **Table 6**. Adjust idle speed in forward gear.

Timing adjustment

1. Connect a suitable timing light to the top, cylinder No.1 spark plug wire.

CAUTION
*Do not run the engine without an adequate water supply and do not exceed 3000 rpm without an adequate load. Refer to **Safety Precautions** at the beginning of this chapter.*

2. Start the engine and allow it to reach normal operating temperature.
3. Reduce the engine speed to idle and shift the gearcase into forward gear.
4. Point the timing light at the flywheel and timing pointer. Note the reading. The idle timing should be within the specification in **Table 6**. If it is not, adjust the idle timing as follows:
   a. Hold the idle stop screw (6, **Figure 21**) against the idle stop.
   b. Adjust the idle timing screw (10, **Figure 21**) to obtain the idle timing in **Table 6**.
   c. Tighten the idle timing screw jam nut. Check the idle timing and readjust as necessary.

5. Point the timing light at the flywheel and timing pointer.

6. Advance the throttle to position the maximum spark advance screw (11, **Figure 21**) against the cylinder block stop. Note the timing reading. The maximum timing advance should be within the specification in **Table 6**. If it is not, adjust the maximum timing advance as follows:

   a. Stop the engine and loosen the jam nut on the maximum spark advance adjusting screw.
   b. Turn the adjustment screw clockwise to retard timing or counterclockwise to advance timing. Securely tighten the jam nut.
   c. Start the engine and recheck the timing. Readjust the timing as necessary.

Oil pump linkage adjustment

Anytime the throttle linkage is adjusted, the oil pump linkage must be synchronized to the throttle linkage.

1A. *Tiller control models*—Rotate the twist grip to the full-idle position.

1B. *Remote control models*—Move the throttle lever to the idle position.

2. With the throttle linkage held in the idle position, adjust the oil pump link rod length to align the stamped mark on the oil pump lever with the stamped mark on the oil pump body. See **Figure 20**.

Shift and throttle cable adjustments (remote control models)

Refer to Chapter Fourteen for additional information on remote control cable adjustments.

Wide-open throttle speed verification

1. Connect an accurate shop tachometer to a spark plug wire.

2. With the engine mounted on a boat, the boat unrestrained in the water and the engine running at wide-open throttle in forward gear, record the maximum engine speed on the tachometer.

3. If the maximum engine speed exceeds the recommended speed range in **Table 6**, check the propeller for damage. Repair or replace the propeller as necessary. If the propeller is in good condition, install a propeller with more pitch or a larger diameter and recheck the wide-open throttle engine speed.

4. If the maximum engine speed does not reach the recommended speed range in **Table 6**, install a propeller with less pitch or a smaller diameter and recheck the wide-open throttle engine speed.

Table 1 GENERAL SPECIFICATIONS: 2.5 AND 3.3 HP MODELS

Rated output	
2.5 hp	2.5 hp (1.9 kW) at 5000 rpm
3.3 hp	3.3 hp (2.5 kW) at 5000 rpm
Number of cylinders	1
Displacement	4.6 cu. in. (75.4 cc)
Standard bore	1.85 in. (47 mm)
Stroke	1.693 in. (43 mm)
Full throttle operating range	
2.5 hp	4000-5000 rpm
3.3 hp	4500-5500 rpm
Idle speed in forward gear	900-1000 rpm
Cranking compression (minimum)	90 psi (620.6 kPa)
Lubrication system	Requires 50:1 premix
Integral fuel tank capacity	0.375 gal. (1.4 L)
Ignition system	CD (capacitor discharge)
Ignition timing	Not adjustable

Table 2 GENERAL SPECIFICATIONS: 4 AND 5 HP MODELS

Rated output	
4 hp	4 hp (2.9 kW) at 5000 rpm
5 hp	5 hp (3.7 kW) at 5000 rpm
Number of cylinders	1
Displacement	6.2 cu. in. (102 cc)
Standard bore	2.165 in. (55 mm)
Stroke	1.693 in. (43 mm)
Full throttle operating range	4500-5500 rpm
Idle speed in forward gear	850 rpm
Cranking compression (minimum)	90 psi (620.6 kPa)
Lubricating system	Requires 50:1 premix
Initial idle mixture adjustment screw	1 1/2 turns out from lightly seated position
Ignition system	CD (capacitor discharge)
Ignition timing	
Idle speed	5° BTDC (not adjustable)
Wide open throttle	28-32° BTDC (not adjustable)

Table 3 GENERAL SPECIFICATIONS: 6-15 HP MODELS

Rated output	
6 hp	6 hp (4.5 kW)
8 hp	8 hp (5.9 kW)
9.9 hp	9.9 hp (7.4 kW)
15 hp	15 hp (11.2 kW)
Number of cylinder	2
Displacement	
6 and 8 hp	12.8 cu. in. (210 cc)
9.9 and 15 hp	16 cu. in. (262 cc)
Standard bore	
6 and 8 hp	2.125 in. (53.975 mm)
9.9 and 15 hp	2.375 in. (60.325 mm)
Stroke	1.8 in. (45.72 mm)
Full throttle operating range	
6 hp	4000-5000 rpm
8 hp	4500-5500 rpm
9.9 and 15 hp	5000-6000 rpm
Idle speed in forward gear	
6 hp	575-725 rpm
8-15 hp	675-775 rpm
Cranking compression	Approximately 100 psi (689.5 kPa)
Lubrication system	Requires 50:1 premix
Initial idle mixture screw adjustment	
6 hp	1–1 1/2 turns out
8-15 hp	1 1/4–1 3/4 turns out
Ignition system type	ADI (alternator driven capacitor discharge)
Ignition timing	
Idle speed	7-9° BTDC
Wide open throttle	36° BTDC

Table 4 GENERAL SPECIFICATIONS: 20 AND 25 HP AND 20 JET MODELS

Rated output	
20 Jet	20 hp (14.9 kW)
20 hp	20 hp (14.9 kW)
25 hp	25 hp (18.7 kW)
Number of cylinder	2
Displacement	24.4 cu. in. (400 cc)
Standard bore	2.562 in. (65.07 mm)
Stroke	2.362 in. (60 mm)

(continued)

Table 4 GENERAL SPECIFICATIONS: 20 AND 25 HP AND 20 JET MODELS (continued)

Recommended full throttle operating range	
20 Jet	5000-6000 rpm
20 hp	4500-5500 rpm
25 hp	5000-6000 rpm
Idle speed in forward gear	700-800 rpm
Fast idle adjustment (neutral gear)	
20 Jet (1998)	1150-1650 rpm
20 Jet (1999-on)	1300-1700 rpm
20 and 25 hp	1300-1700 rpm
Cranking compression	15 psi (103.4 kPa) maximum variation
Lubrication system	Requires 50:1 premix
Initial idle mixture screw adjustment	
20 Jet	1–2 turns out
20 hp	3/4–1 1/4 turns out
25 hp	1–1 1/2 turns out
Ignition system type	ADI (alternator driven capacitor discharge)
Ignition timing	
20 Jet (1998)	
Idle speed	2-6° BTDC (not adjustable)
Wide-open throttle	24-26° BTDC (not adjustable)
20 Jet (1999-on)	
Idle speed	5-7° BTDC
Wide-open throttle	24-26° BTDC
20 and 25 hp (1998-on)	
Idle speed	5-7° BTDC
Wide-open throttle	24-26° BTDC

Table 5 GENERAL SPECIFICATIONS: 30 AND 40 HP MODELS (TWO-CYLINDER)

Rated output	
30 hp	30 hp (22.4 kW) at 5000 rpm
40 hp	40 hp (29.8 kW) at 5250 rpm
Number of cylinder	2
Displacement	39.3 cu. in. (644 cc)
Standard bore	2.993 in. (76 mm)
Stroke	2.796 in. (71 mm)
Recommended full throttle operating range	
30 hp	4500-5500 rpm
40 hp	5000-5500 rpm
Idle speed in forward gear	700-800 rpm
Cranking compression	15 psi (103.4 kPa) maximum variation
Lubrication system	Variable ratio oil injection or 50:1 premix
Initial idle mixture screw adjustment	1 1/4–1 3/4 turns out
Ignition system type	CDM (capacitor discharge module)
Ignition timing	
Idle speed	7-9° BTDC
Wide-open throttle	22-28° BTDC

Table 6 GENERAL SPECIFICATIONS: 50-60 HP MODELS (THREE-CYLINDER)

Rated output	
40 hp	40 hp (29.8 kW)
50 hp	50 hp (37.3 kW)
55 hp	55 hp (41 kW)
60 hp	60 hp (44.7 kW)
Number of cylinder	3
Displacement	39.3 cu. in. (644 cc)
Standard bore	2.993 in. (76 mm)
Stroke	2.796 in. (71 mm)

(continued)

Table 6 GENERAL SPECIFICATIONS: 50-60 HP MODELS (THREE-CYLINDER) (continued)

Recommended full throttle operating range	5000-5500 rpm
Idle speed in forward gear	650-700 rpm
Cranking compression	15 psi (103.4 kPa) maximum variation
Lubrication system	Variable ratio oil injection or 50:1 premix
Initial idle mixture screw adjustment	
40-55 hp (three-cylinder)	1–1 1/2 turns out
60 hp (manual start)	1–1 1/2 turns out
60 hp (electric start)	7/8–1 3/8 turns out
Ignition system type	CDM (capacitor discharge module)
Ignition timing	
Idle timing	
Cranking speed	0° TDC
Running at idle speed	2° ATDC-2° BTDC
Maximum spark advance timing	
Cranking speed	
40 hp, 50 hp and 60 hp	24° BTDC
55 hp	18° BTDC
Running at wide-open throttle	
40 hp, 50 hp and 60 hp	22° BTDC
55 hp	16° BTDC

5

Chapter Six

Fuel System

This chapter includes removal, overhaul, installation and adjustment procedures for fuel pumps, carburetors, reed valves, fuel primer valves, thermal air valves, fuel enrichment valves, bleed (recirculation) systems, portable fuel tanks and connecting lines.

Specific torque values are in **Table 1**. Standard torque values are in **Table 2**. Use the standard torque value for fasteners not in **Table 1**. Carburetor specifications are in **Table 3**. Reed valve specifications are in **Table 4**. **Tables 1-4** are located at the end of this chapter.

NOTE

Metric and U.S. standard fasteners are used on newer model outboards. Always match a replacement fastener to the original. Never run a tap or thread chaser in a hole or over a bolt without first verifying the thread size and pitch. Manufacturer's parts catalogs list every fastener by diameter, length and pitch. Always have the engine model and serial numbers when ordering parts from a dealership.

FUEL PUMP

The 2.5 and 3.3 hp models do not require the use of a fuel pump. They use gravity to supply fuel from the integral fuel tank to the carburetor fuel bowl.

All 4 hp and larger models use a diaphragm-type fuel pump operated by crankcase pressure and vacuum pulses. The fuel pump is an integral part of the carburetor on 4-25 hp models. A remote mounted fuel pump is used on 30-60 hp models.

These types of diaphragm fuel pumps cannot move large quantities of fuel while cranking the engine. Fuel must be transferred to the carburetor by a manually operated primer bulb installed in the fuel supply hose.

Mechanical fuel pumps are operated by crankcase pressure and vacuum pulses created by movement of the piston(s). The pulses reach the fuel pump through either an external hose or internal passages in the crankcase.

Upward piston movement creates low pressure in the crankcase and against the pump diaphragm. This low

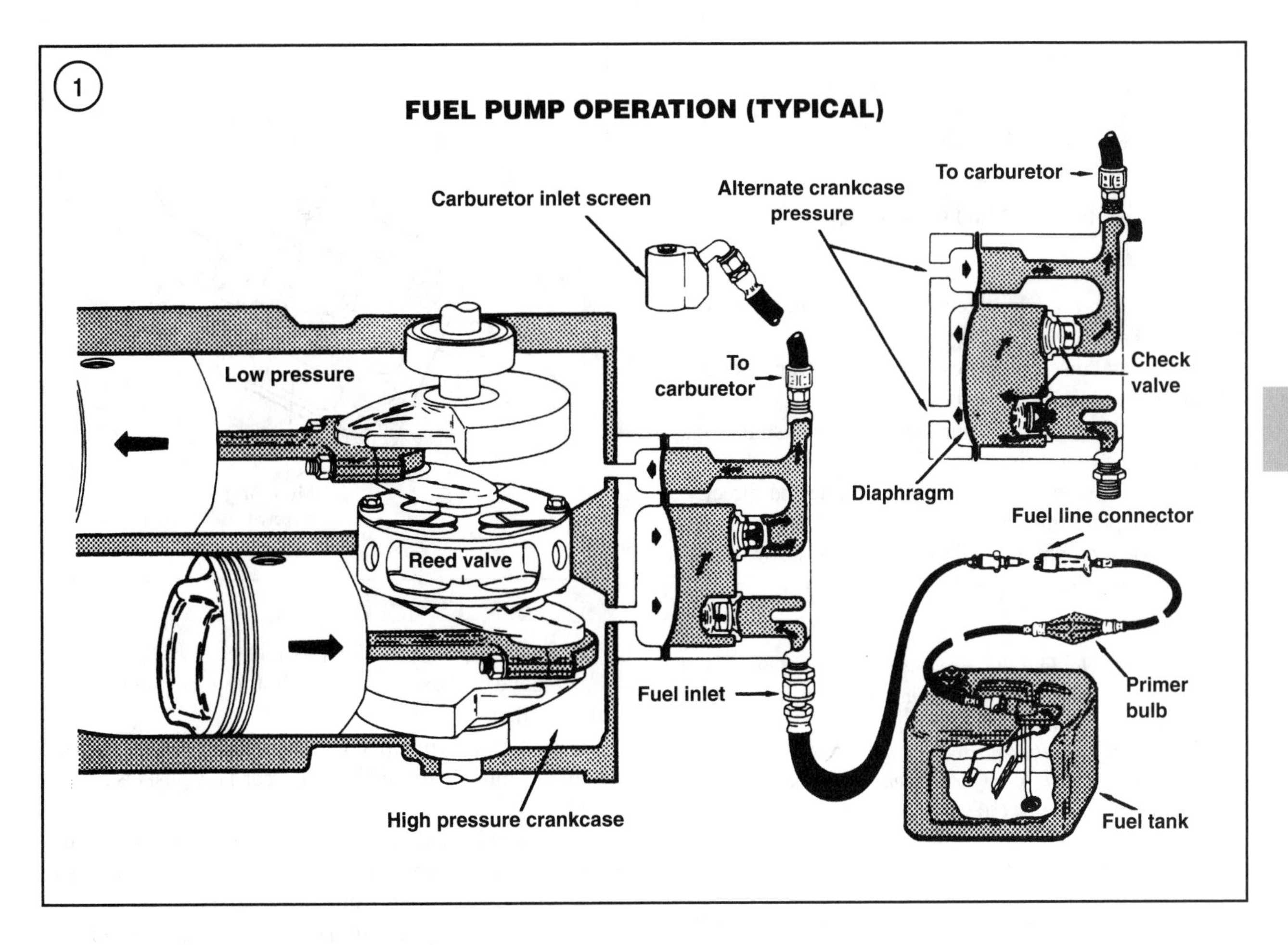

pressure opens the inlet check valve in the pump, drawing fuel from the supply line into the pump.

Downward piston movement creates high pressure in the crankcase and against the pump diaphragm. This pressure closes the inlet check valve and opens the outlet check valve, forcing the fuel out of the fuel pump and into the carburetor(s) or vapor separator. **Figure 1** shows the general operating principles of a pulse driven fuel pump.

Not all Mercury/Mariner diaphragm-type fuel pumps use the pressure/vacuum pulses from two-cylinders as shown in **Figure 1**. Many pumps operate off one cylinder's pressure and vacuum pulses.

NOTE
If the cylinder(s) that supplies crankcase pressure and vacuum to a fuel pump mechanically fails, all of the cylinders starve for fuel. Check the compression of the engine if low fuel pressure occurs and no faults are found with the fuel pump.

Mercury/Mariner fuel pumps have a simple design and operate reliably. Diaphragm failures are the most common problem, although sour fuel, fuel with excessive alcohol or other additives can cause check valve failure. Refer to Chapter Four for fuel recommendations.

Integral fuel pump models (4-25 hp) cannot be tested separately from the carburetor. If fuel is reaching the carburetor inlet and the engine is still suffering a fuel related symptom, disassemble, clean, inspect and rebuild the carburetor and integral fuel pump. Do not service only the carburetor section or only the fuel pump section of these models. Fuel pump service on these models is covered in the carburetor section of this chapter.

Test the remote fuel pump used on 30-60 hp models as described in Chapter Three. Check the fuel delivery hose for restrictions and air leakage by connecting a vacuum gauge and a piece of clear hose to the fuel pump inlet. Check fuel pump output using a pressure gauge connected to the fuel pump outlet.

CAUTION
Fuel pump assemblies and internal fuel pump components vary between models. Use only the correct fuel pump or fuel pump components when replacing or repairing

the fuel pump. An incorrect fuel pump or internal components can cause reduced fuel flow to the engine, resulting in poor performance or power head failure.

Removal/Installation (30-60 hp Models)

Refer to **Figure 2** for this procedure.

1. On 40-60 hp models (three-cylinder), remove the oil reservoir as described in Chapter Twelve.
2. Remove and discard the tie-strap clamps from the fuel delivery hose at the fuel pump.
3. Label the hoses at the pump for correct reinstallation. The inlet and outlet fitting of the fuel pump are marked on the fuel pump cover plate. If a hose is connected directly to the cover plate through a 90° fitting, it is a pulse line from the boost diaphragm. Disconnect all hoses from the pump assembly. See **Figure 2**.

NOTE

If two of the fuel pump screws are slotted or Phillips head, they are the mounting screws. The two screws that hold the fuel pump components together are always hex-head screws. If all four fuel pump screws are hex-head, look at the rear of the pump to determine which two screws pass through the mounting gasket into the power head.

4. Remove the two screws securing the fuel pump assembly to the power head and remove the pump.
5. Carefully clean the fuel pump-to-power head gasket from the power head and fuel pump.
6. Install a new gasket between the fuel pump and power head.
7. Install the pump to the power head and secure it with the two screws. Tighten the pump mounting screws to the specification in **Table 1**.
8. Reconnect the fuel inlet and outlet hoses to the fuel pump. Secure the hoses with new tie-strap clamps.
9. On models so equipped, install and secure the pulse hose to the pump cover fitting using a new tie-strap clamp.
10. On 40-60 hp three-cylinder models, install the oil reservoir and bleed air from the oil injection system as described in Chapter Twelve.

Disassembly (30-60 hp Models)

NOTE

Replace all fuel pump gaskets and diaphragms anytime the fuel pump is disassembled. If the check valves are removed, replace them.

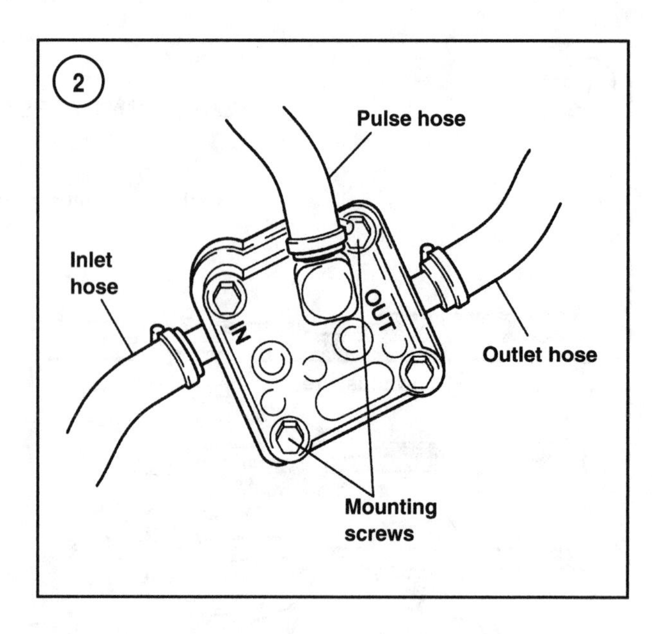

Refer to **Figure 3** for this procedure.

1. Remove the hex-head screws holding the pump assembly together.
2. Separate the pump cover, gaskets and diaphragms from the pump body and base. Discard the gaskets and diaphragms.
3. With needlenose pliers, remove the check valve retainers (4, **Figure 3**) from the pump body. Remove the plastic discs and check valves from the retainers.
4. Remove the cap (9, **Figure 3**) and spring (10) from the pump cover.
5. Remove the cap (5, **Figure 3**) and spring (6) from the pump body.

Cleaning and Inspection (30-60 hp Models)

1. Clean the pump components in a suitable solvent and dry them with compressed air.
2. Inspect the plastic discs for cracks, holes or other damage. Replace as required.
3. Inspect the pump base, pump body and pump cover for cracks, distortion, deterioration or other damage. Replace as required.

Check Valve Installation (30-60 hp Models)

1. Insert a new check valve retainer into the plastic disc, then into a new check valve. See **Figure 4**. Repeat the step for the other check valve assembly.
2. Lubricate the check valve retainers with motor oil. Insert the check valve and retainer assemblies into the pump body. See **Figure 4**.

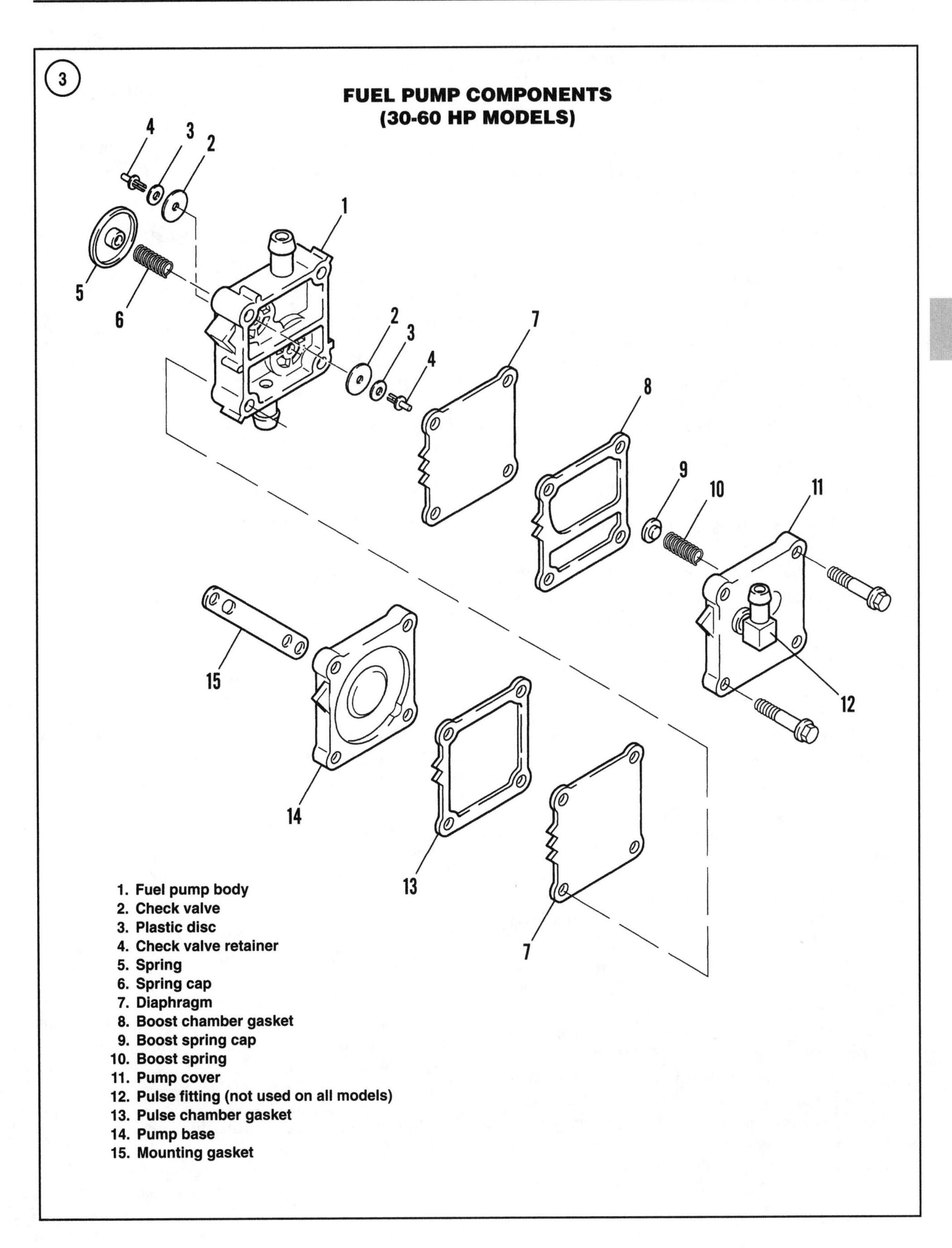
3
FUEL PUMP COMPONENTS
(30-60 HP MODELS)
1
2
3
4
5
6
7
8
9
10
11
12
13
14
15
1. Fuel pump body
2. Check valve
3. Plastic disc
4. Check valve retainer
5. Spring
6. Spring cap
7. Diaphragm
8. Boost chamber gasket
9. Boost spring cap
10. Boost spring
11. Pump cover
12. Pulse fitting (not used on all models)
13. Pulse chamber gasket
14. Pump base
15. Mounting gasket

3. Bend the check valve retainer stem from side to side, until the stem breaks off flush with the retainer cap. Do not discard the stem. See **Figure 5**. Repeat this step for the other retainer.
4. Insert the broken retainer stem into the retainer as shown in **Figure 6**. Use a small hammer and punch to tap the stem into the retainer until it is flush with the retainer cap.

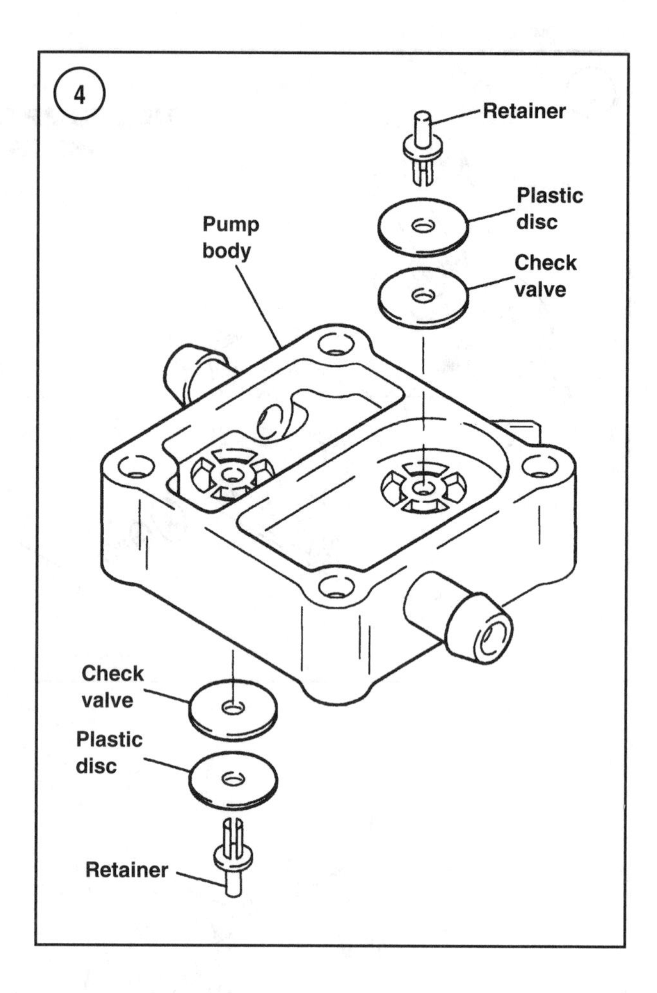

Assembly (30-60 hp Models)

NOTE
The fuel pump covers, body gaskets and diaphragms have one or more V tabs on one side for directional reference during assembly. The V tabs must be aligned. To ensure correct alignment of pump components, use 1/4 in. bolts or dowels as guides. Insert the guides through the pump mounting screw holes.

1. Refer to **Figure 3** and reassemble the pump. Do not use gasket sealer on the pump gaskets or diaphragms. Make sure all pump components are properly aligned, and the springs and spring caps are properly assembled. Spring caps should always push against the diaphragm.
2. Remove one of the alignment bolts or dowels and install a hex-head screw finger-tight. Remove the other alignment bolt or dowel and install the other hex-head screw finger-tight.
3. Tighten both hex-head screws to the specification in **Table 1**.

CARBURETORS

Identification

All Mercury/Mariner carburetors have a carburetor series identification number cast into the carburetor body and a carburetor model identification number stamped into the front or rear flange. The model identification number on 6-60 hp models is in the following format: model number, carburetor location (on multi-carburetor models) and Julian date code. For example, a WMV-7-3-3246 number indicates a WMV series carburetor, model No. 7, mounted in the third carburetor location from the top, built on the 324th day of 1996. Carburetor location refers to the position of the carburetor on the power head. This is not always the same as the cylinder number. Current (1999-2001) two-stroke models use the following carburetors:
1. *2.5 and 3.3 hp models*—One slide valve carburetor.
2. *4 and 5 hp models*—One conventional butterfly valve carburetor with an integral fuel pump.
3. *6-25 hp models*—One WMC series carburetor with an integral fuel pump.
4. *30 and 40 hp two-cylinder models*—Two WME series carburetors.
5. *40-60 hp three-cylinder models*—Three WME series carburetors.

Adjustment (Static)

Refer to **Table 3** for carburetor specifications. A carburetor scale (part No. 91-36392) is recommended for all float adjustments. Refer to *Assembly* in this section for float adjustments.

Carburetor Adjustments (Running)

After servicing or repairing the carburetor(s), perform *Synchronization and Linkage Adjustments* as described in Chapter Five.

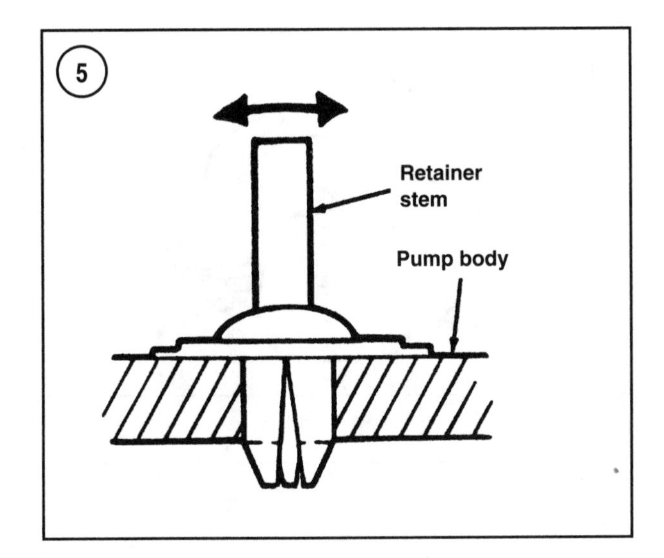

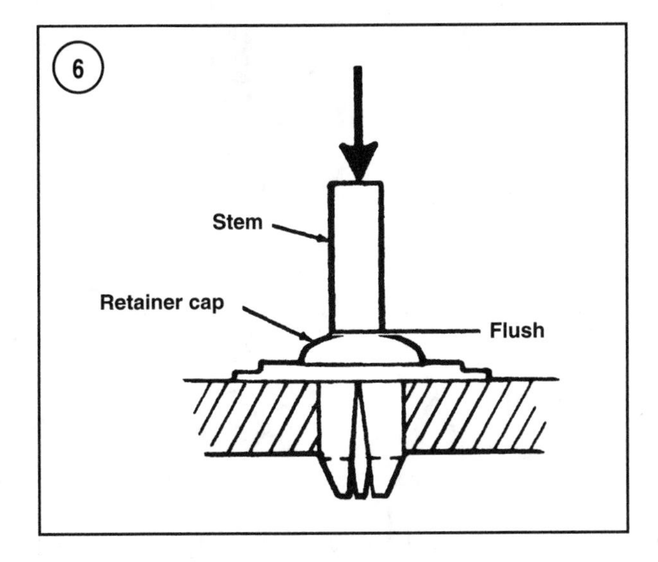

High Altitude Compensation

If an outboard motor is operated at higher altitudes, several factors affect the engine adjustment and operation. As the altitude increases, the air becomes less dense. Since an outboard motor is essentially an air pump, this reduces the efficiency of the engine, reducing the horsepower output proportionally to the air density.

The less dense air affects the engine's carburetor calibration, causing the air/fuel mixture to become richer. Richer mixtures cause the engine to produce less horsepower and lead to fouled spark plugs, reduced fuel economy and accelerated carbon buildup in the combustion chamber. Adjustments to both the carburetor and the propeller may be necessary to bring the engine back to maximum operating efficiency.

CAUTION

High altitude main jets are smaller than the standard main jets and correct the air/fuel ratio by making the mixture leaner. If an engine is rejetted for high altitude operation, it must again be rejetted before operating at a lower altitude. Serious power head damage can occur from operating the engine on an excessively lean air/fuel mixture.

The air/fuel ratio changes any time the air density changes. Adjustments to the carburetor will depend greatly upon the amount of change in air density. All Mercury Mariner carbureted motors come calibrated to operate efficiently between sea level and 2500 ft. (762m). Mercury Mariner recommends rejetting the main jet (high speed jet at altitudes of 5000 ft. (1524m) or higher. While some tables show rejetting specifications beginning at 2500ft. (762m), it is not mandatory to rejet until 5000 ft (1524 m) or higher.

Changes in altitude normally require more than just rejetting. It will require the idle mixture and idle speed to be adjusted, as well as requiring full-throttle engine speed verification (Chapter Five). When unsure of proper adjustments, refer to a reputable dealership familiar with carburetor adjustment and jetting for the altitude at which the engine is to be operated.

CAUTION

Regardless of the altitude, the engine must have a propeller that allows it to operate within, (preferably near the upper limit) of the full-throttle engine speed range. See ***Full-throttle Engine Speed Verification*** *in Chapter five. Change the propeller pitch and diameter as necessary to maintain the full-throttle engine speed capability.*

Cleaning and Inspection (All Models)

CAUTION

Do not remove the throttle plate(s) and throttle shaft unless absolutely necessary and all replacement parts are available. Reinstall each throttle plate in its original bore and orientation. Make sure the throttle plate(s) open and close fully before tightening the screws. Apply Locktite 271 Threadlocking adhesive (part No. 92-809819) to the throttle plate screws.

1. Thoroughly and carefully remove gasket material from all mating surfaces. Do not nick, scratch or damage the mating surfaces.

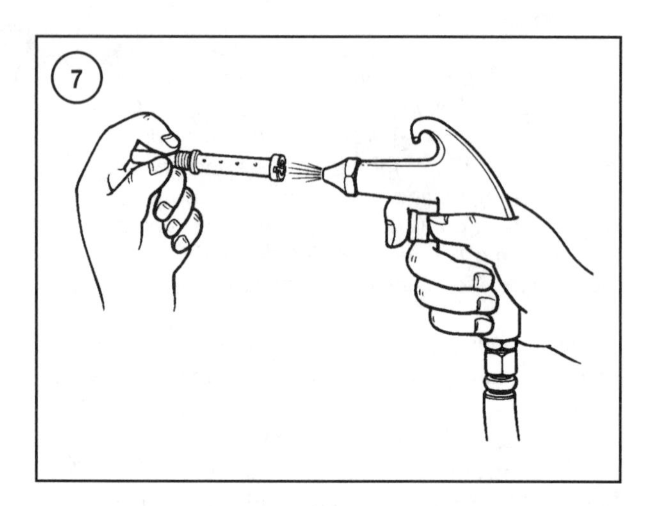

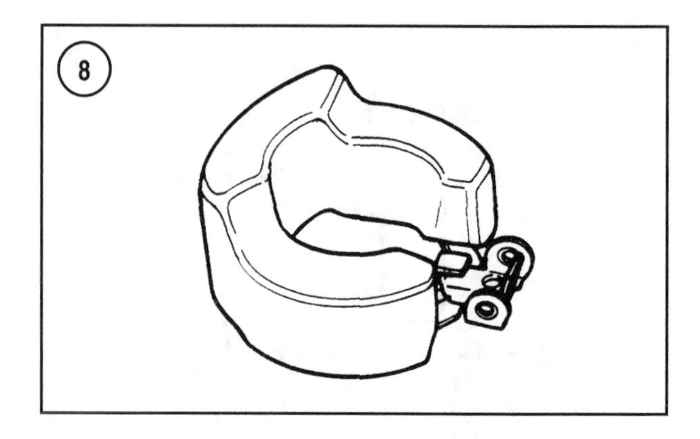

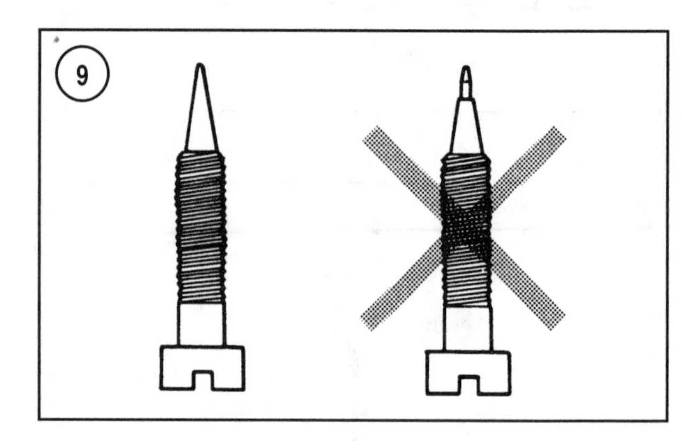

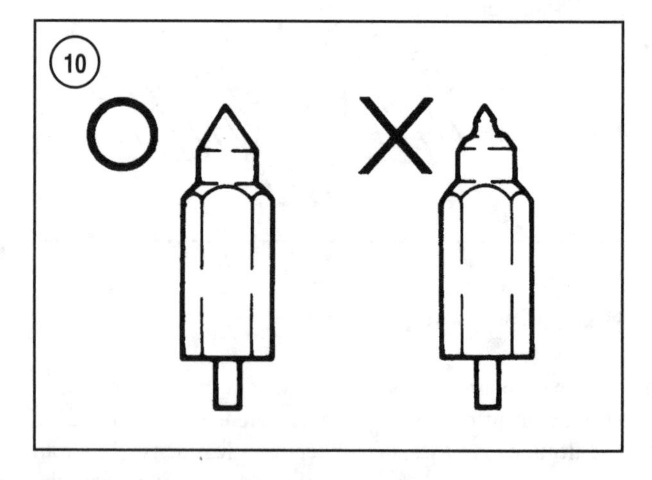

2. Clean the carburetor body and metal parts using an aerosol carburetor and choke cleaning solvent available at automotive parts stores to remove gum, dirt and varnish.
3. Rinse the carburetor components in clean solvent and dry with compressed air (**Figure 7**). Be sure to blow out all orifices, nozzles and passages thoroughly.

CAUTION
Do not use wire or drill bits to clean carburetor passages. This alters calibration and ruins the carburetor.

4. Check the carburetor body casting for stripped threads, cracks or other damage.
5. Check the fuel bowl for distortion, corrosion, cracks, blocked passages or other damage.
6. Check the float (**Figure 8**) for fuel absorption, deterioration or other damage. Check the float arm for wear in the hinge pin and inlet needle contact areas. Replace the float if necessary.
7. Check the idle mixture screw(s) tip (**Figure 9**) for grooving, nicks or other damage. Replace the idle mixture screw(s) if necessary.
8. Check the inlet needle (**Figure 10**) and seat for excessive wear. Some carburetors do not have a replaceable inlet needle seat. Replace the needle and seat as an assembly if both parts are replaceable.

Integral fuel pump models (4-25 hp models)

1. Remove diaphragm or gasket material from the mating surfaces. Do not scratch, nick or damage the mating surfaces.
2. Clean all components in a suitable solvent and dry with compressed air.
3. Inspect the pump body, cover and carburetor mounting flange for nicks, cracks or other damage.
4. Inspect the reed check valves for bends, cracks or other damage. Replace as required.

2.5 and 3.3 hp Models

Refer to **Figure 11** for an exploded view of the carburetor and **Table 3** for specifications. The 2.5 and 3.3 hp carburetor uses gravity to supply fuel from the integral fuel tank to the carburetor fuel bowl.

(11)

CARBURETOR COMPONENTS (2.5 AND 3.3 HP MODELS)

1. Throttle cable
2. Retainer nut
3. Bracket
4. Mixing chamber cover
5. Spring
6. Retainer
7. E-clip
8. Jet needle
9. Throttle valve
10. Inlet valve seat
11. Body
12. Throttle lever
13. Idle speed screw
14. Spring
15. Choke valve
16. Inlet valve needle
17. Gasket
18. Main nozzle
19. Main jet
20. Hinge pin
21. Float arm
22. Float
23. Float bowl
24. Screw
25. Choke lever
26. Gasket
27. Clamp

Removal/installation

1. Remove the knobs from the choke and throttle levers.
2. Remove the front screws and loosen the rear screw securing the manual rewind starter assembly to the power head.
3. Remove the screws from the front intake cover. Lift the front of the manual rewind starter and remove the front intake cover from the carburetor.
4. Close the fuel shutoff valve and disconnect the fuel supply hose from the carburetor.
5. Loosen the carburetor mounting clamp (27, **Figure 11**) and remove the carburetor.

NOTE
It may be necessary to remove the lower cowling to remove the carburetor.

6. Make sure the carburetor gasket (26, **Figure 11**) is properly located inside the carburetor throat.
7. Install the carburetor.
8. Hold the carburetor against the power head and tighten the mounting clamp.
9. Reconnect the fuel delivery hose.
10. Install the front intake cover, then reinstall the choke and throttle lever knobs.
11. Install the front screws on the manual rewind starter. Securely tighten the screws.
12. Refer to Chapter Five for carburetor and linkage adjustment procedures.

Disassembly

Refer to **Figure 11** for the following procedure.

1. Remove the float bowl attaching screws. Separate the float bowl (23, **Figure 11**) from the carburetor. Remove the float bowl gasket.
2. Invert the carburetor and remove the float from the float arm (21, **Figure 11**).
3. Remove the float hinge pin and the float arm (20 and 21, **Figure 11**).
4. Remove the inlet valve needle.
5. Remove the main jet (19, **Figure 11**) using a suitable wide-blade screwdriver.

NOTE
*The main nozzle (18, **Figure 11**) is held in place by the main jet. See **Figure 12**.*

6. Invert the carburetor and catch the main nozzle as it falls out of the carburetor body.
7. Remove the throttle lever.

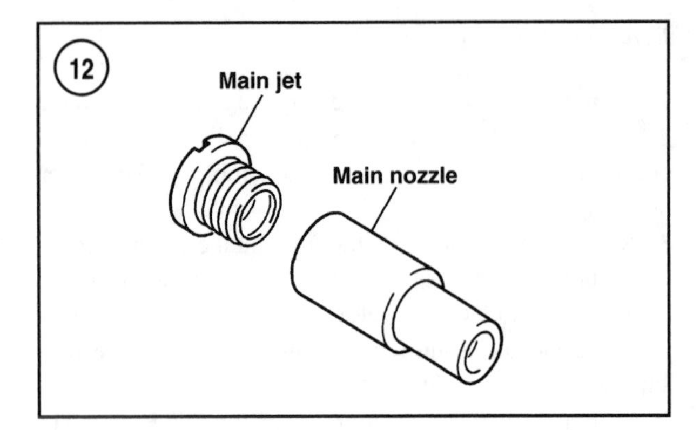

8. Loosen the retainer nut (2, **Figure 11**) several turns, then unscrew the mixing chamber cover (4). Lift the throttle valve assembly (1-9, **Figure 11**) out of the carburetor body.
9. Compress the throttle valve spring (5, **Figure 11**) and disconnect the throttle cable (1) from the throttle valve (9).
10. Remove the jet needle (8, **Figure 11**) and jet retainer (6) from the throttle valve. Do not lose the jet needle E-clip (7, **Figure 11**).
11. Lightly seat the idle speed screw (13, **Figure 11**). Note the turns required for reference during assembly. Remove the idle speed screw and spring to complete disassembly.
12. Refer to *Cleaning and Inspection (All Models)* in this chapter.

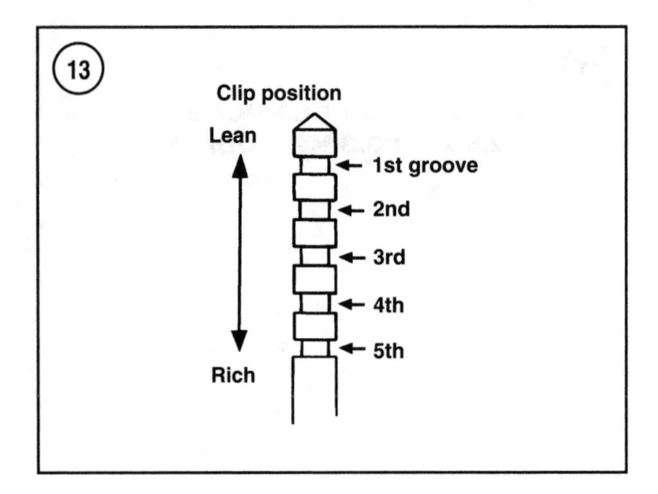

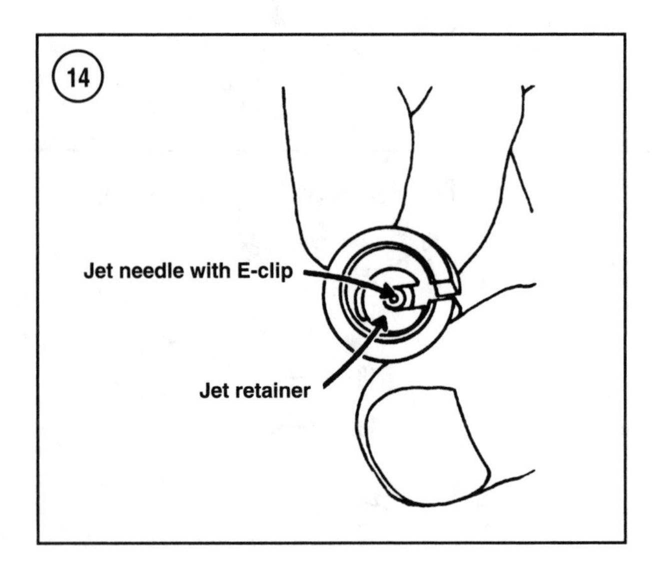

Assembly

Refer to **Figure 11** for the following procedure.
1. Slide the spring onto the idle speed screw. Install the screw into the carburetor body until lightly seated, then back out the number of turns noted during disassembly.
2. Check the jet needle E-clip position. The factory setting is the second groove as shown in **Figure 13**.
3. Insert the jet needle (8, **Figure 11**) into the throttle valve (9). Place the needle retainer (6, **Figure 11**) into the throttle valve over the needle E-clip (7). Align the retainer slot with the slot in the throttle valve as shown in **Figure 14**.
4. Reassemble the throttle valve components as follows:
   a. Place the throttle valve spring (5, **Figure 11**) over the throttle cable (1).
   b. Compress the spring, then slide the throttle cable anchor through the slot and into position in the throttle valve as shown in **Figure 15**.
5. Align the slot in the throttle valve with the alignment pin in the carburetor body. Insert the throttle valve assembly into the body and tighten the mixing chamber cover (4, **Figure 11**). Securely tighten the cover retainer nut (2, **Figure 11**).
6. Reinstall the throttle lever (12, **Figure 11**).
7. Install the main nozzle and main jet (**Figure 12**).
8. Install the inlet valve needle (16, **Figure 11**).
9. Install the float arm and hinge pin (20 and 21, **Figure 11**).
10. With the float bowl removed, invert the carburetor and measure from the mating surface of the carburetor body, with the fuel bowl gasket installed, to the float arm as shown in **Figure 16**. The measurement must equal the float level specification in **Table 3**. If necessary, bend the float arms evenly to obtain the specified measurement.
11. Install the float (22, **Figure 11**).
12. Install the float bowl with a new gasket. Securely tighten the bowl screws.

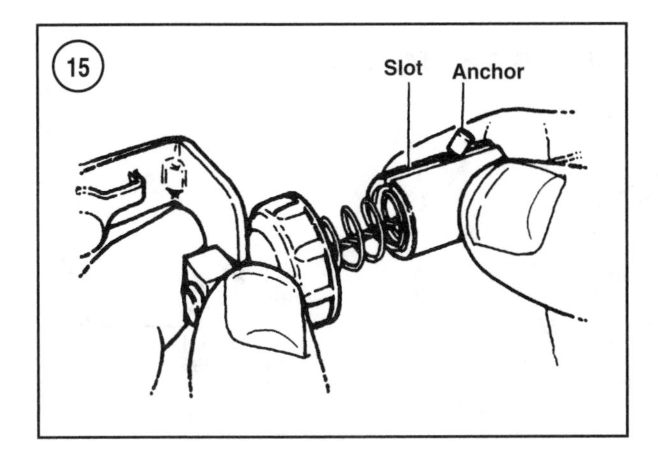

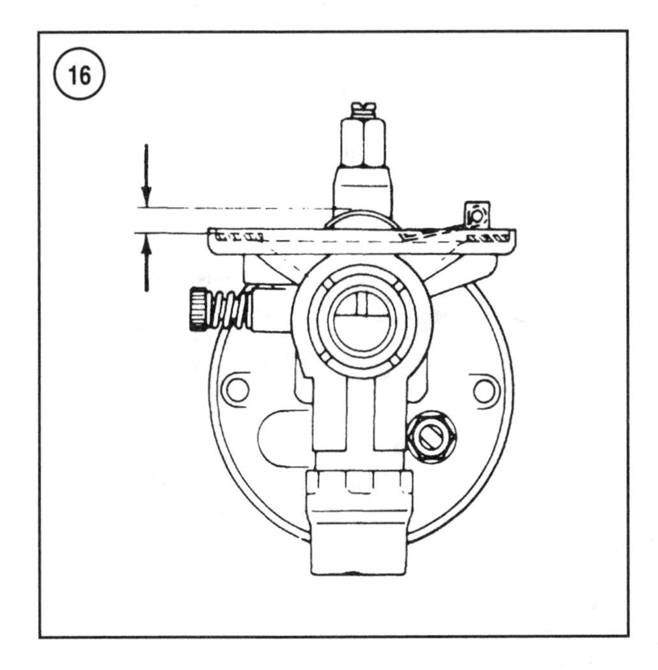

4 and 5 hp Models with Integral Fuel Pump

Although the fuel pump portion of the carburetor can be serviced without removing or disassembling the carburetor, completely disassemble the fuel pump and the carburetor if either needs service. Refer to **Figure 17** for the following procedures.

Removal/Installation

1. Disconnect the fuel hose from the carburetor fuel pump inlet fitting.
2. Loosen the screw (1, **Figure 18**) holding the throttle wire (2) to the carburetor throttle arm. Remove the wire from the throttle arm.
3. Disconnect the choke link rod (3, **Figure 18**) from the carburetor.
4. Remove the screws (36, **Figure 17**) securing the baffle cover (35) to the front of the carburetor. Remove the cover.
5. Remove the bolts (1, **Figure 19**) holding the baffle cover bracket (2) and the carburetor to the crankcase cover. Remove the bracket, carburetor and gasket (31, **Figure 17**). Carefully clean any remaining gasket material from the carburetor and power head.
6. Install the carburetor, baffle cover bracket and a new gasket to the crankcase cover. Tighten the bolts (1, **Figure 19**) evenly and securely.
7. Install the baffle cover (35, **Figure 17**) and secure it with the screw and washers (36 and 37).
8. Connect the choke link rod (3, **Figure 18**) to the carburetor.
9. Insert the throttle wire (2, **Figure 18**) into the hole in the throttle arm retainer. Adjust the throttle wire as follows:
   a. Back the idle speed screw (28, **Figure 17**) off the throttle arm.
   b. Turn the idle speed screw inward until it just contacts the throttle arm, then turn it inward an additional two turns. With the throttle arm in this position, pull upward on the throttle wire to remove slack, then securely tighten the retainer screw (1, **Figure 18**).
10. Refer to Chapter Five for carburetor and linkage adjustment procedures.

Disassembly

Refer to **Figure 17** for the following procedure.

1. Remove the idle mixture screw (26, **Figure 17**) and spring (27).
2. Remove the two screws holding the mixing chamber cover (43, **Figure 17**) to the carburetor. Remove the cover.
3. Remove the plug (44, **Figure 17**) and packing (45) from the carburetor main body mixing chamber pockets.
4. Remove the screws (25, **Figure 17**) securing the fuel pump assembly to the carburetor float bowl. Separate the pump cover from the pump body.
5. Remove the screws (13, **Figure 17**) securing the float bowl (11) to the carburetor body. Remove the float bowl (11, **Figure 17**).
6. Remove the screw (**Figure 20**) holding the float assembly in the carburetor.
7. Remove the hinge pin and inlet needle valve (6 and 9, **Figure 17**) from the float.

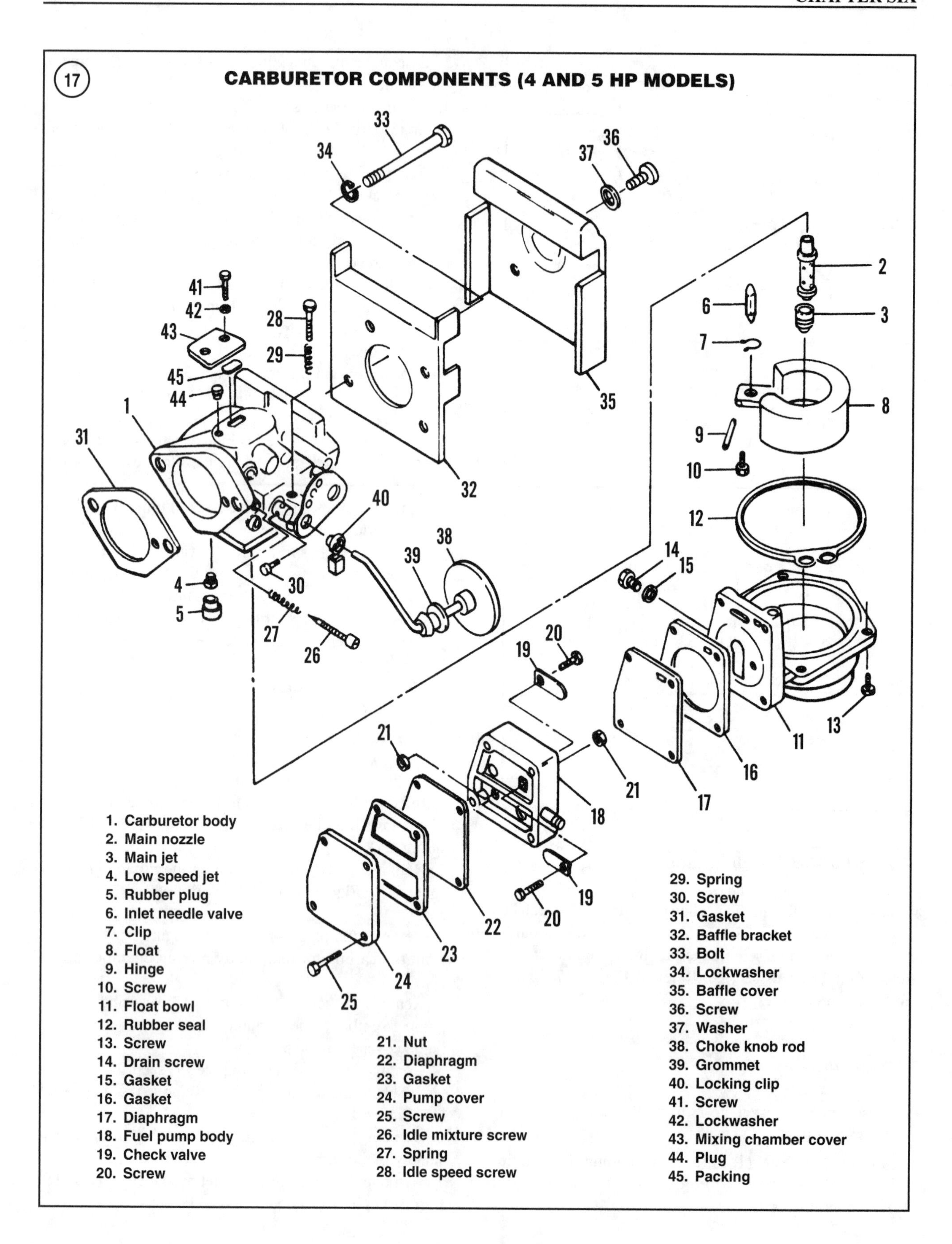
17
CARBURETOR COMPONENTS (4 AND 5 HP MODELS)
1. Carburetor body
2. Main nozzle
3. Main jet
4. Low speed jet
5. Rubber plug
6. Inlet needle valve
7. Clip
8. Float
9. Hinge
10. Screw
11. Float bowl
12. Rubber seal
13. Screw
14. Drain screw
15. Gasket
16. Gasket
17. Diaphragm
18. Fuel pump body
19. Check valve
20. Screw
21. Nut
22. Diaphragm
23. Gasket
24. Pump cover
25. Screw
26. Idle mixture screw
27. Spring
28. Idle speed screw
29. Spring
30. Screw
31. Gasket
32. Baffle bracket
33. Bolt
34. Lockwasher
35. Baffle cover
36. Screw
37. Washer
38. Choke knob rod
39. Grommet
40. Locking clip
41. Screw
42. Lockwasher
43. Mixing chamber cover
44. Plug
45. Packing

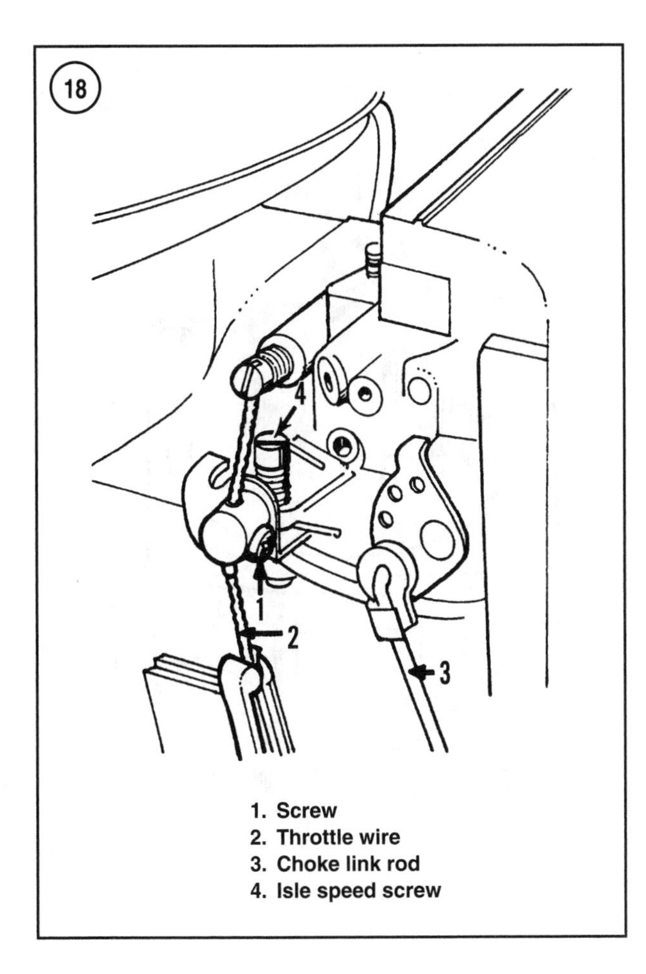

1. Screw
2. Throttle wire
3. Choke link rod
4. Isle speed screw

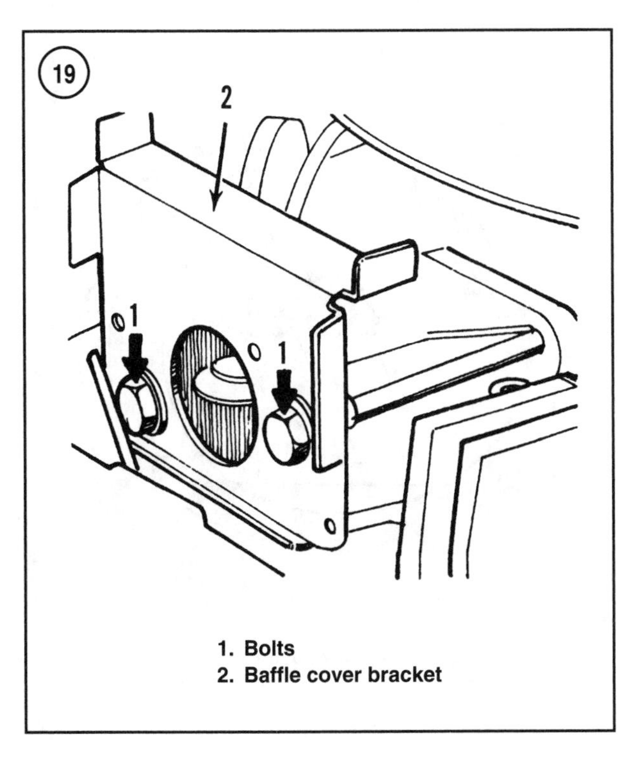

1. Bolts
2. Baffle cover bracket

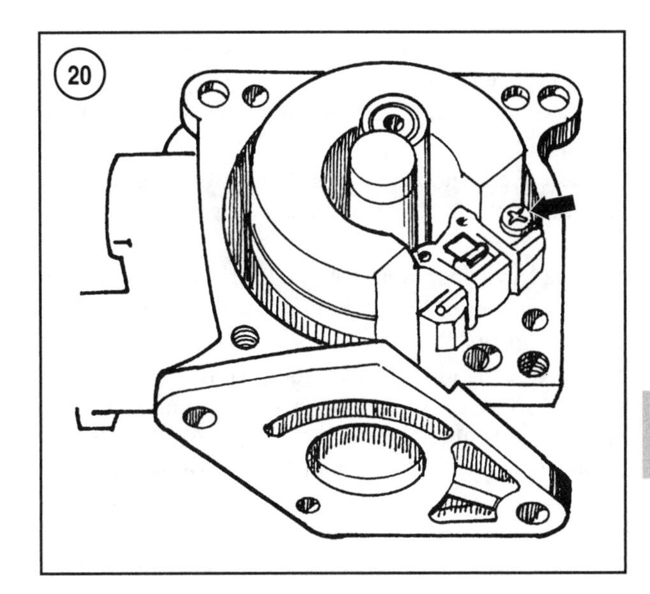

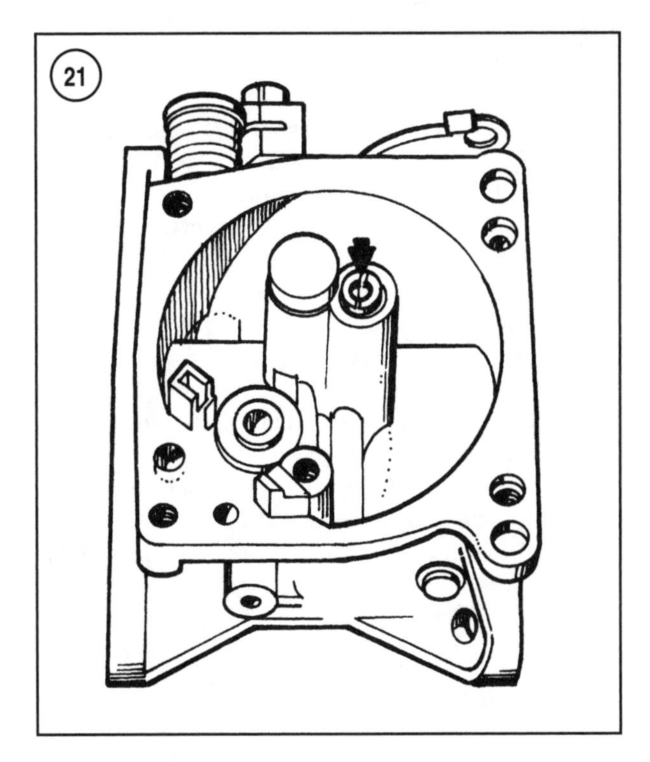

8. Remove the main jet (**Figure 21**). Remove the main nozzle located under the main jet (**Figure 22**).
9. Remove the plug (1, **Figure 23**) and low speed jet (2).
10. Refer to *Cleaning and Inspection (All Models)* in this chapter.

Assembly

Refer to **Figure 17** for the following procedure.

1. To reassemble the carburetor, first install the low speed jet (2, **Figure 23**) and plug (1).
2. Install the main nozzle (**Figure 22**) and the main jet (**Figure 21**).
3. Place the float hinge pin (9, **Figure 17**) into the float arm. Install the inlet needle valve (6, **Figure 17**) onto the float.
4. Install the float, hinge pin and inlet valve assembly onto the carburetor body. Secure the hinge pin with the screw (**Figure 20**).
5. Measure the distance from the float to the float bowl mating surface as shown in **Figure 24**. The measurement must equal the float level specification in **Table 3**. If necessary, adjust by bending the float tang (**Figure 25**).
6. Place a new rubber float bowl seal (12, **Figure 17**) into the bowl groove.
7. Mount the float bowl on the carburetor body. Tighten the screws evenly and securely.
8. Reassemble the fuel pump assembly as follows:
   a. Install new diaphragms on both sides of the pump body.
   b. Install the cover (24, **Figure 17**) onto the pump body (18) using a new gasket (23) and diaphragm (22). Insert the pump mounting screws through the cover and body to maintain component alignment.
   c. Install the pump assembly onto the carburetor float bowl (11, **Figure 17**) using a new gasket (16) and diaphragm (17).
   d. Tighten the screws evenly and securely.
9. Install the idle mixture screw and spring (26 and 27, **Figure 17**). Turn the screw in until lightly seated, then back it out the number of turns in **Table 3**.
10. Install the plug (44, **Figure 17**) and packing (45) into the carburetor main body mixing chamber pockets.
11. Install the mixing chamber cover (43, **Figure 17**) onto the top of the carburetor body. Install the cover screws and tighten them securely.

6-25 hp Models (WMC Series)

This carburetor uses an integral fuel pump and a diaphragm-operated primer system. Remote control models use an electric solenoid to close an air bleed valve, enriching the air/fuel mixture for cold engine starting. The throttle plate must be fully closed for the system to function correctly. Refer to Chapter Three for troubleshooting procedures on the electric primer solenoid.

Although the fuel pump portion of the carburetor can be serviced without removing or disassembling the carburetor, completely disassemble the fuel pump and the carburetor if either needs service. Refer to **Table 3** for

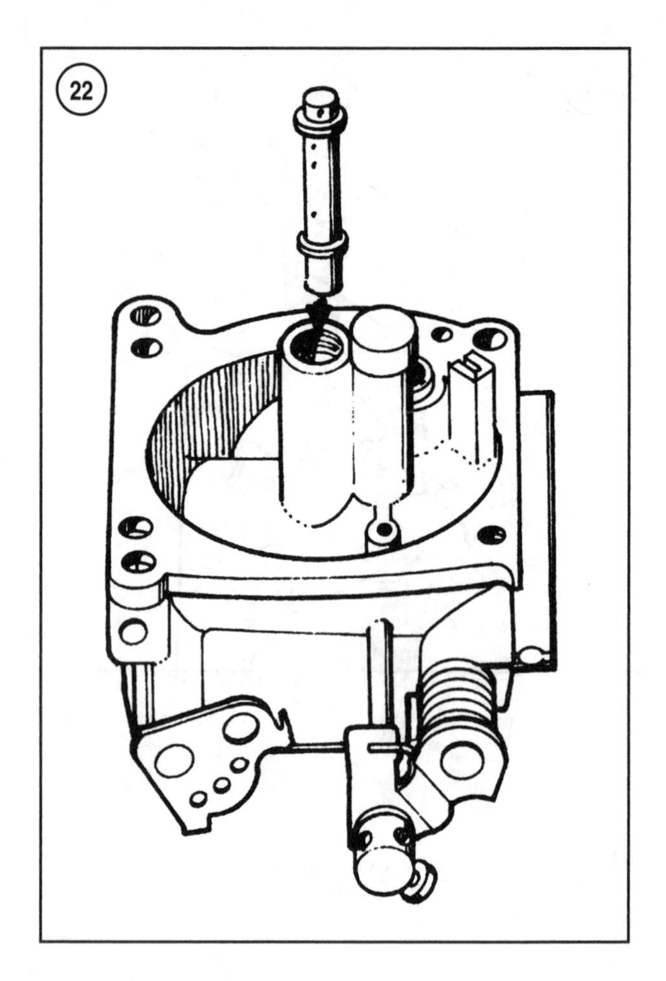

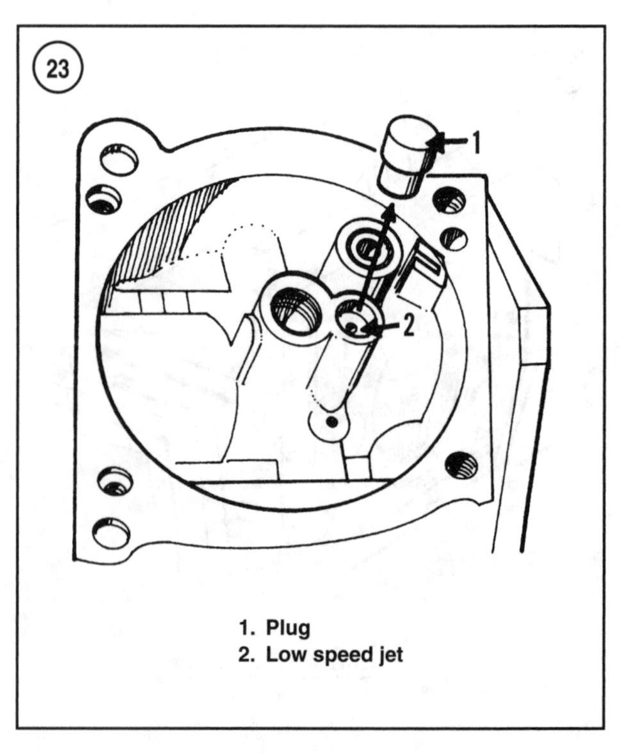

1. Plug
2. Low speed jet

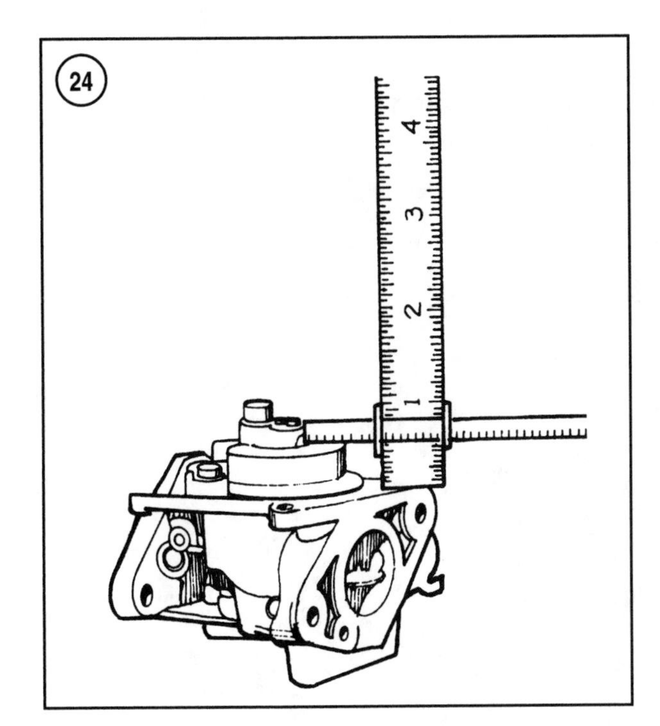

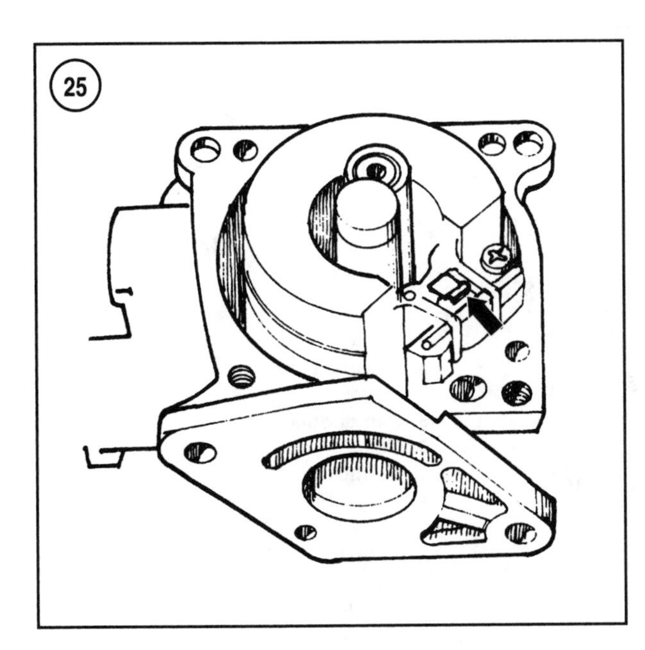

carburetor specifications and **Table 1** for torque specifications.

Figure 26 shows a typical WMC carburetor with the air intake cover removed. The fuel pump cover plate, throttle shaft arm and top cover plate vary from model to model. Remote control models have an electric primer solenoid mounted in place of the top cover. Throttle shafts have either an adjustable idle speed screw, a fixed throttle cam roller or an adjustable throttle cam follower.

Removal/Installation

NOTE
The air intake cover is removed with the carburetor.

1. On electric start models, disconnect the negative battery cable.
2. Disconnect and ground the spark plug leads to the power head.
3. Loosen the primer linkage retaining screw (53, **Figure 26**).
4. Remove the retainer (51, **Figure 26**).
5. Depress the primer lever (30, **Figure 26**) and pull the primer knob, bezel and block from the bottom cowl.
6. Disconnect the wire link from the fast idle cam (29, **Figure 26**).
7. Disconnect the fuel hose from the carburetor. Plug the hose to prevent leakage.
8. On remote control models, disconnect the two primer solenoid wires.
9. Remove the carburetor mounting nuts. Remove the carburetor and gasket (21, **Figure 26**).
10. Disconnect the bleed line from the fitting on the bottom of the carburetor, if so equipped.
11. To install the carburetor, first install the primer bracket and components, except the primer knob and bezel, then install the air intake cover onto the carburetor.
12. Reconnect the bleed hose to the fitting at the bottom of the carburetor.
13. Install the carburetor on the power head using a new base gasket. Tighten the mounting nuts evenly and securely.
14. Attach the fuel hose to the carburetor and secure it with a new tie-strap clamp.
15. On remote control models, reconnect the two primer solenoid wires.
16. Reconnect the fast idle wire link to the fast idle cam (29, **Figure 26**).
17. While pushing down on the primer arm, install the primer knob, bezel and cam block into the primer assembly. Securely tighten the primer linkage retaining screw (53, **Figure 26**).
18. Align the notch in the rear of the bezel with the tab on the lower cowl. Secure the bezel with the retaining clip (51, **Figure 26**).
19. Reconnect the spark plug leads to the spark plugs.
20. On electric start models, reconnect the negative battery cable.
21. Refer to Chapter Five for carburetor and linkage adjustment procedures.

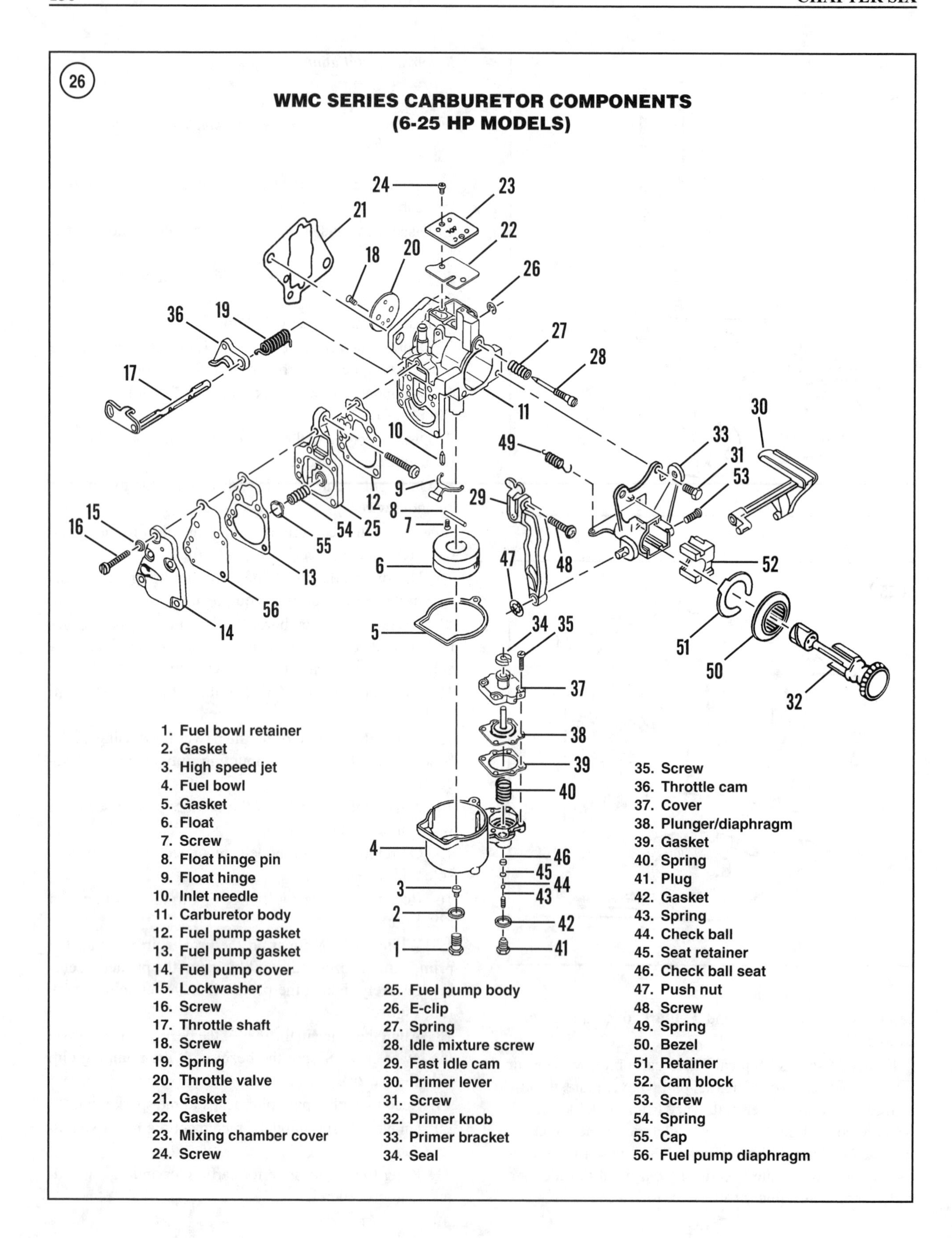

WMC SERIES CARBURETOR COMPONENTS (6-25 HP MODELS)

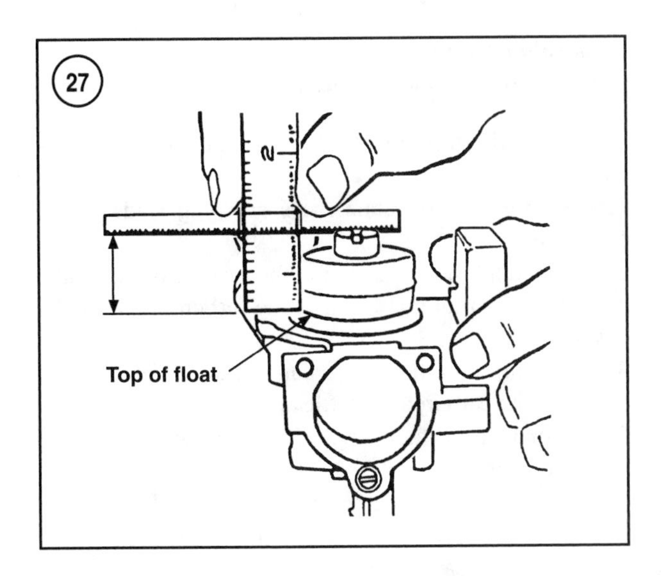

Disassembly

Refer to **Figure 26** for this procedure.

1. Remove the screws retaining the air intake cover and primer bracket assembly to the carburetor body. Remove the primer linkage, fast idle linkage and the air box from the carburetor.

2A. *Tiller control models*—Remove the screws securing the mixing chamber cover (23, **Figure 26**) to the top of the carburetor. Remove the cover and gasket (22 and 23, **Figure 26**).

2B. *Remote control models*—Remove the screws securing the primer solenoid and mixing chamber cover to the top of the carburetor. Remove the primer solenoid assembly.

3. Remove the screws and lockwashers (15 and 16, **Figure 26**) holding the fuel pump to the carburetor. Remove the pump assembly.

4. Separate the pump cover (14, **Figure 26**), diaphragm (56) and pump body (25). Remove the diaphragm from the pump cover or pump body and the gasket from the pump body or carburetor body.

5. Remove the cap and spring (54 and 55, **Figure 26**) from the pump body.

6. Remove the idle speed mixture screw and spring (27 and 28, **Figure 26**).

7. Invert the carburetor, remove the fuel bowl retainer (1, **Figure 26**) and lift the fuel bowl (4) from the carburetor body. Note that the high speed jet is screwed into the fuel bowl retainer.

8. Lift the float (6, **Figure 26**) from the carburetor body.

9. Remove the primer cover screws (35, **Figure 26**). Remove the cover, diaphragm, gasket and spring (37-40, **Figure 26**).

10. Remove the plug (41, **Figure 26**) from the bottom of the primer housing. Remove the gasket, spring and check ball (42-44, **Figure 26**).

NOTE

The inlet needle valve seat is not serviceable. If the seat is worn or damaged, replace the carburetor body.

11. Remove the screw (7, **Figure 26**) securing the float hinge pin (8). Remove the pin and float hinge (9, **Figure 26**), then the inlet needle (10). Remove the fuel bowl gasket (5, **Figure 26**).

12. Refer to *Cleaning and Inspection (All Models)* in this chapter.

Assembly

Refer to **Figure 26** for this procedure.

1. Install the inlet needle (10, **Figure 26**) into the inlet needle seat in the carburetor body. Install the float hinge and float hinge pin (8 and 9, **Figure 26**). Secure the float hinge pin with the screw (7, **Figure 26**). Tighten the screw securely.

2. Install the float onto the carburetor body. Invert the carburetor and measure from the top of the float to the fuel bowl mating surface without the gasket. See **Figure 27**. The measurement must equal the float level specification in **Table 3**. If necessary, adjust the float level by carefully bending the float hinge (9, **Figure 26**).

3. Install the fuel bowl gasket (5, **Figure 26**). Install the fuel bowl onto the carburetor body.

4. If the high-speed jet was removed, reinstall the high-speed jet (3, **Figure 26**) into the fuel bowl retainer screw and tighten the jet securely.

5. Install the fuel bowl retaining screw and tighten it to the specification in **Table 1**.

6. Install the spring and check ball (43 and 44, **Figure 26**) into the bottom of the primer housing. Install the plug (41, **Figure 26**) with a new gasket (42) into the bottom of the primer housing. Tighten the plug to the specification in **Table 1**.

7. Install the primer spring (40, **Figure 26**), a new gasket (39), the diaphragm/plunger (38) and the primer cover (37). Install and evenly tighten the cover screws to the specification in **Table 1**.

8. Install the idle mixture screw and spring. Tighten the screw until it is seated lightly in the carburetor body. Then back out the screw to the specification in **Table 3**.

9. Install the cap and spring (54 and 55, **Figure 26**) into the fuel pump body (25). Install a new diaphragm and gasket (13 and 56, **Figure 26**) between the pump body and

fuel pump cover (14). Install screws at opposite corners of the fuel pump cover to hold the pump body and cover in alignment.

10. Install the fuel pump assembly to the carburetor body with a new gasket (12, **Figure 26**). Install and evenly tighten the fuel pump screws to the specification in **Table 1**.

11A. *Tiller control models*—Install a new gasket and the mixing chamber cover (22 and 23, **Figure 26**) onto the top of the carburetor. Tighten the cover screws to the specification in **Table 1**.

11B. *Remote control models*—Install a new gasket, primer solenoid cover and mixing chamber cover onto the top of the carburetor. Tighten the primer solenoid retaining screws to the specification in **Table 1**.

12. Install the primer linkage, fast idle linkage and the air box to the carburetor. Install and tighten the retaining screws to the specification in **Table 1**.

30-60 hp Models (WME Series)

Remote control models use an electric fuel primer valve to enrich the air/fuel mixture for cold engine starting. The throttle plate(s) must be fully closed for the system to function correctly. Refer to Chapter Three for operational information and troubleshooting procedures on the electric fuel primer valve.

Manual start models use an engine mounted primer bulb. Depressing the primer bulb delivers fuel from the top carburetor's fuel bowl or a T-fitting between the upper two carburetors into one or more fittings on the intake manifold.

A vent jet (back draft) circuit is used on some early model three-cylinder carburetors for improved mid-range fuel economy by controlling the pressure within the float chamber. The outboards covered in this manual do not use vent jets in the carburetor. However, the carburetors on some models are still machined to accept vent jets. Never install plugs or jets in the vent jet opening. Blocking the vent jet leans the air/fuel mixture at mid-range speeds and could cause power head failure.

Figure 28 shows a typical WME series carburetor. Refer to **Table 3** for carburetor specifications and **Table 1** for torque specifications.

NOTE

*Each carburetor must be installed in the correct location according to its identification number. See **Carburetor Identification** at the beginning of this section.*

Removal/installation (30-40 hp two-cylinder models)

1. On electric start models, disconnect the negative battery cable.
2. Disconnect and ground the spark plug leads to the power head.
3. Remove the screws securing the carburetor air intake cover to the carburetor. Remove the carburetor air intake cover.
4. On oil injected models, disconnect the oil pump linkage from the carburetor throttle arm.
5. Disconnect the fuel supply line and fuel primer line from the carburetor.
6. Remove the carburetor mounting bolts and the carburetor air intake cover mounting plate.
7. Remove the carburetor.
8. To install the carburetor, insert the mounting bolts through the air intake cover mounting plate and through the carburetor. Install a new mounting gasket (8, **Figure 28**) over the mounting bolts.
9. Install the carburetor assembly to the intake manifold. Tighten the mounting bolts evenly to the specification in **Table 1**.
10. Reconnect the fuel supply and primer lines. Secure both lines with new tie-straps.
11. On oil injected models, reconnect the oil pump linkage to the carburetor throttle arm.
12. Install the carburetor air intake cover and securely tighten the screws.
13. Reconnect the spark plug leads to the spark plugs.
14. On electric start models, reconnect the negative battery cable.
15. Refer to Chapter Five for carburetor and linkage adjustment procedures.

Removal/installation (40 and 50 hp three-cylinder models)

These models use three carburetors. Do not intermix components. Install each carburetor in its original location.

1. On electric start models, disconnect the negative battery cable.
2. Disconnect and ground the spark plug leads to the power head.
3. Remove the screw securing the top of the carburetor air intake cover. Remove the carburetor air intake cover.
4. On oil injected models, disconnect the oil pump linkage from the carburetor throttle arm.

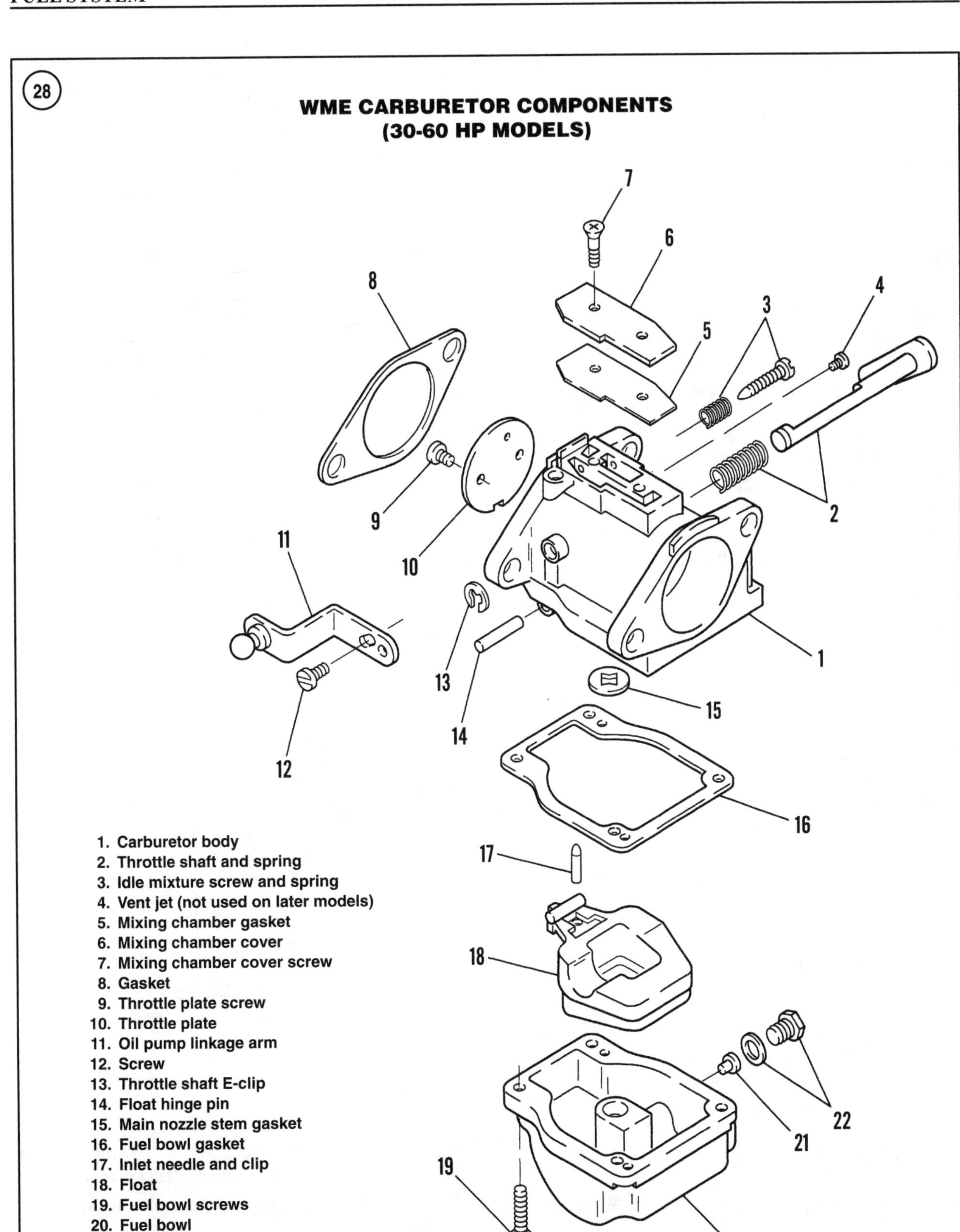
28
WME CARBURETOR COMPONENTS
(30-60 HP MODELS)
1
2
3
4
5
6
7
8
9
10
11
12
13
14
15
16
17
18
19
20
21
22
1. Carburetor body
2. Throttle shaft and spring
3. Idle mixture screw and spring
4. Vent jet (not used on later models)
5. Mixing chamber gasket
6. Mixing chamber cover
7. Mixing chamber cover screw
8. Gasket
9. Throttle plate screw
10. Throttle plate
11. Oil pump linkage arm
12. Screw
13. Throttle shaft E-clip
14. Float hinge pin
15. Main nozzle stem gasket
16. Fuel bowl gasket
17. Inlet needle and clip
18. Float
19. Fuel bowl screws
20. Fuel bowl
21. Main jet
22. Fuel bowl plug and gasket

5. Disconnect the fuel supply line from the top carburetor.
6A. On manual start models, disconnect the primer line from the top carburetor fuel bowl.
6B. On electric start models, disconnect the primer line from the T-fitting between the upper and middle carburetors.
7. Remove the carburetor mounting bolts (1, **Figure 29**) and the carburetor air intake cover mounting plates (2).
8. Remove the carburetors as an assembly.
9. Separate the carburetors by disconnecting the fuel lines and throttle linkage from each carburetor.
10. To install the carburetors, connect the fuel lines and throttle linkage to each carburetor. Secure the fuel lines with a new tie-strap.
11. Insert the mounting bolts through the air intake cover mounting plates on the upper and lower carburetors (2, **Figure 29**). Insert the mounting bolts into the center carburetor. Install a new mounting gasket (8, **Figure 28**) over the mounting bolts of each carburetor.
12. Install the carburetor assembly to the intake manifold. Tighten the mounting bolts (1, **Figure 29**) to the specification in **Table 1**.
13. Reconnect the fuel supply and primer lines. Secure both lines with new tie-straps.
14. On oil injected models, reconnect the oil pump linkage to the carburetor throttle arm.
15. Install the carburetor air intake cover and tighten the screw securely.
16. Reconnect the spark plug leads to the spark plugs.
17. On electric start models, reconnect the negative battery cable.
18. Refer to Chapter Five for carburetor and linkage adjustment procedures.

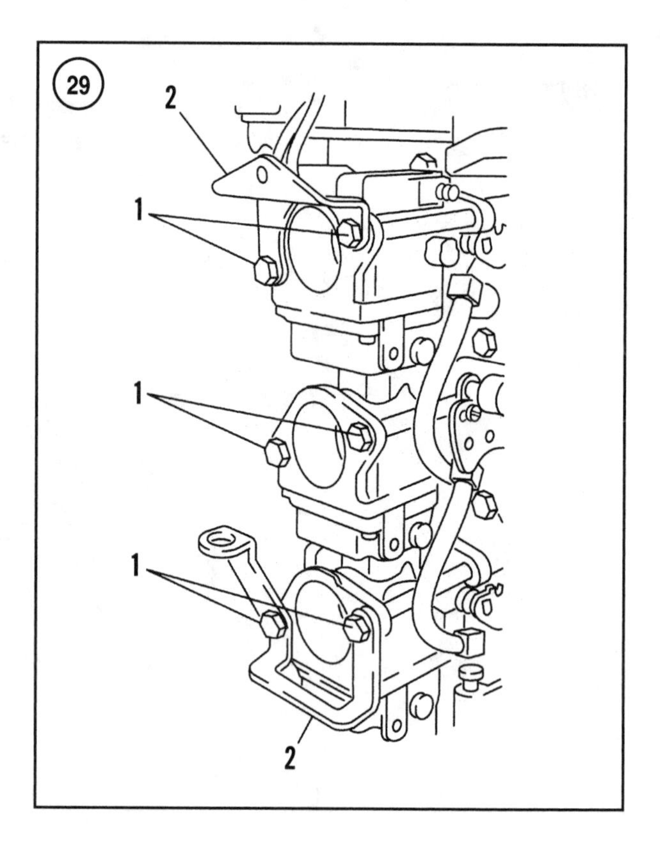

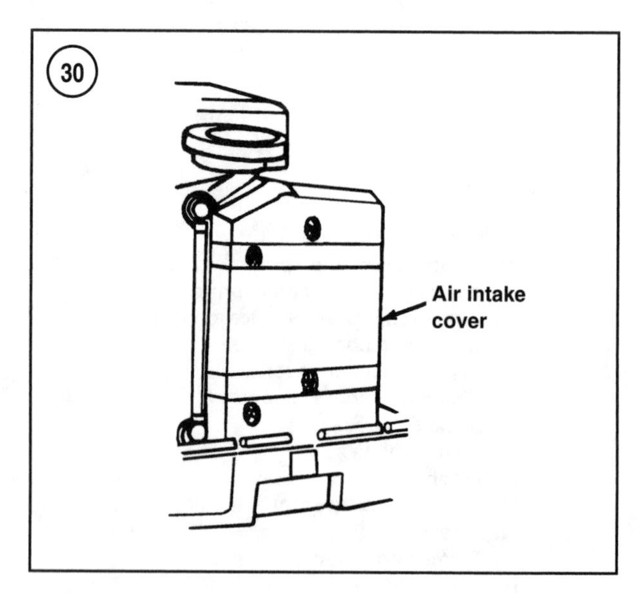

Removal/installation (45 jet, 55 hp and 60 hp models)

These models use three carburetors. Do not intermix components. Install each carburetor in its original location.
1. Disconnect the negative battery cable on electric start models.
2. Disconnect and ground the spark plug leads to the power head.
3. Remove the screws and lift off the air intake cover (**Figure 30**).
4. On oil injected models, disconnect the oil pump linkage from the carburetor throttle arm.
5. Remove the fuel supply line from the top carburetor.
6. Remove the primer line from the top carburetor or T-fitting between the upper and middle carburetors.
7. Remove the carburetor mounting bolts (1, **Figure 31**) and the carburetor air intake cover mounting plates (2).
8. Remove the carburetors as an assembly.
9. Separate the carburetors by disconnecting the fuel lines and throttle linkage from each carburetor.

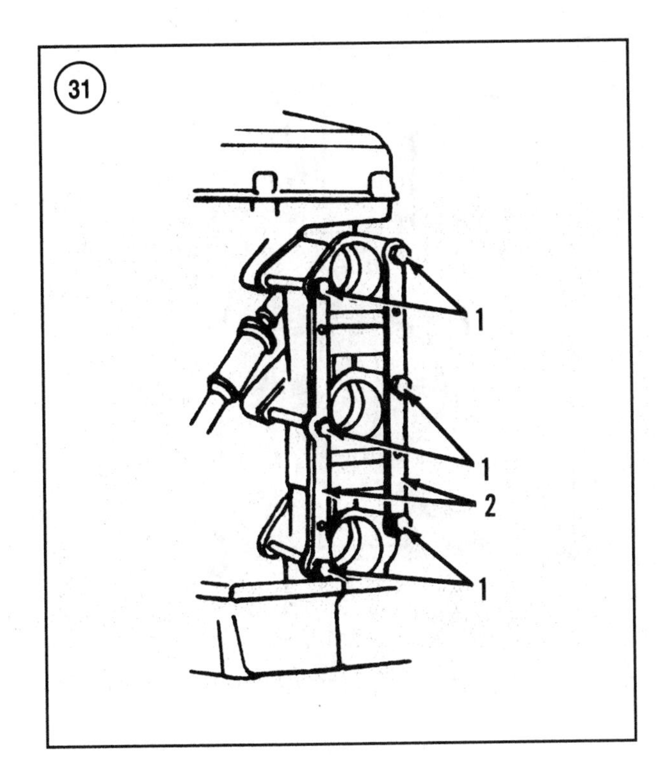

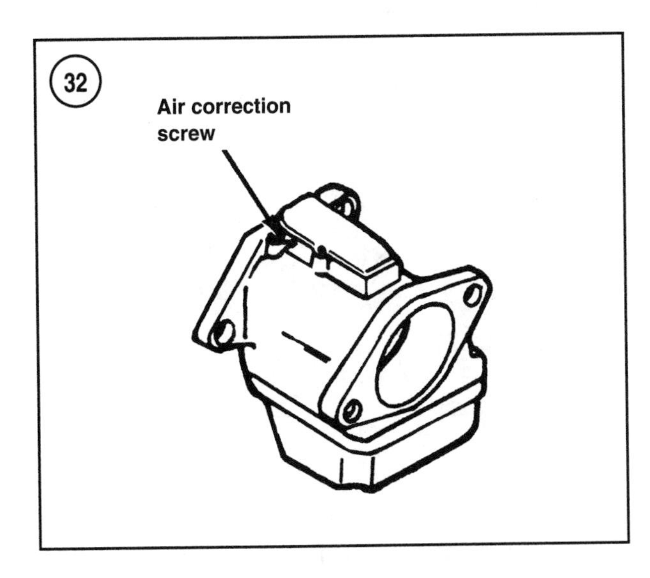

10. To install the carburetors, connect the fuel lines and throttle linkages to each carburetor. Secure the fuel lines with new tie-straps.

11. Insert the mounting bolts through the air intake cover mounting plates and through each carburetor. See **Figure 31**. Install a new mounting gasket (8, **Figure 28**) over the mounting bolts of each carburetor.

12. Install the carburetor assembly to the intake manifold. Tighten the mounting bolts (1, **Figure 31**) evenly to the specification in **Table 1**.

13. Reconnect the fuel supply and primer lines. Secure both lines with new tie-strap clamps.

14. On oil injected models, reconnect the oil pump linkage to the carburetor throttle arm.

15. Install the carburetor air intake cover and securely tighten the screws.

16. Reconnect the spark plug leads to the spark plugs.

17. On electric start models, reconnect the negative battery cable.

18. Refer to Chapter Five for carburetor and linkage adjustment procedures.

Disassembly (all models)

Refer to **Figure 28** for this procedure. On models with multiple carburetors, do not intermix components. Install each carburetor in its original location.

CAUTION
*Do not remove or attempt to adjust the air correction screw (**Figure 32**). This screw is preset by the manufacturer and never requires adjustment. Conventional carburetor cleaning solvent does not affect the brown sealant securing the screw adjustment.*

1. Remove the screws (19, **Figure 28**) securing the fuel bowl (20) to the carburetor body. Remove the fuel bowl.
2. Loosen the screw then remove the float pin (14, **Figure 28**), float (18) and inlet needle (17).
3. Remove the main nozzle stem gasket (15, **Figure 28**).
4. Remove the screws (7, **Figure 28**) securing the mixing chamber cover (6) to the carburetor body.
5. Remove the idle mixture screw and spring (3, **Figure 28**).
6. Remove the main jet plug and gasket from the fuel bowl, then remove the main jet from the fuel bowl. See **Figure 33**.
7. Inspect the throttle shaft and plate (2 and 10, **Figure 28**) for excessive wear and damage. If worn or damaged, remove the shaft and plate as follows:
   a. On oil injected models, remove the screw (12, **Figure 28**) then pull the oil pump linkage arm (11) from the shaft.
   b. Remove the E-clip (13, **Figure 28**) from the end of the throttle shaft.
   c. Remove the screws (9, **Figure 28**) securing the throttle plate (10) to the shaft.
   d. Remove the throttle plate then pull the throttle shaft and spring from the carburetor.

8. Refer to *Cleaning and Inspection (All Models)* in this chapter.

Assembly (all models)

Refer to **Figure 28** for this procedure. On models with multiple carburetors, do not intermix components. Install each carburetor in its original location.

1. If they were removed, insert the throttle shaft and spring (2, **Figure 28**) into the carburetor body. Make sure the throttle shaft spring is properly engaged with the throttle lever and the boss on the carburetor body. Position the throttle plate (10, **Figure 28**) in its original orientation in the throttle bore.
2. Apply Locktite 271 (part No. 92-809819) to the threads of the throttle valve screws (9, **Figure 28**). Install the screws finger-tight.
3. Install the E-clip (13, **Figure 28**) onto the end of the throttle shaft. Check the alignment of the throttle plate to the throttle bore. Make sure the throttle plate opens and closes fully without binding. Adjust the throttle plate position as needed. Tighten the throttle plate screw (9, **Figure 28**) to the specification in **Table 1**.
4. On oil injected models, install the oil pump linkage arm (11, **Figure 28**) and securely tighten the screw (12).
5. Install the main jet into the fuel bowl and tighten it to the specification in **Table 1**. Install the fuel bowl plug (**Figure 33**) and a new gasket. Tighten the fuel bowl plug to the specification in **Table 1**.
6. Install the idle (low speed) mixture screw and spring (3, **Figure 28**). Turn the screw in until lightly seated, then back the screw out the number of turns in **Table 3**. See idle low speed mixture screw adjustment in Chapter Five.
7. Install the mixing chamber cover with a new gasket. Tighten the cover screws (7, **Figure 28**) to the specification in **Table 1**.
8. Install a new main nozzle stem gasket (15, **Figure 28**).
9. Attach the inlet needle wire clip over the float's metal tab. Install the float. Make sure the inlet needle properly enters the inlet valve seat. Install the float pin (14, **Figure 28**) and secure it with the screw.
10. With the fuel bowl removed and the carburetor inverted, measure the distance from the float to the carburetor body as shown in **Figure 34**. The measurement must equal the float level specification in **Table 3**. Carefully bend the float's metal tab as necessary to correct the float level measurement.
11. Install the fuel bowl with a new gasket (16, **Figure 28**). Tighten the screws (19, **Figure 28**) to the specification in **Table 1**.

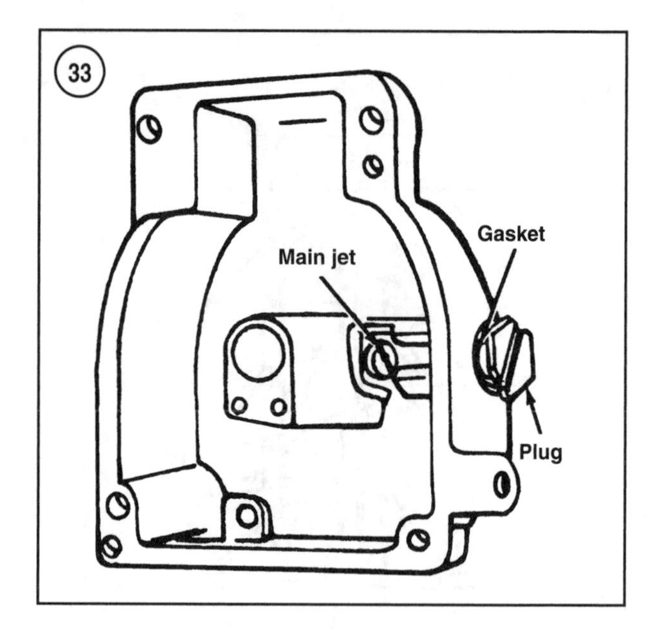

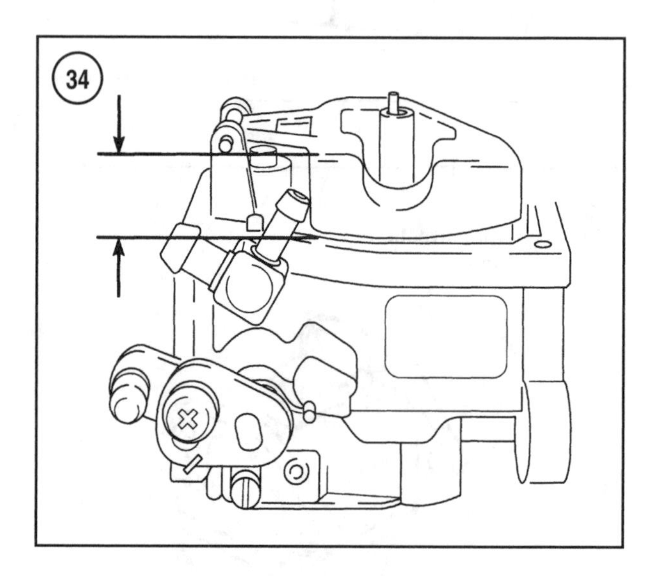

FUEL PRIMER SOLENOID (6-25 HP AND 20 JET REMOTE CONTROL MODELS)

The fuel primer solenoid on 6-25 hp and 20 jet models is a simple spring-loaded, electromechanical plunger solenoid. When activated, the plunger head covers an air bleed valve on the carburetor body. Covering the air bleed valve activates the primer circuit, enriching the air/fuel mixture. The primer circuit works only when the engine is being cranked or run with the throttle plate is closed. The solenoid plunger is exposed to allow for manual activation. Pressing the plunger toward the carburetor activates the primer circuit and the solenoid spring returns the

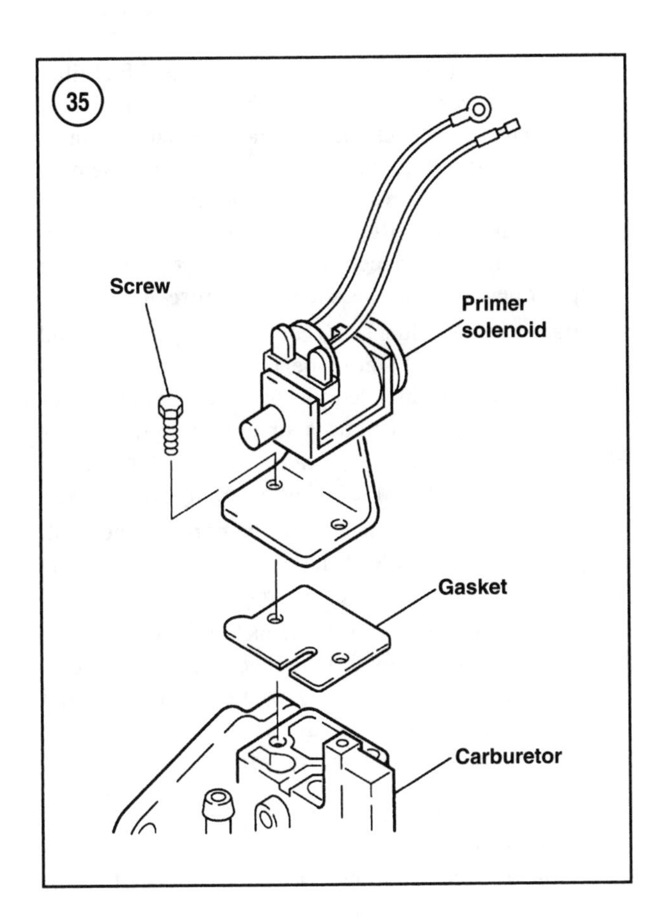

plunger when released. When electricity is applied to the yellow/black lead, the solenoid is energized and moves the plunger toward the carburetor. Refer to Chapter Three for electrical troubleshooting procedures.

Replacement

Refer to **Figure 35** for this procedure.

1. Disconnect the two solenoid electrical leads from the engine wiring harness.
2. Remove the screws (**Figure 35**) securing the solenoid and top cover plate to the carburetor. Remove the solenoid and gasket.
3. Install the solenoid with a new gasket. Tighten the screws to the specification in **Table 1**.
4. Reconnect the two solenoid electrical leads to the engine wiring harness.

FUEL PRIMER VALVE (30-60 HP REMOTE CONTROL MODELS)

The electrically operated fuel primer valve provides additional fuel for easier cold starts. Fuel is gravity fed from the top carburetor or pressure fed from the fuel pump to the fuel primer valve. When the ignition switch is held in the choke or prime position, the valve opens and allows fuel to flow to an intake manifold fitting. The valve can be operated manually by depressing and holding the button located on the valve.

Replacement

1. Disconnect the fuel primer valve yellow/black lead at the engine wiring harness bullet connector.
2. Disconnect the two fuel lines at the valve.
3. Remove the bolt holding the clamp around the valve and the black ground wire. Remove the valve from the clamp.
4. To install the fuel primer valve, position the clamp around the fuel primer valve. Install the bolt through the clamp and ground wire. Secure the clamp, ground wire and valve assembly to the power head. Tighten the bolt securely.
5. Reconnect the two fuel lines to the valve. Secure the fuel lines with new tie-straps.
6. Reconnect the fuel primer valve yellow/black leads to the engine wiring harness bullet connector.

ENGINE MOUNTED PRIMER BULB (30-55 HP MANUAL START MODELS)

The manually operated fuel primer bulb provides additional fuel for easier cold starts. When depressed, the bulb pumps fuel from the top carburetor fuel bowl to a fitting on the intake manifold. The primer bulb is mounted to the front of the lower engine cowl. Replace the bulb as follows:

1. Disconnect the two fuel lines at the valve. Clean up any spilled fuel.
2. Remove the retainer and pull the bulb from the engine cowl.
3. Slip the replacement bulb into its opening in the cowl. Secure the bulb with the retainer.
4. Reconnect the two fuel lines to the bulb. Secure the fuel lines with new tie straps.
5. Check for proper operation of the bulb. Fuel should flow to the intake manifold fitting as the bulb is depressed. If it does not, the fuel lines are improperly connected to the bulb.

ANTISIPHON DEVICES

In accordance with industry safety standards, late model boats equipped with a built-in fuel tank must have some form of antisiphon device installed between the fuel tank outlet and the outboard fuel inlet. The most common method is the antisiphon valve. This device is mounted at the outlet of the fuel tank pickup tube. It is designed to prevent siphoning of the fuel from the fuel tank into the bilge if the fuel line develops a break or leak between the fuel tank and the engine.

Other methods are electrical solenoid and manually operated fuel valves. Often the malfunction of an antisiphon device leads to the replacement of a fuel pump in the mistaken belief that it is defective. Antisiphon devices cause running problems when they restrict the fuel flow to the engine. If the antisiphon device is suspected of causing a fuel starvation problem:

1. The device can be temporarily bypassed. If the fuel starvation problem disappears, the antisiphon device is faulty and must be replaced. Do not return the boat to service with an inoperative antisiphon device.
2. Connect a vacuum gauge to the fuel delivery line using a T-fitting. The vacuum required to draw fuel from the fuel tank must not exceed 4 in. Hg (13.5 kPa) at any engine speed. See Chapter Three for fuel system troubleshooting procedures.
3. Connect a portable fuel tank to the motor. This is the easiest method to determine if the boat's fuel supply system or the antisiphon device is defective. If the motor runs correctly on the portable fuel tank, the problem is in the boat's fuel supply system. Check the easy things first. Check boat mounted fuel filters for blockage. Inspect all fittings, clamps, and fuel delivery and vent lines for secure attachment, leaks and routing problems that could cause a restriction. Inspect the fuel pickup tube filter screen for blockage. Inspect the antisiphon valve for corrosion, mechanical damage or blockage from debris or contamination. Replace any damaged, deteriorated or corroded parts.

FUEL FILTERS

Refer to Chapter Four for standard fuel filter service procedures on all models.

FUEL TANKS

The 2.5 and 3.3 hp models use an integral fuel tank with a tank mounted shutoff valve and fuel filter assembly. The 4 and 5 hp models use an integral tank with a remote shutoff valve and separate inline fuel filter. Some 5 hp models are equipped with a dual fuel supply system. The engine can be run off the integral tank or a remote fuel tank. See **Figure 36**.

Inspect, clean and flush the integral tank and fuel filter at least once a season and during each tune-up or major repair procedure. Inspect the integral fuel tank, shutoff valve, lines, fittings, fuel filter, fill cap and vent assembly for leaks, loose connections, deterioration, corrosion and/or contamination. Repair or replace suspect parts. Secure all fuel line connections with the original spring clamps or new tie-straps as shown in **Figure 36**.

Portable (remote) fuel tanks come in 3.2 gal. (12 L) and 6.6 gal. (25 L) sizes. Later model Quicksilver plastic fuel tanks use a simple threaded pickup tube or threaded retaining nut. The tube or nut is simply unscrewed to remove the pickup tube and clean the filter. Replace the O-ring or seal each time the pickup assembly is removed.

Inspect and clean, if necessary, the portable fuel tank and pickup tube fuel filter at least once a season and during each tune-up or major repair procedure.

Inspect the portable fuel tank, pickup assembly, fuel lines, fittings, connectors and the fill cap and vent assembly for leaks, loose connections, deterioration, corrosion and contamination. Replace suspect parts and secure all fuel line connections with new tie-straps.

NOTE

All integral and portable fuel tanks contain an air vent to allow air into the tank. A plugged or blocked air vent may cause a vacuum to build in the tank and the engine to starve for fuel.

FUEL HOSE AND PRIMER BULB

Figure 37 shows typical fuel hose and primer bulb components. Current fuel line connectors are *Quick Connect* snap lock style connectors. They come in two different hose inside diameter sizes—1/4 in. (6.3 mm) for smaller engines and 5/16 in. (7.9 mm) for larger engines. *Quick Connect* connectors are replaced as assemblies; no internal components are available for the engine or fuel tank ends.

On all engines with permanent mounted fuel tanks, consider connecting the fuel supply line directly to the fuel pump inlet line, eliminating the quick connectors, to minimize possible fuel restrictions and ensure adequate fuel supply to the engine.

Periodically inspect the fuel hose and primer for leaks, deterioration, loose clamps, kinked or pinched lines and other damage. Make sure all fuel hose connections are tight and securely clamped. Replace the fuel supply line and primer bulb as an assembly if there are any doubts as to its integrity.

DUAL FUEL SUPPLY SYSTEM (5 HP MODELS)

1. Fuel cap and vent assembly
2. Grommet
3. Integral fuel tank
4. Fuel shutoff valve
5. T-fitting
6. One-way check valve
7. Fuel filter
8. Remove fuel tank connector
9. Fuel line (to fuel pump and carburetor)

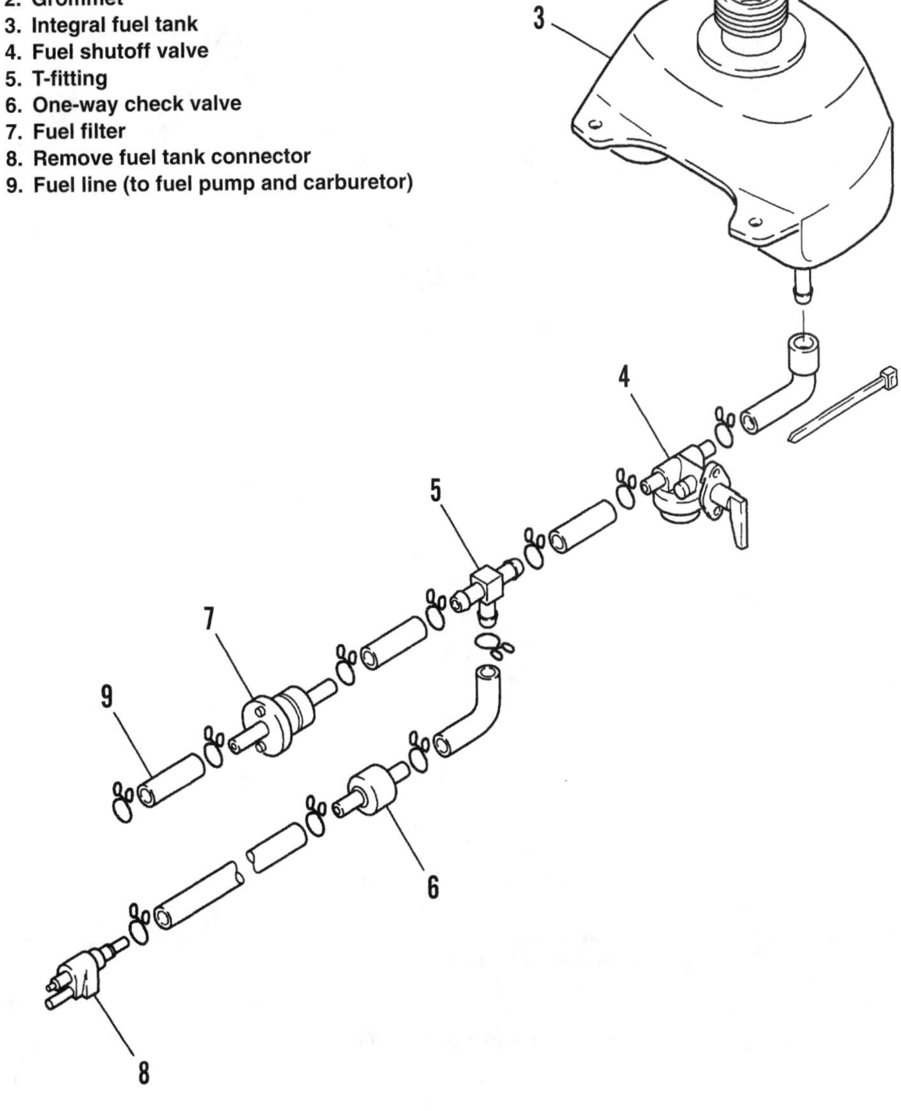

37

FUEL HOSE AND PRIMER BULB

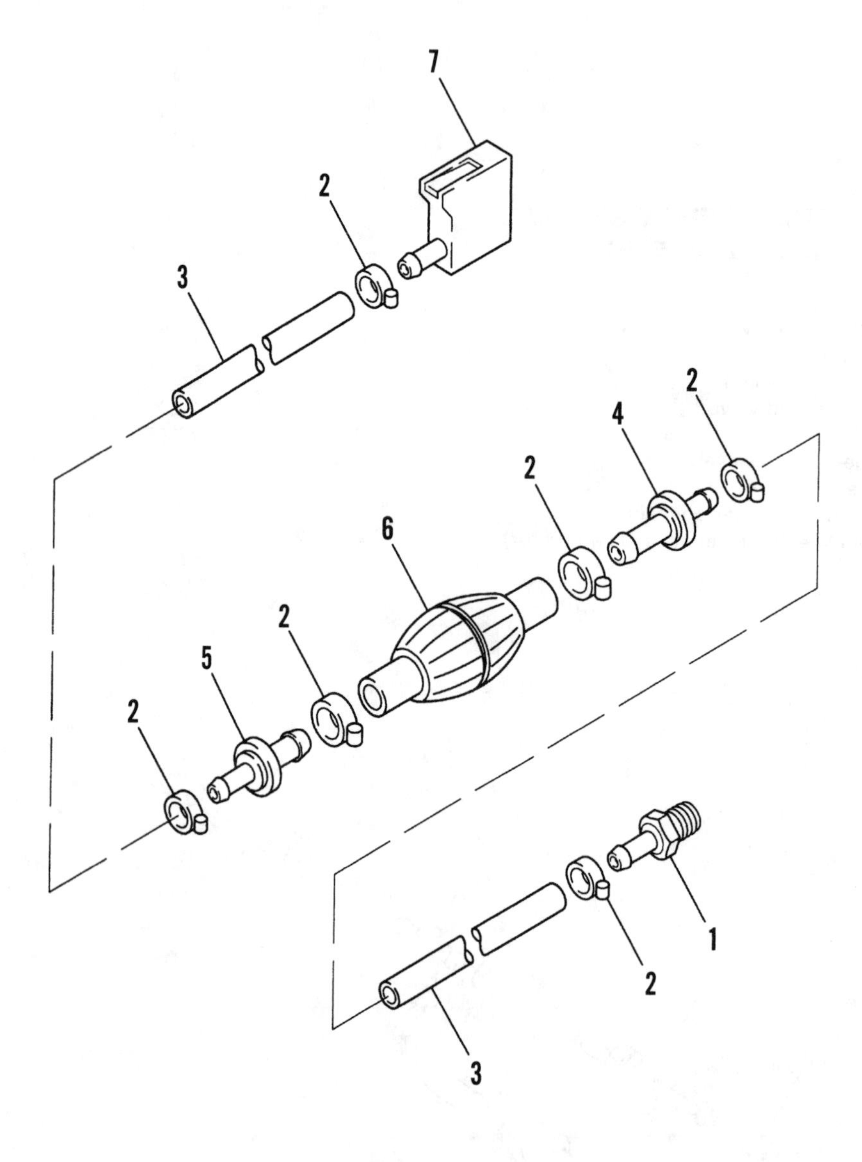

1. Fuel tank fitting
   (pipe thread or quick connect)
2. Clamps
3 Fuel line
4. Inlet check valve
5. Outlet check valve
6. Primer bulb body
7. Engine *Quick Connect* connector

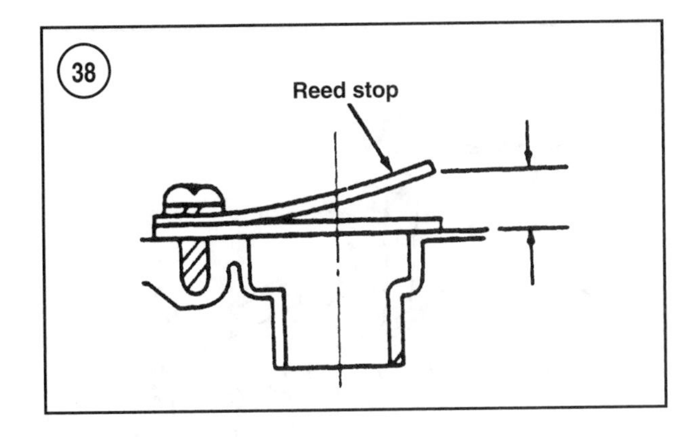

REED VALVE SERVICE

All Mercury/Mariner two-stroke outboard motors are equipped with one set of reed valves per cylinder. The reed valves allow the air/fuel mixture from the carburetor to enter the crankcase, but not exit. They are one-way check valves.

Reed valves are virtually maintenance free and cause very few problems. However, if a reed valve does not seal, the air/fuel mixture escapes from the crankcase and is not transported to the combustion chamber. Some spitting of fuel back out of the carburetor throat at idle can be considered normal, but a substantial discharge of fuel from the carburetor throat indicates reed valve failure.

On most models, the reeds can be inspected with the carburetor removed. A small flashlight and dental mirror can be used to inspect for broken, cracked or chipped reeds. If reed damage is discovered, attempt to locate the missing pieces of the reed petal. The reed petals are made of stainless steel and can damage internal engine components if they pass through the crankcase and combustion chamber.

Anytime the reed valves are removed from the power head, inspect the reeds for excessive stand-open. Reed valve specifications are in **Table 4**. On models with reed stop specifications, measure the reed stop openings. The easiest way to measure the reed stop opening is to use a drill bit of the specified diameter. The shank of the drill bit should just fit between the top surface of the closed reed petal and the inner edge of the reed stop. Some reed stops are adjustable, while some must be replaced if not within specification. The text differentiates between such models.

Never turn over reed petals and reinstall them. This can lead to preloaded reeds. Preloaded reeds require a higher crankcase vacuum level to open. This causes acceleration and carburetor calibration problems. The reed petals should be flush to nearly flush along the entire length of the reed block mating surface with no preload. Most larger engines have a maximum stand-off specification in **Table 4**.

Many new models use rubber coated reed blocks. These reed blocks cushion the impact as the reed closes and improves sealing. Reed block assembly part numbers automatically supersede where applicable. Rubber coated reed blocks are serviced as assemblies only.

2.5-5 hp Models

The reed valves are mounted to the crankcase cover. Power head disassembly is required to service the reeds. Refer to Chapter Eight for power head removal and disassembly procedures. Once the crankcase cover has been removed, inspect the reeds as follows:

Cleaning and inspection

1. Clean the crankcase cover and reed valve assembly thoroughly with solvent.
2. Inspect the reed petals for cracks, chips or evidence of fatigue. The reeds should be flush or nearly flush to the seat along their entire length without being preloaded against the seat. Replace the reeds as necessary.
3. Check for indentation on the face of the seat area in the crankcase cover. If the reeds have worn indentations in the seat, replace the cylinder block and crankcase cover.

NOTE
Do not remove the reeds and reed stop unless replacement is necessary. Never turn reeds over for reuse.

4A. *2.5 and 3.3 hp models*—Measure the reed stop opening at the points indicated in **Figure 38**. If the reed stop opening is not within the specification in **Table 4**, replace the reed stop. The reed stop is not adjustable on 2.5 and 3.3 hp models.

4B. *4 and 5 hp models*—Measure the reed stop opening at the points indicated in **Figure 38**. The reed stop opening should be within the specification in **Table 4**. If necessary, carefully bend the reed stop to obtain the specified measurement. On 4 hp models, one reed stop is flat and holds the reed petal closed.

Replacement

1. Refer to Chapter Eight and remove the crankcase cover.
2. Remove the screws securing the reed stop and reed petals to the crankcase cover. Remove the reed stop and reed petals.

3. Clean and inspect the crankcase cover and reed stop. Check for indentation on the face of the seat area in the crankcase cover. If the reeds have worn indentations in the seat, replace the cylinder block and crankcase cover.
4. Install a new reed petal assembly into the crankcase cover. Place the reed stop over the reed petals.
5. Coat the threads of the reed valve screws with Loctite 271 threadlocking adhesive (part No. 92-809819).
   a. *2.5 and 3.3 hp models*—Install the screws and tighten them securely.
   b. *4 and 5 hp models*—Install the screws and tighten them to the specification in **Table 1**.
6. Check the reed stop setting and reed petals for stand-open and preload as described in *Cleaning and inspection* in the previous section.
7. Reinstall the crankcase cover as described in Chapter Eight.

6-25 hp Models

Two sets of reed petals are mounted to a single reed block assembly. Each set of petals feeds one cylinder. A rubber seal incorporated into the reed block separates the petals.

On 6-15 hp models, the reed block is the intake manifold and the carburetor is bolted directly to the reed block. The reed block is serviced as an assembly and must be replaced if worn or defective.

On 20 and 25 hp models, an intake manifold (carburetor adapter) is bolted between the carburetor and reed block. The reed block assembly is serviceable on these models.

Removal/installation

1. On electric start models, disconnect the negative battery cable.
2. Remove the carburetor as described in this chapter.

3A. *6-15 hp models*—Remove the bolts securing the reed block to the crankcase cover.

3B. *20 and 25 hp models*—Remove the bolts securing the intake manifold and reed block to the power head.

4A. *6-15 hp models*—Carefully remove the reed block assembly from the power head. Do not scratch, warp or gouge the reed block or crankcase cover.

4B. *20 and 25 hp models*—Carefully remove the intake manifold and reed block assembly from the power head. Do not scratch, warp or gouge the reed block, intake manifold or crankcase cover. Separate the reed block from the intake manifold. Remove all gasket material from the reed block, intake manifold and crankcase cover

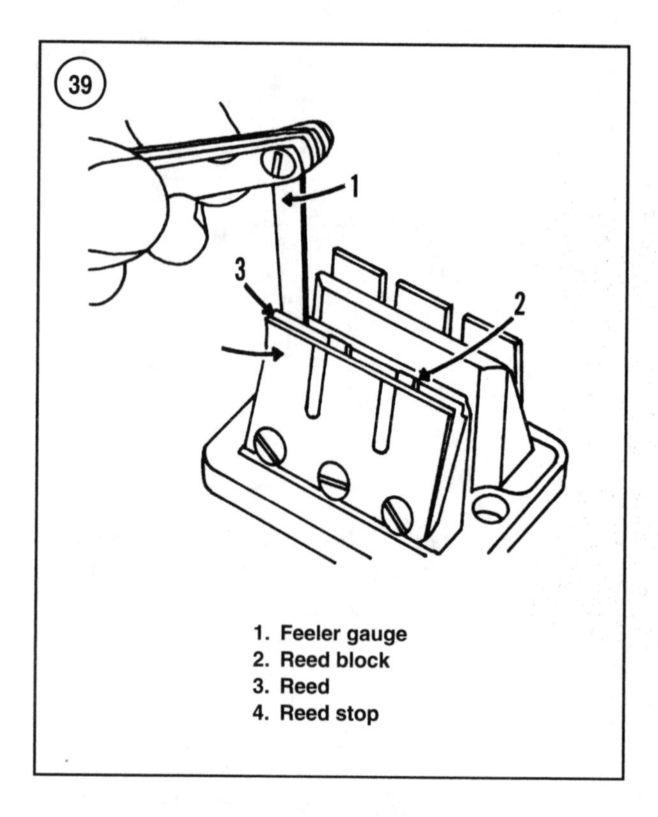

1. Feeler gauge
2. Reed block
3. Reed
4. Reed stop

5. Clean and inspect the reed block assembly as described in the next section.

6A. *6-15 hp models*—Install the reed block with a new gasket over the carburetor mounting studs and seat it against the crankcase cover.

6B. *20 and 25 hp models*—Install the reed block and intake manifold over the carburetor mounting studs and seat them against the crankcase cover.

7A. *6-15 hp models*—Install the three screws and tighten them evenly to the specification in **Table 1**.

7B. *20 and 25 hp models*—Install the three screws and tighten them evenly to the specification in **Table 1**.

8. Install the carburetor as described in this chapter.
9. On electric start models, reconnect the negative battery cable.

Cleaning and inspection

1. Thoroughly clean the reed block assembly in clean solvent.
2. Check for excessive wear, cracks or grooves in the seat area of the reed block. Replace the reed block if it is damaged.
3. Check the reed petals for cracks, chips or evidence of fatigue. Replace the reed petals or reed block assembly if they are damaged.

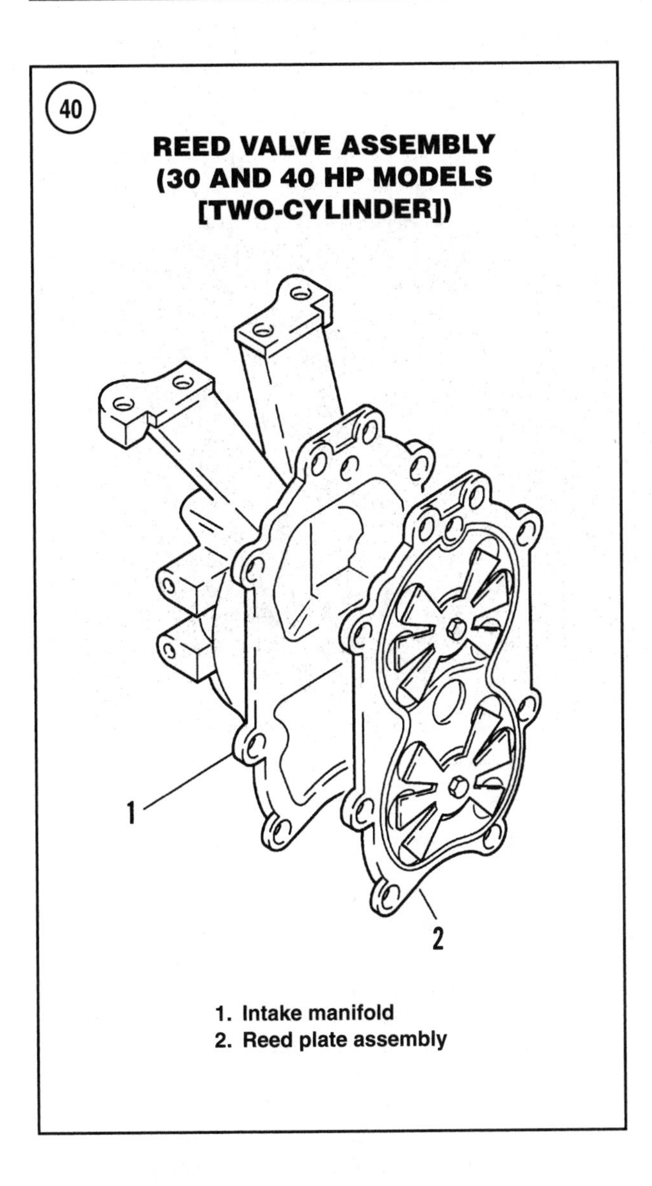

1. Intake manifold
2. Reed plate assembly

CAUTION
Do not remove the reed petal and reed stop unless replacement is necessary. Never turn reed petals over for reuse. Do not disassemble the reed block on 6-15 hp models.

4. Measure the stand-open gap between the reed petals and the reed block mating surface. See **Figure 39**, typical. Replace the reed block assembly if any of the reeds stand open gap exceeds the specification in **Table 4**.
5. To replace the reed petals on 20 and 25 hp models, remove the screws attaching the reed petals and reed stops to the reed block. To reinstall them, apply Loctite 271 threadlocking adhesive (part No. 92-809819) to the threads of the screws. Position the reed petals and reed stops on the reed block. Tighten the screws evenly to the specification in **Table 1**.

30 and 40 hp Two-Cylinder Models (Except 30 Jet)

Two sets of reed valves are mounted to a reed plate. A reed stop is used on 30 hp models, but not on 40 hp models. The reed valves are not serviceable on these models. If they are worn or damaged, replace the reed valve assembly. Refer to **Figure 40** for this procedure.

Removal/installation

1. On electric start models, disconnect the negative battery cable.
2. Remove the carburetor as described previously in this chapter.
3. Remove the bolts securing the intake manifold and reed plate to the power head.
4. Carefully remove the intake manifold and reed plate assembly from the power head. Do not scratch, warp or gouge the reed plate, intake manifold or crankcase cover. Separate the reed plate from the intake manifold. Remove all gasket material from the reed plate, intake manifold and crankcase cover. See **Figure 40**.
5. Clean and inspect the intake manifold or reed plate assembly as described in the next section. Do not remove the reed petals from the intake manifold.
6. Install the reed plate, intake manifold and new gaskets to the power head. Install the bolts finger-tight.
7. Tighten the bolts evenly to the specification in **Table 1** in a circular pattern, starting in the center and working outward.
8. Install the carburetor as described in this chapter.
9. On electric start models, reconnect the negative battery cable.

Cleaning and inspection

1. Thoroughly clean the intake manifold or reed plate assembly in clean solvent.
2. Check for excessive wear, cracks or grooves in the seat area of the intake manifold or reed plate. Replace the reed plate assembly if damaged.
3. Check the reed petals for cracks, chips or evidence of fatigue. Replace the reed plate assembly if it is damaged.

CAUTION
Do not remove the reed petals or stops. The reed petals are not serviceable. Replace the reed block assembly if any of the components are worn or damaged.

4. Check the stand open gap between the reed petals and the intake manifold or reed plate mating surface. See **Figure 41**. Replace the reed plate assembly if the stand open exceeds the specification in **Table 4**.

30 Jet and 40-60 hp Three-Cylinder Models

A single reed plate and single intake manifold is used on 40-60 hp and 45 Jet models. The 40 hp models use a reed stop on each reed plate. Refer to **Table 4** for reed valve specifications.

CAUTION
Do not allow the internal bleed check valves to fall out or become misplaced while the intake manifold is removed.

Removal/installation

1. On electric start models, disconnect the negative battery cable.
2. Remove the carburetors as described previously in this chapter.
3. Remove the bolts securing the intake manifold and reed plate to the power head. See **Figure 42**.
4. Carefully remove the intake manifold and reed plate from the power head. Do not scratch, warp or gouge the reed plate, intake manifold or crankcase cover. Separate the reed plate from the intake manifold. Remove all gasket material from the reed plate, intake manifold and crankcase cover.
5. Clean and inspect the intake manifold and reed plate as described in this chapter. Do not remove the reed plate unless replacement is necessary.

NOTE
Make sure the gaskets are correctly orientated. The gaskets only fit correctly in one direction.

6. Using new gaskets, install the reed plate and intake manifold to the power head. Install the retaining screws finger tight.
7. Tighten the bolts evenly to the specification in **Table 1** and in the pattern shown in **Figure 43**.
8. Install the carburetors as described in this chapter.
9. On electric start models, reconnect the negative battery cable.

41

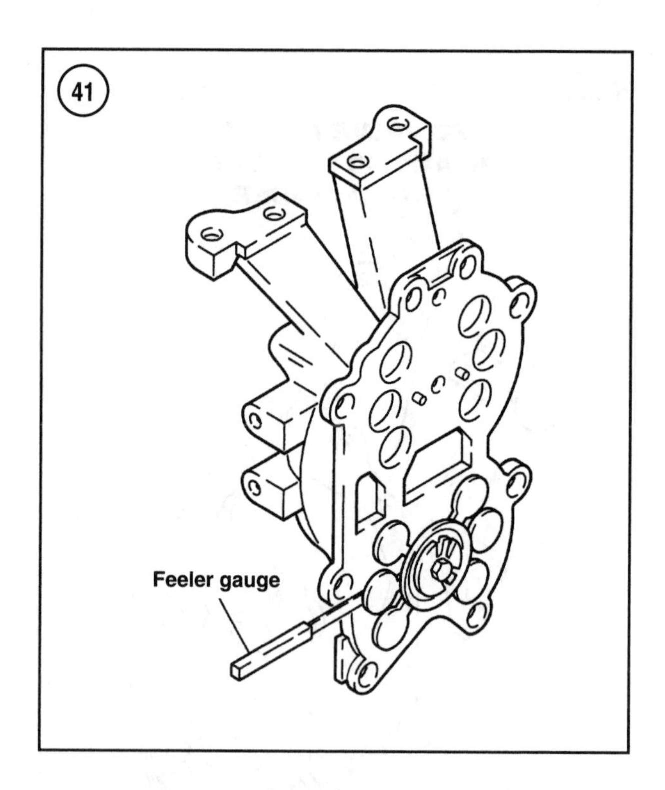

Cleaning and inspection

NOTE
Do not remove the reed petals from the reed plate unless replacement is necessary. Always replace reed petals in complete sets. Never turn a reed over for reuse or attempt to straighten a damaged reed.

1. Thoroughly clean the reed plate using clean solvent.
2. Check for excessive wear, cracks or grooves in the seat area of the reed plate. Replace the reed plate if it is damaged.
3. Check the reed petals for cracks, chips or evidence of fatigue. Replace the reed petals if they are damaged.
4. Check the stand-off gap between the reed petals and the reed plate mating surface. See **Figure 44**, typical. Replace the reed petals if they stand off more than specified in **Table 4**.
5. On 40 hp models, measure from the inside edge of each reed stop to the top of its respective closed reed valve. The reed stop opening must equal the specification in **Table 4**. Carefully bend the reed stop(s) to achieve the correct measurement.
6. To replace the reed petals, bend the lockwasher lock tabs away from the screw heads. Remove the screws attaching the reed petals, reed stop (40 hp) or retaining washer (all other models) to the reed plate. See **Figure 42**, typical.

REED VALVE ASSEMBLY (40-60 HP [THREE-CYLINDER], 30 JET AND 45 JET MODELS)

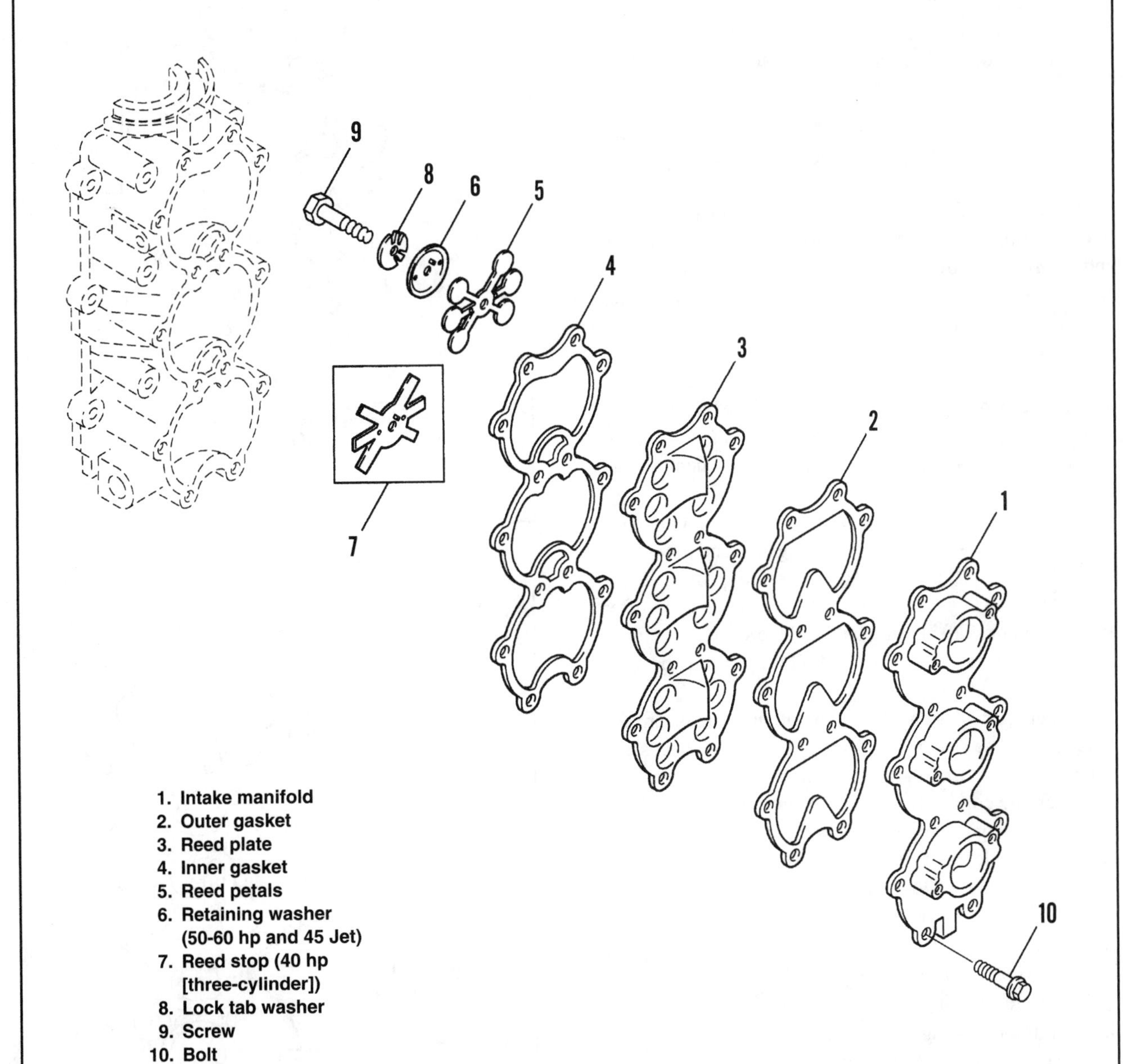

1. Intake manifold
2. Outer gasket
3. Reed plate
4. Inner gasket
5. Reed petals
6. Retaining washer (50-60 hp and 45 Jet)
7. Reed stop (40 hp [three-cylinder])
8. Lock tab washer
9. Screw
10. Bolt

7. To reinstall, apply Loctite 271 threadlocking adhesive (part No. 92-809819) to the threads of the screws. Position the new reed petals, the original reed stops (40 hp) or retaining washers (all other models), and *new* tab washers on the intake manifold. Make sure all components are aligned with the alignment pins in the intake manifold.
8. Tighten the screws to the lower end of the specification in **Table 1** and check the alignment of the screw head to the tab washer. If necessary, continue to tighten the screw to align the screw head with the tab washer. Do not exceed the maximum torque specification in **Table 1**.
9. Bend the lockwasher tabs against each screw head.

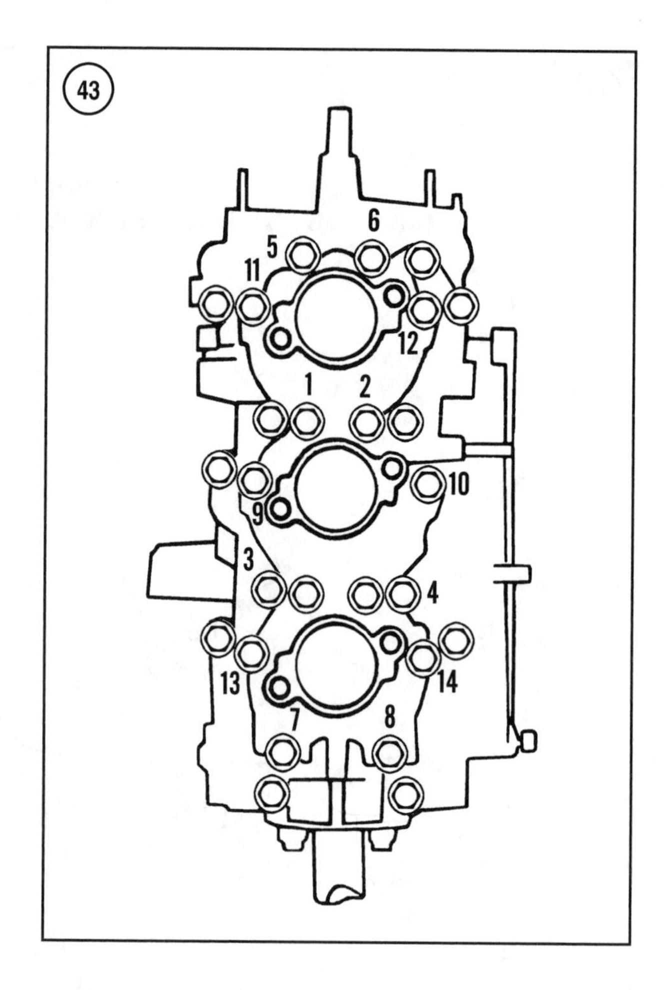

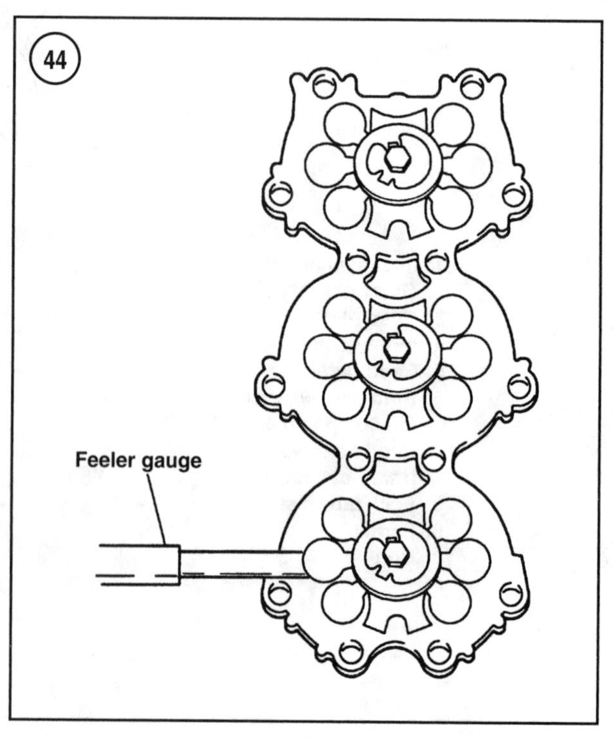

FUEL BLEED (RECIRCULATION) SYSTEM

Multiple cylinder motors are equipped with a fuel bleed (recirculation) system designed to collect unburned fuel and oil from the low spots of the individual crankcase areas. The intake system used by two-stroke engines does not transfer all of the fuel sent through the crankcase to the combustion chamber, especially during low-speed operation. The recirculation system collects the fuel and oil pooled in the low spots of the crankcase. The bleed system pumps the fuel/oil to the intake ports or intake manifold where it can be transferred to the combustion chamber and burned.

Many recirculation systems also collect the fuel and oil pooled in the lower crankshaft bearing area and pump it to the upper crankshaft bearing to ensure proper upper crankshaft bearing lubrication. These models can suffer an upper crankshaft bearing failure if the system malfunctions and does not pump fuel and oil to the upper bearing carrier.

Correct recirculation system operation is important for efficient engine operation. If the system fails, excessive amounts of fuel and oil may puddle in the crankcase and not reach the combustion chamber during low-speed operation, causing a lean mixture. When the engine is accelerated, the puddles of fuel and oil are quickly drawn into the engine causing a temporary excessively rich mixture. This results in the following symptoms:
1. Poor low-speed performance.
2. Poor acceleration.
3. Spark plug fouling.
4. Stalling or missing at idle.
5. Excessive smoke on acceleration.

Recirculation System Service

Refer to **Figure 45** for this procedure.

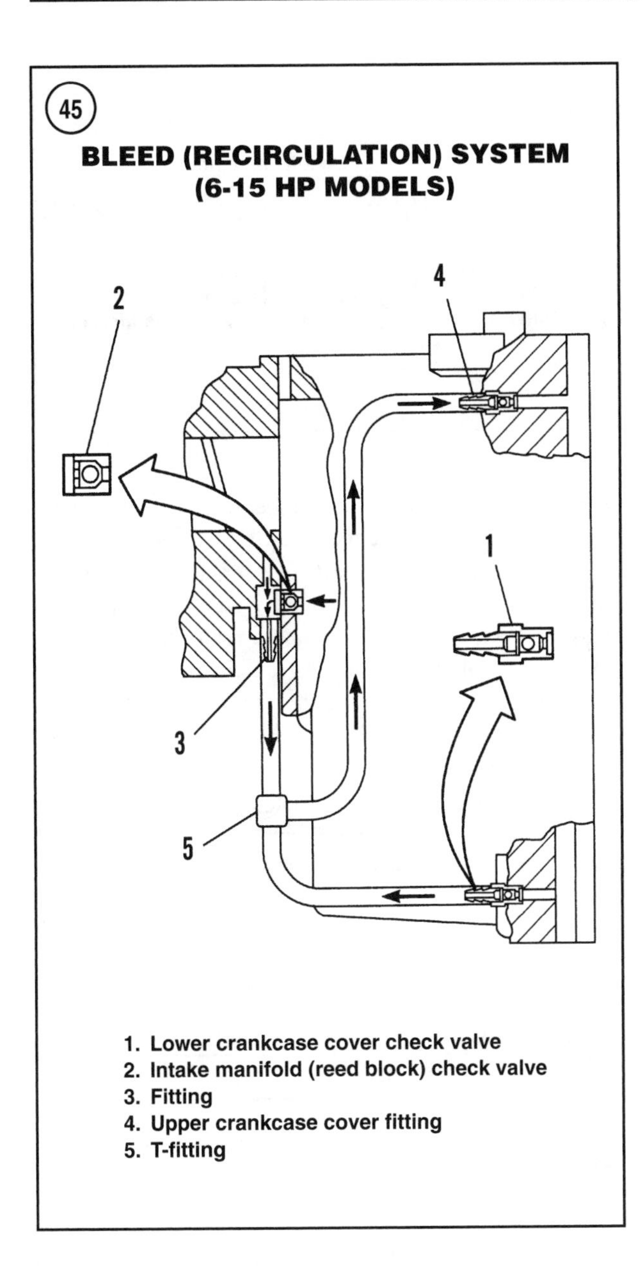

All recirculation systems require a one-way check valve for operation. External check valves are either mounted directly to the crankcase or intake manifold, or mounted in the recirculation lines.

Internal check valves are mounted in the crankcase cover, just behind the intake manifold. Refer to Chapter Eight for additional power head exploded views.

Inspect the internal check valves by removing them from their holder/carrier and looking into the check valve assembly. If light is visible, the nylon check ball is missing or worn and the check valve must be replaced. If light is not visible, insert a fine wire into the check valve and check for slight movement of the check valve. Replace the check valve if the ball does not move.

Replace the check valve holder/carrier if it is burned or otherwise damaged. The end of the check valve with one hole is the inlet side. Fluid must flow into this hole, but not exit. The end of the check valve with two or more holes is the outlet side. Fluid must exit these holes, but not enter.

All check valves must flow in the direction of the arrow (**Figure 45**), but not flow in the opposite direction. Fittings must flow in both directions.

Use a small syringe and a piece of recirculation line to quickly test the system. Push on the syringe plunger to check flow into a fitting or check valve. Pull on the syringe plunger to check flow out of a fitting or check valve. Also, inspect and replace damaged or deteriorated recirculation lines. Clean or replace fittings that will not flow air in both directions. Replace check valves that flow air in both directions or will not flow in either direction.

When replacing fittings or check valves, coat pipe thread fittings with Locktite PST pipe sealant (part No. 92-80822). Press fit external check valves must be coated with Locktite 271 threadlocking adhesive (Part No. 92-809819) prior to installation.

6

2.5 and 3.3 hp models

These models do not have a bleed (recirculation) system.

4 and 5 hp models

These models use a single inline check valve. The recirculation line connects the fitting on the lower port side of the power head, near the crankcase cover parting line and transfers fluid to the fitting on the top cylinder behind the flywheel housing.

6-15 hp models

These models use a check valve mounted at the bottom of the crankcase cover and another check valve mounted in the intake manifold. A T-fitting connects the output from both check valves and transfers it to a fitting at the top of the crankcase to lubricate the upper crankshaft bearing. See **Figure 45**.

20-25 hp models

These models use two check valves mounted at the bottom of the intake manifold and two fittings on the port side of the power head mounted underneath the first and third transfer port covers. The port intake manifold check valve connects to the upper fitting with a piece of flexible hose. The starboard intake manifold check valve connects to the lower fitting with a piece of flexible hose.

30-40 hp two-cylinder models (except 30 jet)

These models use internal and external systems. The first external system uses a check valve mounted at the bottom of the crankcase cover. A flexible line carries the output from the check valve to a fitting mounted on the top of the crankcase cover. This system lubricates the crankshaft upper main bearing.

The second external system is mounted on the port side of the crankcase. An inline check valve mounted in a short piece of hose connects two fittings together. The check valve must be mounted vertically and the check valve arrow must point down. This system removes fuel and oil pooled in the No. 1 cylinder crankcase and transfers it to the intake transfer passages then into the combustion chamber.

The third system is an internal system that consists of a single check valve and carrier (**Figure 46**) mounted in the crankcase cover intake passage of the No. 2 cylinder. The check valve must be installed in the carrier so it moves fluid toward the intake manifold, but not toward the crankcase.

40-60 hp three-cylinder, 30 jet and 45 jet models

These models use internal and external systems. The internal system consists of two check valves and carriers

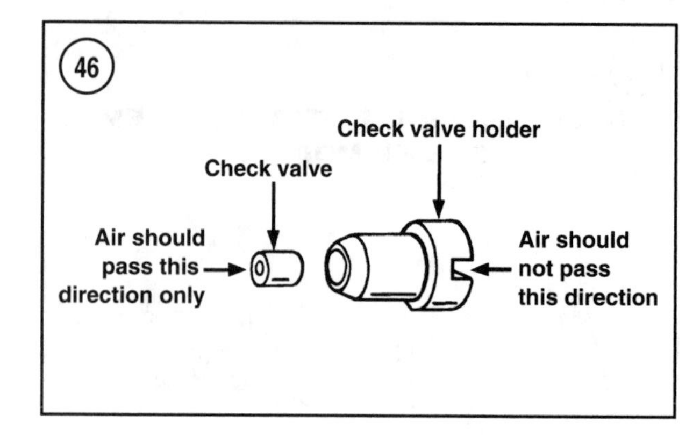

(**Figure 46**) mounted in the crankcase cover intake passages directly behind the reed plate.

The external system consists of a threaded check valve mounted at the bottom of the crankcase cover and a press-in check valve mounted at the top of the crankcase cover. A flexible hose connects the check valve and fitting. This system lubricates the crankshaft upper main bearing. The valve should allow fluid to flow to the upper bearing, but not back down.

A second external system consists of two check valves and two elbow fittings installed into the starboard side of the cylinder block. Each check valve is connected to the elbow fitting with a short loop of hose. An illustration of this system is shown in Chapter Eight. See **Figure 36** (Chapter Eight).

Table 1 TORQUE SPECIFICATIONS

Component	ft.-lb.	in.-lb.	N•m
Fuel pump			
6-25 hp			
Fuel pump to carburetor	–	18	2
30-60 hp			
Mounting screws	–	55	6.2
Cover to pump body screws	–	55	6.2
Carburetor float bowl			
6-25 hp	–	32	3.6
30-60 hp	–	18	2
Fuel primer/cover			
6-25 hp	–	14	1.6
Fuel primer plug	–	32	3.6
Mixing chamber cover			
6-60 hp	–	18	2
Throttle plate screws			
30-60 hp	–	6	0.7
Fuel bowl plug			
30-60 hp	–	22	2.5
Main fuel jet			
30-60 hp	–	14	1.6
Choke solenoid			
6-25 hp	–	18	2

(continued)

Table 1 TORQUE SPECIFICATIONS (continued)

Component	ft.-lb.	in.-lb.	N•m
Air box to carburetor			
6-25 hp	–	32	3.6
Carburetor mounting fasteners			
30-60 hp	–	100	11.3
Reed valve and stop			
4 and 5 hp	–	9	1
20 and 25 hp	–	25	2.8
40-60 hp (three-cylinder)	–	60-100	6.8-11.3
Reed block /intake to power head			
6-15 hp	–	60	6.8
20 and 25 hp	–	80	9
30 and 40 hp (two-cylinder)	16.5	198	22.4
40-60 hp (three-cylinder)	18	–	24.4

Table 2 STANDARD TORQUE SPECIFICATIONS—U.S. STANDARD AND METRIC FASTENERS

Screw or nut size	ft.-lb.	in.-lb.	N•m
U.S. standard fasteners			
6-32	–	9	1
8-32	–	20	2.3
10-24	–	30	3.4
10-32	–	35	4.0
12-24	–	45	5.1
1/4-20	6	72	8.1
1/4-28	7	84	9.5
5/16-18	13	156	17.6
5/16-24	14	168	19
3/8-16	23	270	31.2
3/8-24	25	300	33.9
7/16-14	36	–	48.8
7/16-20	40	–	54
1/2-13	50	–	67.8
1/2-20	60	–	81.3
Metric fasteners			
M5	–	36	4.1
M6	6	72	8.1
M8	13	156	17.6
M10	26	312	35.3
M12	35	–	47.5
M14	60	–	81.3

Table 3 CARBURETOR SPECIFICATIONS

Carburetor ID No.	
6 hp	WMC-57
8 hp	WMC-41A
9.9 hp	WMC-41A
15 hp	WMC-55A
20 hp (except 20 Jet)	WMC-52
20 Jet	WMC-45
25 hp	WMC-53
30 hp	
Manual start	WME-63
Electric start	WME-64
Tiller with oil injection	WME-65

(continued)

Table 3 CARBURETOR SPECIFICATIONS (continued)

Carburetor ID No. (cont.)	
40 hp (two-cylinder)	
Manual start	WME-66
Electric start	WME-67
40 hp (three-cylinder)	
Manual start	WME-69
Electric start	WME-53
50 hp	WME-68
55 hp	WME-57
60 hp	
Manual start	WME-57
Electric start	WME-58
Float level	
2.5 and 3.3 hp (measured from gasket)	0.118 in. (3 mm)
4 and 5 hp	0.5 in. (13 mm)
6-25 hp models	1.0 in. (25.4 mm)
30-60 hp	9/16 in. (14.29 mm)
Needle E-clip setting	
2.5 and 3.3 hp	Second groove from top
Main fuel jet size	
2.5 hp models	No. 96
3.3 hp models	No. 94
4 hp models	
0-2500 ft. elevation (standard jet size)	0.031 in.
2501-5000 ft. elevation	0.029 in.
5001-7500 ft. elevation	0.027 in.
7501-10,000 ft. elevation	0.025 in.
5 hp models	
0-2500 ft. elevation (standard jet size)	0.032 in.
2501-5000 ft. elevation	0.030 in.
5001-7500 ft. elevation	0.028 in
7501-10,000 ft. elevation	0.026 in.
6 hp	
0-5000 ft. elevation (standard jet size)	0.042 in.
5001-7500 ft. elevation	0.040 in.
7501-10,000 ft. elevation	0.038 in.
8 hp	
0-5000 ft. elevation (standard jet size)	0.046 in.
5001-7500 ft. elevation	0.044 in.
7501-10,000 ft. elevation	0.042 in.
9.9 hp	
0-5000 ft. elevation (standard jet size)	0.048 in.
5001-7500 ft. elevation	0.046 in.
7501-10,000 ft. elevation	0.044 in.
15 hp	
0-5000 ft. elevation (standard jet size)	0.072 in.
5001-7500 ft. elevation	0.068 in.
7501-10,000 ft. elevation	0.066 in.
20 hp (except 20 Jet)	
0-5000 ft. elevation (standard jet size)	0.044 in.
5001-7500 ft. elevation	0.042 in.
7501-10,000 ft. elevation	0.040 in.
20 Jet	
0-5000 ft. elevation (standard jet size)	0.076 in.
5001-7500 ft. elevation	0.074 in.
7501-10,000 ft. elevation	0.072 in.
25 hp	
0-5000 ft. elevation (standard jet size)	0.076 in.
5001-7500 ft. elevation	0.074 in.
7501-10,000 ft. elevation	0.072 in.
30 hp (standard jet size)	0.054 in.

(continued)

Table 3 CARBURETOR SPECIFICATIONS (continued)

Main fuel jet size (cont.)	
40 hp (two-cylinder [standard jet size])	0.066 in.
40 hp (three-cylinder [standard jet size])	0.044 in.
50 hp	0.052 in.
55 hp	0.058 in.
60 hp	
Manual start	0.058 in.
Electric start	0.060 in.
Low speed(idle) mixture screw adjustment	
4 and 5 hp models	1 1/2 turns out
6 hp	1–1 1/2 turns out
8-15 hp	1 1/4–1 3/4 turns out
20-hp (except 20 Jet)	3/4–1 1/4 turns out
20 Jet	1-2 turns out
25 hp	1–11/2 turns out
30 and 40 hp (two-cylinder)	1 1/2 turns out
40-55 hp (three-cylinder)	1 1/4 turns out
60 hp (manual start)	1 1/4 turns out
60 hp (electric start)	1 1/8 turns out

Table 4 REED VALVE SPECIFICATIONS

Reed stop opening	
2.5 and 3.3 hp	0.236-0.244 in. (5.99-6.19 mm)
4 and 5 hp	0.240-0.248 in. (6-6.3 mm)
6-15 hp	0.296 in. (7.5 mm)
40 hp (3-cylinder)	0.090 in. (2.29 mm)
Maximum reed tip opening (stand off)	
2.5-5 hp	Not specified
6-25 hp	0.007 in. (0.178 mm)
30-60 hp	0.020 in. (0.5 mm)

Chapter Seven

Ignition and Electrical Systems

This chapter provides service procedures for the battery, starter motor, charging system and ignition system. Wiring diagrams are located at the end of the manual. Torque values are in **Table 1** and **Table 2**. Battery capacity, battery cable size recommendations, battery charge percentage and wire color codes are in **Tables 3-6**. **Tables 1-6** are located at the end of this chapter.

BATTERY

Batteries used in marine applications endure far more rigorous treatment than those used in automotive electrical systems. Marine batteries have a thicker exterior case to cushion the plates during tight turns and rough water operation. Thicker plates are also used with each one individually fastened within the case to prevent premature failure. Spill-proof caps on the battery cells prevent electrolyte from spilling into the bilge.

Automotive batteries should be used in a boat *only* during an emergency situation when a suitable marine battery is not available.

CAUTION
Sealed or maintenance-free batteries are not recommended for use with unregulated charging systems. Excessive charging during continued high-speed operation causes the electrolyte to boil, resulting in loss of electrolyte. Since water cannot be added to sealed batteries, prolonged overcharging will destroy the battery.

Battery Rating Methods

The battery industry has developed specifications and performance standards to evaluate batteries and their energy potential. Several rating methods are available to provide information on battery selection.

Cold cranking amps (CCA)

Cold cranking amps is the amps the battery can deliver for 30 seconds at 0° F (-17.8° C) without dropping below

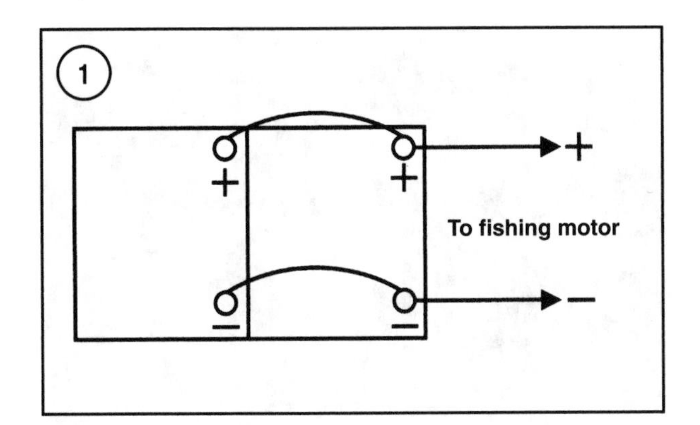

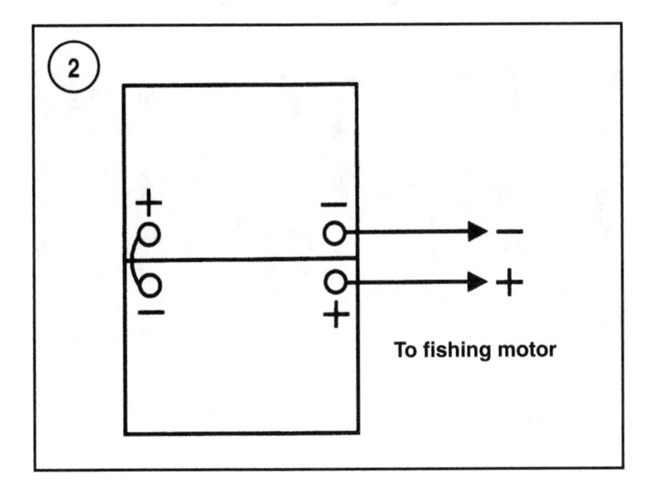

1.2 volts per cell (7.2 volts on a standard 12 volt battery). The higher the number, the more amps the battery can deliver to crank the engine. CCA times 1.3 equals MCA.

Marine cranking amps (MCA)

Marine cranking amps is similar to the CCA, except the test is run at 32° F (0° C) instead of 0° F (-17.8° C). This is closer to actual boat operating environments. MCA times 0.77 equals CCA.

Reserve capacity

Reserve capacity is the number of minutes a fully charged battery at 80°F (26.7° C) can deliver 25 amps, without dropping below 1.75 volts per cell (10.5 volts on a standard 12 volt battery). The reserve capacity rating is how long a typical vehicle can be operated after the charging system fails. This takes into account the power required by the ignition, lighting and other accessories. The higher the reserve capacity rating, the longer the vehicle could be operated after a charging system failure.

Amp-hour rating

The ampere hour rating is also called the 20 hour rating. This rating is the steady current flow the battery delivers for 20 hours while at 80° F (26.7°C) without dropping below 1.75 volts per cell (10.5 volts on a standard 12 volt battery). The rating is actually the steady current flow times 20 hours. For example, a 60 amp-hour battery delivers 3 amps continuously for 20 hours. This rating method has been largely discontinued by the battery industry. Cold cranking amps (or MCA) and reserve capacity ratings are the most common battery ratings.

Battery Recommendations

The manufacturer recommends a battery with a minimum rating of 465 cold cranking amps (CCA) or 350 marine cranking amps (MCA) and 100 minutes reserve capacity for the 6-60 hp models.

Battery Usage

Separate batteries may be used to provide power for accessories such as lighting, fish finders and depth finders. To determine the required capacity of such batteries, calculate the accessory current amperage draw rate and refer to **Table 3**.

Two batteries may be connected in parallel to double the ampere-hour capacity while maintaining the required 12 volts. See **Figure 1**. For accessories which require 24 volts, batteries may be connected in series (**Figure 2**), but only accessories specifically requiring 24 volts should be connected to the system. To charge batteries connected in a parallel or series circuit, disconnect and charge them individually.

Safety concerns

Securely fasten the battery in the boat to prevent it from shifting or moving in the bilge area. Cover the positive battery terminal or the entire top of the battery with a nonconductive shield or boot.

An improperly secured battery may contact the hull or metal fuel tank in rough water or while being transported. If the battery shorts against the metal hull or fuel tank, the resulting short circuit could cause sparks and an electrical fire. An explosion could follow if the fuel tank or battery case are compromised.

If the battery is not properly grounded and the battery contacts the metal hull, the battery shorts to ground through the control cables or the boat's wiring harness.

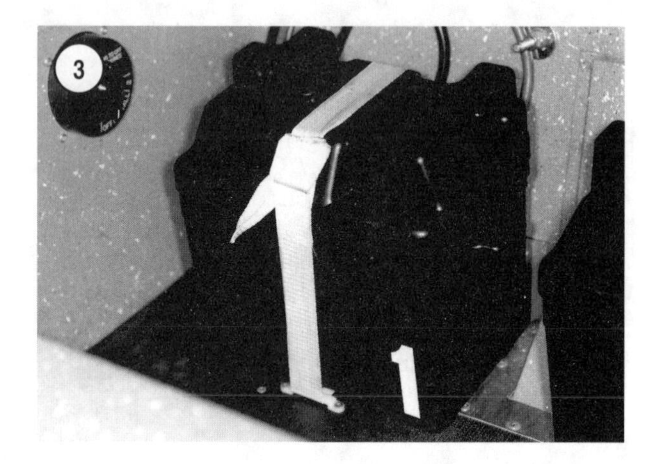

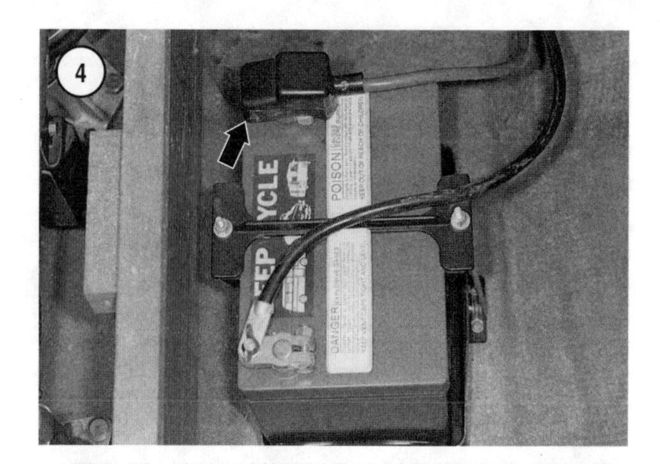

Again, the short circuit could cause sparks and an electrical fire. The control cables and boat wiring harness could be irreparably damaged.

Observe the following preventive steps when installing a battery in a boat, especially a metal boat or a boat with a metal fuel tank.

1. Choose a location as far as practical from the fuel tank while still providing access for maintenance.
2. Secure the battery to the hull with a plastic battery box and tie-down strap (**Figure 3**) or a battery tray with a nonconductive shield or boot covering the positive battery terminal (**Figure 4**).
3. Make sure all battery cable connections, the two at the battery and the two at the engine are clean and tight. Do *not* use wing nuts to secure battery cables. If wing nuts are used, replace them with corrosion resistant hex nuts and lock washers to ensure positive electrical connections. Loose battery connections can cause many problems.
4. Periodically inspect the installation to make sure the battery is secured to the hull and the battery cable connections are clean and tight.

Care and Inspection

1. Remove the battery tray top or battery box cover. See **Figure 3** or **Figure 4**.
2. Disconnect the negative battery cable, then the positive battery cable.

NOTE
*Some batteries have a built-in carry strap (**Figure 5**) for use in Step 3.*

3. Attach a battery carry strap to the terminal posts. Remove the battery from the boat.
4. Inspect the entire battery case for cracks, holes or other damage.

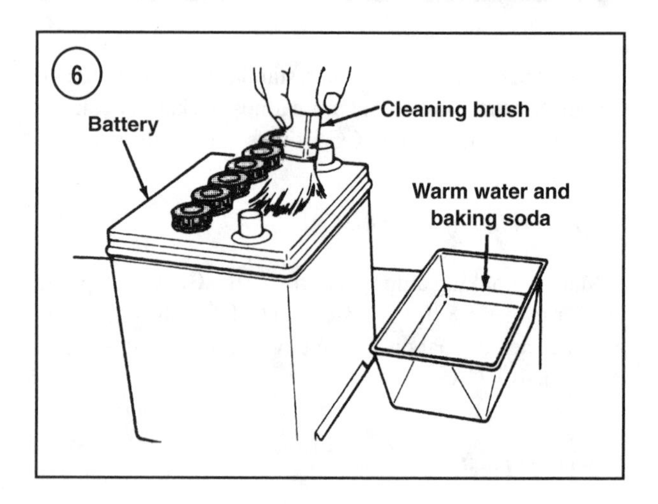

5. Inspect the battery tray or battery box for corrosion or deterioration. Clean as necessary with a solution of baking soda and water.

NOTE
Do not allow the baking soda cleaning solution to enter the battery cells in Step 6 or the electrolyte will be severely weakened.

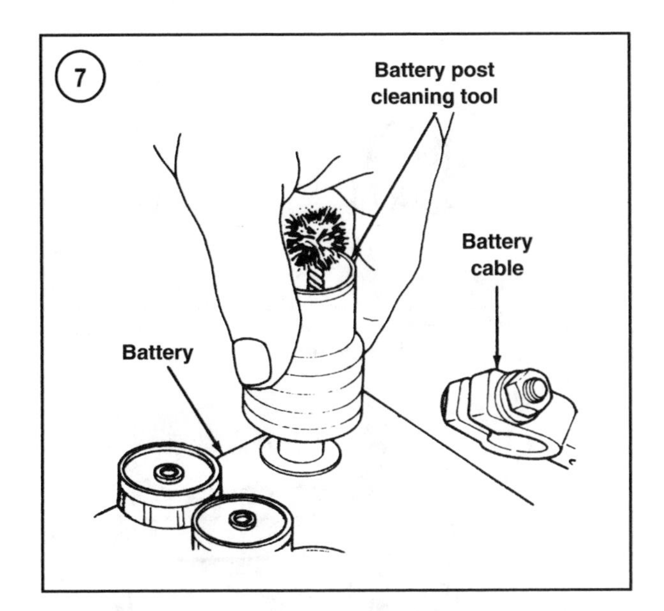

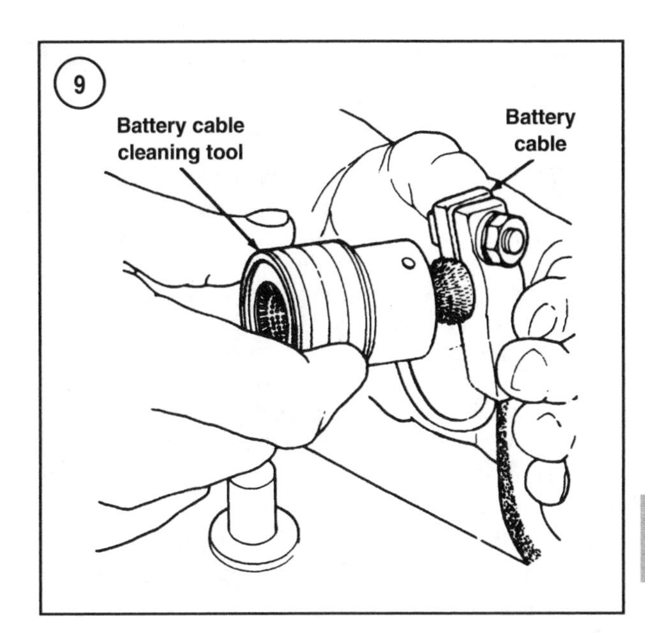

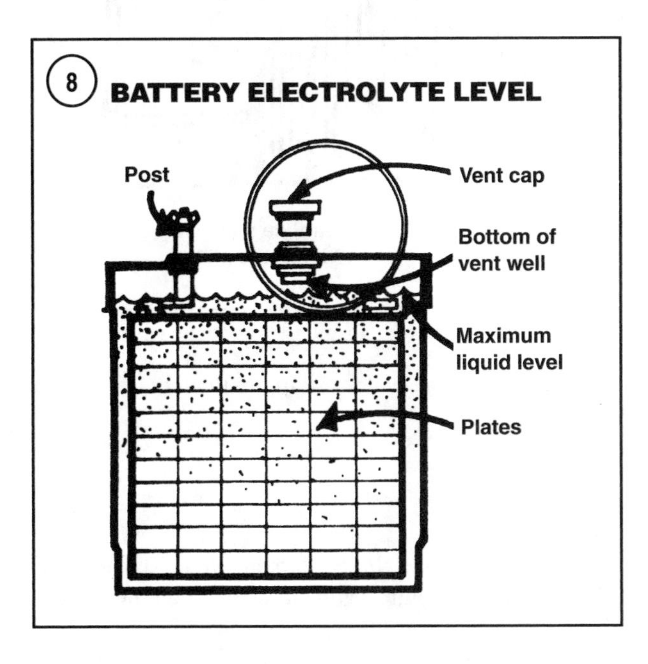

6. Clean the top of the battery with a stiff bristle brush using baking soda and water solution (**Figure 6**). Rinse the battery case with clear water and wipe it dry with a clean cloth or paper towel.
7. Clean the battery terminal posts with a stiff wire brush or battery terminal cleaning tool (**Figure 7**).

NOTE
Do not overfill the battery cells in Step 8. The electrolyte expands due to heat from the charging system and overflows if the level is more than 3/16 in. (4.8 mm) above the battery plates.

8. Remove the filler caps and check the electrolyte level. Add distilled water, if necessary, to bring the level up to 3/16 in. (4.8 mm) above the plates in the battery case. See **Figure 8**.
9. Clean the battery cable clamps with a stiff wire brush (**Figure 9**).
10. Place the battery back into the boat and into the battery tray or battery box. If using a battery tray, install and secure the retaining bracket.
11. Reconnect the positive battery cable first, then the negative cable.

CAUTION
Make sure the battery cables are connected to their proper terminals. Reversing the battery polarity damages the electrical and ignition systems.

12. Securely tighten the battery connections. Coat the connections with petroleum jelly or a light grease to minimize corrosion. If using a battery box, install the cover and secure the assembly with a tie-down strap.

Battery Testing

Hydrometer testing

On batteries with removable vent caps, checking the specific gravity of the electrolyte with a hydrometer is the best method to check the battery state of charge. Use a hydrometer with numbered graduations from 1.100-1.300 rather than one with color-coded bands. To use the hydrometer, squeeze the rubber bulb, insert the tip into a cell, then release the bulb to fill the hydrometer. See **Figure 10**.

NOTE
Do not test specific gravity immediately after adding water to the battery cells, as the water dilutes the electrolyte and lowers the specific gravity. To obtain accurate hydrometer readings, charge the battery after adding water and before testing with a hydrometer.

Draw enough electrolyte to raise the float inside the hydrometer. When using a temperature-compensated hydrometer, discharge the electrolyte back into the battery cell and repeat the process several times to adjust the temperature of the hydrometer to the electrolyte.

Hold the hydrometer upright and note the number on the float even with the surface of the electrolyte (**Figure 11**). This number is the specific gravity for the cell. Discharge the electrolyte into the cell from which it came.

The specific gravity of a cell indicates the cell's state of charge. A fully charged cell reads 1.260 or more at 80°F (26.7° C). A 75 percent charged cell reads from 1.220-1.230 while a 50 percent charged cell reads from 1.170-1.180. Any cell reading 1.120 or less is discharged. All cells should be within 30 points specific gravity of each other. If there is 30 points variation, the battery condition is questionable. Charge the battery and recheck the specific gravity. If 30 points or more variation remains between cells after charging, the battery has failed and must be replaced.

NOTE
If a temperature-compensated hydrometer is not used, add 4 points specific gravity to the actual reading for every 10° above 80° F (26.7°C). Subtract 4 points specific gravity for every 10° below 80° F (26.7°C).

Open-circuit voltage test

On sealed or maintenance-free batteries, check the state of charge by measuring the open-circuit, no load voltage of the battery. Use a digital voltmeter for best results. For the most accurate results, allow the battery to rest for at least 30 minutes to allow the battery to stabilize. Then, observing the correct polarity, connect the voltmeter to the battery and note the meter reading. If the open-circuit voltage is 12.7 volts or higher, the battery is fully charged. A reading of 12.4 volts means the battery is approximately 75 percent charged, a reading of 12.2 means the battery is approximately 50 percent charged and a reading of 12.1 volts means that the battery is approximately 25 percent charged.

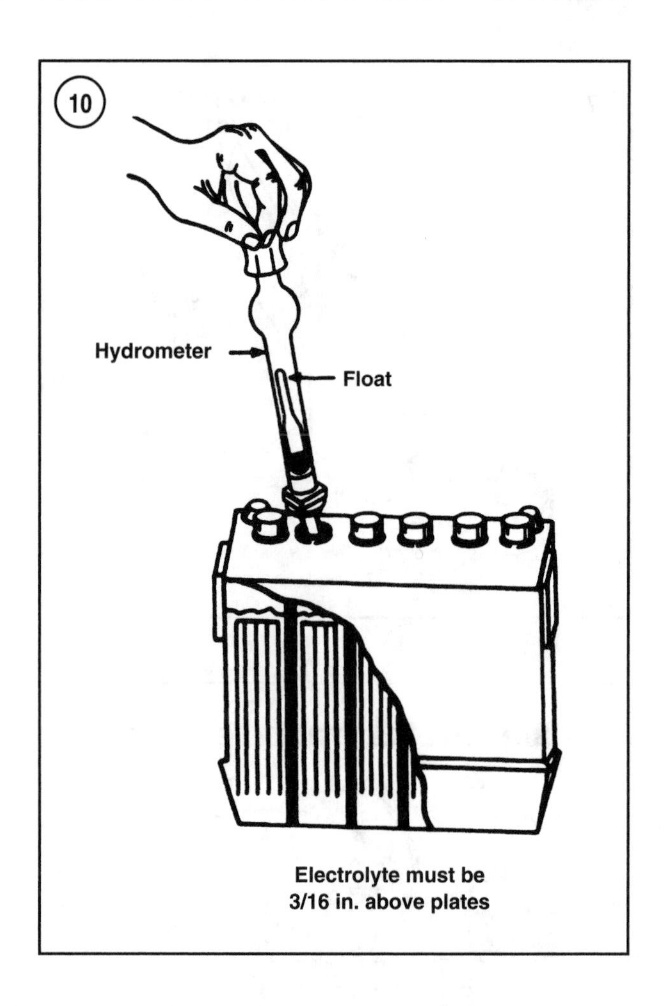

Load testing

Check the battery's ability to maintain the starting system's minimum required voltage while cranking the engine as follows:

1. Attach a voltmeter across the battery as shown in **Figure 12**.
2. Remove and ground the spark plug leads to the power head to prevent accidental starting.
3. Crank the engine for approximately 15 seconds while noting the voltmeter reading.

4A. If the voltage is 9.5 volts or higher at the end of the 15 second period, the battery is sufficiently charged and of sufficient capacity for the outboard motor.

4B. If the voltage is below 9.5 volts at the end of the 15 second period, one of the following conditions is present:

   a. The battery is discharged or defective. Charge the battery and retest.

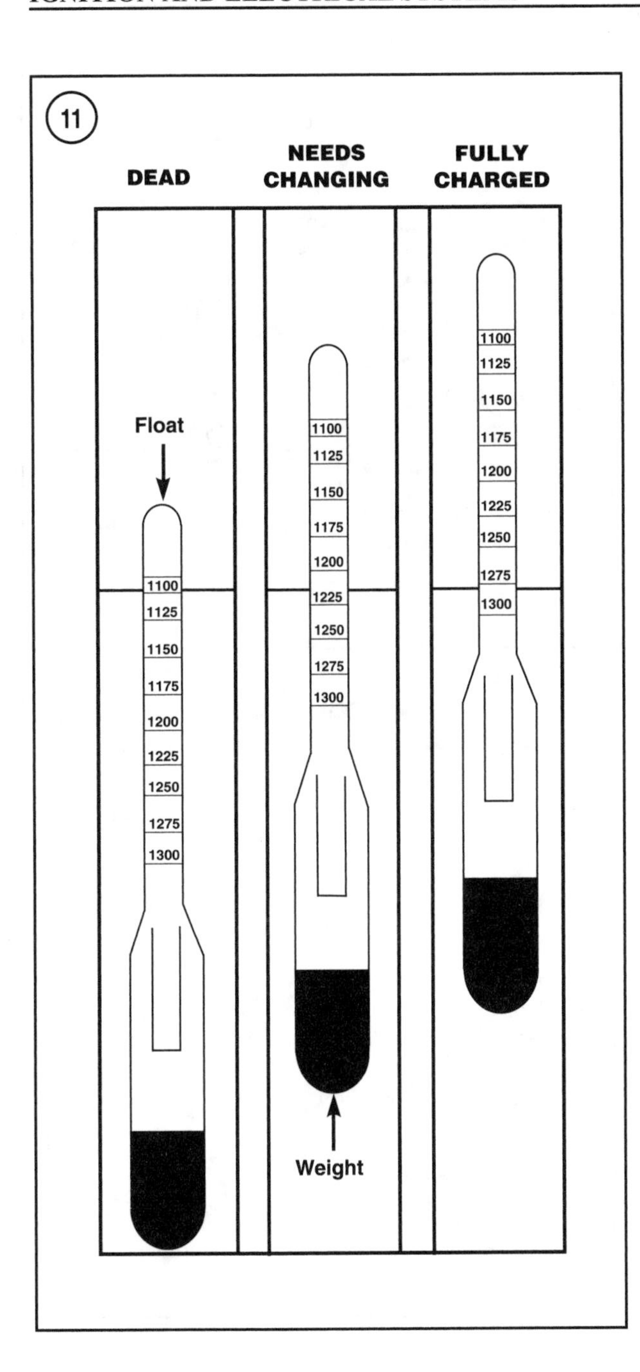

b. The battery is of too small capacity for the outboard motor. Refer to *Battery Recommendation* in this chapter.
c. The starting system is drawing excessive current causing the battery voltage to drop. Refer to Chapter Three for starting system troubleshooting procedures.
d. A mechanical defect is present in the power head or gearcase creating excessive load and current draw on the starting system. Inspect the power head and gearcase for mechanical defects.

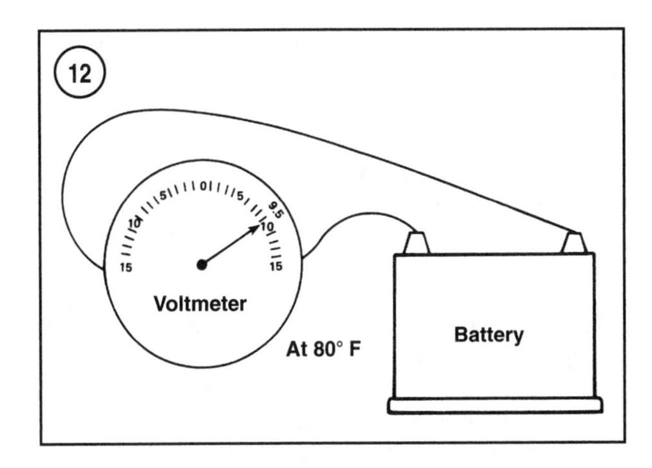

Battery Storage

Wet cell batteries slowly discharge when stored. They discharge faster when warm than when cold. Before storing a battery, clean the case with a solution of baking soda and water. Rinse with clear water and wipe dry. Fully charge the battery then store it in a cool, dry location. Check the electrolyte level and state of charge frequently during storage. If specific gravity falls to 40 points or more below full charge (1.260 volts), or the open circuit voltage falls below 12.4 volts, recharge the battery.

Battery Charging

Maintain a good state of charge in batteries used for starting. Check the state of charge with a hydrometer or digital voltmeter as described in the previous section.

Remove the battery from the boat for charging, since a charging battery releases highly explosive hydrogen gas. In many boats, the area around the battery is not well ventilated and the gas may remain in the area for hours after the charging process has been completed. Sparks or flames near the battery can cause it to explode, spraying battery acid over a wide area.

If the battery cannot be removed for charging, make sure the bilge access hatches, doors or vents are fully open to allow adequate ventilation. Observe the following precautions when charging batteries:

1. Never smoke in close proximity to a battery.
2. Make sure all accessories are turned off before disconnecting the battery cables. Disconnecting electrically active circuits creates a spark that can ignite explosive gas that may be present.
3. Always disconnect the negative battery cable first, then the positive cable.
4. On batteries with removable vent caps, always check the electrolyte level before charging the battery. Maintain

the correct electrolyte level throughout the charging process.

5. Never attempt to charge a frozen battery.

WARNING
Be extremely careful not to create any sparks around the battery when connecting the battery charger.

Connect the charger to the battery. Connect the negative charger lead to the negative battery terminal and the positive charger lead to the positive battery terminal. If the charger output is variable, select a setting of approximately 4 amps. It is better to charge a battery slowly at low amp settings, rather than quickly at high amp settings.

If the charger has a dual voltage setting, set the voltage switch to 12 volts, then switch on the charger.

If the battery is severely discharged, allow it to charge for at least 8 hours. Check the charging process with a hydrometer. The battery is fully charged when the specific gravity of all cells does not increase when checked three times at 1 hour intervals, and all cells are gassing freely.

Jump Starting

If the battery becomes severely discharged, it is possible to jump start the engine from another battery. Jump starting can be dangerous if the proper procedure is not followed.

Check the electrolyte level of the discharged battery before attempting to jump start. If the electrolyte is not visible or if it appears to be frozen, do not jump start the battery.

WARNING
Use extreme caution when connecting the booster battery to the discharged battery. Make sure the jumper cables are connected in the correct polarity.

1. Connect the jumper cables in the order and sequence shown in **Figure 13**.

WARNING
An electrical arc may occur when the final connection is made. This could cause an explosion if it occurs near the battery. Make the final connection to a good engine ground, away from the battery.

2. Make sure jumper cables are out of the way of moving engine parts.

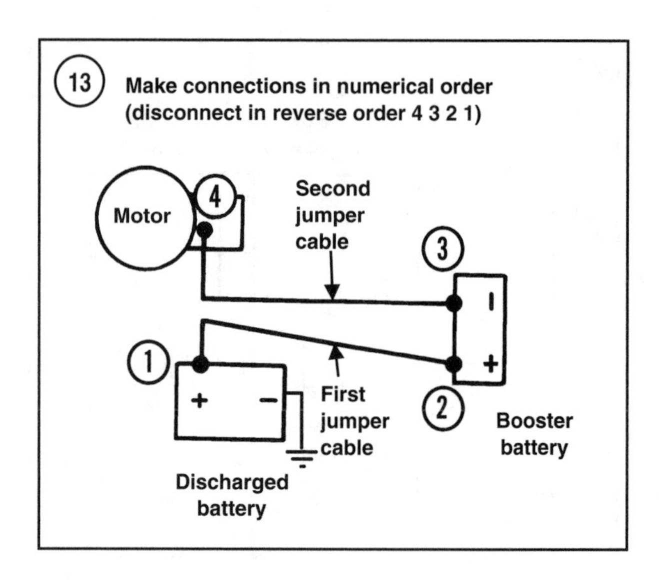

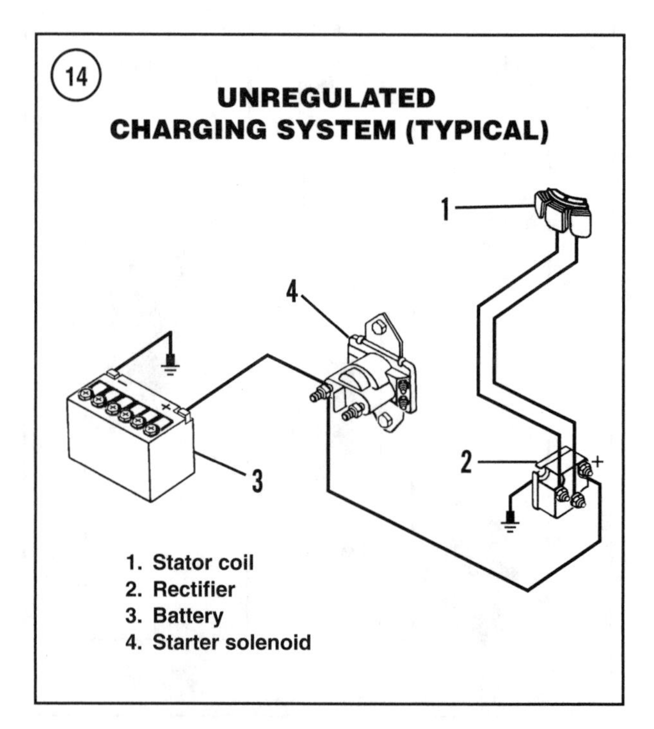

CAUTION
Do not run the engine without an adequate water supply and do not exceed 3000 rpm without an adequate load. Refer to ***Safety Precautions*** *in Chapter Three.*

3. Start the engine. Run the engine at a fast idle.

CAUTION
Running the engine at high speed with a discharged battery can damage the charging system.

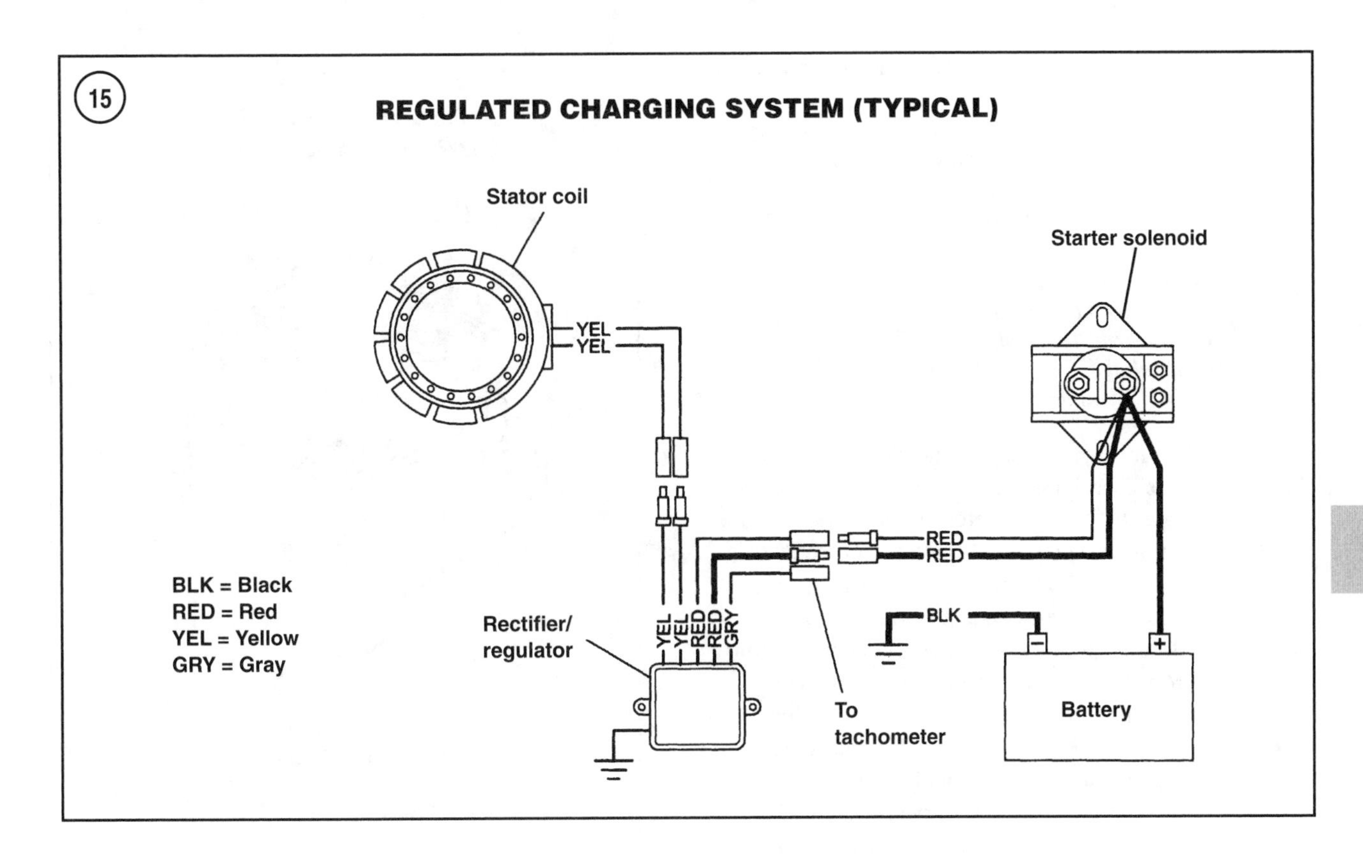

4. Remove the jumper cables in the reverse order shown in **Figure 13**. Remove the cable at point four, then three, then two and finally one.

CHARGING SYSTEM

An alternator charging system is used on all electric start models. The job of the charging system is to keep the battery fully charged and supply current to run accessories. Charging systems can be divided into two basic designs: unregulated and regulated.

Both designs use permanent magnets mounted in the flywheel and a stator coil winding mounted to the power head. As the flywheel rotates, the magnetic fields in the flywheel pass through the stator coil windings, inducing AC (alternating current).

Unregulated systems use a rectifier (a series of four diodes) to change the AC to DC (direct current). See **Figure 14**. The output from an unregulated charging system is directly proportional to engine speed. Because an unregulated system can overcharge the battery during long periods of wide-open throttle operation, do not use maintenance-free batteries. Overcharging a battery causes the electrolyte level to drop, leading to premature battery failure. Vented batteries allow removal of the vent caps and refilling of the electrolyte and have a longer service life.

Regulated systems use the same type flywheel magnets and stator coil windings as the unregulated system, with the rectifier being replaced with a rectifier/regulator. The rectifier portion of the rectifier/regulator changes the AC voltage to DC voltage, while the regulator portion monitors system voltage and controls the charging system output accordingly. Batteries maintained at 13-15 volts stay fully charged without excessive venting. The regulator controls the output of the charging system to keep system voltage at approximately 14.5 volts. The large red lead of the rectifier/regulator is DC output. The small red lead is the sensing terminal which allows the regulator portion to monitor system voltage. See **Figure 15**.

The charging system also provides the signal for the tachometer. The tachometer counts AC voltage pulses coming from the stator before the AC voltage is rectified to DC. Tachometer failure is related to the charging system, not the ignition system. The tachometer connects to one of the stator yellow leads on unregulated systems and connects to the rectifier/regulator gray lead (**Figure 15**) on regulated models.

A malfunction in the charging system generally causes the battery to be undercharged and the tachometer to not register or register erratically. The following conditions damage the charging system.

1. Reversing the battery leads.

2. Disconnecting the battery leads while the engine is running.
3. Loose connections in the charging system circuits, including battery connections and ground circuits.

CAUTION
If a charging system equipped outboard must be operated with the battery removed or disconnected, both stator yellow or yellow and gray leads must be disconnected and insulated (taped or sleeved) at both ends of the connection.

Perform the following visual inspection prior to troubleshooting the charging system. If the visual inspection does not locate the problem, refer to Chapter Three for complete charging system troubleshooting procedures.

1. Make sure the battery cables are connected properly. The positive cable must be connected to the positive battery terminal. If the polarity is reversed, check for a damaged rectifier or rectifier/regulator. See Chapter Three.
2. Inspect the battery terminals for loose or corroded connections. Tighten or clean as necessary. Replace wing nuts with corrosion resistant hex nuts and lock washers.
3. Inspect the physical condition of the battery. Look for bulges or cracks in the case, leaking electrolyte and corrosion buildup. Clean, refill or replace the battery as necessary.
4. Carefully check the wiring between the stator coil and battery for damage or deterioration. Refer to the wiring diagrams at the end of the manual. Repair or replace wires and connectors as necessary.
5. Check all accessory circuits and associated wiring for corroded, loose or disconnected connections. Clean, tighten or reconnect as necessary.
6. Determine if the accessory load on the battery is greater than the charging system's capacity by performing the *Current draw* test. See Chapter Three.

Rectifier Removal/Installation (Unregulated Models 6-25 hp)

1. Disconnect the negative battery cable.
2. Disconnect and ground the spark plug leads to the power head to prevent accidental starting.
3. Locate the rectifier on the power head. On some models, it may be necessary to remove a plastic cover to gain access to the electrical and ignition system components.
4. Note the location and color of the wires attached to the rectifier. See **Figure 16**. Remove the two yellow or one yellow and one gray leads, and one red lead from the rectifier terminal studs.

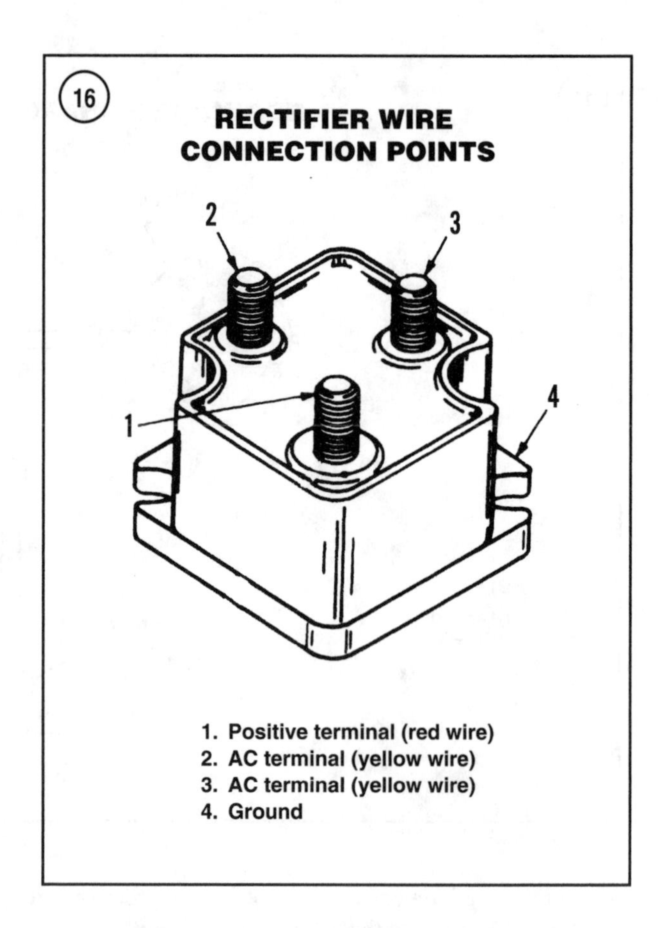

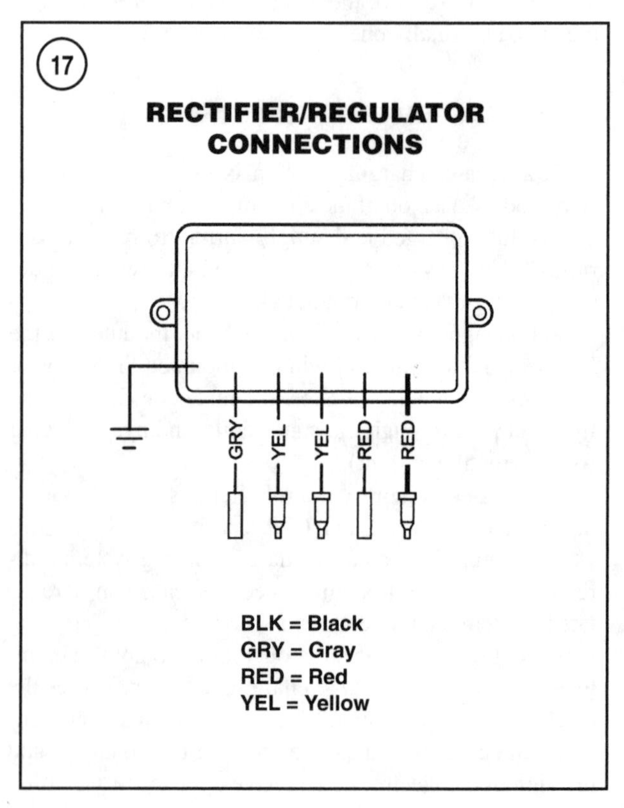

5. Remove the screws securing the rectifier to the engine or electrical bracket. Remove the rectifier from the engine.

6. To install the rectifier, make sure the mounting location is clean and free of corrosion or paint. Clean all corrosion and paint from mounting area as necessary.

7. Secure the rectifier to the engine or electrical bracket with screws. Tighten the screws securely.

8. Connect the two yellow or one yellow and one gray leads, and one red lead to the rectifier as noted during removal or refer to **Figure 16**. Tighten the terminal nuts securely.

9. Coat the terminal studs with Quicksilver Liquid Neoprene (part No. 92-25711-2).

10. Reinstall the electrical and ignition access cover, if so equipped.
11. Reconnect the spark plug leads.
12. Reconnect the negative battery cable.

Rectifier/Regulator Removal/Installation (Regulated Models 30-60 hp)

The rectifier/regulators used on 30-60 hp models are the same basic construction and use the same electrical connections. See **Figure 17**.

Remove and install the rectifier/regulator as follows:

1. Disconnect and ground the spark plug leads to the power head to prevent accidental starting.
2. Disconnect the negative battery cable.
3. Locate the rectifier/regulator(s) on the power head. On some models, it may be necessary to remove a plastic cover (**Figure 18**, typical) to gain access to the electrical and ignition system components. **Figure 19** shows a typical regulator installation.
4. Cut the tie-strap(s) and/or loosen the clamps (B, **Figure 19**) securing the electrical leads and bullet connectors of each regulator to the power head or electrical/ignition bracket.
5. Disconnect the two yellow, two red and one gray lead bullet connecting the regulator to the engine wiring harness. Do not damage the connector terminals or insulating sleeve.
6. Remove the screws (C, **Figure 19**) securing the rectifier/regulator to the power head or electrical/ignition bracket. Remove the rectifier/regulator.
7. To install the rectifier/regulator, position the rectifier/regulator on the power head or electrical/ignition bracket and secure it with the screws. Tighten the screws to the specification in **Table 1**.
8. Connect the rectifier/regulator two yellow, one red and one gray wires to the engine wiring harness. Secure the wires to the power head or electrical/ignition bracket with a new tie-strap or reattach any clamps.
9. Reinstall the electrical and ignition access cover, if so equipped.
10. Reconnect the spark plug leads.
11. Reconnect the negative battery cable.

Stator Removal/Installation

The alternator and ignition windings of the stator are integrated into one assembly on all models except the 4 and 5 hp models and 1998 6-25 hp models. On these models, the charging system stator windings and the ignition system stator windings may be replaced separately.

Removal and installation procedures for all stator windings are covered in *Ignition Systems* later in this chapter.

FUSES

Fuses are designed to protect wire and electrical components from damage due to excessive current flow. A fuse that repeatedly blows indicates a problem with the circuit or component that the fuse is protecting.

Do not install a larger fuse in an attempt to remedy the problem. Refer to Chapter Three and locate the defect causing excessive current flow in the suspect circuit.

A visual inspection can show if a fuse is *bad*, but do not trust a visual inspection to determine if a fuse is *good*. Fuses can be quickly and accurately tested using an ohmmeter. A good fuse indicates a full continuity reading (0 ohms). When testing fuses, do not to touch both ohmmeter probes at the same time with your hands or the reading could be false.

Fuse Locations

A 20 amp blade fuse is located near the rectifier/regulator in a locking fuse holder (D, **Figure 19**).

Fuse Replacement

1. Carefully lift the locking clip and slide the fuse holder out of the protective cover.
2. Pull the defective fuse from the fuse holder.
3. Push a new fuse into the fuse holder.
4. Push the fuse holder into the protective cover until the locking clip snaps into place.

STARTING SYSTEMS

Mercury and Mariner outboards can be equipped with electric start only, manual start only, or both electric and manual starters. Manual starters are covered in detail in Chapter Thirteen.

A typical electric starter system consists of the battery, starter solenoid, neutral safety switch, starter motor, starter or ignition switch and the associated wiring. On tiller models, the neutral safety switch is mounted on the engine shift linkage. On remote control models, the neutral safety switch is mounted in the remote control box. Electric starter system troubleshooting is covered in Chapter Three.

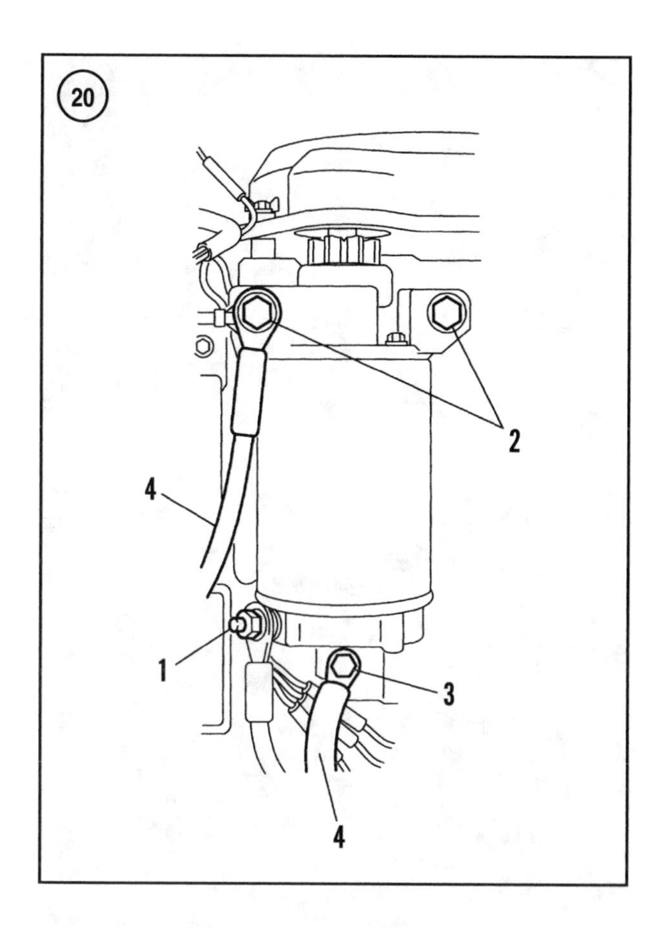

Starter Motor Description

Marine starter motors are similar in design and operation to those found on automotive engines. The starter motors used on outboards covered in this manual have an inertia drive. External spiral splines on the armature shaft mate with internal splines on the drive or bendix assembly.

The starter motor is an intermittent duty electric motor, capable of producing a very high torque, but only for a brief time. The high amperage flow through the starter motor causes the starter motor to overheat very quickly. To prevent overheating, never operate the starter motor continuously for more than 10-15 seconds. Allow the starter motor to cool 2-3 minutes before cranking the engine again.

If the starter motor does not crank the engine, check the battery cables and terminals for loose or corroded connections. If this does not solve the starting problem, refer to Chapter Three for starting system troubleshooting procedures.

CAUTION

Mercury and Mariner electric starter motors use permanent magnets glued to the

(21)

STARTER MOTOR (6-25 HP MODELS)

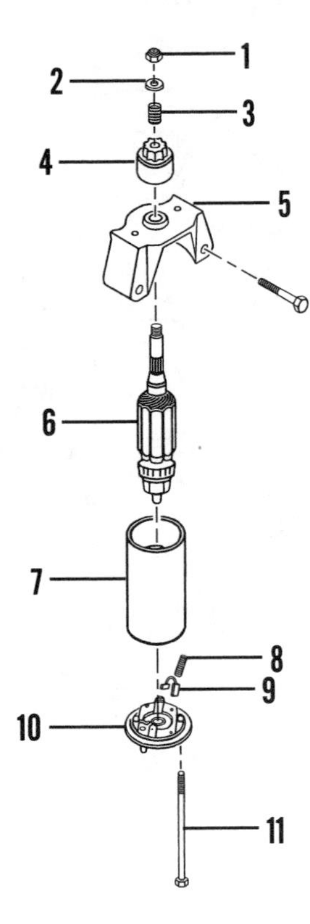

1. Locknut
2. Spacer
3. Spring
4. Drive gear
5. Drive end frame
6. Armature
7. Frame
8. Brush spring
9. Brush
10. End cap
11. Through-bolt

main housing. Never strike a starter as it damages the magnets, leading to starter failure. Inspect the magnets anytime the starter is disassembled. Replace the housing if the magnets are cracked, damaged or loose.

Starter Motor Removal/Installation

1. Disconnect the negative battery cable.
2. Disconnect and ground the spark plug leads to the power head to prevent accidental starting.
3. Remove the starter motor mounting bolts (2, **Figure 20**, typical).
4. Lift the starter motor far enough to remove the electrical cable, which is usually yellow, from the terminal on the bottom of the starter motor (1, **Figure 20** typical). Note the orientation of the cable to the starter motor, then remove the cable.
   a. *30 and 40 hp two-cylinder models*—Remove the bolt (3, **Figure 20**, typical) and large gauge ground wire (4) from the bottom of the starter motor.
   b. *40-60 hp three-cylinder models*—Remove the bolt (3, **Figure 20**) and ground wire (4) from the bottom of the starter motor.
5. Install the starter motor as follows:
   a. *40-60 hp three-cylinder models*—Slip the forward mounting bolt (2, **Figure 20**, typical) through the large gauge wire terminal.
   b. Position the starter motor on the power head. Secure the starter motor with the mounting bolts. Tighten the bolts to the specification in **Table 1**. Reconnect the large gauge ground wires(s). See Step 4.
   c. Secure the electrical cable to the starter motor terminal (1, **Figure 20**, typical). Position the cable to allow installation of the starter motor without pinching the cable between the starter and the power head or lower cowl. Tighten the cable (**Table 1**) once it is positioned.
   d. *30-60 hp models*—Secure the large gauge ground wire (4, **Figure 20**, typical) to the bottom of the starter with the bolt (3). Securely tighten the bolt.

Starter Motor Disassembly/Assembly

6-25 hp models

Refer to **Figure 21** for this procedure. Refer to *Cleaning and Inspection* before assembly.

1. Remove the starter motor as described previously in this chapter.
2. Place match marks (**Figure 22**) on the drive end frame, frame and lower end cap for alignment reference during assembly. Then remove the two throughbolts (11, **Figure 21**).
3. Lightly tap on the drive end frame (5, **Figure 21**) and lower end cap (10) with a rubber mallet until they are loosened. Then remove the end cap taking care not to lose the brush springs.

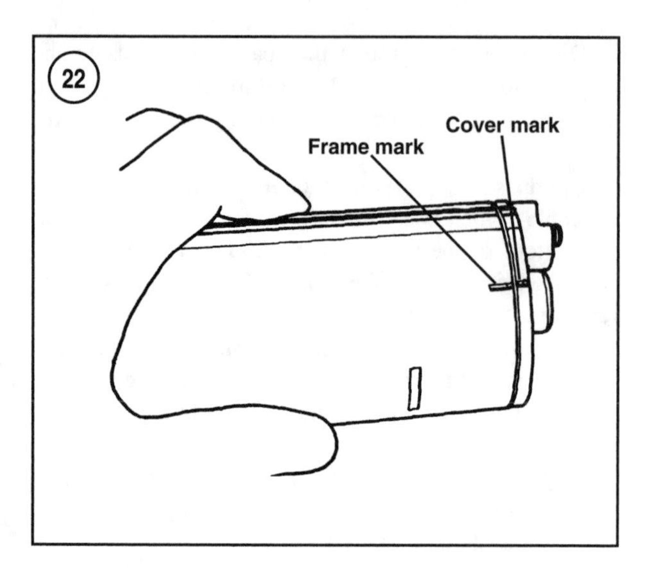

4. Slide the starter housing off the armature and drive end frame assembly.

NOTE
It is not necessary to remove the starter drive assembly in Step 5 unless the drive requires replacement.

5. Secure the armature with a strap wrench (part No. 91-24937A1 or an equivalent). Loosen and remove the starter drive locknut, then remove the drive components (2-4, **Figure 21**) from the armature shaft.
6. To remove the brushes, remove the screw holding the negative brush lead to the end cap (2, **Figure 23**). Remove the negative brush.
7. Remove the hex nut and washers from the positive terminal. Remove the positive terminal and positive brush from the end cap. See **Figure 23**.
8. Install a new positive brush and positive terminal assembly into the end cap. Tighten the terminal nut securely.
9. Install new negative brushes. Tighten the screws securely.

CAUTION
Do not over-lubricate the starter bushings in Step 10. The starter will not operate properly if oil contaminates the commutator or brushes.

10. Lubricate each bushing (**Figure 24**) in the drive end frame and end cap with a single drop of SAE 10W motor oil. *Do not* over-lubricate.
11. Insert the armature (6, **Figure 21**) into the drive end frame (5). Lubricate the armature helical splines with a single drop of SAE 10W motor oil. Install the starter drive gear, spring, spacer and a *new* locknut. Secure the arma-

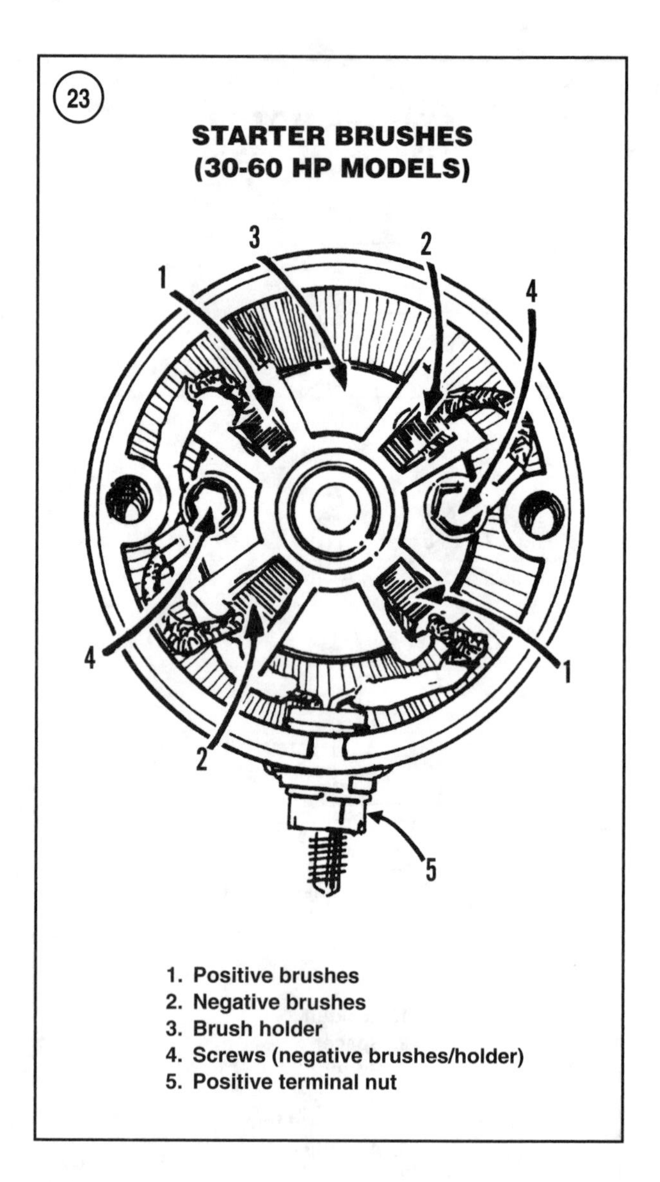

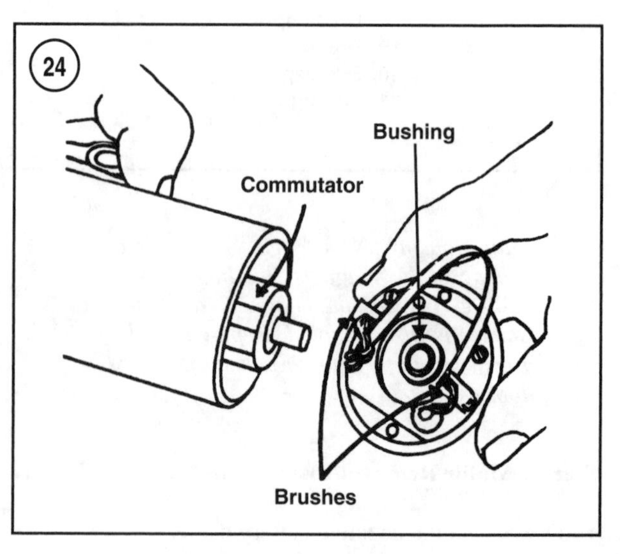

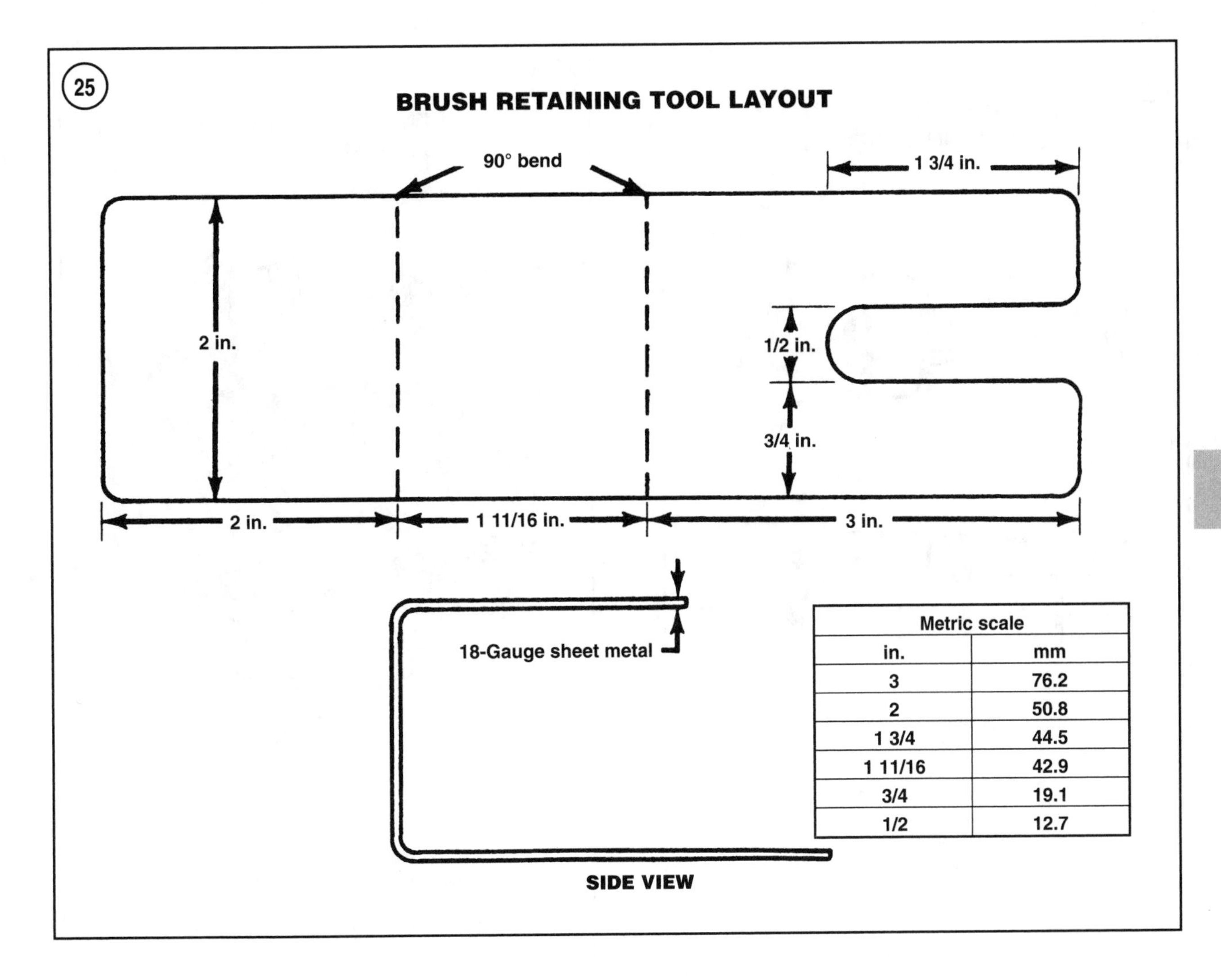

Metric scale	
in.	mm
3	76.2
2	50.8
1 3/4	44.5
1 11/16	42.9
3/4	19.1
1/2	12.7

ture using a strap wrench (part No. 91-24937A1 or an equivalent) and tighten the locknut securely.

12. Place the frame (7, **Figure 21**) over the armature. Make sure the commutator end of the armature is located at the end of the housing with the magnets recessed 1-5/16 in. (33.3 mm). Align the match marks (**Figure 22**) on the housing and the drive end frame.

13. Fit the brush springs and brushes into the brush holders. Press the brushes into the holders and use a strip of flexible metal as shown in **Figure 24** to hold the brushes in place.

14. Push the drive end of the armature shaft into the housing so the commutator end extends out the housing.

15. Install the end cap onto the armature shaft. Remove the brush holding tool, then push the end cap up against the frame assembly.

16. Align the match marks (**Figure 22**) on the end cap and frame. Install the throughbolts and tighten the bolts evenly and securely.

30-60 hp models

Fabricate a brush retaining tool from 18-gauge sheet metal to the dimensions shown in **Figure 25**. This tool is necessary to position the brushes properly and prevent damaging them when reassembling the starter end cap to the housing.

Refer to *Cleaning and Inspection* before assembly.

1. Remove the starter motor as described previously in this chapter.

2. Place match marks (**Figure 22**) on the drive end, frame and lower end cap for alignment reference during assembly.

3. Remove the throughbolts, then lightly tap on the drive end cover and lower end cap with a rubber mallet until they are loosened.

4. Remove the end cap taking care not to lose the brush springs.

7

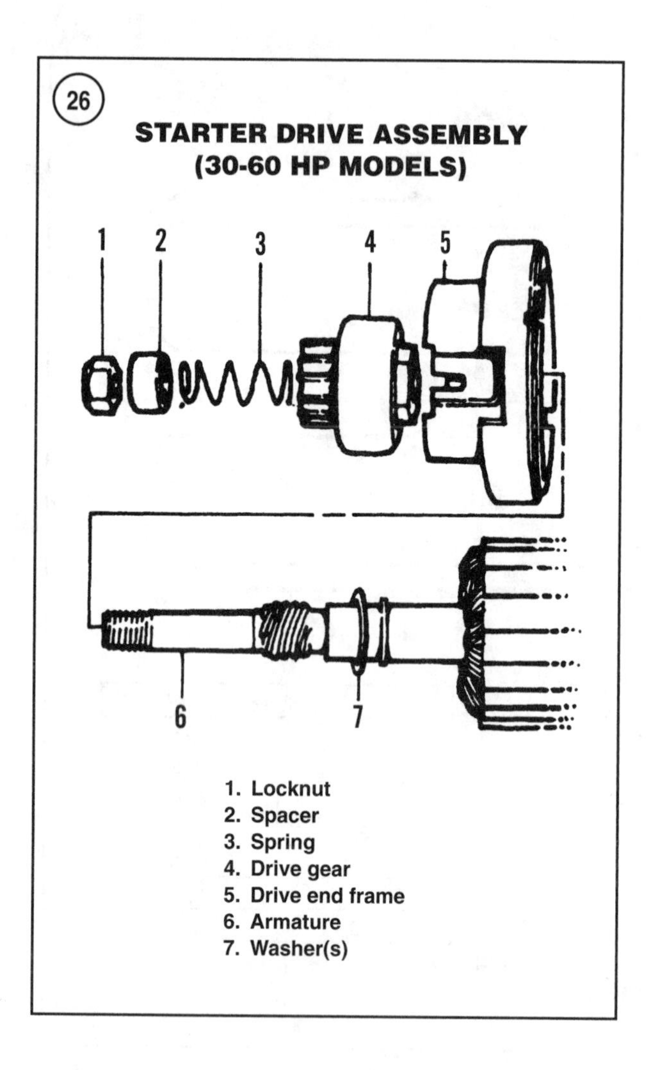

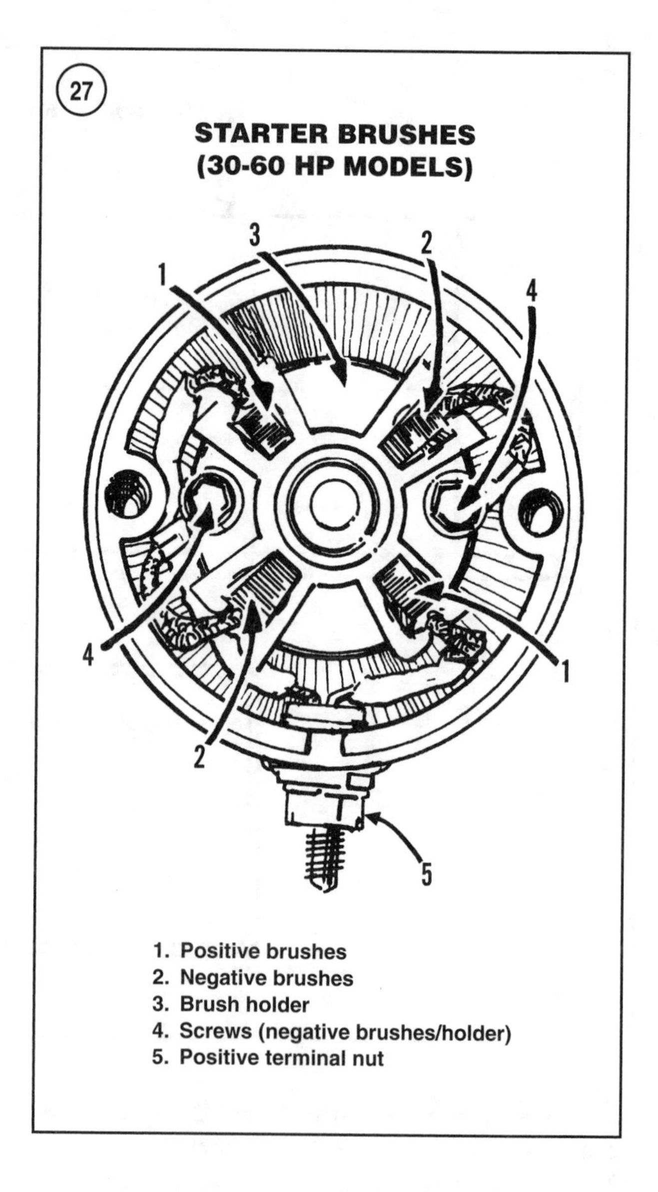

5. Lift the armature and drive end frame assembly from the starter housing.

NOTE
Do not remove the drive assembly in Step 6 and Step 7 unless the drive or end frame requires replacement.

6. Place an appropriate size wrench on the hex area on the back side of the drive gear (4, **Figure 26**).

7. Remove the drive assembly locknut and slide the drive components and drive end frame from the armature shaft.

8. Remove the screws (4, **Figure 27**) securing the brush holder and negative brushes to the end cap. Lift the brush holder from the end cap.

9. Remove the negative brushes from the brush holder.

10. Remove the hex nut and washers from the positive terminal (1, **Figure 28**). Remove the positive terminal and positive brushes from the end cap as an assembly.

11. To reassemble the starter motor, install new positive brushes and terminal assembly into the end cap. Locate the longest brush lead as shown in **Figure 28**.

12. Install the negative brushes into the brush holders. Install the brush holders into the end cap. Tighten the fasteners securely.

13. Fit the springs and brushes into the brush holder. Hold the brushes in place with the brush retaining tool as shown in **Figure 29**.

14. Lubricate the armature shaft splines and the drive end frame bushing with one drop of SAE 10W oil each.

15. Install the drive components onto the armature shaft as shown in **Figure 26**. Tighten the locknut securely while holding the drive gear (4, **Figure 26**) with a wrench.

16. Place the armature and end frame assembly into the starter housing. Make sure the commutator end of the ar-

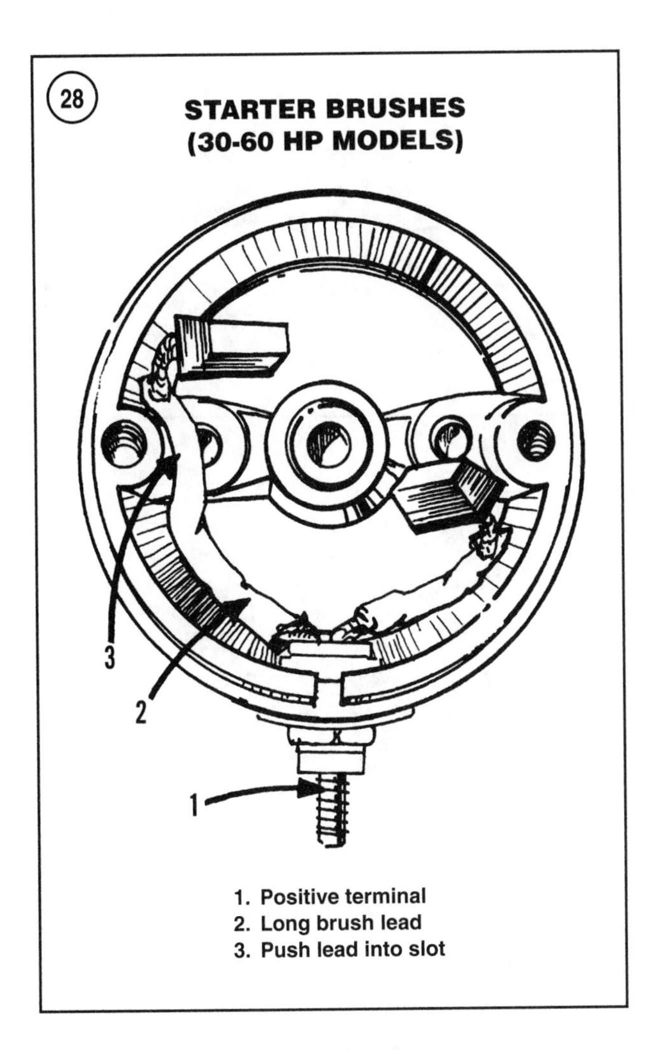

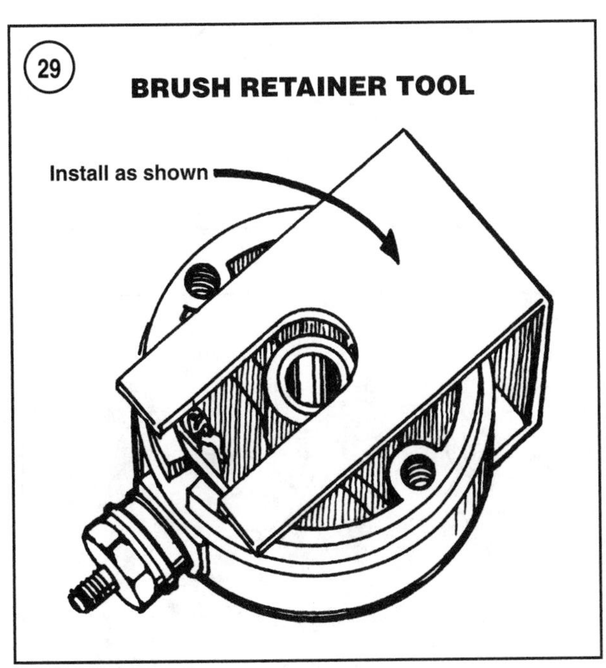

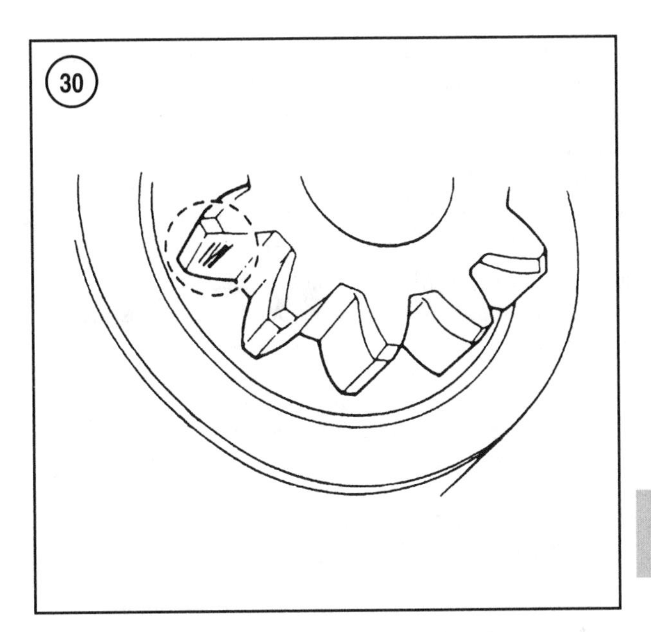

mature is located at the end of the housing with the magnets recessed 1 in. (25.4 mm). Align the match marks on the housing and end frame.

CAUTION

Do not over-lubricate the starter bushings in Step 17. The starter will not operate properly if oil contaminates the commutator or brushes.

17. Lubricate the lower end cap bushing with a single drop of SAE 10W motor oil.
18. Hold the brushes in position with the brush retaining tool (**Figure 29**) and install the end cap onto the armature and up against the frame. Remove the brush retaining tool, align the match marks on the end cap and frame (**Figure 22**), then install the throughbolts. Tighten the throughbolts to the specification in **Table 1**.

Cleaning and Inspection

1. Thoroughly clean all starter motor components with clean solvent, then dry them with compressed air.
2. Check the starter drive gear (**Figure 30**) for chipped teeth, cracks or excessive wear. Replace drive components as necessary.
3. Inspect the starter brushes in the end cap. See **Figure 23** or **Figure 27**. Replace all brushes if they are pitted, chipped, oil soaked or worn to 3/16 in. (4.8 mm) or less.
4. Inspect the armature shaft bushings in the drive end frame and lower end cap for excessive wear or other damage. Replace the end frame or end cap as necessary.

5. Clean the commutator using 00 sandpaper (**Figure 31**). Use a nail file to clean copper particles or other contamination from between the commutator segments (**Figure 32**).

6. If the commutator is pitted, rough or worn unevenly, resurface or replace it. If the armature shows water or overheat damage, check it for shorted windings using an armature growler. Most automotive electrical shops can perform commutator resurfacing and armature testing.

NOTE

*If the armature is resurfaced, the insulation between the commutator segments must be undercut. Undercut the insulation between the commutator segments using a broken hacksaw blade or similar tool. The undercut must be the full width of the insulation and approximately 1/32 in. (0.8 mm) deep (**Figure 33**). Do not damage the commutator segments during the process. Thoroughly clean copper particles from between the segments after undercutting. Clean and smooth the commutator after undercutting using 00 sandpaper.*

7. Use an ohmmeter to check for continuity between each commutator segment and the armature shaft (**Figure 34**). Replace the armature if there is continuity.

8. Use an ohmmeter to check for continuity between each commutator segment (**Figure 35**). There should be continuity between each selected pair of segments. If there is not, the armature is defective and must be replaced.

Starter Solenoid

The large solenoid terminals (**Figure 36**) carry the electrical load from the battery to the starter motor. The large terminals have an open circuit across them when the solenoid is not energized. The large lead from the battery is usually black with red sleeves. The large lead from the solenoid to the starter motor is usually yellow or black with yellow ends (sleeves).

The small terminals (**Figure 36**) are control circuits of the solenoid. When battery voltage is applied to these terminals, the solenoid is energized and the large terminals have a closed circuit across them, allowing electricity to flow from the battery to the starter motor. The polarity of the small terminals is not important as long as one is positive and one is negative. One small lead is always yellow/red and the other small lead is black.

The starter solenoid is always located near the starter motor and, depending on model, may or may not be be-

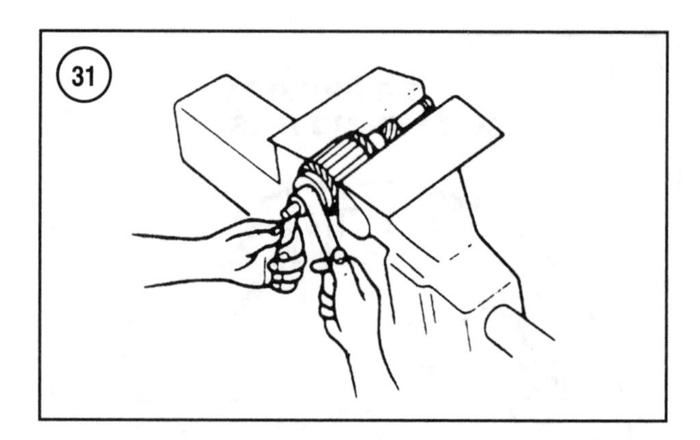

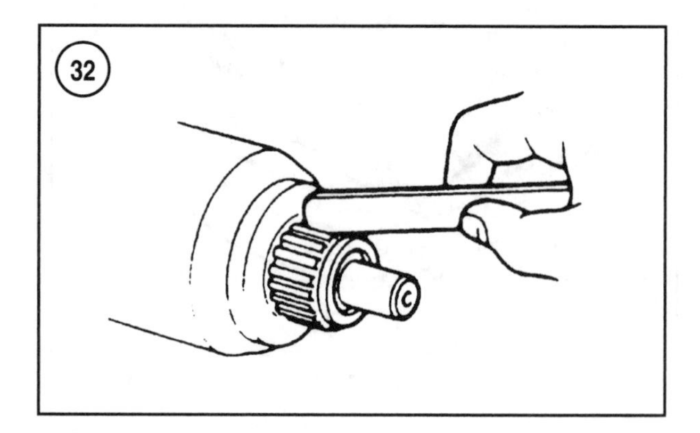

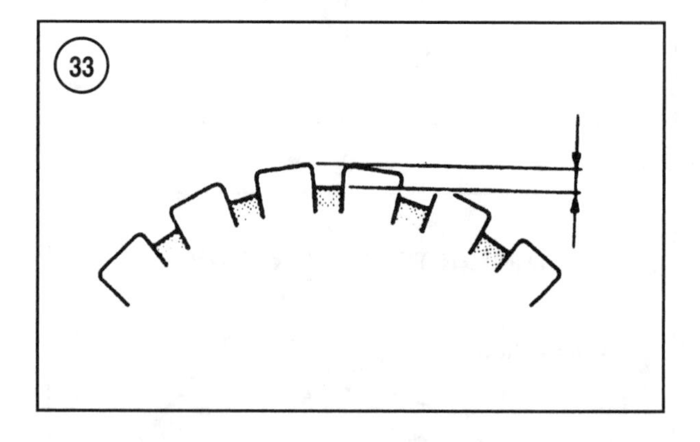

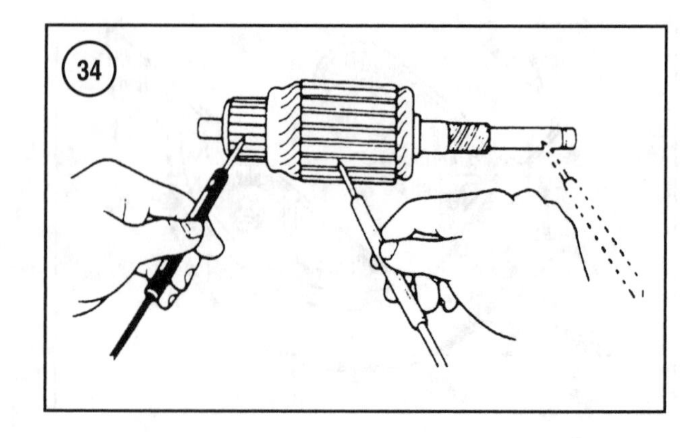

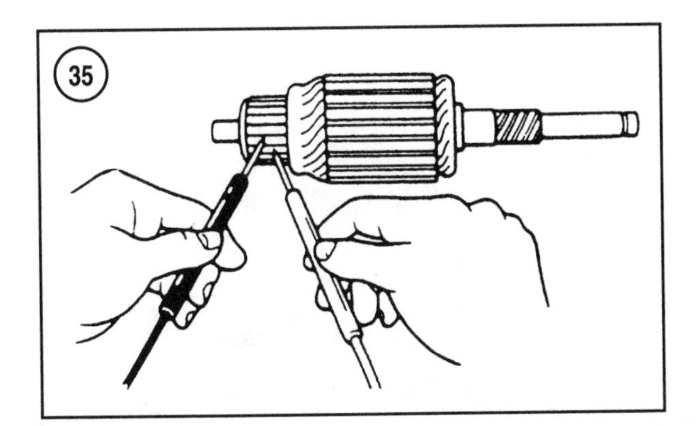

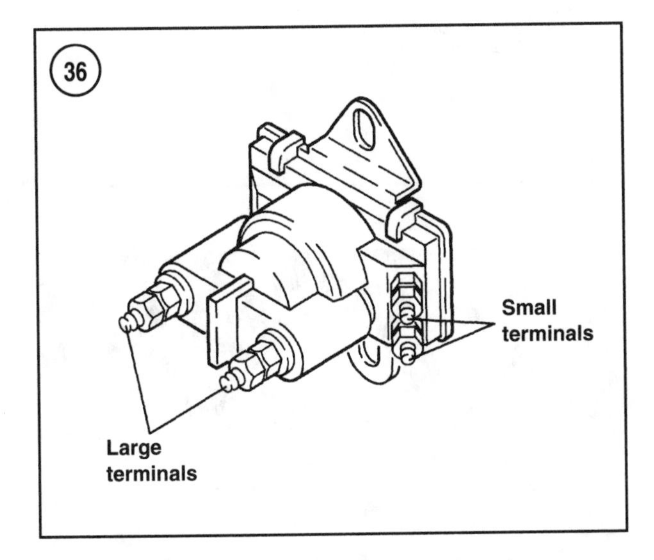

hind an electrical/ignition component access cover. Refer to Chapter Three for troubleshooting procedures.

Removal/installation

Replace the starter solenoid as follows. Refer to **Figure 37**, typical.

1. Disconnect the negative battery cable.
2. Disconnect and ground the spark plug leads to the power head to prevent accidental starting.
3. Locate the starter solenoid on the power head. Remove the electrical/ignition component access cover, if so equipped. Carefully pull the plastic cover from the solenoid, if so equipped.
4. Note the location and position of all wires on the starter solenoid and mounting screws.
5. Remove the nuts, lockwashers and electrical cables from the two large solenoid terminals.
6. Remove the nuts, lockwashers and electrical wires from the two small solenoid terminals.

7. Remove the two solenoid mounting screws, then remove the solenoid from the power head or electrical bracket.
8. To install the solenoid, position the solenoid on the power head or electrical bracket. Make sure any ground wires or ground straps are reconnected to the mounting screws as noted on removal. Tighten both mounting screws to the specification in **Table 1**.
9. Install the small wires, lockwashers and nuts. Tighten the nuts to the specification in **Table 1**.
10. Install the large cables, lockwashers and nuts. Tighten the nuts to the specification in **Table 1**. Make sure the positive battery cable is covered with a protective boot or plastic cover unless the engine is equipped with a plastic electrical/ignition component access cover.
11. Reinstall the electrical/ignition component access cover, if so equipped.
12. Reconnect the spark plug leads. Reconnect the negative battery cable.

Neutral Safety Switch

On tiller models, the neutral safety switch is mounted on the engine shift linkage. On remote control models, the neutral safety switch is mounted in the remote control box. Refer to Chapter Fourteen for Mercury/Mariner remote control box service procedures. If the boat is equipped with an aftermarket control box, consult the control box manufacturer for service procedures.

The neutral safety switch is designed to prevent the engine from starting in gear. The electric starter can only engage when the engine shift linkage is in neutral. A neutral safety switch should have continuity across its two terminals any time the shift linkage is in neutral. The switch should indicate no continuity any time the shift linkage is in gear.

On 20-25 hp models, the switch is mounted to the starboard side of the power head (**Figure 38**). The plunger portion of the switch contacts the shift linkage lever.

On all other models, the switch is mounted to the lower engine cowl (**Figure 39**). The plunger portion of the switch contacts the shift linkage shaft.

Removal/installation (tiller models)

1. Disconnect the negative battery cable.
2. Disconnect and ground the spark plug leads to the power head to prevent accidental starting.
3. Locate the neutral safety switch on the engine shift linkage shaft or power head. The switch typically has two short black or one black and one yellow/red wires with bullet connectors. Disconnect the two wires from the engine wiring harness.
4. Remove the screws (**Figure 38** or **Figure 39**) securing the switch to the shift linkage or power head. Remove the switch.
5. To install the switch, position the switch on the shift linkage or power head. Install the screws and tighten them to the specification in **Table 1**. Do not crush or distort the switch when tightening the screws.
6. Connect the switch wires to the engine harness bullet connectors.
7. Reconnect the spark plug leads.
8. Reconnect the negative battery cable.

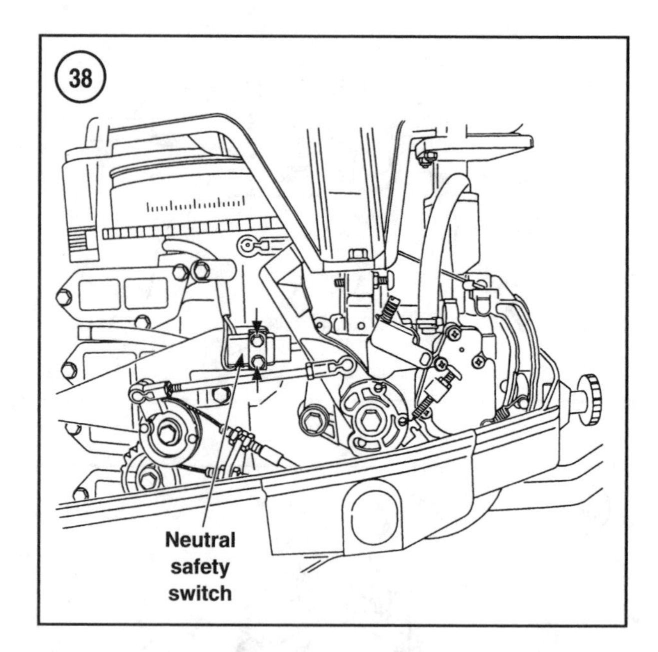

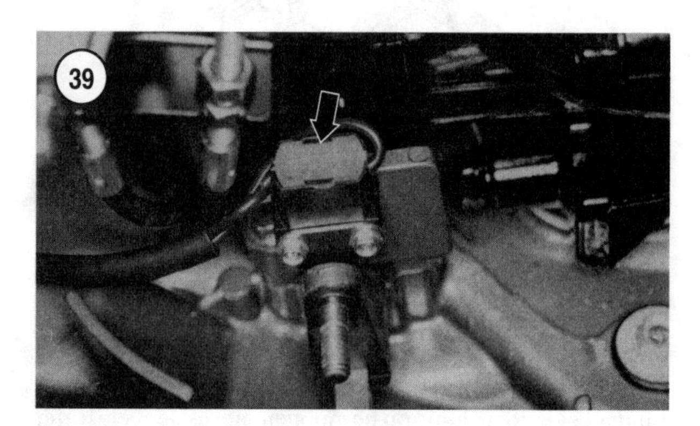

CAPACITOR DISCHARGE IGNITION (2.5-5 HP MODELS)

The CDI (capacitor discharge ignition) system used on 2.5-5 hp models consists of the flywheel, charge coil, trigger coil (4 and 5 hp models), CD module (switch box), ignition coil, spark plug and a combination push button stop/safety lanyard switch.

The 2.5 and 3.3 hp models feature fixed timing with no spark advance or timing adjustments. Because the charge coil and trigger coil are combined, the ignition coil sparks every 180° of crankshaft rotation.

The 4 and 5 hp models incorporate a nonadjustable electronic spark advance built into the CD module (switch box). These models feature a separate trigger coil. The ignition coil sparks once every 360°of crankshaft rotation.

This section includes operating theory and component removal and replacement. Refer to Chapter Three for ignition system troubleshooting procedures and the end of the manual for complete wiring diagrams.

Operation

2.5 and 3.3 hp models have a series of four permanent magnets along the inner diameter of the outer rim of the flywheel. Each magnet is the opposite polarity from the magnet next to it. As the flywheel rotates and a magnet passes the charge coil, voltage is induced in the charge coil windings, transferred to the CD module (switch box) and stored in a capacitor contained inside the switch box. As the flywheel continues to rotate, the next magnet passes the charge coil and induces a voltage pulse of opposite polarity in the charge coil windings. This voltage pulse causes an electronic switch in the switch box called an SCR (silicon controlled rectifier) to close, allowing the stored voltage in the capacitor to discharge into the ignition coil. The ignition coil amplifies this voltage and discharges it into the spark plug wire. This is repeated with each half revolution of the flywheel or the passing of every other magnet. See **Figure 40**.

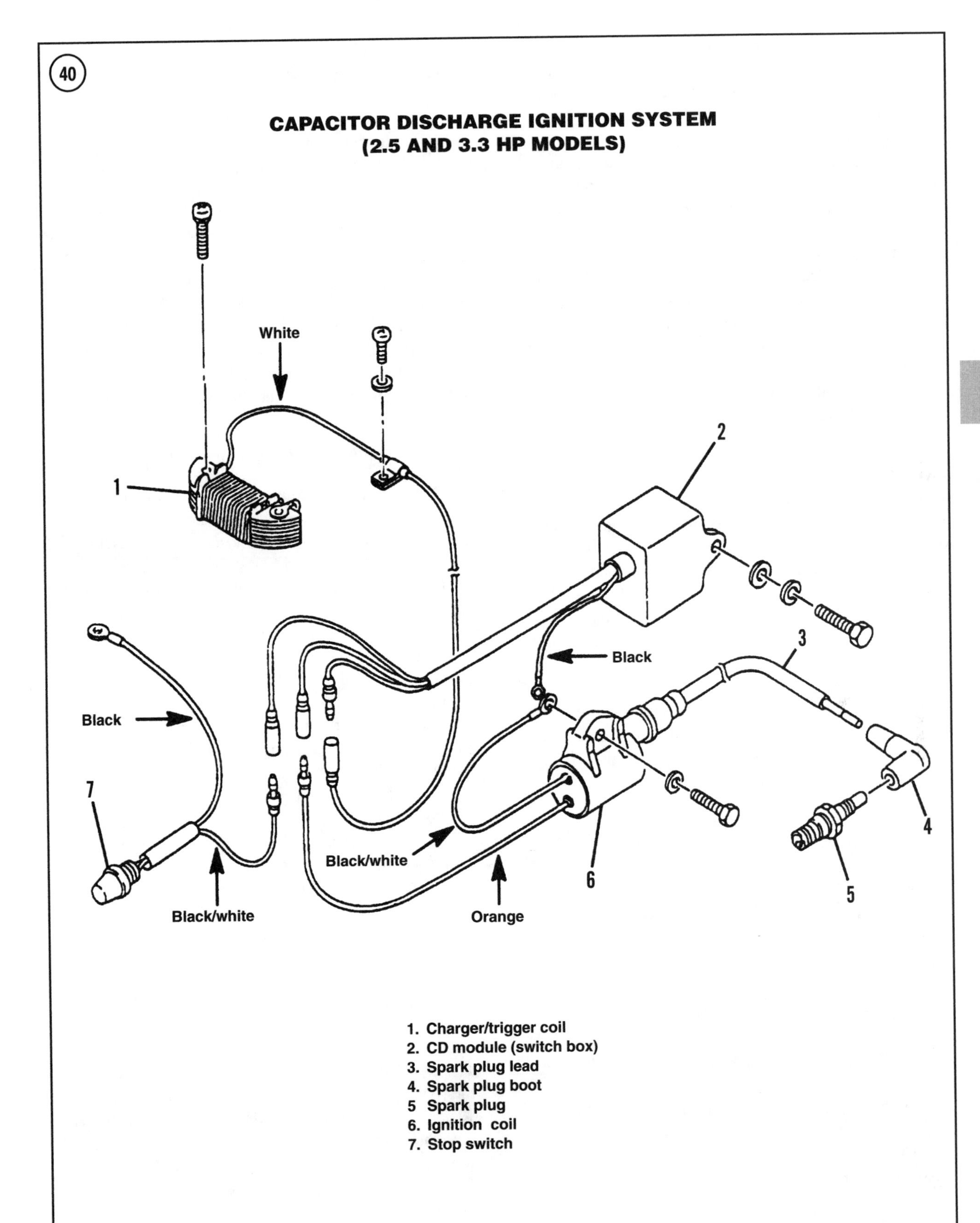

1. Charger/trigger coil
2. CD module (switch box)
3. Spark plug lead
4. Spark plug boot
5 Spark plug
6. Ignition coil
7. Stop switch

4 and 5 hp models have a series of permanent magnets along the inner diameter of the outer rim of the flywheel. As the flywheel rotates, alternating current (AC) is induced into the stator charge coil windings. The AC current flows to the CD module where it is converted (rectified) into direct current (DC) and stored in the capacitor in the CD module. A permanent magnet mounted to the outer diameter of the outer rim of the flywheel rotates past the externally mounted trigger coil assembly and induces a low voltage signal into the trigger coil windings. This voltage pulse causes an electronic switch in the switch box called a SCR (silicon controlled rectifier) to close, allowing the stored voltage in the capacitor to discharge into the ignition coil. The ignition coil amplifies the voltage and discharges it into the spark plug wire. This sequence of events is repeated with each revolution of the flywheel. See **Figure 41**.

Flywheel (2.5-5 hp models)

Since the flywheel contains permanent magnets, never strike the flywheel with a hammer. Striking the flywheel and magnets can cause the magnets to lose their magnetism. Repeatedly striking a flywheel can lead to a weak or erratic spark. Crankshafts are made of hardened steel; striking the flywheel or crankshaft can also permanently damage the crankshaft. Use only the recommended flywheel puller tools or their equivalents.

After removal, inspect the flywheel key and the key slots in the crankshaft and flywheel for wear, cracks and damage. Inspect the flywheel for loose or damaged magnets. Replace all suspect components.

Removal/installation (2.5 and 3.3 hp models)

1. Disconnect and ground the spark plug lead to the power head to prevent accidental starting.
2. Remove the manual rewind starter as described in Chapter Thirteen.
3. Remove the screws securing the rope cup (**Figure 42**) to the flywheel. Remove the rope cup.
4. Hold the flywheel using flywheel holder, part No. 91-83163M or an equivalent, and loosen the flywheel nut. See **Figure 43**, typical.
5. Install the flywheel puller, part No. 91-83164M or an equivalent, on the flywheel as shown in **Figure 44**, typical. Tighten the puller center screw until the flywheel breaks free of the crankshaft taper.
6. Remove the flywheel puller. Lift the flywheel off the crankshaft and remove the key from the crankshaft.
7. Inspect and clean the flywheel, key and crankshaft as described in *Flywheel inspection*.
8. To install the flywheel, make sure the flywheel and crankshaft tapers are clean and dry. Insert the flywheel key into the crankshaft key slot.
9. Lower the flywheel onto the crankshaft, making sure the key slot in the flywheel aligns with the key in the crankshaft.
10. Install the flywheel washer and nut. Hold the flywheel using the flywheel holder (part No. 91-83163M) and tighten the flywheel nut to the specification in **Table 1**.
11. Install the rope cup (**Figure 42**) to the flywheel and secure it with the screws. Tighten the screws securely.
12. Install the manual rewind starter as described in Chapter Thirteen. Reconnect the spark plug lead.

Removal/installation (4 and 5 hp models)

1. Disconnect and ground the spark plug lead to the power head to prevent accidental starting.
2. Remove the manual rewind starter as described in Chapter Thirteen.
3. Hold the flywheel by wrapping a strap wrench (part No. 91-24937A-1) around the starter rope cup (**Figure 42**). Loosen and remove the flywheel nut.
4. Remove the screws securing the rope cup to the flywheel. Remove the cup (**Figure 42**).
5. Install the flywheel puller (part No. 91-83164M or an equivalent) onto the flywheel. Tighten the puller screw until the flywheel is dislodged from the crankshaft taper. See **Figure 44**, typical.
6. Remove the flywheel puller. Lift the flywheel off the crankshaft and remove the key from the crankshaft.
7. Inspect and clean the flywheel, key and crankshaft as described in *Flywheel inspection*.
8. To install the flywheel, make sure the flywheel and crankshaft tapers are clean and dry. Insert the flywheel key into the crankshaft key slot.
9. Lower the flywheel onto the crankshaft, making sure the key slot in the flywheel aligns with the flywheel key.
10. Install the rope cup (**Figure 42**) on the flywheel and secure it with the screws. Tighten the screws to the specification in **Table 1**.
11. Install the flywheel onto the crankshaft. Make sure the flywheel key slot aligns with the flywheel key.
12. Install the flywheel nut. Hold the flywheel by wrapping the strap wrench around the rope cup and tighten the flywheel nut to the specification in **Table 1**.
13. Reinstall the manual rewind starter as described in Chapter Thirteen.
14. Reconnect the spark plug lead.

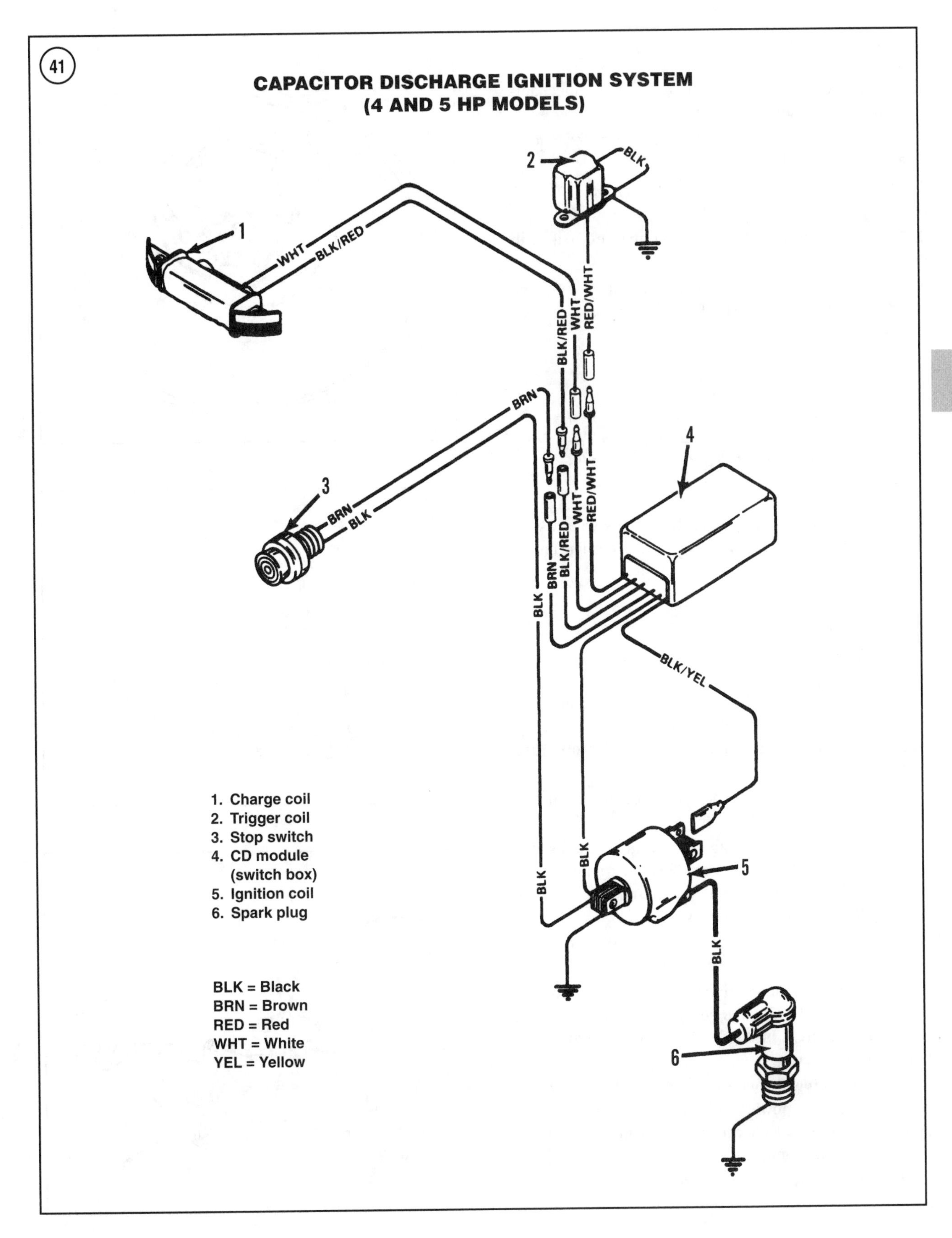
41
CAPACITOR DISCHARGE IGNITION SYSTEM
(4 AND 5 HP MODELS)
1
2
3
4
5
6
WHT
BLK/RED
BLK
RED/WHT
BRN
BLK/YEL
1. Charge coil
2. Trigger coil
3. Stop switch
4. CD module
(switch box)
5. Ignition coil
6. Spark plug
BLK = Black
BRN = Brown
RED = Red
WHT = White
YEL = Yellow

Inspection

1. Inspect the entire flywheel for cracks, chips, mechanical damage, wear and corrosion.
2. Inspect the flywheel and crankshaft tapers for cracks, wear, corrosion and metal transfer.
3. Inspect the flywheel and crankshaft key slots for wear or damage.
4. Inspect the flywheel key. Replace the key if it is in questionable condition.
5. Inspect the flywheel for loose, cracked or damaged magnets. Replace the flywheel if the magnets are loose, cracked or damaged.

WARNING
Replace defective flywheels. A defective flywheel can fly apart at high engine speed, throwing fragments over a large area. Do not attempt to use or repair a defective flywheel.

6. Clean the flywheel and crankshaft tapers with a suitable solvent and blow dry with compressed air. The tapers must be clean and dry.

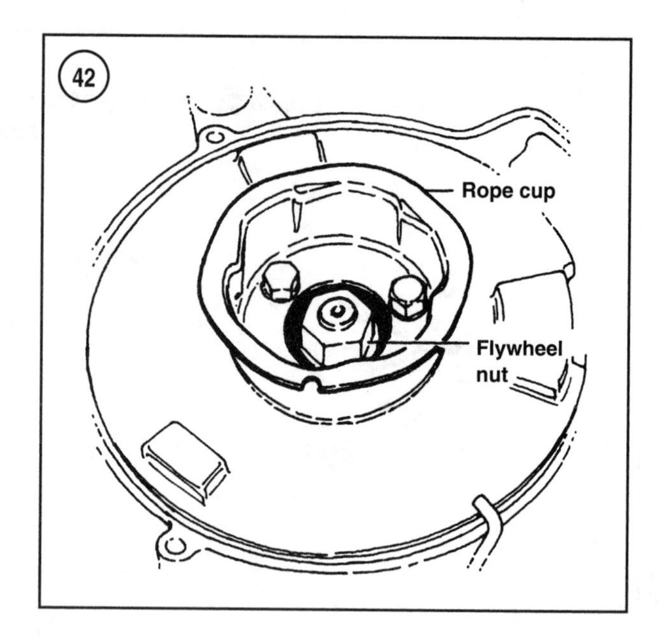

Charge Coil/Trigger Coil Removal/Installation (2.5 and 3.3 hp Models)

The charge coil and trigger coil are a single combined unit on these models. The charge/trigger coil is the only electrical component under the flywheel. Refer to **Figure 40** and **Figure 45** for this procedure.

1. Remove the flywheel as described previously in this section.
2. Disconnect the charge/trigger coil white wire at its bullet connector near the switch box. Note the routing of the white wire. Then remove any screws, clamps or tie-straps securing the white wire to the power head.
3. Remove the screws securing the charge/trigger coil to the stator plate. Remove the coil assembly.
4. To reinstall the charge/trigger coil, mount the coil on the stator plate. Position the coil ground wire under the coil mounting screw. Install the coil mounting screws and tighten them securely.
5. Route the coil white wire as noted on disassembly. Secure the wire to the power head with the original clamp(s) or new tie-strap(s). Connect the coil white wire to the switch box bullet connector.
6. Reinstall the flywheel as described previously in this section.

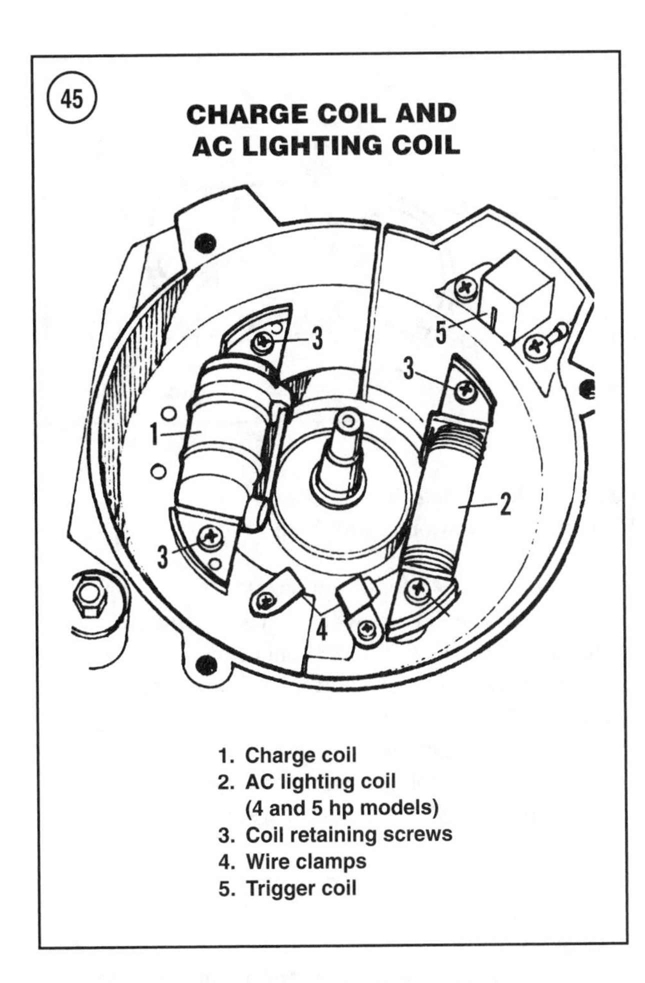

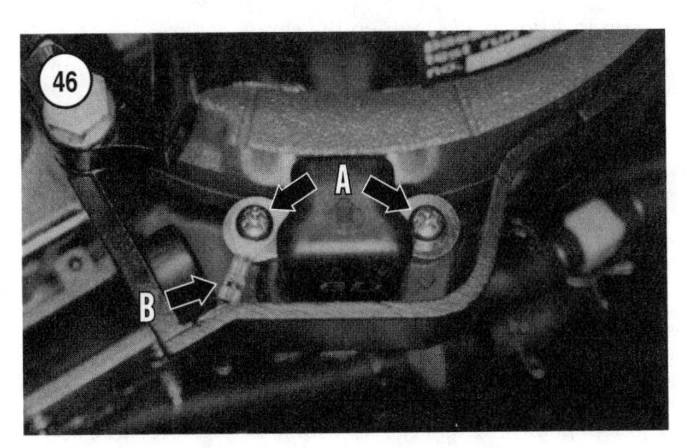

Charge Coil/AC Lighting Coil Removal/Installation (4 and 5 hp Models)

The 4 and 5 hp models may be equipped with an optional AC lighting coil. The charge coil and the AC lighting coil are replaced using the same procedure. Refer to **Figure 41** and **Figure 45** for this procedure.

1. Remove the flywheel as described previously in this section.

2A. *Charge coil*—Disconnect the white and black/red coil wires at the switch box bullet connectors. See **Figure 41**.

2B. *AC lighting coil*—Disconnect the two lighting coil wires from the rectifier, regulator or lighting harness bullet connectors. See **Figure 41**.

3. Remove the screw and clamp (4, **Figure 45**) holding the wires to the power head.
4. Remove the screws holding the coil to the power head. Remove the coil.
5. To install the charge and/or lighting coil, position the coil over the power head mounting bosses.
6. Secure the coil with the screws. Apply Loctite 242 threadlocking adhesive (part No. 92-809821) to the screw threads. Tighten the screws to the specification in **Table 1**.
7. Secure the coil wires with the clamp (4, **Figure 45**). Tighten the clamp screw securely.

8A. *Charge coil*—Connect the white and black/red coil wires to the switch box bullet connectors.

8B. *AC lighting coil*—Connect the two lighting coil wires to the rectifier, regulator or lighting harness bullet connectors.

9. Install the flywheel as described previously in this section.

Trigger Coil Removal/Installation (4 and 5 hp Models)

A single trigger coil is mounted externally on the power head, adjacent to the outer diameter of the flywheel. One red/white wire is connected to the switch box and one ground wire is secured under one of the coil mounting screws. Refer to **Figure 41** and **Figure 46** for this procedure.

1. Disconnect the red/white trigger coil wire from the switch box bullet connector. See **Figure 41**.
2. Remove the screws (A, **Figure 46**) securing the trigger coil to the flywheel housing.
3. Pull the trigger coil from the flywheel housing.
4. Insert the new trigger coil red/white wire through the rubber grommet and connect the lead to the switch box.
5. Seat the trigger coil in the housing and position the black ground wire (B, **Figure 46**) under one coil mounting screw, then tighten the screws securely.

Ignition Coil Removal/Installation (2.5 and 3.3 hp Models)

Refer to **Figure 40** for this procedure.

1. Disconnect the spark plug lead from the spark plug.

2. Disconnect the orange coil primary wire from the switch box bullet connector.
3. Remove the ignition coil mounting fastener and lift the coil off the power head.
4. To install the coil, position the coil on the power head and secure the coil and black ground wire with the mounting screw. Tighten the mounting screw securely.
5. Reconnect the orange coil primary wire to the switch box bullet connector.
6. Reconnect the spark plug wire.

Ignition Coil Removal/Installation (4 and 5 hp Models)

Refer to **Figure 41** and **Figure 47** for this procedure.
1. Disconnect the spark plug wire from the spark plug.
2. Disconnect the black/yellow primary wire from the ignition coil terminal.
3. Remove the screws and washers holding the coil to the power head. Then remove the coil from the power head.
4. To install the coil, position the coil on the power head and secure the coil and the two ground wires to the power head with the mounting screws as shown in **Figure 47**. Tighten the mounting screws securely.
5. Reconnect the black/yellow primary wire to the ignition coil.
6. Reconnect the spark plug wire.

CD Module (Switch Box) Removal/Installation (2.5 and 3.3 hp Models)

It may be necessary to remove the port side lower cowl to gain access to the switch box fastener. Refer to **Figure 40** for this procedure.
1. Remove the ignition coil mounting fastener to free the switch box ground wire.
2. Disconnect the brown, orange and white switch box wires at their bullet connectors.
3. Remove the switch box mounting screw and lift the unit off the power head.
4. To install the switch box, position the switch box on the power head and secure it with the mounting screw. Tighten the mounting screw securely.
5. Position the switch box (black) and ignition coil (black/white) ground wires under the ignition coil mounting screw. Tighten the mounting screw securely.
6. Connect the brown, orange and white wires to the switch box bullet connectors.
7. If the port side lower cowl was removed for access, install it.

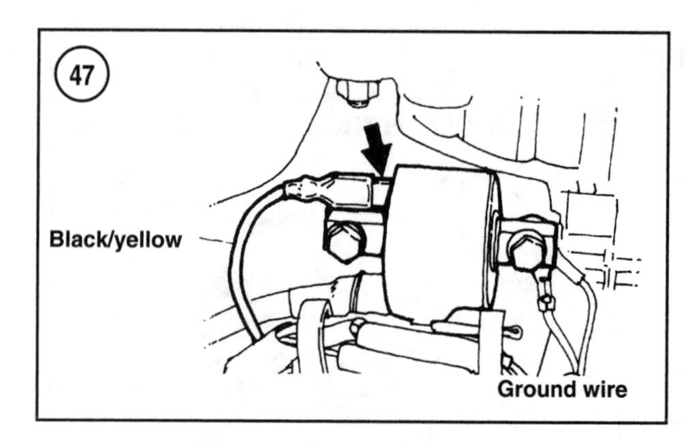

CD Module (Switch Box) Removal/Installation (4 and 5 hp Models)

Refer to **Figure 41** for this procedure.
1. Disconnect the brown, white, black/red and red/white switch box wires at their bullet connectors.
2. Disconnect the black/yellow switch box wire from the ignition coil (**Figure 47**).
3. Remove one ignition coil mounting screw to free the switch box ground (black) wire. See **Figure 47**.
4. Slide the switch box out of its rubber mounting bracket.
5. To install the switch box, slide the switch box into its rubber mounting bracket.
6. Position the switch box ground wire and stop switch ground wire under the ignition coil mounting screw as shown in **Figure 47**. Tighten the screw securely.
7. Connect the black/yellow switch box wire to the ignition coil.
8. Connect the brown, white, black/red and red/white switch box wires to their bullet connectors.

ALTERNATOR DRIVEN CAPACITOR DISCHARGE IGNITION SYSTEM (6-25 HP MODELS)

This section covers all models equipped with alternator driven capacitor discharge ignition (ADI). Refer to Chapter Three for ignition system troubleshooting procedures and the end of the manual for wiring diagrams.

The major components of the ignition system include the flywheel, stator assembly, trigger coil assembly, switch box, ignition coils, spark plugs and related wiring.

Operation (1998 Models)

Two permanent magnets are located along the inner diameter of the outer rim of the flywheel. As the flywheel rotates, alternating current (AC) is induced into the low- and high-speed charge coil windings. The low-speed windings provide the majority of the voltage required for starting and low-speed operation. The high-speed windings provide the majority of the voltage required for high-speed operation. The low- and high-speed winding outputs are combined in the switch box. The charge coil (stator) used on these models is black. The switch box contains a rectifier to convert the AC voltage into direct current (DC) so it can be stored in the switch box capacitor. The capacitor holds this voltage until it is released by a signal from the trigger coil(s).

Another set of permanent magnets is located along the outer diameter of the flywheel inner hub. As the flywheel rotates, low voltage signals are induced in the trigger coil windings. This low voltage pulse is sent to the switch box where it causes an electronic switch in the switch box called an SCR (silicon controlled rectifier) to close, allowing the stored voltage in the capacitor to discharge to the appropriate ignition coil. The ignition coil amplifies the voltage and discharges it into the spark plug wire.

This sequence of events is duplicated for each cylinder of the engine, and is repeated each revolution of the flywheel. One spark occurs for each cylinder for each complete flywheel rotation.

All 1998 20 jet models are equipped with electronic spark advance. The spark advance is controlled by the switch box based on engine speed. The switch box used on jet models utilizes an rpm limit circuit to prevent excessive engine speed if the jet pump ventilates. These circuits switch off ignition to one of the cylinders if the engine speed exceeds approximately 6200 rpm.

All other models use mechanical spark advance. The spark advance is controlled by rotating the position of the trigger coil assembly in relation to the magnets on the flywheel inner hub. The trigger coil rotation is based on throttle lever position.

Operation (1999-On Models)

Four permanent magnets are on the inner diameter of the outer rim of the flywheel. As the flywheel rotates, AC voltage is induced into the charge coil winding. The charge coil (stator) used on these models is red. The AC voltage is directed to the rectifier and voltage regulator circuits in the switch box. These circuits convert the AC voltage to DC so it can be stored in the switch box capacitor. The capacitor holds this voltage until it is released by a signal from the trigger coil(s). The regulator circuits limit the stored DC voltage to approximately 300 volts to prevent damage to the capacitor.

Another set of permanent magnets is located along the outer diameter of the flywheel inner hub. As the flywheel rotates, low voltage signals are induced in the trigger coil windings. This low voltage pulse is sent to the switch box where it causes an electronic switch in the switch box called an SCR (silicon controlled rectifier) to close, allowing the stored voltage in the capacitor to discharge to the appropriate ignition coil. The ignition coil amplifies the voltage and discharges it into the spark plug lead.

This sequence of events is duplicated for each cylinder of the engine and is repeated each revolution of the flywheel. One spark occurs for each cylinder for each complete flywheel rotation.

The switch box used on tiller control and jet models utilizes an rpm limit circuit to prevent excessive engine speed if the jet pump or propeller ventilates. These circuits switch off ignition to one of the cylinders if the engine speed exceeds approximately 6200 rpm.

All 1999-on models use a mechanical spark advance. The spark advance is controlled by rotating the position of the trigger coil assembly in relation to the magnets on the flywheel inner hub. The trigger coil rotation is based on throttle lever position.

Red Stator Upgraded Models

Red stator and adapter module upgrade kits are available for 1998 6-25 hp models. The kit includes a flywheel, stator, switchbox and required wire harnesses. If the stator on the engine is encased in red plastic, the upgrade kit has been installed. If the stator on the engine is encased in black plastic, it is the original stator.

The red stator is generally more reliable than the previous black color stator. The battery charging output from the red stator increased from approximately 4 amps to approximately 6.5 amps.

Component wiring

Electrical wiring is color-coded and the terminals on the components to which each wire connects are embossed with the correct wire color. Used with the correct electrical diagram, incorrect wire connections should be eliminated.

The routing of the wiring harness and individual leads is important to prevent possible electrical interference and/or physical damage to the wiring harnesses from moving engine parts or vibration. Mercury outboards are shipped from the factory with all wiring harnesses and leads properly positioned and secured with the appropriate clamps and tie-straps.

Prior to replacing components, either carefully draw a sketch of the area to be serviced, noting the positioning of all wire harnesses, or take several close-up photographs of the area to be serviced with an instant camera. Reinstall all clamps and new tie-straps where necessary to maintain the correct wire routing.

If wiring harness repairs are required, refer to *Electrical Repairs* at the beginning of Chapter Three.

Flywheel

Since the flywheel contains permanent magnets, never strike the flywheel with a hammer. Striking the flywheel and magnets can cause the magnets to lose their charge. Repeatedly striking a flywheel can lead to a weak or erratic spark. Crankshafts are made of hardened steel, striking the flywheel or crankshaft can also permanently damage the crankshaft. Use only the recommended flywheel puller tools or their equivalents.

Removal/installation

1A. *Electric start models*—Disconnect the negative battery cable.

1B. *Manual start models*—Remove the manual rewind starter as described in Chapter Thirteen.

2. Disconnect the spark plug leads from the spark plugs to prevent accidental starting.

3A. *Electric start models*—Hold the flywheel using the flywheel holder wrench (part No. 91-52344 or an equivalent) to engage the ring gear teeth. See **Figure 48**.

3B. *Manual start models*—Hold the flywheel with strap wrench (part No. 91-24937A-1 or equivalent). See **Figure 49**.

4A. *6-15 hp models*—Loosen and remove the flywheel nut.

4B. *20 and 25 hp models*—Loosen the flywheel bolt. Leave the bolt threaded approximately 3/4 of the way to act as a pressing point for the flywheel puller tool.

CAUTION
Do not thread the puller bolts more than 1/2 in. (12.7 mm) into the flywheel in Step 5 or the stator and trigger coil assemblies will be damaged.

5. Install the flywheel puller (part No. 91-83164M or an equivalent) onto the flywheel. See **Figure 50**. Tighten the puller screw until the flywheel is dislodged from the crankshaft taper. Remove the puller assembly from the flywheel.

6. On 20 and 25 hp models, remove the flywheel bolt.

48

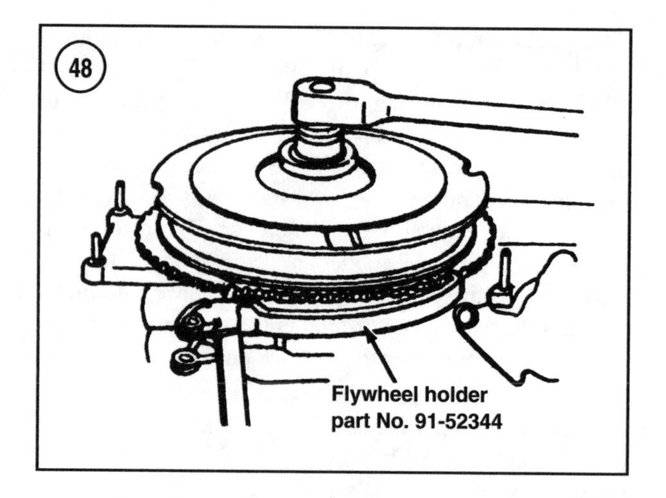

49

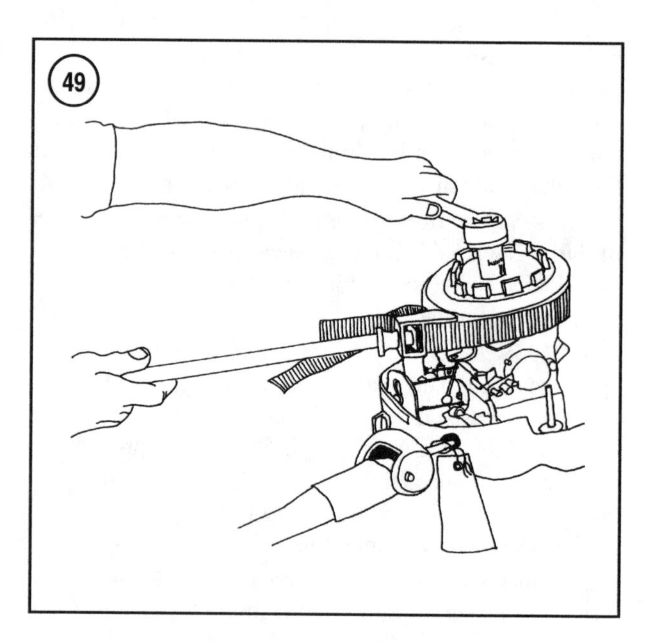

7. Lift the flywheel off the crankshaft. Then remove the flywheel key from the crankshaft slot. Inspect and clean the flywheel, key and crankshaft as described in *Flywheel inspection*.

8. To reinstall the flywheel, insert the flywheel key into the crankshaft key slot.

9. Place the flywheel on the crankshaft. Align the flywheel key with flywheel key slot.

10. Install the flywheel nut or bolt. Hold the flywheel using the strap wrench or flywheel holder. Tighten the nut or bolt to the specification in **Table 1**.

11A. *Electric start models*—Reconnect the negative battery cables.

11B. *Manual start models*—Install the manual rewind starter as described in Chapter Thirteen.

12. Reconnect the spark plug leads.

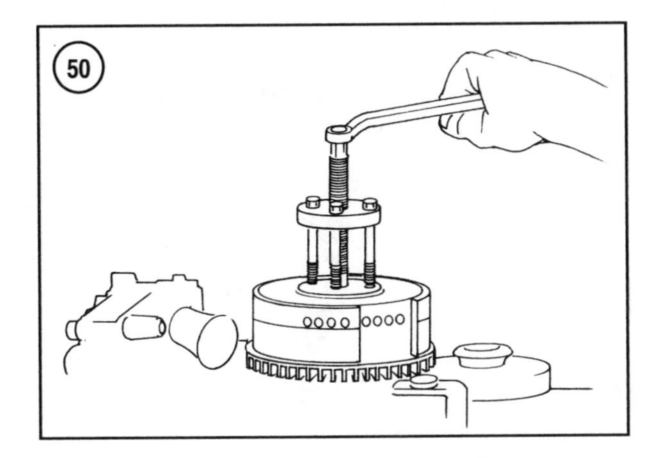

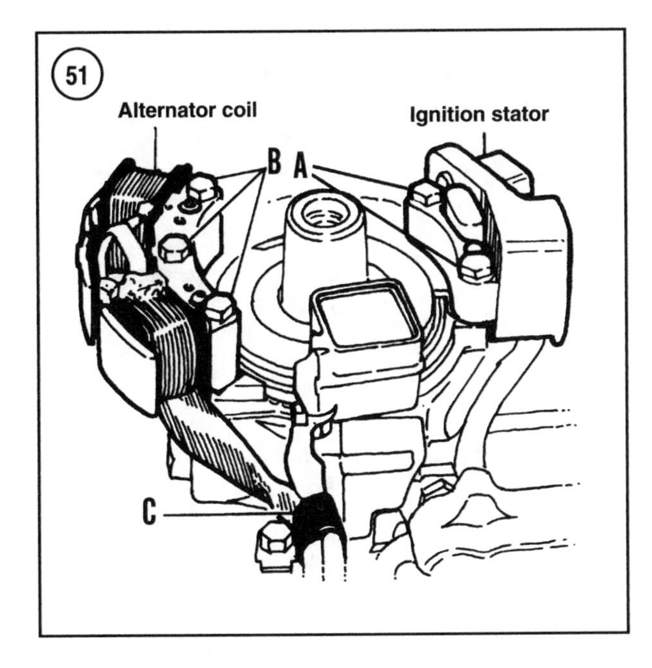

Inspection

1. Inspect the entire flywheel for cracks, chips, mechanical damage, wear and corrosion.
2. Carefully inspect the flywheel and crankshaft tapers for cracks, wear, corrosion and metal transfer.
3. Inspect the flywheel and crankshaft key slots for wear or damage.
4. Carefully inspect the flywheel key. Replace the key if it is in questionable condition.
5. Inspect the flywheel for loose, cracked or damaged magnets. Replace the flywheel if the magnets are loose, cracked or damaged.

WARNING
Replace defective flywheels. A defective flywheel can fly apart at high engine speed throwing fragments over a large area. Do not attempt to use or repair a defective flywheel.

6. Clean the flywheel and crankshaft tapers with a suitable solvent and blow dry with compressed air. The tapers must be clean, dry and free of oil or other contamination.

Stator Removal/Installation (1998 Model)

The ignition stator assembly includes low- and high-speed windings which are not separately serviced. The 6-25 hp models may also be equipped with an optional AC lighting or battery charging coil attached to the stator mounting plate. The stator used on these models is encased in black plastic. This optional coil is called the alternator coil for this procedure. The following procedure covers replacement of both coils.

1. Disconnect and ground the spark plug leads to the power head to prevent accidental starting.
2. On electric start models, disconnect the negative battery cable.
3. Remove the flywheel as described in this chapter.
4A. *Ignition stator*—Remove the screws (A, **Figure 51**) securing the ignition stator assembly to the power head. If the engine is not equipped with an alternator, the stator is mounted with three or four screws.
4B. *Alternator coil*—Remove the screws (B, **Figure 51**) securing the alternator coil to the power head.
5A. *Ignition stator*—Disconnect the stator wires from the switchbox as follows:
  a. *20 jet models*—Disconnect the three-pin connector from the switch box. The wire colors are black, red and light blue.
  b. *All other models (1998)*—Disconnect the black/yellow and black/white wires from the switch box bullet connectors.
5B. *Alternator coil*—Disconnect the alternator coil's two yellow or one yellow and one gray wires from the terminal strip or rectifier.
6. Remove the wire clamp (C, **Figure 51**) and/or tie-straps securing the ignition or alternator wires to the power head.
7. Remove the ignition stator and/or alternator coil from the power head.
8. Position the ignition stator and/or alternator coil on the power head.
9. Install the fasteners and tighten them evenly to the specification in **Table 1**.

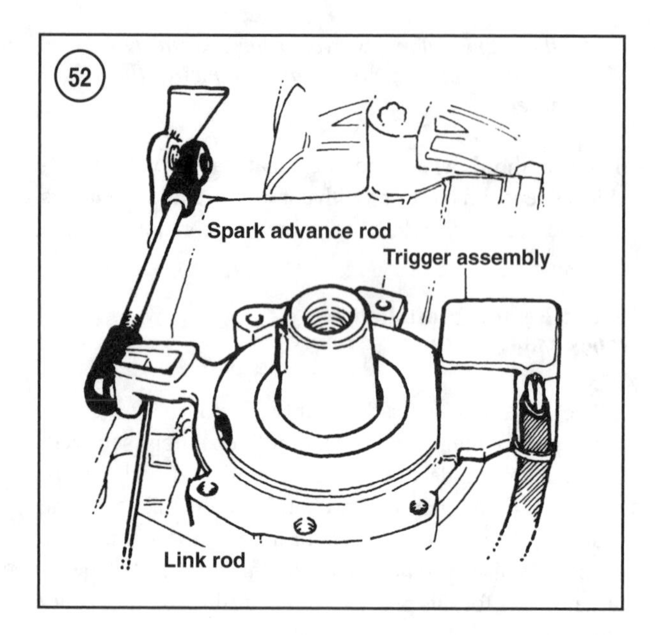

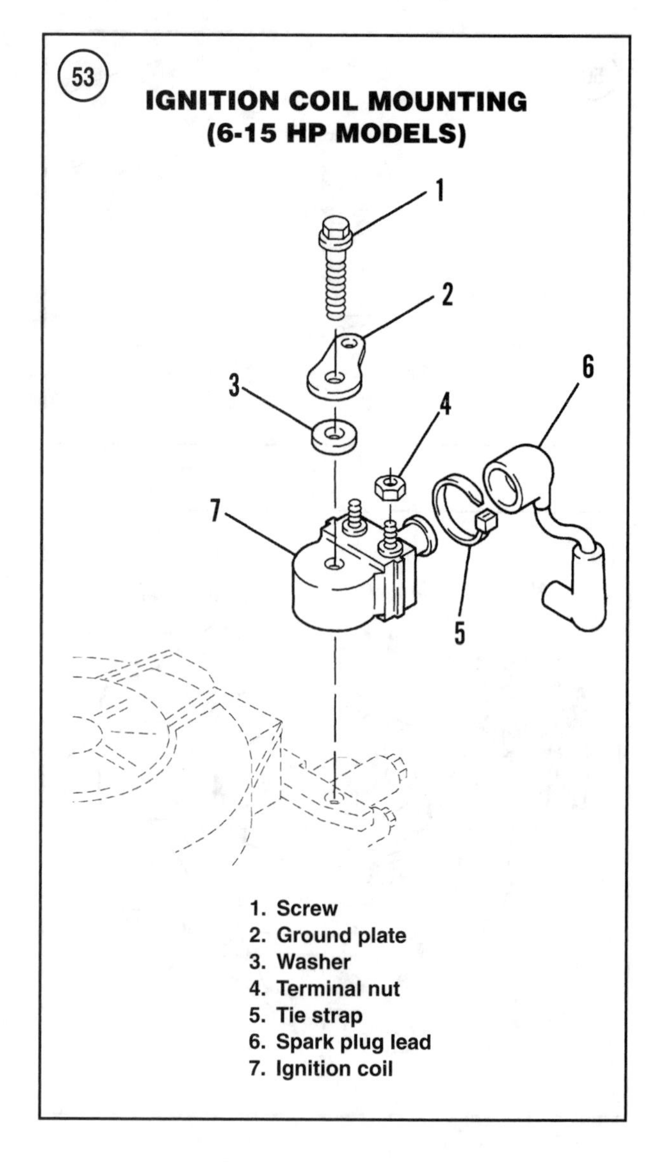

IGNITION COIL MOUNTING (6-15 HP MODELS)

1. Screw
2. Ground plate
3. Washer
4. Terminal nut
5. Tie strap
6. Spark plug lead
7. Ignition coil

10. Secure the wires with the wire clamp (C, **Figure 51**) and/or install new tie-straps to replace any that were removed.

11A. *Ignition stator*—Reconnect the stator leads or connector to the switch box.

11B. *Alternator coil*—Reconnect the two yellow or one yellow and one gray wires to the terminal strip or rectifier.

12. Reinstall the flywheel as described previously in this section.

13. On electric start models, reconnect the negative battery cable.

14. Reconnect the spark plug leads.

Stator Removal/Installation (1999-On Models)

The stator assembly is a one-piece integrated unit, containing both the ignition stator windings and the alternator coils. All stator assemblies are mounted under the flywheel. The stator used on these models is encased in red plastic.

Refer to the wiring diagrams at the end of the manual.

1. Remove the flywheel as described in this chapter.
2. Note the orientation of the stator assembly and stator wires before proceeding.
3. Remove the stator assembly mounting screws.
4. On electric start models, disconnect the gray and yellow wires from the rectifier.
5. Disconnect the white/green and green/white wires from the switch box bullet connectors.
6. Remove the wire clamp (C, **Figure 51**) and/or tie-straps securing the ignition or alternator wires to the power head.
7. Remove the ignition stator and/or alternator coil from the power head.
8. Position the ignition stator and/or alternator coil on the power head.
9. Install the fasteners and tighten them evenly to the specification in **Table 1**.
10. Secure the leads with the wire clamp (C, **Figure 51**) and/or install new tie-straps to replace any that were removed.
11. On electric start models, reconnect the gray and yellow wires to the rectifier.
12. Reconnect the green/white and white/green wires to the switch box bullet connectors.

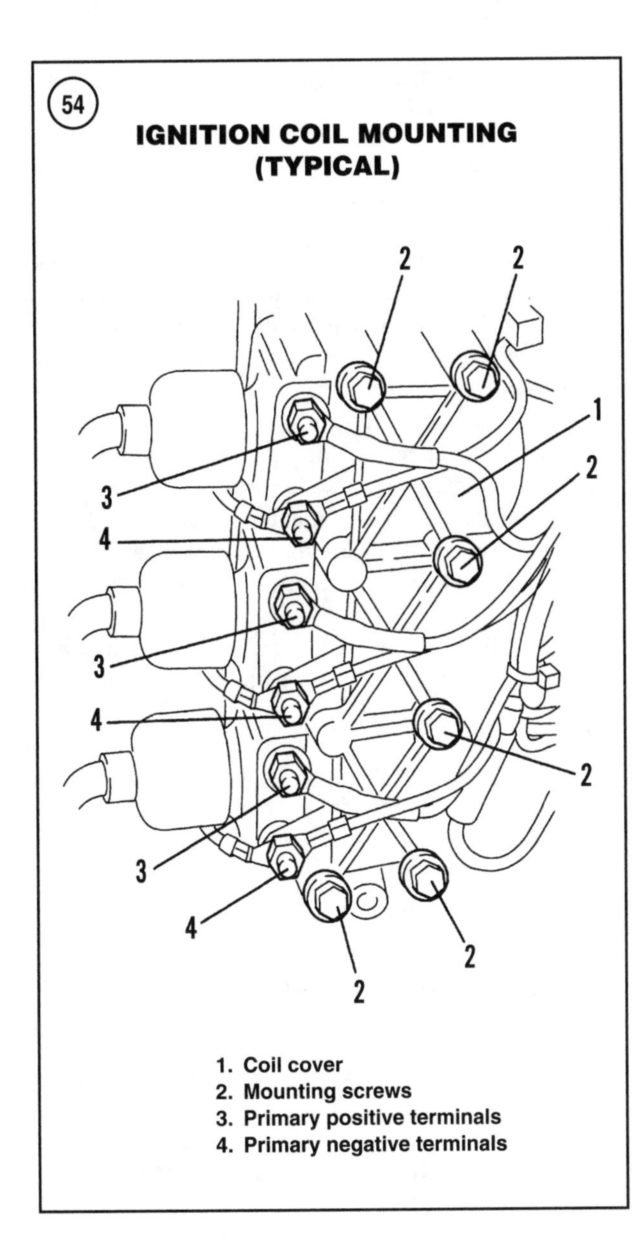

13. Reinstall the flywheel as described in this chapter.
14. On electric start models, reconnect the negative battery cable.
15. Reconnect the spark plug leads.

Trigger Coil Removal/Installation (6-25 hp Models)

A single trigger coil is mounted under the flywheel. Note the trigger wiring harness routing for reference during installation. After trigger coil installation, refer to Chapter Five and perform the synchronization and linkage adjustments.

1. Remove the flywheel as described in this chapter.
2. Remove all stator assembly mounting screws and remove wiring clamps and/or tie-straps, but do not disconnect any stator wires.
3. Remove wire harness clamps securing the trigger wires. Disconnect the trigger wires (brown/white and brown/yellow) from the switch box bullet connectors.
4. Lift the stator assembly upward so the trigger can be removed without interference.
5. Pry the spark advance rod end from the trigger using a screwdriver or similar tool. See **Figure 52**, typical.
6. Lift the trigger assembly off the power head and disconnect the link rod (**Figure 52**), if so equipped.
7. Remove the trigger assembly.

NOTE

*On 20 jet models (1998), the trigger is mounted in a fixed position and will not rotate after the link rod (**Figure 52**) is connected. On all other models, the trigger coil must rotate freely with the linkage after installation is complete.*

8. To install the trigger, lubricate the trigger bearing surfaces with 2-4-C Multi-Lube (part No. 92-825407). Then install the link rod, if so equipped, into the trigger and position the trigger assembly on the power head boss.
9. Install the spark advance rod (**Figure 52**, typical).
10. Route the trigger wire harness as noted during removal. Install harness clamps and new tie-straps to secure the harness properly. On mechanical advance models, the trigger must be able to rotate freely with the control linkage.
11. Connect the trigger wires to their respective switch box bullet connectors.
12. Reinstall the stator assembly as described in this chapter. Make sure all stator wires are clamped and/or tie-strapped in place.
13. Reinstall the flywheel as described in this chapter.
14. Refer to Chapter Five and perform the synchronization and linkage adjustments.

Ignition Coil Removal/Installation

Two types of ignition coils are used. The first type (**Figure 53**) uses a ground strap secured by a through-bolt. This coil is used on all 6-15 hp models.

The second type (**Figure 54**, typical) uses an ignition coil cover to secure the coils. A separate ground wire is used to ground each coil's primary windings. This type of coil is used on 20-25 hp models.

Both types of coils use tie-straps to secure and seal the spark plug leads to the ignition coil body. Refer to **Figure 53** for this procedure.

1. Disconnect the spark plug leads from both spark plugs.

2. Disconnect the switch box primary wires (green/yellow for cylinder No. 1 and green or green/white for cylinder No. 2) at each ignition coil positive terminal.

3. *20 and 25 hp models*—Disconnect the black ground wire from each coil negative terminal.

4. *6-15 hp models*—Remove the terminal nut (4, **Figure 53**) securing the ground plate to each ignition coil negative terminal.

5A. *6-15 hp models*—Remove the throughbolt (1, **Figure 53**) securing each coil to the power head. Remove the coils from the power head.

5B. *20 and 25 hp models*—Remove the coil cover screws (2, **Figure 54**). Remove the coil cover and coils from the engine. Pull each coil from the cover.

6. Cut the tie-strap securing the spark plug lead to each ignition coil and remove the spark plug lead from each coil body.

7. To install the coil(s), connect the spark plug lead to each ignition coil. Use Quicksilver Ignition Coil Insulating Compound (part No. 92-41669-1) to create a water tight seal. Secure the boot to each coil's body with a new tie-strap.

8A. *6-15 hp models*—Position the washer, ground strap and throughbolt on each coil. Position each coil onto the power head and tighten the throughbolt to the specification in **Table 1**. Make sure the ground strap is over each coil's negative terminal.

8B. *20 and 25 hp models*—Install both coils into the coil cover. Position the coil cover onto the power head and secure it with the screws (2, **Figure 54**). Tighten the screws to the specification in **Table 1**.

9A. *6-15 hp models*—Install each coil's negative terminal nut (4, **Figure 53**). Tighten the nut to the specification in **Table 1**.

9B. *20 and 25 hp models*—Connect the ground (black) wire to each coil's negative terminal stud. Secure each wire with a nut. Tighten the nut to the specification in **Table 1**.

10. Connect the green/yellow primary wire to the coil No. 1 positive primary terminal and the green or green/white primary wire to the coil No. 2 positive terminal. Secure each wire with a nut. Tighten each nut to the specification in **Table 1**.

11. Coat the coil primary terminal connections with Quicksilver Liquid Neoprene (part No. 92-25711-2).

12. Reconnect the spark plug leads to the spark plugs.

CD Module (Switch Box) Removal/Installation

All bullet connector switch boxes use wires that are always connected to another lead with the exact same color code. Take note of the wire routing and terminal connections before disconnecting any wires. Correct wire routing is important to prevent insulation and wire damage from heat, vibration or interference with moving parts.

Refer to the wiring diagrams at the end of the manual.

1. Disconnect and ground the spark plug leads to prevent accidental starting.
2. Disconnect the stator, trigger coil, ignition coil and stop circuit wires from the switch box bullet connectors.
3. Remove the switch box mounting screws, then disconnect the ground (black) wire from the power head ground. Once the ground wire is disconnected, remove the switch box.
4. To install the switch box, position the switch box on the power head and secure it with the screws. Tighten the mounting screws securely.
5. Connect the switch box ground wire to the power head ground and tighten the screw securely. The switch box must have a good ground in order to function.
6. Connect the stator, trigger, ignition coil and stop circuit wires to the switch box bullet connectors.
7. Reconnect the spark plug leads.

CAPACITOR DISCHARGE MODULE (CDM) IGNITION (30-60 HP MODELS)

CDM ignition systems are unique in that they combine the switch box and ignition coil functions into one module, called the CDM. There is one CDM for each cylinder. The rectifier in each CDM transforms the ignition stator AC voltage into DC voltage so it can be stored in the CDM capacitor. A voltage regulator circuit in each CDM limits the stored voltage to approximately 300 volt to prevent damage to the capacitor. The capacitor holds the voltage until the SCR (silicon controlled rectifier) releases the voltage to the integral ignition coil primary windings. The SCR is triggered by the trigger coil. The ignition coil transforms the relatively low voltage from the capacitor into voltage high enough (45,000 volts) to jump the spark plug gap and ignite the air/fuel mixture.

Refer to Chapter Three for troubleshooting procedures on all CDM ignition systems. Refer to the end of the manual for wiring diagrams.

The major components of the CDM are the:

1. *Flywheel*—The flywheel inner magnet is for the trigger coil. The outer magnets are for the ignition stator and battery charging stator. See 1, **Figure 55**, typical.

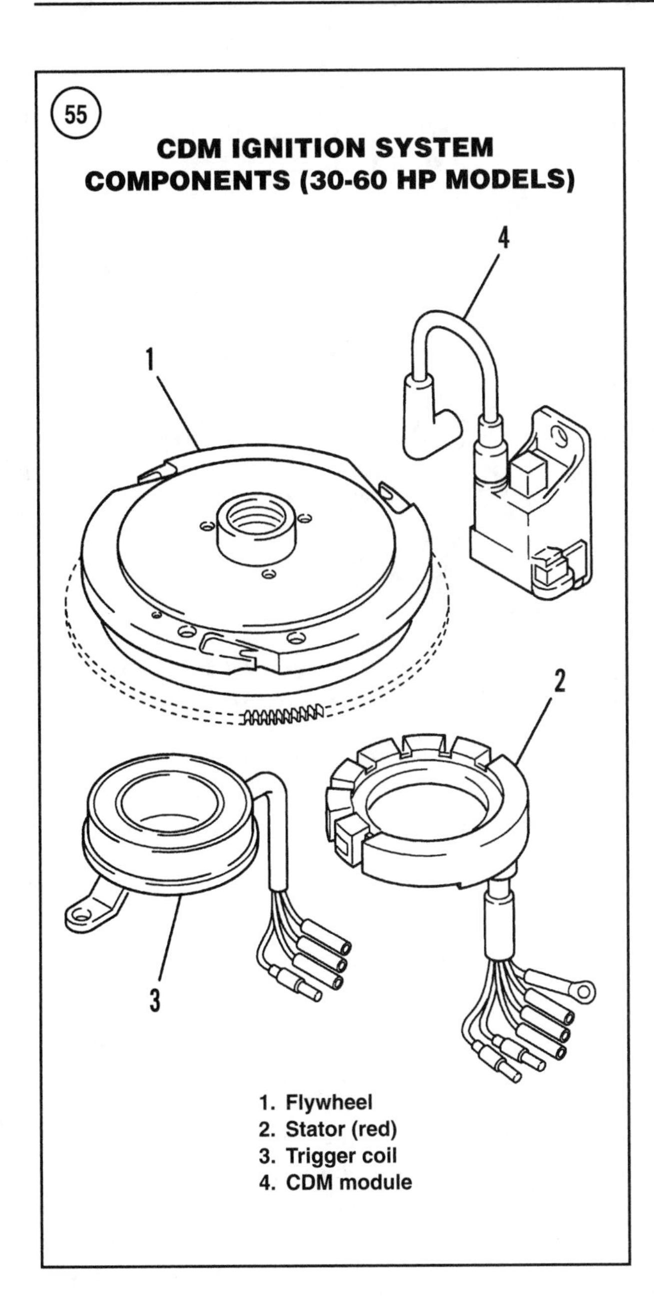

1. Flywheel
2. Stator (red)
3. Trigger coil
4. CDM module

2. *Ignition stator (charge) coils*—The stator coil (2, **Figure 55**, typical) consists of one winding around three bobbins. The ignition stator windings are not grounded to the power head. The ignition stator provides power to the CDM. Stator output is always AC voltage. Manual start models may have an additional winding (light blue/white wire) to power the rpm limit module. This winding is grounded with a black wire.

NOTE
The ignition stator circuits must be complete from the stator to the CDM module and back to the stator through a different CDM module for the system to function. See Chapter Three.

3. *Trigger coil*—The trigger coil (3, **Figure 55**) tells the CDM (4) when to fire. The trigger coil is rotated by mechanical linkage to change the trigger's position relative to the flywheel. This movement advances or retards the ignition spark timing.
4. *CDM, spark plugs and stop circuit*—All of these components function as described in this section.

Spark Plugs

There is one spark plug for each cylinder. Use only the recommended spark plugs or serious engine damage may occur. Resistor or suppressor plugs are designed to reduce RFI (radio frequency interference) emissions that can cause interference with electrical accessories. Use the recommended RFI spark plug if RFI is suspected of causing interference or malfunction of electrical accessories.

Stop Circuit

The stop circuit is connected to one end of the capacitor in each CDM. Whenever the stop circuit is connected to ground, the capacitor is shorted and cannot store electricity. There is no voltage available to send to the ignition coil windings and the ignition system ceases producing spark. The stop circuit must have an open circuit for the engine to run. The stop circuit wires are always color-coded black/yellow.

All models are equipped with an rpm limit module. The rpm limit module is connected to the CDM stop circuit (black/yellow). When engine speed exceeds the preprogrammed limit, the rpm limit module momentarily shorts the black/yellow wire to ground, limiting engine speed.

Component Wiring

Electrical wiring is color coded and the terminals on the components to which each wire connects are embossed with the correct wire color. When used with the correct electrical diagram, incorrect wire connections are eliminated.

The routing of the wiring harness and individual wires is important to prevent possible electrical interference and/or damage to the wiring harnesses from moving engine parts or vibration. Mercury/Mariner outboards come from the factory with all wiring harnesses and wires properly positioned and secured with the appropriate clamps and tie-straps.

Prior to replacing components, either carefully draw a sketch of the area to be serviced, noting the positioning of all wire harnesses, or take several close-up photographs with an instant camera. Reinstall clamps and new tie-straps where necessary to maintain the correct wire routing.

If wiring harness repairs are required, refer to *Electrical Repairs* in Chapter Three.

Flywheel

Since the flywheel contains permanent magnets, never strike the flywheel with a hammer. Striking the flywheel and magnets can cause the magnets to lose their magnetism. Repeatedly striking a flywheel can lead to a weak or erratic spark. Crankshafts are made of hardened steel. Striking the flywheel or crankshaft can also permanently damage the crankshaft. Use only the recommended flywheel puller tools or their equivalents.

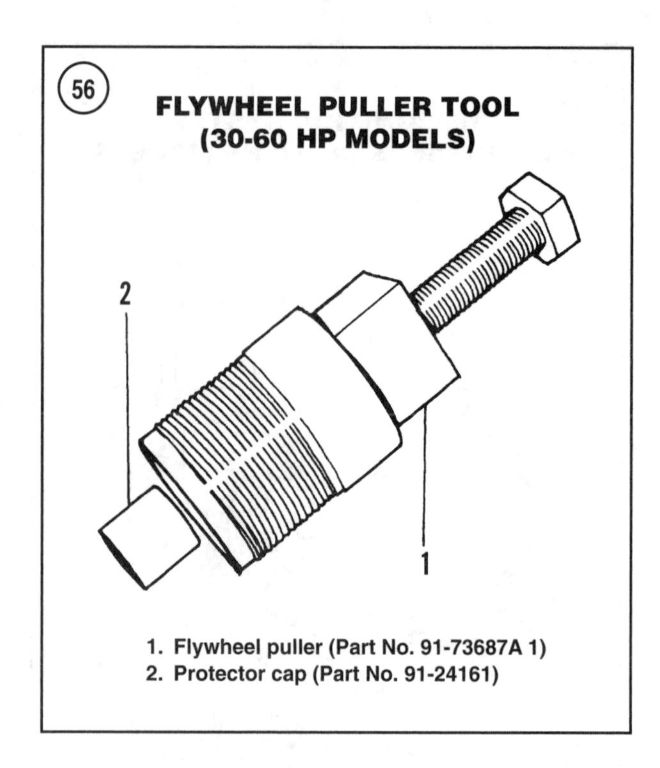

1. Flywheel puller (Part No. 91-73687A 1)
2. Protector cap (Part No. 91-24161)

Removal/installation (30-60 hp models)

1. Disconnect and ground the spark plug lead to the power head to prevent accidental starting.

2A. *Electric start models*—Disconnect the negative battery cable and remove the flywheel cover.

2B. *Manual start models*—Remove the manual rewind starter as described in Chapter Thirteen.

3. Hold the flywheel using a flywheel holder (part No. 91-52344 or an equivalent). See **Figure 48**. If no ring gear teeth are present, hold the flywheel with a strap wrench (part No. 91-24937A-1 or an equivalent). See **Figure 49**, typical. Remove the flywheel nut and washer.

CAUTION

To prevent crankshaft damage, do not remove the flywheel without using a crankshaft protector cap (2, ***Figure 56****).*

4. Install the crankshaft protector cap (part No. 91-24161 or an equivalent) on the end of the crankshaft. Use grease to hold the protector cap in place.

CAUTION

Never apply heat or strike the puller screw with a hammer during flywheel removal. Heat and/or striking can damage the ignition components, flywheel and crankshaft.

5. Thread the flywheel puller (part No. 91-73687A1 or an equivalent) into the flywheel. See **Figure 56**. Hold the puller with a wrench and tighten the puller screw until the flywheel is dislodged from the crankshaft taper. See **Figure 57**.

6. Lift the flywheel off the crankshaft. Remove the flywheel key from the crankshaft slot.

7. Inspect and clean the flywheel, key and crankshaft as described in this chapter.

8. To install the flywheel, insert the flywheel key into the slot in the crankshaft.

9. Align the flywheel key slot with the flywheel key and place the flywheel onto the crankshaft.

10. Install the flywheel nut and washer. Hold the flywheel with the strap wrench or the flywheel holder and tighten the flywheel nut to the specification in **Table 1**.

11A. *Manual start models*—Install the manual rewind starter as described in Chapter Thirteen.

11B. *Electric start models*—Install the flywheel cover and connect the negative battery cable.

12. Connect the spark plug leads to the spark plugs.

Inspection

1. Inspect the entire flywheel for cracks, chips, mechanical damage, wear and corrosion.

2. Carefully inspect the flywheel and crankshaft tapers for cracks, wear, corrosion and metal transfer.

3. Inspect the flywheel and crankshaft key slots for wear or damage.

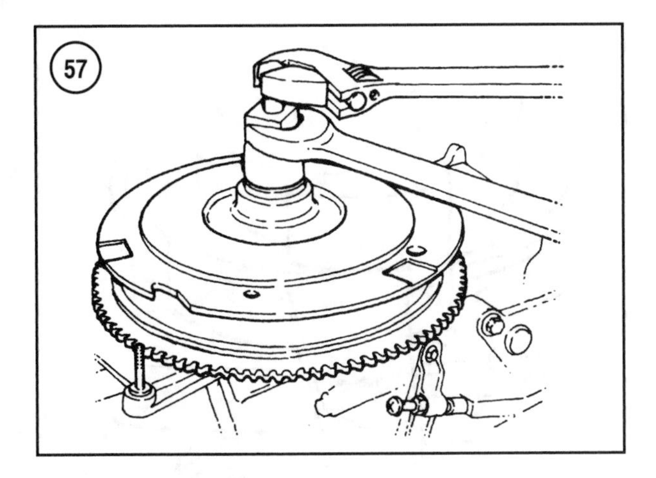

4. Carefully inspect the flywheel key. Replace the key if it is in questionable condition.
5. Inspect the flywheel for loose, cracked or damaged magnets. Replace the flywheel if the magnets are loose, cracked or damaged.

WARNING
Replace a defective flywheel. A defective flywheel can fly apart at high engine speed, throwing fragments over a large area. Do not attempt to use or repair a defective flywheel.

6. Clean the flywheel and crankshaft tapers with a suitable solvent and blow dry with compressed air. The tapers must be clean, dry and free of oil or other contamination.

Stator

The stator assembly (2, **Figure 55**) is a one-piece integrated unit, containing both the ignition stator windings and the alternator coils. The stator assembly is mounted under the flywheel. Refer to the wiring diagrams at the end of the manual to assist with wire routing and connection points.

1. Remove the flywheel as described in this chapter.
2. Note the orientation of the stator assembly and all stator wires before proceeding.
3. Remove the stator assembly mounting screws.
4. Disconnect the green/white and white/green ignition stator wires from wiring harness bullet connectors.

5A. *Electric start models*—Disconnect the two yellow alternator coil wires from the terminal strip, rectifier terminal studs or rectifier/regulator bullet connectors.

5B. *Manual start models*—Disconnect the light blue/white wire from the rpm limit module harness.

6. Remove clamps or tie-straps securing the stator wires to the power head, electrical bracket or wiring harness. Then remove the stator assembly from the engine.

CAUTION
Make sure the stator is positioned in its original location unless instruction or decals included with the replacement stator show otherwise. The stator windings and wire harness must not be between the stator and power head.

7. Position the stator as shown on the stator decal. The appropriate arrow must point aft (rearward). Position the stator wiring harness between 4 and 5 o'clock. Route the stator harness directly through the grommet and into the electrical/ignition components box.
8. Clean the stator mounting screws with Locquic Primer and allow them to air dry. Apply Loctite 271 threadlocking compound (part No. 92-809819) to the threads of the screws.
9. Install the stator screws and tighten them evenly to the specification in **Table 1**.
10. Route the stator wiring harness as noted on removal. Secure the harness to the power head or other harnesses with clamps and/or new tie-straps.
11. Connect all stator wires disconnected during removal to their respective connectors or terminals.
12. Install the electrical/ignition component access cover. Reinstall the flywheel as described in this chapter.

Trigger Coil

The trigger coil assembly is mounted under the flywheel. Note the trigger wiring harness routing for reference during installation. After trigger coil installation, refer to Chapter Five and adjust the ignition timing.

1. Remove the flywheel as described in this chapter.
2. Remove the stator as described in this chapter, but do not disconnect the stator electrical wires. Lift the stator from the power head and set it to one side.
3. Disconnect the trigger link rod ball and socket connector (**Figure 58**) from the ball on the power head.
4. Disconnect the trigger wire harness wires from the CDM harness connectors.
5. Remove the trigger coil from the power head.
6. To install the trigger, lubricate the trigger bearing surfaces with 2-4-C Multi-Lube (part No. 92-825407). Then position the trigger assembly on the power head boss.
7. Connect the trigger link rod ball and socket connector to the ball on the power head. See **Figure 58**.

8. Route the trigger wires as noted during removal. Install the necessary clamps and/or tie-straps to secure the harness.
9. Connect the trigger wire harness to the CDM harness connectors.
10. Reinstall the stator assembly as described in this chapter. Make sure all stator wires are clamped and/or tie-strapped in place.
11. Reinstall the flywheel as described in this chapter.
12. Refer to Chapter Five and check engine timing.

CDM Removal/Installation

1. Disconnect and ground the spark plug leads to the power head to prevent accidental starting.
2. Disconnect the four-pin connector from each CDM to be removed.
3. Remove the CDM mounting bolts from each module. Then remove the module(s) from the engine.
4. To install the CDM(s), position each module in its mounted position and secure it with the screws. Tighten the mounting screws to the specification in **Table 1**.
5. Reconnect the four-pin connector to each module.
6. Reconnect the spark plug leads.

RPM Limit Module Removal/Installation

All 30-60 hp models are equipped with an rpm limit module. The rpm limit module is connected to the CDM stop circuit (black/yellow). When engine speed exceeds the preprogrammed limit, the rpm limit module momentarily shorts the black/yellow wire to ground, limiting engine speed. There are four wires on the rpm limit module. The purple wire is power for the module from the key switch or stator on manual start models. The brown wire is connected to the brown trigger coil wire and is an engine speed signal for the module. The black/yellow wire is connected to the CDM stop circuit and is shorted to ground by the module to control engine speed by switching the ignition system on and off. The black wire is the ground path for the module.

Locate the rpm limit module on the power head. The module is typically mounted on the electrical/ignition

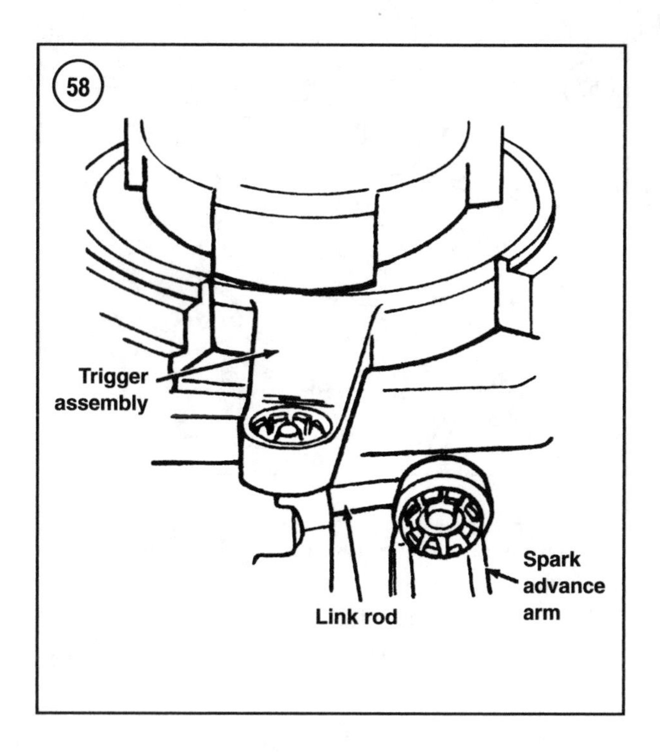

component plate. Verify the location by matching the colors of the wires.
1. Disconnect the negative battery cable.
2. Disconnect and ground the spark plug leads to the power head to prevent accidental starting.
3. Disconnect the brown, purple and black/yellow wires at the module bullet connectors. If necessary, cut tie-straps securing the module wires.
4. Remove the screw securing the black ground wire to the power head.
5. Remove the module mounting screws, then remove the module from the power head or electrical bracket.
6. To install the module, position the module on the power head or electrical bracket and install the screws. Tighten the screws to the specification in **Table 1**.
7. Connect the module ground wire to the power head. Tighten the ground screw securely.
8. Connect the brown, purple and black/yellow wires to the module bullet connectors. If necessary, secure the wires with a new tie-strap
9. Reconnect the spark plug leads and the negative battery cable.

Table 1 TORQUE SPECIFICATIONS

Component	ft.-lb.	in.-lb.	N•m
Flywheel nut/bolt			
2.5 and 3.3 hp	30	–	40.6
4 and 5 hp	40	–	54.2
6-15 hp	50	–	67.8
20 and 25 hp	58	–	78.6
30 and 40 hp (two-cylinder)	95	–	129
40-60 hp (three-cylinder)	125	–	169
Flywheel ring gear			
6-25 hp	–	100	11.3
Starter cup			
4 and 5 hp	–	70	7.9
Ignition charge coil			
4 and 5 hp	–	14	1.6
Battery charge/alternator coil			
4 and 5 hp	–	14	1.6
6-15 hp (black stator)	–	40	4.5
20 and 25 hp (black stator)	–	80	9
Rectifier/regulator mounting			
30-60 hp	–	60	6.8
Rectifier/regulator terminals			
6-25 hp	–	25	2.8
Stator/ignition charge coil			
6-15 hp	–	40	4.5
20 and 25 hp	–	80	9
30 and 40 hp (two-cylinder)	–	50	5.6
40-50 hp (three-cylinder)	–	60	6.8
Ignition coil			
6-15 hp	–	35	4.9
20 and 25 hp	–	80	9
Capacitor discharge module			
30-60 hp	–	60	6.8
RPM limit module			
30 and 40 hp (two-cylinder)	–	30	3.4
40-60 hp (three-cylinder)	–	40	4.5
Coil terminals			
6-15 hp	–	20	2.3
20 and 25 hp	–	25	2.8
Starter mounting			
6-20 hp	18	216	24.4
30-60 hp	17	204	23
Starter terminals			
30-60 hp	–	60	6.8
Starter through-bolts			
30-60 hp	–	70	7.9
Starter solenoid mounting			
6-15 hp	–	70	7.9
20 and 25 hp	–	40	4.5
30 and 40 hp (two-cylinder)	–	40	4.5
40-60 hp (three-cylinder)	–	60	6.8
Starter solenoid terminals			
6-15 hp			
Large terminals	–	30	3.4
Small terminals	–	15	1.7
20 and 25 hp			
Large terminals	–	30	3.4
Small terminals	–	25	2.8
30 and 40 hp (two-cylinder)			
Large terminals	–	50	5.6

(continued)

Table 1 TORQUE SPECIFICATIONS (continued)

Component	ft.-lb.	in.-lb.	N•m
Starter solenoid terminals (cont.)			
40-60 hp (three-cylinder)			
Large terminals	–	30	3.4
Small terminals	–	20	2.3
Neutral safety switch			
6-25 hp	–	5	0.6

Table 2 STANDARD TORQUE SPECIFICATIONS—U.S. STANDARD AND METRIC FASTENERS

Screw or nut size	ft.-lb.	in.-lb.	N•m
U.S. standard fasteners			
6-32	–	9	1
8-32	–	20	2.3
10-24	–	30	3.4
10-32	–	35	4.0
12-24	–	45	5.1
1/4-20	6	72	8.1
1/4-28	7	84	9.5
5/16-18	13	156	17.6
5/16-24	14	168	19
3/8-16	23	270	31.2
3/8-24	25	300	33.9
7/16-14	36	–	48.8
7/16-20	40	–	54
1/2-13	50	–	67.8
1/2-20	60	–	81.3
Metric fasteners			
M5	–	36	4.1
M6	6	72	8.1
M8	13	156	17.6
M10	26	312	35.3
M12	35	–	47.5
M14	60	–	81.3

Table 3 BATTERY CAPACITY (HOURS)

Accessory draw	Provides continuous power for:	Approximate recharge time
80 amp-hour battery		
5 amps	13.5 hours	16 hours
15 amps	3.5 hours	13 hours
25 amps	1.6 hours	12 hours
105 amp-hour battery		
5 amps	15.8 hours	16 hours
15 amps	4.2 hours	13 hours
25 amps	2.4 hours	12 hours

Table 4 BATTERY CABLE RECOMMENDATIONS

Cable length	Minimum cable gauge size (AWG)
To 3 1/2 ft.	4
3 1/2 to 6 ft.	2
6 to 7 1/2 ft.	1
7 1/2 to 9 1/2 ft.	0
9 1/2 to 12 ft.	00
12 to 15 ft.	000
15 to 19 ft.	0000

Table 5 BATTERY STATE OF CHARGE

Specific Gravity Reading	Percentage of Charge Remaining
1.120-1.140	0
1.135-1.155	10
1.150-1.170	20
1.160-1.180	30
1.175-1.195	40
1.190-1.210	50
1.205-1.225	60
1.215-1.235	70
1.230-1.250	80
1.245-1.265	90
1.260-1.280	100

Table 6 BOAT WIRING HARNESS STANDARD WIRE COLORS

Circuit	Wire color
Starter engagement	Yellow/red
Tachometer	Gray
Stop circuit (ignition side)	Black/yellow
Stop circuit (ground side)	Black
Choke or primer	Yellow/black
Overheat/low oil warning	Tan/blue
Switched B+	Purple
Protected B+	Red/purple
Temperature gauge	Tan
Fuel sender circuit	Pink
Grounds	Black
Trim motor up	Blue
Trim motor down	Green
Trim switching circuit up	Blue/white
Trim switching circuit down	Green/white
Trim switching circuit B+	Red/purple
Trim trailer circuit	Purple/white
Trim sender circuit	Brown/white
Trim system grounds	Black

Chapter Eight

Power Head

This chapter provides power head removal/installation, disassembly/reassembly, and cleaning and inspection procedures for all models. The power head can be removed from the outboard motor without removing the entire outboard from the boat.

The power heads from different model groups differ in construction and require different service procedures. Whenever possible, engines with similar service procedures have been grouped together.

The components shown in the accompanying illustrations are generally from the most common models. While the components shown in the illustrations may not be identical to those being serviced, the step-by-step procedures cover each model. Exploded illustrations, typical of each power head model group are in the appropriate *Disassembly* section and are helpful references for many service procedures.

This chapter is arranged in a normal disassembly/assembly sequence. If only a partial repair is required, follow the procedure(s) to the point where the faulty parts can be replaced, then jump to reassembly. Many procedures require the use of manufacturer recommended special tools, which can be purchased from a Mercury or Mariner outboard dealership.

Power head work stands and holding fixtures are available from specialty shops or marine and industrial product distributors.

Make sure the work bench, work station, engine stand or holding fixture is sufficient size and strength to support the size and weight of the power head.

SERVICE CONSIDERATIONS

Performing internal service procedures on the power head requires considerable mechanical ability. Carefully consider the expertise necessary before attempting any operation involving major disassembly of the engine.

If power head disassembly or repair will be made by a dealership, consider separating the power head from the outboard motor and removing the fuel, ignition and electrical systems and all accessories before taking the basic power head to the dealership for the overhaul or major repair.

Since most marine dealerships often have lengthy waiting lists for service especially during the spring and summer, this practice can reduce the time the unit is in the shop. If preliminary work is done beforehand, repairs can be scheduled and performed much more quickly. Always discuss options with the dealership before taking them a disassembled engine. Dealerships will often want to install, adjust and test run the engine in order to be comfortable providing warranty coverage for the overhaul or repair.

No matter who is doing the work, repair will be quicker and easier if the motor is clean before any service procedure is started. There are many special cleaners available from automotive supply stores. Most of these cleaners are simply sprayed on, then rinsed off with a garden hose after the recommended time period. Always follow all instructions provided by the manufacturer. Never spray cleaning solvent onto electrical and ignition components or spray into the induction system.

WARNING

Never use gasoline as a cleaning agent. Gasoline presents an extreme fire and explosion hazard. Work in a well-ventilated area when using cleaning solvent. Keep a large fire extinguisher rated for gasoline and oil fires nearby in case of an emergency.

Thoroughly read this chapter and understand what is involved in completing the repair satisfactorily before doing service work. Make arrangements to buy or rent the necessary special tools and obtain a source for replacement parts *before* starting. It is frustrating and time-consuming to start a major repair, then be unable to finish because the necessary tools or parts are not available.

NOTE

A series of at least five photographs, taken from the front, rear, top and both sides of the power head before removal are very useful during reassembly and installation. The photographs are especially useful when trying to route electrical harnesses, fuel, primer and recirculation lines. They are also helpful during the installation of accessories, control linkages and brackets.

Before beginning the job, review Chapter One and Chapter Two of this manual.

Table 1 lists specific torque specifications for most power head fasteners. **Table 2** list standard torque specifications for both American standard and Metric fasteners. Use the standard torque specification with fasteners not in **Table 1**. **Tables 3-6** list power head dimensional specifications. **Table 7** list model number codes. **Tables 1-7** are located at the end of this chapter.

MODEL IDENTIFICATION

All Mercury/Mariner outboard models use an individual, unique serial number for the primary means of identification. The serial number plate is on the midsection (**Figure 1**) of the outboard motor and is usually attached to the swivel bracket or the stern bracket. Serial numbers are never duplicated. If the engine is still equipped with its original power head, the serial number is also stamped on a welch plug affixed to the power head (**Figure 2**, typical).

The serial number plate contains a coded model number and model information. See **Figure 3**. Model number codes are in **Table 7**. Shaft length relates to the transom height of the boat. A long shaft (20 in. [50.8 cm]) motor is designed to fit a transom that measures approximately 20 in. (50.8 cm). If no shaft length code is specified, the motor is a standard (15 in. [38.1 cm]) shaft model.

When ordering replacement parts, provide the serial number to the Mercury/Mariner dealership. On engines equipped with coded model numbers and model information, make sure all of this data is also provided to the dealership.

POWER HEAD BREAK-IN

Whenever a power head has been rebuilt or replaced, or if any new internal parts have been installed, treat it as a new engine. Run the engine on the specified fuel/oil mixture and operate according to the recommended break-in procedure as described in Chapter Four.

CAUTION

Failure to follow the recommended break-in procedure may result in premature power head failure.

SERVICE RECOMMENDATIONS

If the power head fails determine the cause of the failure. Refer to the *Engine* section in Chapter Three for troubleshooting procedures.

Many failures are caused simply from using incorrect or stale fuel and incorrect lubricating oil. Refer to Chapter Four for fuel and oil recommendations.

When rebuilding or performing a major repair on the power head, consider performing the following steps to prevent the failure from reoccurring.

1. Service the water pump. Replace the impeller and all seals and gaskets. See Chapter Nine.

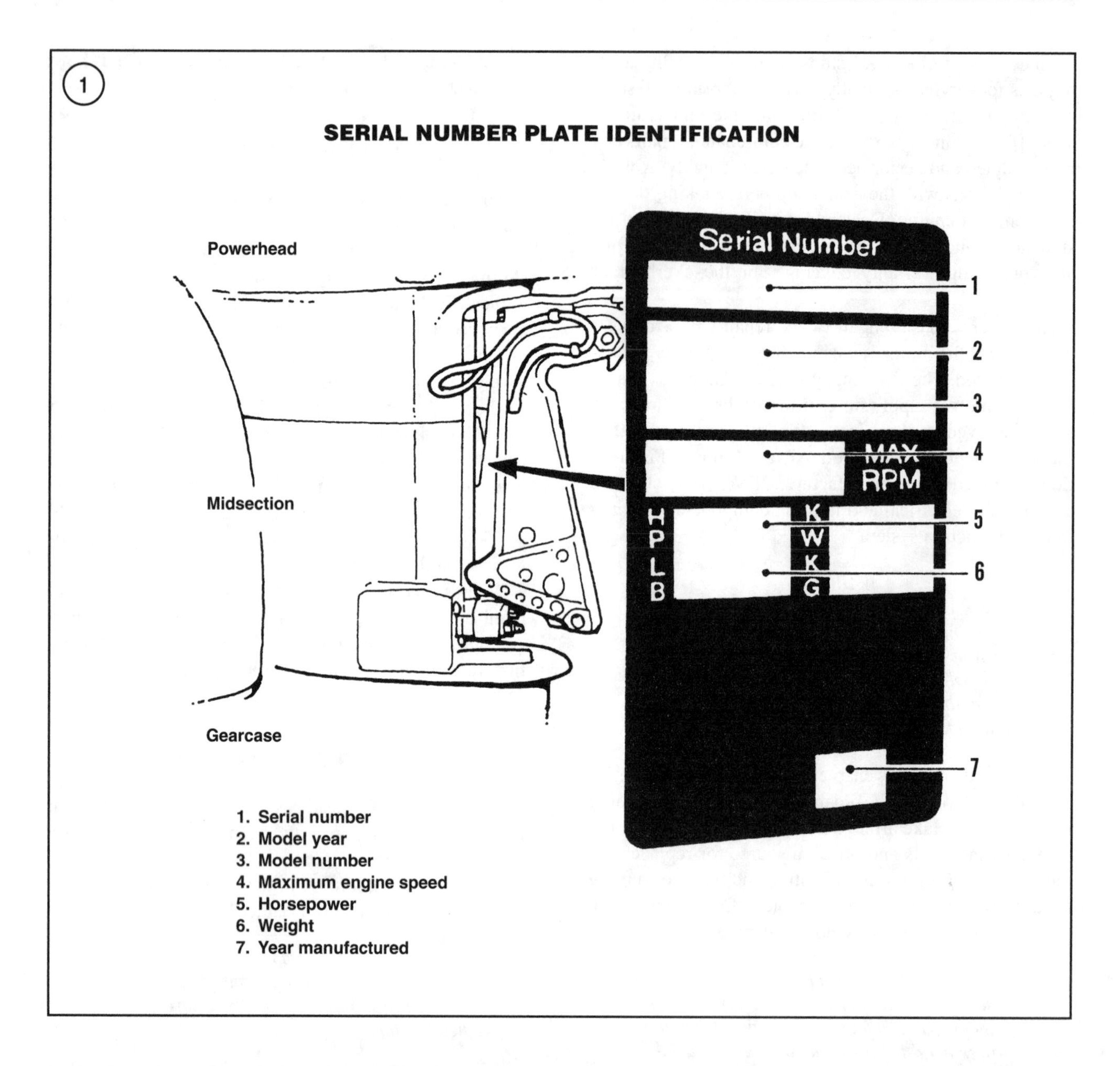

2. Replace the thermostat(s), and remove and inspect the poppet valve assembly, on models so equipped, as described in this chapter. Replace suspect components.
3. Drain the fuel tank(s) and dispose of the old fuel in an approved manner.
4. Fill the fuel tank with fresh fuel and add the recommended oil to the fuel tank at the *break-in* ratio as described in Chapter Four.
5. Replace or clean all fuel filters. See Chapter Four.
6. Clean and adjust the carburetors. See Chapter Six.
7. On oil-injected models, drain and clean the oil reservoir(s). Dispose of the old oil in an approved manner. Then refill the oil system with the specified oil (Chapter Four) and bleed the oil system as described in Chapter Twelve.
8. Install new spark plugs as described in Chapter Four. Use only the recommended spark plugs.
9. Perform *all* of the synchronization and linkage adjustments as described in Chapter Five before returning the motor to service.

LUBRICANTS, SEALANTS AND ADHESIVES

The Mercury (Quicksilver) part numbers for the lubricants, sealants and adhesives are specified in this section.

Equivalent products are acceptable for use as long as they meet or exceed the original manufacturer's specifications.

During power head assembly, lubricate all internal engine components with Quicksilver two-cycle (TCW-3) outboard motor oil. Do not assemble any components *dry*. Lubricate all seal lips and O-rings with Quicksilver 2-4-C Multi-Lube grease (part No. 92-825407). Lubricate and hold all needle and roller bearings in place with Quicksilver Needle Bearing Assembly Grease (part No. 92-825265A 1).

To efficiently remove the carbon from the pistons and combustion chambers, use Quicksilver Power Tune Engine Cleaner (part No. 92-15104). Allow ample time for the cleaner to soak into and soften carbon deposits.

When no other sealant or adhesive is specified, coat all gaskets with Quicksilver Perfect Seal (part No. 92-34227-1). Coat the threads of all external fasteners when no other sealant or adhesive is specified with Quicksilver Perfect Seal to help prevent corrosion and ease future removal.

When sealing the crankcase cover/cylinder block, both mating surfaces must be free of all sealant residue, dirt, oil or other contamination. Locquic Primer (part No. 92-809824), lacquer thinner, acetone or similar solvents work well when used in conjunction with a plastic scraper. Do not use solvents with an oil, wax or petroleum base.

Loctite Master Gasket Sealant (part No. 92-12564-2) is the only recommended sealant for sealing the crankcase cover-to-cylinder block mating surfaces on models without a gasket. The sealant comes in a kit that includes a special primer. Follow the instructions included in the kit for preparing the surfaces and applying the sealant. Apply the sealant bead to the inside (crankshaft side) of all crankcase cover screw holes.

Apply Loctite 271 threadlocking adhesive (part No. 92-809819) to the outer diameter of all seals before pressing the seals into place. Also apply this adhesive to the threads of all internal fasteners if no other adhesive is specified.

Whenever a Loctite product is called for, always clean the surfaces or threads with Locquic Primer (part No. 92-809824). Locquic Primer cleans and primes the surface and ensures a quick secure bond by leaving a thin film of catalyst on the surface or threads. The primer must be allowed to air dry, as blow drying will remove the catalyst.

SEALING SURFACES

Clean all sealing surfaces carefully to prevent nicks and gouges. Often a shop towel soaked in solvent can be used to rub gasket material and/or sealant from a mating surface. If scrapers must be used, try using a plastic scraper, such as a household electrical outlet cover, or a piece of Lucite with one edge ground to a 45° angle to prevent damage to the sealing surfaces. Nick gouges usually prevent the sealant from curing.

NOTE

Use plate glass or a machinist's surface plate or straightedge for surface checking. Ordinary window glass does not have a uniform surface and will give false readings. Plate glass has a very uniform surface flatness.

Once the surfaces are clean, check the component for warp by placing the component onto a piece of plate glass or a machinist's surface plate. Apply uniform downward pressure and try to insert a selection of feeler gauges between the plate and the component. Specifications for maximum cylinder head warpage are in **Table 5**.

CAUTION
Do not lap the cylinder block-to-crankcase cover.

To remove minor warp, minor nicks or scratches, or traces of sealant or gasket material, place a large sheet of 320-400 grit wet sandpaper onto the plate glass or surface plate. Apply light downward pressure and move the component in a figure-eight pattern as shown in **Figure 4**. Use a light oil, such as WD-40, to keep the sandpaper from loading up. Remove the component from the sandpaper and recheck the sealing surface. Use a machinist's straightedge to check areas that cannot be accessed using the glass or surface plate. See **Figure 5**.

It may be necessary to repeat the lapping process several times to achieve the desired results. Never remove more material than is absolutely necessary. Make sure the component is thoroughly washed to remove all grit before reassembly.

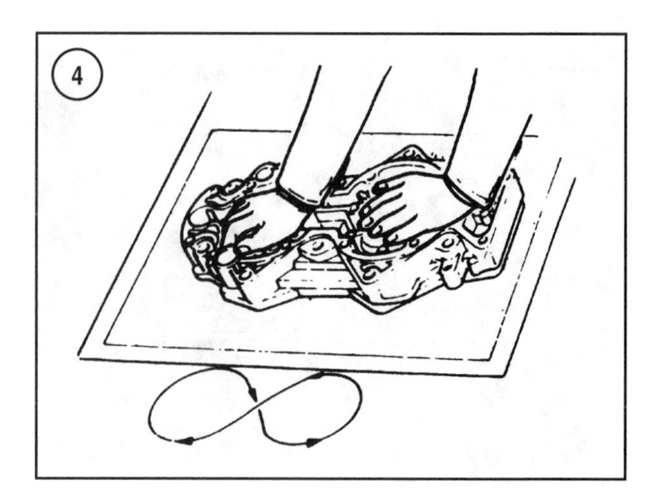

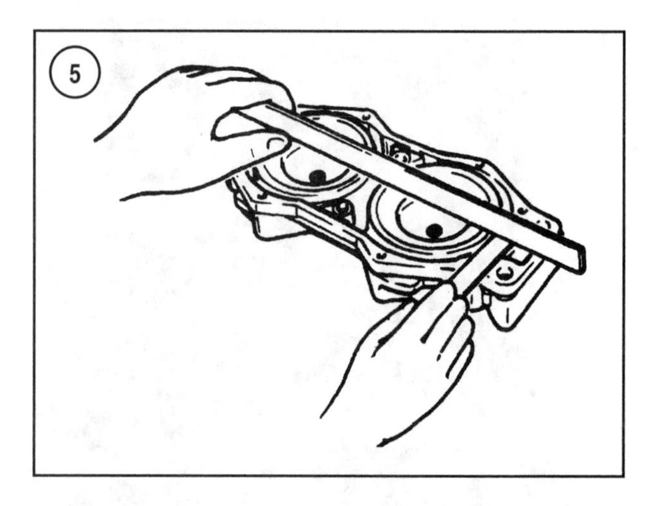

FASTENERS AND TORQUE

Always replace a worn or damaged fastener with one of equal size, type and torque requirement. Power head torque values are in **Table 1**. If a specification is not provided for a given fastener, use the standard torque values in **Table 2** according to fastener size.

Determine the fastener size by measuring the shank of the screw or bolt as shown in **Figure 6**. Determine the thread pitch using the appropriate metric or American thread pitch gauge as shown in **Figure 7**.

Damaged threads in components and castings may be repaired using a Heli-Coil, or an equivalent, stainless steel threaded insert (**Figure 8**, typical). Heli-coil kits are available at automotive or marine and industrial supply stores. Never run a thread tap or thread chaser into a hole equipped with a Heli-coil. Replace damaged Heli-coils by gripping the outermost coil with needlenose pliers and unthreading the coil from the hole. Do not pull the coil straight out or the threads in the hole will be damaged.

CAUTION
American and Metric fasteners are used on these engines. Always match a replacement fastener to the original. Do not run a tap or thread chaser into a hole or over a bolt without first verifying the thread size and pitch. Newer model manufacturer's parts catalogs list each standard fastener by diameter, length and pitch. Always have the engine model and serial numbers when ordering parts from a dealership.

Unless otherwise specified, components secured by more than one fastener must be tightened in a *minimum* of three steps. Evenly tighten all fasteners hand-tight as a first step. Then evenly tighten all fasteners to 50% of the specified torque value. Finally, evenly tighten all fasteners to 100% of the specified torque value.

Follow torque sequences as directed. If no tightening sequence is specified, start at the center of the component and tighten in a circular pattern, working outward. All torque sequences are listed in the appropriate assembly section of this chapter.

CAUTION
Many models use a new torque process for the rod bolts, cylinder head bolts and crankcase cover screws. This procedure is called torque and turn. Follow the new procedure as outlined in this chapter to prevent dam-

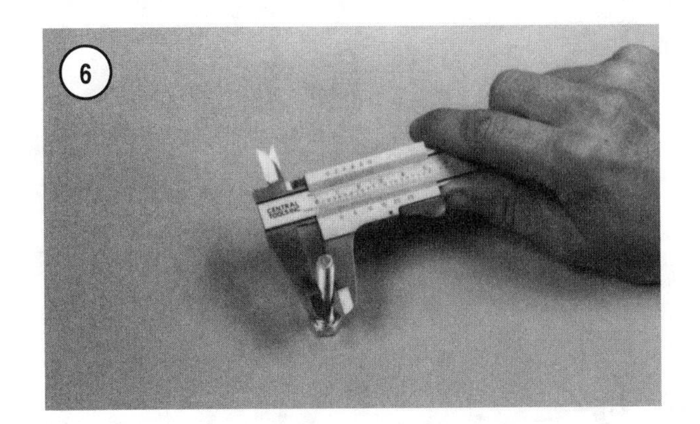

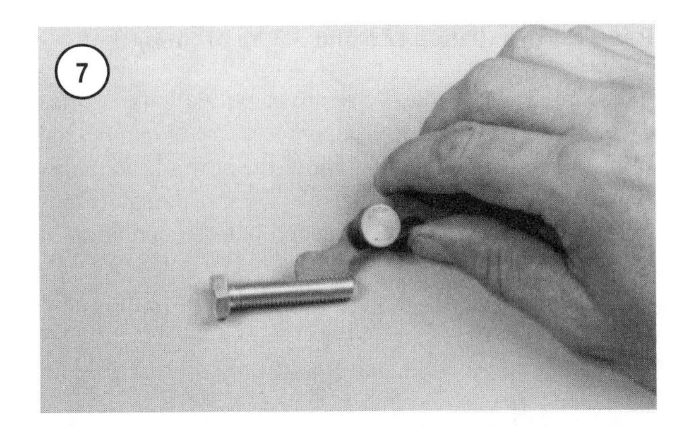

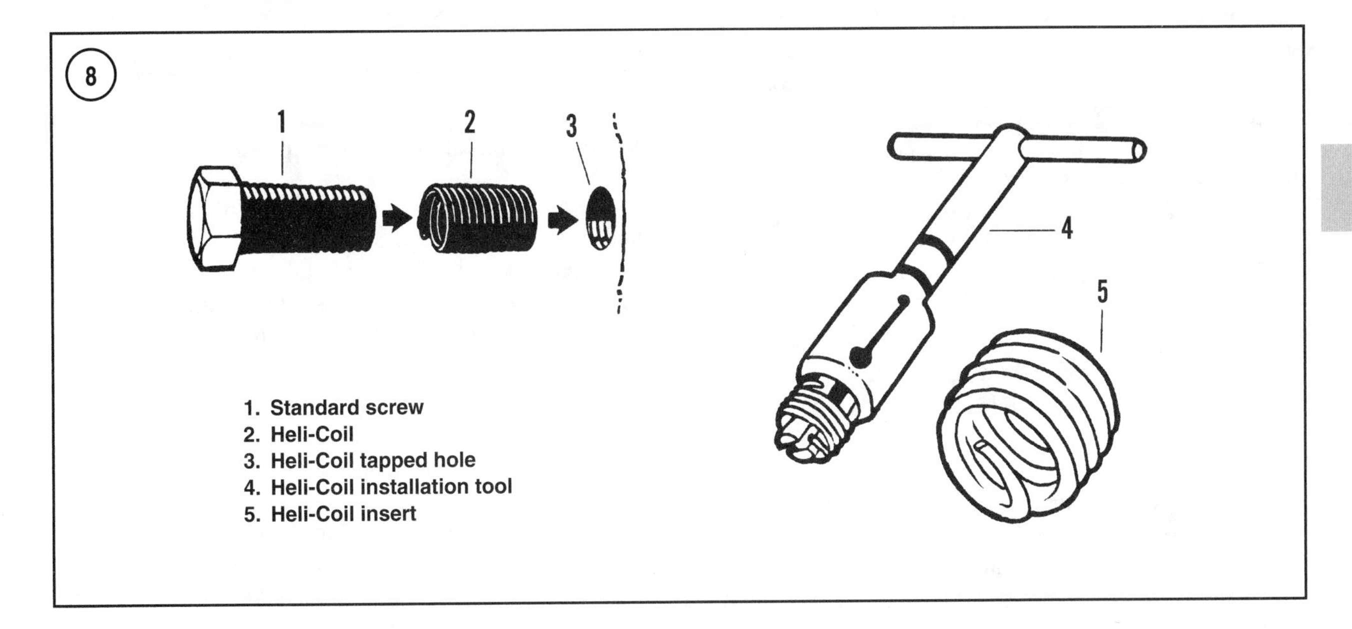

1. Standard screw
2. Heli-Coil
3. Heli-Coil tapped hole
4. Heli-Coil installation tool
5. Heli-Coil insert

aging components. Never retorque a fastener secured by the torque and turn method.

On models *not* equipped with torque and turn fasteners, retorque the cylinder head screws after the engine has reached operating temperature and been allowed to cool. To retorque the cylinder head screws and any other fasteners, loosen each fastener approximately one turn, then retighten it to the specified torque value. Repeat the process until all of the fasteners are retorqued.

Retorque spark plugs to ensure proper heat transfer and to prevent preignition and detonation (Chapter Three). Retorque spark plugs after the engine has reached operating temperature and been allowed to cool. Do not loosen the spark plug, simply retighten the plug to the specified torque value.

When no other sealant or adhesive is specified, coat the threads of all external fasteners with Quicksilver Perfect Seal (part No. 92-34227-1) to help prevent corrosion and ease future removal.

POWER HEAD REMOVAL/INSTALLATION

The removal and installation procedures in this chapter are presented in the most efficient sequence for removing the power head while preparing for complete disassembly. If complete disassembly is not necessary, stop disassembly at the appropriate point, then begin reassembly where disassembly stopped. Remove the power head as an assembly if major repair must be performed. Power head removal is not required for certain service procedures such as cylinder head removal, intake and exhaust cover removal, if so equipped, ignition component replacement, fuel system component replacement and reed block/intake manifold removal.

Removal/Installation (2.5 and 3.3 hp Models)

The power head may be removed with all accessories and systems installed. It may be necessary to remove the split lower cowl in order to remove the power head in this fashion. If Steps 2-4 are skipped, make sure the safety lanyard switch leads are disconnected before continuing at Step 5.

1. Disconnect the spark plug lead and remove the spark plug.
2. Remove the rewind starter assembly as described in Chapter Thirteen.
3. Remove the flywheel, charge/trigger coil, CD module (switch box) and ignition coil as described in Chapter Seven.
4. Remove the fuel tank and carburetor as described in Chapter Six.
5. Remove the screws securing the power head to the drive shaft housing (**Figure 9**).
6. Lift the power head off the drive shaft housing and place it on a clean bench. It may be necessary to tap the power head with a plastic or rubber mallet to help break the base gasket free.
7. Thoroughly clean all gasket material from the drive shaft housing and power head mating surfaces.
8. To install the power head, install a new gasket on the drive shaft housing.
9. Lubricate the drive shaft with Quicksilver 2-4-C Multi-Lube grease. Wipe excess grease from the top surface of the drive shaft.
10. Install the power head onto the drive shaft housing. Rotate the crankshaft as necessary to align the drive shaft splines.
11. Apply Quicksilver Perfect Seal to the threads of the power head mounting screws. Install the screws and evenly tighten them to the specification in **Table 1**.
12. Install the fuel tank and carburetor as described in Chapter Six.
13. Install the charge/trigger coil, CD module (switch box), ignition coil and flywheel as described in Chapter Seven.
14. Install the rewind starter as described in Chapter Thirteen.
15. Install the spark plug and torque it to the specification in **Table 1**. Reconnect the spark plug lead.
16. Make sure all cable clamps are reinstalled in their original positions and *new* tie-straps are installed to replace any that were removed.
17. Refer to Chapter Four for fuel and oil recommendations and break-in procedures as needed. Then refer to Chapter Five and perform all synchronization and linkage adjustments.

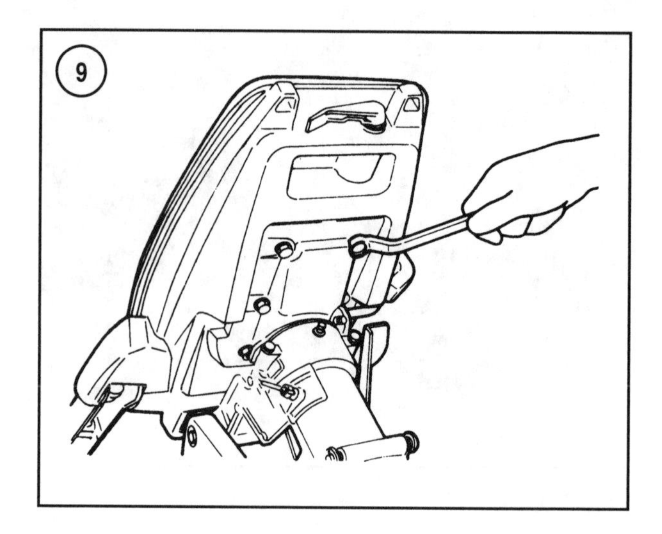

Removal/Installation (4 and 5 hp Models)

The power head may be removed with all accessories and systems installed. If Steps 2-4 are skipped, make sure the safety lanyard switch leads, all applicable fuel supply lines, carburetor throttle control cable, choke valve linkage and the rewind starter interlock rod are disconnected before continuing at Step 5.

1. Disconnect the spark plug lead and remove the spark plug.
2. Remove the rewind starter as described in Chapter Thirteen.
3. Remove the flywheel, charge and trigger coils, CD module (switch box), and ignition coil as described in Chapter Seven. If the engine is equipped with an AC lighting coil, remove it at this time.
4. Remove the fuel tank and carburetor as described in Chapter Six.
5. Remove the six screws securing the power head to the drive shaft housing (**Figure 9**).
6. Lift the power head off the drive shaft housing and place it on a clean bench. It may be necessary to tap the power head with a plastic or rubber mallet to help break the base gasket free.
7. Thoroughly clean all gasket material from the power head and drive shaft housing mating surfaces.
8. To install the power head, lubricate the drive shaft splines with Quicksilver 2-4-C Multi-Lube. Wipe excess lubricant from the top of the drive shaft.
9. Place a new gasket on the drive shaft housing.
10. Install the power head onto the drive shaft housing. Rotate the crankshaft as necessary to align the drive shaft splines.
11. Coat the threads of the six mounting screws with Quicksilver Perfect Seal. Then install the six power head

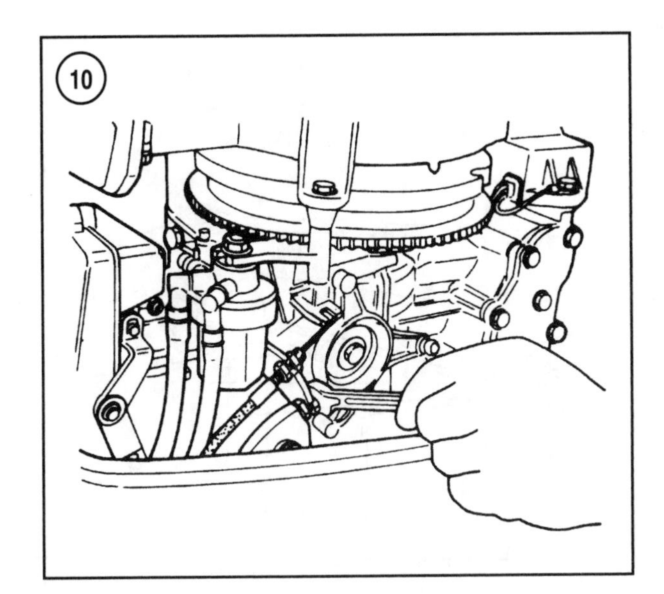
10

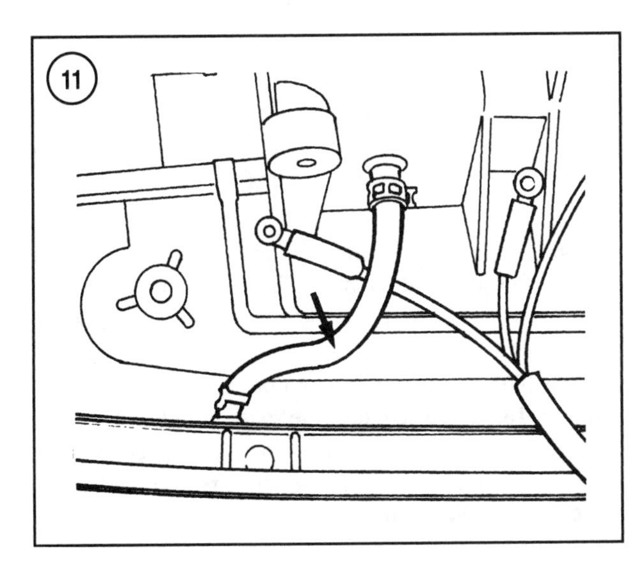
11

mounting bolts and evenly tighten them to the specification in **Table 1**.

12. Install the fuel tank and carburetor as described in Chapter Six.

13. Install the ignition coil, CD module (switch box), charge and trigger coils, ignition coil, and flywheel as described in Chapter Seven. If the engine is equipped with an AC lighting coil, install it before installing the flywheel.

14. Install the rewind starter as described in Chapter Thirteen.

15. Install the spark plug and reconnect the spark plug lead. Tighten the spark plug to the specification in **Table 1**.

16. Make sure all cable clamps are reinstalled in their original positions and new tie-straps are installed to replace any that were removed.

17. Refer to Chapter Four for fuel and oil recommendations and break-in procedures as needed. Then refer to Chapter Five and perform all synchronization and linkage adjustments.

Removal/Installation (6-15 hp Models)

The power head may be removed with most of the accessories and systems installed. If Steps 2-4 are skipped, make sure the fuel supply line at the fuel pump and the safety lanyard switch leads are disconnected before continuing at Step 5.

1. Disconnect the spark plug leads and remove the spark plugs.
2. Remove the rewind starter as described in Chapter Thirteen.
3. Remove the entire ignition system. This includes the flywheel, ignition stator, trigger coil, switch box and both ignition coils. See Chapter Seven. If the engine is equipped with an AC lighting coil or battery charging coil and rectifier, remove them at this time.
4. Remove the carburetor/fuel pump assembly as described in Chapter Six.

NOTE
The control linkages must be disconnected or removed as described in Steps 5-7 to remove the power head. Additional removal of linkages and brackets is not necessary unless they interfere with power head disassembly.

5. Disconnect the shift rod bellcrank arm from the lower gearcase shift rod as described in *Gearcase Removal* in Chapter Nine.

6A. *Tiller control models*—Remove the throttle cables as follows:

  a. Note the routing of the throttle cables on the throttle pulley/control arm. Mark the upper cable with a felt-tipped marker or a piece of tape for reassembly purposes.
  b. Loosen the jam nuts on both throttle cables (**Figure 10**). Then remove both throttle cables from the throttle pulley/control arm.

6B. *Remote control models*—Disconnect and remove the throttle and shift cables as described in Chapter Fourteen.

7. *Side-shift tiller control*—Remove the screw and washer securing the shift lever to the shift shaft. Remove the shift lever by pulling it from the shift shaft.

8. Disconnect the water discharge hose from the lower cowl. See **Figure 11**.

9. Remove the screws and nuts holding the power head to the drive shaft housing (**Figure 12**).

CAUTION
At this point, there should be no hoses, wires or linkages connecting the power head to the drive shaft housing. Make sure nothing will interfere with power head removal before continuing.

10. Rock the power head to break the gasket seal between the drive shaft housing and power head. Remove the power head and mount it on a power head stand (part No. 91-13662A-1). The power head stand must be securely clamped in a suitable vise. If this stand is not available, mount the power head to a power head work stand and holding fixture (**Figure 13**).
11. Thoroughly clean all gasket material from the drive shaft housing and power head mating surfaces.
12. To install the power head, lubricate the drive shaft splines with Quicksilver 2-4-C Multi-Lube. Wipe excess lubricant from the top of the drive shaft.
13. Place a new gasket on the drive shaft housing.
14. Install the power head on the drive shaft housing. Rotate the crankshaft as necessary to align the drive shaft splines.
15. Apply Loctite 271 threadlocking adhesive to the threads of the power head mounting screws and studs. Install the screws and nuts, and evenly tighten the fasteners to the specification in **Table 1**. See **Figure 12**.
16. Reconnect the water discharge hose to the lower cowl.
17. Reconnect the shift rod bellcrank arm to the lower gearcase shift rod as described in *Gearcase installation* in Chapter Nine.
18. *Side-shift tiller control*—Position the shift lever onto the shift shaft and secure it with the screw and washer. Tighten the screw to the specification in **Table 1**.
19A. *Tiller control models*—Install the throttle cables as follows:
   a. Route the throttle cables on the throttle pulley/control arm as noted during disassembly.
   b. Adjust the jam nuts on both throttle cables as described in Chapter Five.

19B. *Remote control models*—Reconnect and adjust the throttle and shift cables as described in Chapter Fourteen.
20. Install the carburetor/fuel pump assembly as described in Chapter Six. Secure the fuel line with a new tie-strap.
21. Install the entire ignition system. This includes the ignition stator, trigger coil, switch box, both ignition coils and the flywheel. See Chapter Seven. If the engine is equipped with an AC lighting coil or battery charging coil and rectifier, install them before installing the flywheel.

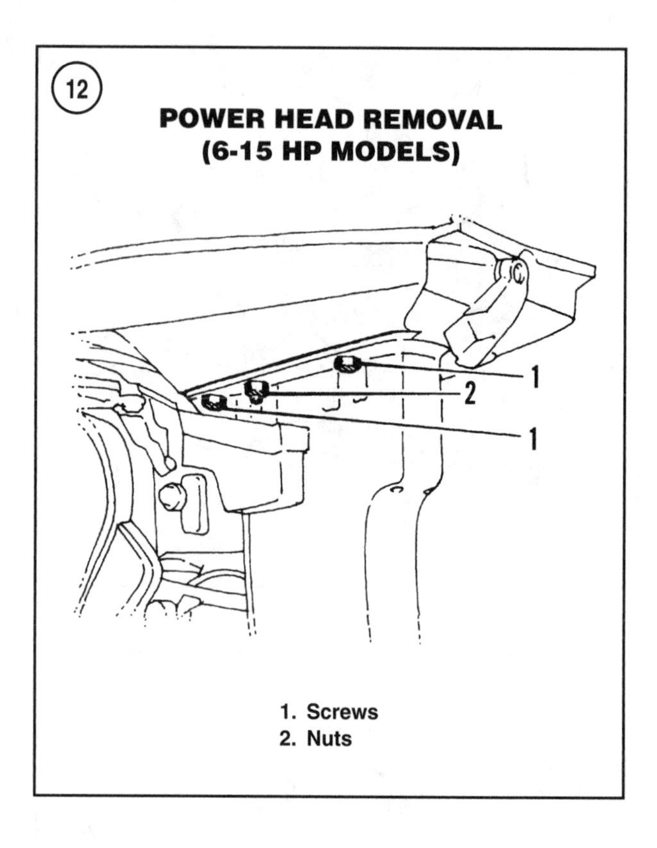

22. Install the rewind starter as described in Chapter Thirteen.
23. Install the spark plugs and torque them to the specification is **Table 1**. Reconnect the spark plug leads.
24. Make sure all cable clamps are reinstalled in their original positions and new tie-straps are installed to replace any that were removed.
25. Refer to Chapter Four for fuel and oil recommendations and break-in procedures as needed. Then refer to Chapter Five and perform all synchronization and linkage adjustments.

Removal/Installation (20-25 hp and 20 Jet Models)

The power head may be removed with most of the accessories and systems installed. If Steps 3-6 are skipped, make sure the fuel supply line at the fuel pump and the safety lanyard switch leads are disconnected before continuing at Step 7.

1. Disconnect the spark plug leads and remove the spark plugs to prevent accidental starting.
2. On electric start models, disconnect both battery cables from the battery. Then remove the cables from the

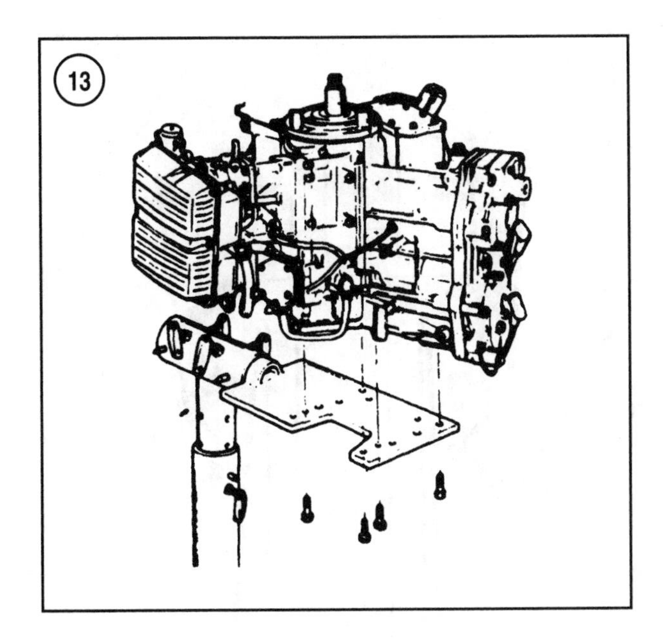

power head. Note each connection's location and the cable routing for reassembly.

3. Remove the rewind starter as described in Chapter Thirteen.

4. Remove the entire ignition system. This includes the flywheel, ignition stator, trigger coil, switch box and both ignition coils. See Chapter Seven. If the engine is equipped with an AC lighting coil or battery charging coil and rectifier, remove them at this time.

5. On electric start models, remove the electric starter, starter solenoid and wiring harness. See Chapter Seven.

6. Remove the carburetor/fuel pump assembly as described in Chapter Six.

NOTE

The control linkages must be disconnected or removed as described in Steps 7-9 to remove the power head. Additional removal of linkages and brackets is not necessary unless they interfere with power head disassembly. The 20 jet model does not require disconnecting the shift cable or linkage for power head removal.

7. On 20-25 hp models except jet, disconnect the shift rod bellcrank arm from the lower gearcase shift rod as described in *Gearcase Removal* in Chapter Nine.

8A. *Tiller control models*—Remove the throttle cables as follows:

a. Note the routing of the throttle cables on the throttle pulley/control arm. Mark the upper cable with a felt-tipped marker or a piece of tape for reassembly purposes.

b. Loosen the jam nuts on both throttle cables (**Figure 10**). Then remove both throttle cables from the throttle pulley/control arm.

8B. *Remote control models*—Disconnect and remove the throttle and shift cables as described in Chapter Fourteen.

9. On side-shift tiller control models, remove the screw and washer securing the shift lever to the shift shaft. Remove the shift lever by pulling it from the shift shaft.

10. Disconnect the water discharge hose (**Figure 11**) from the lower cowl.

11. Remove the six screws and washers securing the power head to the drive shaft housing. There are three screws on each side of the drive shaft housing. See **Figure 14**, typical.

CAUTION

At this point, there should be no hoses, wires or linkages connecting the power head to the drive shaft housing. Make sure nothing will interfere with power head removal before continuing.

12. Rock the power head to break the gasket seal between the drive shaft housing and power head. Remove the power head and mount it on a power head stand (part No. 91-25821A-1). The power head stand must be securely clamped in a vise. If this stand is not available, mount the power head to a power head work stand and holding fixture (**Figure 13**).

13. Thoroughly clean all gasket material from the drive shaft housing and power head mating surfaces.

14. To install the power head, lubricate the drive shaft splines with Quicksilver 2-4-C Multi-Lube. Wipe excess lubricant from the top of the drive shaft.

15. Place a new gasket on the drive shaft housing.

16. Install the power head onto the drive shaft housing. Rotate the crankshaft as necessary to align the drive shaft splines.

17. Apply Loctite 271 threadlocking adhesive to the threads of the power head mounting screws. Install the screws and washers (**Figure 14**). Evenly tighten the fasteners to specification in **Table 1**.

18. Reconnect the water discharge hose to the lower cowl.

19. Reconnect the shift rod bellcrank arm to the lower gearcase shift rod as described in *Gearcase Installation* in Chapter Nine.

20. On side-shift tiller control models, position the shift lever on the shift shaft and secure it with the screw and washer. Tighten the screw to the specification in **Table 1**.

21A. *Tiller control models*—Install the throttle cables as follows:

a. Route the throttle cables on the throttle pulley/control arm as noted during disassembly.
b. Adjust the jam nuts on both throttle cables as described in Chapter Five.

21B. *Remote control models*—Reconnect and adjust the throttle and shift cables as described in Chapter Fourteen.

22. Install the carburetor/fuel pump assembly as described in Chapter Six. Secure the fuel line with a new tie-strap.

23. Install the entire ignition system. This includes the ignition stator, trigger coil, switch box, both ignition coils and the flywheel. See Chapter Seven. If the engine is equipped with an AC lighting coil or battery charging coil and rectifier, install these items before installing the flywheel.

24A. On electric start models, install the electric starter, starter solenoid and wiring harness. See Chapter Seven.

24B. On electric start models, connect the battery cables to the power head as noted on removal, then connect the cables to the battery. Tighten all connections securely.

25. Install the rewind starter as described in Chapter Thirteen.

26. Install the spark plugs and torque them to the specification in **Table 1**. Reconnect the spark plug leads.

27. Make sure all cable clamps are reinstalled in their original positions and new tie-straps are installed to replace any that were removed.

28. Refer to Chapter Four for fuel and oil recommendations and break-in procedures as needed. Then refer to Chapter Five and perform all synchronization and linkage adjustments.

Removal/Installation (30 and 40 hp Two-cylinder Models)

The power head is best removed with most of the accessories and systems left installed. These items will be removed after the power head is separated from the drive shaft housing. Refer to the wiring diagrams at the end of the manual. Make sure all cable clamps are reinstalled in their original positions and new tie-straps are installed to replace any that were removed.

If the oil tank is interfering with access to other components or connections, refer to Chapter Twelve for removal and installation procedures. Bleed the oil pump after oil tank installation as described in Chapter Twelve.

1. Disconnect the spark plug leads and remove both spark plugs.

2. On electric start models, disconnect both battery cables at the battery, then remove the cables from the power head. Note each cable's location and routing for reassembly.

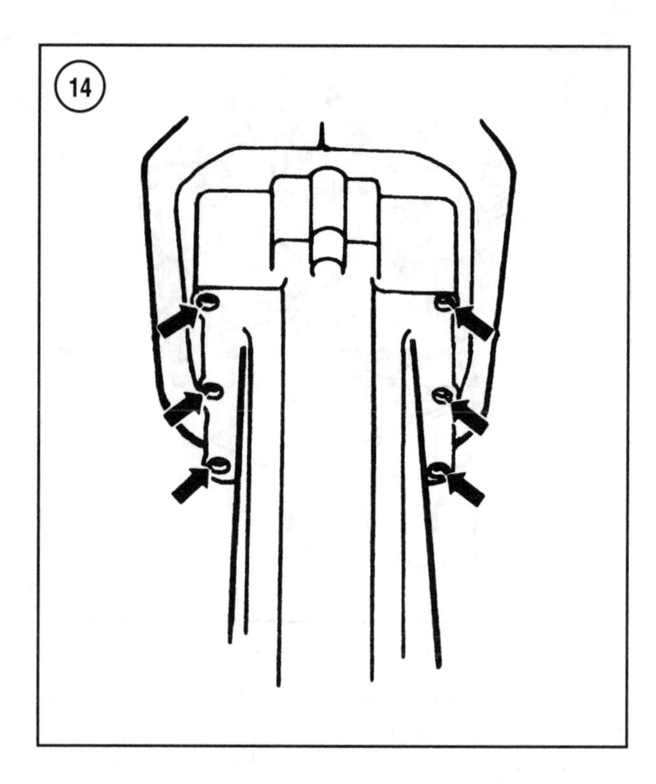

3A. *Manual start models*—Disconnect the safety lanyard switch, stop button switch and warning horn from their bullet connectors.

3B. *Electric start models*—Disconnect the remote control harness or key switch harness from the engine harness at the main wiring harness connector. Disconnect any additional remote control harness leads from their bullet connectors. For models equipped with trim and tilt, disconnect the lower cowl mounted switch leads at their bullet connectors.

4. Disconnect the neutral safety switch leads from their bullet connectors.

5. On manual start models, disconnect the primer bulb lines from the intake manifold and carburetor fittings.

6A. *Manual start models*—Remove the rewind starter as described in Chapter Thirteen.

6B. *Electric start models*—Remove the flywheel cover if it is not already removed.

7. Disconnect the water discharge hose (**Figure 11**) from the power head fitting.

8A. *Tiller control models*—Remove the throttle cables as follows:

a. Note the routing of the throttle cables on the throttle pulley/control arm. Mark the upper cable with a felt-tipped marker or a piece of tape for reassembly reference.

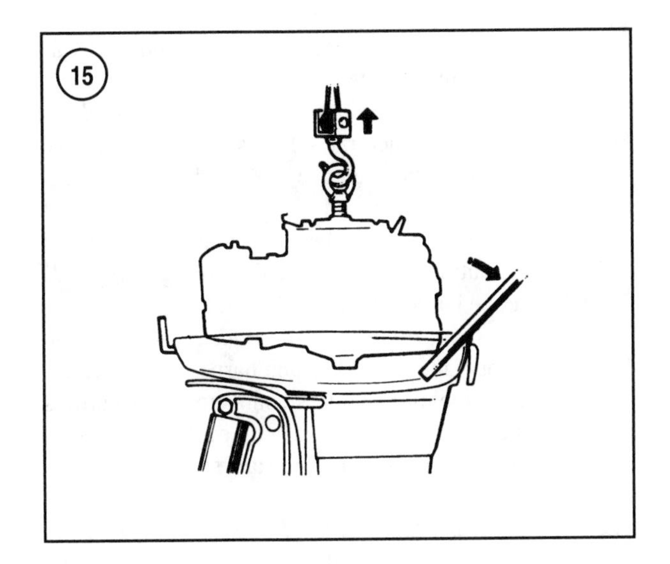

b. Loosen the jam nuts on both throttle cables (**Figure 10**). Then remove both throttle cables from the throttle pulley/control arm.

8B. *Remote control models*—Remove the throttle and shift cables as described in Chapter Fourteen.

9. Disconnect the fuel supply line connector from the lower cowl or disconnect the fuel line from the connector or the fuel filter inlet fitting.

10. Remove the screw, washer and locknut securing the trim cover to the drive shaft housing. The screw is located at the front of the cover, close to the swivel bracket. Spread the front of the cover and slide it rearward from the drive shaft housing to gain access to the power head mounting screws.

11. Remove the six screws securing the power head to the drive shaft housing. There are three screws on each side of the drive shaft housing. See **Figure 14**, typical.

12. Remove the plastic protective cap from the center of the flywheel and install the lifting eye (part No. 91-90455 or an equivalent) into the flywheel a minimum of five full turns.

CAUTION

At this point, there should be no hoses, wires or linkages connecting the power head to the drive shaft housing. Make sure nothing will interfere with power head removal before continuing.

13. Rock the power head to break the gasket seal between the drive shaft housing and power head. Then lift the power head from the drive shaft housing with the aid of an assistant or a suitable hoist. See **Figure 15**.

14. Mount the power head onto a power head (crankshaft) stand (part No. 91-827001A-1). The power head stand must be securely clamped in a vise. If the stand is not available, mount the power head to a power head work stand and holding fixture (**Figure 13**).

15. Thoroughly clean all gasket material from the drive shaft housing and power head mating surfaces.

16. To install the power head, lubricate the drive shaft splines with Quicksilver Special Lube 101. Wipe excess lubricant from the top of the drive shaft.

17. Place a new gasket on the drive shaft housing.

18. Install the power head onto the drive shaft housing. Rotate the crankshaft as necessary to align the drive shaft splines.

19. Apply Quicksilver Perfect Seal to the threads of the power head mounting screws. Install and evenly tighten the screws (**Figure 14**, typical) to the specification in **Table 1**.

20. Install the trim cover over the drive shaft housing and against the lower cowl. Secure the cover with the screw, washer and locknut. Tighten the screw securely.

21. Reconnect the water discharge hose from the power head fitting (**Figure 11**). Secure the connection with a new tie-strap.

22. Install the fuel supply line connector to the lower cowl or reconnect the fuel line to the connector or the fuel filter inlet fitting. Secure the connection with a new tie-strap.

23A. *Tiller control models*—Install the throttle cables as follows:

a. Route the throttle cables onto the throttle pulley/control arm as noted during disassembly.
b. Adjust the jam nuts on both throttle cables as described in Chapter Five.

23B. *Remote control models*—Reconnect and adjust the throttle and shift cables as described in Chapter Fourteen.

24A. *Manual start models*—Install the rewind starter as described in Chapter Thirteen.

24B. *Electric start models*—Install the flywheel cover.

25. On manual start models, connect the primer bulb lines to the intake manifold and carburetor fittings. Secure the connections with new tie-straps.

26. Connect the neutral safety switch leads to the engine wiring harness bullet connectors.

27A. *Manual start models*—Connect the safety lanyard switch, stop button switch and warning horn to the appropriate engine wiring harness bullet connectors.

27B. *Electric start models*—Connect the remote control harness or key switch harness to the engine at the main wiring harness connector. Reconnect any additional remote control harness leads, such as trim/tilt leads, to their bullet connectors.

28. On electric start models, connect both battery cables to the power head, then connect the cables to the battery. Tighten all connections securely.
29. Install the spark plugs and torque them to the specification in **Table 1**. Reconnect the spark plug leads.
30. Bleed the oil system as described in Chapter Twelve.
31. Refer to Chapter Four for fuel and oil recommendations and break-in procedures as needed. Then refer to Chapter Five and perform all synchronization and linkage adjustments.

Removal/Installation (40-60 hp, 30 Jet and 45 Jet Three-cylinder Models)

The power head is best removed with most of the accessories and systems installed. These items are removed after the power head is separated from the drive shaft housing. Refer to the wiring diagrams at the end of the manual.

If the oil tank is hindering access to other components or connections, refer to Chapter Twelve for removal and installation procedures. Bleed the oil pump after oil tank installation as described in Chapter Twelve.

1. Disconnect the spark plug leads and remove all spark plugs.
2. On electric start models, disconnect both battery cables from the battery, then remove the cables from the power head. Note each cable's location and routing for reassembly.

3A. *40 and 50 hp tiller control models*—Disconnect the throttle cable as follows:

a. Note the routing of the throttle cables on the throttle pulley/control arm. Mark the upper cable with a felt-tipped marker or a piece of tape for reassembly purposes.
b. Loosen the jam nuts on both throttle cables (**Figure 10**). Then remove both throttle cables from the throttle pulley/control arm.

3B. *40 and 50 hp remote control models*—Disconnect and remove the throttle and shift cables. See Chapter Fourteen.

3C. *55 and 60 hp models*—Disconnect and remove the throttle and shift cables. See Chapter Thirteen. Cable removal procedures are the same for tiller handle and remote control models.

4. Remove the screw securing the fuel connector to the lower cowl.
5. Disconnect the water discharge hose (**Figure 11**) from the starboard rear corner of the lower cowl. Then disconnect the black ground lead from the port rear corner of the lower cowl.

6A. *Manual start models*—Remove the rewind starter as described in Chapter Thirteen.

6B. *Electric start tiller control models (40 and 50 hp)*—Disconnect the neutral safety switch leads at the bullet connectors. Then remove the flywheel cover if it has not already been removed.

7A. *Manual start models*—Disconnect the safety lanyard switch, push button stop switch and warning horn leads from the engine harness at the bullet connectors.

7B. *Electric start tiller control models (40 and 50 hp)*—Disconnect the control switch harness from the engine harness at the main harness connector. This harness connects the safety lanyard switch, push button stop switch, warning horn, primer button and start button to the main harness.

7C. *Electric start tiller control models (55 and 60 hp)*—Disconnect the key switch harness from the engine harness at the main harness connector. Then disconnect the blue/white and green/white power trim/tilt switch leads at their bullet connectors. There are two switches, one on the tiller handle and one on the lower cowl.

7D. *Electric start remote control models*—Disconnect the remote control harness from the engine harness at the main harness connector. On models equipped with trim and tilt, disconnect the blue/white and green/white power trim/tilt leads at the remote control harness bullet connectors. Disconnect the lower cowl mounted trim/tilt switch leads in the same manner.

8. Remove the screw, washer and locknut securing the trim cover to the drive shaft housing. The screw is located at the front of the cover, close to the swivel bracket. Spread the front of the cover and slide it rearward from the drive shaft housing to gain access to the power head mounting screws.
9. Remove the six screws securing the power head to the drive shaft housing. There are three screws on each side of the drive shaft housing. See **Figure 14**, typical.
10. Remove the plastic protective cap from the center of the flywheel and install a lifting eye (part No. 91-90455 or an equivalent) into the flywheel a minimum of five full turns. On electric start models, remove the flywheel cover if it is not already removed to gain access to the flywheel.

CAUTION
At this point, there should be no hoses, wires or linkage connecting the power head to the drive shaft housing. Make sure nothing will interfere with power head removal before continuing.

11. Rock the power head to break the gasket seal between the drive shaft housing and power head. Then lift the power head from the drive shaft housing with a suitable

hoist. See **Figure 15**. On 55 and 60 hp models, make sure the shift slide is pushed off the end of the shift rail *before* the power head is lifted far from the drive shaft housing.

12. Remove the power head and mount it on a power head (crankshaft) stand (part No. 91-25821A-1). The power head stand must be securely clamped in a vise. If this stand is not available, mount the power head to a power head work stand and holding fixture (**Figure 13**).
13. Thoroughly clean all gasket material from the drive shaft housing and power head mating surfaces.
14. To install the power head, lubricate the drive shaft splines with Quicksilver Special Lube 101 or 2-4-C Multi-Lube. Wipe excess lubricant from the top of the drive shaft.
15. Place a new gasket on the drive shaft housing.
16. Thread the lifting eye (part No. 91-90455) into the flywheel a minimum of five full turns. Support the power head with a suitable hoist.
17. Position the power head over the drive shaft housing and lower it into position. Rotate the crankshaft as necessary to align the drive shaft splines. On 55 and 60 hp models, make sure the shift slide is piloted over the end of the shift rail as the power head is lowered into place.
18. Apply Quicksilver Perfect Seal to the threads of the power head mounting screws. Install and evenly tighten the screws (**Figure 14**, typical) to the specification in **Table 1**.
19. Install the trim cover over the drive shaft housing and against the lower cowl. Secure the cover with the screw, washer and locknut. Tighten the screw to the specification in **Table 1**.
20. Connect the water discharge hose (**Figure 11**) to the power head fitting. Secure the connection with a new tie-strap. Then connect the black ground lead to the port rear corner of the lower cowl.
21. Position the fuel line connector in the lower cowl and secure it with one screw. Tighten the screw securely.

22A. *Manual start models*—Connect the safety lanyard switch, push button stop switch and warning horn leads to the appropriate engine harness bullet connectors.

22B. *Electric start tiller control models (40 and 50 hp)*—Connect the control switch harness to the engine harness at the main harness connector. This harness connects the safety lanyard switch, push button stop switch, warning horn, primer button and start button to the main harness.

22C. *Electric start tiller control models (55 and 60 hp)*—Connect the key switch harness to the engine harness at the main harness connector. Then connect the blue/white and green/white power trim/tilt switch leads to the appropriate bullet connectors. There are two switches, one on the tiller handle and one on the lower cowl.

22D. *Electric start remote control models*—Connect the remote control harness to the engine harness at the main harness connector. On models equipped with trim and tilt, reconnect the blue/white and green/white power trim/tilt leads to the appropriate engine harness bullet connectors, then connect the lower cowl mounted trim/tilt switch leads in the same manner.

23A. *Tiller control models (40 and 50 hp)*—Install the throttle cables as follows:

   a. Route the throttle cables onto the throttle pulley/control arm as noted on disassembly.
   b. Adjust the jam nuts on both throttle cables as described in Chapter Five.

23B. *55 and 60 hp models*—Install the throttle and shift cables as described in Chapter Fourteen. Cable installation procedures are the same for tiller handle and remote control models.

24A. *Manual start models*—Install the rewind starter as described in Chapter Thirteen

24B. *Electric start tiller control models (40 and 50 hp)*—Connect the neutral safety switch leads to the appropriate engine harness bullet connectors.

25. *Electric start models*—Install the flywheel cover if it is not already installed. Then, connect both battery cables to the power head and finally connect the cables to the battery. Tighten all connections securely.
26. Install the spark plugs and torque them to the specification in **Table 1**. Reconnect the spark plug leads.
27. Bleed the oil system as described in Chapter Twelve.
28. Make sure all cable clamps are reinstalled in their original positions and new tie-straps are installed to replace any that were removed.
29. Refer to Chapter Four for fuel and oil recommendations and break-in procedures as needed. Then refer to Chapter Five and perform all synchronization and linkage adjustments.

POWER HEAD DISASSEMBLY

Power head overhaul gasket sets are available for 6 hp and larger models. It is more economical and simpler to order the gasket set instead of each component individually. Replace *every* gasket, seal and O-ring during power head reassembly. Replace the piston rings if the piston(s) are taken out of the cylinder bore.

Dowels are used to position the crankcase halves to each other and to position some crankshaft bearings. The dowel pins do not require removal if they are securely seated in their bores on either crankcase half or in the bearing. However, do not lose them during disassembly and reassembly. If a dowel pin can be easily removed from its bore, remove it and store it with the other internal components until reassembly.

The connecting rods on 6 hp and larger models are fractured cap design. The cap is broken from the rod during the manufacturing process, leaving a jagged mating surface that mates perfectly when installed in its original orientation. If the cap is installed reversed and the rod screws are tightened, the rod will be distorted and must be discarded. While alignment marks are provided, always mark the rod and cap with a permanent marker for redundancy. See **Figure 16**. Correct orientation is obvious if the time is taken to examine the mating surfaces of the rod and cap.

Clean and inspect all power head components before reassembly. If the power head has experienced a serious failure involving numerous broken components, it may be more economical to replace the basic power head as an assembly.

The part Nos. of all manufacturer recommended special tools are listed in the repair procedures. Replacing parts damaged by the incorrect tool can often be more expensive than the original cost of the tool.

A large number of fasteners of different lengths and sizes are used in a power head. Plastic sandwich bags and/or cupcake tins are excellent methods of keeping small parts organized. Tag all larger internal parts for location and orientation. A felt-tipped permanent marker can be used to mark components after they have been cleaned. Avoid scribing or stamping internal components as the marking process may damage or weaken the component.

NOTE

A series of at least five photographs taken from the front, rear, top and both sides of the power head before removal are useful during reassembly and installation. The photographs are useful to route electrical harnesses, fuel primer and recirculation lines. They are also helpful during the installation of accessories, control linkages and brackets.

Disassembly (2.5 and 3.3 hp Models)

Refer to **Figure 17** for this procedure. The 3.3 hp (shifting gearcase) model has an extended lower end cap (27, **Figure 17**) that uses a gasket and two internal seals. The 2.5 hp (non-shifting gearcase) model uses a simple plate (28, **Figure 17**) to hold the lower crankshaft seal in place.

1A. *2.5 hp models*—Remove the screws and washers securing the lower end cap (28, **Figure 17**) to the power head. Remove the end cap from the power head.

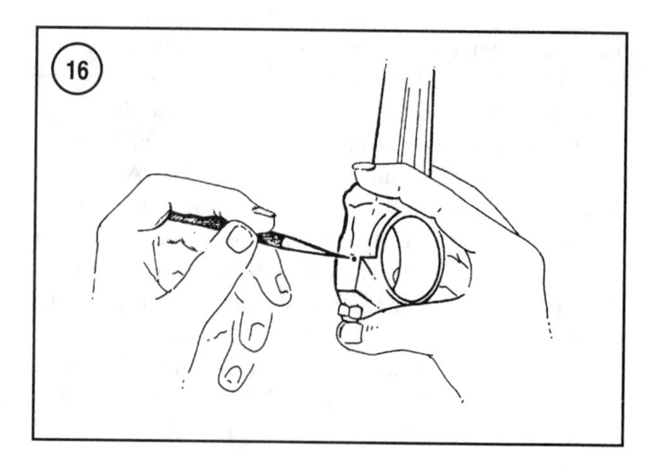

1B. *3.3 hp*—Remove the screws and washers securing the extended lower end cap (27, **Figure 17**) to the power head. Tap the end cap with a soft-faced plastic or rubber hammer to break it free from the power head. Remove the gasket.

2. On 3.3 hp models, remove the two seals (26, **Figure 17**) in the extended lower end cap using a suitable seal puller. Do not damage the lower cap seal bore during the removal process.

3. Remove the screws and washers (18, **Figure 17**) securing the cylinder head (17) to the power head. Tap the head with a soft-faced hammer to break it free from the power head. Remove the gasket.

CAUTION

The crankcase cover and cylinder block are a matched, align-bored unit. Never install the crankcase cover from one cylinder block to a different cylinder block. Do not scratch, nick or damage the machined mating surfaces.

4. Remove the six crankcase cover screws, then remove the crankcase cover (2, **Figure 17**) from the cylinder block. Tap the cover with a soft-faced hammer to break it free from the cylinder block.

5. Lift the crankshaft, connecting rod and piston assembly from the cylinder block (**Figure 18**). Then locate and secure the two hollow dowel pins (10, **Figure 17**) as necessary.

6. Slide the crankshaft seals (**Figure 19**) and half-moon retainer rings (6, **Figure 17**) from each end of the crankshaft.

WARNING

Wear suitable eye protection for piston ring and piston pin lock ring removal procedures.

(17)

POWER HEAD COMPONENTS (2.5 AND 3.3 HP MODELS)

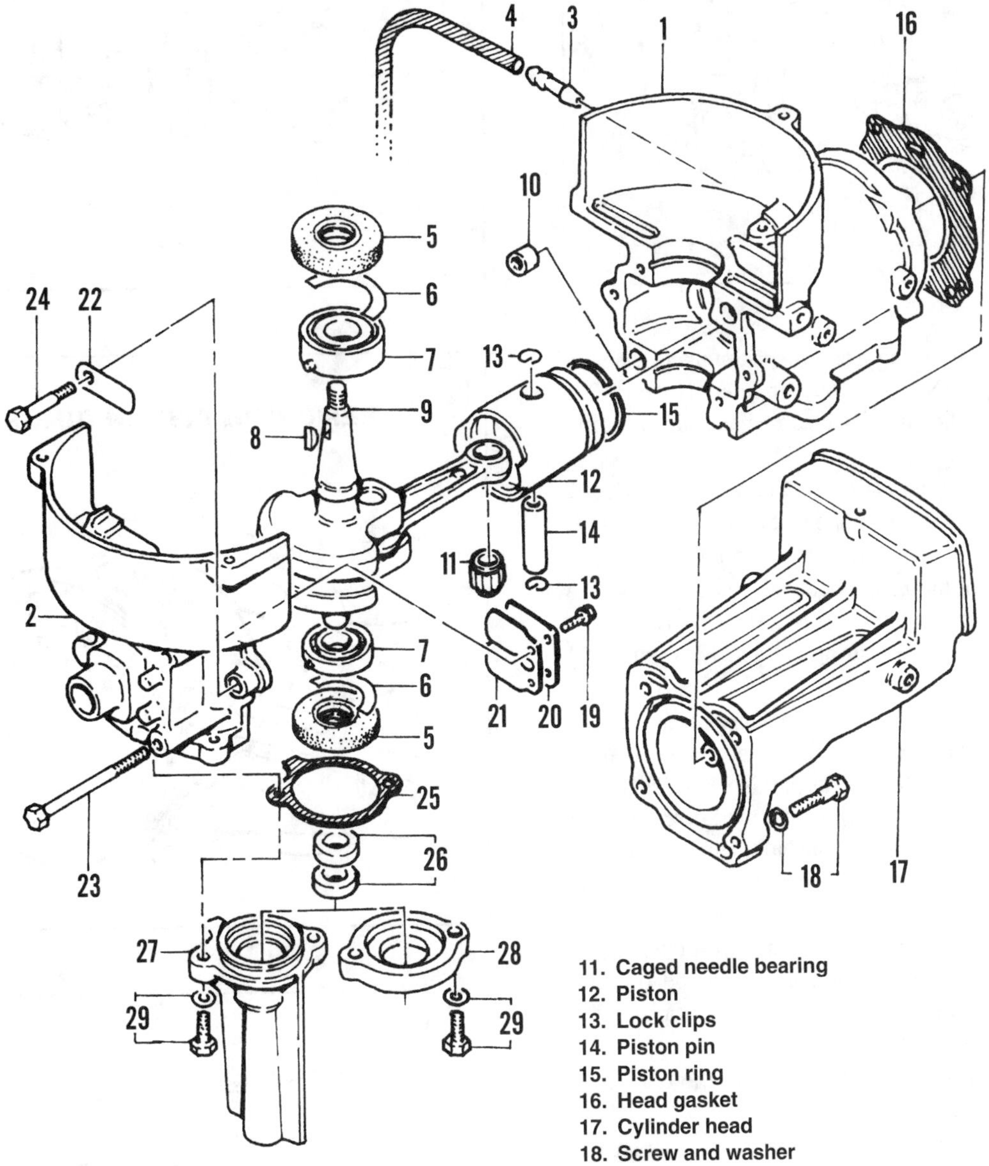

1. Cylinder block
2. Crankcase cover
3. Fitting
4. Water discharge (tell-tale) hose
5. Crankshaft seal
6. Half-moon retainer clips
7. Ball bearings
8. Flywheel key
9. Crankshaft assembly
10. Dowel
11. Caged needle bearing
12. Piston
13. Lock clips
14. Piston pin
15. Piston ring
16. Head gasket
17. Cylinder head
18. Screw and washer
19. Screw
20. Reed stop
21. Reed petals
22. Clamp plate
23. Long screw
24. Short screw
25. Gasket (3.3 hp)
26. Seals (3.3 hp)
27. Extended lower end cap (3.3 hp)
28. Lower end cap (2.5 hp)
29. Screw and washer

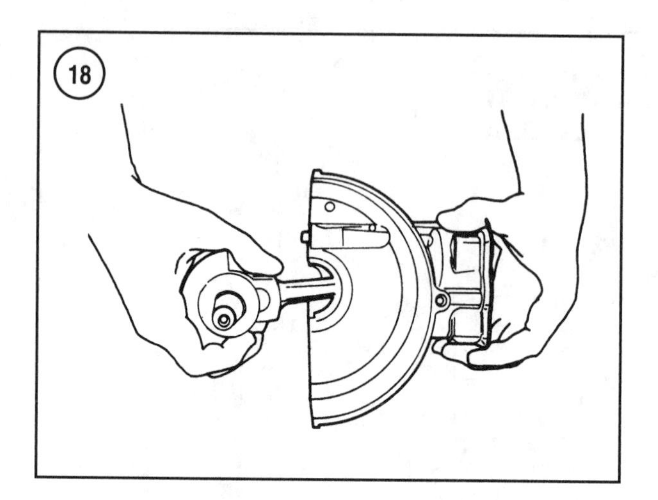

18

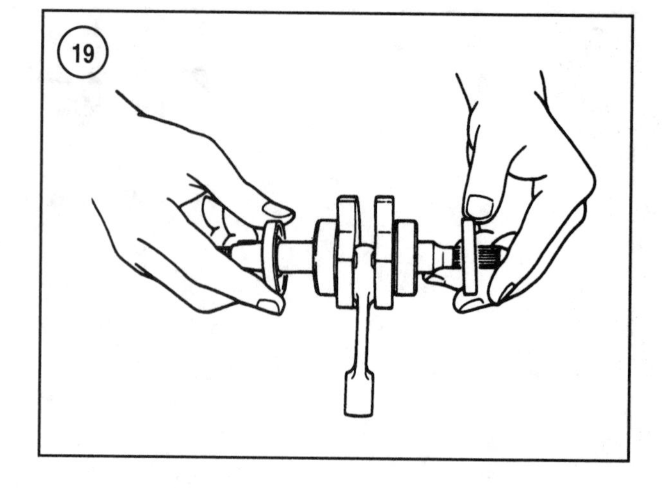

19

7. Remove the piston ring from the piston using a piston ring expander (part No. 91-24697 or an equivalent). See **Figure 20**. Keep the ring for cleaning the piston's ring groove.

8. Use needlenose pliers (**Figure 21**) or a small screwdriver (**Figure 22**) to remove the piston pin lock clip (13, **Figure 17**) from each end of the piston pin bore.

9. Use a section of tubing or socket extension to push the piston pin out of the piston (**Figure 23**). Remove the piston and the caged needle bearing (11 and 12, **Figure 17**) from the connecting rod.

10. If the ball bearings of the crankshaft must be replaced, remove each bearing by supporting it in a knife-edged bearing plate, such as part No. 91-37241. See **Figure 24**. Press against the crankshaft (**Figure 25**) until the bearing is free from the crankshaft. Then discard the bearing(s). When removing the upper bearing, install the flywheel nut to protect the crankshaft threads.

11. The connecting rod is not removable from the crankshaft. If the connecting rod or connecting rod big end bearing is excessively worn or damaged, replace the crankshaft as an assembly.

12. Clean and inspect the reed valves in the crankcase cover as described in Chapter Six.

13. Refer to *Cleaning and Inspection* in this chapter before beginning the reassembly procedure.

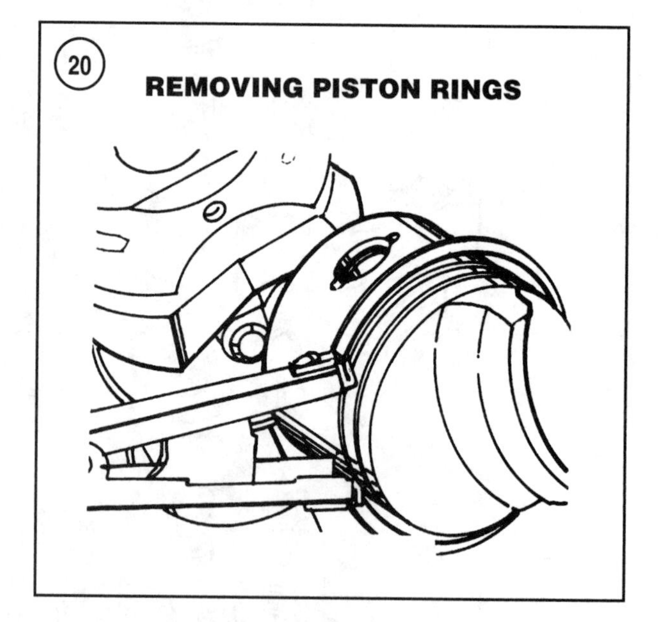

20

REMOVING PISTON RINGS

Disassembly (4 and 5 hp Models)

Refer to **Figure 26** for this procedure.

1. Remove the screws and washers (30, **Figure 26**) securing the extended lower end cap to the power head. Tap the end cap with a soft plastic or rubber hammer to break it free from the power head. Remove the gasket (26, **Figure 26**).

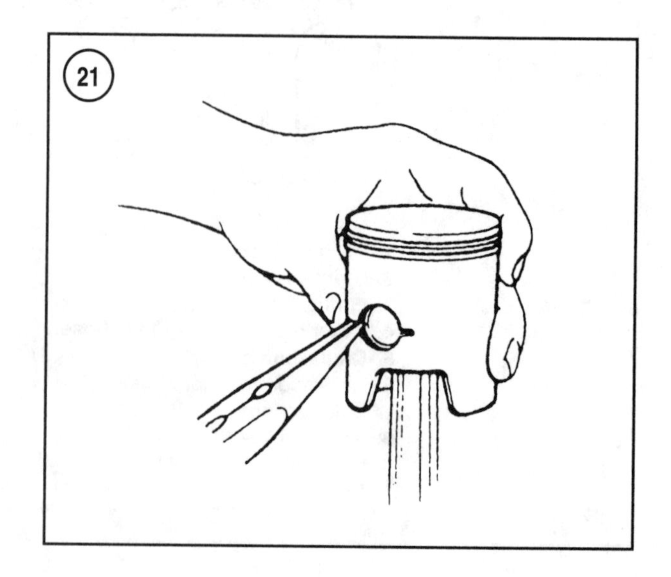

21

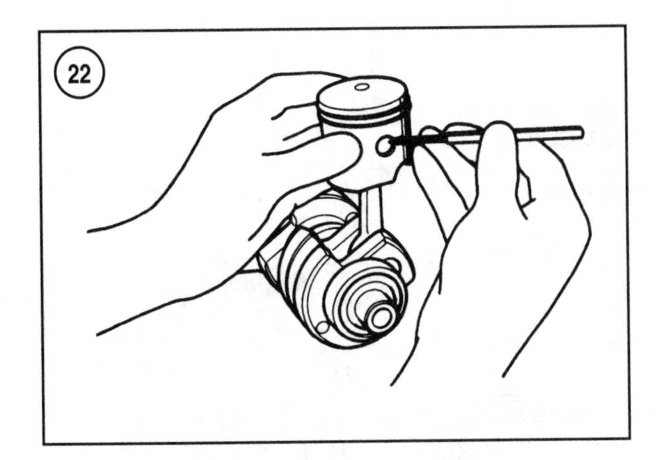

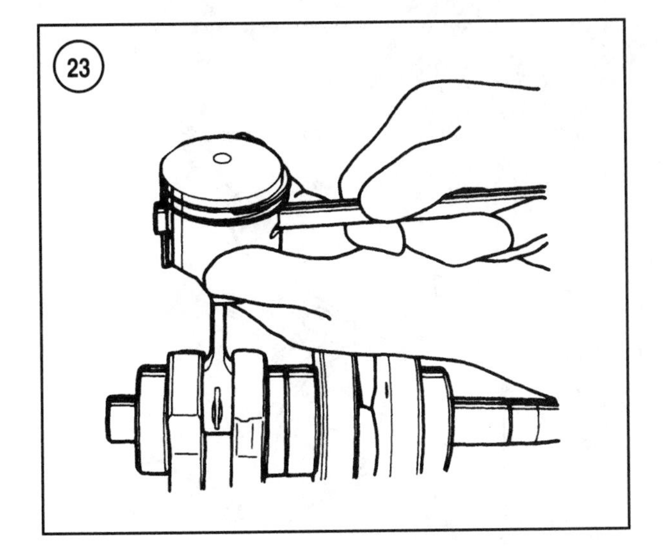

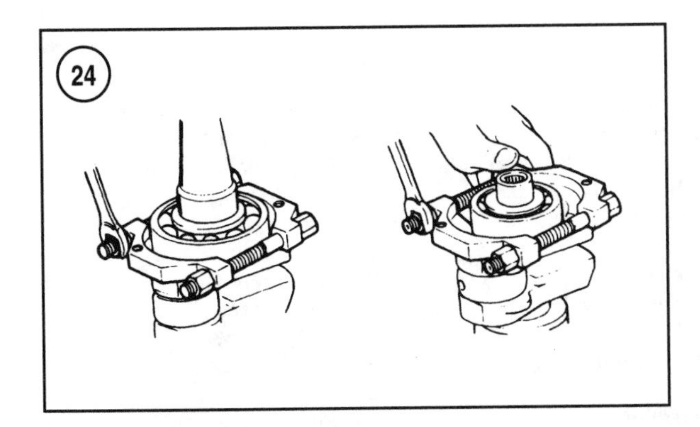

2. Remove the lower end cap seals (27, **Figure 26**) using a suitable seal puller. Do not damage the lower cap seal bore during the removal process. Retain the seal spacer (28, **Figure 26**) for reassembly.

3. If the lower end cap bushing (31, **Figure 26**) must be replaced, clamp the end cap in a vise with protective jaws

and drive the bushing from the end cap with a suitable punch and hammer.

4. Remove the screws and washers (17, **Figure 26**) securing the cylinder head (16) to the power head. Tap the head with a soft-faced hammer to break it free from the power head. Remove the gasket.

CAUTION
The crankcase cover and cylinder block are a matched, align-bored unit. Never install the crankcase cover from one cylinder block to a different cylinder block. Do not scratch, nick or damage the machined mating surfaces.

5. Remove the crankcase cover screws and washers (24 and 25, **Figure 26**).

6. Insert a screwdriver blade under the tabbed locations on each side of the crankcase cover (**Figure 27**). Carefully pry at the tab points to break the crankcase seal, then remove the crankcase cover from the cylinder block. If necessary, gently tap the cover with a soft-faced hammer to break it free.

7. Locate and secure the two hollow dowel pins (9, **Figure 26**) as necessary. Then lift the crankshaft, connecting rod and piston assembly from the cylinder block (**Figure 18**).

8. Slide the crankshaft seals and locating washers from each end of the crankshaft (**Figure 19**).

WARNING
Wear suitable eye protection for piston ring and piston pin lock ring removal procedures.

9. Remove both piston rings from the piston using piston ring expander (part No. 91-24697 or an equivalent). See **Figure 20**. Retain the ring for cleaning the piston's ring groove.

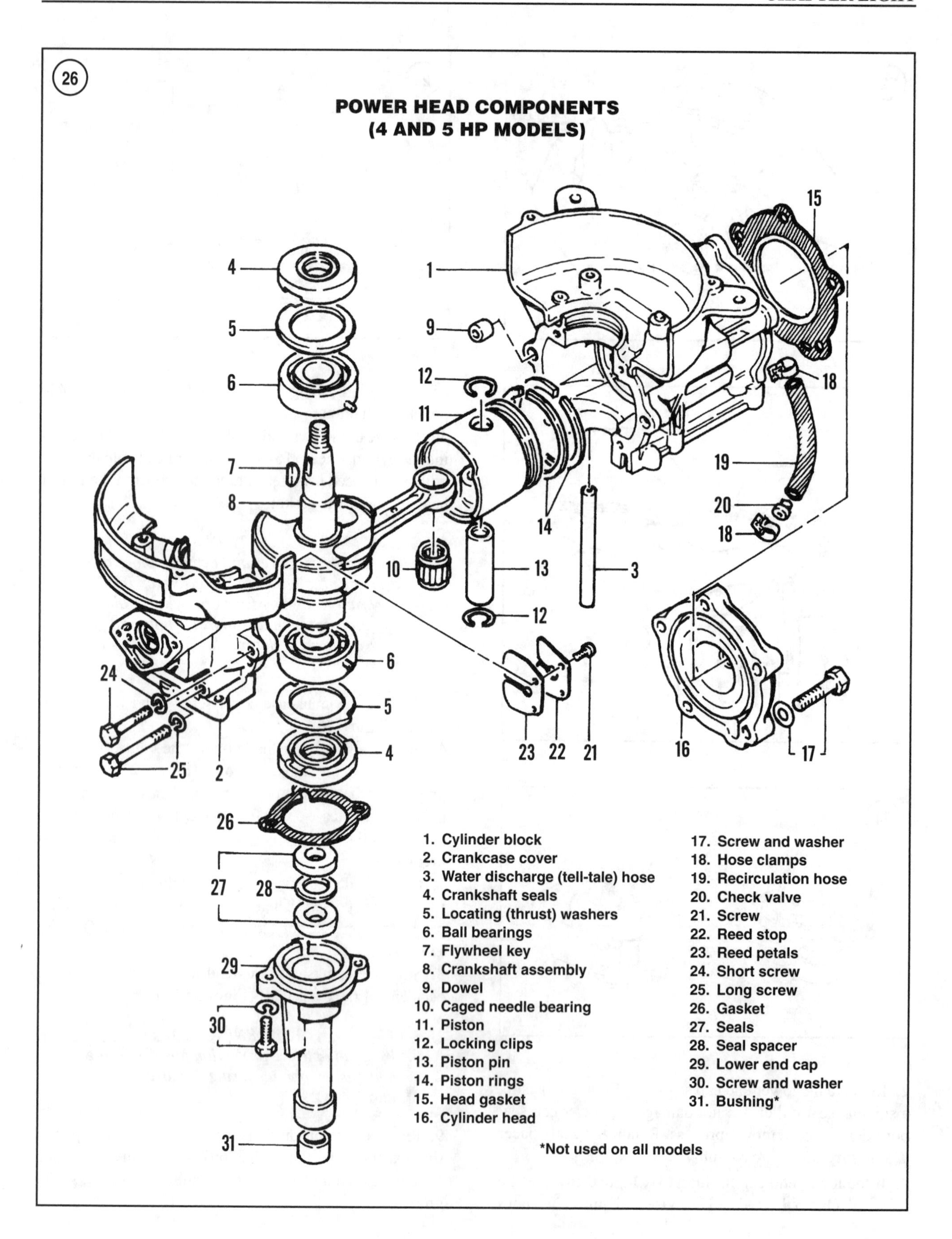
26
POWER HEAD COMPONENTS
(4 AND 5 HP MODELS)
1
2
3
4
5
6
7
8
9
10
11
12
13
14
15
16
17
18
19
20
21
22
23
24
25
26
27
28
29
30
31
1. Cylinder block
2. Crankcase cover
3. Water discharge (tell-tale) hose
4. Crankshaft seals
5. Locating (thrust) washers
6. Ball bearings
7. Flywheel key
8. Crankshaft assembly
9. Dowel
10. Caged needle bearing
11. Piston
12. Locking clips
13. Piston pin
14. Piston rings
15. Head gasket
16. Cylinder head
17. Screw and washer
18. Hose clamps
19. Recirculation hose
20. Check valve
21. Screw
22. Reed stop
23. Reed petals
24. Short screw
25. Long screw
26. Gasket
27. Seals
28. Seal spacer
29. Lower end cap
30. Screw and washer
31. Bushing*
*Not used on all models

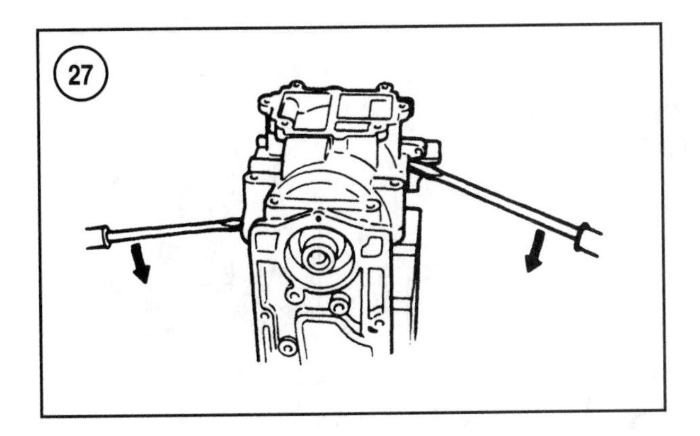

10. Use a suitable awl and carefully pry the piston pin lock ring (12, **Figure 26**) from each end of the piston pin bore (**Figure 22**).

11. Use a section of tubing or socket extension to push the piston pin out of the piston (**Figure 23**). Remove the piston and the caged needle bearing (10 and 11, **Figure 26**) from the connecting rod.

12. If the ball bearings of the crankshaft must be replaced, remove each bearing by supporting it in a knife-edged bearing plate, such as part No. 91-37241. See **Figure 24**. Press against the crankshaft (**Figure 25**) until the bearing is free from the crankshaft. When removing the upper bearing, install the flywheel nut or place the protector cap (part No. 91-24161) over the end of the crankshaft to protect the crankshaft threads.

13. The connecting rod cannot be removed from the crankshaft. If the connecting rod or connecting rod big end bearing is excessively worn or damaged, replace the crankshaft as an assembly.

14. Remove the two hose clamps (18, **Figure 26**) securing the recirculation hose (19) to the cylinder block. Remove the hose from the block. Test the check valve in the hose as described in Chapter Six, under *Fuel Bleed (Recirculation) systems*.

15. Clean and inspect the reed valves in the crankcase cover as described in Chapter Six.

16. Refer to *Cleaning and Inspection* in this chapter before beginning the reassembly procedure.

Disassembly (6-15 hp Models)

A 12.8 cu. in. (210 cc) power head is used on 6 and 8 hp models. A 16.0 cu. in. (262 cc) power head is used on 9.9 and 15 hp models. The 12.8 cu. in. (210 cc) power head uses stamped split bearing races (20, **Figure 28**) on the center main bearing. The 16.0 cu. in. (262 cc) power head uses machined bearing liners and a retainer ring (27, **Figure 28**) on the center main bearing. Refer to **Figure 28** for this procedure.

1. Remove the intake manifold, reed block and the fuel bleed (recirculation) system as described in Chapter 6.

NOTE

A thermostat is standard equipment on 9.9-15 hp models and optional on 6-8 hp models.

2. Remove the screws securing the thermostat cover to the cylinder block. Remove the cover and thermostat, if equipped. Remove the cover gasket and the thermostat grommet.

3. Remove the screws holding the cylinder block cover (3, **Figure 28**) to the power head. Remove the cover.

4. Remove the screws securing the intake cover (8, **Figure 28**) to the cylinder block. Remove the cover, then remove the gasket or two O-rings.

5. Remove the screws holding the exhaust cover (5, **Figure 28**) to the cylinder block. If necessary, tap the assembly gently with a soft-faced mallet to break the gasket seal. Separate the exhaust cover and baffle plate. Do not warp or damage the plate or cover. Remove the gaskets.

CAUTION

The crankcase cover and cylinder block are a matched, align-bored unit. Never install the crankcase cover from one cylinder block to a different cylinder block. Do not scratch, nick or damage the machined mating surfaces.

6. Remove the screws holding the crankcase cover to the cylinder block. Carefully tap the crankcase cover with a soft-face mallet to break the gasket seal. Remove the crankcase cover from the cylinder block. Locate and secure the two locating dowel pins as necessary.

7. Lift the crankshaft assembly straight up and out of the cylinder block. Mount the crankshaft vertically into the power head (crankshaft) stand (part No. 91-13662A-1 or an equivalent). Make sure the stand is securely clamped in a vise.

8A. *6 and 8 hp models*—Collect the center main caged roller bearing halves (19, **Figure 28**), then remove the stamped center main bearing liners (20, **Figure 28**) from the cylinder block and crankcase cover.

8B. *9.9 and 15 hp models*—Remove the retaining ring (**Figure 29**) from the center main bearing. Then remove the outer race halves and caged roller bearing halves. Locate and secure the center main bearing locating pin as necessary.

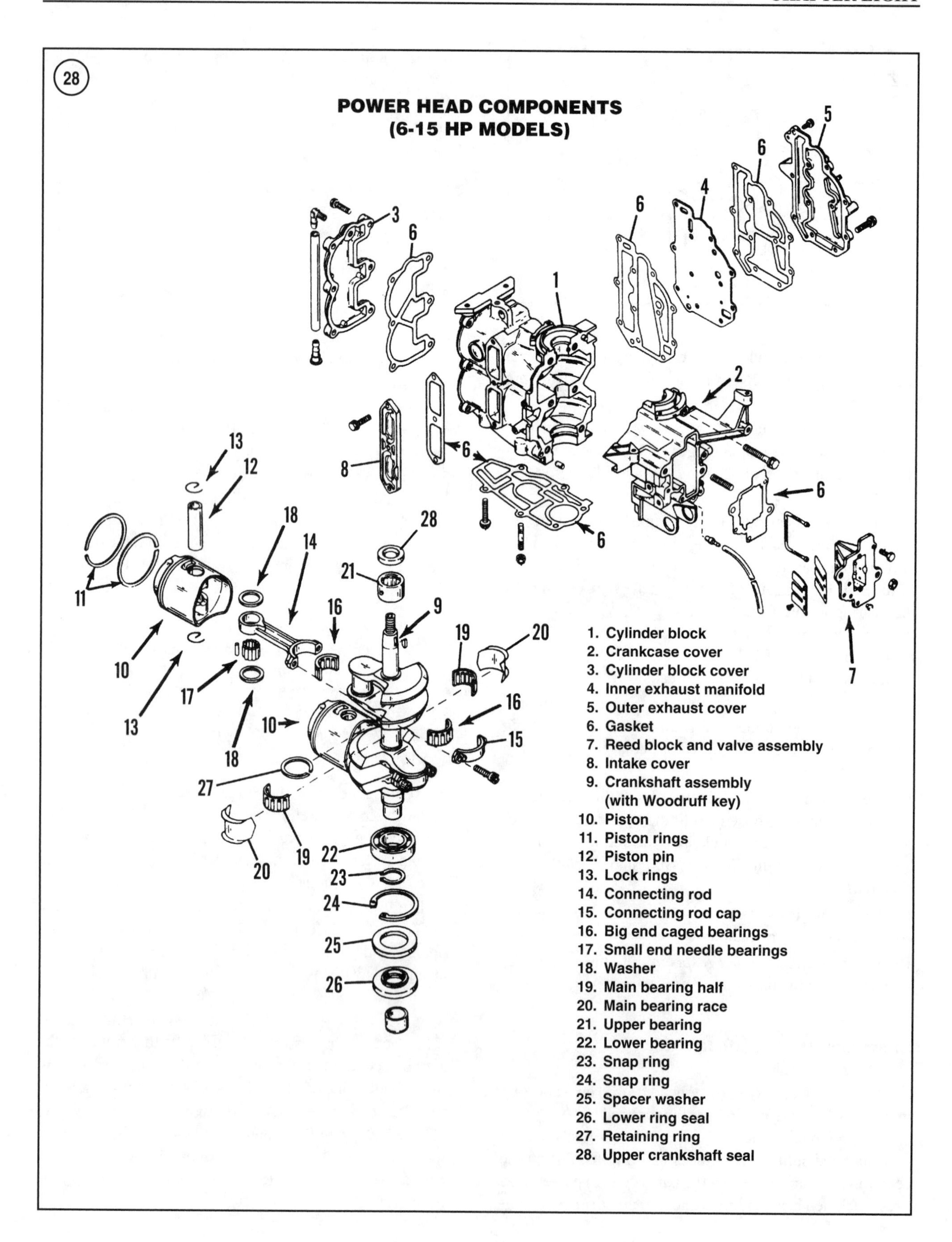
28
POWER HEAD COMPONENTS
(6-15 HP MODELS)
1. Cylinder block
2. Crankcase cover
3. Cylinder block cover
4. Inner exhaust manifold
5. Outer exhaust cover
6. Gasket
7. Reed block and valve assembly
8. Intake cover
9. Crankshaft assembly (with Woodruff key)
10. Piston
11. Piston rings
12. Piston pin
13. Lock rings
14. Connecting rod
15. Connecting rod cap
16. Big end caged bearings
17. Small end needle bearings
18. Washer
19. Main bearing half
20. Main bearing race
21. Upper bearing
22. Lower bearing
23. Snap ring
24. Snap ring
25. Spacer washer
26. Lower ring seal
27. Retaining ring
28. Upper crankshaft seal

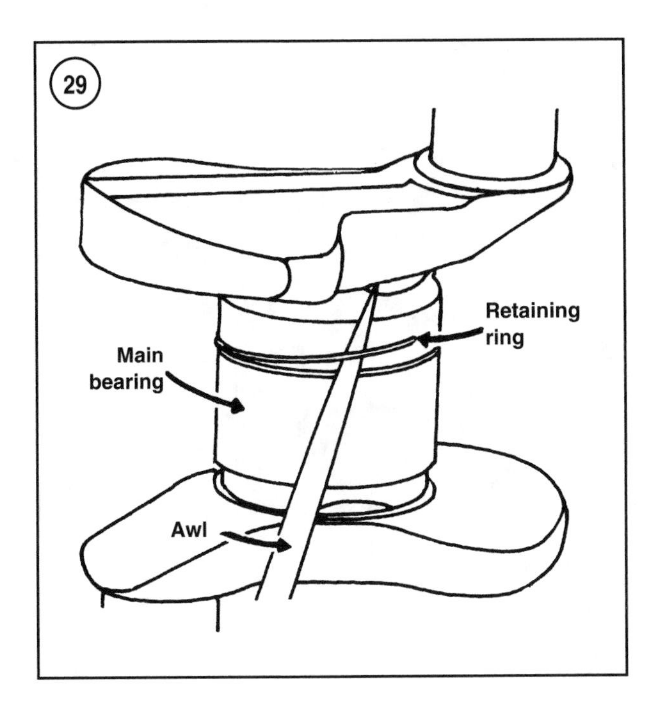

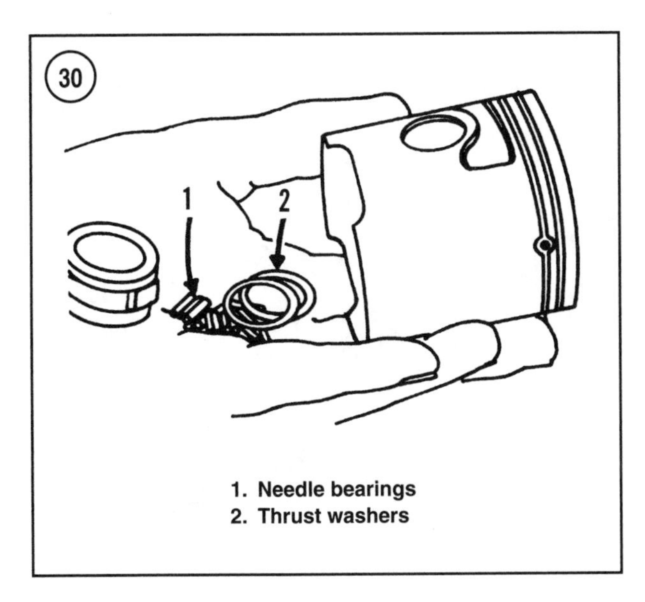

1. Needle bearings
2. Thrust washers

9. Remove the sealing ring from the groove above the center main bearing journal.

10. Slide the upper crankshaft seal and bearing (21 and 28, **Figure 28**) off the crankshaft.

11. Remove the lower crankshaft seal, spacer washer and retaining ring (25-27, **Figure 28**).

12. Remove the wear sleeve and O-ring from the bottom of the crankshaft using a suitable puller, such as collet (part No. CG 40-15) and expander rod (part No. CG 40-4) from Snap-on tools, attached to a slide hammer (part No. 91-34569A-1).

13. Mark the corresponding cylinder number on the pistons and connecting rods with a felt-tipped permanent marker. See **Figure 16**. Mark the connecting rods and rod caps so the rod caps can be reinstalled in their original orientation.

WARNING
Wear suitable eye protection for piston ring and piston pin lock ring removal procedures.

CAUTION
Always store all components from each connecting rod and piston assembly together. They must be reinstalled in their original locations.

14. Remove both piston rings from both pistons using piston ring expander (part No. 91-24697 or an equivalent). See **Figure 20**. Keep the rings for cleaning the piston's ring grooves.

15. Remove each connecting rod and piston assembly as follows:

a. Remove the connecting rod screws from the upper, cylinder No. 1 connecting rod. Alternately loosen each screw a small amount until all tension is off both screws.
b. Tap the rod cap with a soft rubber or plastic mallet to separate the cap from the rod.
c. Remove the cap and rod from the crankshaft, then remove the caged roller bearing halves from the crankshaft.
d. Reinstall the rod cap to the connecting rod in its original orientation. Tighten the rod cap screws finger-tight.
e. Store the caged roller bearing halves in clean numbered containers, corresponding to the cylinder number.
f. Repeat this procedure to remove the lower, cylinder No. 2 connecting rod assembly.

16. Use needlenose pliers to remove the piston pin lock ring from each end of both piston pin bores. See **Figure 21**.

17. Place the piston pin tool (part No. 91-13663A-1 or equivalent) into one end of the cylinder No. 1 piston pin bore. Support the bottom of the piston with one hand and drive the pin tool and pin from the piston. See **Figure 23**.

18. Remove the piston from the connecting rod. Remove the thrust washers and the 24 loose needle bearings (**Figure 30**). Store the components in clean, numbered containers corresponding to the cylinder number.

19. Repeat Steps 17 and 18 to separate the cylinder No. 2 piston from its connecting rod.

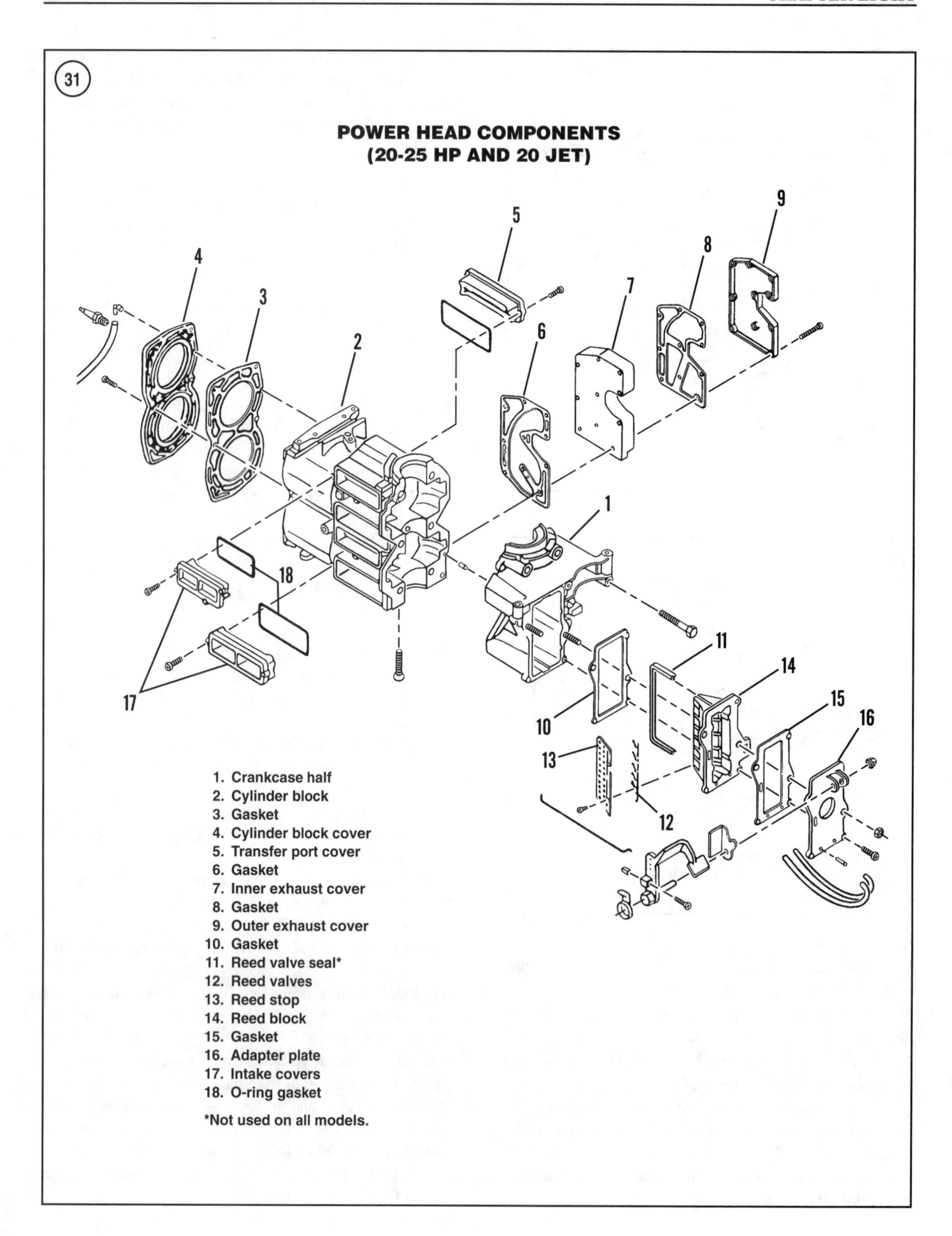
31
POWER HEAD COMPONENTS
(20-25 HP AND 20 JET)
1
2
3
4
5
6
7
8
9
10
11
12
13
14
15
16
17
18
1. Crankcase half
2. Cylinder block
3. Gasket
4. Cylinder block cover
5. Transfer port cover
6. Gasket
7. Inner exhaust cover
8. Gasket
9. Outer exhaust cover
10. Gasket
11. Reed valve seal*
12. Reed valves
13. Reed stop
14. Reed block
15. Gasket
16. Adapter plate
17. Intake covers
18. O-ring gasket
*Not used on all models.

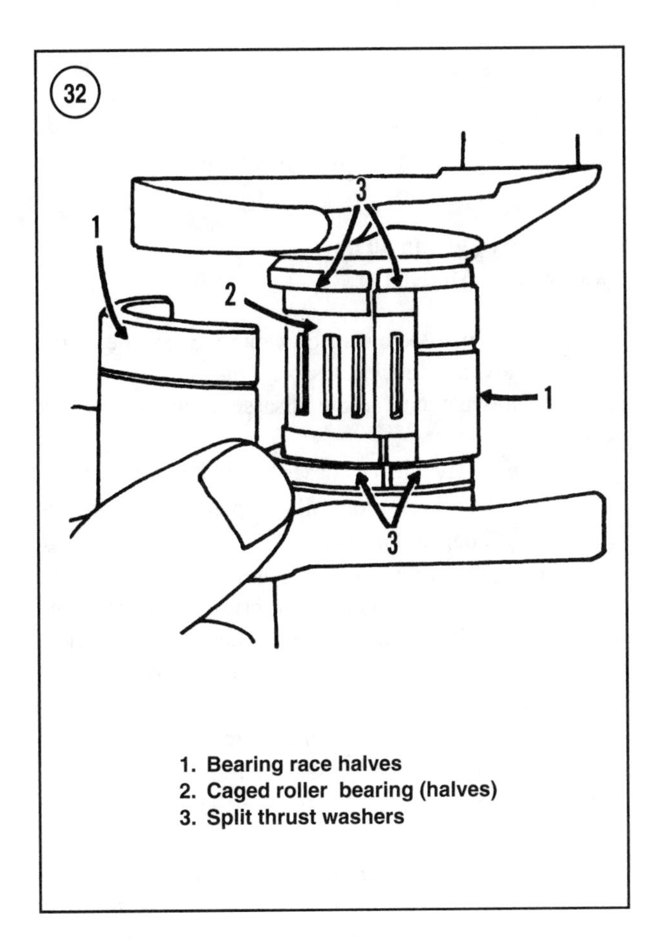

1. Bearing race halves
2. Caged roller bearing (halves)
3. Split thrust washers

20. If the ball bearings of the crankshaft must be replaced, remove each bearing by supporting it in a knife-edged bearing plate (part No. 91-37241). See **Figure 24**. Press against the crankshaft (**Figure 25**) until the bearing is free from the crankshaft. When removing the upper bearing, install the flywheel nut over the end of the crankshaft to protect the crankshaft threads.
21. Refer to *Cleaning and Inspection* in this chapter before beginning reassembly procedures.

Disassembly (20-25 hp and 20 Jet Models)

Refer to **Figure 31** for this procedure.
1. Remove the intake manifold, reed block and fuel bleed (recirculation) lines as described in Chapter Six.
2. Remove the screws securing the thermostat cover to the cylinder block. Remove the cover and thermostat. Remove the cover gasket and the thermostat grommet.
3. Remove the remaining screws holding the cylinder block cover (4, **Figure 31**) to the power head. Remove the cover.
4. Remove the two intake covers (17, **Figure 31**) from the port side of the power head. There are two screws securing each cover. Note the orientation of each cover as it is removed, then remove each cover's O-ring gasket.

NOTE
A special screw with a ball head for the control linkage is used on one of the starboard side intake covers. Note the position of the screw before removal.

5. Remove the four intake covers from the starboard side of the power head. Note the orientation of the screws and cover as each are removed, then remove each cover's O-ring.
6. Remove the screws holding the exhaust cover and exhaust manifold (7 and 9, **Figure 31**) to the cylinder block. If necessary, tap the assembly gently with a soft-faced mallet to break the gasket seal. Separate the exhaust cover and the exhaust manifold. Do not damage the plate or manifold. Remove the gaskets.

8

CAUTION
The crankcase cover and cylinder block are a matched, align-bored unit. Never install the crankcase cover from one cylinder block to a different cylinder block. Do not scratch, nick or damage the machined mating surfaces.

7. Remove the six screws holding the crankcase cover (1, **Figure 31**) to the cylinder block. One screw is internally located behind the reed block mounting cavity. Carefully tap the crankcase cover with a soft-faced mallet to break the gasket seal. Remove the crankcase cover from the cylinder block. Locate and secure the two locating dowel pins as necessary.
8. Lift the crankshaft assembly straight up and out of the cylinder block and set it onto a clean workbench. Locate and secure the crankshaft center main bearing locating pin as necessary.
9. Slide the upper and lower crankshaft seals and caged roller bearings from the crankshaft.
10. Mount the crankshaft vertically on a power head (crankshaft) stand (part No. 91-25821A-1 or an equivalent). Make sure the stand is securely clamped in a vise.
11. Remove the retaining ring from the center main bearing (**Figure 29**). Then remove the outer race halves, caged roller bearing halves and four split thrust washers. See **Figure 32**.
12. Mark the corresponding cylinder number on the pistons and connecting rods with a felt-tipped permanent marker. Mark the connecting rods and rod caps so

the rod caps can be reinstalled in their original orientation.

CAUTION
Wear suitable eye protection for piston ring and piston pin lock ring removal procedures.

NOTE
Always store the components from each connecting rod and piston assembly together. They must be reinstalled in their original locations.

13. Remove both piston rings from both pistons using a piston ring expander (part No. 91-24697 or an equivalent). See **Figure 20**. Keep the rings for cleaning the piston ring grooves.
14. Remove each connecting rod and piston assembly as follows:
   a. Remove the connecting rod screws from the upper, cylinder No. 1 connecting rod. Alternately loosen each screw a small amount until all tension is off both screws.
   b. Tap the rod cap with a soft rubber or plastic mallet to separate the cap from the rod.
   c. Remove the cap and rod from the crankshaft, then remove the 24 loose needle bearings and two bearing cages from the crankshaft.
   d. Reinstall the rod cap to the connecting rod in its original orientation. Tighten the rod cap screws finger-tight.
   e. Store the loose roller bearing and bearing cages in clean numbered containers, corresponding to the cylinder number.
   f. Repeat this procedure to remove the lower, cylinder No. 2 connecting rod assembly.
15. Use needlenose pliers to remove the piston pin lock ring from each end of both piston's pin bore. See **Figure 21**.
16. Place the piston pin tool (part No. 91-76160A-2 or an equivalent) into one end of the cylinder No. 1 piston pin bore. Support the bottom of the piston with one hand and drive the pin tool and pin from the piston. See **Figure 23**.
17. Remove the piston from the connecting rod. Remove the locating washers and the 27 loose needle bearings (**Figure 30**). Store the components in clean, numbered containers corresponding to the cylinder number.
18. Repeat Steps 16 and 17 to separate the cylinder No. 2 piston from its connecting rod.
19. Remove the wear sleeve and O-ring from the bottom of the crankshaft using a suitable puller or pair of pliers. Do not damage the crankshaft surface during the removal process.
20. Refer to *Cleaning and Inspection* in this chapter before beginning the reassembly procedure.

Disassembly (30 and 40 hp Two-cylinder Models)

Refer to **Figure 33** for this procedure. Since the power head is removed with all accessories installed, it is necessary to remove them before disassembling the power head.

1. Remove the flywheel as described in Chapter Seven.
2. Remove the oil reservoir, oil pump and all oil lines as described in Chapter Twelve.
3. Remove the carburetor, fuel pump, fuel filter, intake manifold, reed block, and all fuel, primer and fuel bleed (recirculation) lines as described in Chapter Six. On electric start models, remove the fuel primer valve with the other components.
4. Remove the internal fuel check (recirculation) valve and carrier (8 and 9, **Figure 33**) from the crankcase cover.
5. On electric start models, remove the electric starter motor as described in Chapter Seven.
6. Remove the remaining ignition and electrical components as an assembly. This includes the stator assembly, trigger coil, CDM modules and the electrical/ignition plate. On electric start models, this also includes the voltage regulator and the starter solenoid. Do not disconnect electrical components from each other unless absolutely necessary. Remove the mounting screws, cable clamps and tie-straps. See Chapter Seven.
7. Remove any remaining control linkages that will interfere with power head disassembly.
8. Remove the screws (11, **Figure 33**) securing the thermostat cover to the port upper side of the cylinder block. Remove the cover and gasket. Then remove the thermostat and grommet, and the poppet valve and spring assembly.
9. Remove the screw securing the engine temperature switch (21, **Figure 33**) to the cylinder block. Remove the switch assembly.
10. Remove the screws (28, **Figure 33**) securing the lower end cap (27) to the power head. Do not remove the lower end cap at this time.

CAUTION
The crankcase cover and cylinder block are a matched, align-bored unit. Never install the crankcase cover from one cylinder block to a different cylinder block. Do not scratch,

(33)

POWER HEAD COMPONENTS (30 AND 40 HP MODELS [TWO-CYLINDER])

1. Cylinder block
2. Crankcase cover
3. Bearing locating pin
4. Dowel pin
5. Gaskets
6. Long screw
7. Short screw
8. Check valve
9. Check valve carrier
10. Roll pin
11. Thermostat screw
12. Thermostat housing
13. Thermostat grommet
14. Thermostat
15. Spring
16. Screw
17. Cup
18. Diaphragm
19. Poppet valve
20. Brass pipe plug
21. Engine temperature switch
22. Screw and washer
23. Water hose (tell-tale) fitting
24. O-ring
25. Large crankshaft seal
26. Small crankshaft seal
27. Lower end cap
28. Screw
29. Crankshaft upper seal
30. Caged needle bearing
31. Flywheel key
32. Crankshaft
33. Piston pin retainers
34. Piston pin
35. Piston
36. Piston rings
37. Connecting rod and rod cap
38. Screw
39. Thrust washer
40. Loose needle bearings
41. Caged roller bearing halves
42. Loose needle bearings
43. Bearing race halves
44. Bearing retainer ring
45. Crankshaft seal ring
46. Drive key
47. Oil pump drive gear
48. Ball bearing
49. Retaining ring

nick or damage the machined mating surfaces.

11. Remove the six long and four short screws securing the crankcase cover to the cylinder block. Tap the crankcase cover with a soft-faced rubber or plastic hammer to break the crankcase seal. Remove the crankcase cover. Locate and secure the two locating dowel pins as necessary.
12. Lift the crankshaft assembly straight up and out of the cylinder block, and set it on a clean workbench. Then pull the lower end cap from the crankshaft. Remove the O-ring and two seals. Do not damage the end cap when removing the seals.
13. Mount the crankshaft vertically in a power head (crankshaft) stand (part No. 91-827001A-1 or an equivalent). Make sure the stand is securely clamped in a vise.
14. Locate and secure the crankshaft center main bearing locating pin (3, **Figure 33**) as necessary.
15. Slide the upper crankshaft seal and caged roller bearing (29 and 30, **Figure 33**) from the crankshaft.
16. Remove the retaining ring from the center main bearing (**Figure 29**). Then remove the outer race halves, 14 loose rollers and the crankshaft seal ring.
17. Mark the corresponding cylinder number on the pistons and connecting rods with a felt-tipped permanent marker. See **Figure 16**. Mark the connecting rods and rod caps so the rod caps can be reinstalled in their original position.

WARNING

Wear suitable eye protection for piston ring and piston pin lock ring removal procedures.

CAUTION

Always store the components from each connecting rod and piston assembly together. They must be reinstalled in their original locations or engine damage will occur.

18. Remove both piston rings from both pistons using a piston ring expander (part No. 91-24697 or an equivalent). See **Figure 20**. Keep the rings for cleaning the piston's ring grooves.
19. Remove each connecting rod and piston assembly as follows:
    a. Remove the connecting rod screws from the upper, cylinder No. 1 connecting rod. Alternately loosen each screw a small amount until all tension is off both screws.
    b. Tap the rod cap with a soft rubber or plastic mallet to separate the cap from the rod.

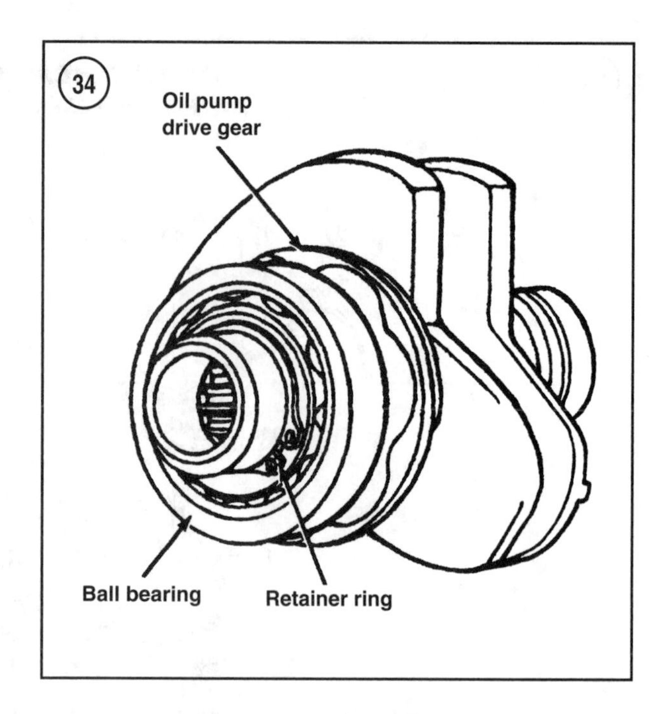

    c. Remove the cap and rod from the crankshaft, then remove the two caged roller bearing halves from the crankshaft.
    d. Reinstall the rod cap to the connecting rod in its original orientation. Tighten the rod cap screws finger-tight.
    e. Store the caged roller bearing assemblies in clean, numbered containers, corresponding to the cylinder number.
    f. Repeat this procedure to remove the lower, cylinder No. 2 connecting rod assembly.
20. Use the lock ring remover (part No. 91-52952A-1) or a suitable awl (**Figure 22**) to remove the piston pin lock ring from each end of both piston's pin bore.
21. Place the piston pin tool (part No. 91-76160A-2 or an equivalent) into one end of the cylinder No. 1 piston pin bore. Support the bottom of the piston with one hand and drive the pin tool and pin from the piston. See **Figure 23**.
22. Remove the piston from the connecting rod. Remove the thrust washers and the 29 loose needle bearings (**Figure 30**). Store the components in clean, numbered containers corresponding to the cylinder number.
23. Repeat Steps 21 and 22 to separate the cylinder No. 2 piston from its connecting rod.
24. If the crankshaft ball bearing and/or the oil pump drive gear requires replacement, proceed as follows:
    a. Remove the retainer ring (**Figure 34**) with a suitable pair of snap ring pliers.

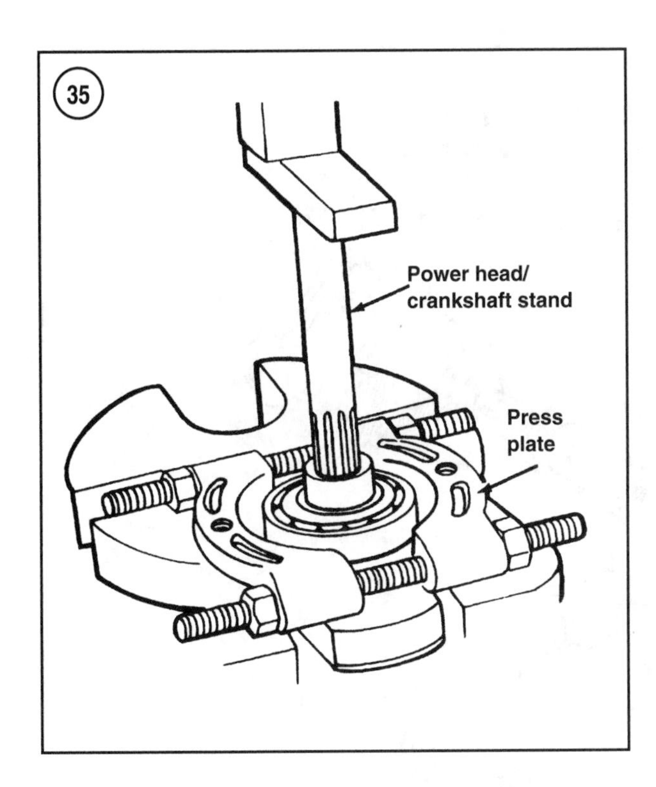

b. Support the ball bearing in a knife-edged press plate and press against the crankshaft using the power head stand as a mandrel until the bearing is free from the crankshaft. See **Figure 35**.

c. If necessary, pull the oil pump gear from the crankshaft. Then locate and secure the gear drive key (46, **Figure 33**).

25. Refer to *Cleaning and Inspection* in this chapter before beginning reassembly procedures.

Disassembly (40-60 hp, 30 Jet and 45 Jet Three-cylinder Models)

Refer to **Figure 36** and **Figure 37** for this procedure. Since the power head is removed with most of the accessories installed, it is necessary to remove them before disassembling the power head.

1. Remove the flywheel as described in Chapter Seven.
2. Remove the oil reservoir if it is not already removed, oil pump and all oil lines as described in Chapter Twelve.
3. Remove the carburetors, fuel pump, fuel filter, intake manifold, and reed block, and all fuel, primer and fuel bleed (recirculation) lines as described in Chapter Six. On electric start models, remove the fuel primer valve with the other components.
4. Remove the internal fuel bleed (recirculation) valves and carriers (8 and 9, **Figure 36**) from the crankcase cover. Remove the hoses and check valves (13 and 15, **Figure 36**) from the cylinder block.
5. On electric start models, remove the electric starter motor as described in Chapter Seven.
6. Remove the remaining ignition and electrical components as an assembly. This includes the stator assembly, trigger coil, and the electrical/ignition plate containing the CDM modules. On electric start models, this includes the voltage regulator and the starter solenoid. If equipped with an rpm limit module, remove it with the other components. Do not disconnect electrical components from each other unless absolutely necessary. Only remove the mounting screws, cable clamps and tie-straps. See Chapter Seven.
7. Remove any remaining control linkages that will interfere with power head disassembly.
8. Remove the screws securing the thermostat cover (22, **Figure 36**) to the port upper side of the cylinder block. Remove the cover and gasket. Remove the thermostat, O-ring and thermostat carrier (18-20, **Figure 36**).
9. Remove the screw and washer securing the engine temperature switch (16, **Figure 36**) to the cylinder block. Remove the switch assembly.
10. Remove the screws securing the lower end cap (24, **Figure 37**) to the power head. Do not remove the lower end cap at this time.

CAUTION

The crankcase cover and cylinder block are a matched, align-bored unit. Never install the crankcase cover from one cylinder block to a different cylinder block. Do not scratch, nick or damage the machined mating surfaces.

11. Remove the eight long and six short screws securing the crankcase cover to the cylinder block. Tap the crankcase cover with a soft-faced rubber or plastic hammer to break the crankcase seal. Remove the crankcase cover. Locate and secure the locating dowel pins as necessary.
12. Lift the crankshaft assembly straight up and out of the cylinder block (**Figure 38**) and set it on a clean workbench. Then pull the lower end cap (24, **Figure 37**) from the crankshaft. Remove the O-ring and two seals (21-23, **Figure 37**). Do not damage the end cap while removing the seals.
13. Mount the crankshaft vertically on a power head (crankshaft) stand (part No. 91-25281A-1 or an equivalent). Make sure the stand is securely clamped in a vise.
14. Locate and secure the crankshaft center main bearing locating pins (3, **Figure 36**) as necessary.
15. Slide the upper crankshaft seal (1, **Figure 37**) and caged roller bearing (2) from the crankshaft.

36

CYLINDER BLOCK COMPONENTS (40-60 HP, 30 JET AND 45 JET MODELS)

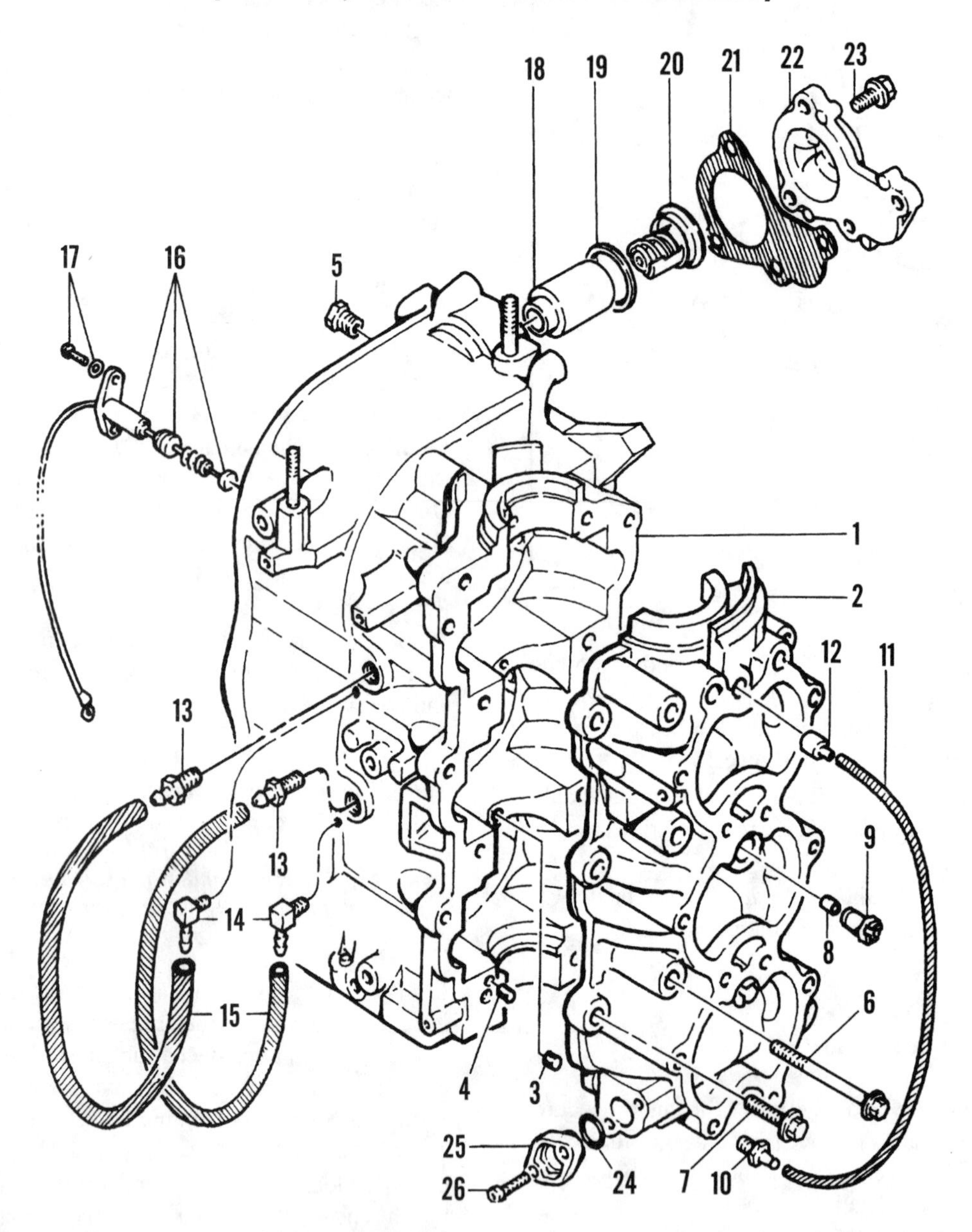

1. Cylinder block
2. Crankcase cover
3. Bearing locating pin
4. Dowel pin
5. Brass pipe plug
6. Main bearing (long) screw
7. Outer (short) screw
8. Check valve
9. Check valve carrier
10. Check valve
11. Fuel bleed (recirculation) hose
12. Check valve
13. Check valves
14. Fittings
15. Fuel bleed (recirculation) hoses
16. Engine temperature switch
17. Screw and washer
18. Thermostat carrier
19. O-ring
20. Thermostat
21. Gasket
22. Thermostat housing
23. Screw
24. O-ring
25. Cap (except oil injected models)
26. Screw

CRANKSHAFT ASSEMBLY (40-60 HP, 30 JET AND 45 JET MODELS)

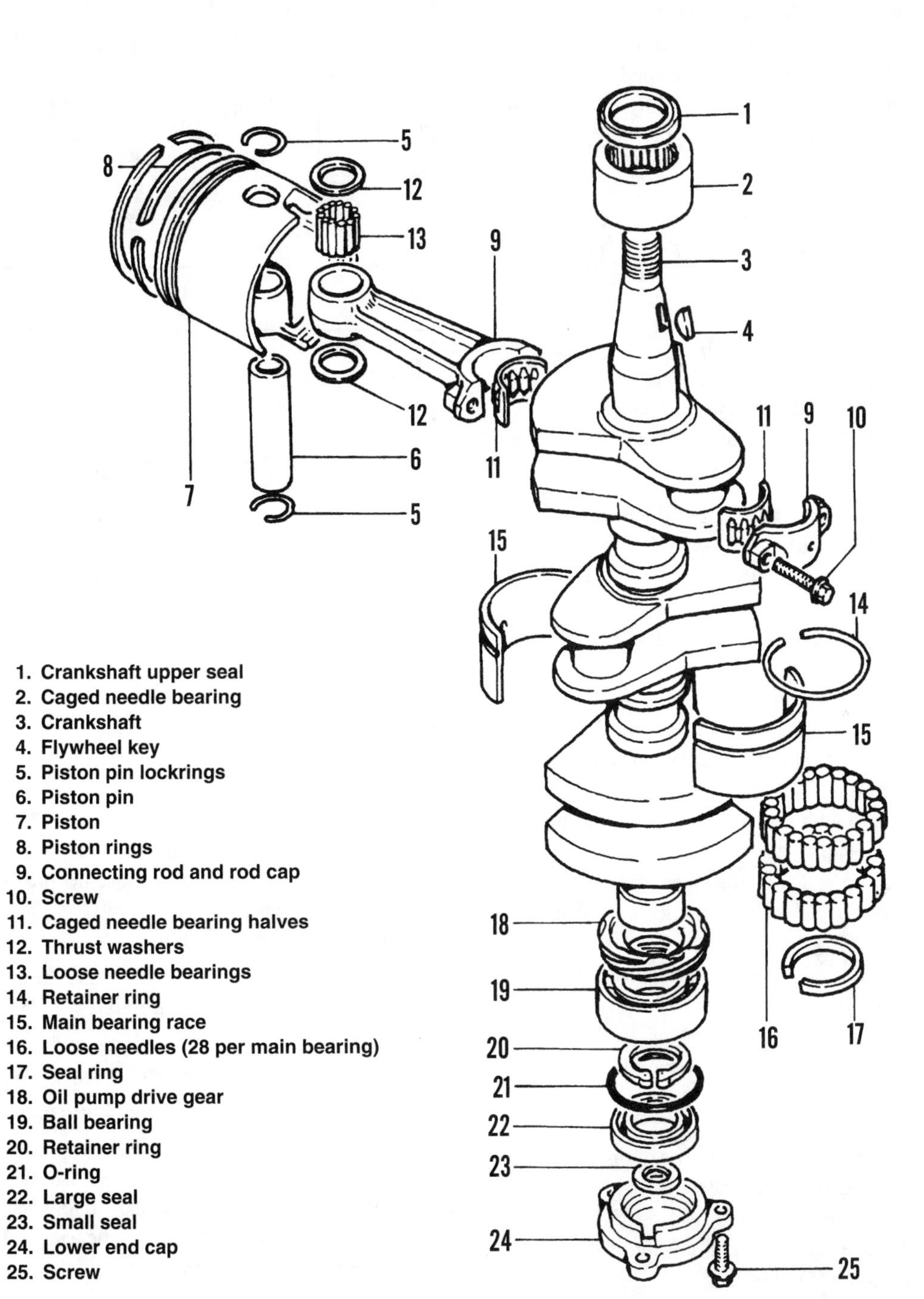

1. Crankshaft upper seal
2. Caged needle bearing
3. Crankshaft
4. Flywheel key
5. Piston pin lockrings
6. Piston pin
7. Piston
8. Piston rings
9. Connecting rod and rod cap
10. Screw
11. Caged needle bearing halves
12. Thrust washers
13. Loose needle bearings
14. Retainer ring
15. Main bearing race
16. Loose needles (28 per main bearing)
17. Seal ring
18. Oil pump drive gear
19. Ball bearing
20. Retainer ring
21. O-ring
22. Large seal
23. Small seal
24. Lower end cap
25. Screw

16. Remove the retaining ring (14, **Figure 37**) from the center main bearing. See **Figure 32**. Then remove the outer race halves, 28 loose rollers and the crankshaft seal ring. Store the bearing components in marked containers to allow reassembly in their original locations. Repeat the procedure for the other center main bearing.
17. Mark the corresponding cylinder number on the pistons and connecting rods with a felt-tipped permanent marker. See **Figure 16**. Mark the connecting rods and rod caps so the rod caps can be reinstalled in their original orientation.

WARNING
Wear suitable eye protection for piston ring and piston pin lockring removal procedures.

CAUTION
Always store the components from each connecting rod and piston assembly together. They must be reinstalled in their original locations or engine damage will occur.

18. Remove the piston rings from all of the pistons using a piston ring expander (part No. 91-24697 or an equivalent). See **Figure 20**, typical. Keep the rings for cleaning the piston's ring grooves.
19. Remove each connecting rod and piston assembly as follows:
   a. Remove the connecting rod screws from the cylinder No. 1 (top cylinder) connecting rod. Alternately loosen each screw a small amount until all tension is off both screws.
   b. Tap the rod cap with a soft rubber or plastic mallet to separate the cap from the rod.
   c. Remove the cap and rod from the crankshaft, then remove the two caged roller bearing halves from the crankshaft.
   d. Reinstall the rod cap to the connecting rod in its original orientation. Tighten the rod cap screws finger-tight.
   e. Store the caged roller bearing assemblies in clean containers numbered according to the cylinder number.
   f. Repeat this procedure to remove the cylinders No. 2 and 3 connecting rod assemblies.
20. Use a lockring remover (part No. 91-52952A-1) or a suitable awl (**Figure 22**) to remove the piston pin lockrings from all piston pin bores.
21. Place the piston pin tool (part No. 91-76160A-2 or an equivalent) into one end of the cylinder No. 1 piston pin bore. Support the bottom of the piston with one hand and drive the pin tool and pin from the piston. See **Figure 23**.

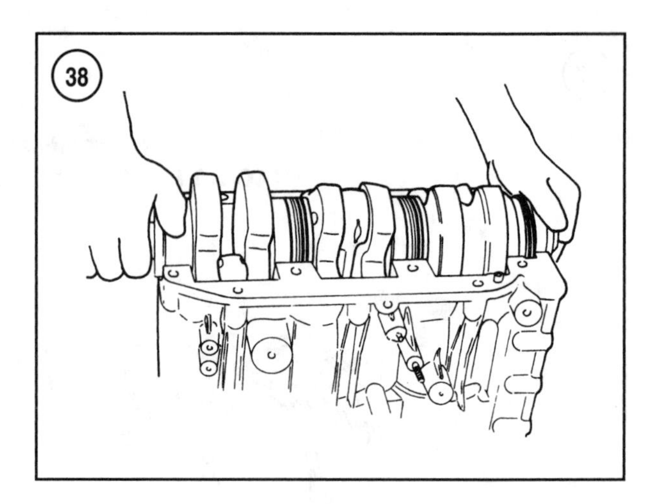

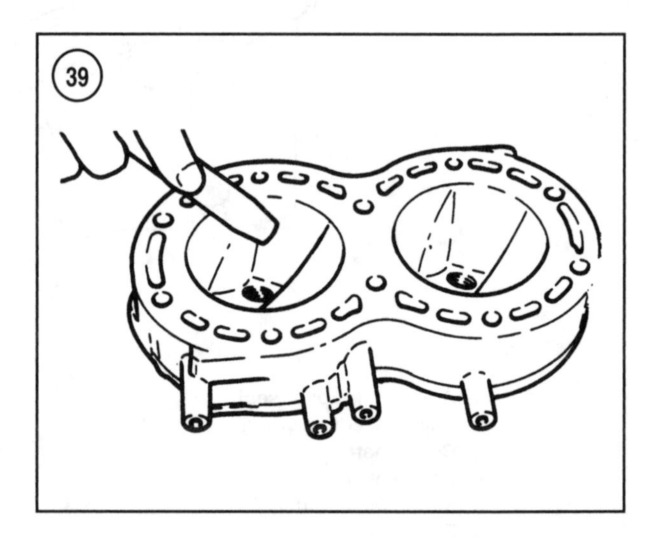

22. Remove the piston from the connecting rod. Remove the thrust washers and the 29 loose needle bearings (**Figure 30**). Store the components in clean containers numbered according to the cylinder number.
23. Repeat Steps 21 and 22 to separate the cylinders No. 2 and 3 pistons and connecting rods.
24. If the crankshaft ball bearing and/or the oil pump drive gear requires replacement, proceed as follows:
   a. Remove the retainer ring (**Figure 34**) with a suitable pair of snap ring pliers.
   b. Support the ball bearing in a knife-edged plate and press against the crankshaft using the power head stand as a mandrel until the bearing is free from the crankshaft. See **Figure 35**.
   c. If necessary, pull the oil pump gear from the crankshaft.
25. Refer to *Cleaning and Inspection* in this chapter before beginning reassembly procedures.

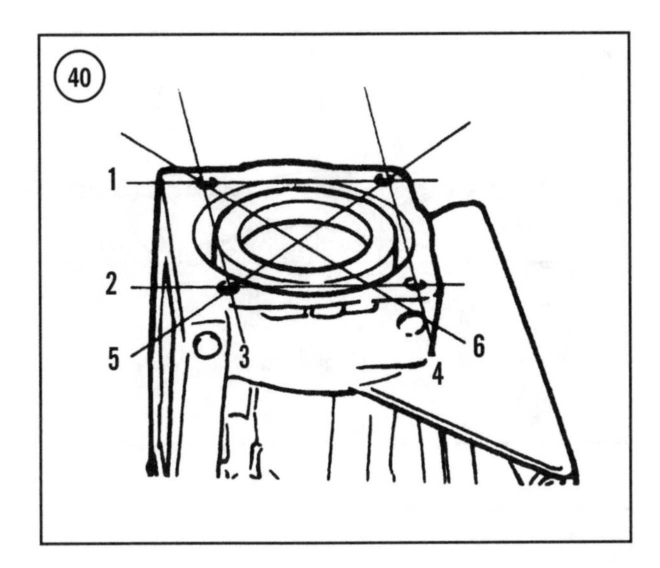

CLEANING AND INSPECTION

Refer to Chapter Six and clean and inspect the reed blocks and the fuel bleed (recirculation) system. Test all check valves in the fuel bleed system for correct function as described in Chapter Six.

Review *Lubricants, Sealants and Adhesives*, *Sealing Surfaces*, and *Fasteners and Torque* in this chapter before assembling the power head.

Replace all seals, O-rings, gaskets, connecting rod screws, piston pin lock rings, piston rings and all bearings any time a power head is disassembled.

Clean and inspect the components before assembling the power head.

Cylinder Block/Crankcase (All Models)

Mercury outboard cylinder blocks and crankcase covers are matched, align-bored assemblies. Do not attempt to assemble an engine with parts salvaged from other blocks. If the block or cover requires replacement, replace the cylinder block and crankcase cover as an assembly.

NOTE
Remove all fuel bleed hoses, T-fittings, threaded fittings, check valves and check valve carriers before submerging the block or cover in a strong cleaning solution. See Chapter Six.

1. Clean the cylinder block and crankcase cover thoroughly with clean solvent and a parts washing brush. Carefully remove all gasket and sealant material from mating surfaces.
2. Remove all carbon and varnish deposits from the combustion chambers, exhaust ports and exhaust cavities with a carbon removing solvent, such as Quicksilver Power Tune (part No. 92-15104). A hardwood dowel or plastic scraper (**Figure 39**) can be used to remove stubborn deposits. Do not scratch, nick or gouge the combustion chambers or exhaust ports.

WARNING
Use suitable hand and eye protection when using muriatic acid products. Avoid breathing the vapors. Only use them in a well-ventilated area.

CAUTION
Do not allow muriatic acid to contact the aluminum surfaces of the cylinder block. Never use muriatic acid to clean Mercosil cylinders (20-25 hp and 20 Jet models) or the cylinders will be destroyed.

NOTE
Mercosil (high-silicon) treated aluminum bores are used on 20-25 hp and 20 Jet models. All other models use cast iron bores.

3A. *Cast-iron cylinder bores*—If the cylinder bore(s) has aluminum transfer from the piston(s), clean loose deposits using a stiff bristle brush. Apply a *small* amount of diluted muriatic acid to the aluminum deposits. Bubbling indicates the aluminum is being dissolved. Wait 1-2 minutes, then thoroughly wash the cylinder with hot water and detergent. Repeat this procedure until the aluminum deposits are removed. Lightly oil the cylinder wall to prevent rusting.
3B. *Mercosil cylinder bores (20-25 hp and 20 jet models*—Do not use muriatic acid to clean Mercosil cylinder bores as the bores will be destroyed by the acid. Mercosil bores can be bored or honed the same as cast-iron cylinders.
4. Check the cylinder block and crankcase cover for cracks, fractures stripped threads or other damage.
5. Inspect gasket mating surfaces for nicks, grooves, cracks or distortion. Any of these defects will cause leaks. Check the surfaces for distortion as described in *Sealing Surfaces* in this chapter. Replace the component if distortion is more than 0.004 in. (0.1 mm), unless otherwise specified. Smaller imperfections can be removed by lapping the component as described under *Sealing Surfaces*. **Figure 40-42** show typical directions in which to check for warp on the cylinder head and exhaust cover/manifold surfaces.
6. Check all water, oil and fuel bleed passages in the block and cover for obstructions. Make sure all pipe plugs

are installed tightly. Seal pipe plugs with Loctite 567 PST pipe sealant (part No. 92-809822).

Cylinder bore inspection

1. *Cast iron cylinder bores*—Inspect the cylinder bores for scoring, scuffing, grooving, cracks or bulging, and other mechanical damage. Inspect the cylinder block (casting) and cast-iron liner for separation or delamination. There must be no gaps or voids between the aluminum casting and the liner. Remove aluminum deposits as described in this section. If the cylinders are in visually acceptable condition, hone the cylinders as described in *Cylinder wall honing.* If the cylinders are in unacceptable condition, rebore the defective cylinder bore(s) or replace the cylinder block and crankcase cover as an assembly.
2. *Mercosil cylinder bores*—Inspect the cylinder bores for scoring, scuffing, grooving, cracks or bulging, and other mechanical damage. If the cylinder walls are in visually acceptable condition, hone the cylinders as described in *Cylinder wall honing.* If the cylinders are in unacceptable condition, rebore the defective cylinder bore(s) or replace the cylinder block and crankcase cover as an assembly.

NOTE
It is not necessary to rebore all cylinders in a cylinder block. Rebore only the cylinders that are defective. It is acceptable to have a mix of standard and oversize cylinders in a power head as long as the correct piston (standard or oversize) is used to match each bore. Always check the manufacturer's parts catalog for oversize piston availability and bore sizes, before boring the cylinder(s).

Cylinder bore honing

The manufacturer recommends using only a rigid type cylinder hone to deglaze the bore to aid in the seating of new piston rings. If the cylinder has been bored oversize, the rigid hone will be used in two steps: a rough, deburring hone to remove the machining marks and a finish (final) hone to establish the correct cross-hatch pattern in the cylinder bore.

Flex (ball type) hones and spring-loaded hones are not acceptable as they will not produce a true bore.

NOTE
If proficiency with the correct use of a rigid cylinder hone is in doubt, have the honing performed at a qualified machine shop or reputable dealership. The manufacturer recommends rigid hones.

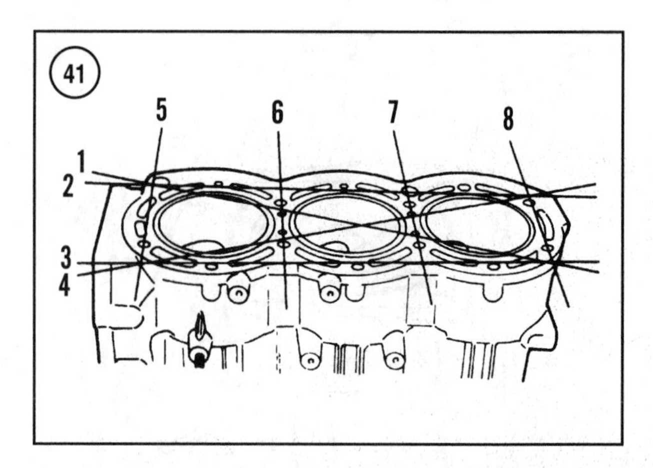

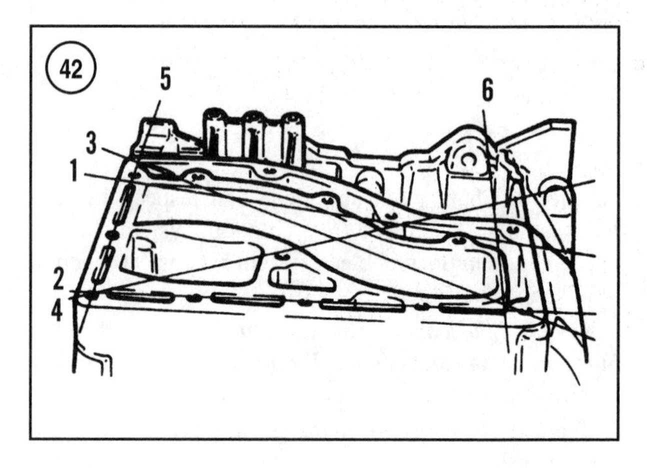

If the cylinders are in visually acceptable condition, prepare the cylinder bore for new piston rings and remove any glazing, light scoring and/or scuffing by lightly honing the cylinders as follows:

1. Follow the rigid hone manufacturer's instructions when using the hone. Make sure the correct stones for the bore (cast-iron or aluminum) are installed on the hone.
2. A continuous flow of honing oil must be pumped into the bore during the honing operation. If an oil pumping system is not available, have an assistant with an oil can keep the cylinder walls flushed with honing oil.
3. If the hone loads (slows down) at one location in the bore, this is the narrowest portion of the bore. Remove stock in this location until the hone maintains the same speed and load throughout the entire bore.
4. Frequently remove the hone from the cylinder bore and inspect the bore. Do not remove more material than necessary.

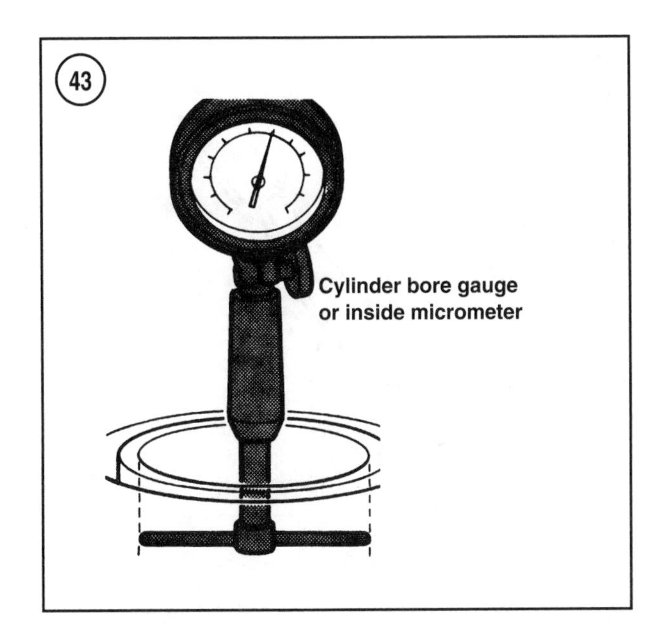

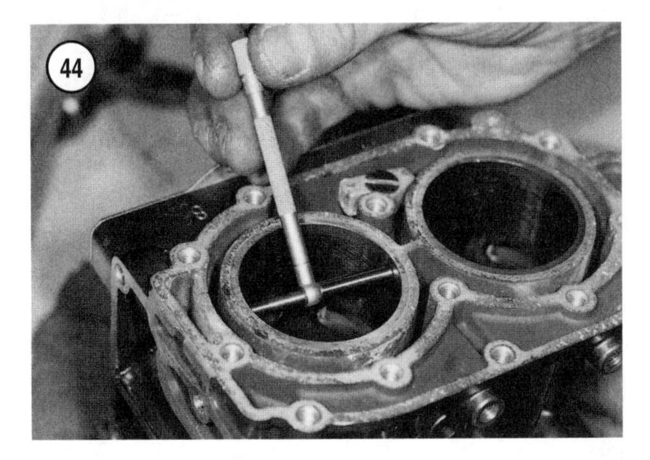

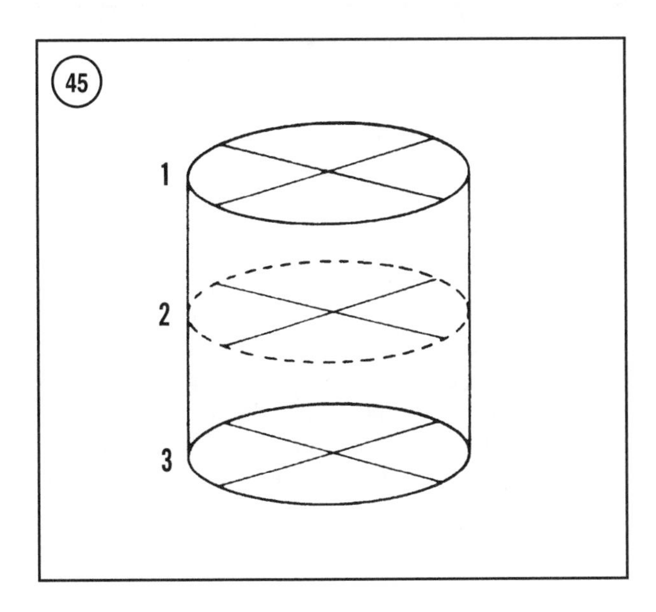

5. Attempt to achieve a stroke rate of approximately 30 cycles per minute, adjusting the speed of the hone to achieve a cross-hatch pattern with an intersecting angle of approximately 30°. Do not exceed a cross-hatch of more than 45°.
6. After honing, thoroughly clean the cylinder block with hot water, detergent and a stiff bristle brush. Make sure all abrasive material from the honing process is removed. After washing and flushing, coat the cylinder walls with a film of outboard motor oil to prevent rust.
7. Proceed to *Cylinder bore measurements* to determine if the cylinder bores are within the manufacturer's specifications for wear, taper and out-of-round.

Cylinder bore measurements

Measure each cylinder bore as follows. Oversize bore specifications are the standard bore specification *plus* the oversize dimension. Check the parts catalog for available oversize dimensions. All standard bore specifications, maximum taper and out-of-round specifications are in **Table 5**. The maximum wear limit on a given cylinder is equal to the standard bore plus the maximum taper specification.

Use a cylinder bore gauge (**Figure 43**), inside micrometer, or a telescoping gauge (**Figure 44**) and a regular micrometer to measure the entire area of ring travel in the cylinder bore. Three sets of readings must be taken at the top, middle and bottom of the ring travel area (**Figure 45**).
1. Take the first reading at the top of the ring travel area (approximately 1/2 in. [12.7 mm] from the top of the cylinder bore) with the gauge aligned with the crankshaft centerline. Record the reading. Then turn the gauge 90° to the crankshaft centerline and record another reading.
2. The difference between the two readings is the cylinder out-of-round. Cylinder out-of-round cannot exceed the specification in **Table 5**.
3. Take a second set of readings at the midpoint of the ring travel area (just above the ports) using the same alignment points described in Step 1. Record the readings. Calculate the cylinder out-of-round by determining the difference between the two readings. Cylinder out-of-round cannot exceed the specification in **Table 5**.
4. Take a third set of readings at the bottom of the ring travel area (near the bottom of the cylinder bore) using the same alignment points described in Step 1. Record the readings. Calculate the cylinder out-of-round by determining the difference between the two readings. Cylinder out-of-round cannot exceed the specification in **Table 5**.
5. To determine the cylinder taper, subtract the readings taken at the top of the cylinder bore (Step 1) from the readings taken at the bottom of the cylinder bore (Step 4). The

difference is the cylinder taper. Cylinder taper cannot exceed the specification in **Table 5**.

NOTE
If the cylinder has already been bored oversize, add the oversize dimension to the standard bore dimension and the maximum out-of-round specification in the next step.

6. To determine if the cylinder is excessively worn, add the maximum taper specification (**Table 5**) to the standard cylinder bore. If any of the readings taken in Steps 1-4 exceed this figure, the cylinder is excessively worn.
7. Repeat Steps 1-6 for each remaining cylinder.
8. If any cylinder exhibits excessive out-of-round, taper or is excessively worn, bore the cylinder(s) oversize or replace the cylinder block and crankcase cover as an assembly.

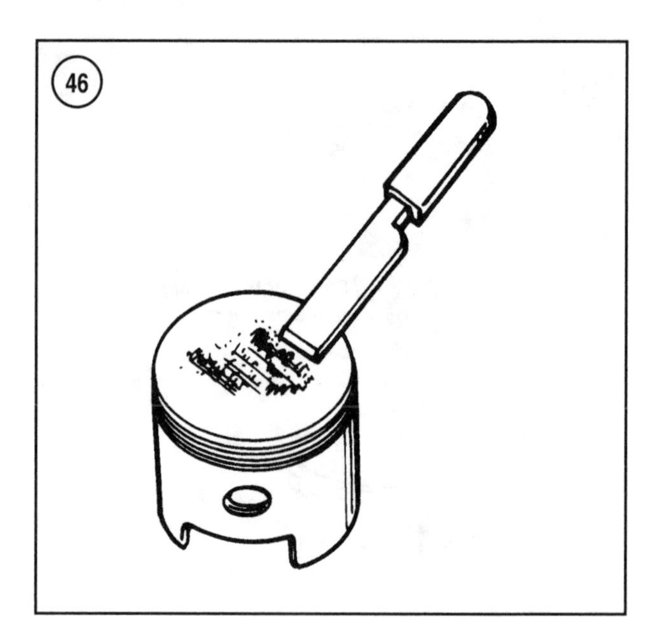

Piston

The piston and piston pin are serviced as an assembly. If either is damaged, replace them together. Install the piston pin only in its original piston.

CAUTION
Do not use an automotive ring groove cleaning tool as it can damage the ring grooves and loosen the ring locating pins.

1. Clean the piston(s), piston pin(s) thrust washers and the piston pin needle bearing assemblies thoroughly with clean solvent and a parts washing brush. Do not use a wire brush on the piston as metal from the wire wheel may become imbedded in the piston. This can lead to preignition and detonation.
2. Remove all carbon and varnish deposits from the top of the piston, piston ring groove(s) and under the piston crown with a carbon removing solvent, such as Quicksilver Power Tune (part No. 92-15104). Use a piece of hardwood or a plastic scraper to remove stubborn deposits (**Figure 46**). Do not scratch, nick or gouge any part of the piston. Do not remove stamped or cast identification marks.
3. Clean stubborn deposits from the ring groove(s) as follows:
   a. Fashion a ring cleaning tool from the original piston ring(s). Rings are a different shape for each ring groove. Use the correct original ring for each ring groove.
   b. Break off approximately 1/3 of the original ring. Grind a beveled edge on the broken end of the ring.

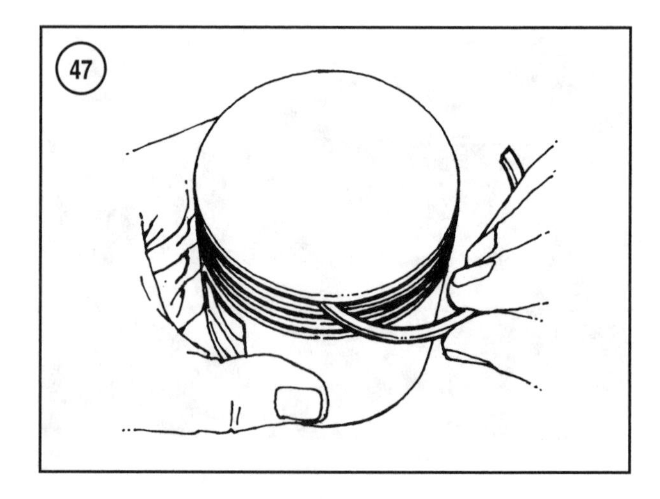

NOTE
On keystone and semi-keystone shaped rings, grind off enough of the ring taper to allow the inside edge of the broken ring to reach the inside diameter of the ring groove.

   c. Use the ground end of the ring to gently scrape the ring groove clean (**Figure 47**). Be careful to only remove the carbon. Do not gouge the metal and do not damage or loosen the piston ring locating pin(s).
4. Polish nicks, burrs or sharp edges on and about the piston skirt with crocus cloth or 320 grit carborundum cloth. Do not remove cast or stamped identification markings. Wash the piston thoroughly to remove all abrasive grit.
5. Inspect the piston(s) overall condition for scoring, cracks, worn or cracked piston pin bosses and other mechanical damage. Carefully inspect the crown and the top

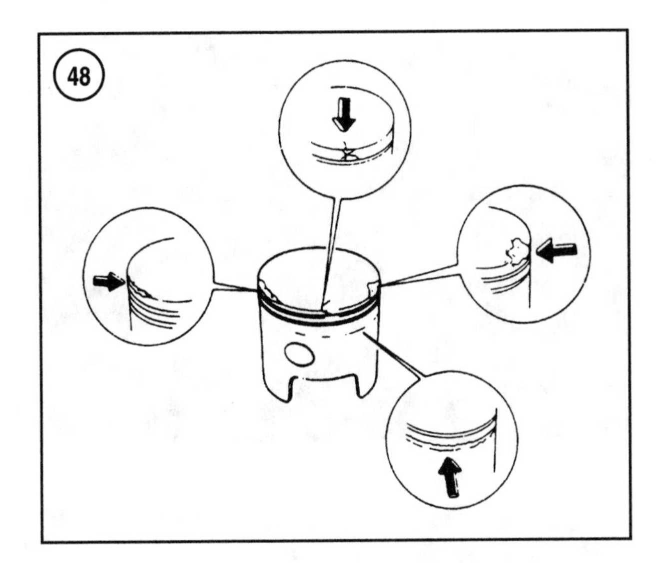

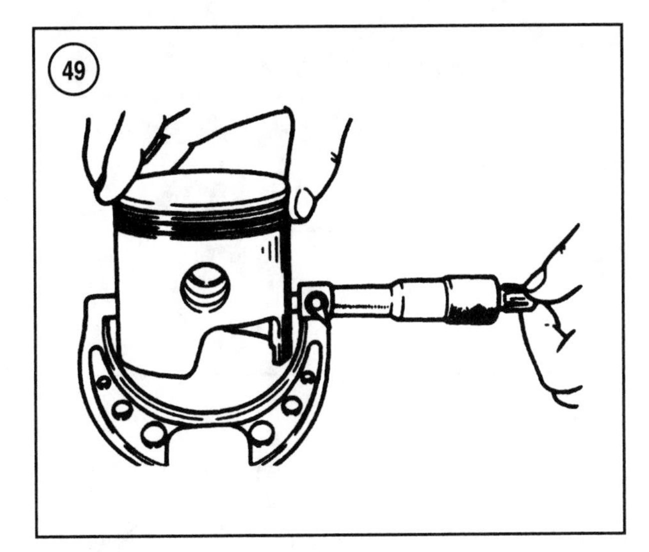

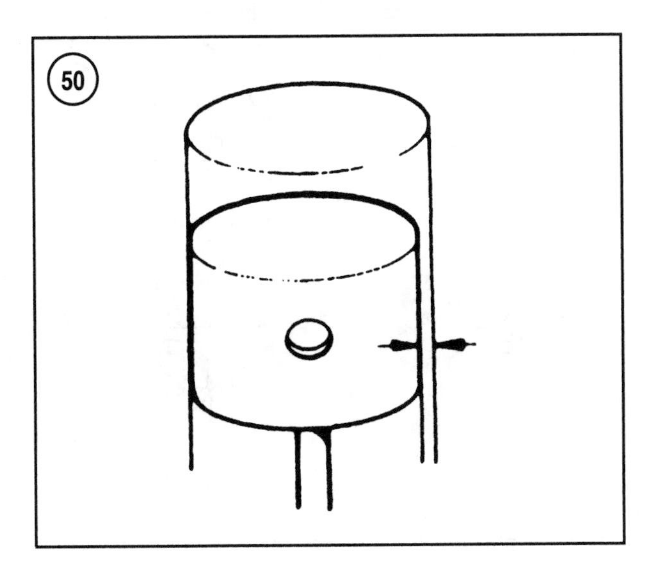

outer diameter for burning, erosion, evidence of ring migration and mechanical damage (**Figure 48**). Replace the piston and pin as necessary.
6. Check piston ring grooves for wear, erosion, distortion and loose ring locating pins.
7. Inspect the piston pin for water etching, pitting, scoring, heat discoloration, excessive wear, distortion and mechanical damage. Roll the pin across a machinist's surface plate to check the pin for distortion.
8. Inspect the thrust washers and needle bearings for water damage, pitting, scoring, overheating, wear and mechanical damage.

Piston measurements

Most pistons used in these engines are cam shaped. The piston is intentionally manufactured out-of-round. The piston is engineered to fit the bore perfectly when at operating temperature and fully expanded. This makes the engine run more quietly and efficiently. The piston must be measured at the specified point(s) for accurate readings.

Measure each piston skirt with a micrometer as described in the following text and compare the readings to the specifications in **Table 6**.

To calculate the specified skirt dimension on oversize pistons, add the oversize dimension to the standard skirt diameter in **Table 6**.

1. *2.5 and 3.3 hp models*—There is no specification for the piston skirt dimension. Use a micrometer to measure the piston skirt diameter at a point even with and at a 90° angle to the piston pin bore. See **Figure 49**. Record the reading. Subtract this reading from the cylinder bore dimension to determine the piston-to-cylinder clearance (**Figure 50**). The clearance should be within the specification in **Table 6**. If it is not, compare the cylinder bore dimension to specification. If the cylinder is within specification, replace the piston.
2. *4 and 5 hp models*—Use a micrometer to measure the piston skirt diameter at a point 5/8 in. (15.9 mm) up from the bottom of the skirt and at a 90° angle to the piston pin bore. See **Figure 49**, typical. Record the reading. Subtract this reading from the cylinder bore dimension to determine the piston-to-cylinder clearance (**Figure 50**). The clearance should be within the specification in **Table 6**. If it is not, compare the piston skirt diameter to the specification in **Table 6**. Replace the piston and/or bore the cylinder oversize as necessary.
3. *6-15 hp models*—Use a micrometer to measure each piston's skirt diameter at a point 0.10 in. (2.54 mm) up from the bottom of the skirt and at a 90° angle to the piston pin bore. See **Figure 49**, typical. Record the reading. Subtract this reading from the appropriate cylinder bore di-

mension to determine the piston-to-cylinder clearance. The clearance should be within the specification in **Table 6**. If it is not, compare the piston skirt diameter to the specification **Table 6**. Replace the piston(s) and/or bore the cylinder(s) oversize as necessary.

4. *20-25 hp and 20 jet models*—Use a micrometer to measure each piston skirt diameter at a point 0.50 in. (12.7 mm) from the bottom of the skirt and at a 90° angle to the piston pin bore. See **Figure 49**, typical. Record the reading. Compare the reading to the specification in **Table 6**. Each piston dimension should be as specified. If it is not, replace the piston(s).

5. *30-60 hp models*—Use a micrometer to measure each piston skirt diameter at a point 0.50 in. (12.7 mm) up from the bottom of the skirt and at a 90° angle to the piston pin bore. See **Figure 49**, typical. Record the reading. Each piston diameter should be within the specification in **Table 6**. If it is not, replace the piston(s).

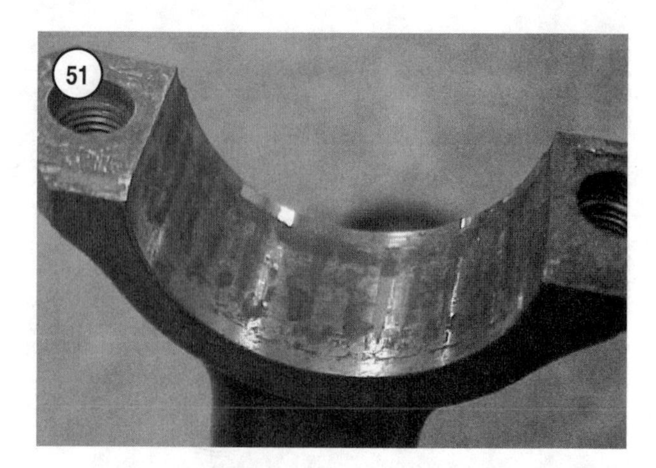

Connecting Rods (6-60 hp and 30-45 Jet)

All connecting rods on 6-60 hp models are the fractured cap design. The cap is broken from the rod during the manufacturing process, leaving a jagged mating surface that will mate perfectly if installed in its original orientation. If the cap is installed reversed and the rod screws are tightened, the rod will be distorted and must be replaced. Though alignment marks are provided, always mark the rod and cap with a felt-tipped permanent marker for easy identification. Correct orientation is obvious if the time is taken to examine the mating surfaces of the rod and cap. Hold the connecting rod cap firmly in position as the screws are installed.

New connecting rod screws must be installed upon final assembly, but the old screws can be used for cleaning and inspection. All connecting rod screw torque specifications are in **Table 1**.

Some 40-60 hp, 30 jet and 45 jet three-cylinder models use a *Torque and Turn* process to tighten the connecting rod screws. In this process, oil is applied to the screw threads and under the screw head. The screws are tightened to an initial torque (**Table 1**), and the alignment of the rod cap is verified. Then the screws are tightened to a second, higher torque (**Table 1**), and finally both screws are turned an additional 90°.

Clean and inspect the connecting rods as follows:

1. Clean the connecting rods thoroughly with clean solvent and a parts washing brush.

2. Check the connecting rod big end (**Figure 51**) and small end bearing surfaces (**Figure 52**) for rust, water damage, pitting, spalling, chatter marks, heat discoloration and excessive or uneven wear. If the defect can be

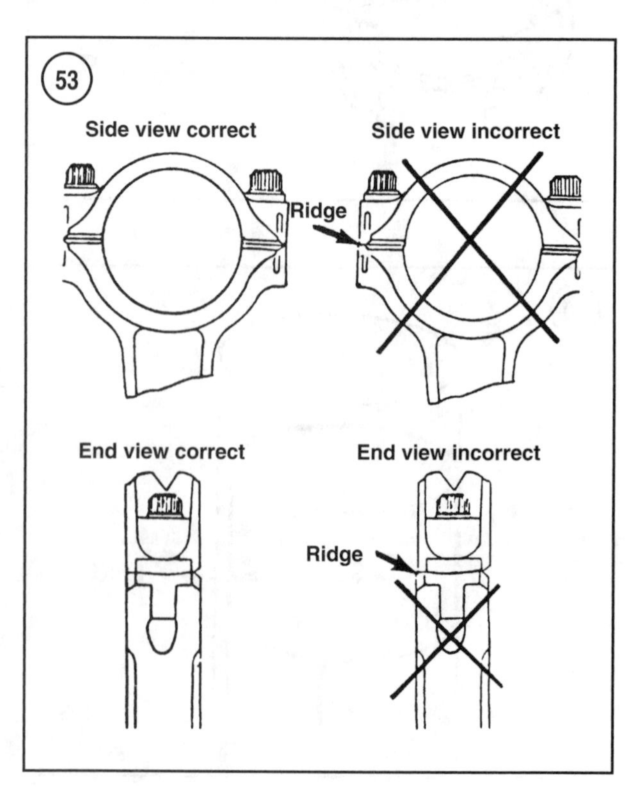

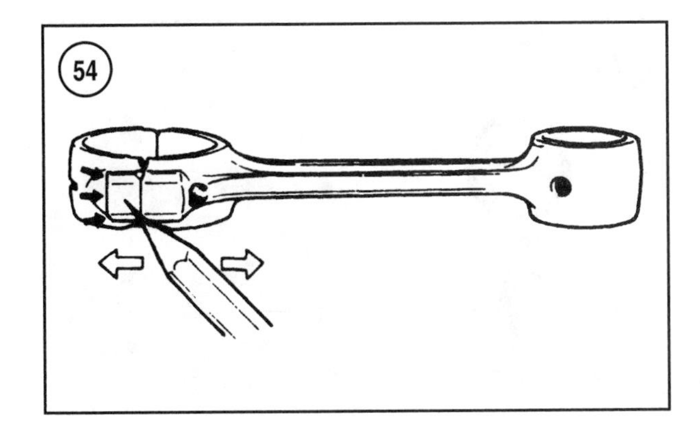

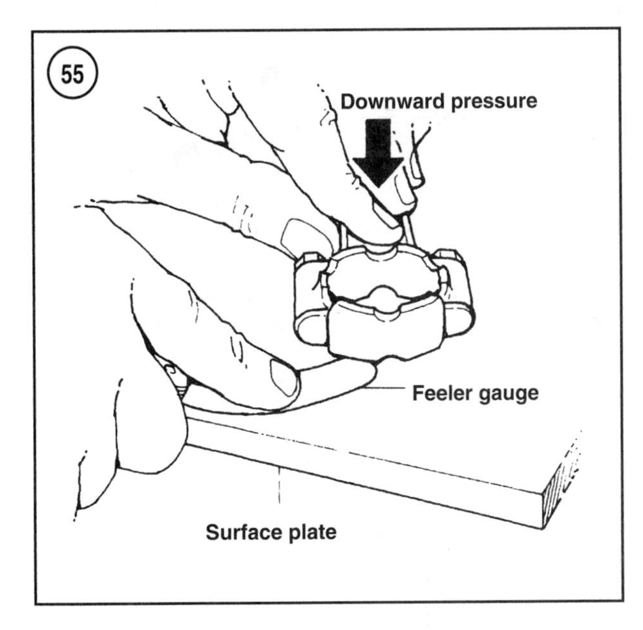

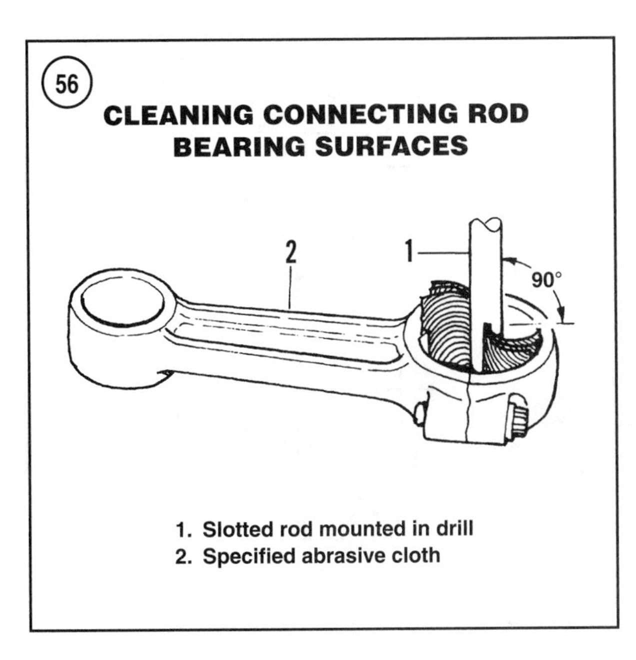

1. Slotted rod mounted in drill
2. Specified abrasive cloth

felt by dragging a pencil lead or a fingernail over it, replace the rod. Stains or marks that cannot be felt can be removed by polishing the bearing surface as described in later steps.

3. Assemble the rod cap to the connecting rod as follows:
   a. Clamp the cylinder No. 1 connecting rod securely in a vise with protective jaws.
   b. Install the matching connecting rod cap to the connecting rod in its original orientation. Carefully observe fracture and alignment marks to ensure correct installation (**Figure 53**).
   c. Lubricate the screw threads and underside of the screw head with outboard lubricant. Then hold the cap firmly in position, and install the connecting rod screws and thread them fully into the rod.
   d. Tighten each screw to 15 in.-lb. (1.7 N•m). Run a fingernail or pencil lead over each edge of the rod to cap joint (**Figure 54**). No ridge should be seen or felt. Loosen, realign and retorque the cap as necessary.
   e. Once the alignment is correct, finish torquing the rod screws as specified in **Table 1**. Make a final check of alignment after the final torque is applied.
   f. Repeat this procedure for each remaining rod.
4. Check the connecting rods for straightness. Place each rod/cap assembly on a machinist's surface plate and press downward on the rod beam. The rod must not wobble under pressure. hold the rod against the plate, and attempt to insert a 0.002 in. feeler gauge between the machined surfaces of the rod and the plate (**Figure 55**). If the feeler gauge can be inserted between any machined surface of the rod and the surface plate, the rod is bent and must be replaced.

CAUTION

Only use crocus cloth or 320 carborundum cloth to clean the connecting rod bearing surfaces in the following step.

5. If the connecting rod has passed all inspections to this point, slight defects noted in Step 2 in either bearing surface may be cleaned up as follows:
   a. Fabricate a holder by cutting a 1 in. (25.4 mm) notch lengthwise into a 4 in. (102 mm) long, 5/16 in. (8 mm) wide shank, rod or bolt with a hacksaw.
   b. Clean the bearing surface with a strip of 320 Carborundum cloth or a strip of crocus cloth mounted in the fabricated holder. Mount the holder in a drill. Spin the cloth using the drill as shown in **Figure 56** until the surface is polished. Maintain a 90° angle as shown and do not remove more material than necessary.

c. Wash the connecting rod thoroughly in clean solvent to remove abrasive grit, then inspect the bearing surfaces. Replace any connecting rod assembly that does not clean up properly.
d. Remove and replace the rod cap screws. Wash the rod and cap again in clean solvent. Retag the rod and cap for identification. Lightly oil the bearing surfaces with outboard lubricant to prevent rust.
e. Repeat this process for each remaining connecting rod.

6. On 6-40 hp two-cylinder models, measure the inside diameters of the bearing surfaces (**Figure 57**) and compare to the specification in **Table 3**. Replace any connecting rod assembly that is excessively worn.

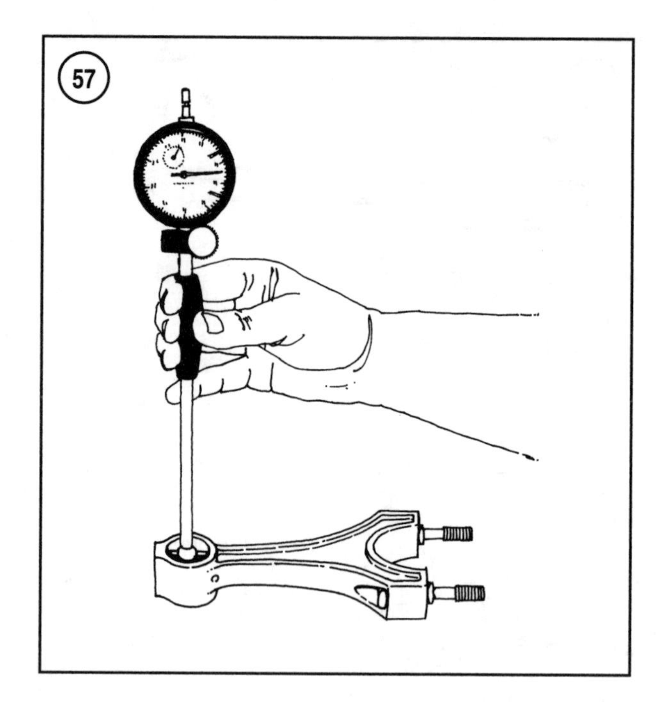

Crankshaft (2.5-5 hp Models)

1. Thoroughly wash the crankshaft assembly and all bearings in clean solvent with a parts washing brush.

2A. *2.5 and 3.3 hp models*—Inspect the square drive end of the crankshaft for corrosion, excessive wear or other damage.

2B. *4 and 5 hp models*—Inspect the drive shaft splines in the lower end of the crankshaft for wear, corrosion or other damage.

3. Inspect the flywheel taper, flywheel key groove and threads for corrosion, wear, cracks and mechanical damage.

4. Inspect the seal surfaces for excessive grooving, pitting, nicks or burrs. Polish seal surfaces with crocus cloth as necessary.

5. Inspect the connecting rod large end bearing by rotating the rod around its journal. The bearing must rotate smoothly without excessive play or noise.

6A. If the ball bearings have been removed, check the upper and lower ball bearing journals for overheating, corrosion and evidence of a bearing race spinning on the journal. The bearing's inner races must be a press fit to the crankshaft. The journals may be polished with crocus cloth as necessary.

6B. If the ball bearings are installed, rotate the ball bearings. The bearing should rotate smoothly without excessive play or noise. There should be no discernible end or axial play between the inner and outer races of the bearings (**Figure 58**). If the bearing shows visible wear, corrosion or deterioration, replace the bearing.

7. Clean the connecting rod small end bearing surfaces as necessary with crocus cloth. Use the same procedure as described under *Connecting rods* in this chapter.

8. Inspect the connecting rod small end needle bearing assembly for water damage, wear, pitting, overheating

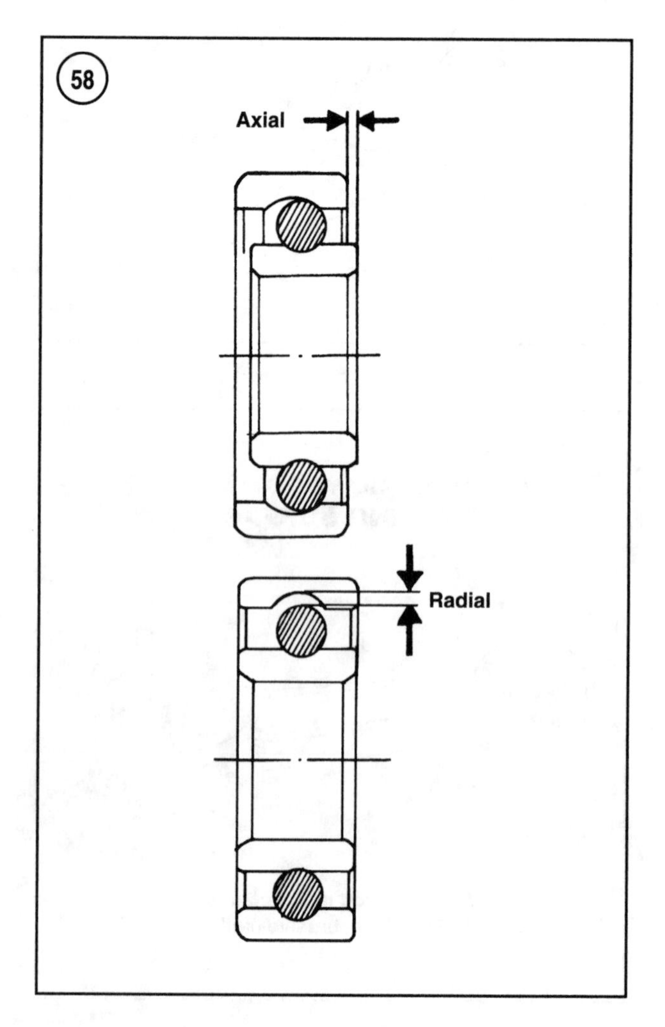

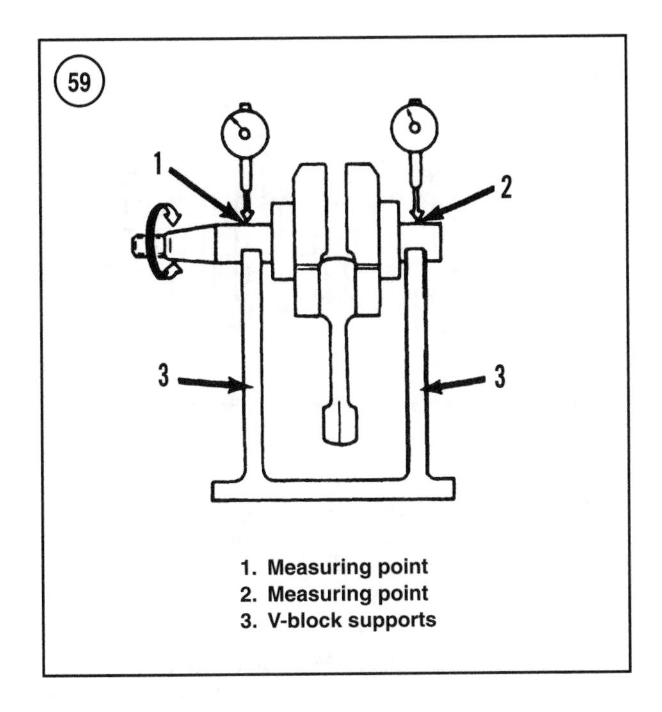

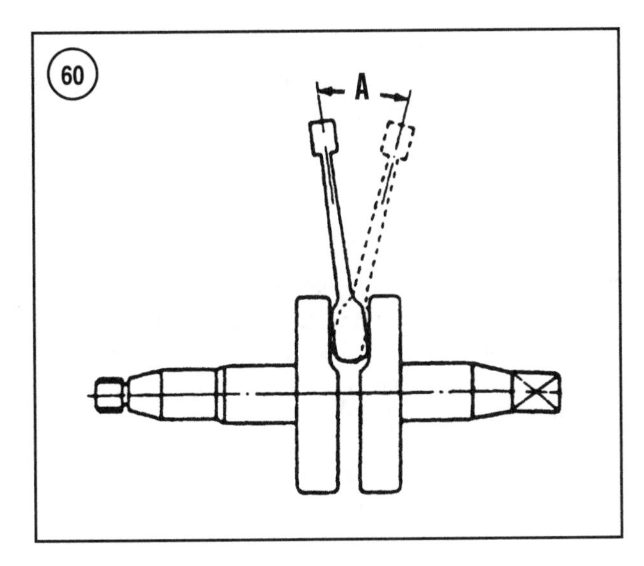

and mechanical damage. See **Figure 52**. Replace the bearing if there is any doubt about its condition.

9. Support the crankshaft assembly as shown in **Figure 59**. Mount a dial indicator at the measuring points shown, then rotate the crankshaft assembly and check the runout. Replace the crankshaft if runout exceeds the specification in **Table 4**.

10A. *2.5 and 3.3 hp models*—Determine connecting rod, crankpin and crankpin bearing wear by measuring the connecting rod deflecting at the small end (A, **Figure 60**). Replace the crankshaft assembly if the deflection exceeds the specification in **Table 3**.

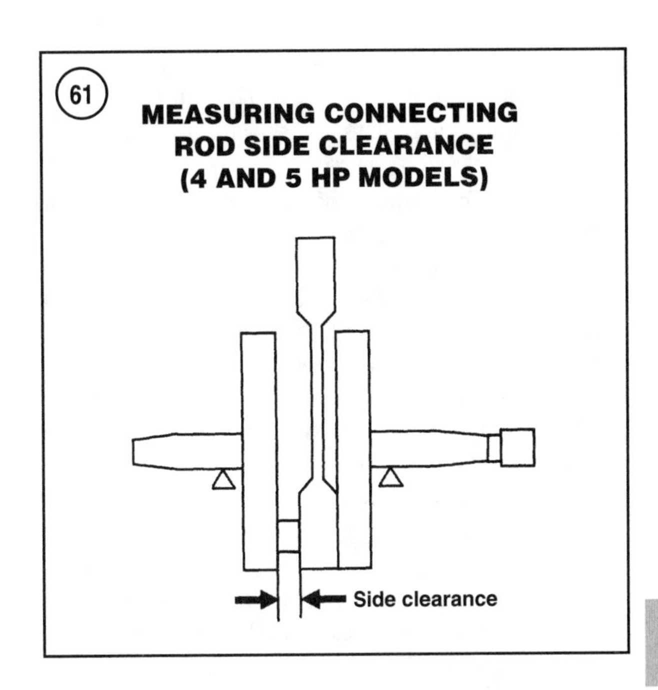

10B. *4 and 5 hp models*—Measure the connecting rod side clearance (**Figure 61**) by inserting a feeler gauge between the rod big end and crankshaft counterweights. Replace the crankshaft assembly if the side clearance exceeds the specification in **Table 3**.

Crankshaft (6-60 hp Models)

Crankshaft dimensions are in **Table 4**. Measure all surfaces where specifications are given. Replace any crankshaft that is excessively worn.

1. Thoroughly wash the crankshaft and the main and connecting rod bearing assemblies with clean solvent and a parts washing brush.
2. Inspect the drive shaft splines, flywheel taper, flywheel key groove or splines and flywheel nut or bolt hole threads for corrosion, cracks, excessive wear and mechanical damage.
3. Inspect the upper and lower seal surfaces for excessive grooving, pitting, nicks or burrs. The seal surfaces may be polished with crocus cloth as necessary. If the crankshaft is equipped with a wear sleeve on the drive shaft end, replace it if it is damaged.
4. Inspect the seal ring groove at each center main journal location for wear and mechanical damage.
5. Check the crankshaft bearing surfaces for rust, water damage, pitting, spalling, chatter marks, heat discoloration and excessive or uneven wear. If a defect can be felt by dragging a pencil lead or fingernail over the bearing surfaces, replace the crankshaft. Stains or marks that cannot be felt can be removed by polishing the bearing sur-

8

face with a strip of 320 grit carborundum cloth or a strip of crocus cloth. Work the cloth back and forth evenly over the entire journal until the surface is polished. Do not remove more material than necessary.

6. Thoroughly clean the crankshaft again in clean solvent and recheck the crankshaft surfaces. Replace the crankshaft if it cannot be properly cleaned. If the crankshaft is in visually acceptable condition, lightly oil the crankshaft to prevent rust.

7. On oil injected models, inspect the oil pump drive gear for worn or chipped teeth, heat damage or any other damage. Replace the oil pump drive gear if it is damaged. The drive gear is on the bottom end of the crankshaft.

8. Inspect the bearings as follows:
   a. *Ball bearings*—The ball bearing must rotate without rough spots, catches or noise. There must be no discernible end or axial play (**Figure 58**) between the inner and outer races of the bearing. Replace the bearing if it shows visible signs of wear, corrosion or deterioration.
   b. *Roller/needle bearings*—Inspect the rollers and/or needles for water etchings, pitting, chatter marks, heat discoloration and excessive or uneven wear. Inspect the cages for wear and mechanical damage. Replace bearings as an assembly. Do not attempt to replace individual rollers or needles.

CAUTION
Some bearing cages are designed to retain the rollers and others are not. If some rollers fall out of the cage and others do not, the bearings and/or cage(s) are worn excessively and must be replaced.

9. Support the crankshaft assembly at the upper and lower main bearing journals as shown in **Figure 62**, typical. Rotate the crankshaft assembly and check the runout at each main journal with a dial indicator. Replace the crankshaft if runout exceeds the specification in **Table 4**.

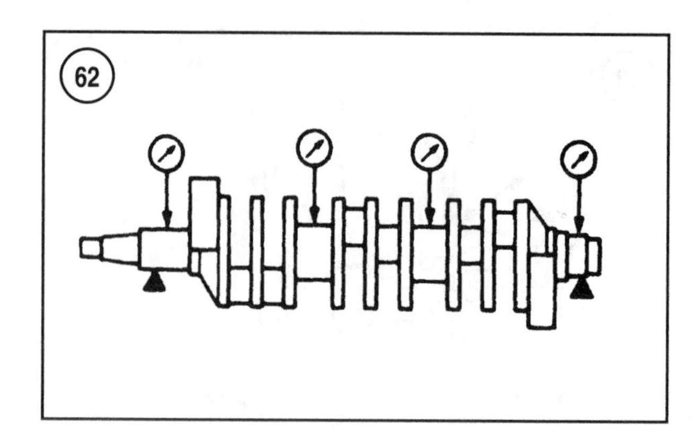

Cylinder Head(s) or Cylinder Block Cover

NOTE
The cylinder head on 4 and 5 hp models can be lapped or resurfaced to remove warpage up to 0.010 in. (0.254 mm). Do not remove too much material.

1. Clean the cylinder head(s) or block cover thoroughly with solvent and a parts washing brush. Carefully remove all gasket and sealant material from mating surfaces.

2. Remove all carbon and varnish deposits from the combustion chambers with a carbon removing solvent, such as Quicksilver Power Tune (part No. 92-15104). A sharpened hardwood dowel or plastic scraper can be used to remove stubborn deposits (**Figure 63**, typical). Do not scratch, nick or gouge the combustion chambers.

3. Check the cylinder head(s) and block cover for cracks, fractures, distortion or other damage. Check the cylinder head(s) for stripped or damaged threads. Refer to *Sealing Surfaces* at the beginning of this chapter and check the cylinder head(s) for warp. Remove minor imperfections by lapping the cylinder head as described in *Sealing Surfaces*.

4. Inspect all gasket surfaces or O-ring and water seal grooves for nicks, grooves, cracks, corrosion or distortion. Replace the cylinder head or block cover if a defect is severe enough to cause leakage.

5. Check all water, oil and fuel bleed passages in the head(s) for obstructions. Make sure all pipe plugs are installed tightly. Seal pipe plugs with Loctite 567 PST pipe sealant (part No. 92-80822).

Exhaust Cover/Manifold/Plate

1. Clean the exhaust cover, manifold and plate thoroughly with solvent and a parts washing brush. Carefully remove all gasket and sealant material from mating surfaces.

2. Remove all carbon and varnish deposits with a carbon removing solvent, such as Quicksilver Power Tune (part No. 92-15104). A hardwood dowel or plastic scraper can be used to remove stubborn deposits. Do not scratch, nick or gouge the mating surfaces.

3. Inspect the component and all gasket surfaces for nicks, grooves, cracks, corrosion or distortion. Replace the cover/manifold/plate if a defect is severe enough to cause leakage.

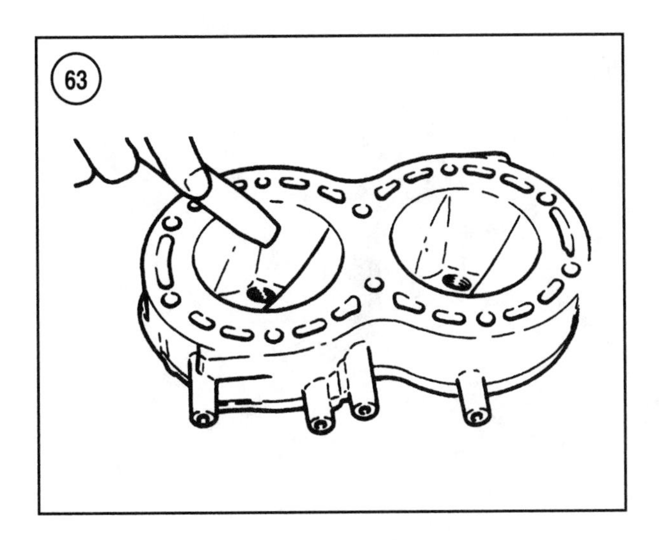

Intake Cover(s)

1. Clean the intake cover(s) thoroughly with clean solvent and a parts washing brush. Carefully remove all gasket and sealant material from mating surfaces.
2. Inspect the cover and all gasket surfaces or O-ring grooves for nicks, grooves, cracks, corrosion or distortion. Replace the cover if a defect could cause leakage.

End Caps

1. Clean the end cap(s) thoroughly with clean solvent and a parts washing brush. Carefully remove all sealant material from mating surfaces.
2. Inspect the seal bore(s) for nicks, gouges or corrosion that would cause the seal to leak around its outer diameter. Replace the end cap if the seal bore is damaged.
3. Inspect the end cap mating surface and O-ring groove for nicks, grooves, cracks, corrosion or distortion. Replace the end cap(s) if a defect could cause leakage.
4. If the end cap contains a bearing, inspect the bearing as described under *Crankshaft (6-60 hp models)* in the *Cleaning and Inspection* section in this chapter.

Thermostat and Poppet Valve Assembly (Models So Equipped)

The thermostat regulates the water leaving the power head. Replace the thermostat during major disassembly or repair. Correct thermostat operation is vital to engine break-in, spark plug life, smooth consistent idling, and maximum performance and durability.

The poppet valve assembly, if equipped, is controlled by water pressure in the block. At higher engine speeds, the water pump pressure forces the poppet valve open. When the poppet valve opens, an additional exit for heated cooling water is provided. This increased flow of heated water leaving the cylinder head in addition to the thermostat flow, allows additional cold inlet water to enter the block, lowering the operating water temperature.

The block stays warm enough at low speeds under thermostat control to maintain smooth idle and keep the plugs from fouling, but cools enough at high speeds to prevent preignition and detonation. Install a new poppet valve diaphragm, poppet and grommet seat during major disassembly or repair.

Refer to the appropriate power head disassembly procedure for illustrations of the thermostat and poppet valve assemblies specific to the engine model.

1. Carefully clean all gasket material from the thermostat housing, poppet valve inner and outer plates, and their mating surfaces. Most thermostats are sealed by a grommet around the thermostat and/or a gasket between the thermostat and the housing.
2. Check thermostat covers and poppet valve inner and outer plates for cracks, corrosion or distortion. Replace as necessary. All poppet valves use a similar construction.
3. If the thermostat must be reused, refer to *Engine Temperature* in Chapter Three for cleaning, inspection and testing procedures.
4. Inspect the poppet valve diaphragm for cracks, pin holes or deterioration. Replace the diaphragm if there is doubt about its condition.
5. Inspect the plastic poppet valve for melting and mechanical damage. Replace the poppet valve if there is doubt about its condition.

POWER HEAD ASSEMBLY

Before beginning assembly, complete all applicable sections of the *Cleaning and Inspection* in this chapter.

Review *Sealing Surfaces, Fasteners and Torque*, and *Sealants, Lubricants and Adhesives* at the beginning of this chapter.

Replace all seals, O-rings, gaskets, connecting rod screws, piston pin lockrings, piston rings and all bearings any time a power head is disassembled. If the original bearings must be reused, reinstall them in their original positions.

CAUTION

Install new connecting rod screws for final assembly.

Any identification mark on a piston ring must face up when installed. Some pistons use a combination of ring styles. Rings may be rectangular, semi-keystone or full keystone.

Rectangular and full keystone rings fit their grooves in either direction, but must be installed with their identification mark facing up.

Semi-keystone rings are beveled 7-10° on the upper surface only. These rings will not fit their groove correctly if installed upside down. Carefully examine the construction of the rings and look for identification marks before installation. The beveled side must face up, matching the ring groove. The beveled side is identified by the mark on the upper surface.

Lubricate the needle and roller bearings with Quicksilver Needle Bearing Assembly Grease (part No. 92-825265A1). This grease holds the needles, rollers and cages in position during assembly. Lubricate all other internal components with Quicksilver Two-Cycle TC-W3 outboard oil. Do not use any lubricant inside the power head that is not gasoline soluble.

A selection of torque wrenches is essential for correct assembly and to ensure maximum longevity of the power head assembly. Failing to torque items as specified may result in a premature power head failure.

All power head torque values are in **Table 1**. Standard torque values are in **Table 2**.

Mating surfaces must be absolutely free of gasket material, sealant residue, dirt, oil, grease or any other contaminant. Lacquer thinner, acetone, isopropyl alcohol and similar solvents are excellent oil, petroleum and wax-free solvents to use for the final preparation of mating surfaces.

All power head specifications are listed in **Tables 3-6**. All tables are located at the end of the chapter.

Piston Ring End Gap (All Models)

Check the piston ring end gap and adjust it if necessary before installing the piston rings onto the pistons.

Insufficient end gap causes the piston to stick in the cylinder bore when the engine is hot. There must be adequate end gap to allow for heat expansion.

Excessive end gap causes an excessive amount of combustion gases to leak past the gap between the ring ends. This causes a reduction in performance and can lead to excessive carbon buildup in the ring grooves and on the piston skirt.

Once the end gap has been set, tag the rings for correct installation in the bore in which they were checked. The 2.5 and 3.3 hp models have one piston ring. All other engines use two rings on each piston. Both rings must be checked and fitted.

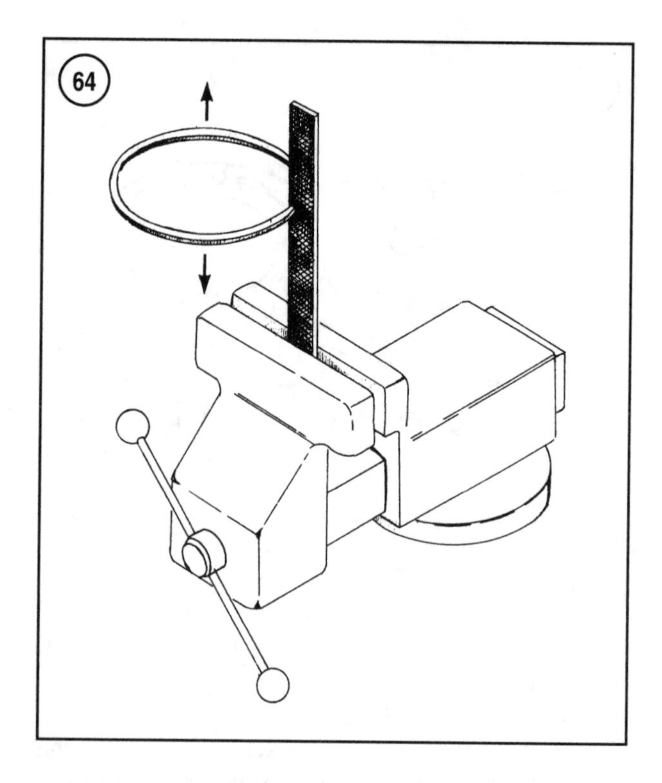

Excessive ring end gap can be caused by a worn or oversize bore. Recheck the cylinder bore as described in this chapter and/or make sure the correct piston rings are being used for the bore size.

On 2.5 and 3.3 hp models, insufficient ring end gap can be corrected by carefully filing the ring ends until the correct end gap (**Table 6**) is achieved. See **Figure 64**. Precision ring end gap grinders are available.

On 4-60 hp models, insufficient ring gap can only be corrected by trying a new ring in the suspect bore. The cylinder bore is undersize or the rings are oversize if insufficient ring gap occurs with multiple rings.

Check the piston ring end gap as follows. Refer to **Figure 65**.

1. Select a piston ring and place it into the cylinder No. 1 bore. Push the ring squarely into the bore using the piston. The ring must be square in the bore.
2. Measure the ring end gap with a feeler gauge as shown in **Figure 65**. If the ring gap is not within the specification in **Table 6**, repeat the measurement with the same ring in the cylinder No. 2 bore. Repeat the process until a bore is found that the ring fits correctly or there are no more cylinder bores to check.

3A. *Excessive end gap in all bores*—The ring is defective or the cylinder is oversize. Measure the cylinder bore and recheck the piston ring part number. If the bore is within specification and the correct ring is being used, the ring is defective and must be replaced with another new ring.

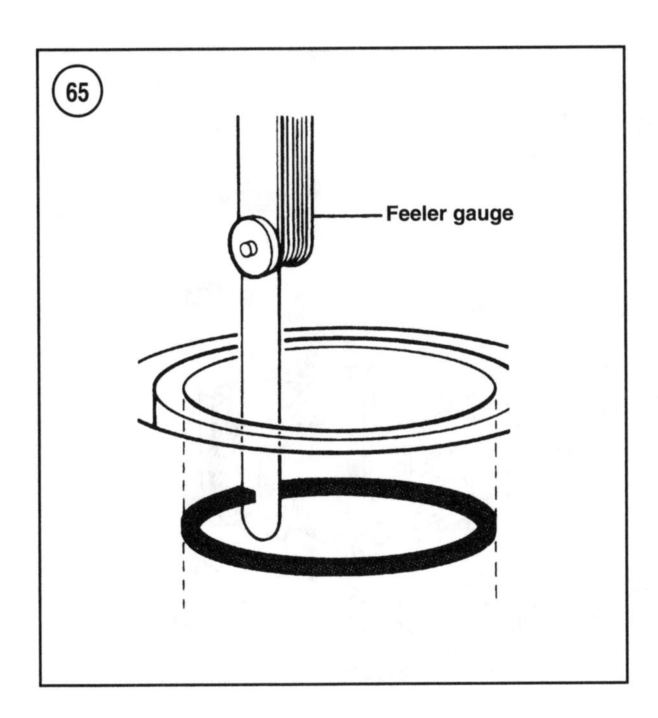

3B. *Insufficient end gap in all bores*—Measure the cylinder bore and recheck the piston ring part number.

a. *2.5 and 3.3 hp models*—If the bore is within specification and the correct ring is being used, carefully file the ends of the ring as shown in **Figure 64**. Keep the file at a 90° angle to the ring. Do not remove more material than necessary and do not create any burrs on the ring ends.

b. *4-60 hp models*—If the bore is within specification and the correct ring is being used, the ring is defective and must be replaced with another new ring.

4. Once a ring correctly fits in a cylinder or has been fitted to a cylinder, tag the ring with the cylinder number so it can be installed onto the correct piston during power head assembly.

5. Repeat this process until all piston rings are fitted to a specific cylinder bore and properly tagged for identification.

2.5 and 3.3 hp Models

Refer to **Figure 66** for this procedure.

1. Check the new piston ring's end gap as described in *Piston Ring End Gap* in this chapter.

2. If the crankshaft ball bearings were removed, install new bearings as follows:

a. Lubricate the new bearings and the crankshaft journals with outboard oil.

b. Slide the upper bearing over the flywheel end of the crankshaft. Position the locating pin in the bearing's outer race toward the bottom of the crankshaft.

c. Support the crankshaft under the upper counterweight in a press. Use a suitable socket or section of tubing to press against the inner race of the bearing until it is seated onto the crankshaft. See **Figure 67**.

d. Slide the lower bearing over the drive shaft end of the crankshaft. Position the locating pin in the bearing's outer race toward the bottom of the crankshaft.

e. Support the crankshaft under the lower counterweight in a press. Use a suitable socket or section of tubing to press against the inner race of the bearing until it is seated onto the crankshaft. See **Figure 68**.

3. Lubricate the piston pin caged needle bearing with needle bearing grease, then install the bearing into the connecting rod.

NOTE

Install the piston onto the connecting rod with the arrow on the piston crown facing downward toward the exhaust port.

4. Position the piston on the connecting rod with the arrow on the piston crown facing down. Lubricate the piston pin with outboard oil and push the pin through the piston, bearing and connecting rod.

5. Secure the piston pin with two new lockrings (13, **Figure 66**). Install the lockrings with needlenose pliers. See **Figure 69**. Make sure the lockrings are completely seated in their grooves at each end of the piston pin.

6. Use a ring expander (part No. 91-24697 or an equivalent) to install the single piston ring onto the piston. See **Figure 70**. Install the ring with its grooved ends facing up toward piston crown as shown in **Figure 71**. The ring end gap must straddle the piston ring locating pin inside the ring groove.

7. Use a feeler gauge to measure the clearance between the piston ring and the side of the ring groove. See **Figure 72**, typical. The ring side clearance must be within the specification in **Table 6**. If the clearance exceeds the specification, replace the piston.

8. Lubricate the seal lips of new upper and lower crankshaft seals with Quicksilver 2-4-C Multi-Lube. Install the upper and lower seals onto the crankshaft. The lips of both seals must face downward toward the gearcase when installed.

9. Install the crankshaft bearing half moon retainers (6, **Figure 66**) into the cylinder block grooves.

10. Lubricate the piston ring, piston and cylinder bore with outboard oil.

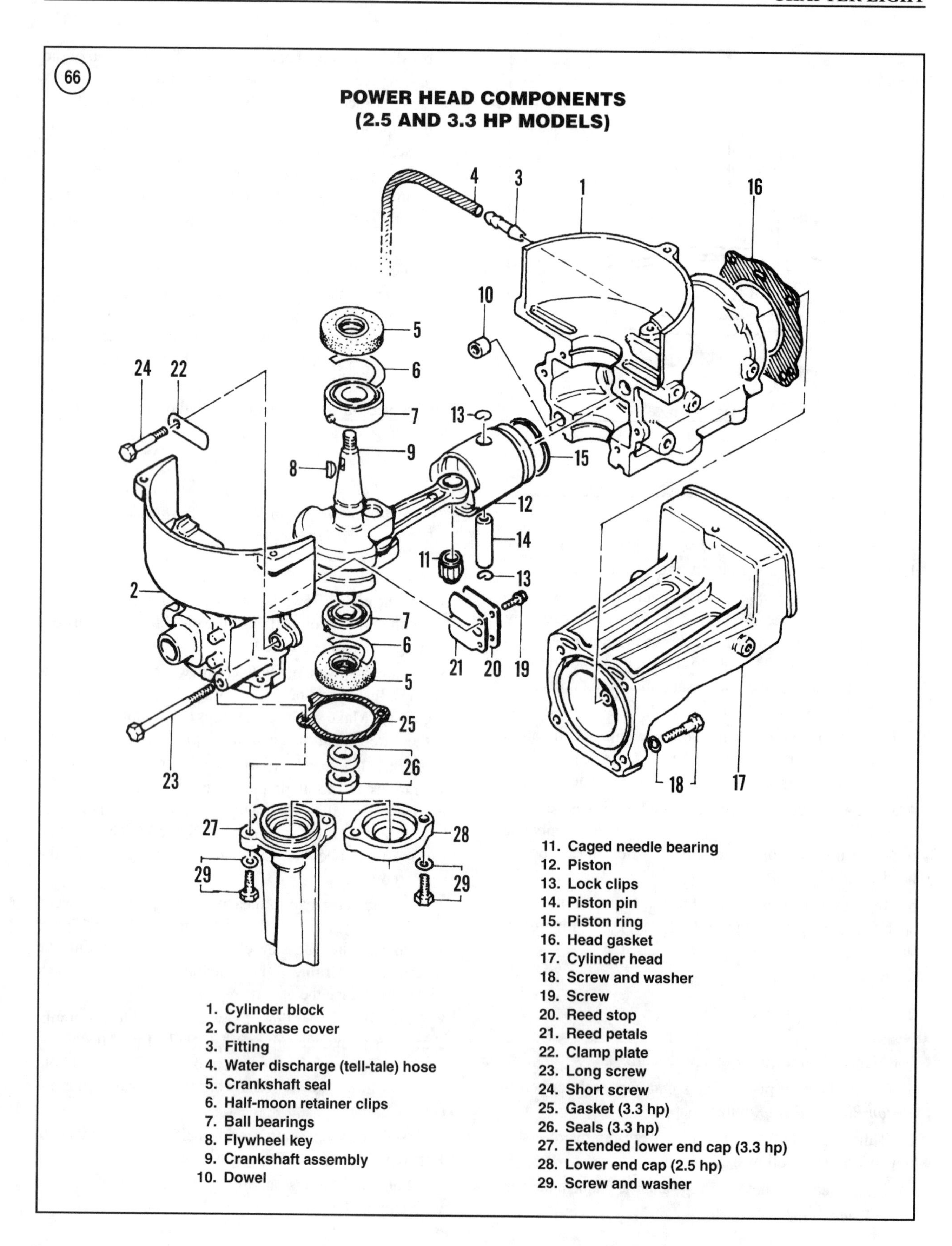
66
POWER HEAD COMPONENTS
(2.5 AND 3.3 HP MODELS)
1
2
3
4
5
6
7
8
9
10
11
12
13
14
15
16
17
18
19
20
21
22
23
24
25
26
27
28
29
1. Cylinder block
2. Crankcase cover
3. Fitting
4. Water discharge (tell-tale) hose
5. Crankshaft seal
6. Half-moon retainer clips
7. Ball bearings
8. Flywheel key
9. Crankshaft assembly
10. Dowel
11. Caged needle bearing
12. Piston
13. Lock clips
14. Piston pin
15. Piston ring
16. Head gasket
17. Cylinder head
18. Screw and washer
19. Screw
20. Reed stop
21. Reed petals
22. Clamp plate
23. Long screw
24. Short screw
25. Gasket (3.3 hp)
26. Seals (3.3 hp)
27. Extended lower end cap (3.3 hp)
28. Lower end cap (2.5 hp)
29. Screw and washer

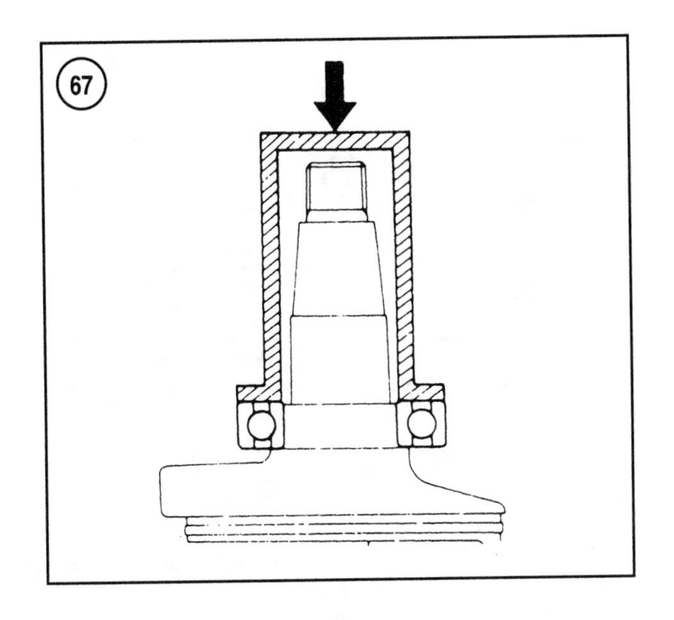
67

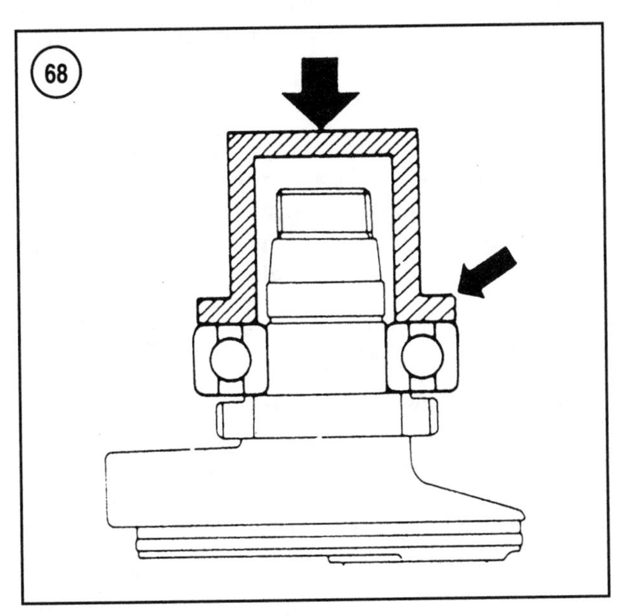
68

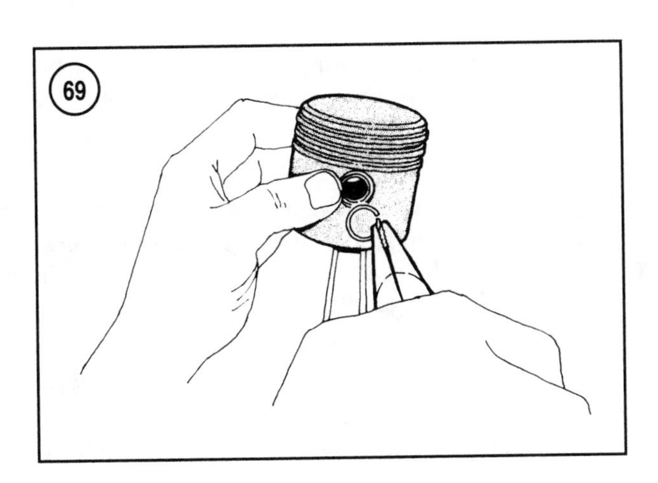
69

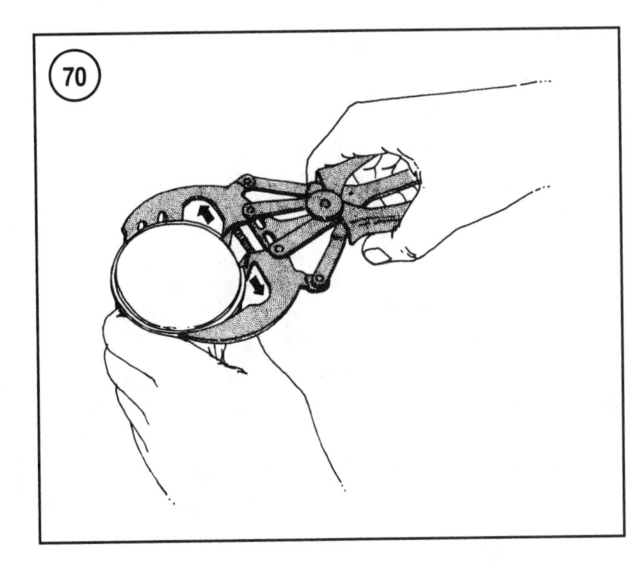
70

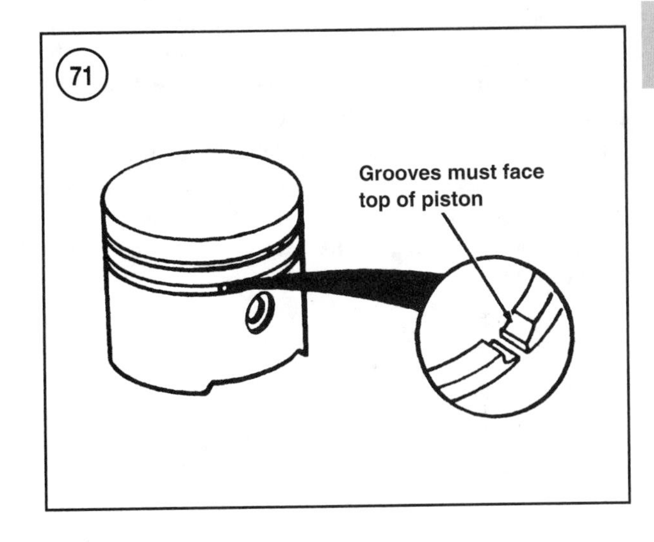
71
Grooves must face
top of piston

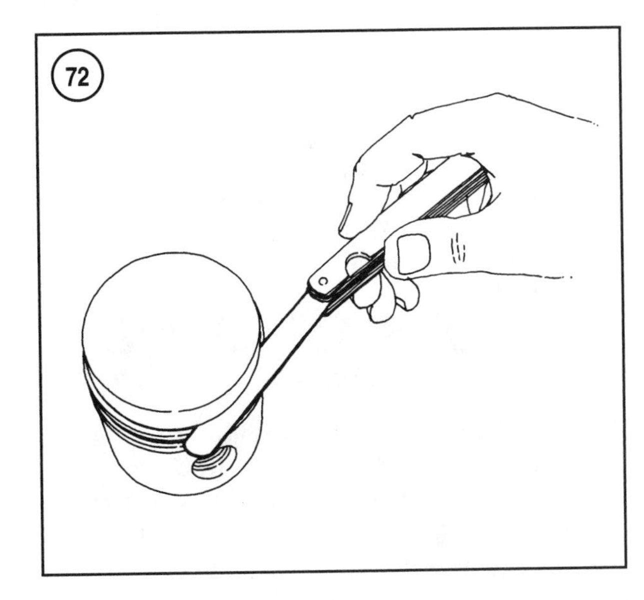
72

11. Make sure the ring can be rotated freely in its groove, then position the end gap of the piston ring to straddle the ring locating pin in its groove.

NOTE
A ring compressor is not required as the cylinder bore has a tapered entrance at the bottom of the bore.

12. Position the cylinder block so it is sitting on the cylinder head mating surface. Position the crankshaft and piston assembly over the cylinder block. Slowly lower the assembly toward the block while feeding the piston into the cylinder bore. Keep the crankshaft perfectly horizontal. Rock the piston slightly to help it enter the cylinder bore, making sure the piston ring does not rotate or catch while entering the bore.
13. Once the piston ring has entered the cylinder bore, seat the crankshaft ball bearings into the cylinder block. Rotate the crankshaft bearings as necessary to position the bearing locating pins in the cylinder block notches in the port side of the mating surface.
14. Insert a thin screwdriver blade into the exhaust port and carefully depress the piston ring (**Figure 73**). If the ring fails to spring back when released, it was probably broken during installation. Replace the ring if it is damaged or broken.
15. If the reed valve assembly was removed, install new reed valves into the crankcase cover as described in Chapter Six.
16. Use an oil and wax free solvent, such as acetone or lacquer thinner to clean the cylinder block and crankcase cover mating surfaces.
17. Install the two hollow dowel pins (10, **Figure 66**) into the cylinder block or crankcase cover, if they are not already installed.

CAUTION
Loctite Master Gasket Sealant (part No. 92-12564-2) is the only sealant recommended to seal the crankcase cover-to-cylinder block mating surfaces. The sealant comes in a kit that includes a special primer. Follow the instructions included in the kit for preparing the surfaces and applying the sealant. Apply the sealant bead to the inner, crankshaft side, of all crankcase cover screw and dowel holes.

18. Following the instructions supplied with the sealer, apply a continuous bead of Loctite Master Gasket Sealant to the mating surface of the cylinder block (**Figure 74**). Run the sealant bead along the inside of all bolt holes. The bead must be continuous.

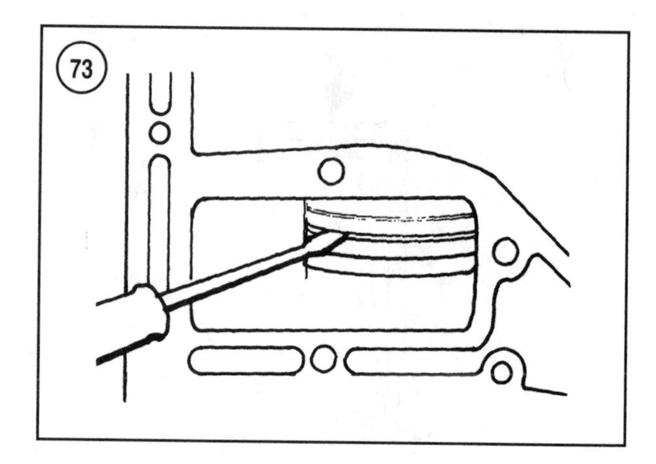

19. Install the crankcase cover into position on the cylinder block. Seat the cover to the block with hand pressure. Coat the threads of the crankcase cover screws with Loctite 242 threadlocking adhesive (part No. 92-809821), then install the cover screws. Tighten the cover screws evenly in a crossing pattern starting with the center screws and working outward to the specification in **Table 1**.

20A. *2.5 hp models*—Install the lower end cap (28, **Figure 66**) with the recessed side facing the power head. Coat the threads of the two screws with Loctite 242 threadlocking adhesive, then install and tighten the screws to the specification in **Table 1**.

20B. *3.3 hp models*—Reassemble and install the extended lower end cap as follows:

a. Coat the outer diameter of two new end cap seals (26, **Figure 66**) with Loctite 271 threadlocking adhesive. Press both seals into the end cap with a suitable mandrel. The lips of both seals must face downward towards the gearcase.
b. Lubricate the seal lips with Quicksilver 2-4-C Multi-Lube. Then position a new gasket over the end cap and align the screw holes.
c. Install the lower end cap to the power head. Coat the threads of the two screws with Loctite 242 threadlocking adhesive. Install and tighten the screws and washers to the specification in **Table 1**.

21. Rotate the crankshaft several revolutions to check for binding or unusual noise. If binding or noise occurs, disassemble the power head, and locate and correct the cause of the defect before proceeding.

22. Install the cylinder head using a new gasket. Coat the threads of the four screws with Quicksilver Perfect Seal, then install the screws and washers. Evenly tighten the screws in a crossing pattern to the specification in **Table 1**.

74

4 and 5 hp Models

Refer to **Figure 75** for this procedure.

1. Check the end gap of both new piston rings as described in *Piston Ring End Gap* in this chapter. The ring end gap must be within the specification in **Table 1**.
2. If the crankshaft ball bearings were removed, install new bearings as follows:
   a. Lubricate the new bearings and the crankshaft journals with outboard oil.
   b. Slide the upper bearing over the flywheel end of the crankshaft with the numbered side of the bearing facing up away from the closest counterweight.
   c. Support the crankshaft under the lower counterweight in a press. Use a suitable socket or section of tubing to press against the inner race of the bearing until it is seated on the crankshaft. See **Figure 67**.
   d. Slide the lower bearing over the drive shaft end of the crankshaft. Position the locating pin in the bearing's outer race toward the bottom of the crankshaft.
   e. Use a suitable socket or section of tubing to press against the inner race of the bearing until it is seated on the crankshaft. See **Figure 68**.
3. Lubricate the piston pin caged needle bearing with needle bearing grease, then install the bearing into the connecting rod.
4. Install the piston onto the connecting rod with the *UP* mark facing toward the flywheel.
5. Position the piston on the connecting rod with the *UP* mark facing up (**Figure 76** typical). Lubricate the piston pin with outboard oil and push the pin through the piston, bearing and connecting rod. See **Figure 77**.
6. Secure the piston pin with two new piston pin lockrings (12, **Figure 75**). Install the lockrings with needlenose pliers. See **Figure 69**. Make sure the lockrings are completely seated in their grooves at each end of the piston pin.
7. Use a ring expander (part No. 91-24697 or an equivalent) to install the piston rings. See **Figure 70**. Install the lower ring first, then the upper ring with the grooved ends facing up toward piston crown as shown in **Figure 71**. The ring end gap must straddle the piston ring locating pin inside the ring groove.
8. Use a feeler gauge to measure the clearance between the piston rings and the side of the ring grooves. See **Figure 72**, typical. The ring side clearances should be within the specifications in **Table 6**. If either of the clearances exceeds the specification, replace the piston.
9. Lubricate the piston rings, piston and cylinder bore with outboard oil.
10. Make sure each ring can be rotated freely in its groove, then position the end gap of each piston ring so it straddles the locating pin in its groove.

NOTE

A ring compressor is not required as the cylinder bore has a tapered entrance at the bottom of the bore.

11. Position the cylinder block so it is sitting on the cylinder head mating surface. Position the crankshaft and piston assembly over the cylinder block. Slowly lower the assembly toward the block while feeding the piston into the cylinder bore. The crankshaft must be kept perfectly horizontal. Rock the piston slightly to help it enter the cylinder bore. Make sure the piston rings do not rotate or catch while entering the bore.
12. Once the piston rings enter the cylinder bore, seat the crankshaft ball bearings into the cylinder block. Rotate the crankshaft bearings as necessary to position the bearing locating pins in the cylinder block notches in the port side of the mating surface.
13. Insert a thin screwdriver blade into the exhaust port and carefully depress each piston ring (**Figure 73**). If either ring fails to spring back when released, it was probably broken during installation. Replace broken or damaged rings.
14. Install the crankshaft bearing thrust washers (5, **Figure 75**) over each end of the crankshaft. Seat the larger diameter half of each washer into the cylinder block grooves.
15. Lubricate the seal lips of new upper and lower crankshaft seals with Quicksilver 2-4-C Multi-Lube. Install the upper and lower seals onto the crankshaft. Install the lip of the upper seal so it faces downward toward the gearcase and install the lip of the lower seal so it faces up toward the flywheel.
16. If the reed valve assembly was removed, install new reed valves into the crankcase cover as described in Chapter Six.

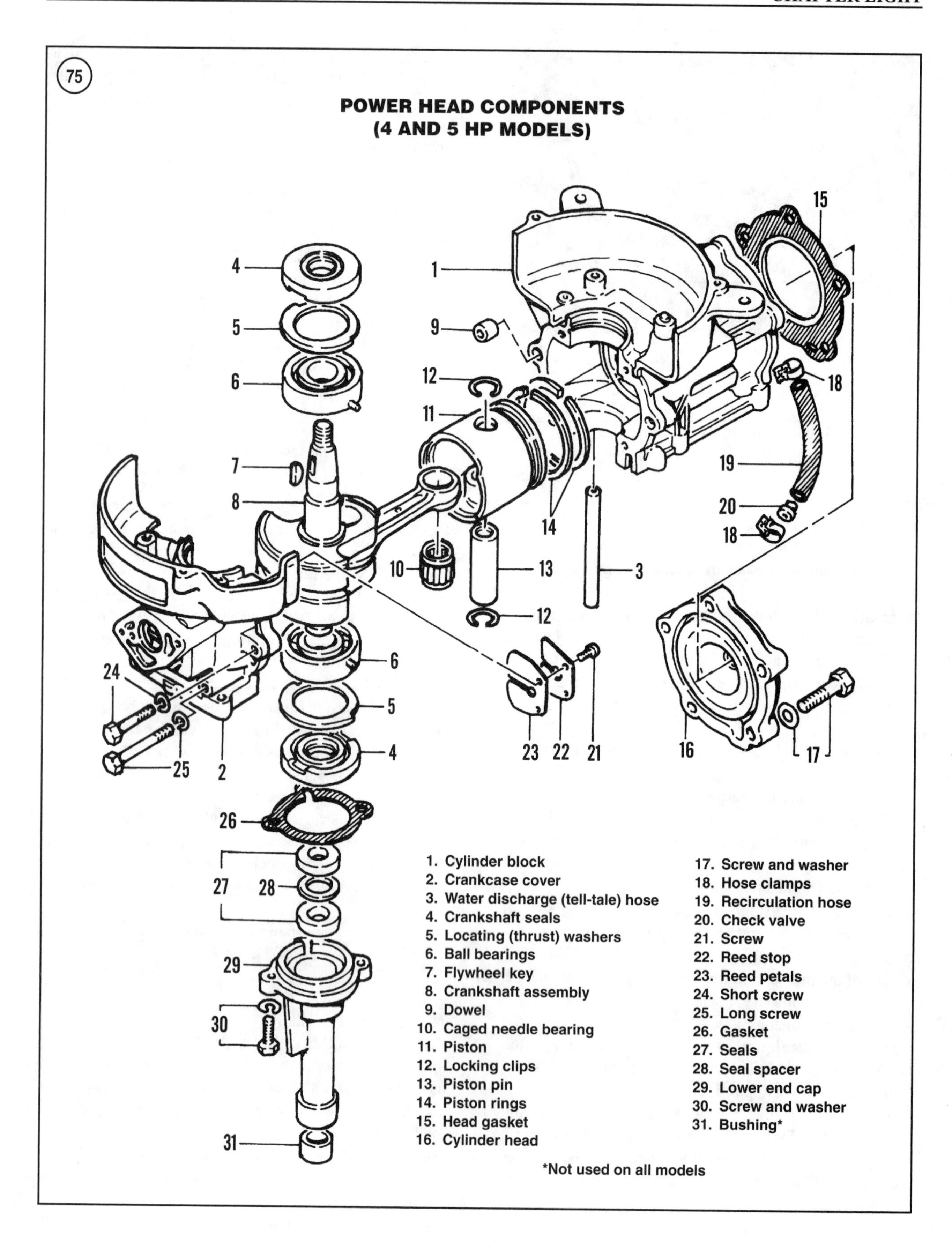
75
POWER HEAD COMPONENTS
(4 AND 5 HP MODELS)
1
2
3
4
5
6
7
8
9
10
11
12
13
14
15
16
17
18
19
20
21
22
23
24
25
26
27
28
29
30
31
1. Cylinder block
2. Crankcase cover
3. Water discharge (tell-tale) hose
4. Crankshaft seals
5. Locating (thrust) washers
6. Ball bearings
7. Flywheel key
8. Crankshaft assembly
9. Dowel
10. Caged needle bearing
11. Piston
12. Locking clips
13. Piston pin
14. Piston rings
15. Head gasket
16. Cylinder head
17. Screw and washer
18. Hose clamps
19. Recirculation hose
20. Check valve
21. Screw
22. Reed stop
23. Reed petals
24. Short screw
25. Long screw
26. Gasket
27. Seals
28. Seal spacer
29. Lower end cap
30. Screw and washer
31. Bushing*
*Not used on all models

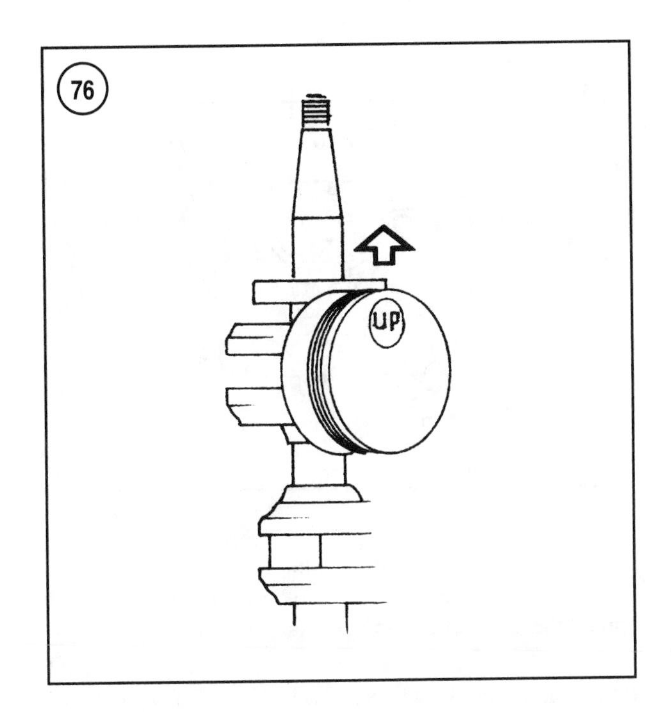

17. Use an oil and wax free solvent, such as acetone or lacquer thinner, to clean the cylinder block and crankcase cover mating surfaces.

18. Install the two hollow dowel pins (9, **Figure 75**) into the cylinder block or crankcase cover if they were not already installed.

CAUTION

Loctite Master Gasket Sealant (part No. 92-12564-2) is the only sealant recommended to seal the crankcase cover-to-cylinder block mating surfaces. The sealant comes in a kit that includes a special primer. Follow the instructions included in the kit for preparing the surfaces and applying the sealant.

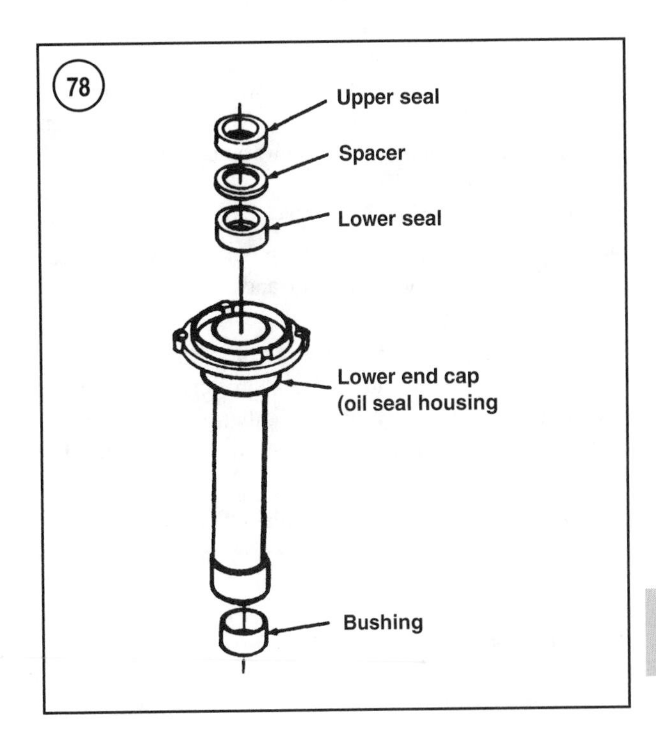

19. Apply a continuous bead of Loctite Master Gasket Sealant to the mating surface of the cylinder block (**Figure 74**). Run the sealant bead along the inside of all bolt holes. The bead must be continuous.

20. Install the crankcase cover onto the cylinder block. Seat the cover to the block with hand pressure. Coat the threads of the crankcase cover screws with Loctite 242 threadlocking adhesive (part No. 92-809821), then install the cover screws. Tighten the cover screws evenly and in a crossing pattern starting with the center screws and working outward to the specification in **Table 1**.

21. Refer to **Figure 78** and reassemble and install the extended lower end cap as follows:

   a. Lubricate a new bushing with Quicksilver 2-4-C Multi-Lube. Press the bushing into the lower end of the end cap with a suitable mandrel.

   b. Coat the outer diameter of the two new end cap seals with Loctite 271 threadlocking adhesive (part No. 92-809819). Press the lower seal into the end cap with a suitable mandrel. The seal lip must face downward toward the gearcase.

   c. Install the seal spacer onto the top of the lower seal, then press the upper seal into the end cap with a suitable mandrel. The seal lip must face downward toward the gearcase.

   d. Lubricate the seal lips with Quicksilver 2-4-C Multi-Lube. Then position a new gasket over the end cap and align the screw holes.

8

e. Install the lower end cap onto the power head. Coat the threads of the two screws with Loctite 242 threadlocking adhesive (part No. 92-809821), then install and tighten the screws and washers to the specification in **Table 1**.

22. Rotate the crankshaft several revolutions to check for binding or unusual noise. If binding or noise occurs, disassemble the power head, and locate and correct the cause of the defect before proceeding.

23. Install the cylinder head using a new gasket. Coat the threads of the screws with Quicksilver Perfect Seal, then install the screws and washers. Evenly tighten the screws to the specification in **Table 1** in the pattern shown in **Figure 79**.

24. Install the fuel bleed (recirculation) hose and check valve. See Chapter Six.

6-15 hp Models

Refer to **Figure 80** for the following procedures.

Crankshaft and pistons

1. Check the end gap of the new piston rings for each cylinder as described under *Piston Ring End Gap* in this chapter. The end gap must be within the specification in **Table 6**.

2. If the crankshaft ball bearing (22, **Figure 80**) was removed, install a new bearing as follows:
   a. Lubricate the new bearing and the crankshaft journal with outboard oil.
   b. Slide the bearing over the drive shaft end of the crankshaft with the open side of the bearing cage facing the crankshaft.
   c. Support the crankshaft under the lower counterweight in a press. Press against the inner race of the bearing using the power head (crankshaft stand) and its sleeve (part No. 91-13662A-1) until the bearing seats on the crankshaft. See **Figure 68**.

3. Install a new wear sleeve and O-ring onto the lower end of the crankshaft as follows:
   a. Grease the O-ring and the inner diameter of the wear sleeve with Quicksilver 2-4-C Multi-Lube.
   b. Position the O-ring against the lower end of the crankshaft, then pilot the wear sleeve over the O-ring and crankshaft.
   c. Press the wear sleeve into position using the power head stand *without* its sleeve. Press it in until the tool bottoms on the crankshaft.

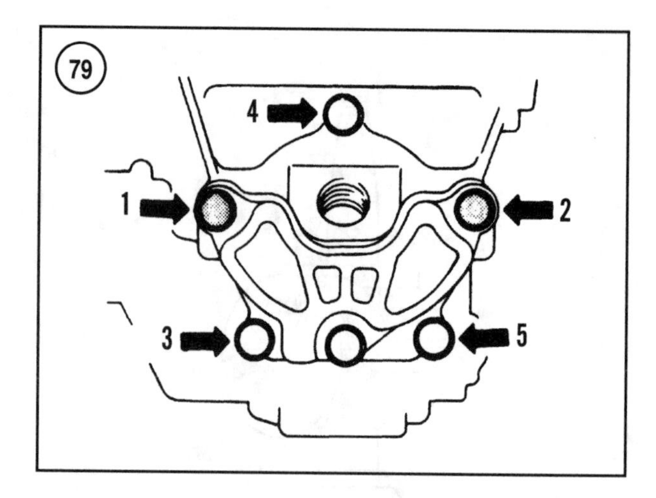

4. Install a new crankshaft sealing ring in the groove just above the crankshaft center main bearing journal. Do not spread the seal more than necessary to install.

5. Mount the crankshaft vertically on the power head (crankshaft) stand (part No. 91-13662A-1 or an equivalent).

CAUTION
If reusing the original bearings, install them in their original locations.

6. Assemble the connecting rods to the pistons as follows:
   a. Grease the sleeve portion of the piston pin installation tool (part No. 91-13663A-1) with needle bearing assembly grease.
   b. Position the cylinder No. 1 connecting rod so the beveled edge of the fractured cap mating surface is facing upward. When the cap is installed, this edge will form a V-notch with the beveled edge of the cap.
   c. Hold the lower thrust washer (18, **Figure 80**) under the connecting rod small end, then insert the greased sleeve into the small end bore.
   d. Lubricate the 24 needles with needle bearing assembly grease and insert them into the small end around the sleeve as shown in **Figure 81**.
   e. Position the upper thrust washer (18, **Figure 80**) on top of the needles. Carefully slide the cylinder No. 1 piston over the rod with the steep intake deflector facing the rod's beveled edge and align the piston pin bores. Then insert the main body of piston pin tool (part No. 91-13663A-1) into the piston pin bore and through the connecting rod, pushing the sleeve out the other side of the piston pin bore. Remove the sleeve.

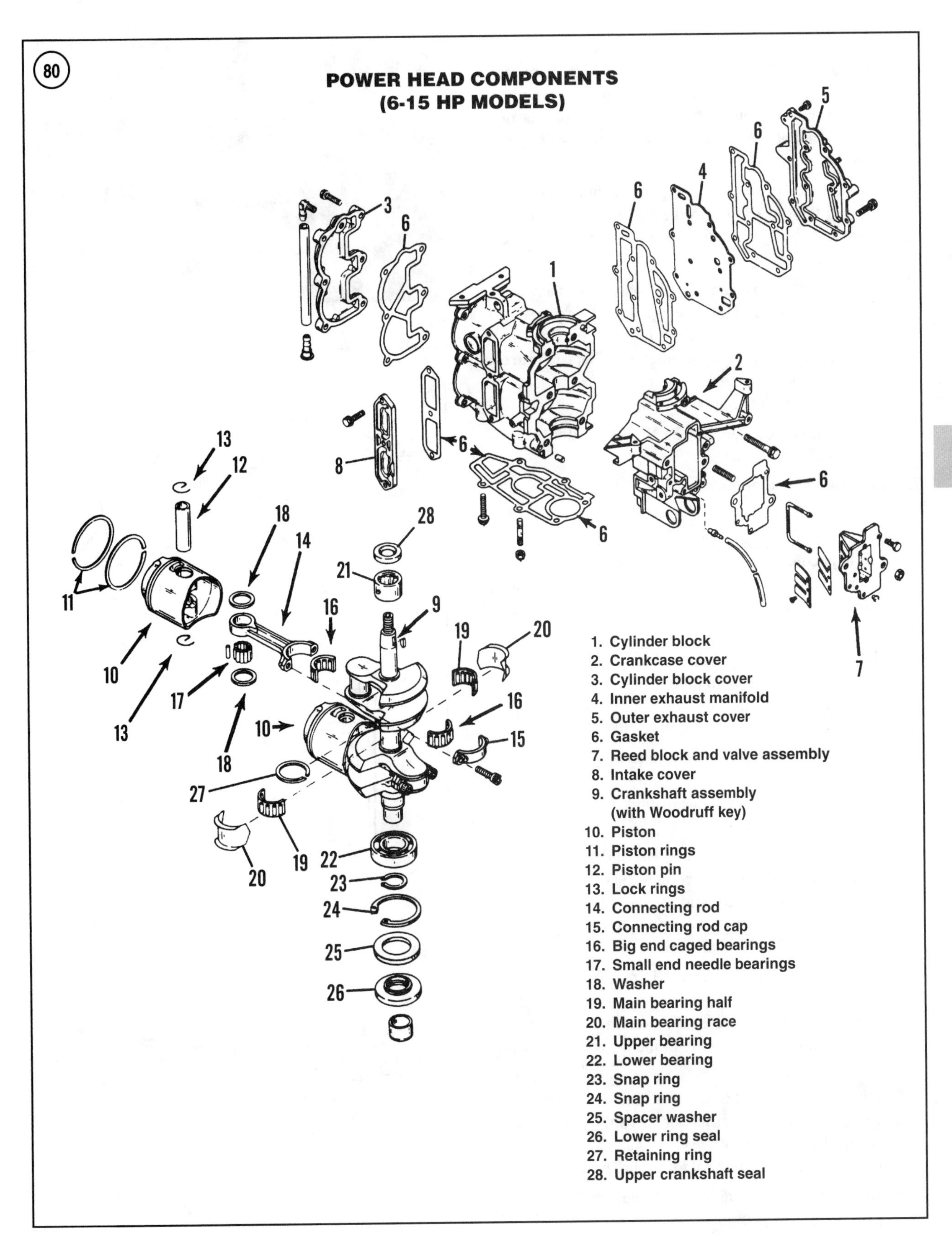
80
POWER HEAD COMPONENTS
(6-15 HP MODELS)
1. Cylinder block
2. Crankcase cover
3. Cylinder block cover
4. Inner exhaust manifold
5. Outer exhaust cover
6. Gasket
7. Reed block and valve assembly
8. Intake cover
9. Crankshaft assembly (with Woodruff key)
10. Piston
11. Piston rings
12. Piston pin
13. Lock rings
14. Connecting rod
15. Connecting rod cap
16. Big end caged bearings
17. Small end needle bearings
18. Washer
19. Main bearing half
20. Main bearing race
21. Upper bearing
22. Lower bearing
23. Snap ring
24. Snap ring
25. Spacer washer
26. Lower ring seal
27. Retaining ring
28. Upper crankshaft seal

f. Lubricate the piston pin with outboard oil and pilot it into the open end of the piston pin bore. Support the piston and tool with one hand and drive the piston pin into the piston with a soft-faced rubber or plastic mallet. Allow the pin tool to exit as the piston pin is driven in.

g. Remove the piston pin tool from the bottom of the piston, then insert it into the top of the piston pin bore and gently tap it until the pin is centered in the pin bore.

h. Make sure all needles and both locating washers are properly positioned, then secure the piston pin with two new piston pin lockrings (13, **Figure 80**). Install the lockrings with needlenose pliers (**Figure 69**). Make sure the lockrings are completely seated in their grooves at each end of the piston pin.

i. Repeat this procedure for the No. 2 cylinder piston and connecting rod.

7. Install the rod and piston assemblies on the crankshaft as follows:

a. Grease the connecting rod big end bearing halves with needle bearing assembly grease and install them to the crankshaft journals. If the original bearings are reused, install them in their original locations.

b. Position the No. 1 cylinder rod and piston assembly over the No. 1 crankpin journal and bearing halves. When looking at the crown of the piston, the steep intake deflector must be on the right.

c. Install the connecting rod cap in its original orientation. Carefully observe fracture and alignment marks to ensure correct installation (**Figure 82**, typical).

d. Lubricate the screw threads and underside of the screw heads of *new* connecting rod screws with outboard lubricant. Then hold the cap firmly in position and install the connecting rod screws and thread them fully into the rod.

e. Tighten each screw to 15 in.-lb. (1.7 N•m). Run a fingernail or pencil lead over each edge of the rod-to-cap joint (**Figure 83**). No ridge should be seen or felt. Realign and retorque the cap as necessary.

f. Once the alignment is verified, finish torquing the rod screws to the specification in **Table 1** in a minimum of two progressive steps. Make a final check of alignment after the final torque has been applied.

g. Repeat this procedure to install the No. 2 cylinder piston and rod assembly.

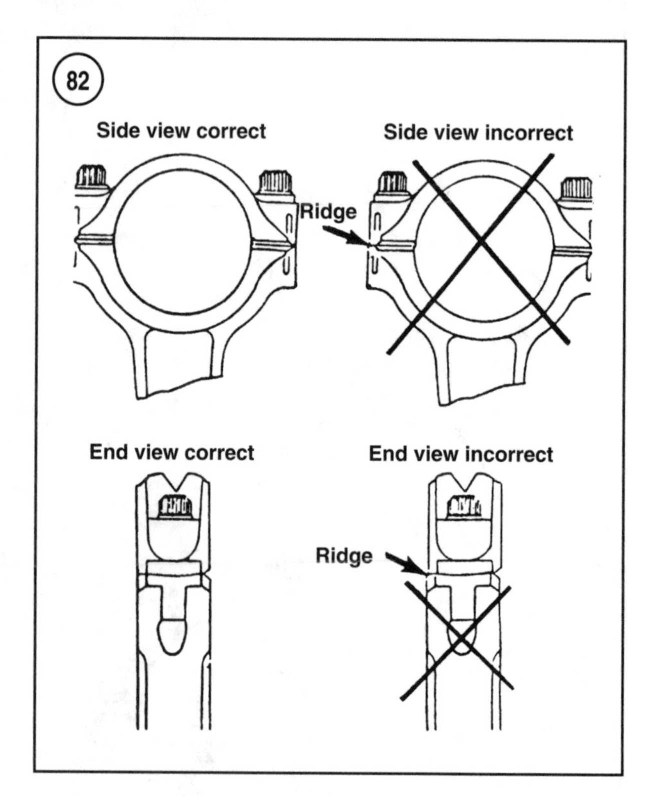

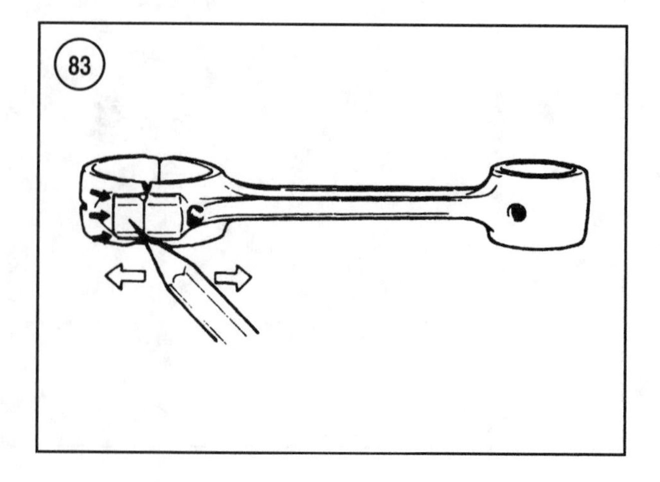

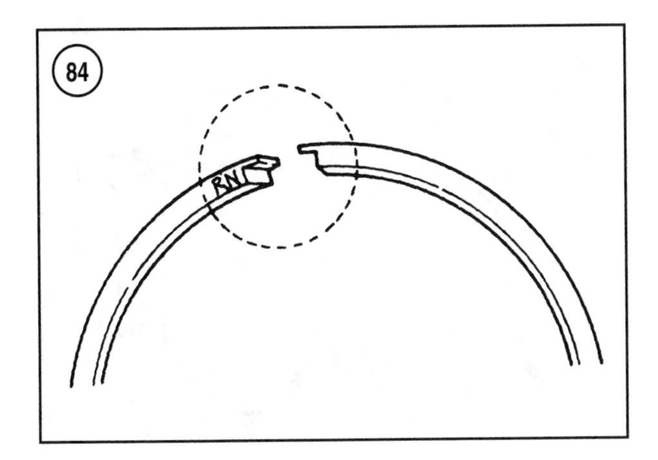

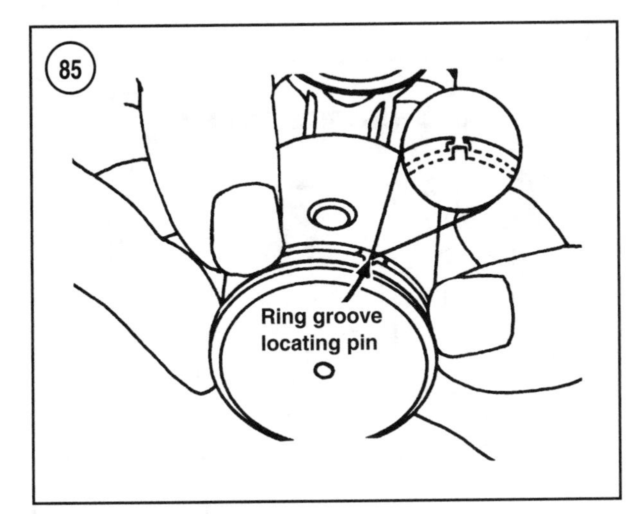

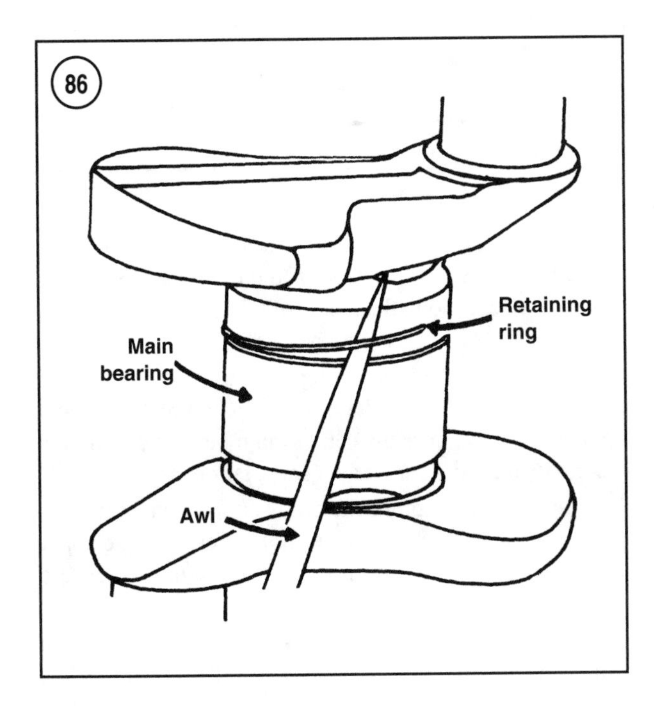

CAUTION
Install the piston rings onto the pistons that match the cylinder bore for which the rings were fitted.

8. Install the piston rings onto each piston using a ring expander (part No. 91-24697 or an equivalent). See **Figure 70**. Install the bottom, rectangular ring first, then the top, semi-keystone ring. Expand each one just enough to slip over the piston. The beveled edge of the top, semi-keystone ring must face upward. If the ring has an identification mark, position the ring so the mark (**Figure 84** typical) faces up.

9. Make sure each ring can be rotated freely in its groove, then position the end gap of each piston ring to straddle the locating pin in its groove. See **Figure 85**.

10. Apply a thick coat of needle bearing assembly grease to the crankshaft center main bearing journal. Position the caged roller bearing halves (19, **Figure 80**) on the journal.

11A. *6 and 8 hp models*—Position one of the split bearing liner halves in the cylinder block. Make sure the liner is protruding above the mating surface equally on both sides.

11B. *9.9 and 15 hp models*—Position the outer race halves over the bearing halves. The retaining ring groove must face up toward the flywheel. Carefully align the fracture lines, then install the retainer ring. Position the retainer ring (**Figure 86**) to cover as much of both fracture lines as possible.

12. Lubricate the upper main bearing with outboard oil and install it onto the crankshaft. The bearing locating pin must face downward toward the drive shaft.

13. Lubricate the lip of a new upper seal with Quicksilver 2-4-C Multi-Lube, then slide it over the flywheel end of the crankshaft. The seal lip must face downward toward the drive shaft.

NOTE
Lift the crankshaft from the power head stand if necessary, to install the following components.

14. Install the snap ring (23, **Figure 80**) over the drive shaft end of the crankshaft with the beveled edge facing away from the ball bearing. Then install the spacer washer (25, **Figure 80**) against the retaining ring.

15. Lubricate the lips of a new lower seal (26, **Figure 80**) with Quicksilver 2-4-C Multi-Lube, then slide it over the crankshaft and against the spacer washer. The protruding lip of the seal must face up toward the flywheel.

Power head assembly

1. Lubricate the piston rings, pistons and cylinder bores with outboard oil.
2. Make sure each piston ring end gap is still straddling the locating pin in its ring groove (**Figure 85**) and the center main bearing locating pin is installed in the cylinder block.

NOTE
A ring compressor is not required as the cylinder bore has a tapered entrance at the bottom of the bore.

3. Position the cylinder block so it is sitting on the cylinder block cover mating surface. Position the crankshaft assembly over the cylinder block. Slowly lower the assembly toward the block while feeding the pistons into each cylinder bore. Keep the crankshaft perfectly horizontal. Rock the pistons slightly to help each enter its cylinder bore. Make sure the piston rings do not rotate or catch while entering the bore.
4. Once the piston rings have entered the cylinder bore, seat the crankshaft assembly in the cylinder block. Rotate the crankshaft upper bearing as necessary to position the locating pin in the cylinder block notch in the port side of the mating surface. Carefully push inward on the crankshaft seals to ensure they are properly seated.

5A. *6 and 8 hp models*—Install the second half of the center main bearing liner over the center main bearing journal. Match the ends of the liner with the liner already installed in the block.

5B. *9.9 and 15 hp models*—Rotate the center main bearing assembly as necessary to align the hole in the bearing race with the bearing locating pin.

6. Insert a thin screwdriver blade into both cylinder's exhaust ports and carefully depress each piston ring (**Figure 73**). If a ring fails to spring back when released, it was probably broken during installation. Replace broken or damaged rings.
7. Use an oil and wax free solvent, such as acetone or lacquer thinner, to clean the cylinder block and crankcase cover mating surfaces.
8. Install the two dowel pins into the cylinder block or crankcase cover if they are not already installed.

CAUTION
Use Loctite Master Gasket Sealant (part No. 92-12564-2) to seal the crankcase cover-to-cylinder block mating surfaces. This sealant comes in a kit that includes a special primer. Follow the instruction included in the kit for preparing the surfaces and applying the sealant.

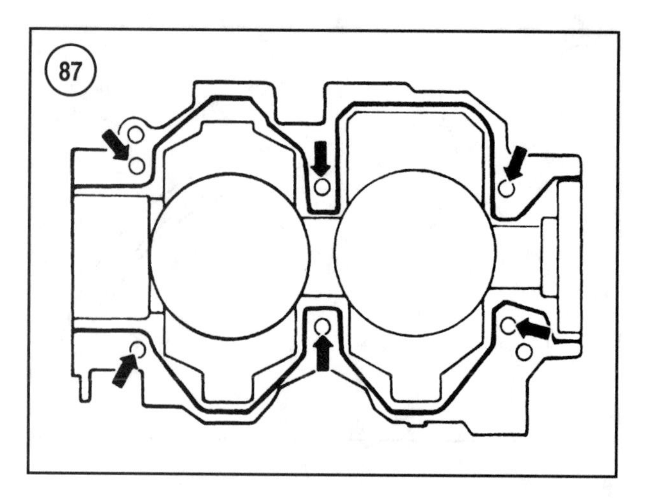

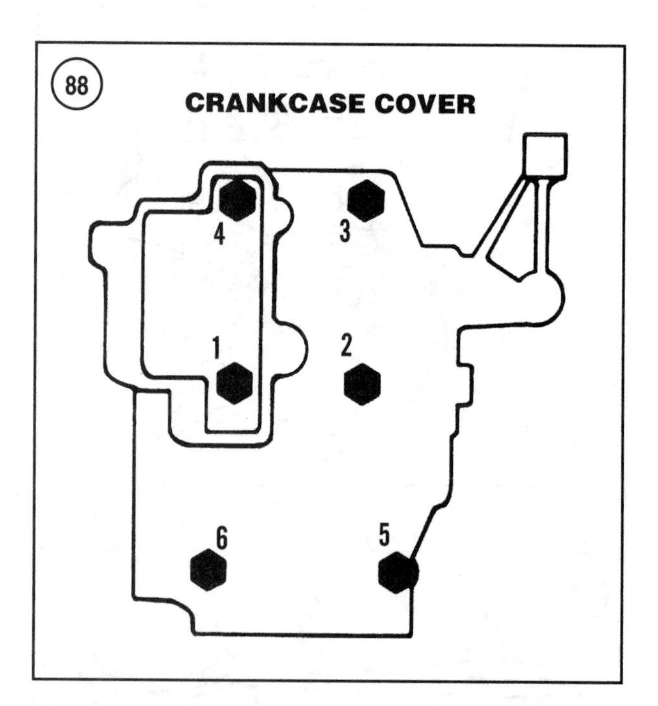

9. Apply a continuous bead of Loctite Master Gasket Sealant to the mating surface of the cylinder block. Run the sealant bead along the inside of all bolt holes as shown in **Figure 87**. The bead must be continuous.
10. Install the crankcase cover onto the cylinder block. Seat the cover to the block with hand pressure. On 6 and 8 hp models, make sure the center main bearing split liners are not displaced during cover installation.
11. Tap the lower end of the crankshaft with a soft hammer to seat the ball bearing in its bore, then push the upper and lower seals firmly into their bores until they are seated.
12. Coat the threads of the crankcase cover screws with Loctite 242 threadlocking adhesive (part No. 92-809821),

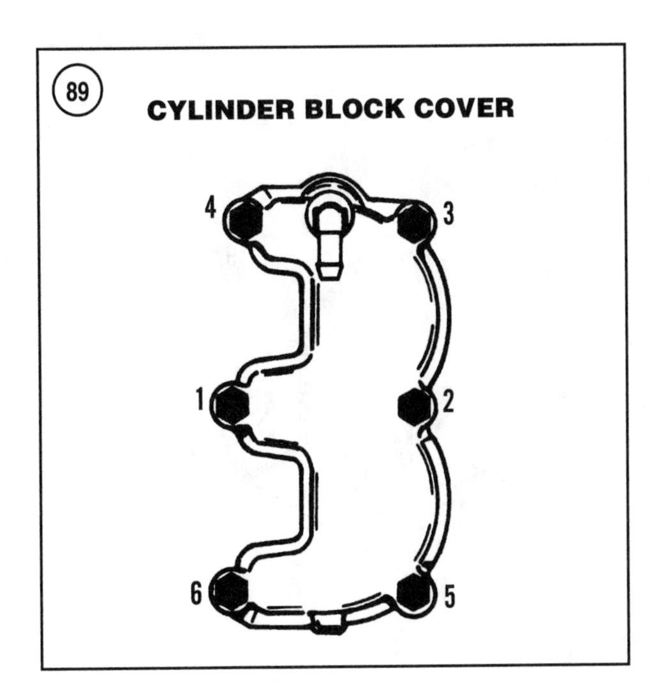

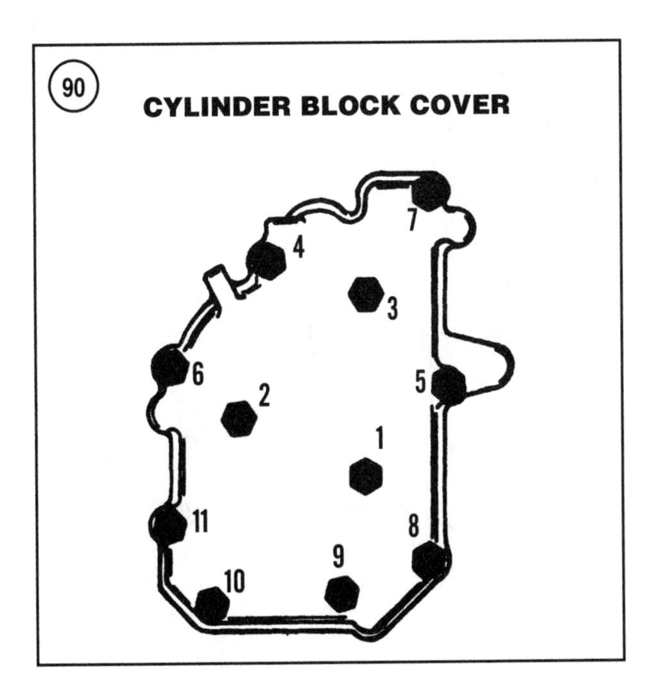

then install the cover screws. Tighten the cover screws evenly to the specification in **Table 1** in the pattern shown in **Figure 88**.

13. Rotate the crankshaft several revolutions to check for binding or unusual noise. If binding or noise occurs, disassemble the power head, and locate and correct the cause of the defect before proceeding.

14. Install the intake cover (8, **Figure 80**) with a new gasket or two O-rings. Coat the gasket or O-rings and the threads of the screws with Quicksilver Perfect Seal. Install and tighten the mounting screws to the specification in **Table 1**.

15. Install the cylinder block cover (3, **Figure 80**) with a new gasket (6). Tighten the mounting screws evenly to the specification in **Table 1** in the pattern shown in **Figure 89**.

16. Install a new grommet around the thermostat, if equipped. Then install the thermostat with the sensing pellet facing the power head. Install the thermostat cover and a new gasket. Coat the screw threads with Quicksilver Perfect Seal. Tighten the screws evenly to the specification in **Table 1**.

17. Assemble the exhaust plate and cover (4 and 5, **Figure 80**) with two new gaskets. Sandwich the exhaust plate between the gaskets, then position the cover over the outer gasket and plate. Position the assembly on the power head and secure it. Coat the screw threads with Loctite 242 threadlocking adhesive and tighten them to the specification in **Table 1** in the pattern shown in **Figure 90**.

18. Install the reed block, intake manifold and fuel bleed (recirculation) system as described in Chapter Six.

20-25 hp and 20 Jet Models

Refer to **Figure 91** for the following procedures.

Crankshaft and pistons

1. Check the end gap of the new piston rings for each cylinder as described under *Piston Ring End Gap* in this chapter. The ring gap must be within the specification in **Table 6**.

CAUTION
The wear sleeve is made of very thin material and is easy to crush or distort during installation.

2. Install a new wear sleeve and O-ring onto the lower end of the crankshaft as follows:
   a. Coat the wear sleeve contact area of the crankshaft with Loctite 271 threadlocking adhesive (part No. 92-809819).
   b. Position the O-ring against the lower end of the crankshaft, then guide the wear sleeve over the O-ring and crankshaft.
   c. Using a suitable block of wood and a hammer, drive the wear sleeve onto the crankshaft until the sleeve seats on the crankshaft shoulder. Make sure the sleeve is driven straight onto the crankshaft.
   d. Grease the O-ring with Quicksilver 2-4-C Multi-Lube.

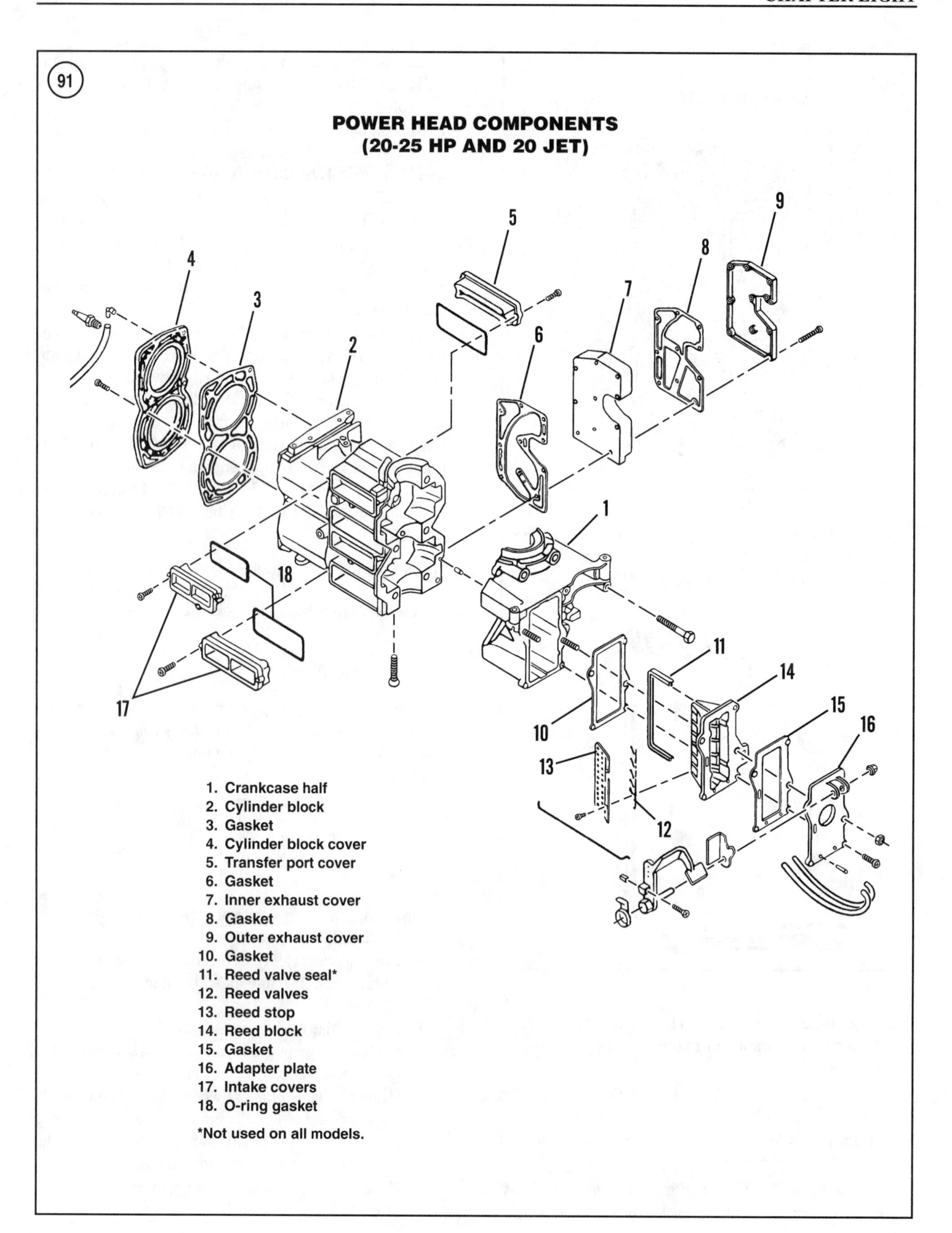
91
POWER HEAD COMPONENTS
(20-25 HP AND 20 JET)
1
2
3
4
5
6
7
8
9
10
11
12
13
14
15
16
17
18
1. Crankcase half
2. Cylinder block
3. Gasket
4. Cylinder block cover
5. Transfer port cover
6. Gasket
7. Inner exhaust cover
8. Gasket
9. Outer exhaust cover
10. Gasket
11. Reed valve seal*
12. Reed valves
13. Reed stop
14. Reed block
15. Gasket
16. Adapter plate
17. Intake covers
18. O-ring gasket
*Not used on all models.

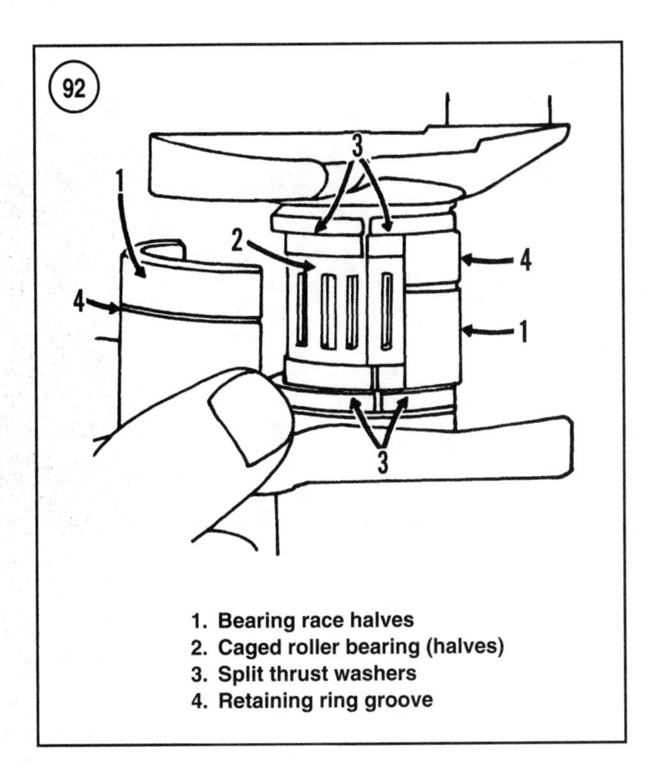

1. **Bearing race halves**
2. **Caged roller bearing (halves)**
3. **Split thrust washers**
4. **Retaining ring groove**

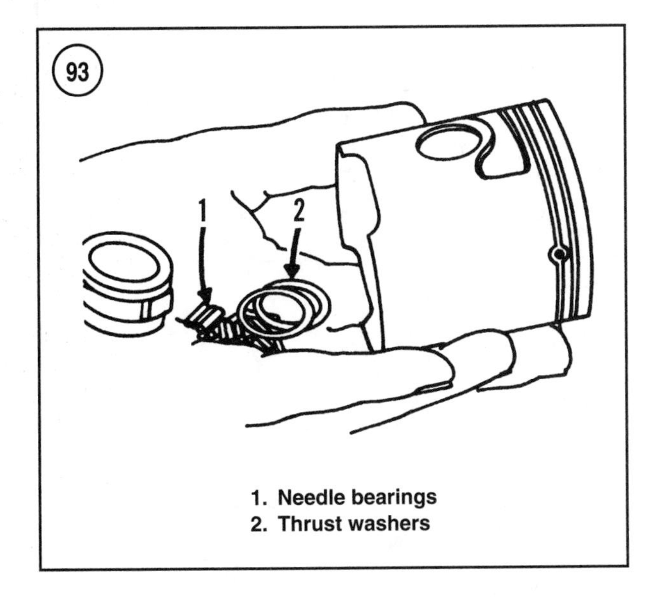

1. **Needle bearings**
2. **Thrust washers**

3. Mount the crankshaft vertically on a power head (crankshaft) stand (part No. 91-25821A-1 or an equivalent). Clamp the power head stand securely in a vise.

CAUTION
If reusing the original bearings, install them in their original locations.

4. Refer to **Figure 92** and assemble the center main bearing as follows:

a. Apply a thick coat of needle bearing assembly grease (part No. 92-825265A 1) to the crankshaft center main bearing journal.

b. Position the split thrust washers (3, **Figure 92**) and caged roller bearing halves (2) on the journal.

c. Position the outer race halves over the bearing halves. Position the retaining ring groove up toward the flywheel. Carefully align the fracture lines, then install the retainer ring. Position the retainer ring to cover as much of both fracture lines as possible.

d. Push the bearing assembly toward one end of the crankshaft. Check the clearance between the crankshaft shoulder and thrust washers with a feeler gauge. If the clearance exceeds 0.030 in. (0.76 mm), install new thrust washers. The recommended bearing end play is 0.004-0.025 in. (0.10-0.64 mm).

5. Assemble the connecting rods to the pistons as follows:

a. Grease the sleeve portion of piston pin installation tool (part No. 91-76160A-2) with needle bearing assembly grease.

b. Position the No. 1 cylinder connecting rod so the beveled edge of the fractured cap mating surface is facing upward. When the cap is installed, this edge will form a V-notch with the beveled edge of the cap.

c. Hold the lower thrust washer (2, **Figure 93**) under the connecting rod small end, then insert the greased sleeve into the small end bore.

d. Lubricate the 27 needles (1, **Figure 93**) with needle bearing assembly grease and insert them into the small end around the sleeve as shown in **Figure 81**.

e. Position the upper thrust washer (2, **Figure 93**) on top of the needles. Carefully slide the cylinder No. 1 piston over the rod with the cast or stamped *UP* mark (inside the skirt or on bottom of crown) facing up and align the piston pin bores. Then insert the main body of piston pin tool (part No. 91-76160A-2) into the piston pin bore and through the connecting rod. Push the sleeve out the other side of the piston pin bore. Remove the sleeve.

f. Lubricate the piston pin with outboard oil and pilot it into the open end of the piston pin bore. Support the piston and tool with one hand and drive the piston pin into the piston with a soft-faced rubber or plastic mallet. Allow the pin tool to exit as the piston pin is driven in.

g. Remove the piston pin tool from the bottom of the piston, then insert it into the top of the piston pin bore and gently tap it until the pin is centered in the pin bore.

h. Make sure none of the needles or thrust washers were displaced, then secure the piston pin with two new piston pin lockrings. Install the lockrings with needlenose pliers (**Figure 94**). Make sure the lockrings are completely seated in their grooves at each end of the piston pin.
i. Repeat this procedure for the No. 2 cylinder piston and connecting rod.

6. Install the rod and piston assemblies to the crankshaft as follows:
   a. Grease the crankpin journals with a thick coat of needle bearing assembly grease. Install the bearing cage halves and the 12 needle bearings. If the original bearings are reused, install them in their original position.
   b. Position the cylinder No. 1 rod and piston assembly over the No. 1 crankpin journal and bearings. The cast or stamped *UP* mark must face upward toward the flywheel. See **Figure 95**. If the *UP* mark is not present, position the piston with the raised ridge on the piston pin boss (bottom side of piston) facing upward.
   c. Install the matching connecting rod cap in its original orientation. Carefully observe fracture and alignment marks to ensure correct installation (**Figure 82**, typical).
   d. Lubricate the screw threads and underside of the screw heads of *new* connecting rod screws with outboard lubricant. Then hold the cap firmly in position, and install the connecting rod screws and thread them fully into the rod.
   e. Tighten each screw to 15 in.-lb. (1.7 N•m). Run a fingernail or pencil lead over each edge of the rod-to-cap joint (**Figure 83**). No ridge should be seen or felt. Realign and retorque the cap as necessary.
   f. Once the alignment is verified, finish torquing the rod screws in a minimum of three progressive steps to the specification in **Table 1**. Make a final check of alignment after the final torque has been applied.
   g. Repeat this procedure to install the No. 2 cylinder piston and rod assembly.

CAUTION

Install the piston rings onto the pistons that match the cylinder bore for which the rings were fitted.

7. Install the piston rings onto each piston with a ring expander (part No. 91-24697 or an equivalent). See **Figure 96**. Install the bottom rectangular ring (3, **Figure 97**) first, then the top semi-keystone ring (2). Expand each ring just enough to slip over the piston. Position each ring's identification mark (4, **Figure 97**) facing up.

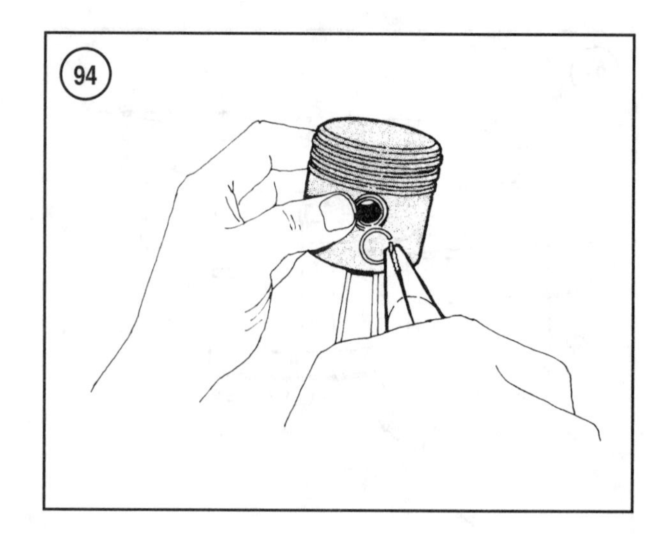

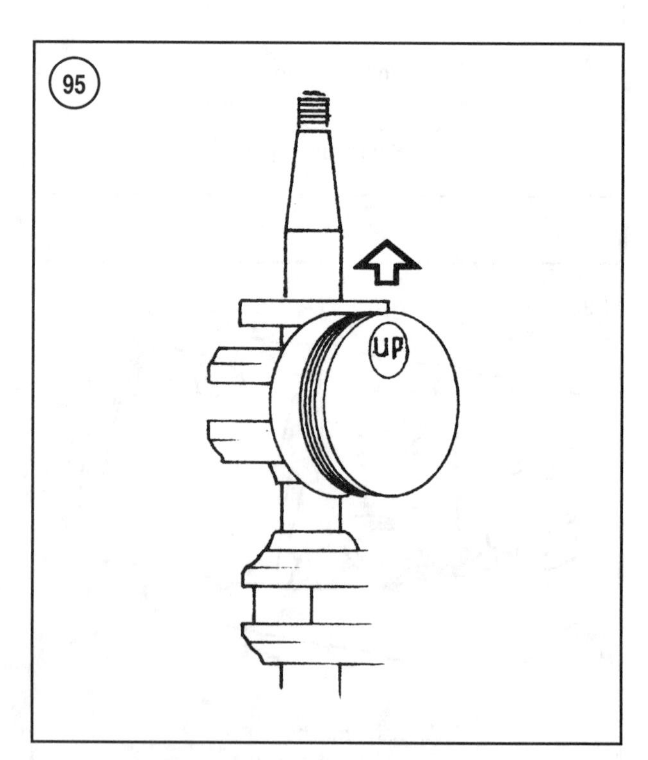

8. Make sure each ring can be rotated freely in its groove, then position the end gap of each piston ring to straddle the ring locating pin in its groove (**Figure 85**).
9. Lubricate the upper main bearing with outboard oil and install it onto the crankshaft.
10. Lubricate the lip of a new upper seal with Quicksilver 2-4-C Multi-Lube, then slide it over the flywheel end of the crankshaft. The seal lip must face downward toward the drive shaft.

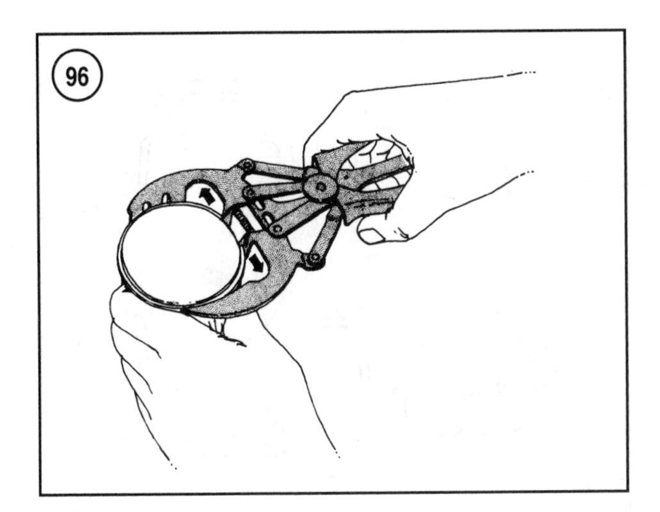

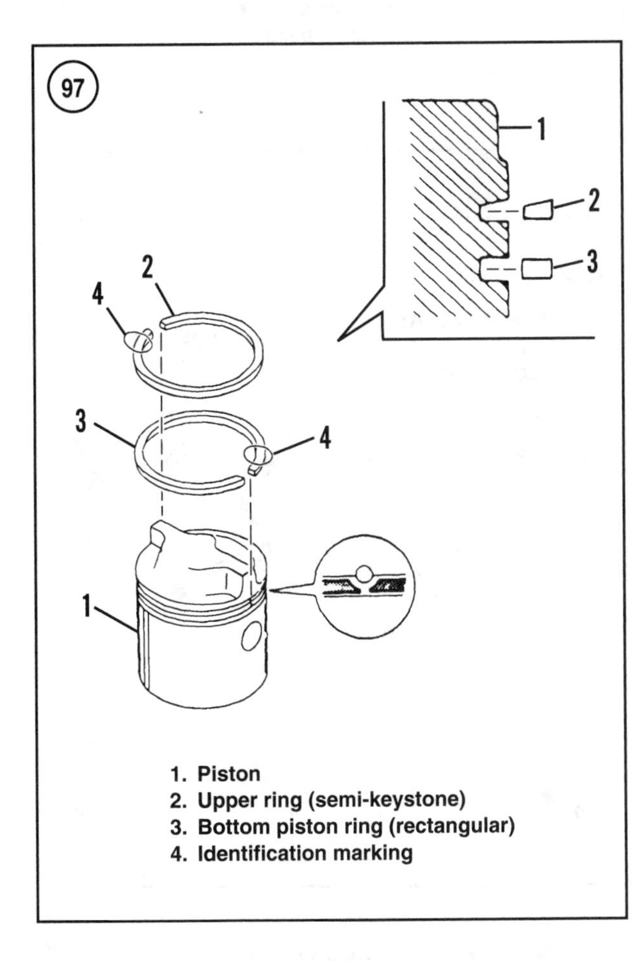

1. Piston
2. Upper ring (semi-keystone)
3. Bottom piston ring (rectangular)
4. Identification marking

NOTE
Lift the crankshaft from the power head stand if necessary to install the following components.

11. Lubricate the lower main bearing with outboard oil and install it onto the crankshaft. Then lubricate the lip of a new lower seal with Quicksilver 2-4-C Multi-Lube and slide it over the drive shaft end of the crankshaft. The seal lip must face downward toward the drive shaft.

Power head assembly

1. Lubricate the piston rings, pistons and cylinder bores with outboard oil.
2. Make sure each piston ring end gap is still straddling the locating pin in its ring groove (**Figure 85**) and the center main bearing locating pin is installed in the cylinder block correctly.

NOTE
A ring compressor is not required as the cylinder bore has a tapered entrance at the bottom of the bore.

3. Position the cylinder block so it is sitting on the cylinder block cover mating surface. Position the crankshaft assembly over the cylinder block. Slowly lower the assembly toward the block while feeding the pistons into each cylinder bore. Keep the crankshaft perfectly horizontal. Rock the pistons slightly to help each enter its cylinder bore. Make sure the piston rings do not rotate or catch while entering the bore.
4. Once the piston rings enter the cylinder bore, seat the crankshaft assembly into the cylinder block. Rotate the crankshaft upper and lower bearings as necessary to position the locating pin into the cylinder block notches in the port side of the mating surface. Rotate the center main bearing assembly as necessary to align the hole in the bearing race with the bearing locating pin. Then carefully push inward on the crankshaft seals to ensure they are properly seated.
5. Insert a thin screwdriver blade into both cylinder's exhaust ports and carefully depress each piston ring (**Figure 73**). If a ring fails to spring back when released, it was probably broken during installation. Replace broken or damaged rings.
6. Use an oil and wax free solvent, such as acetone or lacquer thinner, to clean the cylinder block and crankcase cover mating surfaces.
7. Install the two dowel pins into the cylinder block or crankcase cover if they are not already installed.

CAUTION
Loctite Master Gasket Sealant (part No. 92-12564-2) is the only sealant recommended to seal the crankcase cover-to-cylinder block mating surfaces. The sealant comes in a kit that includes a special primer. Follow the instructions included in the kit

for preparing the surfaces and applying the sealant.

8. Apply a continuous bead of Loctite Master Gasket Sealant to the mating surface of the cylinder block. Run the sealant bead along the inside of all bolt holes as shown in **Figure 98**. The bead must be continuous.
9. Install the crankcase cover onto the cylinder block. Seat the cover to the block with hand pressure. Press against the upper and lower seals to make sure they are seated in their bores.
10. Coat the threads of the crankcase cover screws with Loctite 242 threadlocking adhesive (part No. 92-809821), then install the cover screws. Tighten the cover screws evenly to the specification in **Table 1** in the pattern shown in **Figure 88**.
11. Rotate the crankshaft several revolutions to check for binding or unusual noise. If binding or noise occurs, disassemble the power head, and locate and correct the cause of the defect before proceeding.
12. Install the two starboard intake covers (17, **Figure 91**) with new O-rings. Coat the O-rings with Quicksilver 2-4-C Multi-Lube and coat the screw threads with Quicksilver Perfect Seal. Install and tighten the mounting screws to the specification in **Table 1**.
13. Install the port intake cover (5, **Figure 91**) with new O-ring. Coat the O-ring with Quicksilver 2-4-C Multi-Lube and coat the screws with Quicksilver Perfect Seal. Install the cover with the extended edge facing upward toward the flywheel. Install and tighten the mounting screws to the specification in **Table 1**.
14. Install the cylinder block cover (4, **Figure 91**) with a new gasket (3). Coat the threads of the cover screws with Quicksilver Perfect Seal. Install the cover screws and tighten them finger-tight at this time.
15. Install a new grommet around the thermostat. Then install the thermostat with the sensing pellet facing the power head. Install the thermostat cover and gasket, and secure the cover with two screws. Coat the screw threads with Quicksilver Perfect Seal. Tighten the thermostat housing and cylinder block screws evenly to the specification in **Table 1** in the pattern shown in **Figure 99**.
16. Assemble the exhaust manifold and cover (7 and 9, **Figure 91**) with two new gaskets. Sandwich the exhaust plate between the gaskets, then position the cover over the outer gasket and manifold. Position the assembly on the power head and install the screws. Coat the screw threads with Loctite 242 threadlocking adhesive and tighten them to the specification in **Table 1** in the pattern shown in **Figure 99**.
17. Install the reed block, intake manifold and fuel bleed (recirculation) system as described in Chapter Six.

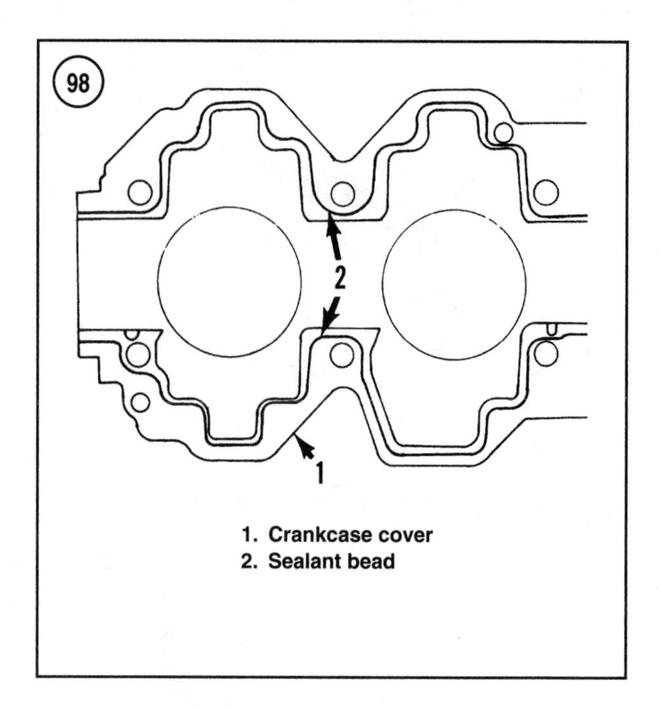

1. Crankcase cover
2. Sealant bead

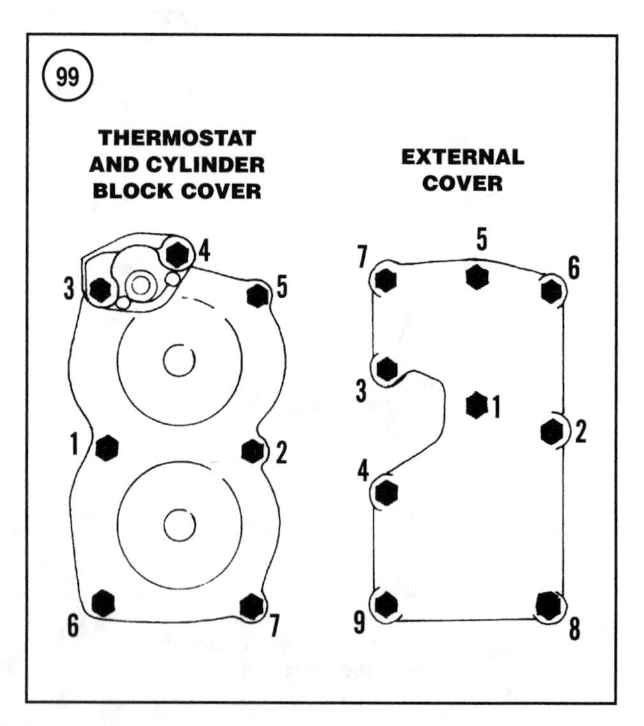

30 and 40 hp Two-Cylinder Models

Refer to **Figure 100** for the following procedure.

1. Check the end gap of the new piston rings for each cylinder as described under *Piston Ring End Gap* in this chapter. The ring end gap must be within the specification in **Table 6**.

100

POWER HEAD COMPONENTS (30 AND 40 HP MODELS [TWO-CYLINDER])

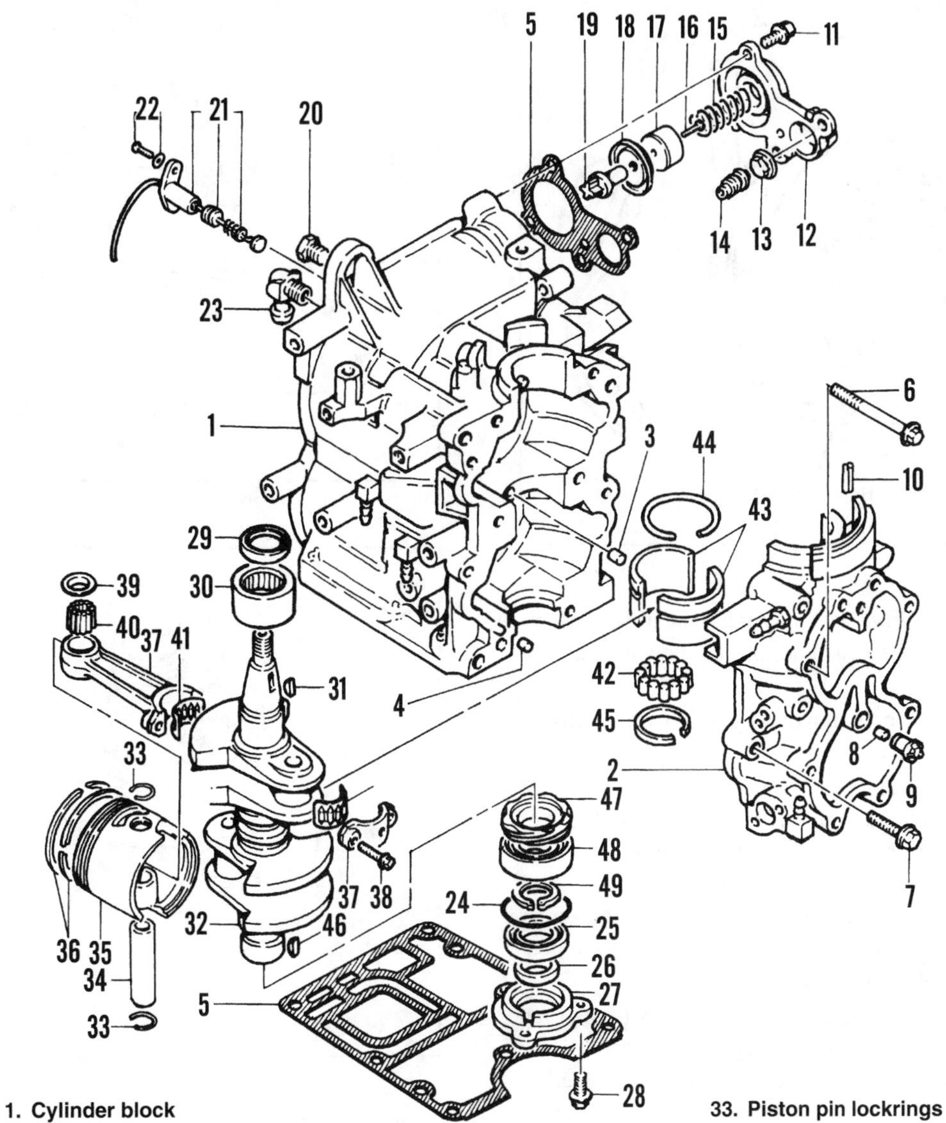

1. Cylinder block
2. Crankcase cover
3. Bearing locating pin
4. Dowel pin
5. Gaskets
6. Long screw
7. Short screw
8. Check valve
9. Check valve carrier
10. Roll pin
11. Thermostat housing
12. Thermostat grommet
13. Thermostat grommet
14. Thermostat
15. Spring
16. Screw
17. Cup
18. Diaphragm
19. Poppet valve
20. Brass pipe plug
21. Engine temperature switch
22. Screw and washer
23. Water hose (tell-tale) fitting
24. O-ring
25. Large crankshaft seal
26. Small crankshaft seal
27. Lower end cap
28. Screw
29. Crankshaft upper seal
30. Caged needle bearing
31. Flywheel key
32. Crankshaft
33. Piston pin lockrings
34. Piston pin
35. Piston
36. Piston rings
37. Connecting rod and rod cap
38. Screw
39. Thrust washer
40. Loose needle bearings
41. Caged roller bearing halves
42. Loose needle bearings
43. Bearing race halves
44. Bearing retainer ring
45. Crankshaft seal ring
46. Drive key
47. Oil pump drive gear
48. Ball bearing
49. Retaining ring

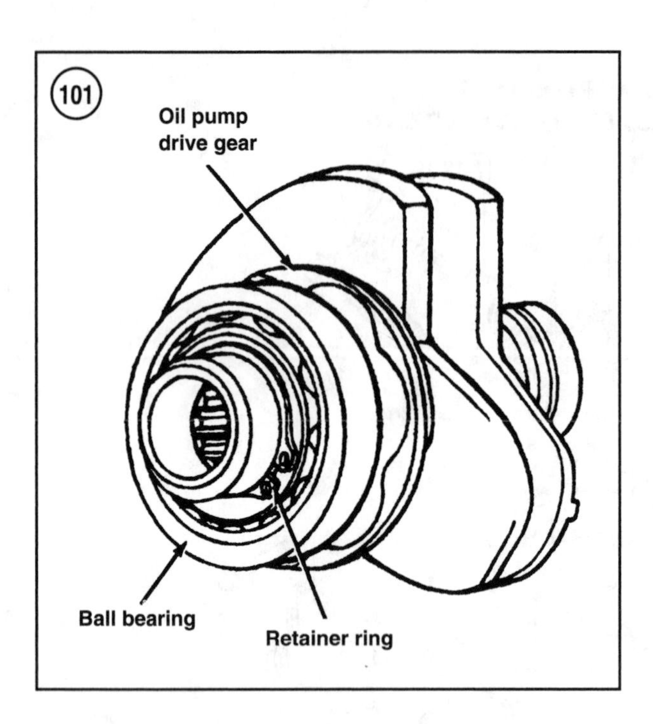

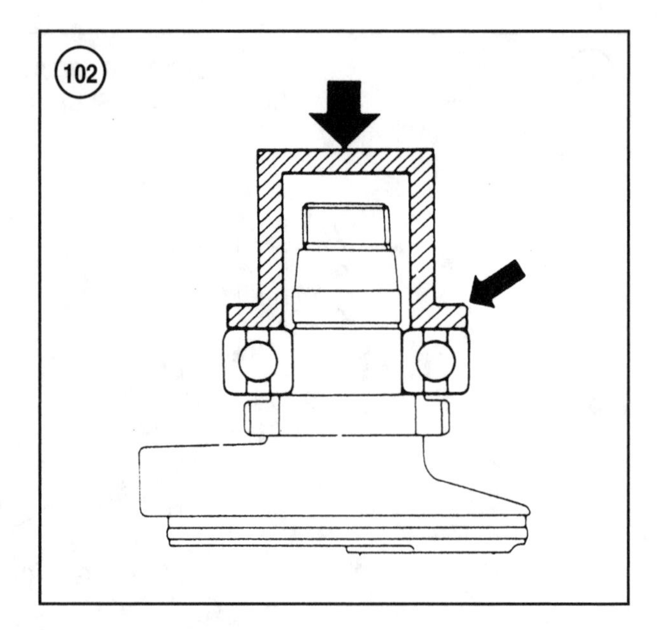

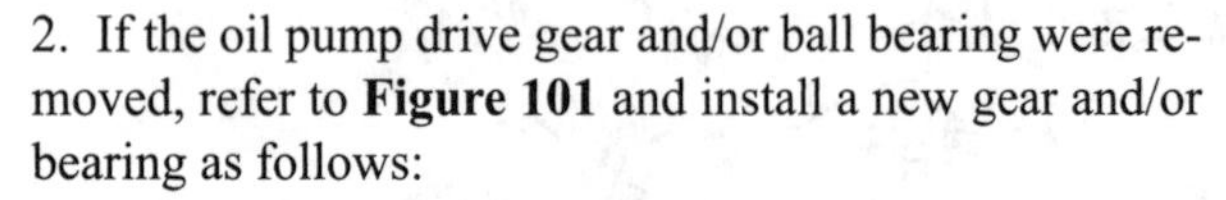

2. If the oil pump drive gear and/or ball bearing were removed, refer to **Figure 101** and install a new gear and/or bearing as follows:

  a. Make sure the oil pump drive gear key is installed in the crankshaft.
  b. Lubricate the crankshaft and a new oil pump drive gear with outboard oil. Position the recessed side of the gear toward the crankshaft counterweight, align the keyway and seat the gear against the crankshaft shoulder. If necessary, support the crankshaft under the lower counterweight and press against the gear with a suitable mandrel.
  c. Lubricate a new ball bearing with outboard oil, then slide the bearing over the drive shaft end of the crankshaft with the numbered side of the bearing facing away from the crankshaft.
  d. Support the crankshaft (under the lower counterweight) in a press. Press against the inner race of the bearing using a suitable mandrel (**Figure 102**) until the bearing seats on the crankshaft.
  e. Install the retainer ring (49, **Figure 100**) with a suitable pair of snap ring pliers (**Figure 103**). Make sure the ring is fully seated in the crankshaft groove.

3. Mount the crankshaft vertically on a power head (crankshaft) stand (part No. 91-24697 or an equivalent). Securely clamp the power head stand in a vise.

CAUTION

If reusing the original bearings, install them in their original locations.

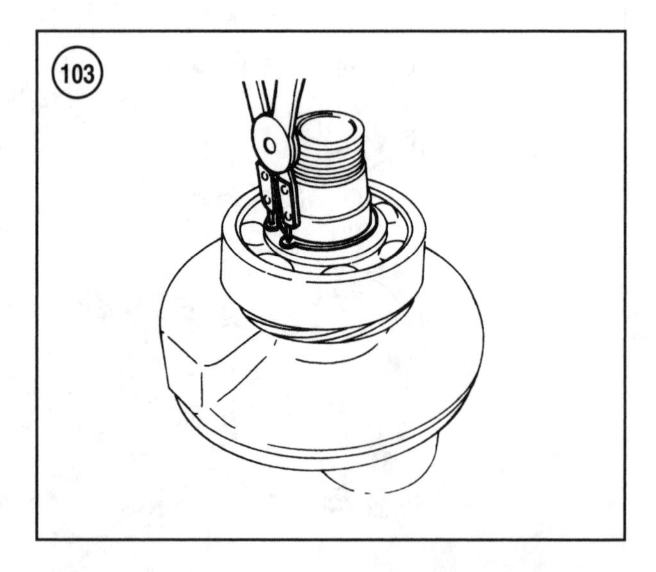

4. Refer to **Figure 104** and assemble the center main bearing as follows:

  a. Install a new sealing ring into the center main bearing journal seal groove. Do not expand the ring any further than necessary.
  b. Apply a thick coat of needle bearing assembly grease (part No. 92-825265A 1) to the crankshaft center journal bearing surface. Then install the 14 loose rollers to the bearing surface.
  c. Position the outer race halves over the bearings. The retaining ring groove must be positioned up toward the flywheel.
  d. Carefully align the fracture lines, then install the retainer ring (44, **Figure 100**). Position the retainer

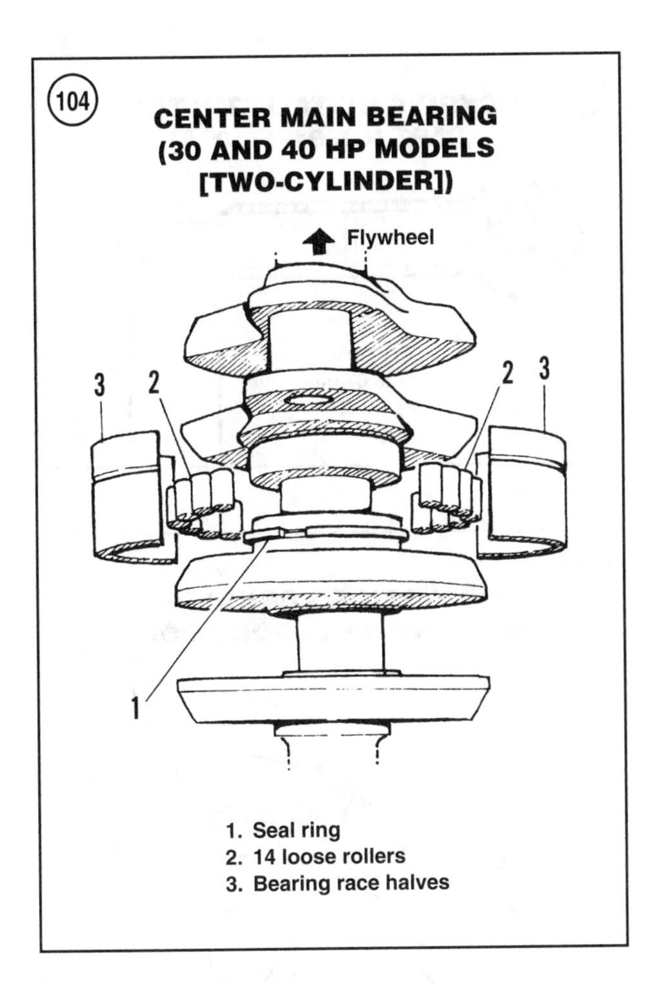

ring to cover as much of both fracture lines as possible.

5. Lubricate the upper main bearing (30, **Figure 100**) with outboard oil, then install the bearing onto the crankshaft.

6. Lubricate the lip of a new upper seal with Quicksilver 2-4-C Multi-Lube, then slide the seal (29, **Figure 100**) over the flywheel end of the crankshaft. The lip of the seal must face downward toward the drive shaft.

7. Begin assembly of the connecting rods to the pistons by greasing the sleeve portion of the piston pin installation tool (part No. 91-76160A-2) with needle bearing assembly grease.

8. Position the cylinder No. 1 connecting rod in its original orientation as marked during disassembly.

9. Hold the lower thrust washer (2, **Figure 93**) under the connecting rod small end, then insert the greased sleeve into the small end bore.

10. Lubricate the 29 needles (1, **Figure 93**) with needle bearing assembly grease and insert them into the small end around the sleeve as shown in **Figure 105**.

11. Position the upper thrust washer (2, **Figure 93**) on top of the needles. Carefully slide the No. 1 cylinder piston over the rod with the *UP* mark facing upward (**Figure 95**) and align the piston pin bores. Then insert the main body of a piston pin tool (part No. 91-76160A-2) into the piston pin bore through the connecting rod, pushing the sleeve out the other side of the piston pin bore. Remove the sleeve.

12. Lubricate the piston pin with outboard oil and guide it into the open end of the piston pin bore. Support the piston and tool with one hand and drive the piston pin into the piston with a soft-faced rubber or plastic mallet. Allow the pin tool to exit as the piston pin is driven in.

13. Remove the piston pin tool from the bottom of the piston, then insert it into the top of the piston pin bore and gently tap it until the pin is centered in the pin bore.

14. Make sure no needles or locating washers were displaced, then secure the piston pin with two new piston pin lockrings (33, **Figure 100**) using a lockring installation tool (part No. 91-77109A-2) as follows:

NOTE
*The small shaft of the drive handle of the lockring installation tool (part No. 91-77109A-1) must be modified to the dimension shown in **Figure 106**. The A-2 version of the tool may not need modification. Check the instructions included with the tool.*

a. Position a new lockring into the stepped, open end of the tool's sleeve (**Figure 107**).
b. Insert the drive handle into the opposite end of the sleeve.
c. Pilot the stepped end of the sleeve into either end of the No. 1 cylinder piston pin bore.
d. While holding the sleeve to the pin bore, press the drive handle quickly and firmly to install the ring.

8

e. Remove the handle and sleeve. Make sure the lockring is completely seated in its groove in the piston pin bore.
f. Install the second lockring in the opposite end of the piston pin bore in the same manner.

15. Repeat Steps 7-14 for the No. 2 cylinder piston and connecting rod.

CAUTION
Install the piston rings onto the pistons that match the cylinder bore for which the rings were fitted.

16. Install the two semi-keystone piston rings onto each piston using a ring expander (part No. 91-24697 or equivalent). See **Figure 96**. Install the bottom ring first, then the top ring. Expand each ring just enough to slip over the piston. Each ring's identification dot or letter (**Figure 108**) must face upward.
17. Make sure each ring can be rotated freely in its groove, then position the end gap of each piston ring to straddle the locating pin in its groove (**Figure 109**).
18. Assemble the lower end cap as follows:
   a. Coat the outer diameter of two new seals with Loctite 271 threadlocking adhesive (part No. 92-809819).
   b. Press the small diameter seal into the end cap with a suitable mandrel. The seal lip must face downward toward the drive shaft.
   c. Press the large diameter seal into the end cap with a suitable mandrel. The seal lip must face up toward the flywheel.
   d. Grease a new O-ring with Quicksilver 2-4-C Multi-Lube and install it onto the end cap.
19. Lubricate the piston rings, pistons and cylinder bores with outboard oil. Then make sure each piston ring end gap is still straddling the locating pin in each ring groove (**Figure 109**, typical).

NOTE
A ring compressor is not required as the cylinder bore has a tapered entrance at the bottom of the bore.

20. Install each piston into its appropriate cylinder bore. Make sure the *UP* marks are facing the flywheel end of the cylinder block and the connecting rods are aligned with the crankshaft throws. Rock each piston slightly to help it enter its cylinder bore, making sure the piston rings do not rotate or catch and break entering the bore. Seat each piston at the bottom of its bore.
21. Make sure the center and upper main bearing locating pins (3, **Figure 100**) are installed in the cylinder block.

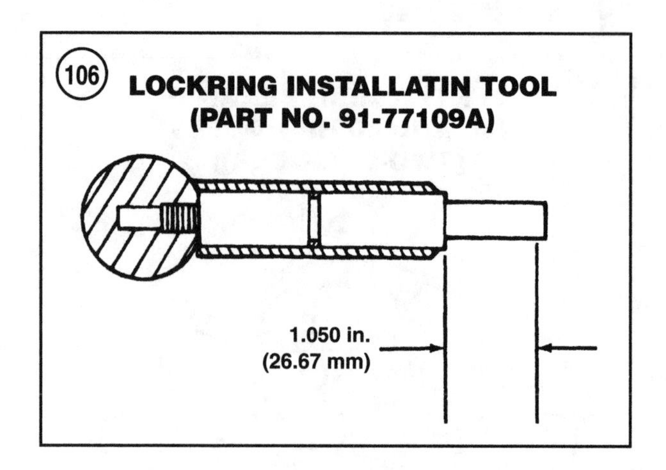

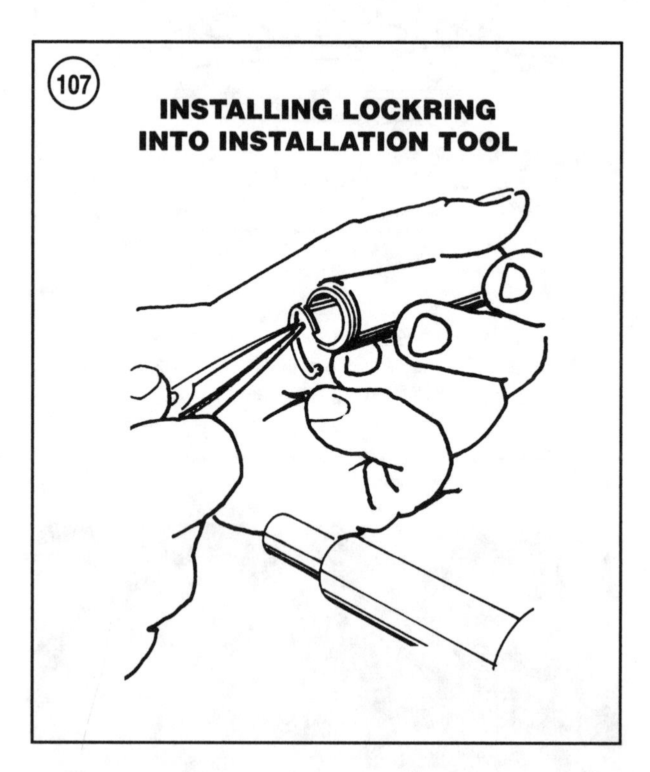

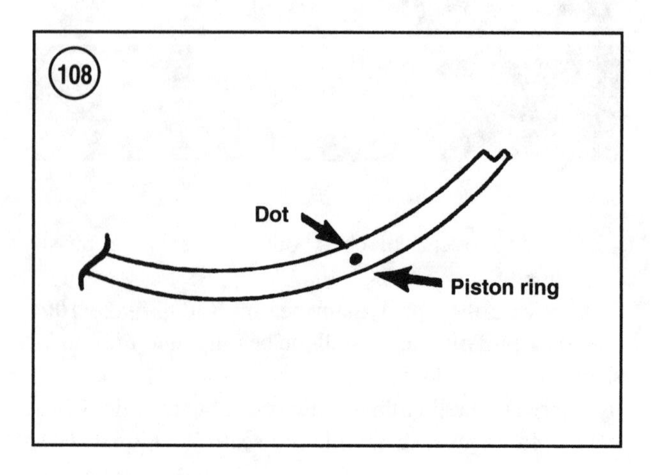

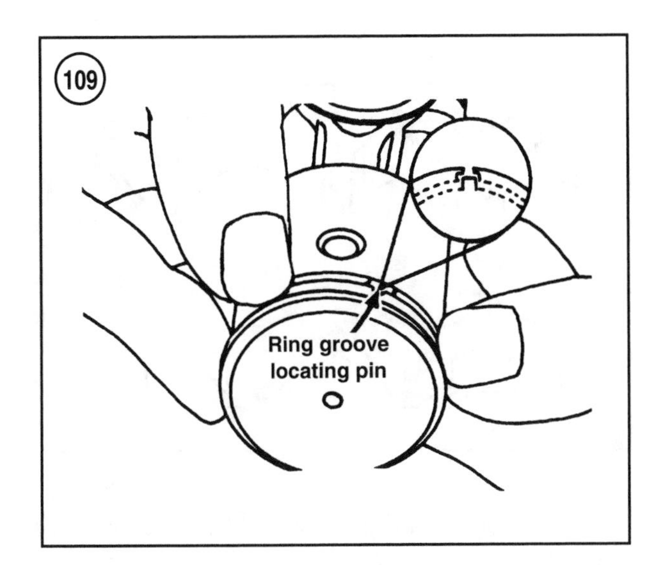

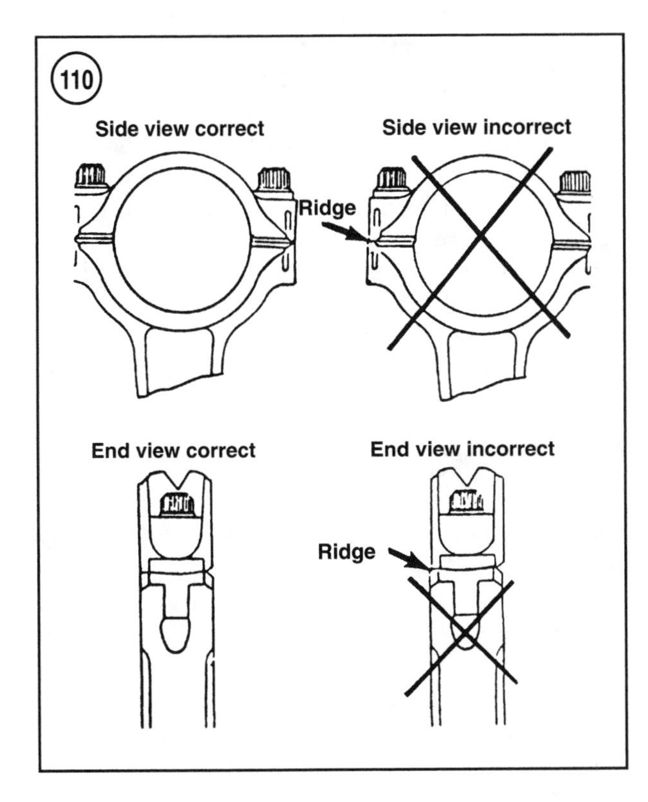

Then, position the cylinder block so the block-to-crankcase cover mating surface is pointing upward. Move both connecting rods toward one side on the cylinder block.

22. Slowly lower the crankshaft assembly into the block. Rotate the crankshaft center and upper main bearing assemblies as necessary to align the hole in the bearing races with the bearing locating pins (3, **Figure 100**). Then carefully tap the lower end of the crankshaft with a soft hammer to seat the ball bearing in its bore.

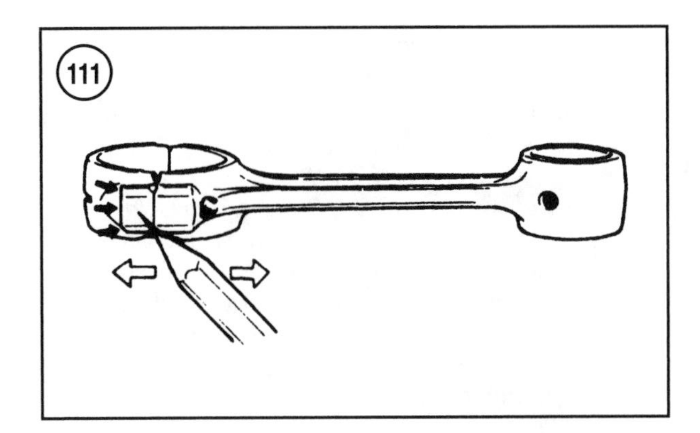

23. Install the connecting rods to the crankshaft journals as follows:
   a. Grease the crankpin journals with a thick coat of needle bearing assembly grease. Install the caged needle bearing halves to the journals. If the original bearings are reused, install them in their original position.
   b. Pull the No. 1 cylinder rod and piston assembly up to the No. 1 crankpin journal and bearings. Rotate the crankshaft as necessary to allow mating of the rod and journal.
   c. Install the matching connecting rod cap in its original orientation. Carefully observe fracture and alignment marks to ensure correct installation (**Figure 110**, typical).
   d. Lubricate the screw threads and underside of the screw heads of the *new* connecting rod screws with outboard lubricant. Then, while holding the cap firmly in position, install the connecting rod screws and thread them fully into the rod.
   e. Tighten each screw to 15 in.-lb. (1.7 N•m). Run a fingernail or pencil lead over each edge of the rod-to-cap joint (**Figure 111**). No ridge should be seen or felt. Realign and retorque the cap as necessary.
   f. Once the alignment is verified, finish torquing the rod screws in a minimum of three progressive steps to the specification in **Table 1**. Make a final check of alignment after the final torque has been applied.
   g. Repeat this procedure to install the cylinder No. 2 connecting rod to its crankshaft journal.

24. Use an oil and wax free solvent, such as acetone or lacquer thinner, to clean the cylinder block and crankcase cover mating surfaces.

25. Install the dowel pin (4, **Figure 100**) into the cylinder block or crankcase cover if it is not already installed.

26. Install the lower end cap over the crankshaft. Coat the mating surfaces of the end cap with Quicksilver Perfect

Seal. Align the screw hole in the cylinder block and seat the cap to the block. Install a screw finger-tight to hold the end cap in position.

CAUTION

Loctite Master Gasket Sealant (part No. 92-12564-2) is the only sealant recommended to seal the crankcase cover-to-cylinder block mating surfaces. The sealant comes in a kit that includes a special primer. Follow the instruction included in the kit for preparing the surfaces and applying the sealant.

27. Apply a continuous bead of Loctite Master Gasket Sealant to the mating surface of the cylinder block. Run the sealant bead along the inside of all bolt holes as shown in **Figure 112**. The bead must be continuous.
28. Install the crankcase cover into position on the cylinder block. Seat the cover to the block with hand pressure. Tap the lower end of the crankshaft with a soft hammer to seat the ball bearing in its bore, then push against the upper seal until it is seated in its bore.
29. Coat the threads of the six long main bearing and four short outer crankcase cover screws with Loctite 242 threadlocking adhesive (part No. 92-809821), then install the cover screws. Tighten the cover screws evenly to the specification in **Table 1** in the pattern shown in **Figure 113**.
30. Rotate the crankshaft several revolutions to check for binding or unusual noise. If binding or noise occurs, disassemble the power head, and locate and correct the cause of the defect before proceeding.
31. Remove the end cap screw installed previously. Coat the threads of the end cap screws with Loctite 242 threadlocking adhesive. Install and evenly tighten the screws to the specification in **Table 1**.
32. Install a new grommet around the thermostat. Then install the thermostat with the sensing pellet facing the power head. Install the poppet valve assembly as shown in **Figure 100**. Install the cover with a new gasket. Coat the screw threads with Quicksilver Perfect Seal. Evenly tighten the screws to the specification in **Table 1**.
33. Install the engine temperature switch into the cylinder block. Secure the switch with a screw and washer. Tighten the screw securely.
34. Install any control linkage removed during power head disassembly.
35. Install the ignition and electrical components as an assembly. This includes the stator assembly, trigger coil, CDMs and the electrical/ignition plate. On electric start models, this will also include the voltage regulator and the starter solenoid. Secure all cables and harnesses with the original clamps and/or new tie-straps. See Chapter Seven.

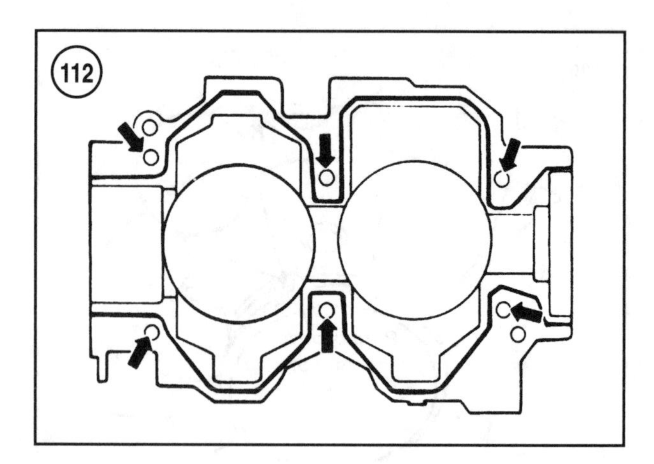

36. On electric start models, install the electric starter motor as described in Chapter Seven
37. Install the internal fuel bleed (recirculation) valve and carrier (8 and 9, **Figure 100**) into the crankcase cover. See Chapter Six for fuel bleed system information.
38. Install the intake manifold and reed block, carburetor, fuel pump, fuel filter, and all fuel, primer and fuel bleed (recirculation) lines as described in Chapter Six. On electric start models, this will include the fuel primer valve.
39. Install the oil pump, oil reservoir and all oil lines as described in Chapter Twelve.
40. Install the flywheel as described in Chapter Seven.

40-60 hp, 30 Jet and 45 Jet Three-Cylinder Models

Refer to **Figure 114** and **Figure 115** for the following procedure.
1. Check the end gap of the new piston rings as described under *Piston Ring End Gap* in this chapter. The ring end gap must be within the specification in **Table 6**.
2. If the oil pump drive gear and/or ball bearing were removed, refer to **Figure 101** and install a new gear and/or bearing as follows:
   a. Make sure the oil pump drive gear key is installed in the crankshaft.
   b. Lubricate the crankshaft and a new oil pump drive gear with outboard oil. Position the recessed side of the gear toward the crankshaft counterweight, align the keyway and seat the gear against the crankshaft shoulder. If necessary, support the crankshaft under the lower counterweight and press against the gear with a suitable mandrel.
   c. Lubricate a new ball bearing with outboard oil, then slide the bearing over the drive shaft end of the

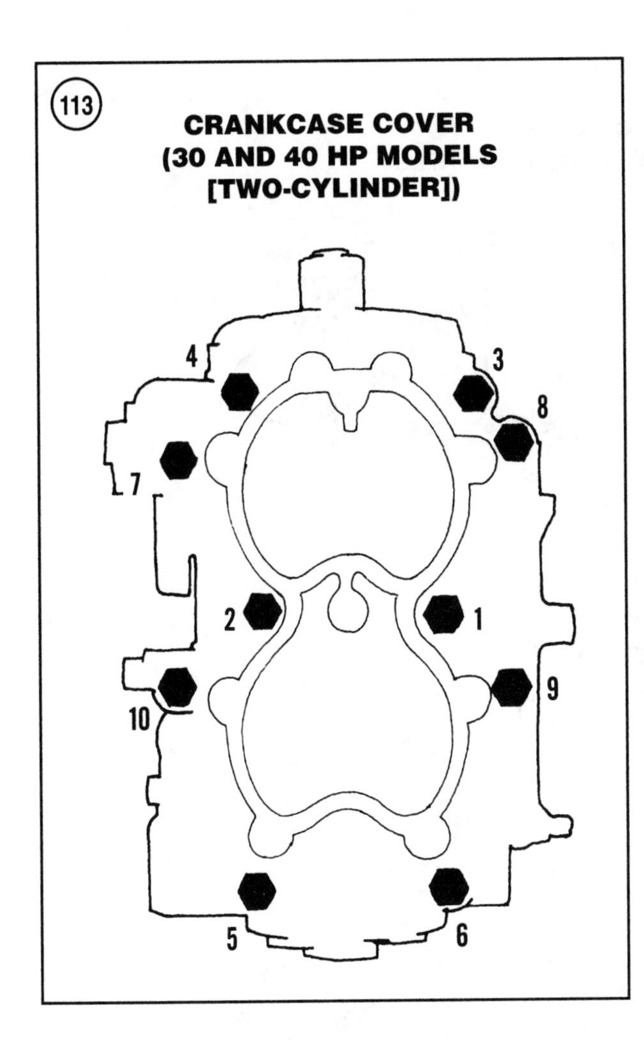

crankshaft with the numbered side of the bearing facing away from the crankshaft.

d. Support the crankshaft under the lower counterweight in a press. Press against the inner race of the bearing using a suitable mandrel (**Figure 102**) until the bearing seats on the crankshaft.

e. Install the retainer ring (20, **Figure 115**) with a suitable pair of snap ring pliers (**Figure 103**). Make sure the ring is fully seated in the crankshaft groove.

3. Mount the crankshaft vertically on a power head (crankshaft) stand (part No. 91-25821A 1 or an equivalent). Securely clamp the power head stand in a vise.

CAUTION
If reusing the original bearings, install them in their original locations.

4. Assemble the two center main bearings as follows:

a. Install a new sealing ring (14, **Figure 115**) into the upper center main bearing journal's seal groove. Do not expand the ring any further than necessary to install.

b. Apply a thick coat of needle bearing assembly grease (part No. 92-825265A 1) to the journal's twin bearing surfaces. Then install the 14 loose rollers (16, **Figure 115**) to each bearing surface (28 rollers for each bearing assembly).

c. Position the outer race halves over the bearing rollers and seal ring. The retaining ring groove must be positioned up toward the flywheel.

d. Carefully align the fracture lines, then install the retainer ring (14, **Figure 115**). Position the retainer ring to cover as much of both fracture lines as possible.

e. Repeat this procedure to assemble the lower center main bearing.

5. Lubricate the upper main bearing (2, **Figure 115**) with outboard oil, then install the bearing onto the crankshaft.

6. Lubricate the lip of a new upper seal with Quicksilver 2-4-C Multi-Lube, then slide the seal (1, **Figure 115**) over the flywheel end of the crankshaft. The lip of the seal must face downward toward the drive shaft.

7. Begin assembly of the connecting rods to the pistons by greasing the sleeve portion of the piston pin installation tool (part No. 91-76160A-2) with needle bearing assembly grease.

8. Position the cylinder No. 1 connecting rod in its original orientation as marked during disassembly.

9. Hold the lower thrust washer (2, **Figure 93**) under the connecting rod small end, then insert the greased sleeve into the small end bore.

10. Lubricate the 29 needles (1, **Figure 93**) with needle bearing assembly grease and insert them into the small end around the sleeve as shown in **Figure 105**.

11. Position the upper thrust washer (2, **Figure 93**) on top of the needles. Carefully slide the No. 1 cylinder piston over the rod with the *UP* mark facing up (**Figure 95**) and align the piston pin bores. Then insert the main body of a piston pin tool (part No. 91-76160A-2) into the piston pin bore through the connecting rod, pushing the sleeve out the other side of the piston pin bore. Remove the sleeve.

12. Lubricate the piston pin with outboard oil and guide it into the open end of the piston pin bore. Support the piston and tool with one hand and drive the piston pin into the piston with a soft-faced rubber or plastic mallet. Allow the pin tool to exit as the piston pin is driven inward.

13. Remove the piston pin tool from the bottom of the piston, then insert it into the top of the piston pin bore and gently tap it until the pin is centered in the piston pin bore.

14. Make sure no needles or locating washers were displaced, then secure the piston pin with two new piston pin

(114)

CYLINDER BLOCK COMPONENTS (40-60 HP, 30 JET AND 45 JET MODELS)

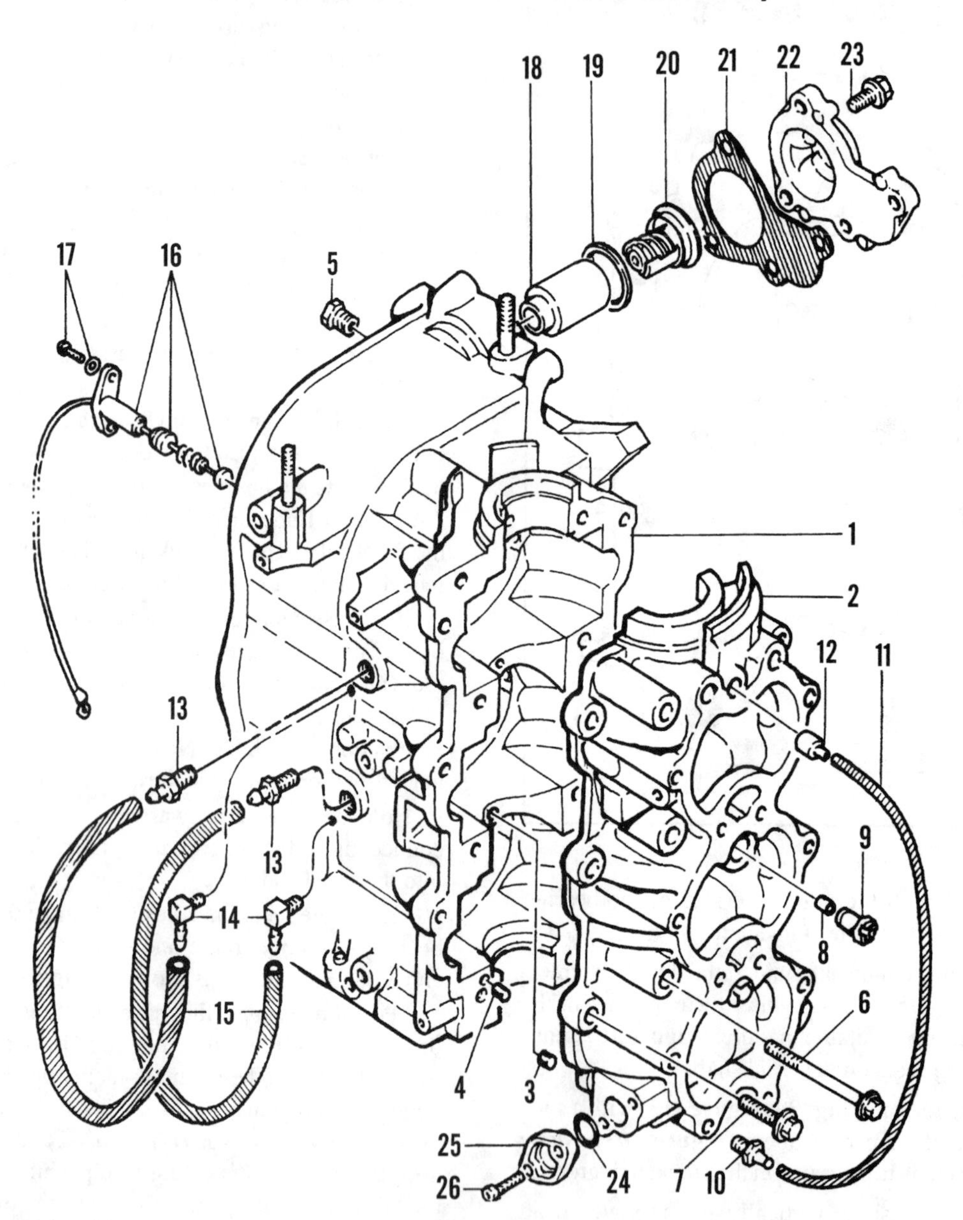

1. Cylinder block
2. Crankcase cover
3. Bearing locating pin
4. Dowel pin
5. Brass pipe plug
6. Main bearing (long) screw
7. Outer (short) screw
8. Check valve
9. Check valve carrier
10. Check valve
11. Fuel bleed (recirculation) hose
12. Check valve
13. Check valves
14. Fittings
15. Fuel bleed (recirculation) hoses
16. Engine temperature switch
17. Screw and washer
18. Thermostat carrier
19. O-ring
20. Thermostat
21. Gasket
22. Thermostat housing
23. Screw
24. O-ring
25. Cap (except oil injected models)
26. Screw

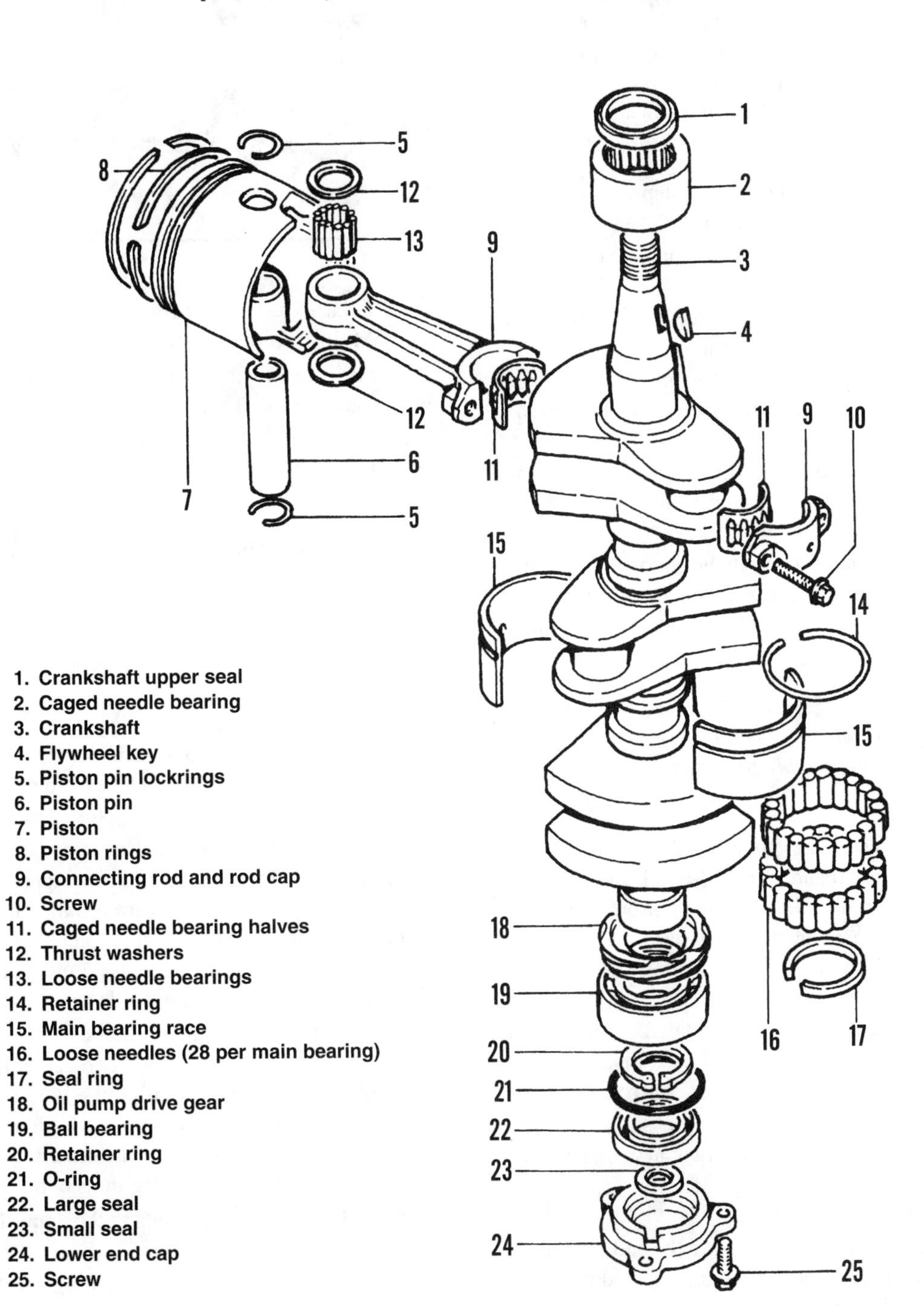
CRANKSHAFT ASSEMBLY
(40-60 HP, 30 JET AND 45 JET MODELS)
1
2
3
4
5
12
13
9
8
7
6
11
10
14
15
16
17
18
19
20
21
22
23
24
25
1. Crankshaft upper seal
2. Caged needle bearing
3. Crankshaft
4. Flywheel key
5. Piston pin lockrings
6. Piston pin
7. Piston
8. Piston rings
9. Connecting rod and rod cap
10. Screw
11. Caged needle bearing halves
12. Thrust washers
13. Loose needle bearings
14. Retainer ring
15. Main bearing race
16. Loose needles (28 per main bearing)
17. Seal ring
18. Oil pump drive gear
19. Ball bearing
20. Retainer ring
21. O-ring
22. Large seal
23. Small seal
24. Lower end cap
25. Screw

8

lockrings (5, **Figure 115**) using a lockring installation tool (part No. 91-77109A-2) as follows:

NOTE
*The small shaft of the drive handle of the lockring installation tool (part No. 91-77109A-1) must be modified to the dimension shown in **Figure 106**. The A-2 version of the tool may not need modification. Check the instructions included with the tool.*

a. Position a new lockring into the stepped, open end of the tool's sleeve (**Figure 107**).
b. Insert the drive handle into the opposite end of the sleeve.
c. Guide the stepped end of the sleeve into either end of the No. 1 cylinder piston pin bore.
d. While holding the sleeve to the pin bore, press the drive handle quickly and firmly to install the lockring.
e. Remove the handle and sleeve. Make sure the lockring is completely seated in its groove in the piston pin bore.
f. Install the second lockring in the opposite end of the piston pin bore in the same manner.

15. Repeat Steps 7-14 for the No. 2 and No. 3 cylinder piston and connecting rod.

CAUTION
Install the piston rings onto the pistons that match the cylinder bore for which the rings were fitted.

16. Install the two semi-keystone piston rings onto each piston using a ring expander (part No. 91-24697 or an equivalent). See **Figure 96**. Install the bottom ring first, then the top ring, expanding each ring just enough to slip over the piston. Each ring's identification dot or letter (**Figure 108**) must face upward.

17. Make sure each ring can be rotated freely in its groove, then position the end gap of each piston ring to straddle the locating pin in its groove (**Figure 109**).

18. Assemble the lower end cap as follows:

a. Coat the outer diameter of two new seals with Loctite 271 threadlocking adhesive (part No. 92-809819).
b. Press the small diameter seal into the end cap with a suitable mandrel. The seal lip must face downward toward the drive shaft.
c. Press the large diameter seal into the end cap with a suitable mandrel. The seal lip must face up toward the flywheel.

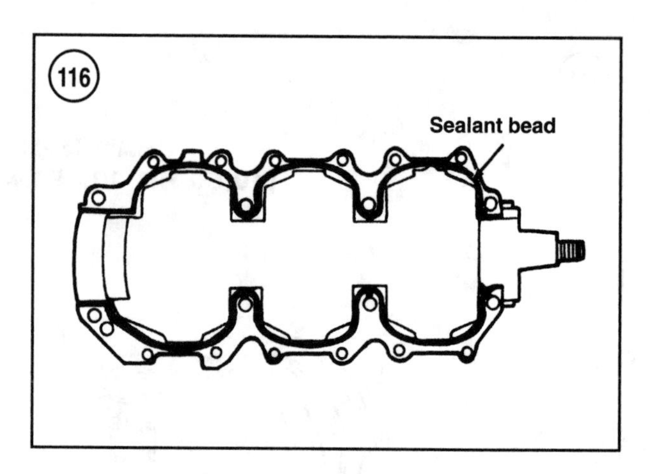

d. Grease a new O-ring with Quicksilver 2-4-C Multi-Lube and install it onto the end cap.

19. Lubricate the piston rings, pistons and cylinder bores with outboard oil. Then make sure each piston ring end gap is still straddling the locating pin in each ring groove (**Figure 109**, typical).

NOTE
A ring compressor is not required as the cylinder bore has a tapered entrance at the bottom of the bore.

20. Install each piston into its appropriate cylinder bore. Make sure the *UP* marks are facing the flywheel end of the cylinder block and the connecting rods are aligned with the crankshaft throws. Rock each piston slightly to help it enter its cylinder bore, making sure the piston rings do not rotate or catch and break entering the bore. Seat each piston at the bottom of its bore.

21. Make sure the two center and one upper main bearing locating pins (3, **Figure 114**) are installed in the cylinder block. Then position the cylinder block so the block-to-crankcase cover mating surface is pointing upward. Make all connecting rods toward one side on the cylinder block.

22. Slowly lower the crankshaft assembly into the block. Rotate the crankshaft upper and two center main bearing assemblies as necessary to align the hole in the bearing races with the bearing locating pins (3, **Figure 114**). Then carefully tap the lower end of the crankshaft with a soft hammer to seat the ball bearing in its bore.

23. Install the connecting rods to the crankshaft journals as follows:

a. Grease the crankpin journals with a thick coat of needle bearing assembly grease. Install the caged needle bearing halves to the journals. If the original bearings are reused, install them in their original position.

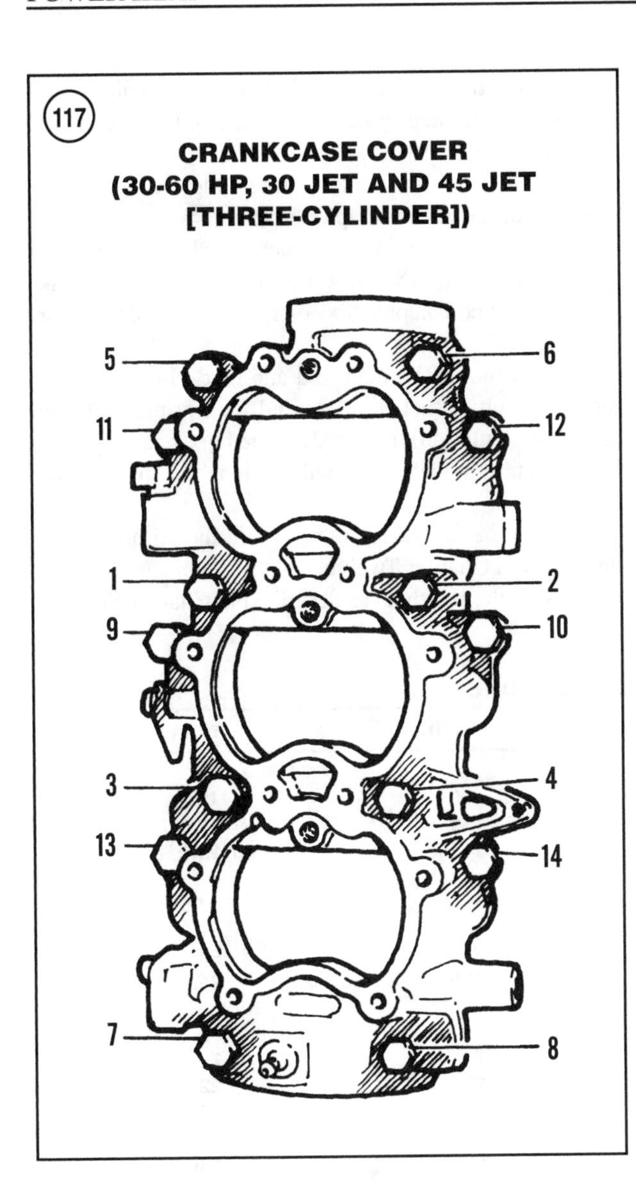

b. Pull the No. 1 cylinder rod and piston assembly up to the No. 1 crankpin journal and bearings. Rotate the crankshaft as necessary to allow mating of the rod and journal.

c. Install the matching connecting rod cap in its original orientation. Carefully observe fracture and alignment marks to ensure correct installation (**Figure 110**, typical).

d. Lubricate the screw threads and underside of the screw heads of *new* connecting rod screws with outboard lubricant. Then, while holding the cap firmly in position, install the connecting rod screws and thread them fully into the rod.

e. Tighten each screw to 15 in.-lb. (1.7 N•m). Run a fingernail or pencil lead over each edge of the rod-to-cap joint (**Figure 111**). No ridge should be seen or felt. Realign and retorque the cap as necessary.

f. Once the alignment is verified, finish torquing the rod bolt to the specification in **Table 1**. Tighten each rod bolt and additional 90°. Make a final check of alignment after the final torque has been applied.

g. Repeat this procedure to install the cylinder No. 2 and 3 connecting rods to their respective crankshaft journals.

24. Use an oil and wax free solvent, such as acetone or lacquer thinner, to clean the cylinder block and crankcase cover mating surfaces.

25. Install the dowel pin (4, **Figure 114**) into the cylinder block or crankcase cover if it is not already installed.

26. Install the lower end cap over the crankshaft. Coat the mating surfaces with Quicksilver Perfect Seal. Align the screw hole in the cylinder block and seat the end cap to the block. Install a bolt finger-tight to hold the end cap in position.

CAUTION

Loctite Master Gasket Sealant (part No. 92-12564-2) is the only sealant recommended to seal the crankcase cover-to-cylinder block mating surfaces. The sealant comes in a kit that includes a special primer. Follow the instructions included in the kit for preparing the surfaces and applying the sealant.

27. Apply a continuous bead of Loctite Master Gasket Sealant to the mating surface of the cylinder block. Run the sealant bead along the inside of all bolt holes as shown in **Figure 116**. The bead must be continuous.

28. Install the crankcase cover on the cylinder block. Seat the cover to the block with hand pressure. Tap the lower end of the crankshaft with a soft hammer to seat the ball bearing in its bore, then push against the upper seal until it is seated in its bore.

29. Coat the threads of the eight long, main bearing, and six short, outer, crankcase cover bolts with Loctite 242 threadlocking adhesive (part No. 92-809821), then install the cover bolts. Tighten the cover bolts evenly to the specification in **Table 1** in the pattern shown in **Figure 117**.

30. Rotate the crankshaft several revolutions to check for binding or unusual noise. If binding or noise occurs, disassemble the power head, and locate and correct the cause of the defect before proceeding.

31. Remove the end cap bolt installed previously. Coat the threads of the three end cap bolts with Loctite 242 threadlocking adhesive. Install and evenly tighten the bolts to the specification in **Table 1**.

32. Install the carrier and O-ring (18 and 19, **Figure 114**) over the thermostat (20). Then, install the carrier and thermostat assembly into the power head as shown in **Figure 114**. Install the cover using a new gasket and secure with screws. Coat the screw threads with Quicksilver Perfect Seal. Evenly tighten the screws to the specification in **Table 1**.
33. Install the engine temperature switch (16, **Figure 114**) into the cylinder block. Secure the switch with a screw and washer. Tighten the screw securely.
34. Install any control linkage removed during power head disassembly.
35. Install the ignition and electrical components that were removed as an assembly. This includes the stator assembly, trigger coil and the electrical/ignition plate. On electric start models, this will also include the voltage regulator and the starter solenoid. Install the rpm limit module at this time, if equipped. Secure all cables and harnesses with the original clamps and/or new tie-straps. See Chapter Seven.
36. On electric start models, install the electric starter motor as described in Chapter Seven.
37. Install the two internal fuel bleed (recirculation) valves and carriers (8 and 9, **Figure 114**) into the crankcase cover. See Chapter Six for fuel bleed system information.
38. Install the intake manifold and reed block, carburetors, fuel pump, fuel filter, and all fuel, primer and fuel bleed (recirculation) lines as described in Chapter Six. On electric start models, this will include the fuel primer valve.
39. Install the oil pump, oil reservoir and all oil lines as described in Chapter Twelve.
40. Install the flywheel as described in Chapter Seven.

Table 1 TORQUE SPECIFICATIONS

Component	ft.-lb.	in.-lb.	N•m
Connecting rod			
6-15 hp	–	100	11.3
20-25 hp	–	150	16.9
30 and 40 hp (two-cylinder)	16	192	21.7
40-60 hp (three-cylinder)			
Initial torque	–	120*	13.6*
Cylinder head bolt			
2.5 and 3.3 hp	–	85	9.6
4 and 5 hp	18	216	24.4
Lower end cap to crankcase			
2.5 and 3.3 hp	–	50	5.6
4 and 5 hp	–	70	7.9
30 and 40 hp (two-cylinder)	16.5	198	22.4
40-60 hp (three-cylinder)	18	216	24.4
Crankcase cover			
2.5 and 3.3 hp	–	50	5.6
4 and 5 hp	–	90	10.2
6-15 hp	16.6	200	22.6
20 and 25	30	–	40.7
30 and 40 hp (two-cylinder)	16.5	198	22.4
40-60 hp (three-cylinder)	18	–	24.4
Cylinder block (water cover)			
6-15 hp	–	60	6.8
20 and 25 hp	–	140	15.8
Exhaust cover			
6-15 hp	–	60	6.8
20 and 25 hp	–	140	15.8
Flywheel nut/bolt			
2.5 and 3.3 hp	30	–	40.6
4 and 5 hp	40	–	54.2
6-15 hp	50	–	67.8
20 and 25 hp	58	–	78
30 and 40 hp (two-cylinder)	95	–	129
40-60 hp (three-cylinder)	125	–	169
Flywheel ring gear screws			
20 and 25 hp	–	100	11.3

(continued)

Table 1 TORQUE SPECIFICATIONS (continued)

Component	ft.-lb.	in.-lb.	N•m
Intake (transfer) port covers screws			
6-15 hp	–	60	6.8
20 and 25	–	40	4.5
Power head mounting hardware			
2.5 and 3.3 hp	–	50	5.6
4 and 5 hp	–	70	7.9
6-15 hp	–	100	11.3
20 and 25 hp	20	–	27.1
30 and 40 hp (two-cylinder)	29	–	39.3
40-60 hp (three-cylinder)	28	–	38.0
Spark plug			
2.5 and 3.3 hp	20	–	27.1
4 and 5 hp	14	168	19
6-60 hp	20	–	27.1
Thermostat cover			
6-15 hp	–	60	6.8
20 and 25	–	140	15.8
30 and 40 hp (two-cylinder)	16.5	198	22.4
40-60 hp (three-cylinder)	18	216	24.4
Side shift handle			
6-15 hp	–	35	4.0
20 and 25 hp	–	50	5.6
Trim cover (lower cowl)			
30 and 40 hp (two-cylinder)	–	85	9.6
40-60 hp (three-cylinder)	–	80	9.0

*Initial torque specification only. Tighten the fastener using the *Torque and Turn* method as described in the assembly procedures.

8

Table 2 STANDARD TORQUE SPECIFICATIONS—U.S. STANDARD AND METRIC FASTENERS

Screw or nut size	ft.-lb.	in.-lb.	N•m
U.S. standard fasteners			
6-32	–	9	1
8-32	–	20	2.3
10-24	–	30	3.4
10-32	–	35	4.0
12-24	–	45	5.1
1/4-20	6	72	8.1
1/4-28	7	84	9.5
5/16-18	13	156	17.6
5/16-24	14	168	19
3/8-16	23	270	31.2
3/8-24	25	300	33.9
7/16-14	36	–	48.8
7/16-20	40	–	54
1/2-13	50	–	67.8
1/2-20	60	–	81.3
Metric fasteners			
M5	–	36	4.1
M6	6	70	8.1
M8	13	156	17.6
M10	26	312	35.3
M12	35	–	47.5
M14	60	–	81.3

Table 3 CONNECTING ROD SERVICE SPECIFICATIONS

Measurement	Specification in. (mm)
Maximum misalignment (on a surface plate)	
6-40 hp (two-cylinder)	0.002 (0.051)
Deflection at small end	
2.5 and 3.3 hp	0.022-0.056 (0.559-1.422)
Side clearance at big end	
4 and 5 hp	0.005-0.015 (0.127-0.381)
Piston end inside diameter	
6-15 hp	0.8195 (20.815)
20 and 25 hp	0.897 (22.784)
30 and 40 hp (two-cylinder)	0.957 (24.308)
Crankshaft end inside diameter	
6-15 hp	1.0635 (27.013)
20 and 25 hp	1.196 (30.378)
30 and 40 hp (two-cylinder)	1.499 (38.075)

Table 4 CRANKSHAFT SERVICE SPECIFICATIONS

Measurement	Specification in. (mm)
Crankpin diameter	
6-15 hp	0.8125 (20.638)
20 and 25 hp	0.883 (22.43)
30 and 40 hp (two-cylinder)	1.181 (29.997)
Center main journal diameter	
6-15 hp	0.8108 (20.594)
20 and 25 hp	1.0 (25.4)
30 and 40 hp (two-cylinder)	1.216 (30.886)
Top main journal diameter	
6-15 hp	0.7517 (19.093)
20 and 25 hp	1.251 (31.775)
30 and 40 hp (two-cylinder)	1.375 (34.925)
Bottom main journal diameter	
6-15 hp	0.7880 (20.015)
20 and 25 hp	1.125 (28.575)
30 and 40 hp (two-cylinder)	1.385 (35.179)
End play	
20 and 25 hp	0.004-0.019 (0.102-0.483)
Maximum crankshaft runout	
2.5 and 3.3 hp	0.001 (0.025)
4 and 5 hp	0.002 (0.051)
6-25 hp	0.003 (0.076)
30 and 40 hp (two-cylinder)	0.003 (0.076)

Table 5 CYLINDER BLOCK SERVICE SPECIFICATIONS

Measurement	Specification in. (mm)
Bore diameter-standard	
2.5 and 3.3 hp	1.850-1.852 (47.00-47.041)
4 and 5 hp	2.165 (54.991)
6 and 8 hp	2.125 (53.98)
9.9 and 15 hp	2.375 (60.33)
20 and 25 hp	2.562 (65.07)
30-60 hp	2.993 (76.02)

(continued)

Table 5 CYLINDER BLOCK SERVICE SPECIFICATIONS (continued)

Measurement	Specification in. (mm)
Maximum cylinder out-of-round	
2.5 and 3.3 hp	0.002 (0.051)
4 and 5 hp	0.003 (0.076)
6-15 hp	0.004 (0.102)
20-60 hp	0.003 (0.076)
Maximum cylinder taper	
2.5 and 3.3 hp	0.002 (0.051)
4 and 5 hp	0.003 (0.076)
6-15 hp	0.004 (0.102)
20-60 hp	0.003 (0.076)
Maximum cylinder head warpage	
2.5-5 hp	0.002 (0.051)

Table 6 PISTON SERVICE SPECIFICATIONS

Measurement	Specification in. (mm)
Piston-to-cylinder clearance	
2.5 and 3.3 hp	0.002-0.005 (0.051-0.127)
4 and 5 hp	0.0012-0.0024 (0.030-0.061)
6-15 hp	0.002-0.005 (0.051-0.127)
20 and 25 hp	0.003-0.004 (0.076-0.102)
Skirt diameter	
4 and 5 hp	2.164 (54.97)
6 and 8 hp	2.123 (53.92)
9.9 and 15 hp	2.373 (60.27)
20 and 25 hp	2.5583-2.5593 (64.981-65.006)
30-60 hp	2.988 (75.895)
Piston ring end gap	
2.5 and 3.3 hp	0.006-0.012 (0.15-0.30)
4 and 5 hp	0.008-0.016(0.20-0.41)
6-15 hp	0.010-0.018 (0.25-0.46)
20 and 25 hp	0.011-0.025 (0.28-0.64)
30-60 hp	0.010-0.018 (0.25-0.46)
Piston ring side clearance	
2.5 and 3.3 hp	0.0003-0.001 (0.0076-0.0254)
4 and 5 hp	
Top ring	0.0012-0.0028 (0.03-0.07)
Second ring	0.0008-0.0024 (0.02-0.06)

Table 7 MODEL NUMBER CODES

Code	Definition
E	Electric start
H	Handle (tiller steering handle)
M	Manual start (rope recoil starter)
O	Oil injection
PT	Power trim and tilt
L	Long shaft (20 in.)
XL	Extra long shaft (25 in.)

Chapter Nine

Gearcase

This chapter provides lower gearcase removal/installation, rebuilding and resealing procedures for all models. **Table 1** lists specific torque specifications for most gearcase fasteners. **Table 2** lists standard torque specifications. Use the standard torque specifications for fasteners not in **Table 1**. **Table 3** lists the gear ratio and lubricant capacity for all models. **Table 4** lists service specifications. **Tables 1-4** are at the end of the chapter.

Exploded illustrations of each lower gearcase are in the appropriate *Disassembly* section and are helpful references for many service procedures.

The lower gearcase can be removed from the outboard motor without removing the entire outboard from the boat.

The gearcases covered in this chapter differ in construction and require different service procedures. The chapter is arranged in a normal disassembly/assembly sequence. When only a partial repair is required, follow the procedure(s) to the point where the faulty parts can be replaced, then jump ahead to reassemble the unit.

Since this chapter covers a large range of models, the gearcases shown in the accompanying illustrations are the most common models. While the components shown in the pictures may not be identical to components being serviced, the step-by-step procedures covers each model in this manual.

GEARCASE OPERATION

A drive shaft transfers engine torque from the engine crankshaft to the lower gearcase (**Figure 1**). A pinion (drive) gear on the drive shaft is in constant mesh with forward and reverse (driven) gears in the lower gearcase housing. These gears are spiral bevel cut to change the vertical power flow into the horizontal flow required by the propeller shaft. The spiral bevel design also provides for smooth, quiet operation.

All models, except 2.5 and 3.3 hp, have full shifting capability. A sliding clutch, splined to the propeller shaft, engages the spinning forward or reverse gear. This creates a direct coupling of the drive shaft to the propeller shaft. Since this is a straight mechanical engagement, shift the engine only at idle speed. Shifting at higher speeds will cause premature gearcase failure.

The 2.5 hp gearcase only has a pinion and forward gear. The propeller shaft is directly connected to the forward

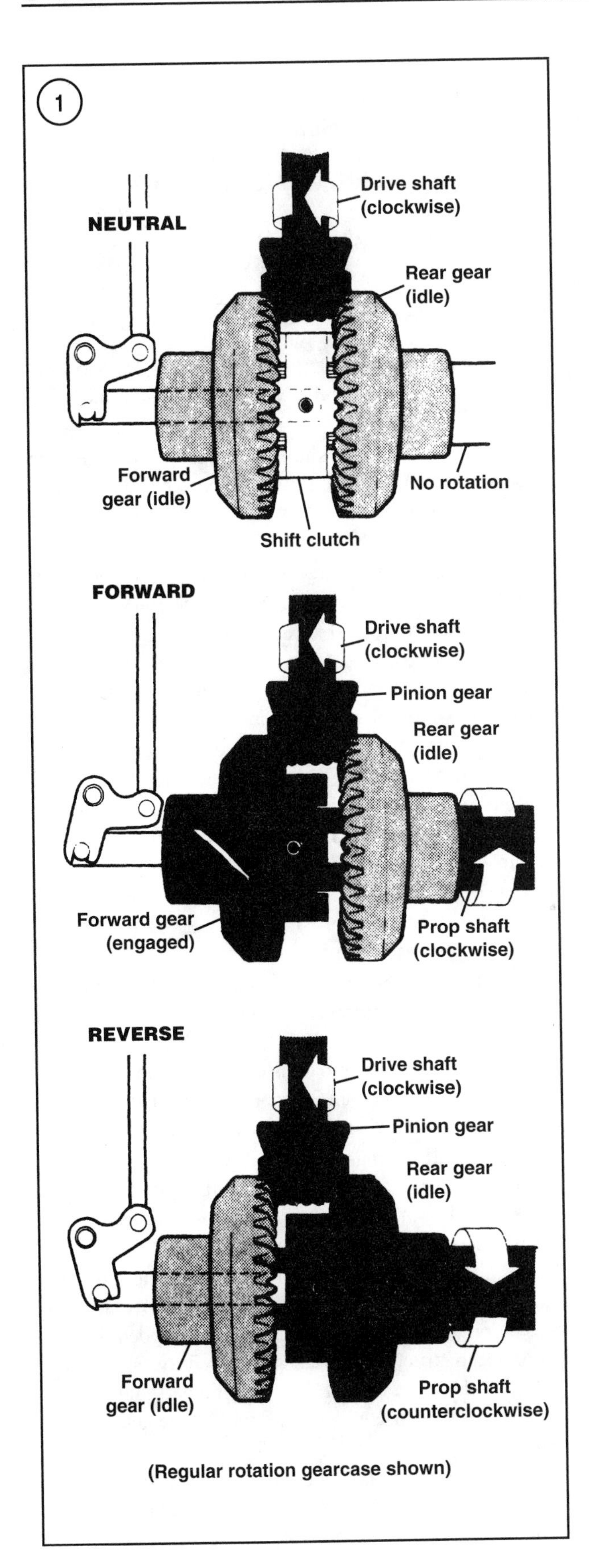

(Regular rotation gearcase shown)

gear at all times. This is a direct drive gearcase, meaning that anytime the engine is cranked or running the propeller shaft is turning.

The 3.3 hp gearcase also only has a pinion and forward gear, but the propeller shaft has a sliding clutch to allow *neutral* and *forward* operation. There is no reverse.

All lower gearcases incorporate a water pump to supply cooling water to the power head. The water pump can be changed on 2.5 hp models without removing the gearcase from the engine. All other models require gearcase removal to service the water pump. Water pump removal and installation procedures are in this chapter.

Larger gearcases use precision shimmed gears. The gears are precisely located in the gear housing by very thin metal spacers, called shims. After assembly, check correct shimming of the gears by measuring the *gear lash*, also called *backlash*. Gear lash is the measurement of the clearance or air gap between a tooth on the pinion gear and two teeth on the forward or reverse gear.

Excessive gear lash indicates that the gear teeth are too far apart. This will cause excessive gear noise and a reduction in gear strength and durability.

Insufficient gear lash indicates that the gear teeth are too close together. This can result in gear failure since there is not enough clearance to maintain a film of lubricant. Heat expansion compounds the problem.

GEAR RATIO

The gear ratio refers to the amount of gear reduction between the crankshaft and the propeller shaft provided by the lower gearcase. Gear ratios range from as low as 2.30:1 to as high as 1.64:1. A gear ratio of 2.30:1 means the crankshaft turns 2.3 times for every one turn of the propeller shaft. Higher number ratios are easier for the engine to turn.

If the gear ratio is suspected as being incorrect, the gear ratio can be determined by two different methods. The first method does not require removal of the gearcase. Mark the flywheel and a propeller blade for counting purposes. Manually shift the gearcase into *forward* gear. Turn the flywheel in the normal direction of rotation (clockwise as viewed from the top of the flywheel) until the propeller shaft has made exactly ten turns while counting the number of flywheel rotations. Divide the number of flywheel rotations by ten and compare the result with the list of gear ratios in **Table 3**. Round the result to the nearest ratio listed.

The second method of determining gear ratio involves counting the actual number of teeth on the gears. This method requires at least partial disassembly of the gearcase. To determine the gear ratio, divide the driven gear tooth count (forward or reverse gear) by the drive gear (pinion or drive shaft gear) tooth count.

For example, on a gearcase with a 15:28 (drive to driven) tooth count, divide 28 (driven) by 15 (drive) which is a 1.87 ratio.

Only operate the engine with the factory recommended gear ratio. Running the engine with an incorrect gear ratio can cause poor performance and poor fuel economy, and make it difficult or impossible to obtain the correct wide-open throttle engine speed which may lead to power head failure.

Some horsepower groups that use the same gearcase housing have several factory ratios, depending on the exact horsepower of the engine within that group. Refer to **Table 3** for gear ratio specifications for each specific engine.

Regardless of gear ratio, only operate engine within the recommended speed range at wide-open throttle. Change propeller pitch and diameter as necessary to adjust engine speed. Increasing pitch or diameter increases the load on the engine and reduces the wide-open throttle speed. Decreasing the pitch or diameter reduces the load on the engine and increases the wide-open throttle speed. Use an accurate shop tachometer for wide-open throttle engine speed verification.

HIGH-ALTITUDE OPERATION

On certain models, a high-altitude gear ratio set is available. The factory recommended gear ratio is adequate for altitudes to 5000 ft. (1524 m). At higher altitudes, change the gear ratio to a higher number ratio to compensate for the loss of horsepower caused by the thinner air.

The propeller may also need to be changed to maintain the recommended wide-open throttle speed range.

If the boat is returned to lower altitudes, change the gear ratio back to the factory recommended ratio and adjust the wide-open throttle speed with propeller changes as necessary.

At high altitude, the lower density air affects the engine's carburetor calibration, causing the engine's air/fuel mixture to become richer. Richer mixtures cause the engine to produce less horsepower and lead to fouled spark plugs, reduced fuel economy and accelerated carbon build-up in the combustion chamber.

All Mercury/Mariner outboards come from the factory calibrated to operate efficiently between sea level and 2500 ft. (762 m). Mercury Marine recommends rejetting the carburetor for operation at altitudes of 5000 ft. (1524 m) or higher.

NOTE

If the boat is operated in a high-altitude environment temporarily, only change the propeller to achieve the correct wide-open throttle speed while at high altitude. Change back to the original propeller when the boat is returned to its normal altitude.

SERVICE PRECAUTIONS

When working on a gearcase, keep in mind the following precautions to make the work easier, faster and more accurate.

1. Never use elastic locknuts more than twice. Replace older nuts when they are removed. Never use an elastic locknut that can be turned by hand without the aid of a wrench.
2. Use special tools where noted. Makeshift tools can damage components and cause serious personal injury.
3. Use the appropriate fixtures to hold the gearcase housing whenever possible. Use a vise with protective jaws to hold smaller housings or individual components. If protective jaws are not available, insert blocks of wood or similar padding on each side of the housing or component before clamping.
4. Remove and install pressed-on parts with an appropriate mandrel, support and arbor or hydraulic press. Do not attempt to pry or hammer press-fit components on or off.
5. Refer to **Table 1** and **Table 2** for torque values. Proper torque is essential to ensure long life and satisfactory service from gearcase components.
6. To help reduce corrosion, especially in saltwater areas, apply Quicksilver Perfect Seal (part No. 92-34227-1 or an equivalent) to all external surfaces of bearing carriers, housing mating surfaces and fasteners when no other sealant, adhesive or lubricant is recommended. Do not apply sealing compound where it can get into gears or bearings.
7. Remove all O-rings, seals and gaskets during disassembly. Apply Quicksilver 2-4-C Multi-Lube grease (part No. 92-825407 or an equivalent) to new O-rings and seal lips to provide initial lubrication.
8. Tag all shims with the location and thickness of each shim as it is removed from the gearcase. Shims are reusable as long as they are not damaged or corroded. Follow shimming instructions closely and carefully. Shims control gear location and/or bearing preload. Incorrectly shimming a gearcase will cause failure of the gears and/or bearings.
9. Work in an area with good lighting and sufficient space for component storage. Keep an ample number of clean containers available for parts storage. Cover parts and as-

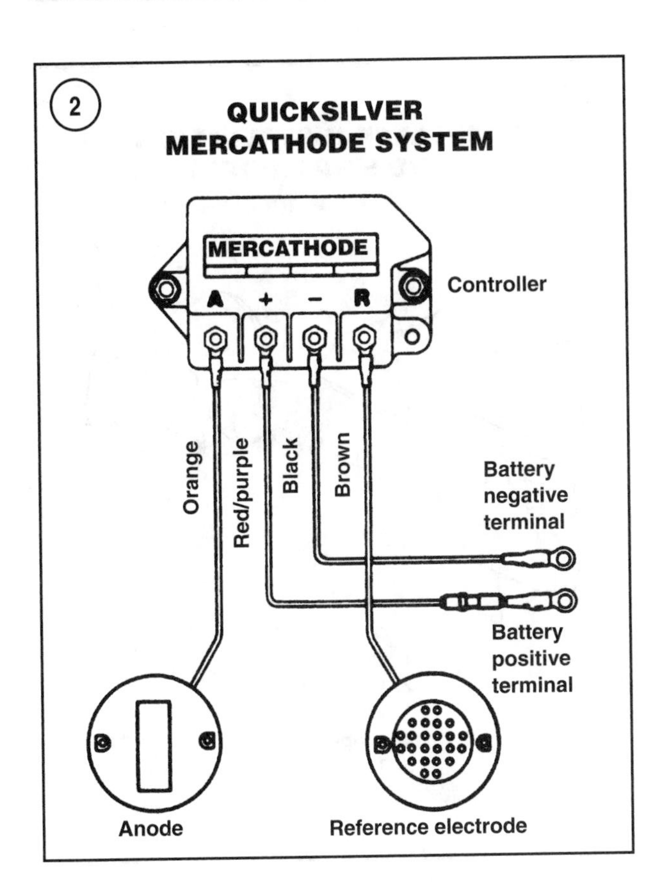

semblies with clean shop towels or plastic bags when they are not being worked on.

10. Whenever a threadlocking adhesive is specified, first spray the threads of the threaded hole or nut and the screw with Locquic Primer. Allow the primer to air dry before proceeding. Locquic primer cleans the surfaces and allows better adhesion. Locquic primer also accelerates the cure rate of threadlocking adhesives from an hour or longer, to 15-20 minutes.

CAUTION

Metric and American fasteners are used on Mercury/Mariner gearcases. Always match a replacement fastener to the original. Do not run a tap or thread chaser into an opening or over a bolt without first verifying the thread size and pitch. Check all threaded openings for Heli-Coil stainless steel locking thread inserts. Never run a tap or thread chaser into a Heli-Coil equipped opening. Heli-Coil inserts are replaceable if damaged.

CORROSION CONTROL

Sacrificial zinc or aluminum anodes are standard equipment on all models. The anodes must have good electrical continuity to ground or they will not function. Anodes are inspected visually and tested electrically. Anodes must not be painted or coated with any material.

The most common location for the anode is the anodic trim tab, but some newer models use a painted trim tab with an anode mounted at the rear of the gearcase, above the antiventilation plate. On 30 hp and larger engines, an anode is mounted across the bottom of the clamp brackets. Refer to the illustrations of the lower gearcase in the appropriate *Disassembly* section for exact anode location(s) for a specific gearcase.

If the unit is operated exclusively in freshwater, magnesium anodes are available from Quicksilver Parts and Accessories. Magnesium anodes provide better protection in freshwater, but must *not* be used in saltwater. Magnesium anodes overprotect the unit in saltwater and cause the paint to blister and peel.

Electronic corrosion control, or the MerCathode system, is also available from Quicksilver Parts and Accessories. The controller module is mounted inside the bilge and the reference electrode and anode are mounted on the transom, below the waterline (**Figure 2**).

Sacrificial Anode Visual Inspection

Check for loose mounting hardware, make sure the anodes are not painted and check the amount of deterioration present. Replace anodes if they are half their original size. Test the electrical continuity of each anode after installation as described in the next section.

Sacrificial Anode Electrical Testing

1. Set the ohmmeter on the R × 1 scale.
2. Connect one ohmmeter lead to the anode being tested. Connect the other ohmmeter lead to a good ground point on the gearcase on which the anode is mounted. The meter should indicate continuity.
3. If the meter indicates an open circuit, remove the anode and clean the mounting surfaces of the anode, gearcase and mounting hardware. Reinstall the anode and retest continuity.
4. Test the continuity of the gearcase to the engine and negative battery post by connecting one ohmmeter lead to the negative battery cable and the other ohmmeter lead to a good ground point on the lower gearcase. The meter should indicate continuity.
5. If the meter indicates an open circuit, check the electrical continuity of the lower gearcase to the drive shaft housing, the upper drive shaft housing to the power head and the power head to the negative battery terminal.

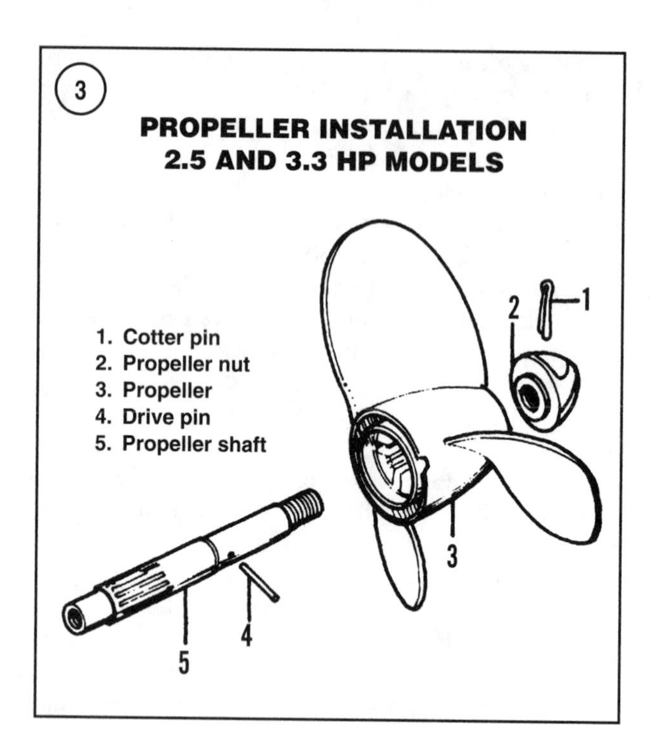

Check for loose mounting hardware, broken or missing ground straps, or excessive corrosion. Repair as necessary to establish a good electrical ground path.

GEARCASE LUBRICATION

Lubricate the gearcase periodically to ensure maximum performance and durability. Change the gearcase lubricant every 100 hours of operation or once each season.

The recommended lubricant for all models is Quicksilver Premium Blend Gear Lubricant. If the gearcase is subjected to severe duty, consider using Quicksilver High Performance Gear Lubricant.

Refer to Chapter Four to change the lower gearcase lubricant.

PROPELLER

The outboard motors covered in this manual use many variations of propeller attachments.

2.5 and 3.3 hp models use a propeller drive pin that engages notches in the propeller hub to a hole in the propeller shaft. See **Figure 3**. The drive pin is secured by the propeller, which is secured by a cotter pin. The drive pin is designed to absorb the propeller thrust and shearing loads.

4 and 5 hp models use a splined propeller shaft and propeller hub to transfer engine torque. A thrust washer transfers propeller thrust to a stepped muster on the propeller

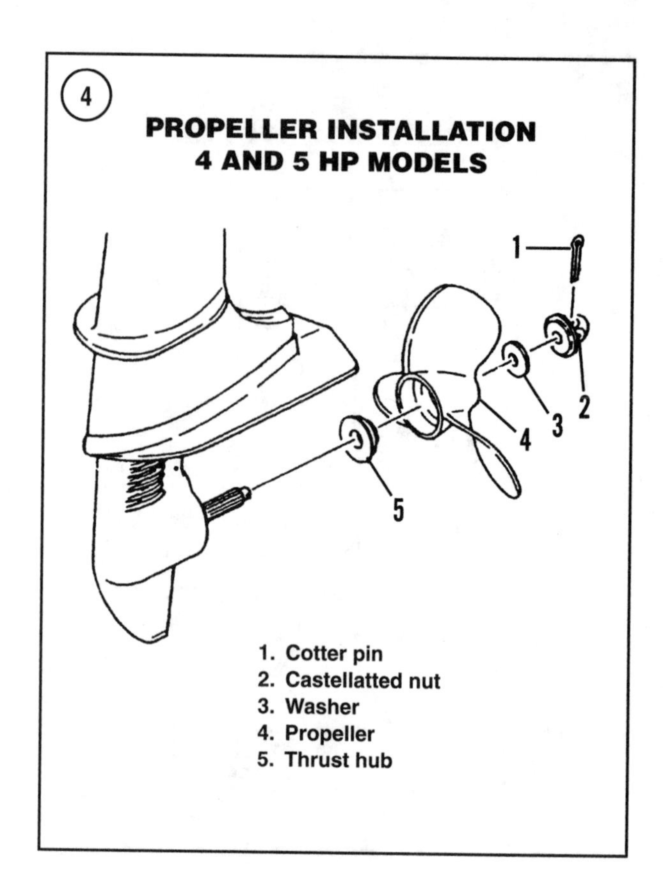

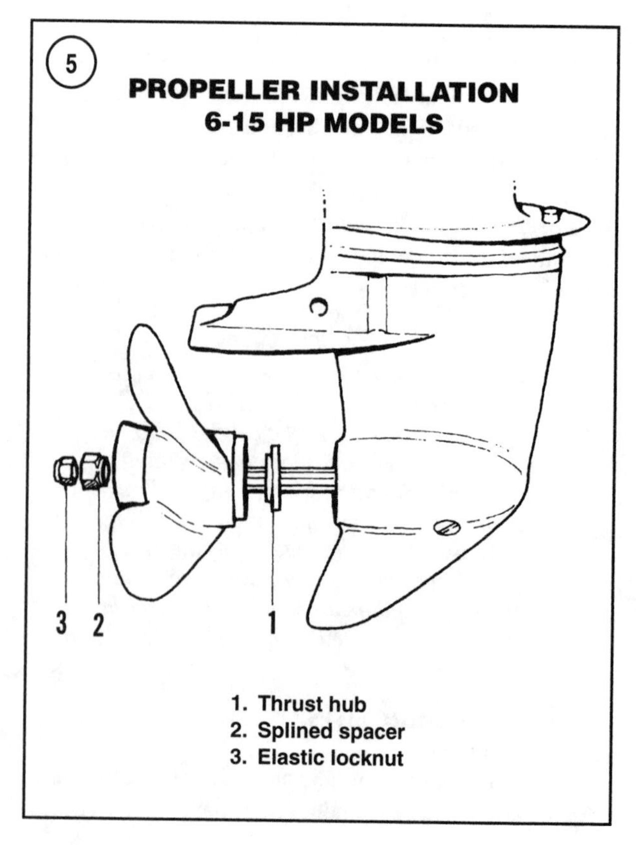

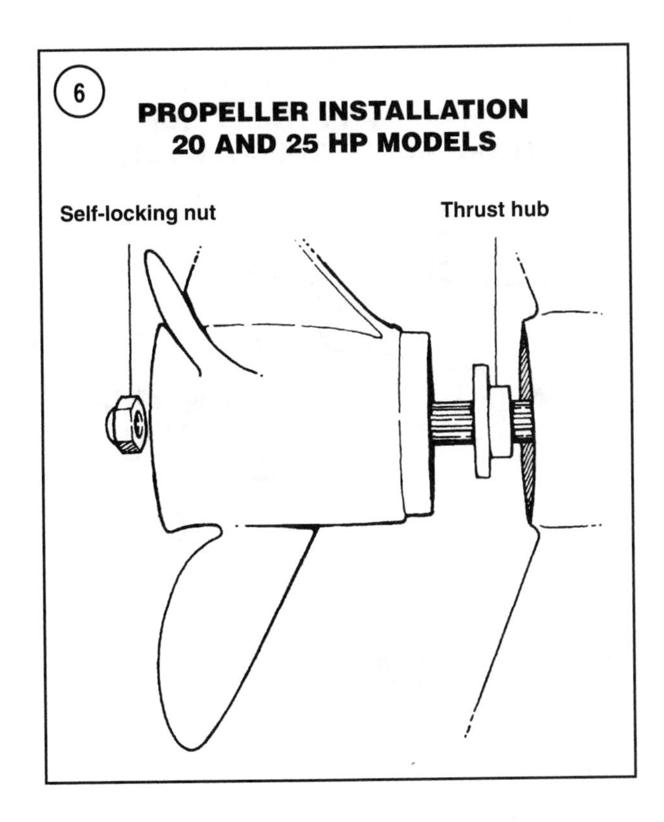

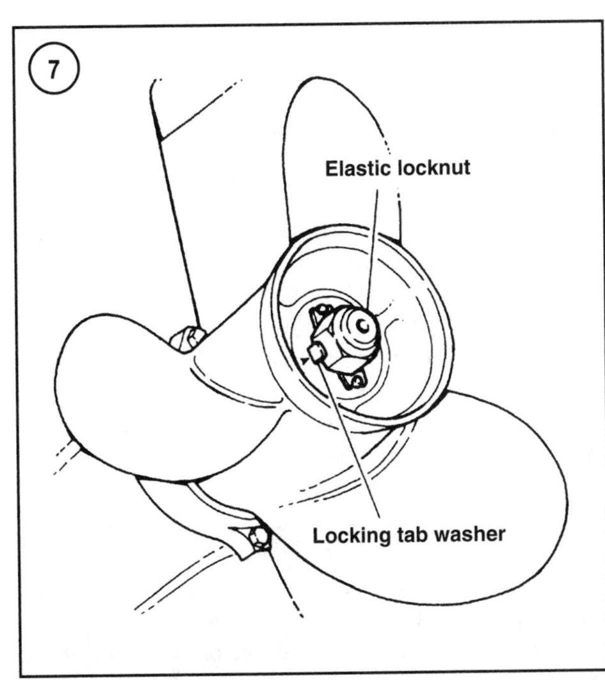

shaft. The propeller is retained by a washer, castellated nut and cotter pin. See **Figure 4**.

6-15 hp models use a splined propeller shaft and propeller hub to transfer engine torque. A thrust washer transfers propeller thrust to a stepped muster on the propeller shaft.

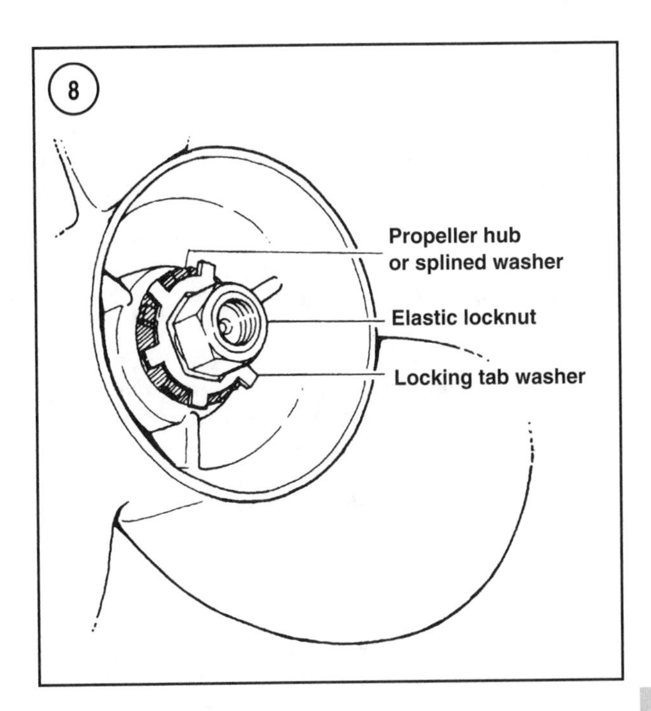

The propeller is retained by a splined spacer and a self-locking propeller nut. See **Figure 5**.

20 and 25 hp models use a propeller that requires a thrust hub that rides against a stepped muster on the propeller shaft. The small diameter end of the thrust hub must face the gearcase and the large diameter end must face the propeller. The propeller is retained by an elastic locknut without a washer. See **Figure 6**.

30-60 hp models use propellers that push against a thrust washer that rides against a tapered step on the propeller shaft. The propellers are retained by an elastic locknut and a lock tab washer that engages two protrusions on the propeller hub, or by a spline washer (built into propeller on some models), a lock tab washer and an elastic locknut. After the elastic locknut is tightened, the locking tabs are either bent up against the nut (**Figure 7**) or driven down into the propeller hub or spline washer (**Figure 8**).

All propellers use a shock absorbing rubber or Delrin hub which absorbs the shock loads produced from shifting the unit into gear. When a hub fails, it generally slips at higher throttle settings, but still allows the boater to return to port at reduced throttle. On 2.5-25 hp models, the propeller must be replaced if the hub fails. On 30-60 hp engines, the defective rubber hub can be removed and a new hub pressed into the propeller using a hydraulic or arbor press (generally at a propeller repair station), or the propeller can be replaced.

Late model Quicksilver or Mercury Marine Propeller Company propellers incorporate the Flo-Torq II square

Delrin drive hub that can be replaced by the operator without the use of a hydraulic or arbor press. See **Figure 9**. If the propeller was not originally equipped with a square rubber hub, it may be possible to upgrade the propeller to the Flo-Torq II style hub. Consult a Mercury or Mariner dealership.

Propeller Removal and Installation

WARNING
To prevent accidental engine starting during propeller service, disconnect and ground all spark plug leads to the power head. Remove the ignition key and safely lanyard from models so equipped.

2.5 and 3.3 hp models

Refer to **Figure 3** for this procedure.

1. Remove the cotter pin (1, **Figure 3**) securing the propeller to the propeller shaft. Slide the propeller off the shaft.
2. Remove the drive pin (4, **Figure 3**) from the propeller shaft. Inspect the pin for damage and replace it if it is distorted, bent or worn.
3. Clean the propeller shaft and propeller hub thoroughly. Inspect the pin engagement hole for elongation, wear or cracks. Rotate the propeller shaft to check for a bent propeller shaft. Replace damaged parts.
4. Liberally coat the propeller shaft and propeller hub bore with Quicksilver 2-4-C Multi-Lube grease (part No. 92-825407) or Special Lube 101 (part No. 92-13872A1).
5. Install the drive pin into the propeller shaft hole, then slide the propeller over the propeller shaft and seat it against the drive pin.
6. Secure the propeller with a new stainless steel cotter pin. Bend both prongs of the cotter pin for a secure attachment.

4-25 hp models

NOTE
The 6-25 hp models do not use a locking tab washer. The elastic locknut is the only means of retention. Replace the nut if the locking feature is diminished.

1. On 4 and 5 hp models, remove the cotter pin (1, **Figure 4**) from the castellated nut (2).
2. Place a block of wood between the propeller blades and antiventilation plate to prevent the propeller from turning. See **Figure 10**, typical.

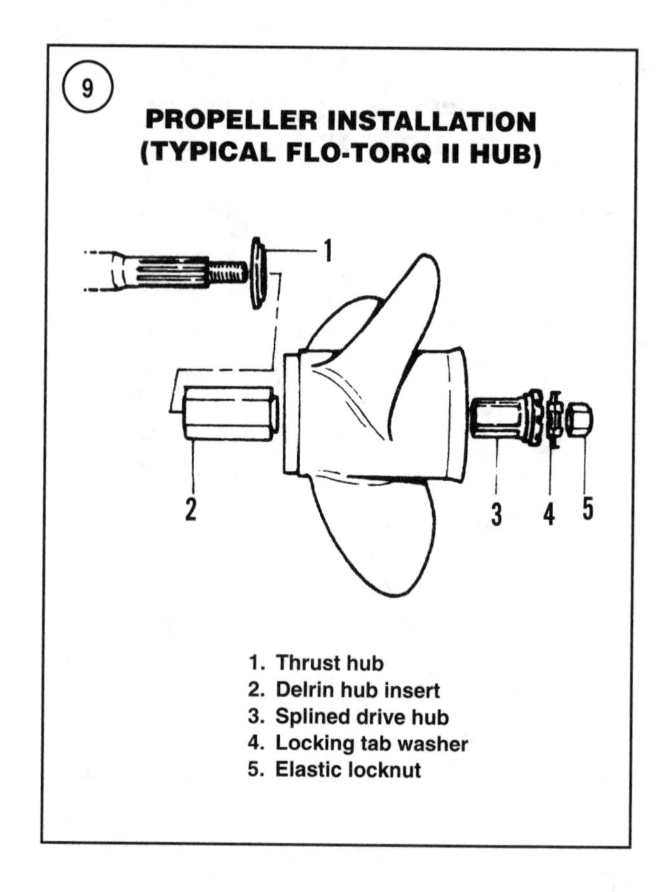

3. Remove the propeller nut and washer or splined spacer. Then remove the propeller.
4. Remove the thrust washer (4-15 hp) or thrust hub (20-25 hp).
5. Clean the propeller shaft thoroughly. Inspect the propeller shaft for cracks, wear or damage. Rotate the propeller shaft to check for a bent propeller shaft. Inspect the propeller thrust washer or thrust hub and rear washer or splined spacer for wear or damage. Replace damaged parts.
6. Lubricate the propeller shaft liberally with 2-4-C Multi-Lube grease (part No. 92-825407) or Special Lube 101 (part No. 92-13872A1).

7A. *4-15 hp models*—Slide the propeller thrust washer onto the propeller shaft. The large diameter of the thrust washer must face the gearcase. Then align the splines and seat the propeller against the thrust hub.

7B. *20 and 25 hp models*—Slide the propeller thrust hub (**Figure 6**) onto the propeller shaft. The small diameter of the thrust washer must face the gearcase. Then align the splines and seat the propeller against the thrust washer.

8A. *4 and 5 hp models*—Install the washer and castellated nut (2 and 3, **Figure 4**). Place a block of wood between a propeller blade and the antiventilation plate (**Figure 10**). Then tighten the nut to the specification in

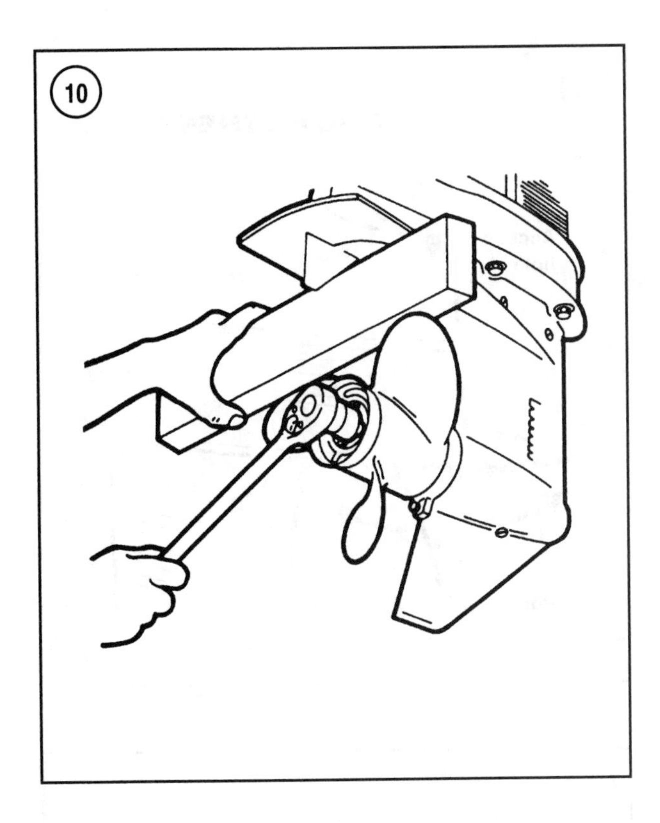

Table 1. Secure the nut with a new stainless steel cotter pin. Bend both prongs of the cotter pin for a secure attachment.

8B. *6-15 hp models*—Install the splined spacer and elastic locknut (2 and 3, **Figure 5**). Place a block of wood between a propeller blade and the antiventilation plate (**Figure 10**). Then tighten the nut to the specification in **Table 1**.

8C. *20 and 25 hp models*—Install the elastic locknut (**Figure 6**). Place a block of wood between a propeller blade and the antiventilation plate (**Figure 10**). Then tighten the nut to the specification in **Table 1**.

30-60 hp models

All models have the rear washer built into the propeller or drive hub on Flo-Torq II models. All models use a locking tab washer and elastic locknut.

1. Pry the lock tab(s) up from the propeller or rear splined washer, or down from the elastic stop nut as required with the appropriate tool.
2. Place a suitable block of wood between a propeller blade and the antiventilation plate to prevent propeller rotation. See **Figure 10**.
3. Remove the propeller elastic stop nut with a socket. Replace the nut if it can be unthreaded by hand.
4. Slide the propeller and all related hardware from the propeller shaft.
5. Clean the propeller shaft thoroughly. Inspect the propeller shaft for cracks, wear or damage. Rotate the propeller shaft to check for a bent propeller shaft. Inspect the propeller thrust washer and rear washer(s) for wear or damage. Replace damaged parts.
6. On Flo-Torq II models, inspect the Delrin hub (2, **Figure 9**) for wear, deterioration, damage or failure.
7. Lubricate the propeller shaft liberally with 2-4-C Multi-Lube (part No. 91-825407) or Special Lube 101 (part No. 92-13872A1).
8. Slide the propeller thrust washer onto the propeller shaft.

9A. *Rubber hub models*—Align the splines and seat the propeller against the thrust washer.

9B. *Flo-Torq II models*—Assemble the Delrin hub, propeller and drive hub as shown in **Figure 9**. Align the splines and seat the propeller and drive hub against the thrust washer.

10. Install the continuity washer (if equipped), splined washer (if equipped), locking tab washer and the elastic locknut.
11. Place a suitable block of wood between a propeller blade and the antiventilation plate (**Figure 10**) to prevent propeller rotation and tighten the propeller nut to the specification in **Table 1**.
12. Secure the propeller nut in one of the following ways:
    a. Bend both lock tabs securely against the appropriate flats of the elastic stop nut. See **Figure 7**. If necessary, tighten the propeller nut slightly to align the tabs.
    b. Select three lock tabs that align with the notches in the propeller hub or rear splined washer. Drive the lock tabs into the notches with a hammer and punch. See **Figure 8**. If necessary, tighten the propeller nut slightly to align the tabs.

TRIM TAB ADJUSTMENT

Adjust the trim tab so the steering wheel turns with equal ease in each direction at the normal cruising speed and trim angle. The trim tab can only provide neutral steering effort for the speed and trim angle for which it was set. Trimming the outboard out (up) or in (down), or changing engine speed changes the torque load on the propeller and the steering effort.

If adjustment is desired, run the boat at the speed and trim angle desired. If the boat turns more easily to starboard than port, loosen the trim tab retaining screw and move the tab trailing edge slightly to starboard. If the boat turns more easily to port, move the tab slightly to port. See

Figure 11, typical. Tighten the trim tab retaining screw to the specification in **Table 1** after adjustment and before water testing.

On models equipped with an anodic trim tab, deterioration of the anode reduces the effectiveness of the trim tab. Replace the anodic trim tab as necessary.

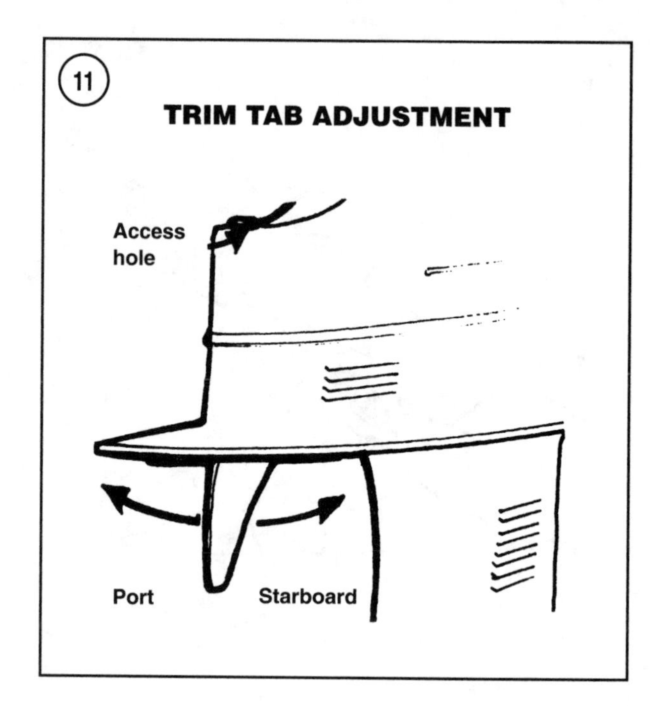

GEAR HOUSING

Removal/Installation (2.5 hp Models)

The water pump can be serviced without gear housing removal on 2.5 hp models. Refer to *Water Pump* in this chapter.

The square drive shaft runs inside a removable support tube in the drive shaft housing. The tube must seal into the lower crankshaft seal.

1. Disconnect and ground the spark plug lead to the power head to prevent accidental starting.
2. Close the fuel shutoff valve.
3. Tilt the outboard to the fully *up* position.
4. Remove the upper mounting bolt, lockwasher and flat washer at the leading edge of the gearcase, and the lower mounting bolt, lockwasher and flat washer directly above the propeller. Then pull the gear housing from the drive shaft housing and place the gearcase on a clean workbench.
5. If the square drive shaft and support tube does not come out with the gearcase, remove the drive shaft and support tube from the drive shaft housing.
6. Remove the water tube seal from the gearcase.
7. To install the gearcase, tilt the outboard to the fully up position.
8. Apply Quicksilver 2-4-C Marine Lubricant (part No. 92-825407) to the outside diameter of the drive shaft support tube (**Figure 12**), then insert the tube into the drive shaft housing until it is seated in the lower crankshaft seal.
9. Apply Quicksilver 2-4-C Multi-Lube grease to the inside diameter of both ends of the upper square drive shaft. Insert the shaft into the tube (**Figure 12**), then engage the shaft to the crankshaft.
10. Glue a new water tube seal into the lower gearcase with Quicksilver Bellows Adhesive (part No. 92-86166). Then apply Quicksilver 2-4-C Multi-Lube grease to the inside diameter of the water tube seal.
11. Apply Quicksilver Perfect Seal (part No. 92-34227-1) to the threads of the two gear housing mounting bolts.
12. Install the gear housing while aligning the water tube and the upper square to the lower drive shaft. Rotate the propeller shaft clockwise to engage the drive shafts. Install the gear housing mounting bolt, lockwashers and flat washers. Tighten both bolts to the specification in **Table 1**.

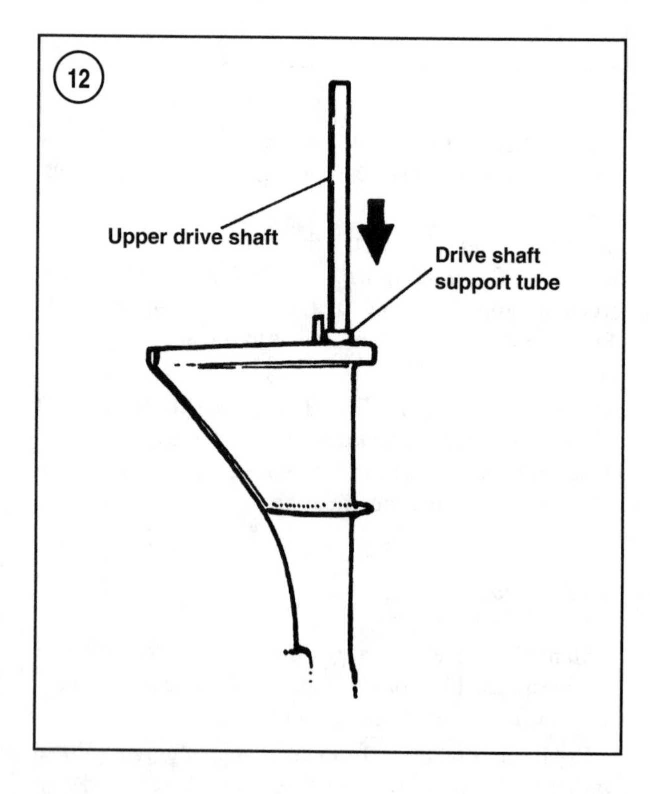

13. Reconnect the spark plug lead.
14. Check the lubricant level or refill the gearcase with the recommended lubricant as described in Chapter Four.

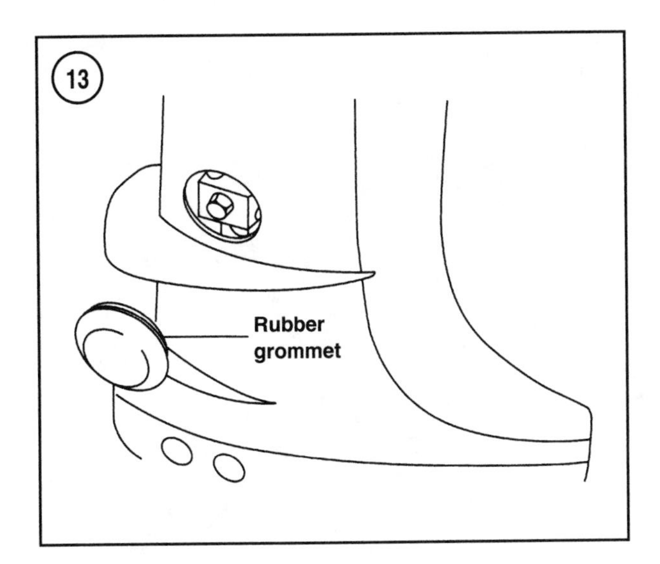

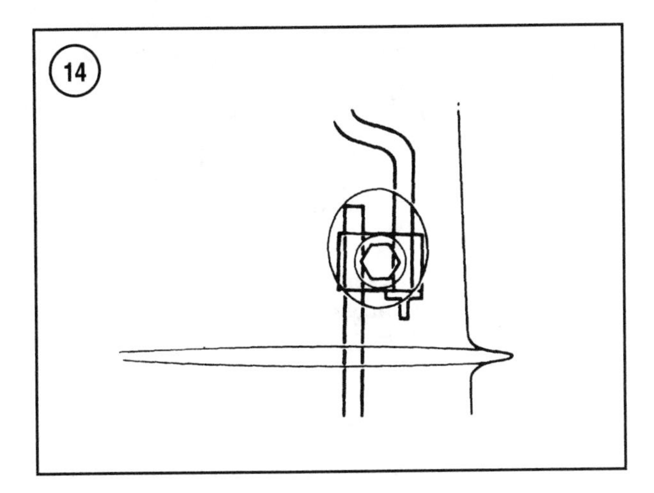

Removal/Installation (3.3 hp Models)

1. Close the fuel shutoff valve.
2. Disconnect and ground the spark plug lead to the power head to prevent accidental starting.
3. Tilt the outboard to the fully up position.
4. Shift the outboard into neutral gear.
5. Remove the rubber grommet on the port side of the drive shaft housing (**Figure 13**). Insert a socket into the hole and loosen, but do not remove, the shift rod clamp screw (**Figure 14**) enough to allow the shift rods to separate.
6. Remove the gearcase mounting bolts, lockwashers and flat washers. Then pull the gearcase straight down and away from the drive shaft housing. It is not necessary to remove the anode screw. Place the gearcase on a clean workbench.
7. Remove the water tube grommet from the water pump housing.
8. To install the gearcase, glue a new water tube grommet into the water pump housing with Quicksilver Bellows Adhesive (part No. 92-86166).

CAUTION
Do not apply lubricant to the top of the drive shaft in the next step. Excess lubricant between the top of the drive shaft and the engine crankshaft can create a hydraulic lock, preventing the drive shaft from fully engaging the crankshaft.

9. Clean the drive shaft splines as necessary, then coat the splines with Quicksilver 2-4-C Multi-Lube grease (part No. 92-825407).
10. Place the gearcase and the outboard shift lever in neutral.
11. Apply Quicksilver 2-4-C Multi-Lube to the inside diameter of the water tube seal and apply Quicksilver Perfect Seal (part No. 92-343227-1) to the threads of the two gear housing mounting bolts.
12. Install the gear housing to the drive shaft housing, guiding the water tube and shift rod into place. The end of the lower shift rod must be properly positioned in the upper shift rod clamp. If necessary, rotate the flywheel clockwise to align the crankshaft and drive shaft splines. When all components are correctly aligned, seat the gearcase to the drive shaft housing.
13. Install the two mounting bolts, lockwashers and flat washers. Tighten the bolts to the specification in **Table 1**.
14. Securely tighten the shift rod clamp bolt.
15. Verify shift rod adjustment by placing the shift lever in neutral. The propeller shaft should turn freely in both directions. Place the shift lever in forward. The propeller shaft should turn approximately 90° in either direction, then stop. If it does not, loosen the shift rod clamp bolt and adjust the rods as necessary.
16. Install the shift rod clamp bolt and rubber access grommet after the shift adjustment is verified.
17. Check the lubricant level or refill the gearcase with the recommended lubricant as described in Chapter Four.
18. Reconnect the spark plug lead.

Removal/Installation (4 and 5 hp Models)

1. Disconnect and ground the spark plug lead to the power head to prevent accidental starting.
2. Shift the outboard into reverse gear.
3. Tilt the outboard to the fully up position and engage the tilt lock lever.
4. Remove the propeller as described in this chapter.

5. Remove the rubber grommet from the starboard side of the drive shaft housing (**Figure 15**). Insert a socket into the hole and loosen the shift shaft clamp bolt enough to allow the shift rods to separate from the clamp.
6. Remove the two bolts and flat washers from the front and rear of the gearcase, just below the antiventilation plate.
7. Carefully pull the gearcase straight down and away from the drive shaft housing.
8. Mount the gearcase in a suitable holding fixture or place the gearcase on a clean workbench.
9. Remove the water tube grommet from the water pump housing.
10. To install the gearcase, install a new water tube grommet into the water pump housing. Make sure the raised tabs align with the holes in the water pump housing.
11. Place the outboard shift lever into reverse. While rotating the propeller shaft clockwise, push downward on the gear housing shift shaft to engage reverse gear.

CAUTION
Do not apply lubricant to the top of the drive shaft in the next step. Excess lubricant between the top of the drive shaft and the engine crankshaft can create a hydraulic lock, preventing the drive shaft from fully engaging the crankshaft.

12. Clean the drive shaft splines as necessary, then coat the splines with Quicksilver 2-4-C Multi-Lube grease (part No. 92-825407).
13. Apply Quicksilver 2-4-C Multi-Lube to the inside diameter of the water tube seal and apply Quicksilver Perfect Seal (part No. 92-34227-1) to the threads of the two gear housing mounting bolts.
14. Install the gear housing to the drive shaft housing, guiding the water tube and shift rod into place. The end of the lower shift rod must be properly positioned in the upper shift rod clamp (**Figure 14**). If necessary, rotate the flywheel clockwise to align the crankshaft and drive shaft splines. When all components are correctly aligned, seat the gearcase to the drive shaft housing.
15. Install the two mounting bolts and flat washers. Tighten the bolts to the specification in **Table 1**.
16. Securely tighten the shift rod clamp bolt (**Figure 16**).
17. Verify shift rod adjustment by placing the shift lever in neutral. The propeller shaft should turn freely in both directions. Place the shift lever in forward. The propeller shaft should turn approximately 90° in either direction, then stop. Place the shift lever in reverse. The propeller shaft should turn approximately 90° in either direction, then stop. If it does not, loosen the shift rod clamp bolt and adjust the rods as necessary.

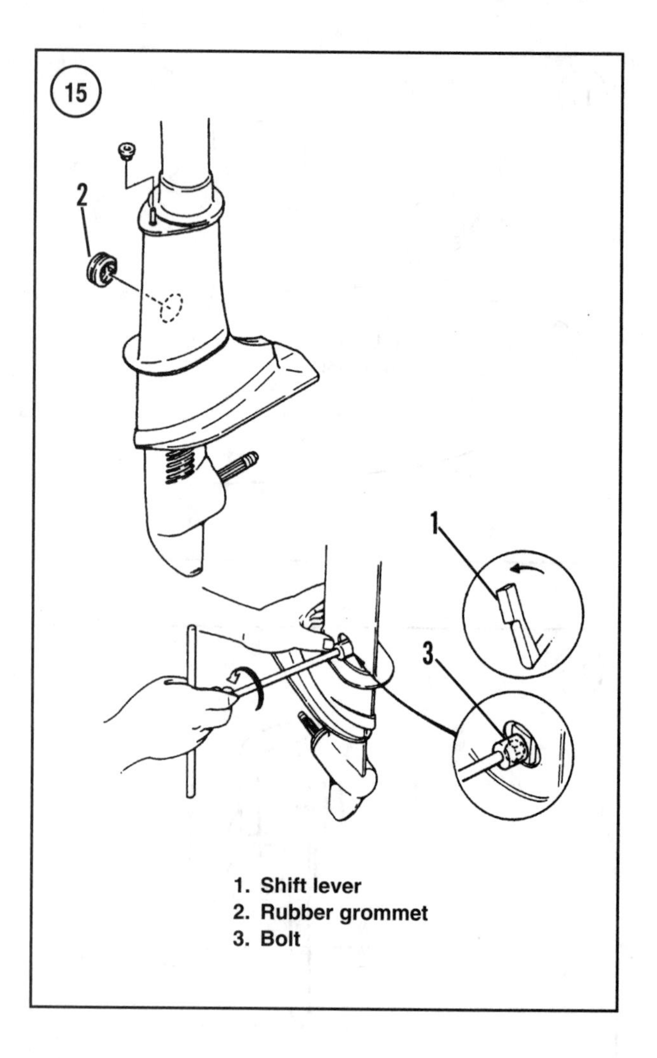

1. Shift lever
2. Rubber grommet
3. Bolt

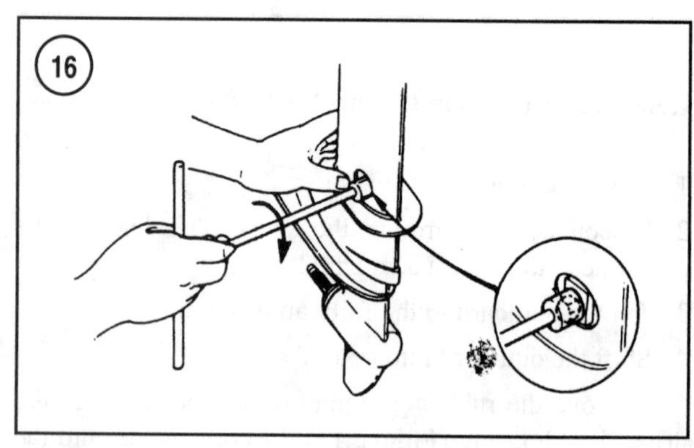

18. Install the shift rod clamp bolt and rubber access plug after the shift adjustment is verified.
19. Reinstall the propeller as described in this chapter.
20. Release the tilt stop/lock and return the outboard to the normal operating position.
21. Check the lubricant level or refill the gearcase with the recommended lubricant as described in Chapter Four.

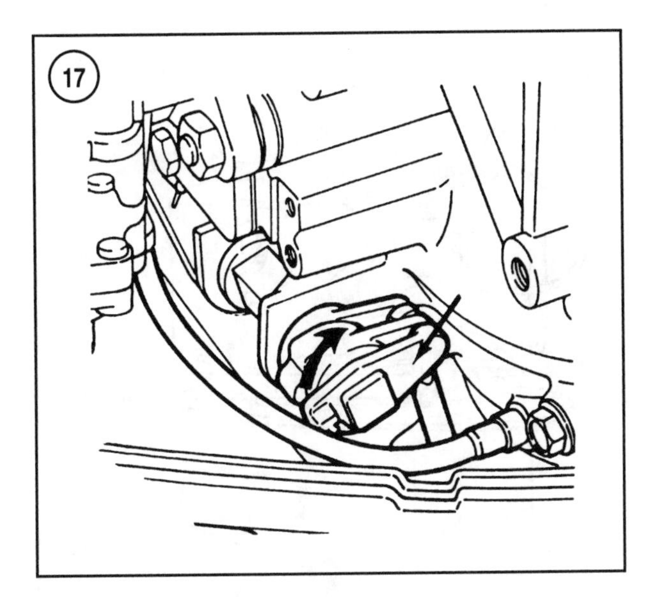

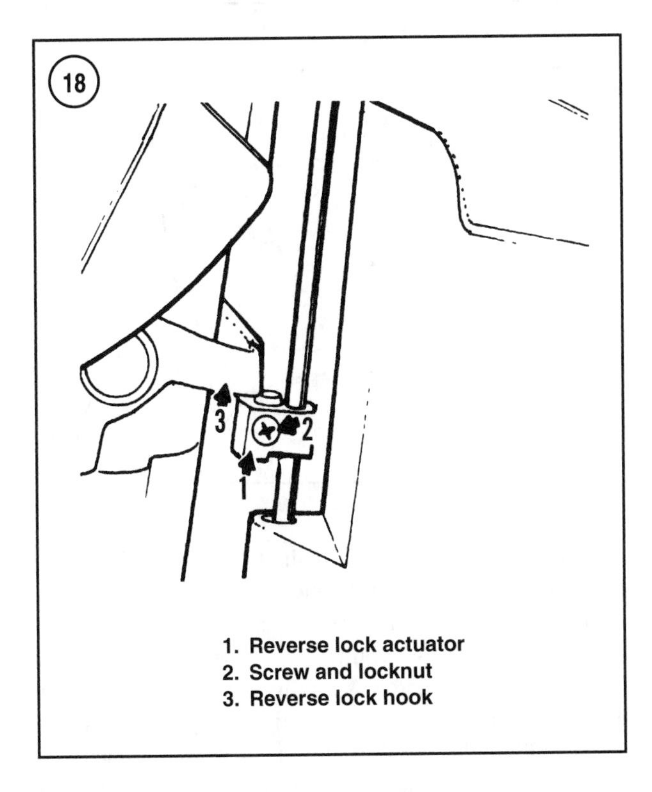

1. Reverse lock actuator
2. Screw and locknut
3. Reverse lock hook

22. Reconnect the spark plug lead.

Removal/Installation (6-15 hp Models)

1. Disconnect and ground the spark plug leads to the power head to prevent accidental starting.
2. Remove the propeller as described in this chapter.
3. Shift the outboard into forward gear.
4. Tilt the outboard to the fully up position, then engage the tilt lock lever.
5. Remove the shift shaft retainer (**Figure 17**) located under the cowl. Rotate the tabbed end of the retainer upward, then slide the retainer from the shift shaft and bellcrank.
6. Remove the reverse lock actuator (1, **Figure 18**) from the shift shaft.
7. Remove the two lower attaching bolts, one on each side of the rear of the gear housing. Then remove the top bolt at the leading edge of the gearcase.
8. Carefully pull the gearcase straight down and away from the drive shaft housing.
9. Mount the gearcase in a suitable holding fixture or place the gearcase on a clean workbench.
10. Remove the water guide tube, then remove the water tube grommet from the water pump housing.
11. To install the gearcase, install a new water tube grommet into the water pump housing. Make sure the raised tabs align with the holes in the water pump housing.
12. Verify the shift shaft height adjustment by measuring from the water pump base to the center of the shift shaft attachment hole with the gearcase in neutral. The propeller should rotate freely in either direction when in neutral. Rotate the shift shaft as necessary to adjust the dimension to the specification in **Table 4**.
13. Once the shift shaft height is verified, shift the gearcase into reverse gear by pulling the shift shaft upward while rotating the propeller.
14. Run a 1/4 in. (6.4 mm) diameter bead of RTV sealant (part No. 92-809826) along the water dam at the rear of the water pump base.

CAUTION
Do not apply lubricant to the top of the drive shaft in the next step. Excess lubricant between the top of the drive shaft and the engine crankshaft can create a hydraulic lock, preventing the drive shaft from fully engaging the crankshaft.

15. Clean the drive shaft splines as necessary, then coat the splines with Quicksilver 2-4-C Multi-Lube grease (part No. 92-825407).
16. Apply Quicksilver 2-4-C Multi-Lube to the inside diameter of the water tube seal and apply Quicksilver Perfect Seal (part No. 92-34227-1) to the threads of the three gear housing mounting bolts.
17. Install the gear housing to the drive shaft housing, guiding the water tube and shift shaft into place. The shift shaft must enter the coupling yoke through the lower cowl opening. If necessary, rotate the flywheel clockwise to align the crankshaft and drive shaft splines. When all

components are correctly aligned, seat the gearcase to the drive shaft housing.

18. Install the three mounting bolts. Tighten the bolts evenly to the specification in **Table 1**.

19. Align the shift shaft with the shift linkage bellcrank. Secure the shift shaft to the bellcrank with the retainer (**Figure 17**). Rotate the retainer down until it locks into place.

20. With the shift linkage in forward gear, the propeller shaft should not turn counterclockwise. Shift the outboard into neutral. The propeller shaft should rotate freely in either direction. Shift the outboard into reverse gear. The propeller shaft should not turn clockwise. Adjust the shift linkage as necessary to ensure correct gear shift operation.

21. Place the shift lever into neutral and loosely install the reverse lock actuator. Position the actuator so it just touches the reverse lock hook and tighten the screw. See **Figure 18**.

22. Reinstall the propeller as described in this chapter.

23. Release the tilt stop/lock and return the outboard to the normal operating position.

24. Check the lubricant level or refill the gearcase with the recommended lubricant as described in Chapter Four.

25. Reconnect the spark plug leads.

Removal/Installation (20 and 25 hp Models)

1. Disconnect and ground the spark plug leads to prevent accidental starting. Disconnect the negative battery cable on electric start models.

2. Remove the shift shaft retainer (**Figure 17**) securing the shift shaft to the shift linkage bellcrank arm. Rotate the tabbed end of the retainer upward, then slide the retainer from the shift shaft and bellcrank.

3. Tilt the outboard to the full up position and engage the tilt lock lever.

4. Remove the propeller as described previously in this chapter.

NOTE

A flat washer is installed near the top of the shift shaft. The washer may slide from the shaft during gearcase removal. Do not lose the washer, as it must be reinstalled in the same location.

5. Remove the four screws (two on each side) and washers securing the gearcase to the drive shaft housing. Pull the gearcase straight down and away from the drive shaft housing.

6. Mount the gearcase in a suitable holding fixture or place the gearcase on a clean workbench.

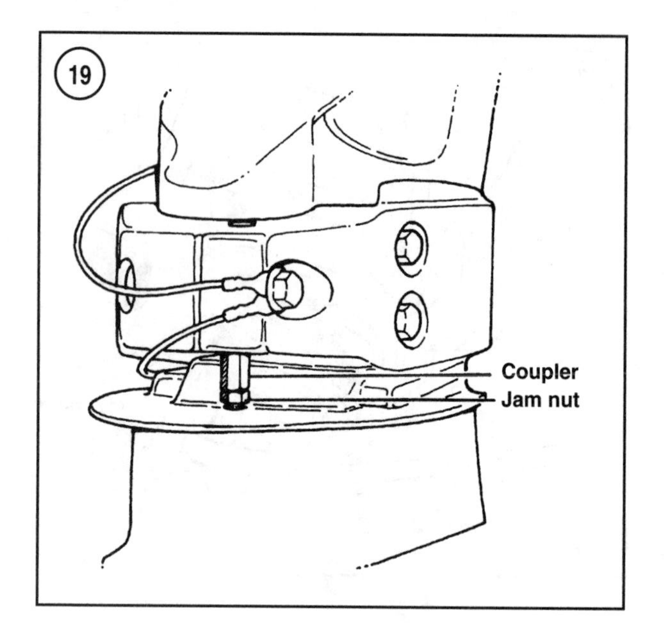

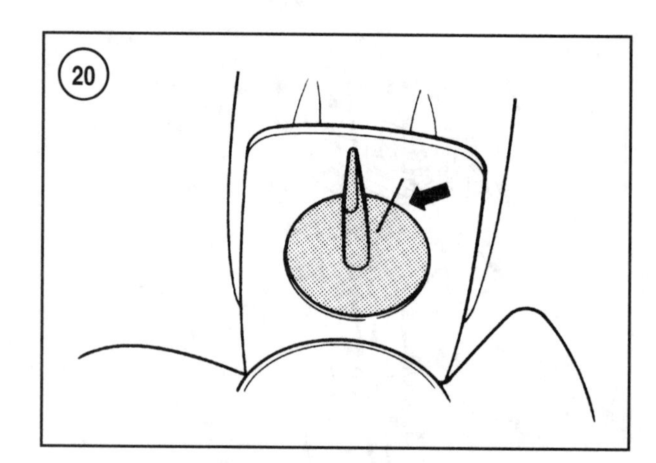

7. Locate and secure the shift shaft flat washer. Then remove the drive shaft O-ring and the water tube grommet from the water pump housing.

8. If the water tube was pulled out of the drive shaft housing when the gearcase was removed, lubricate the upper end with Quicksilver 2-4-C Multi-Lube grease (part No. 92-825407) and reinstall it into the drive shaft housing.

9. To install the gearcase, install a new drive shaft O-ring and water tube grommet into the water pump. Lubricate the O-ring and grommet with Quicksilver 2-4-C Multi-Lube grease.

10. Rotate the drive shaft clockwise while pulling up on the shift shaft until the gearcase is fully engaged in forward gear.

11. Install the flat washer onto the shift shaft. Then make sure the water tube is installed in the drive shaft housing.

12. Place the outboard shift lever into forward gear.

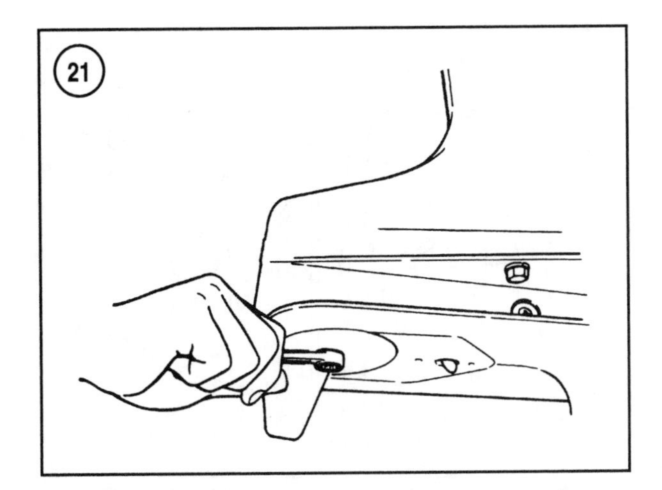

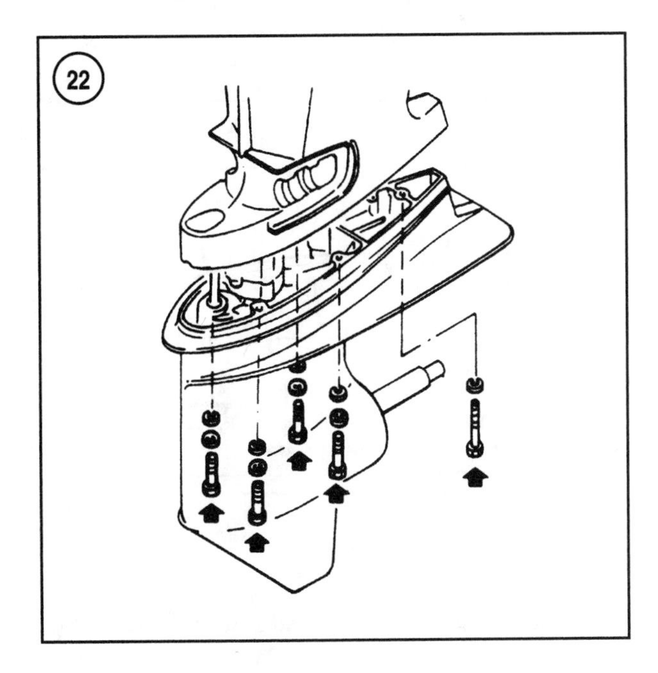

CAUTION

Do not apply lubricant to the top of the drive shaft in the next step. Excess lubricant between the top of the drive shaft and the engine crankshaft can create a hydraulic lock, preventing the drive shaft from fully engaging the crankshaft.

13. Clean the drive shaft splines as necessary, then coat the splines with Quicksilver 2-4-C Multi-Lube grease.
14. Apply Loctite 271 threadlocking adhesive (part No. 92-809819) to the threads of the gearcase mounting bolts.
15. Install the gear housing to the drive shaft housing, guiding the water tube and shift shaft into place. The shift shaft must pass through the reverse lock yoke and align with the bellcrank arm. If necessary, rotate the flywheel clockwise to align the crankshaft and drive shaft splines. When all components are correctly aligned, seat the gearcase to the drive shaft housing.
16. Install the mounting bolts and flat washers. Tighten the bolts evenly to the specification in **Table 1**.
17. Align the shift shaft with the shift linkage bellcrank. Secure the shift shaft to the bellcrank with the retainer. Rotate the retainer down until it locks into place.
18A. *Tiller control models*—With the shift linkage in forward gear, the propeller shaft should not turn counterclockwise. Shift the outboard into neutral. The propeller shaft should rotate freely in either direction. Shift the outboard into reverse gear. The propeller shaft should not turn clockwise. Adjust the shift linkage as necessary to ensure correct gear shift operation.
18B. *Remote control models*—Refer to Chapter Fourteen for shift cable adjustment procedures.
19. Reinstall the propeller as described in this chapter.
20. Release the tilt lock and return the outboard to the normal operating position.
21. Check the lubricant level or refill the gearcase with the recommended lubricant as described in Chapter Four.
22. Reconnect the spark plug leads. Reconnect the negative battery cable on electric start models.

Removal/Installation (30-50 hp Models)

1. Disconnect and ground the spark plug leads to the power head to prevent accidental starting.
2. Tilt the outboard to the full up position and engage the tilt lock.
3. Remove the propeller as described in this chapter.
4. Shift the gearcase into neutral.
5. Loosen the jam nut above the shift coupler. See **Figure 19**. Then turn the shift rod coupler until the upper and lower shift rods are separated.
6. Mark the trim tab position with a felt tip marker (**Figure 20**) and remove the trim tab retaining screw (**Figure 21**). Remove the trim tab.
7. Remove the four screws (two on each side) and washers, and one locknut and washer in the trim tab cavity securing the gearcase to the drive shaft housing. (See **Figure 22**). Pull the gearcase straight down and away from the drive shaft housing.
8. Place the gearcase in a suitable holding fixture or on a clean workbench.
9. If the water tube guide and seal remained on the water tube in the drive shaft housing, remove the water tube guide and seal from the water tube. Inspect the guide and seal, and replace it if it is damaged.
10. To install the gearcase, make sure the water tube guide and seal are securely attached to the water pump

housing. If the guide and seal are loose, glue the guide and seal to the water pump housing with Loctite 405 adhesive.

CAUTION
Do not apply lubricant to the top of the drive shaft in the next step. Excess lubricant between the top of the drive shaft and the engine crankshaft can create a hydraulic lock, preventing the drive shaft from fully engaging the crankshaft.

11. Clean the drive shaft splines as necessary, then coat the splines with Quicksilver 2-4-C Multi-Lube grease (part No. 92-825407). Coat the inner diameter of the water tube seal with the same grease.
12. Shift the control box and gearcase into neutral gear.
13. Position the gearcase under the drive shaft housing. Align the water tube in the water pump and drive shaft with crankshaft splines.

CAUTION
Do not rotate the flywheel counterclockwise in Step 14 or water pump impeller damage can occur.

14. Push the gearcase towards the drive shaft housing, rotating the flywheel clockwise as required to align the drive shaft and crankshaft splines.
15. Make sure the water tube seats in the water pump guide and seal, then push the gearcase against the drive shaft housing. Install the four screws and washers, and one elastic locknut and washer (**Figure 22**). Evenly tighten all fasteners to the specification in **Table 1**.
16. Align the upper and lower shift rods. Reconnect the shift rods by turning the shift rod coupler (**Figure 19**).
17. Adjust the shift linkage. See *Gear shift linkage adjustment* in this chapter.
18. Reinstall the propeller as described previously in this chapter.
19. Install the trim tab and secure it with the screw and washer. Align the marks made during removal (**Figure 20**) and tighten the trim tab screw to the specification in **Table 1**.
20. Check the lubricant level or refill the gearcase with the recommended lubricant as described in Chapter Four.
21. Release the tilt lock and return the outboard to the normal operating position.
22. Reconnect the spark plug leads.

Gear shift linkage adjustment (30-50 hp models)

1. Disconnect the remote control shift cable from the engine (Chapter Fourteen).

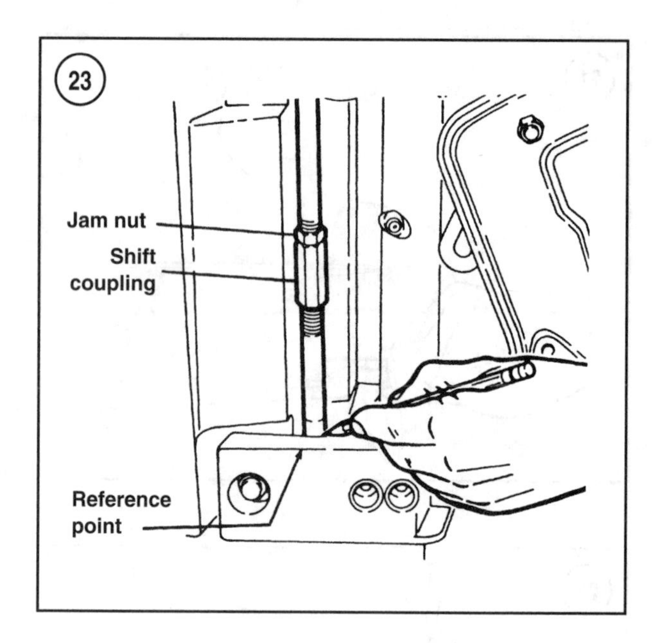

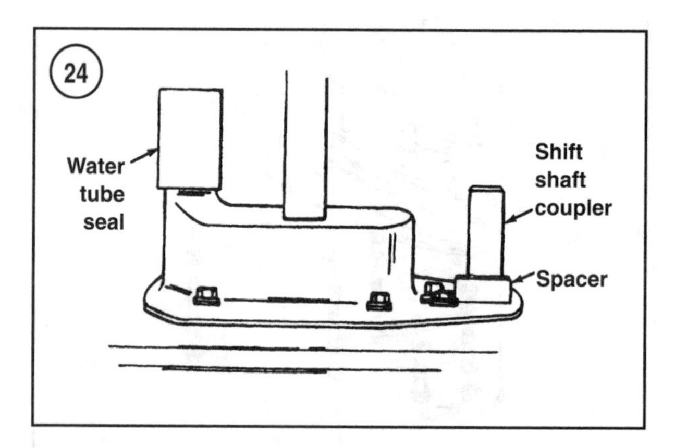

2. Manually move the power head gear shift lever to forward gear while spinning the propeller to ensure full clutch engagement.
3. Mark an accessible part of the shift rod, referenced from a drive shaft housing component. See **Figure 23**, typical.
4. Manually move the power head gear shift lever to reverse gear while spinning the propeller to ensure full clutch engagement.
5. Make another mark on the shift rod from the same drive shaft housing reference point as Step 3.
6. Make a final mark on the shift rod at the midpoint of the two previous marks. This mark represents true mechanical neutral for the gearcase.
7. Manually move the power head gear shift lever to align the center mark with the reference point on the drive shaft housing. Note the position of the power head gear shift lever in neutral.

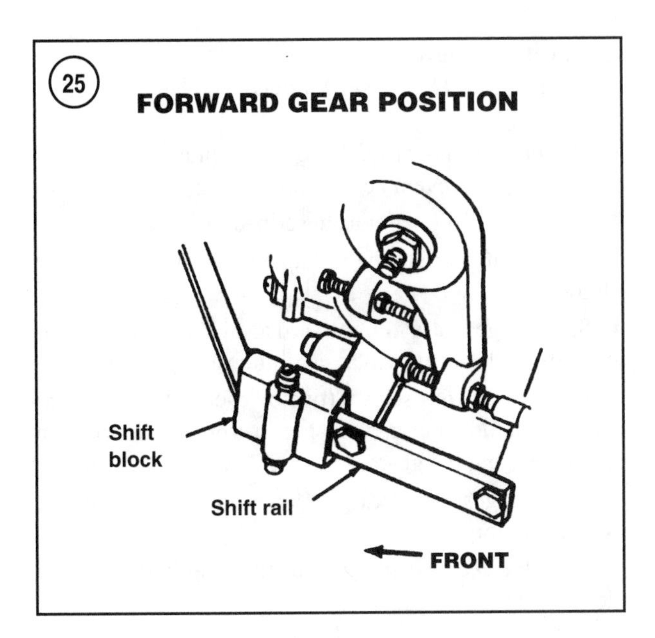

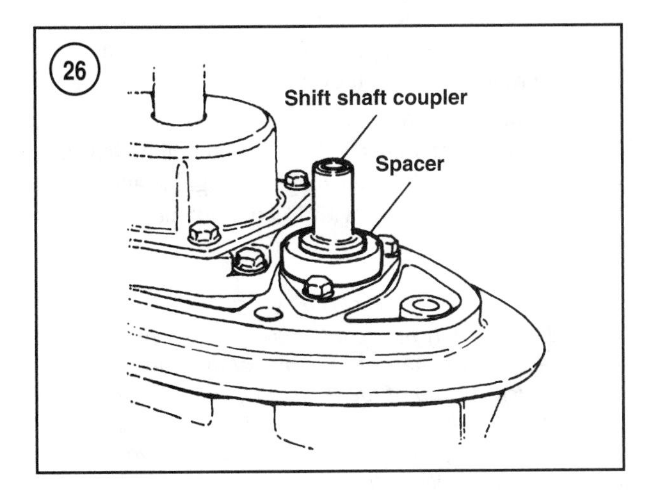

8. Manually shift into each gear while rotating the propeller. Note the travel required to engage each gear as referenced from the neutral position in Step 7.
9. Gear shift lever travel at the power head must be equal in each gear. If the shift lever moves further to engage one gear than it moves for the other, adjust the shift rod coupler to provide equal shift travel into each gear. Retighten the jam nut securely when finished.
10. Reconnect the remote control shift cable (Chapter Fourteen).

Removal/Installation (55 and 60 hp [Standard Gearcase] Models)

1. Disconnect and ground the spark plug leads to the power head to prevent accidental starting.
2. Remove the propeller as described in this chapter.
3. Tilt the outboard to the fully up position and engage the tilt lock.
4. Shift the gearcase into forward gear. Rotate the propeller shaft while shifting to assist full gear engagement.
5. Mark the trim tab position with a felt tip marker (**Figure 20**) and remove the trim tab retaining screw (**Figure 21**). Remove the trim tab.
6. Remove the four screws (two on each side) and washers, and one locknut and washer in the trim tab cavity securing the gearcase to the drive shaft housing. See **Figure 22**.
7. Pull the gearcase straight down and away from the drive shaft housing.
8. Place the gearcase in a suitable holding fixture or on a clean workbench.
9. If the water tube guide and seal remains on the water tube in the drive shaft housing, remove the water tube guide and seal from the water tube. Inspect the guide and seal, and replace them if they are damaged.
10. To install the gearcase, make sure the water tube guide and seal are securely attached to the water pump housing. If the guide and seal are loose, glue the guide and seal to the water pump housing with Loctite 405 adhesive. See **Figure 24**.

CAUTION
Do not apply lubricant to the top of the drive shaft in the next step. Excess lubricant between the top of the drive shaft and the engine crankshaft can create a hydraulic lock, preventing the drive shaft from fully engaging the crankshaft.

11. Clean the drive shaft splines as necessary, then coat the splines with Quicksilver 2-4-C Multi-Lube grease (part No. 92-825407). Coat the shift shaft splines and inner diameter of the water tube seal in the water pump housing with the same grease. See **Figure 24**.
12. Shift the gearcase into forward gear. Rotate the propeller shaft to assist gear engagement. When forward gear is engaged and the propeller shaft is turned clockwise, the sliding clutch will ratchet.
13. Position the shift block on the power head in the full forward position. If the remote control cable is attached, make sure the block is traveling to the full forward position. See **Figure 25**.
14. Install the nylon spacer over the shift shaft (**Figure 26**). Fit the shift shaft coupler onto the shift shaft as shown in **Figure 26**.
15. Position the gearcase under the drive shaft housing. Align the water tube in the water pump, the drive shaft

with crankshaft splines and the shift shaft with the shift shaft coupler.

CAUTION
Do not rotate the flywheel counterclockwise in Step 16 or water pump impeller damage can occur.

16. Push the gearcase toward the drive shaft housing, rotating the flywheel clockwise as required to align the drive shaft and crankshaft splines. In addition, move the shift block *slightly* on the power head as necessary to align the shift shaft splines.
17. Make sure the water tube seats in the water pump seal and the shift rod splines engage, then push the gearcase against the drive shaft housing.
18. Apply Loctite 271 threadlocking adhesive (part No. 92-809819) to the threads of the four mounting bolts.
19. Secure the gearcase to the drive shaft housing with the four mounting bolts and one locknut with a washer. See **Figure 22**. Tighten the fasteners hand-tight at this time.

NOTE
If the gearcase does not shift as described in the next step, the shift shafts are incorrectly indexed. Remove the gearcase and repeat Steps 12-19.

20. Shift the outboard into forward. The propeller shaft should ratchet when turned clockwise. Shift into neutral. The propeller shaft should turn freely in both directions. Shift into reverse gear. The propeller shaft should not turn in either direction. If shift operation is not as specified, remove the gearcase from the drive shaft housing and re-index the upper shift shaft with the lower shift shaft coupler.
21. Once shift function is correct, evenly tighten the gearcase mounting bolts and locknut to the specification in **Table 1**.
22. Install the trim tab and secure it with the screw and washer. Align the marks made during removal (**Figure 20**) and tighten the trim tab screw to the specification in **Table 1**.
23. Install the propeller as described in this chapter.
24. Release the tilt lock and return the outboard to the normal operating position.
25. Check the lubricant level or refill the gearcase with the recommended lubricant as described in Chapter Four.
26. If remote control shift cable adjustment is needed, refer to Chapter Fourteen.
27. Reconnect the spark plug leads.

Removal/Installation (60 hp [Bigfoot Gearcase] Models)

1. Disconnect and ground the spark plug leads to the power head to prevent accidental starting.
2. Remove the propeller as described in this chapter.
3. Tilt the outboard to the full up position and engage the tilt lock.
4. Shift the gearcase into forward gear. Rotate the propeller shaft while shifting to assist full gear engagement.
5. Remove the four screws (two on each side) and washers, and one locknut and washer in front of the trim tab cavity securing the gearcase to the drive shaft housing.
6. Pull the gearcase straight down and away from the drive shaft housing.
7. Place the gearcase in a suitable holding fixture or on a clean workbench.
8. If the water tube guide and seal remains on the water tube in the drive shaft housing, remove the water tube guide and seal from the water tube. Inspect the guide and seal, and replace them if they are damaged.
9. To install the gearcase, make sure the water tube guide and seal are securely attached to the water pump housing. If the guide and seal are loose, glue the guide and seal to the water pump housing with Loctite 405 adhesive. See **Figure 24**.

CAUTION
Do not apply lubricant to the top of the drive shaft in the next step. Excess lubricant between the top of the drive shaft and the engine crankshaft can create a hydraulic lock, preventing the drive shaft from fully engaging the crankshaft.

10. Clean the drive shaft splines as necessary, then coat the splines with Quicksilver 2-4-C Multi-Lube grease (part No. 92-825407). Coat the shift shaft splines and inner diameter of the water tube seal in the water pump housing with the same grease. See **Figure 24**.
11. Shift the gearcase into forward gear. Rotate the propeller shaft to assist gear engagement. When forward gear is engaged and the propeller shaft is turned clockwise, the sliding clutch will ratchet.
12. Position the outboard shift block in the forward gear (**Figure 25**). When in the proper position, the shift block extends 1/8 in. (3.2 mm) past the front of the shift rail.
13. Install the nylon spacer over the shift shaft (**Figure 26**). Fit the shift shaft coupler onto the shift shaft as shown in **Figure 26**. Install the shift shaft spacer and coupler shaft to the gearcase shift shaft.

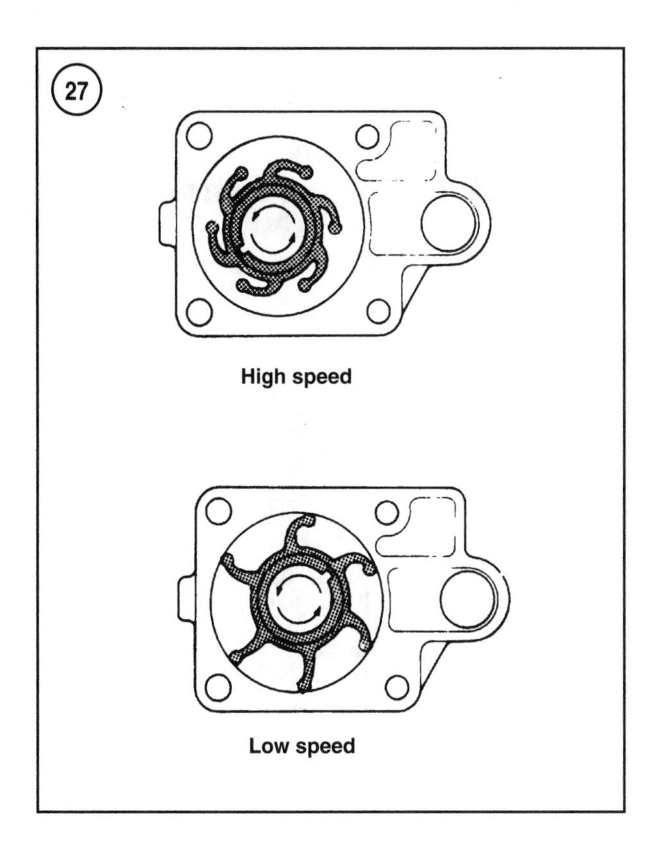

14. Run a 1/4 in. (6.4 mm) bead of RTV sealant (part No. 92-990113-2) along the water dam at the rear of the water pump base.
15. Position the gearcase under the drive shaft housing. Align the water tube in the water pump with the coupling, the drive shaft with the crankshaft splines and the shift shaft with the shift shaft coupler.

CAUTION
Do not rotate the flywheel counterclockwise in Step 16 or water pump impeller damage can occur.

16. Push the gearcase toward the drive shaft housing, rotating the flywheel clockwise as required to align the drive shaft and crankshaft splines. Also, move the shift block *slightly* on the power head as necessary to align the shift shaft splines.
17. Make sure the water tube seats in the water pump seal and the shift rod splines are engaged, then push the gearcase against the drive shaft housing.
18. Apply Loctite 271 threadlocking adhesive (part No. 92-809819 to the threads of the four mounting bolts.
19. Secure the gearcase to the drive shaft housing with the four mounting bolts and one locknut with a washer. Tighten the fasteners hand-tight at this time.

NOTE
If the gearcase does not shift as described in the next step, the shift shafts are incorrectly indexed. Remove the gearcase and repeat Steps 11-19.

20. Shift the outboard into forward gear. The propeller shaft should ratchet when turned clockwise. Shift into neutral. The propeller shaft should turn freely in both directions. Shift into reverse gear. The propeller shaft should not turn in either direction. If shift operation is not as specified, remove the gearcase from the drive shaft housing and re-index the upper shift shaft with the lower shift shaft coupler.
21. Once shift function is correct, evenly tighten the gearcase mounting bolts and locknut to the specification in **Table 1**.
22. Install the propeller as described in this chapter.
23. Release the tilt lock and return the outboard to the normal operating position.
24. Check the lubricant level or refill the gearcase with the recommended lubricant as described in Chapter Four.
25. If remote control shift cable adjustment is needed, refer to Chapter Fourteen.
26. Reconnect the spark plug leads.

WATER PUMP

All outboard motors in this manual use an offset-center pump housing that causes the vanes of the impeller to flex during rotation. At low speed, the pump operates as a positive displacement pump. At high speed, water resistance causes the impeller vanes to flex inward, causing the pump to operate as a centrifugal pump. See **Figure 27**.

The pump draws water into the intake port(s) as the vanes expand and pumps water out of the discharge port(s) as the vanes compress as shown in **Figure 28**.

Two basic designs of water pumps are used: the high-pressure pump (**Figure 29**, typical) and the high-volume pump (**Figure 30**, typical). The high-pressure pump develops more pressure than volume, while the high-volume pump delivers more volume than pressure.

On 2.5 hp models, the pump is at the rear of the gearcase and is driven by a pin in the propeller shaft. On all other models, the pump is on the gearcase upper deck and is driven by a key in the drive shaft.

The impeller rotates in a clockwise rotation with the drive shaft or propeller shaft and is held in a flexed position at all times. Over time, this causes the impeller to take a *set* in one direction. Turning an impeller over and attempting to turn it against its natural set will cause premature impeller failure and power head damage from

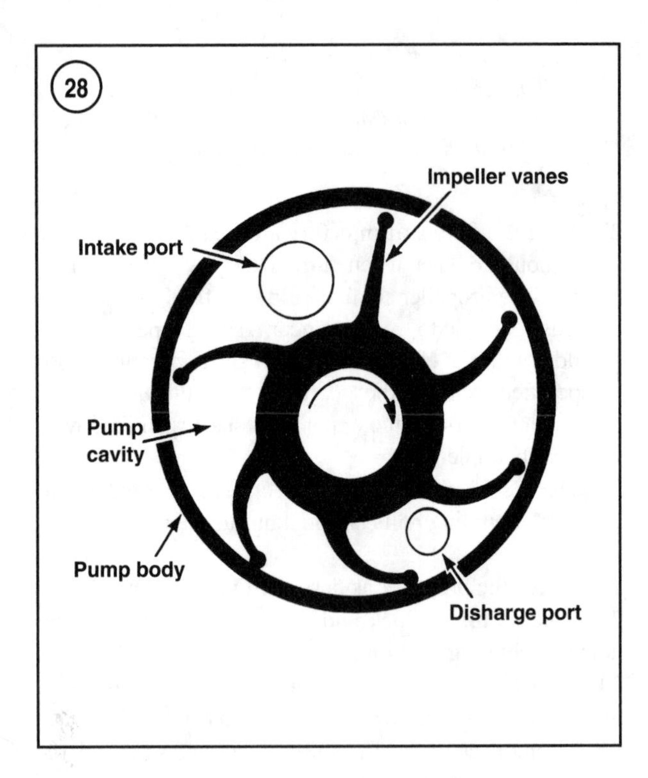

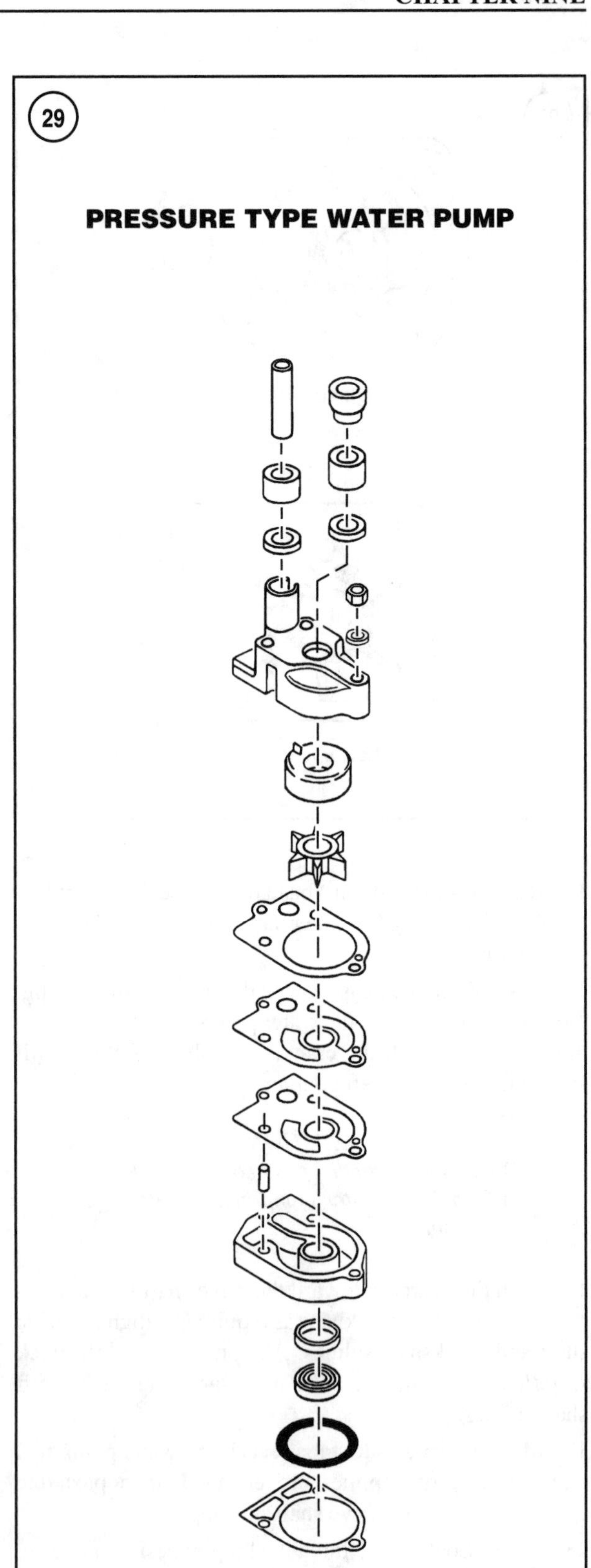

overheating. Replace the impeller each time the water pump is disassembled. Only reuse the impeller if there is no other option. If the impeller must be reused, reinstall the impeller in its original position.

Overheating and extensive power head damage can result from a faulty water pump. Replace the water pump impeller, seals and gaskets as follows:

1. *High pressure pump*—Once a year or every 100 hours of operation. The 2.5-5 hp models use a high-pressure pump. An exploded view of a typical high-pressure pump is shown in **Figure 29**.
2. *High-volume pump*—Every three years or 300 hours of operation. All 6-60 hp models use a high-volume pump. An exploded view of a typical high-volume pump is shown in **Figure 30**.

Individual operating conditions may dictate that the pump requires service more often. Also service the water pump any time the lower gearcase or jet pump unit is removed for service.

Removal and Disassembly

Replace all seals and gaskets whenever the water pump is disassembled. Since the drive shaft seals operate at crankshaft speed and are under motion when the engine is running, remove the water pump base, on models so equipped, and replace the drive shaft seals any time the water pump is disassembled. The manufacturer also rec-

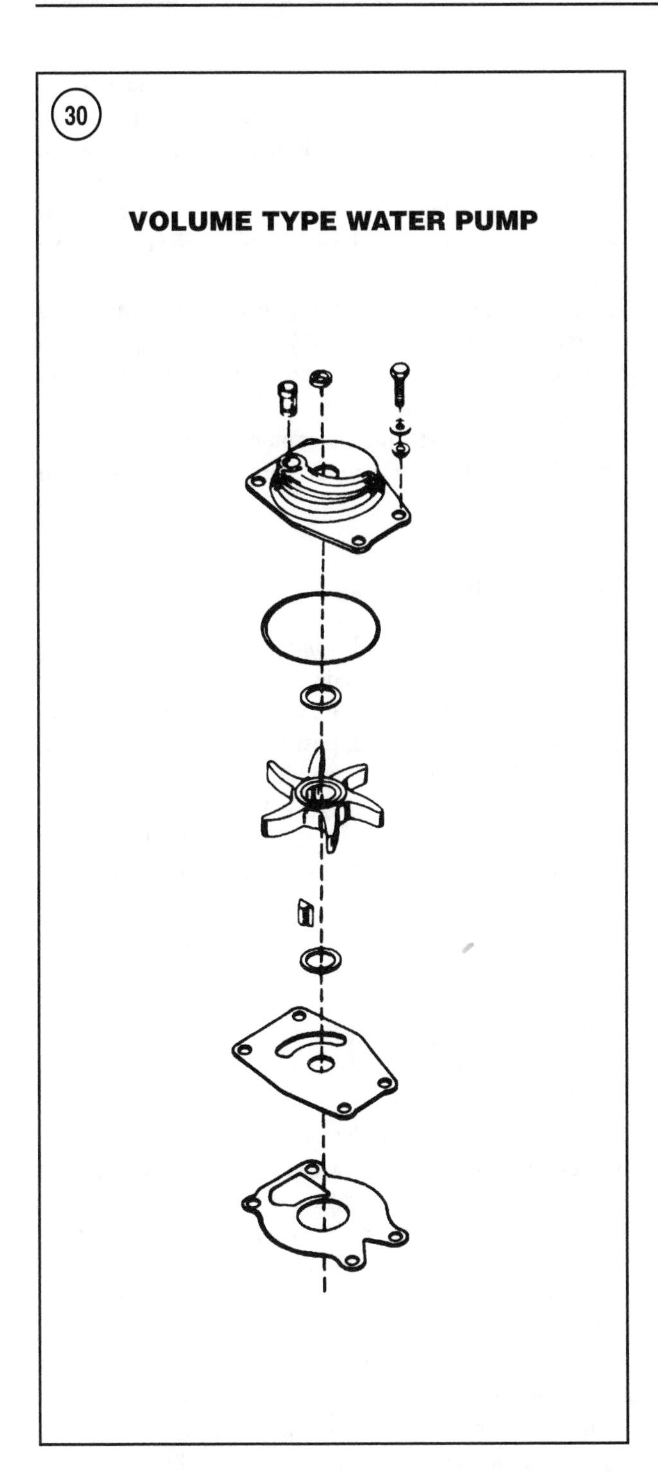

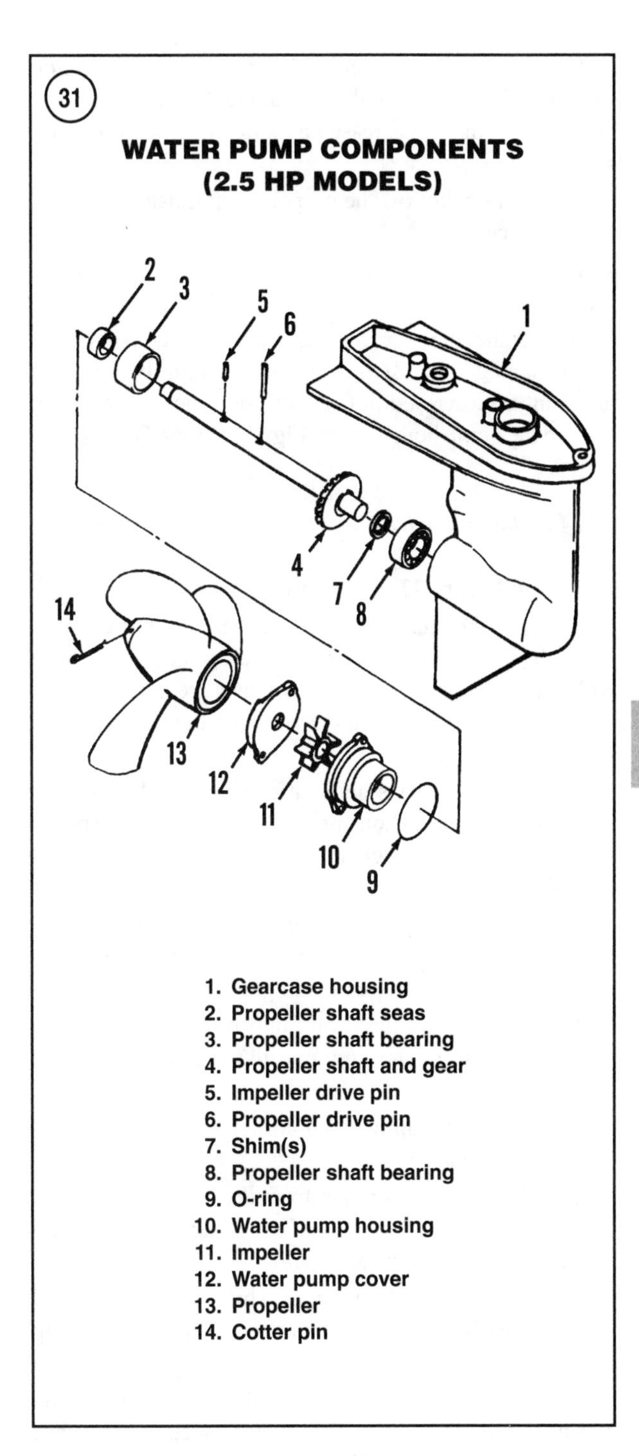

ommends replacing the water pump impeller anytime the pump is disassembled.

2.5 hp models

A rubber impeller water pump is located at the rear of the gearcase, just in front of the propeller. The impeller is mounted on and driven by the propeller shaft. The water pump can be serviced without removing the gearcase from the outboard motor. Refer to **Figure 31** for this procedure.

1. Remove the propeller and propeller drive pin as described in this chapter. Drain the gearcase lubricant as described in Chapter Four.

2. Remove the screws and washers securing the pump cover (12, **Figure 31**) to the rear of the gearcase. Separate the cover from the gearcase and slide it off the propeller shaft.

3. Pry the impeller off the propeller shaft using a screwdriver or needlenose pliers.

4. Remove the impeller drive pin from the propeller shaft.

5. Inspect the impeller wear surface in the pump housing. If it is damaged or excessively worn, drain the gearcase lubricant as described in Chapter Four, then remove and discard the pump housing (10, **Figure 31**) and O-ring (9).

3.3, 4 and 5 hp models

Refer to **Figure 32** for this procedure.

1. Remove the gearcase as described in this chapter. Place the gearcase in a suitable holding fixture or a vise with protective jaws. If protective jaws are not available, position the unit upright in the vise with the skeg between wooden blocks.

2. Remove the screws and flat washers (1, **Figure 32**) from the water pump housing (3). Carefully pry the pump housing up enough to grasp it, then slide the housing up and off the drive shaft.

3. On 4 and 5 hp models, remove the water supply tube from the pump housing. Remove the water supply tube grommets from the gearcase and pump housing.

4. If the impeller remains on the drive shaft, slide it up and off the shaft. Then remove the impeller drive key from the drive shaft.

5. Remove the pump housing gasket, impeller plate and impeller plate gasket.

6. Remove the impeller liner from the pump housing. Inspect the liner for wear and damage. Replace the liner if it is excessively worn or damaged.

7A. *3.3 hp models*—Remove the shift shaft bushing and two O-rings from the shift rod bore in the water pump housing.

7B. *4 and 5 hp models*—Remove the screw, washer and retainer at the front of the pump base securing the shift shaft and bushing assembly in place. Remove the shift shaft and bushing assembly. Remove the shift shaft bushing outer O-ring.

NOTE

If the drive shaft is pulled loose from the pinion gear in Step 8, the propeller shaft and bearing carrier must be removed from the gearcase to reinstall the drive shaft.

8. Carefully pry the pump base upward and slide it off the drive shaft while holding the drive shaft down. Do not allow the drive shaft to be lifted during pump base removal. Remove the pump base gasket. On 4 and 5 hp models, make sure any shims stuck to the pump base are reinstalled into the drive shaft bearing bore on top of the drive shaft bearing.

9. Support the pump base in a vise with protective jaws and remove the drive shaft seal with a suitable puller and appropriate jaws. Do not distort or damage the pump base during seal removal.

10. Clean and inspect the water pump components as described under *Water Pump Cleaning and Inspection* in this chapter.

6-15 hp models

Refer to **Figure 33** for this procedure.

1. Remove the gearcase as described in this chapter. Secure the housing in a vise with protective jaws. If protective jaws are not available, position the unit upright in a vise with the skeg between wooden blocks.

2. Remove the screws (3, **Figure 33**) and washers securing the water pump housing. Carefully pry the housing up enough to grasp it, then slide the housing up and off the drive shaft.

3. If the impeller remains on the drive shaft, slide the impeller and the upper impeller nylon washer up and off the shaft.

4. Remove the impeller drive key from the flat on the drive shaft. Then remove the lower impeller nylon washer.

5. Remove the impeller plate and plate gasket.

6. Remove the screw and washer (12, **Figure 33**) from the front of the water pump base. Lift the pump base and shift shaft assembly from the gearcase, disengaging the water tube from its seal at the rear of the gearcase.

7. Remove the pump base gasket and the gearcase water tube seal.

8. Remove the screw and retainer (14 and 15, **Figure 33**) holding the water tube to the pump base, then pull the water tube from the pump base. Remove the water tube seal.

9. Remove the shift shaft lock clip and unthread the shift shaft cam. Then pull the shift shaft from the water pump base. Remove the shift shaft quad ring seal from the pump base.

10. Remove the two pump base seals with a suitable punch or screwdriver. Do not damage or distort the pump base during the removal process.

11. Remove the pump housing drive shaft and water tube seal assembly (2, **Figure 33**).

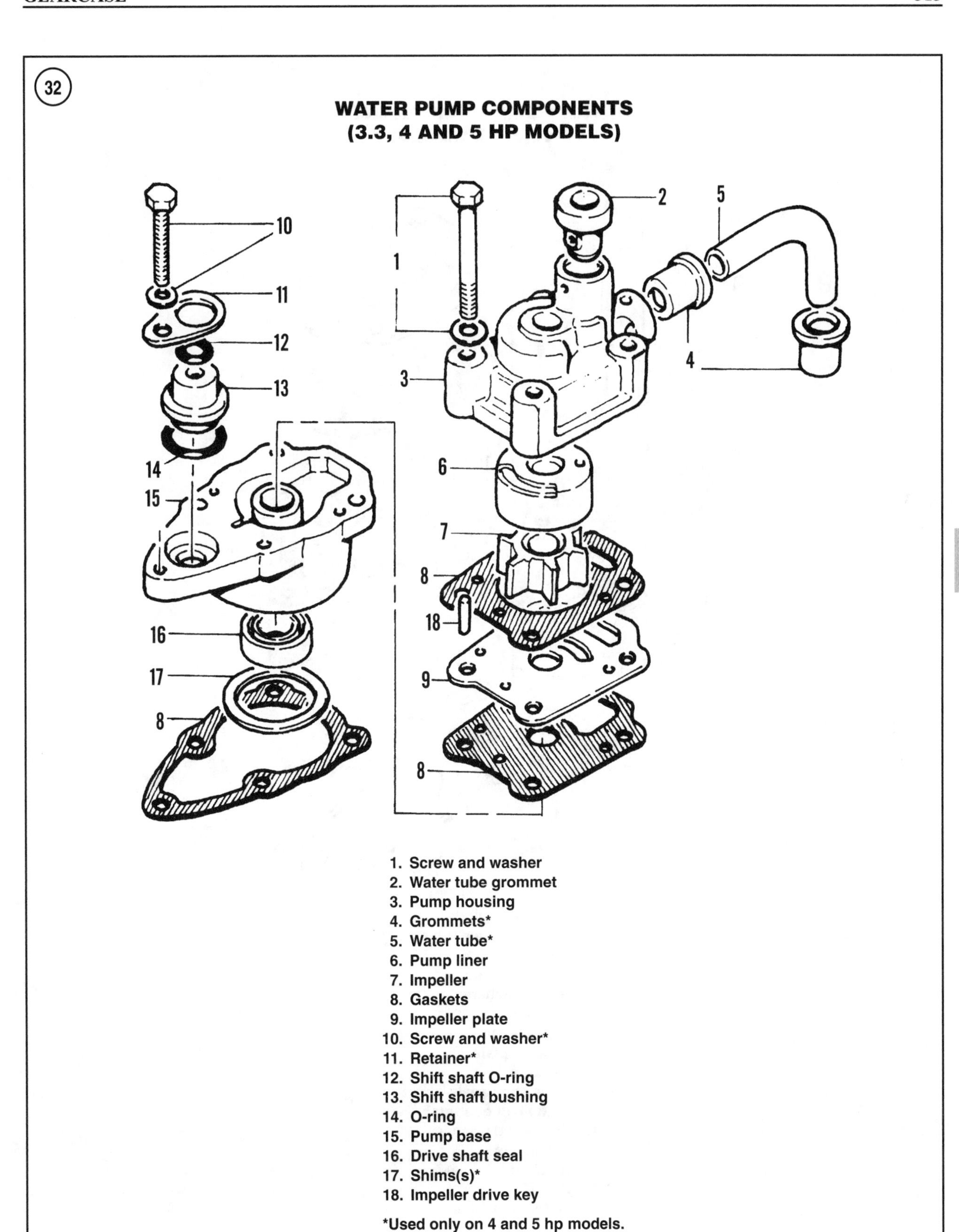

1. Screw and washer
2. Water tube grommet
3. Pump housing
4. Grommets*
5. Water tube*
6. Pump liner
7. Impeller
8. Gaskets
9. Impeller plate
10. Screw and washer*
11. Retainer*
12. Shift shaft O-ring
13. Shift shaft bushing
14. O-ring
15. Pump base
16. Drive shaft seal
17. Shims(s)*
18. Impeller drive key

*Used only on 4 and 5 hp models.

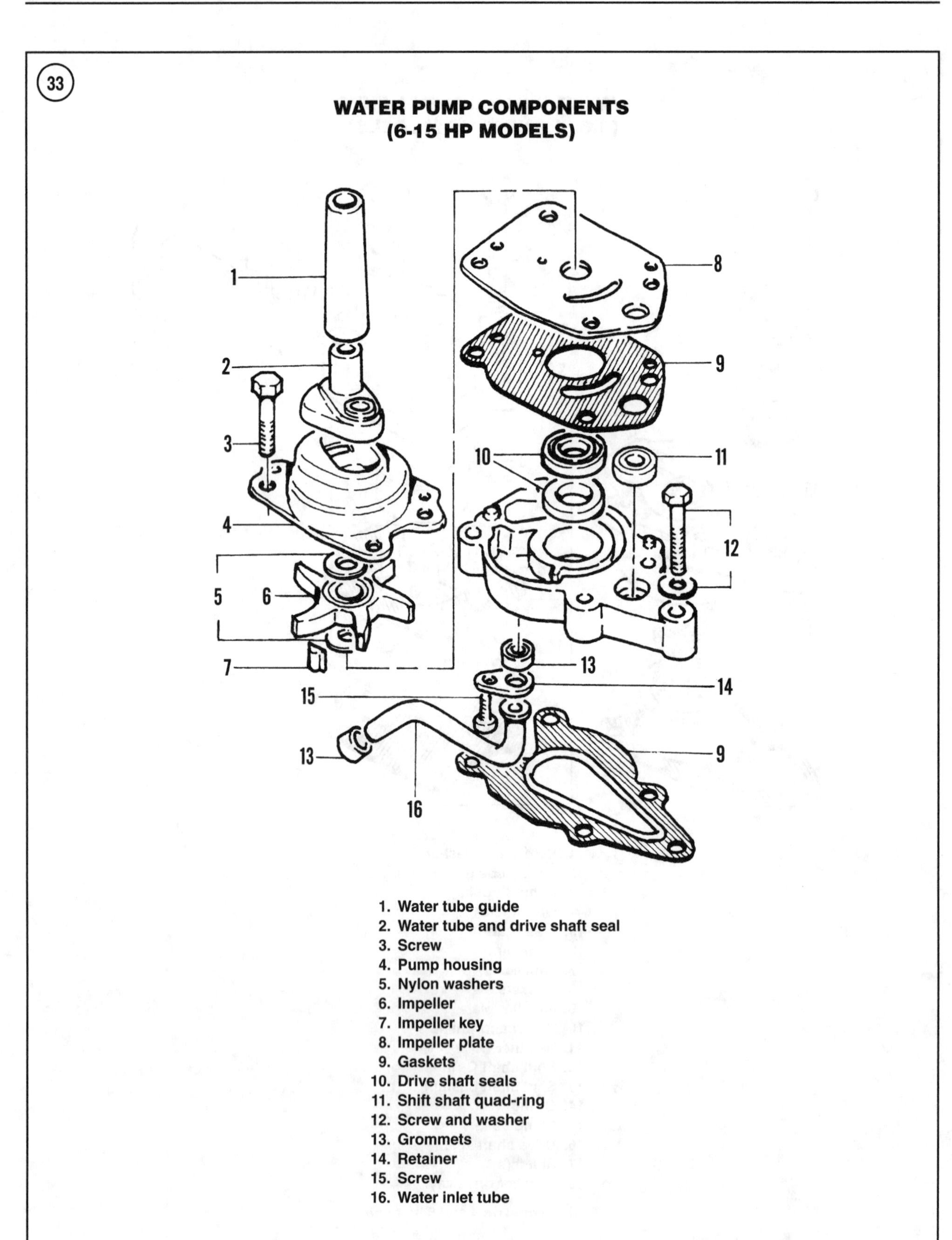
33
WATER PUMP COMPONENTS
(6-15 HP MODELS)
1
2
3
4
5
6
7
8
9
10
11
12
13
14
15
13
9
16
1. Water tube guide
2. Water tube and drive shaft seal
3. Screw
4. Pump housing
5. Nylon washers
6. Impeller
7. Impeller key
8. Impeller plate
9. Gaskets
10. Drive shaft seals
11. Shift shaft quad-ring
12. Screw and washer
13. Grommets
14. Retainer
15. Screw
16. Water inlet tube

(34)

WATER PUMP COMPONENTS (20 HP, 25 HP AND 20 JET MODELS)

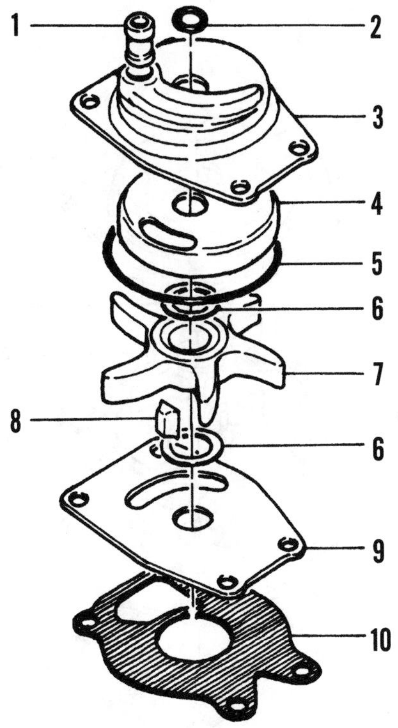

1. Water tube seal
2. Drive shaft O-ring
3. Pump housing
4. Pump liner*
5. O-ring
6. Washers
7. Impeller
8. Impeller key
9. Impeller plate
10. Gasket

*Not used on all models

12. Clean and inspect the water pump components as described in *Water Pump Cleaning and Inspection* in this chapter.

20 hp, 25 hp and 20 jet models

Refer to **Figure 34** for this procedure.

NOTE

Drive shaft seal replacement requires disassembly of the gearcase and removal of the drive shaft. Removal of the drive shaft is necessary to remove the drive shaft seal carrier. If seal replacement is required, refer to the appropriate gearcase disassembly section in this chapter.

1. Remove the gearcase as described in this chapter. Secure the housing in a suitable holding fixture or a vise with protective jaws. If protective jaws are not available, position the unit upright in a vise with the skeg between wooden blocks.
2. Remove the screws holding the pump cover to the gearcase.
3. Slide the pump housing up and off the shaft. Remove the housing O-ring (5, **Figure 34**).

NOTE

The impeller and upper washer may come off with the pump housing. If so, remove them from the housing.

4. Remove the upper washer, impeller, impeller key and lower washer from the gearcase.
5. Remove the pump liner (4, **Figure 34**) from the pump housing if it is worn or damaged.
6. Carefully loosen the impeller plate from the gearcase using a putty knife. Slide the plate up and off the drive shaft. Remove the gasket.
7. Clean and inspect the water pump components as described in *Water Pump Cleaning and Inspection* in this chapter.

30-50 hp (except 45 jet) models

The water pump housing is replaced as an assembly on all models. Refer to **Figure 35** for this procedure.

NOTE

Drive shaft seal replacement requires disassembly of the gearcase and removal of the drive shaft. Removal of the drive shaft is necessary to remove the drive shaft seal carrier. If seal replacement is required, refer to the appropriate gearcase disassembly section in this chapter.

1. Remove the gearcase as described in this chapter. Secure the housing in a suitable holding fixture or a vise with protective jaws. If protective jaws are not available, position the unit upright in a vise with the skeg between wooden blocks.

2. Remove the screws holding the pump cover to the gearcase.
3. Slide the pump housing up and off the shaft. Remove the housing gasket (6, **Figure 35**).

NOTE
The pump impeller and upper washer may come off with the pump housing. If so, remove them from the housing.

4. Remove the upper washer, impeller, impeller key and lower washer from the gearcase.
5. Carefully loosen the impeller plate (7, **Figure 35**) from the gearcase using a putty knife. Slide the plate up and off the drive shaft. Remove the gasket (8, **Figure 35**).
6. Clean and inspect the water pump components as described in *Water Pump Cleaning and Inspection* in this chapter.

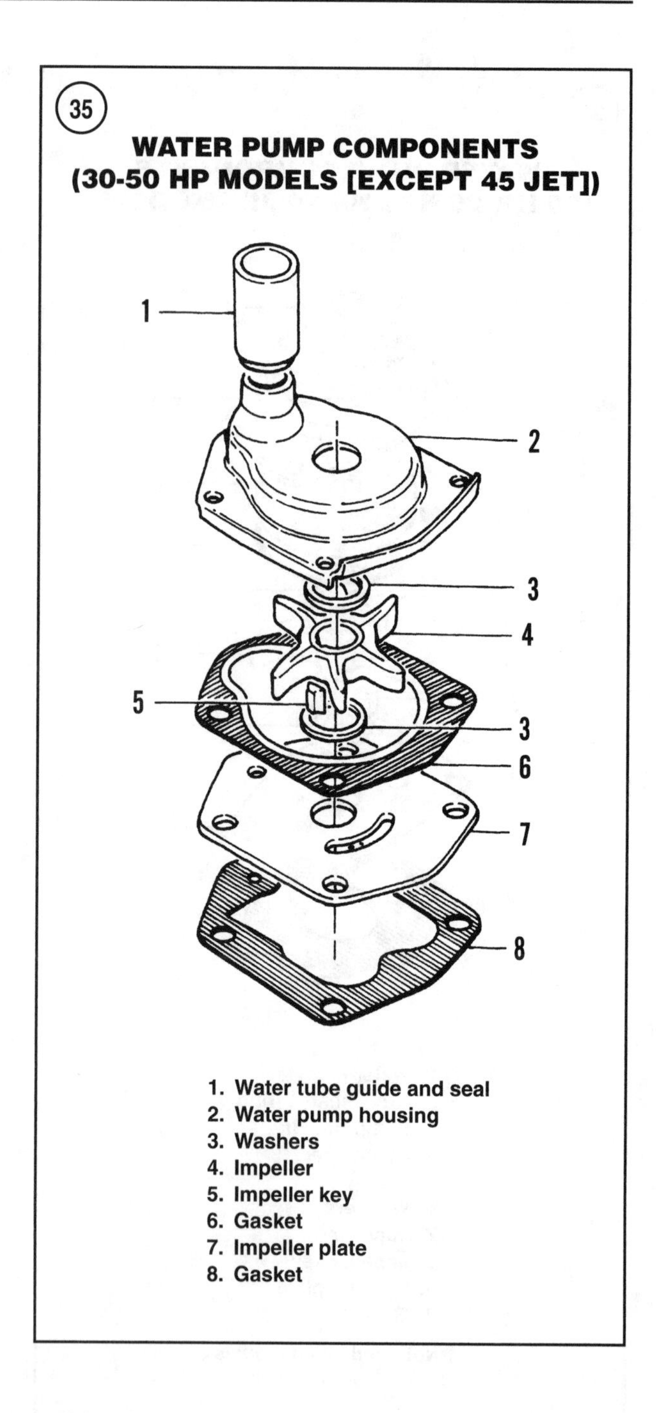

1. Water tube guide and seal
2. Water pump housing
3. Washers
4. Impeller
5. Impeller key
6. Gasket
7. Impeller plate
8. Gasket

45 jet, 55 hp and 60 hp standard gearcase models

Refer to **Figure 36** for this procedure.
1. Remove the gearcase as described in this chapter. Secure the gearcase in a suitable holding fixture or a vise with protective jaws. If protective jaws are not available, position the unit upright in a vise with the skeg between wooden blocks.
2. Remove the water tube guide and seal from the pump housing.
3. Remove the pump housing screws (and washers and insulators if used). Lift the pump housing up and off the drive shaft. If necessary, carefully pry at each end with a suitable tool.
4. Remove the impeller from the drive shaft.
5. Remove the impeller drive key from the drive shaft flat.
6. Remove the screws (and washers and insulators if used) at the front of the impeller plate. Remove the impeller plate and the upper and lower gaskets.
7. Remove the pump base from the gearcase. Remove the pump base O-ring and seal(s). Do not damage or distort the pump base when removing the seals.
8. Inspect the water pump components as described in *Water Pump Cleaning and Inspection* in this chapter.

60 hp Bigfoot gearcase models

Refer to **Figure 37** for this procedure.
1. Remove the gearcase as described in this chapter. Secure the gearcase in a suitable holding fixture or a vise with protective jaws. If protective jaws are not available, position the unit upright in a vise with the skeg between wooden blocks.
2. Remove the water tube guide and seal from the pump housing.
3. Remove the pump housing screws. Lift the pump housing up and off the drive shaft. If necessary, carefully pry at each end with a suitable tool.
4. Remove the impeller from the drive shaft.
5. Remove the impeller drive key from the drive shaft flat.

36 WATER PUMP COMPONENTS (45 JET, 55 HP AND 60 HP MODELS [STANDARD GEARCASE])

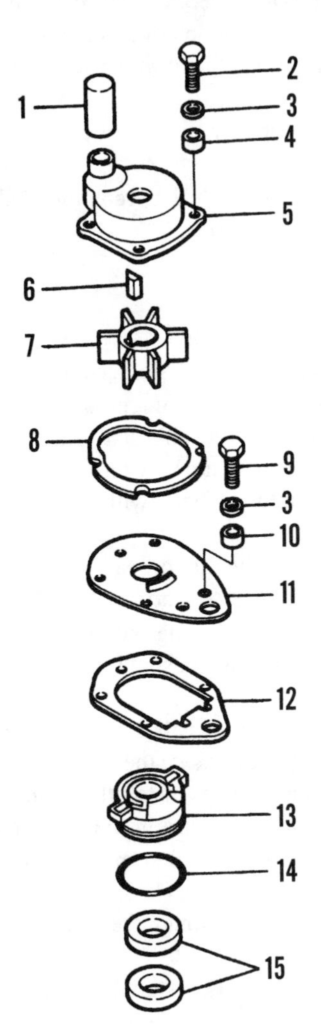

1. Water tube guide and seal
2. Screw
3. Washer*
4. Insulator*
5. Pump housing
6. Impeller key
7. Impeller
8. Gasket
9. Screw
10. Insulator*
11. Impeller plate
12. Gasket
13. Pump base
14. O-ring
15. Drive shaft seals

*Not used on all models

37 WATER PUMP COMPONENTS 60 HP WITH BIGFOOT GEARCASE

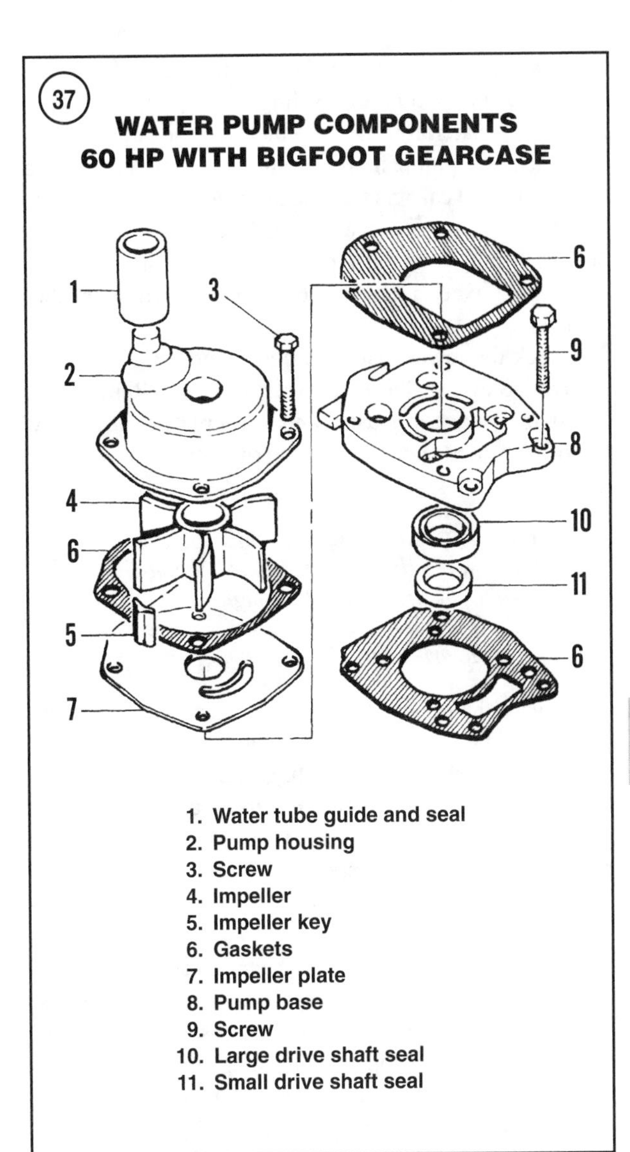

1. Water tube guide and seal
2. Pump housing
3. Screw
4. Impeller
5. Impeller key
6. Gaskets
7. Impeller plate
8. Pump base
9. Screw
10. Large drive shaft seal
11. Small drive shaft seal

6. Remove the impeller plate and the upper and lower gaskets.

7. Remove the screws securing the pump base to the gearcase. Carefully pry the base (8, **Figure 37**) loose from the gearcase. Lift the base up and off the drive shaft. Remove the base-to-gearcase gasket. Carefully pry the drive shaft seal from the pump base.

8. Inspect the water pump components as described in *Water Pump Cleaning and Inspection* in this chapter.

Water Pump Cleaning and Inspection

1. Clean all metal parts in solvent and dry with compressed air.

2. Clean all gasket material from all mating surfaces. Do not gouge or distort gasket sealing surfaces and do not allow gasket material to fall into the gearcase housing.
3. Check the pump housings, pump bases and plastic drive shaft seal carriers for cracks or distortion from overheating or improper service procedures.
4. Check the pump liner or impeller bore in the pump housing for excessive wear, corrosion, distortion or other damage.
5. Check the lower plate and pump liner or housing for grooves, rough surfaces or excessive wear. Replace the pump liner insert or housing and lower plate as necessary. Grooves from the impeller sealing rings are not a concern.

NOTE
The water pump impeller must be able to float on the drive shaft. Clean the impeller area of the drive shaft thoroughly using emery cloth. The impeller must slide easily over the drive shaft.

6. Replace the impeller any time it is removed. If the impeller must be reused, check the bonding of the rubber to the impeller hub for separation. Check the side seal surfaces and blade ends for cracks, tears, excessive wear or a glazed or melted appearance. If any of these defects are noted, do *not* reuse the impeller.
7. On 30-50 hp models (except 45 jet), perform the following:
   a. Measure the thickness of the pump cover at the discharge slots. Replace the cover if the metal thickness is 0.060 in. (1.52 mm) or less.
   b. Inspect the pump housing and impeller plate for grooves or excessive wear. Replace the housing and/or the impeller plate if grooves, except the impeller seal ring grooves, exceed 0.030 in. (0.76 mm) deep.

Assembly and Installation

2.5 hp models

Refer to **Figure 31** for this procedure.
1. If the pump housing is replaced, coat a new O-ring with Quicksilver 2-4-C Multi-Lube (part No. 92-825407) grease and install it over the new water pump housing assembly. Lubricate the housing, flange and propeller shaft seal lip with the same grease.
2. Install the pump housing into the gearcase.
3. Insert the impeller drive pin into the propeller shaft.
4. Lubricate the impeller with Quicksilver 2-4-C Multi-Lube grease. Slide the impeller onto the propeller shaft. Install the impeller into the pump housing while rotating the impeller in a clockwise direction. Make sure the impeller properly engages the drive pin.
5. Install the pump cover over the propeller shaft and seat it against the pump housing.
6. Apply Quicksilver Perfect Seal (part No. 92-34227-1) to the threads of the pump cover retaining screws. Install the screws and tighten them to the specification in **Table 1**.
7. Install the propeller as described previously in this chapter.
8. Fill the gearcase with the recommended lubricant as described in Chapter Four.

3.3, 4 and 5 hp models

Refer to **Figure 32** for this procedure.
1. Coat the outer diameter of a new drive shaft seal with Loctite 271 threadlocking adhesive (part No. 92-809819). Then place the water pump base on a flat work surface and press the seal into its bore with a mandrel (part No. 91-84530M) until it is seated.
2. Lubricate the drive shaft seal lips with Quicksilver 2-4-C Multi-Lube grease (part No. 92-825407).
3. Install a new outer O-ring onto the shift rod bushing. Lubricate the O-ring and bushing with Quicksilver 2-4-C Multi-Lube grease.
4. On 3.3 hp models, install a new inner O-ring in the shift shaft bushing. Lubricate the O-ring with Quicksilver 2-4-C Multi-Lube grease. Install the bushing and O-ring assembly into the pump base from the bottom side.
5. With a new gasket, install the pump base over the drive shaft and onto the gearcase. On 3.3. hp models, be careful to guide the shift shaft through the shift shaft bushing and O-ring.
6. On 4 and 5 hp models, install the shift rod and bushing assembly into the water pump base. Make sure the outer O-ring is correctly positioned between the pump base and the shift shaft bushing. Then install the retainer plate, washer and screw. Tighten the screw to the specification in **Table 1**.
7. Place new upper and lower gaskets onto the water pump plate. Install the plate and gaskets over the drive shaft, and shift rod on 3.3 hp models, then seat the plate and gaskets against the pump base.
8. If the pump liner was removed from the housing, align the liner tab with its respective hole or recess in the housing. Press the insert into the housing until it is seated.
9. Install the water discharge tube seal into the pump housing. Make sure the rubber tabs align with the housing holes. Lubricate the inside diameter of the seal with Quicksilver 2-4-C Multi-Lube grease.

10. On 4 and 5 hp models, install the pump pickup tube seals into the pump housing and the gearcase housing. Lubricate the inside diameter of the seals with Quicksilver 2-4-C Multi-Lube grease. Then install the water pickup tube into the pump housing seal.
11. Lubricate the impeller drive key with 2-4-C Marine Lubricant. Fit the key on the flat of the drive shaft, then slide the impeller over the drive shaft and drive key. Lubricate the impeller blades with Quicksilver 2-4-C Multi-Lube grease.
12. Slide the pump housing over the drive shaft and into position over the impeller. On 4 and 5 hp models, insert the water pickup tube into the gearcase seal.
13. Turn the drive shaft clockwise and press the pump housing down to feed the impeller into the housing. Seat the pump body against the plate. Make sure the water pickup tube on 4 and 5 hp models is fully seated into the water pump and gearcase seals.
14. Install the housing screws and washers. Tighten the screws to the specification in **Table 1**.
15. Install the gearcase on the outboard motor as described in this chapter.

6-15 hp models

Refer to **Figure 33** for this procedure.
1. Apply Loctite 271 Threadlocking adhesive (part No. 92-809819) to the outer diameter of two new drive shaft seals. Install the first seal with the spring lip side facing the pump base using seal installer (part No. 91-13655 or an equivalent). Press the seal into its bore until seated. Install the second seal with its spring lip side facing away from the pump base using the same installer. Press the second seal into its bore until it is seated against the first seal. Lubricate the seal lips with Special Lube 101 (part No. 92-13872A1).
2. Lubricate the water supply tube seal with Quicksilver 2-4-C Multi-Lube grease (92-825407). Install the seal onto the water supply tube. Insert the seal end of the tube into the water pump base and secure it with the retainer and screw. Tighten the screw securely.
3. Lubricate a new shift shaft quad ring with Quicksilver Special Lube 101. Install the quad ring into the pump base. Then insert the shift shaft through the pump base and install the lock clip into the shift shaft groove and the shift cam onto the shift shaft threads.
4. Lubricate the gearcase water supply tube seal with Quicksilver Special Lube 101. Install the tube into the gearcase with the tapered end facing upward.
5. Install a new water pump base gasket. Make sure the drain hole in the gasket is positioned on the starboard side of the housing.
6. Install the water pump base and shift shaft assembly into the housing, guiding the water supply tube into the tapered end of the gearcase seal. Make sure the ramp side of the shift cam faces the propeller shaft. Seat the water pump base against the gearcase.
7. Apply Loctite 271 Threadlocking adhesive to the threads of the pump base screw. Install the screw and washer, and tighten them to the specification in **Table 1**.
8. With a new gasket, install the impeller plate. If the shift shaft quad ring has moved from its bore, seat the quad ring back into its bore before installing the gasket and impeller plate.
9. Lubricate a new water tube and drive shaft seal with Quicksilver 2-4-C Multi-Lube grease. Press the seal onto the pump cover. Install the water tube guide onto the water tube seal.
10. Install an impeller washer over the drive shaft and position it against the plate. Grease the impeller key with 2-4-C Multi-Lube grease or an equivalent and position it on the drive shaft.
11. Lubricate the impeller with Quicksilver 2-4-C Multi-Lube grease. Slide the impeller onto the drive shaft and engage the key. Install the other impeller washer over the drive shaft and into position on top of the impeller.
12. Slide the pump housing over the drive shaft and into position over the impeller. Turn the drive shaft clockwise and press the housing down to feed the impeller into the body. Seat the pump housing against the plate.
13. Apply Loctite 271 Threadlocking adhesive to the threads of the pump housing screws. Install the screws and tighten them in a crossing pattern to the specification in **Table 1**.
14. Install the gearcase to the outboard motor as described previously in this chapter. Make sure the shift shaft height is set as specified in the gearcase installation procedures.

20 hp, 25 hp and 20 jet models

Refer to **Figure 34** for this procedure.
1. Place a new gasket onto the gearcase. Slide the impeller plate over the drive shaft and into position on the gasket.
2. Install the pump liner into the pump housing if it was removed. Glue a new O-ring into the pump body groove with a suitable contact adhesive.
3. Install an impeller washer over the drive shaft and position it against the plate. Grease the impeller key with 2-4-C Multi-Lube grease (part No. 92-825407) and position it on the drive shaft.
4. Lubricate the impeller with Quicksilver 2-4-C Multi-Lube grease. Slide the impeller onto the drive shaft

and engage the key. Install the other impeller washer over the drive shaft and into position on top of the impeller.

5. Slide the pump housing over the drive shaft and into position over the impeller. Turn the drive shaft clockwise and press the housing down to feed the impeller into the housing. Seat the pump housing against the plate.

6. Coat the threads of the housing screws with Loctite 271 Threadlocking adhesive (part No. 92-809819). Install and evenly tighten the screws to the specification in **Table 1**.

7. Install the water tube seal into the pump housing and the drive shaft O-ring into the drive shaft groove. Lubricate the seal and O-ring with Quicksilver 2-4-C Multi-Lube grease.

8. Install the gearcase to the outboard motor as described previously in this chapter.

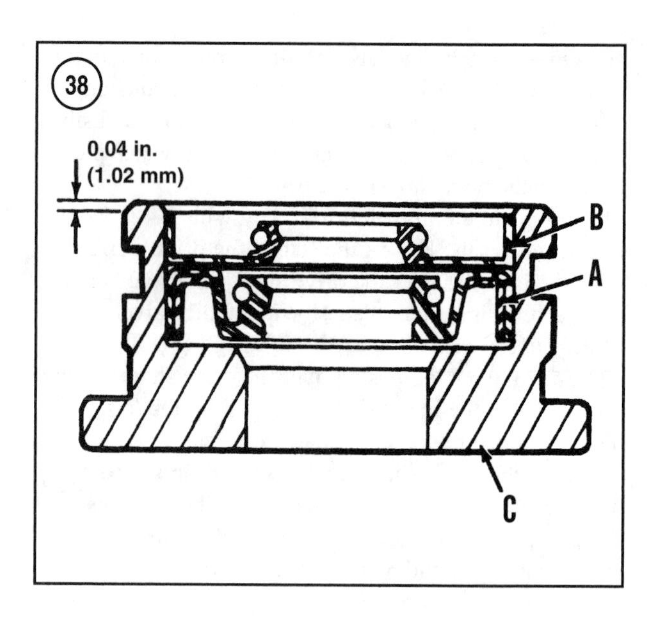

30-50 hp (except 45 jet) models

Refer to **Figure 35** for this procedure.

1. Place a new gasket onto the gearcase. Slide the impeller plate over the drive shaft and into position on the gasket.

2. Install a new gasket, with the raised rib up, on top of the water pump plate. Make sure the exhaust deflector plate is installed in its recess, just behind the water pump base.

3. Install an impeller washer over the drive shaft and position it against the plate. Grease the impeller key with 2-4-C Multi-Lube grease (part No. 92-825407) and position it on the drive shaft.

4. Lubricate the impeller with Quicksilver 2-4-C Multi-Lube grease. Slide the impeller onto the drive shaft and engage the key. Install the other impeller washer over the drive shaft and into position on top of the impeller.

5. Slide the pump housing over the drive shaft and into position over the impeller. Turn the drive shaft clockwise and press the housing down to feed the impeller into the housing. Seat the pump housing against the plate.

6. Coat the threads of the housing screws with Loctite 271 Threadlocking adhesive (part No. 92-809819). Install and evenly tighten the screws to the specification in **Table 1**.

7. If the water tube guide and seal was removed, reattach or replace it. Secure the water tube guide and seal with Loctite 405 adhesive.

8. Install the gearcase to the outboard motor as described previously in this chapter.

45 jet, 55 hp and 60 hp standard gearcase models

Refer to **Figure 36** for this procedure.

1. Coat the outer diameter of the two new drive shaft seals with Loctite 271 Threadlocking adhesive (part No. 92-809819). Use the longer muster of a seal installer (part No. 91-817006 or an equivalent) to press the thicker seal (A, **Figure 38**) into the water pump base. With the spring side facing the gearcase, press the seal until the installer bottoms on the base.

2. Use the short side of the seal installer to press the thinner seal (B, **Figure 38**) into the water pump base with the spring side again facing the gearcase until the installer bottoms on the base. When correctly installed, the top of the second seal is 0.040 in. (1.02 mm) below the surface of the base as shown (C, **Figure 38**).

3. Lubricate a new O-ring with Quicksilver Special Lube 101 (part No. 92-13872A1). Install the O-ring into the pump base groove.

4. Carefully install the pump base over the drive shaft and seat it into the gearcase bore. Do not damage or cut the drive shaft seals on the drive shaft splines.

5. If the exhaust deflector was removed, coat both ends of the deflector with RTV sealant (part No. 92-809826) and install it into the gearcase recess just behind the pump base.

6. With a new gasket, install the impeller plate over the drive shaft and into position on the gearcase. Secure the impeller plate with two screws (and washers and insulators on early models). Tighten the screws finger-tight at this time.

7. Lubricate the impeller key with Quicksilver 2-4-C Multi-Lube grease (part No. 92-825407). Position the impeller key on the drive shaft flat.
8. Lubricate the impeller with Quicksilver 2-4-C Multi-Lube grease. Slide the impeller over the drive shaft and engage the impeller key.
9. Install a new pump housing gasket on the impeller plate with the sealing bead facing upward.
10. Slide the pump housing assembly over the drive shaft and into position over the impeller. Turn the drive shaft clockwise and press the housing down to feed the impeller into the housing. Seat the pump housing against the plate.
11. Secure the pump housing with four screws (and washers and insulators on early models). Tighten the screws finger-tight at this time.
12. Tighten the pump housing screws in a crossing pattern, then tighten the plate screws. Tighten all screws to the specification in **Table 1**.
13. If the water tube guide and seal was removed, glue the guide and seal to the water pump housing with Loctite 405 adhesive. Then lubricate the water tube seal with Quicksilver 2-4-C Multi-Lube grease.
14. Install the gearcase to the outboard motor as described in this chapter.

60 hp Bigfoot gearcase models

Refer to **Figure 37** for this procedure.

1. Apply Loctite 271 Threadlocking adhesive (part No. 92-809819) to the outer diameter of the new drive shaft seals. Set the water pump base with the stepped side facing up into a press. Install the Teflon coated seal (flat brown/black color) with the spring facing the power head onto the longer stepped side of the seal installer (part No. 91-13949 or an equivalent). Press the seal into the water pump base until the tool bottoms. Install the non-Teflon coated seal (glossy black color) with the spring facing the gearcase onto the short stepped side of the same installer. Press the seal into the water pump base until the tool bottoms. Coat the seal lips with 2-4-C Multi-Lube grease (part No. 92-825407).
2. Place a new gasket onto the gearcase. Slide the water pump base over the drive shaft and into position on the gasket. Be careful not to cut or damage the seals on the drive shaft splines. Make sure the water pump base is piloted into the gearcase bore and has not pinched the gasket in the gearcase bore.
3. Coat the water pump base screws with Loctite 271 threadlocking adhesive. Install the screws and washers. Evenly tighten the screws to the specification in **Table 1**.
4. Install a new gasket on top of the water pump base. Slide the impeller plate over the drive shaft and into position on the gasket. Then install a new housing gasket on the plate.
5. Grease the impeller key with 2-4-C Multi-Lube grease and position it on the drive shaft flat. Slide the impeller onto the drive shaft and engage the key.
6. Slide the pump housing over the drive shaft and into position over the impeller. Turn the drive shaft clockwise and press the housing down to feed the impeller into the body. Seat the pump housing against the plate.
7. Coat the threads of the housing screws with Loctite 271 Threadlocking adhesive. Install and evenly tighten the screws to the specification in **Table 1**.
8. If the water tube guide and seal were removed, glue the guide and seal to the water pump housing with Loctite 405 adhesive. Then lubricate the water tube seal with Quicksilver 2-4-C Multi-Lube grease.
9. Install the gearcase to the outboard motor as described previously in this chapter.

GEARCASE DISASSEMBLY/ASSEMBLY

This section covers disassembly and reassembly procedures for each lower gearcase covered in this manual. Once the gearcase is disassembled, refer to *Gearcase Cleaning and Inspection* before assembly.

Larger models require the gears to be shimmed, and the gear lash (clearance) between the forward gear and the pinion gear to be verified before continuing with assembly. The assembly procedure refers to *Gearcase Shimming* at the proper time.

Disassembly (2.5 hp Models)

Refer to **Figure 39** for this procedure.

NOTE

Do not remove the ball bearings (6, ***Figure 39****) unless they must be replaced. Remove the impeller housing ball bearing to access the propeller shaft seal.*

1. Remove the water pump components as described in this chapter.
2. Remove the water pump impeller housing. If removal is difficult, tap the housing with a soft plastic or lead hammer to break it free. Remove the housing O-ring.
3. Using a suitable tool, reach into the gearcase cavity and remove the E-clip securing the pinion gear to the lower drive shaft (**Figure 40**). Lift the drive shaft out the top of the gearcase, then remove the pinion gear from the gear cavity.

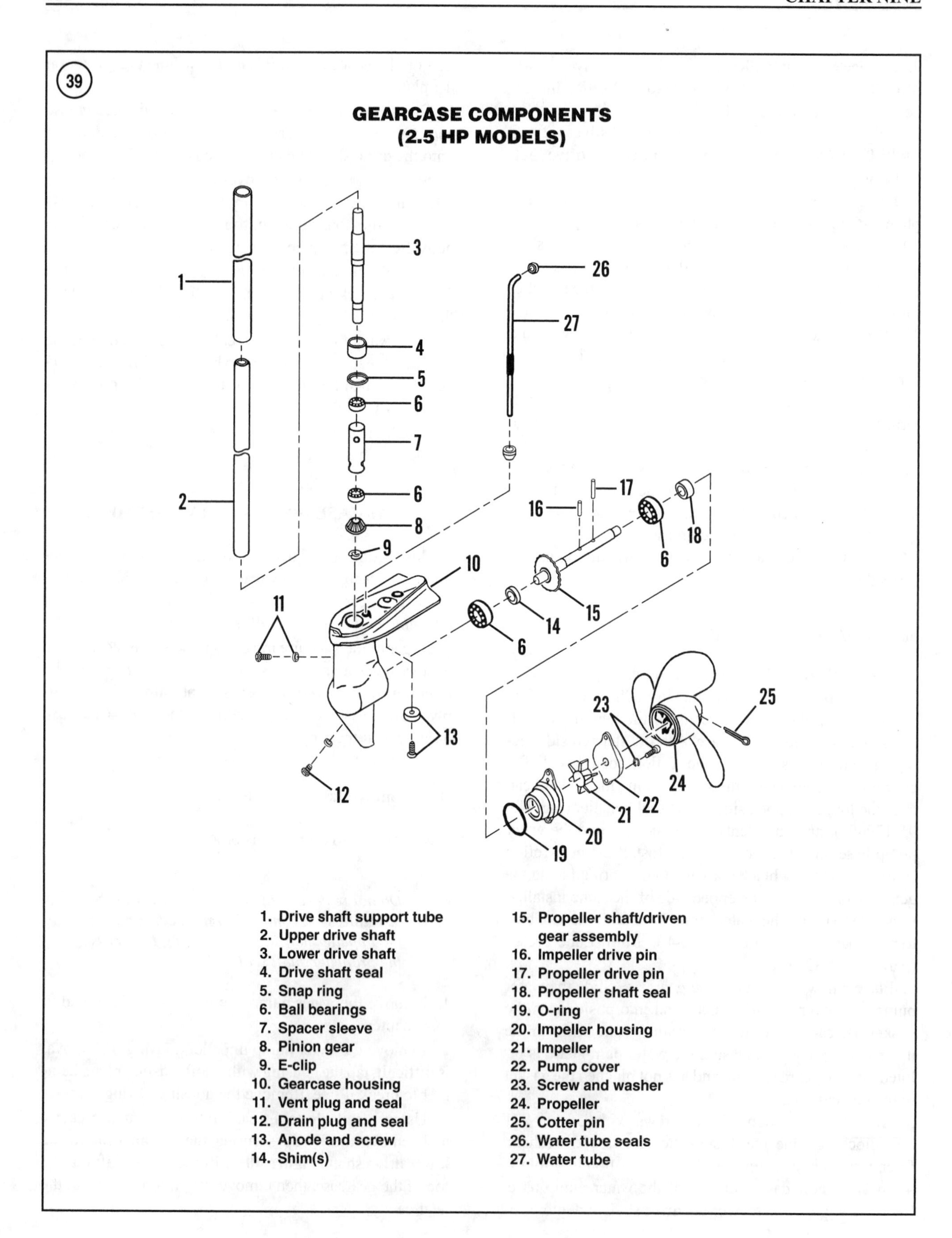
39
GEARCASE COMPONENTS
(2.5 HP MODELS)
1
2
3
4
5
6
7
6
8
9
10
11
12
13
14
15
16
17
18
6
6
19
20
21
22
23
24
25
26
27
1. Drive shaft support tube
2. Upper drive shaft
3. Lower drive shaft
4. Drive shaft seal
5. Snap ring
6. Ball bearings
7. Spacer sleeve
8. Pinion gear
9. E-clip
10. Gearcase housing
11. Vent plug and seal
12. Drain plug and seal
13. Anode and screw
14. Shim(s)
15. Propeller shaft/driven gear assembly
16. Impeller drive pin
17. Propeller drive pin
18. Propeller shaft seal
19. O-ring
20. Impeller housing
21. Impeller
22. Pump cover
23. Screw and washer
24. Propeller
25. Cotter pin
26. Water tube seals
27. Water tube

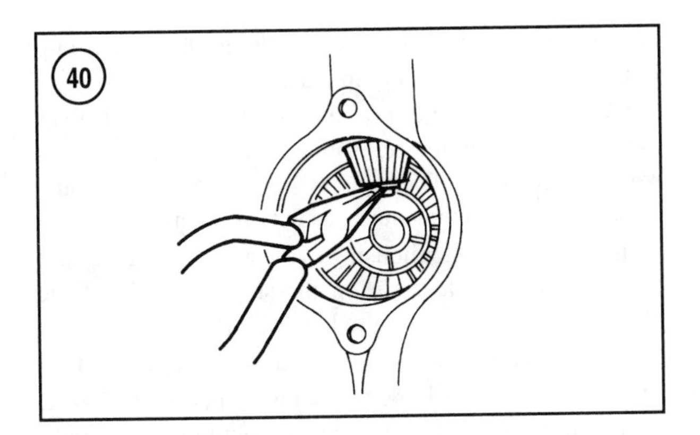

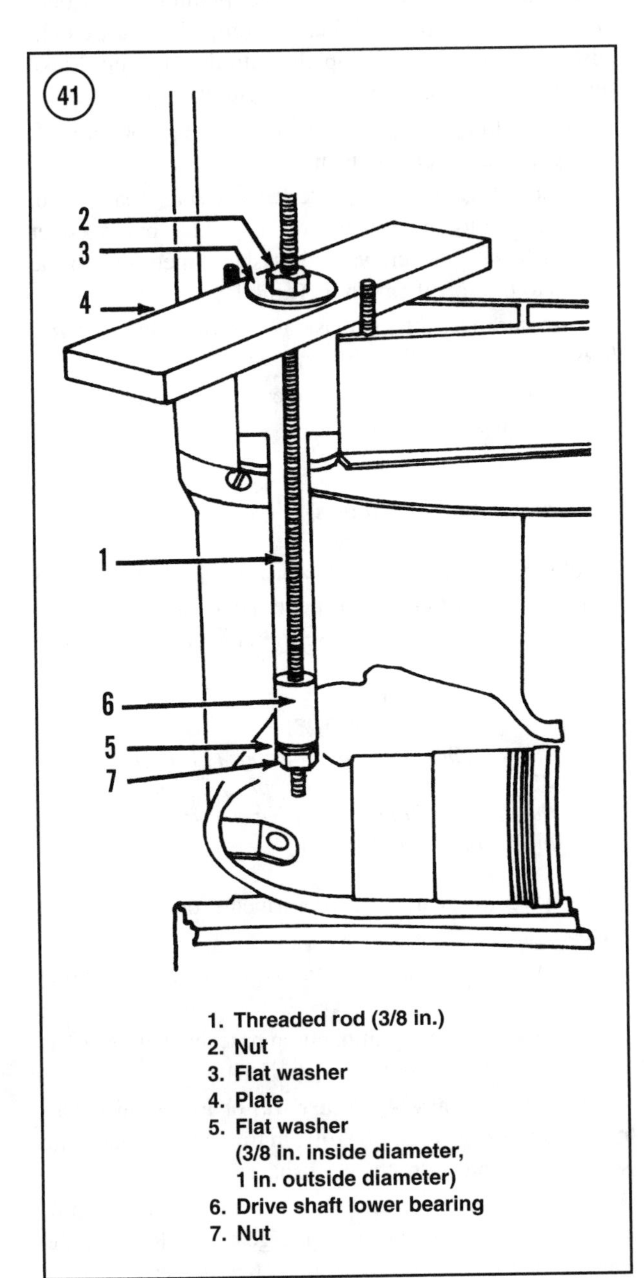

1. Threaded rod (3/8 in.)
2. Nut
3. Flat washer
4. Plate
5. Flat washer (3/8 in. inside diameter, 1 in. outside diameter)
6. Drive shaft lower bearing
7. Nut

4. Pull the propeller shaft and driven gear assembly from the gearcase. Be sure to retrieve the shim(s) on the end of the shaft. See 14, **Figure 39**.
5. Remove the drive shaft and water tube seals from the gearcase deck using a screwdriver or suitable tool. Do not damage the seal bore during removal.
6. Replace the propeller shaft seal and ball bearing as follows:
   a. Remove the impeller housing ball bearing using a suitable bearing puller, such as a collet and expander rod attached to a slide hammer.
   b. Set the impeller housing on a workbench with the bearing bore facing downward.
   c. Drive the seal from the housing with a suitable punch and hammer. Do not damage the seal bore during removal.
7. If the forward propeller shaft ball bearing must be replaced, pull the bearing from the gearcase with a suitable bearing puller, such as a collet and expander rod, attached to a slide hammer (part No. 91-34569A-1).
8. If the drive shaft ball bearings are to be replaced, replace them as follows:
   a. Remove the snap ring securing the drive shaft upper bearing to the gearcase.
   b. Remove the drive shaft upper ball bearing using a suitable bearing puller, such as a collet and expander rod, attached to a slide hammer (part No. 91-34569A-1).
   c. Reach into the drive shaft bore and remove the bearing spacer sleeve (7, **Figure 39**) from the gearcase.
   d. Use a suitable mandrel and driver rod to drive the lower ball bearing into the propeller shaft cavity.
9. Refer to *Gearcase Cleaning and Inspection*. Clean and inspect all components as described before beginning the reassembly procedure.

Assembly (2.5 hp Models)

Lubricate all internal components with Quicksilver Premium Blend gear oil (part No. 92-13783 or an equivalent). Do not assemble components *dry*. Refer to **Figure 39** for this procedure.

1. If the drive shaft ball bearings were removed, install new bearings as follows:
   a. Assemble a 3/8 in. threaded rod with a plate and washers as shown in **Figure 41**. The outer diameter of the lower washer (5, **Figure 41**) must be 1 in. (25.4 mm).
   b. Lubricate the bearing and the gearcase bearing bore.
   c. Place the drive shaft lower bearing onto the rod, with the numbered side facing up, on top of the 1 in.

(25.4 mm) washer as shown. Tighten the top nut (2, **Figure 41**) until the washer contacts the gearcase housing and the bearing seats flush in the gearcase.
d. Remove the threaded rod assembly
e. Insert the spacer sleeve into the drive shaft bore.
f. Lubricate a new upper ball bearing and the gearcase bearing bore. Set the bearing into the gearcase bore with the numbered side facing up.
g. Use a suitable mandrel to drive the bearing into the housing until seated. Drive only on the bearing outer race.
h. Install the bearing snap ring. Make sure the snap ring is fully expanded into its groove.

2. If the propeller shaft front ball bearing was removed, install a new bearing as follows:
a. Lubricate the bearing and gearcase bearing bore.
b. Set the bearing into its bore with the numbered side facing out.
c. Drive the bearing into the housing until it is seated by installing a washer (part No. 12-37249) over the gear end of the propeller shaft and using the propeller shaft as a driving tool. Use a soft plastic or lead hammer to prevent damage to the propeller shaft.

3. Install a new seal and bearing into the impeller housing as follows:
a. Coat the outer diameter of a new seal with Loctite 271 threadlocking adhesive (part No. 92-809819).
b. Position the seal into the impeller housing with the spring side facing upward.
c. Use a suitable mandrel to press the seal into the housing until it is seated.
d. Lubricate a new ball bearing and the impeller housing bearing bore.
e. Position the bearing into the housing with the numbered side facing upward.
f. Use a suitable mandrel to press the bearing into the housing until it is seated. Press only on the outer bearing race.

4. Apply Quicksilver Bellows Adhesive (part No. 92-86166) to the outer diameter of new drive shaft and water tube seals. Install the water tube seal and drive shaft seals into the gearcase. Then lubricate the inner diameter of both seals with Quicksilver 2-4-C Multi-Lube grease (part No. 92-825407).

5. Install the original shim(s) onto the front of the driven gear and propeller shaft assembly.

6. Install the drive gear and propeller shaft assembly into the gearcase. Make sure the shaft is fully engaged to the front ball bearing.

7. Install the lower drive shaft and pinion gear into the gearcase. Secure the pinion gear by installing the E-clip into the drive shaft groove (**Figure 40**). Replace the E-clip if it does not lock securely in place.

8. Coat a new impeller housing O-ring with Quicksilver 2-4-C Multi-Lube grease. Install the O-ring onto the impeller housing. Then coat the impeller housing-to-gearcase mating surfaces with the same grease.

9. Install the impeller housing to the gearcase. Make sure the water intake slots are on the starboard side of the gearcase. Seat the housing to the gearcase.

10. Install the impeller drive pin into the propeller shaft.

11. Lubricate the impeller with Quicksilver 2-4-C Multi-Lube grease. Install the water pump impeller onto the shaft. Rotate the impeller and propeller shaft clockwise while pushing the impeller into the housing. Make sure the impeller is fully engaged with the drive pin.

12. Install the water pump cover over the propeller shaft and against the pump housing.

13. Apply Quicksilver Perfect Seal (part No. 92-34227-1) to the threads of the water pump cover screws. Install the screws and washers, then tighten the screws to the specification in **Table 1**.

14. Pressure test the gearcase as described in *Gearcase Pressure Testing* in this chapter.

15. Fill the gearcase with the recommended lubricant as described in Chapter Four.

Disassembly (3.3 hp Models)

Refer to **Figure 42** for this procedure. The propeller shaft bearing carrier and propeller shaft can be removed without removing the gearcase from the drive shaft housing.

NOTE
*Do not remove the ball bearings (14, **Figure 42**) unless they must be replaced. Remove the impeller housing ball bearing to access the propeller shaft seal.*

1. Remove the gearcase as described previously in this chapter.

2. Drain the gearcase lubricant as described in Chapter Four.

3. Remove the water pump, pump base and shift shaft as described in this chapter.

4. If the shift cam (34, **Figure 42**) or lower shift shaft need replacement, drive the roll pin from the shift cam and separate the cam from the shift shaft.

5. Remove the screws and flat washers securing the propeller shaft bearing carrier to the gearcase. Remove the carrier. If it is difficult to remove, tap the carrier with a

soft plastic or lead hammer to break it free. Remove the carrier O-ring.

6. Pull the propeller shaft from the gearcase.
7. Pull the drive shaft assembly from the gearcase. Reach into the gearcase bore and remove the spacer sleeve.
8. Remove the pinion and forward gears from the gearcase.
9. Locate and secure the forward gear shim(s) (20, **Figure 42**).
10. Remove the propeller shaft seal and rear ball bearing as follows:
    a. Remove the propeller shaft bearing carrier ball bearing using a suitable bearing puller, such as a collet and expander rod, attached to a slide hammer (part No. 91-34569A-1).
    b. Secure the carrier in a vise with protective jaws or support the carrier on blocks with the bearing bore facing downward.
    c. Drive the seal from the carrier using a suitable punch and hammer. Do not damage the seal bore during removal.
11. If the front propeller shaft ball bearing needs replacement, remove the bearing using a bearing puller (part No. 91-27780 or an equivalent) slide hammer. See **Figure 43**.
12. If the drive shaft ball bearings need replacement, remove both bearings as follows:
    a. Support the drive shaft upper ball bearing with a bearing plate (part No. 91-37241 or an equivalent). Install a protector cap (part No. 91-24161 or an equivalent) over the pinion gear splines. Press against the protector cap until the bearing is removed from the drive shaft.
    b. Use a suitable mandrel and driver rod to drive the lower ball bearing into the propeller shaft cavity.
13. If the propeller shaft requires disassembly, disassemble it as follows:
    a. Pull the shift cam follower out of the forward end of the propeller shaft.
    b. Use a small screwdriver, awl or similar tool to compress the clutch spring towards the aft end of the propeller shaft far enough to pull the sliding clutch out of the propeller shaft slot.
    c. Remove the spring (23, **Figure 42**) from the shaft.
14. Refer to *Gearcase Cleaning and Inspection* in this chapter. Clean and inspect all components as described before beginning the reassembly procedures.

Assembly (3.3 hp Models)

Lubricate all internal components with Quicksilver Premium Blend gear oil (part No. 92-13783 or an equivalent). Do not assemble components *dry*. Refer to **Figure 42** for this procedure.

1. If the drive shaft ball bearings were removed, install new bearings as follows:
    a. Lubricate the lower bearing and the entire gearcase drive shaft bearing bore.
    b. Place the bearing into the gearcase bore from the gearcase deck with the numbered side facing upward.
    c. Use a suitable mandrel and driver rod to drive the bearing into the bearing bore until the bottom of the bearing outer race is approximately 1/8 in. (3.2 mm) from the bottom of the bore as shown in **Figure 44**. Drive only on the outer bearing race.

NOTE

*The drive shaft upper bearing must be installed to a point 3-5/8 in. (92.1 mm) from the pinion gear end of the drive shaft as shown in **Figure 45**. To ease installation, fabricate an installation tool from a section of pipe with a 13/32 in. (10.3 mm) inside diameter and a length of 3-5/8 in. (92.1 mm).*

    d. Lubricate the ball bearing and the lower 4 in. (102 mm) of the drive shaft.
    e. Place the drive shaft with the crankshaft end down on a padded surface. Place the upper bearing on the pinion gear end of the drive shaft with the numbered side of the bearing facing upward.
    f. Place the fabricated piece of pipe over the end of the drive shaft and install the bearing by pressing on the pipe until it is even with the end of the drive shaft. When installed, the bearing must be 3-5/8 in. (92.1 mm) from the pinion gear end of the drive shaft as shown in **Figure 45**.
2. If the propeller shaft front ball bearing was removed, install a new bearing as follows:
    a. Lubricate the bearing and gearcase bearing bore.
    b. Set the bearing into its bore with the numbered side facing out.
    c. Drive the bearing into the housing until it is seated using a mandrel (part No. 91-38628) and driver rod (part No. 91-37323) or an equivalent. See **Figure 46**.
3. Install a new seal and bearing into the propeller shaft bearing carrier as follows:
    a. Coat the outer diameter of a new seal with Loctite 271 threadlocking adhesive (part No. 92-809819).
    b. Position the seal into the bearing carrier with the flat side facing downward.
    c. Use a suitable mandrel to press the seal into the carrier until it is seated.

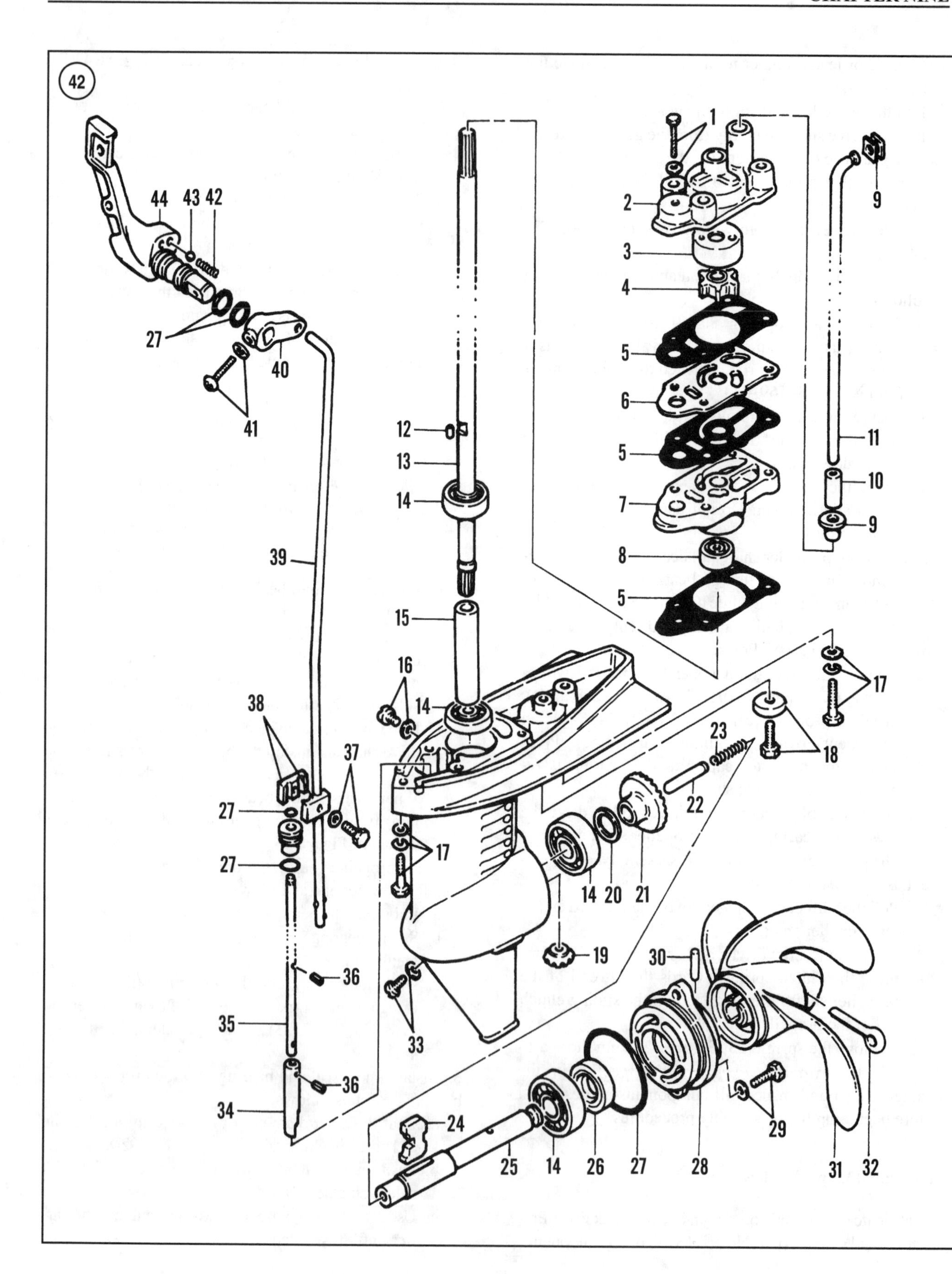
42
1
2
3
4
5
6
5
7
8
5
9
11
10
9
12
13
14
15
16
14
17
18
23
22
21
20
14
19
17
38
37
27
27
39
40
41
44
43
42
27
36
35
36
34
33
24
25
14
26
27
28
30
29
31
32

GEARCASE COMPONENTS (3.3 HP MODELS)

1. Screw and washer
2. Water pump housing
3. Pump liner
4. Impeller
5. Gaskets
6. Impeller plate
7. Pump base
8. Drive shaft seal
9. Water tube seals
10. Water tube guide
11. Water tube
12. Impeller drive key
13. Drive shaft
14. Ball bearings
15. Spacer sleeve
16. Vent plug and seal
17. Screw, lockwasher and washer
18. Anode and screw
19. Pinion gear
20. Shim(s)
21. Forward gear
22. Shift cam follower
23. Spring
24. Sliding clutch
25. Propeller shaft
26. Propeller shaft seal
27. O-rings
28. Propeller shaft bearing carrier
29. Screw and washer
30. Propeller drive pin
31. Propeller
32. Cotter pin
33. Fill/drain plug and seal
34. Shift cam
35. Lower shift shaft
36. Roll pins
37. Screw and washer
38. Clamp
39. Upper shift shaft
40. Shift lever
41. Screw and washer
42. Spring
43. Detent ball
44. Shift control lever

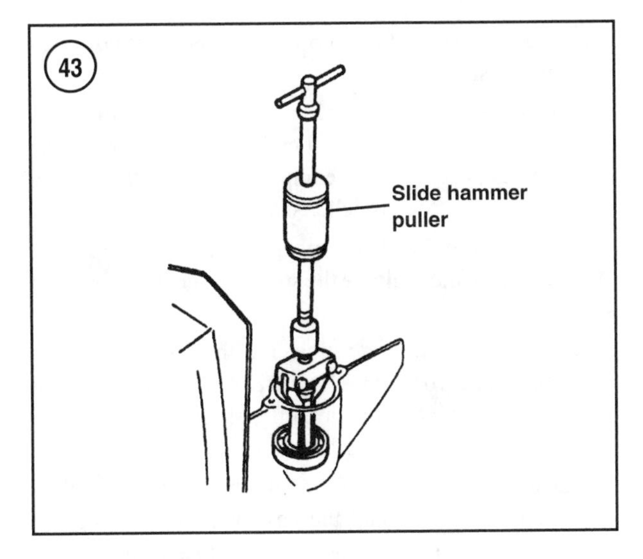

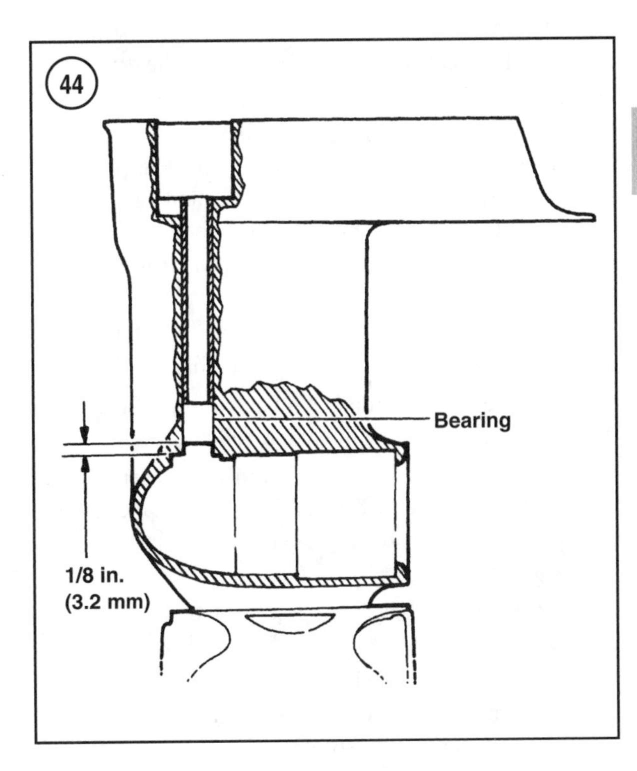

d. Lubricate a new ball bearing and the carrier bearing bore.

e. Position the bearing into the carrier with the numbered side facing upward.

f. Use a mandrel (part No. 91-37312) and driver rod (part No. 91-37323) or an equivalent to press the bearing into the housing until it is seated. Press only on the outer bearing race.

4. If the propeller shaft was disassembled, reassemble it as follows:

a. Lubricate the clutch spring and insert it into the propeller shaft.
b. Use a screwdriver, awl or similar tool to compress the spring toward the aft end of the shaft. When the spring is sufficiently compressed, insert the sliding clutch into the propeller shaft slot.
c. Make sure the clutch engagement pawls are facing forward, then release the spring to secure the sliding clutch.
d. Coat the shift cam follower with Quicksilver Special Lube 101 (part No. 92-13872A1) and insert it into the front of the propeller shaft. Make sure the rounded tip is facing forward.

5. Place the original shim(s) on the forward gear. Lubricate the gear, then install the gear and shim(s) into the gearcase. Make sure the gear and shim(s) are fully seated in the ball bearing.

6. Insert the drive shaft spacer sleeve into the drive shaft bore.

7. Position the pinion gear under the drive shaft lower ball bearing. While holding the pinion gear in place, install the drive shaft into the gearcase and through the spacer sleeve and lower ball bearing. Rotate the shaft as necessary to engage the pinion gear, then seat it fully in the gearcase bore.

8. Install the propeller shaft into the gearcase and guide it into the forward gear and bearing assembly.

9. Install the shift cam on the lower shift rod if it was removed. Secure the cam with a new roll pin.

10. Install the water pump base, shift shaft assembly and the water pump as described in this chapter. Make sure the shift cam's ramp is properly positioned against the shift cam follower.

11. Check the gear lash as described under *Gearcase Shimming* in this chapter. Once the gear lash is verified, proceed with the next step.

12. Lubricate a new O-ring with Quicksilver 2-4-C Multi-Lube grease (part No. 92-825407) and install it in the propeller shaft bearing carrier groove. Carefully slide the propeller shaft bearing carrier over the propeller shaft and seat it against the gearcase.

13. Coat the threads of the bearing carrier screws with Quicksilver Perfect Seal (part No. 92-34227-1). Install the screws and washers, and tighten them to the specification in **Table 1**.

14. Pressure test the gearcase as described in *Gearcase Pressure Testing* in this chapter.

15. Fill the gearcase with the recommended lubricant as described in Chapter Four.

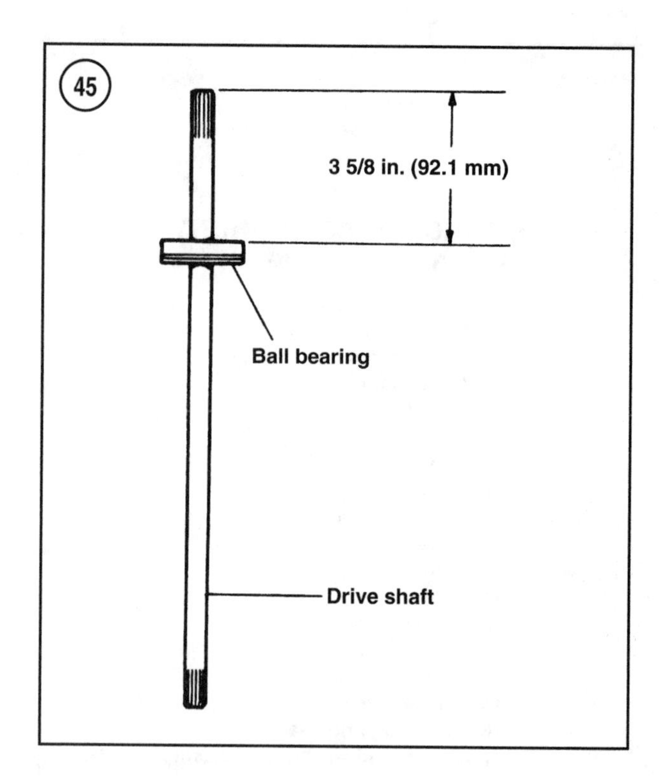

Disassembly (4 and 5 hp Models)

Refer to **Figure 47** for this procedure. The propeller shaft bearing carrier and propeller shaft can be removed without removing the gearcase from the drive shaft housing.

NOTE
Do not remove the ball bearings (25, 29 and 40, ***Figure 47****) unless they must be replaced. Remove the propeller shaft bearing carrier ball bearing (40,* ***Figure 47****) to access the propeller shaft seal (41).*

1. Remove the gearcase as described in this chapter.

2. Drain the gearcase lubricant as described in Chapter Four.

3. Remove the water pump, pump base and shift shaft as described previously in this chapter.

4. Slide the shift shaft bushing from the lower shift shaft. Remove the shift shaft bushing inner O-ring.

5. If the shift cam (20, **Figure 47**) or lower shift shaft require replacement, drive the roll pin from the shift cam and separate the cam from the shift shaft.

6. Remove the screws and washers holding the propeller shaft bearing carrier to the gearcase.

7. Clamp the propeller shaft into a vise with protective jaws.

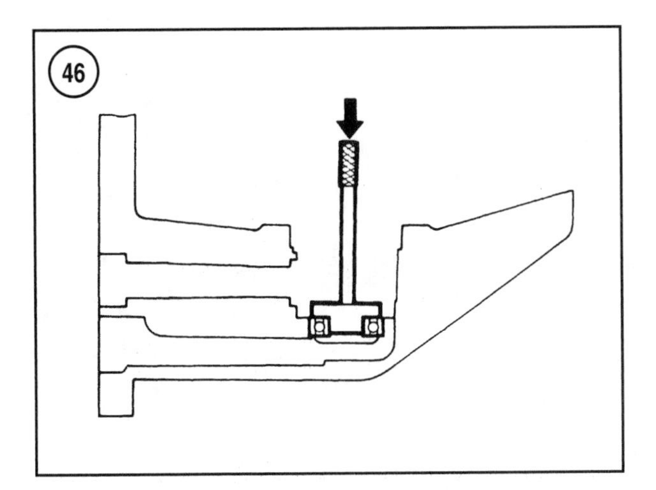

8. Tap the gearcase at a point midway between the antiventilation plate and the propeller shaft with a soft plastic or rawhide mallet while pulling the gearcase from the propeller shaft assembly.

9. Remove the propeller shaft and bearing carrier assembly from the vise. Make sure the cam follower (32, **Figure 47**) came out with the propeller shaft. If it did not, retrieve it from the gearcase cavity and temporarily reinstall it into the propeller shaft.

10. Remove the propeller shaft and reverse gear assembly from the propeller shaft bearing carrier. Remove the carrier O-ring (42, **Figure 47**).

11. Slide the reverse gear and thrust washer (38 and 39, **Figure 47**) from the propeller shaft.

12. Pull the drive shaft from the gearcase. Locate and secure the upper ball bearing shim(s) (17, **Figure 47**). Then remove the pinion gear from the propeller shaft cavity.

13. Remove the forward gear and shim by tapping the open end of the gearcase propeller shaft cavity bore against a block of wood. Locate and secure the forward gear shim(s) (30, **Figure 47**).

14. Clamp the propeller shaft bearing carrier in a vise with protective jaws with the ball bearing pointing up. Use a suitable two-jaw puller set, such as part No. 91-83165M, to remove the ball bearing and seal. Do not damage the seal bore during the removal process.

15. If the forward gear ball bearing requires replacement, remove the bearing using a bearing puller (part No. 91-27780 or an equivalent) slide hammer (**Figure 43**).

16. If the lower drive shaft needle bearing requires replacement, remove the bearing by driving the bearing from the housing using a remover/installer (part No. 91-17351 or an equivalent). See **Figure 48**. Make sure the tool collar is installed on the driver. Drive the bearing down into the gear cavity.

17. If the drive shaft ball bearing (25, **Figure 47**), drive shaft sleeve (26) or bearing sleeve (27) requires replacement, replace the drive shaft as an assembly.

18. If the propeller shaft requires disassembly, disassemble it as follows:

   a. Remove the shift cam follower and spring guide from the front end of the shaft.
   b. Place the shaft on a suitable support, such as the open, padded jaws of a vise, and carefully drive the retaining pin (35, **Figure 47**) from the sliding clutch with a suitable punch and hammer. Do not remove the punch from the sliding clutch at this time.
   c. Position the propeller shaft upright with the cam follower end resting on a solid surface. Slowly withdraw the punch from the clutch and carefully release the spring compression. Remove the spring from the end of the shaft, then slide the clutch from the propeller shaft.

19. Refer to *Gearcase Cleaning and Inspection*. Clean and inspect all components as described before beginning the reassembly procedure.

Assembly (4 and 5 hp Models)

Lubricate all internal components with Quicksilver Premium Blend gear oil (part No. 92-13783 or an equivalent). Do not assemble components *dry*. Refer to **Figure 47** for this procedure.

1. If the lower drive shaft needle bearing was removed, install a new bearing as follows:

   a. Lubricate the bearing and the gearcase bearing bore.
   b. Position the bearing on the installer tool (part No. 91-17351 or an equivalent). The numbered side of the bearing must face up, against the tool. Remove the tool's collar for bearing installation.
   c. Position the bearing and installer tool assembly in the gearcase bearing bore and drive the bearing into the gearcase until the driver head contacts the guide bushing.

2. If the forward gear ball bearing was removed, install a new bearing as follows:

   a. Lubricate the bearing and the gearcase bore.
   b. Assemble the bearing installer (part No. 91-84532M) to the driver rod (part No. 91-84529M).
   c. Position the bearing on the installer with the numbered side of the bearing facing the installer.
   d. Drive the bearing into the gearcase bore until the bearing seats in the bore. See **Figure 46**.

3. Reassemble the propeller shaft as follows:

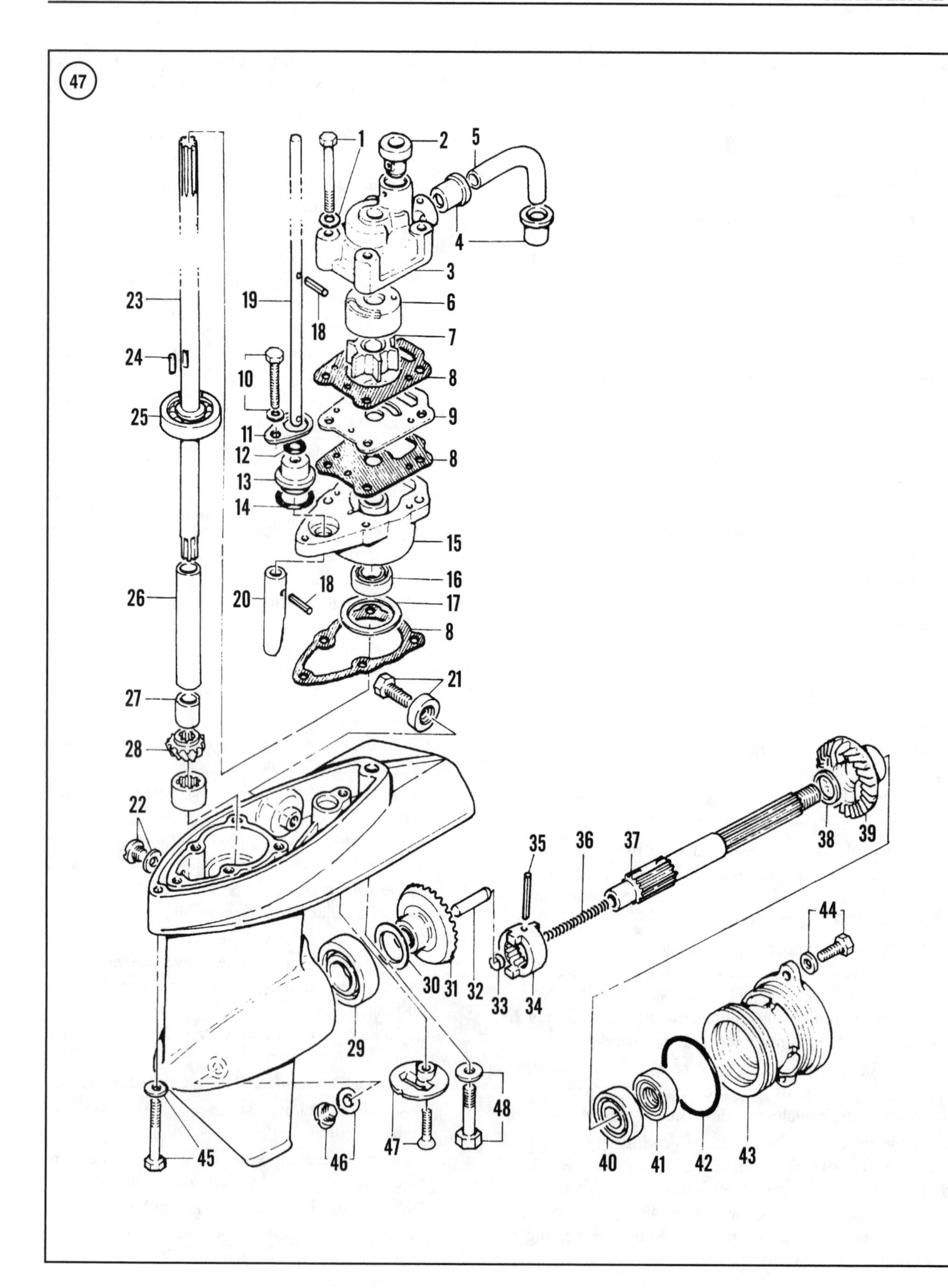
47
1
2
5
3
4
23
19
18
6
7
24
10
8
25
9
11
12
8
13
14
15
16
26
20
18
17
8
21
27
28
22
35
36
37
38
39
44
30
31
32
33
34
29
48
47
46
45
40
41
42
43

GEARCASE COMPONENTS (4 AND 5 HP MODELS)

1. Screw and washer
2. Water tube grommet
3. Pump housing
4. Grommets
5. Water tube
6. Pump liner
7. Impeller
8. Gaskets
9. Impeller plate
10. Screw and washer
11. Retainer
12. Shift shaft O-ring
13. Shift shaft bushing
14. Bushing O-ring
15. Pump base
16. Drive shaft seal
17. Shim(s)
18. Roll pins
19. Lower shift shaft
20. Shift cam
21. Anode and screw
22. Vent plug and seal
23. Drive shaft
24. Impeller drive key
25. Ball bearing
26. Drive shaft sleeve
27. Bearing sleeve
28. Pinion gear
29. Ball bearing
30. Shim(s)
31. Forward gear
32. Shift cam follower
33. Spring guide
34. Sliding clutch
35. Retaining pin
36. Spring
37. Propeller shaft
38. Thrust washer
39. Reverse gear
40. Ball bearing
41. Propeller shaft seal
42. O-ring
43. Propeller shaft bearing carrier
44. Screw and washer
45. Screw and washer
46. Drain/fill plug and seal
47. Water screen and screw
48. Screw and washer

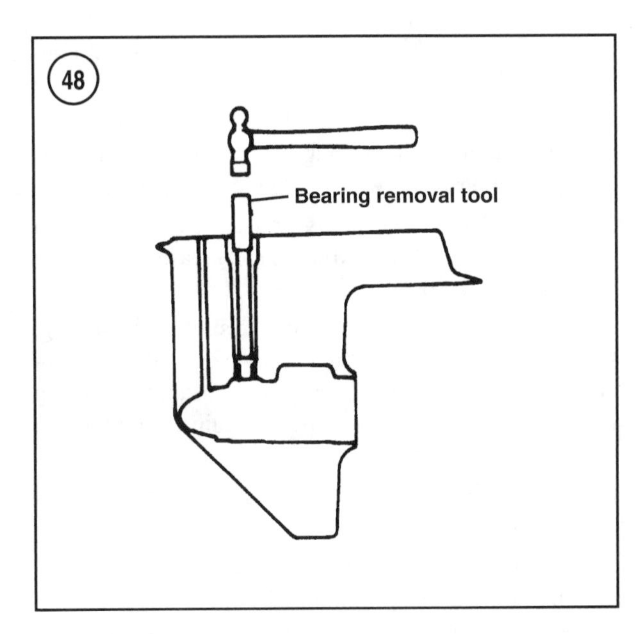

a. Install the sliding clutch onto the propeller shaft. Align the retaining pin hole in the clutch with the slot in the shaft.
b. Install the spring into the shift cam follower end of the propeller shaft.
c. Clamp a suitable pin punch horizontally in a vise. Position the shaft so the punch fits inside the cam follower bore and compresses the spring.
d. Once the spring is compressed beyond the clutch retaining pin hole, insert another pin punch through the hole in the sliding clutch to hold the spring compressed.
e. Install a new retaining pin into the clutch from the opposite side of the pin punch. Tap the retaining pin into the clutch until it contacts the pin punch, then carefully drive the retaining pin into the clutch, pushing out the pin punch. When properly installed, the cross pin is centered in the sliding clutch.
f. Lubricate the shift cam follower and spring guide with Quicksilver 2-4-C Multi-Lube grease (part No. 92-825407). Install the spring guide, then the cam follower, into the propeller shaft bore.

4. Install a new seal and bearing into the propeller shaft bearing carrier as follows:
   a. Coat the outer diameter of a new seal with Loctite 271 threadlocking adhesive (part No. 92-809819).
   b. Position the seal into the bearing carrier.
   c. Use a seal installer (part No. 91-83147M) and driver rod (part No. 91-84529M) or an equivalent to press the seal into the carrier until it is seated.
   d. Lubricate a new ball bearing and the carrier bearing bore.

e. Position the bearing into the carrier with the numbered side facing upward.

f. Use a mandrel (part No. 91-84536M or an equivalent) to press the bearing into the housing until it is seated. Press only on the outer bearing race.

g. Lubricate the seal lips and a new O-ring with Quicksilver 2-4-C Multi-Lube grease. Install the O-ring into the carrier groove.

5. Install a new inner O-ring into the shift shaft bushing. Lubricate the O-ring, bushing and lower shift shaft with Quicksilver 2-4-C Multi-Lube grease. Install the bushing over the shift shaft and into position against the roll pin at the center of the shaft.

6. Reinstall the shift cam onto the lower shift shaft and secure it with a new roll pin.

7. Lubricate the forward gear hub. Then install the original shim(s) onto the forward gear.

8. Position the forward gear and shim(s) on the propeller shaft and insert the assembly into the gear cavity. Guide the gear and shim(s) into the forward bearing, then remove the propeller shaft.

9. Position the pinion gear under the drive shaft needle bearing. While holding the pinion gear in place, install the drive shaft into the gearcase bore and through the needle bearing. Rotate the shaft as necessary to engage the pinion gear, then seat it fully in the gearcase bore.

10. Reinstall the original shim(s) on top of the drive shaft ball bearing. A small amount of grease may be used to hold the shim(s) in position.

11. Install the thrust washer and reverse gear onto the propeller shaft and up against the sliding clutch.

12. Slide the propeller shaft bearing carrier assembly over the propeller shaft splines and seat the carrier against the reverse gear.

13. While holding the propeller shaft splines, insert the carrier and shaft assembly into the gearcase until the carrier flange seats against the gearcase housing.

14. Coat the threads of the bearing carrier screws with Quicksilver Perfect Seal (part No. 92-343227-1). Install the screws and washers, and tighten them to the specification in **Table 1**.

15. Install the water pump base, shift shaft assembly and the water pump as described in this chapter. Make sure the shift cam's ramp is properly positioned against the shift cam follower.

16. Pressure test the gearcase as described under *Gearcase Pressure Testing*.

17. Fill the gearcase with the recommended lubricant as described in Chapter Four.

Disassembly (6-15 hp Models)

Refer to **Figure 49** for this procedure. The propeller shaft bearing carrier and propeller shaft can be removed without removing the gearcase from the drive shaft housing.

NOTE
If the forward gear roller bearing requires replacement, replace the bearing rollers and race as an assembly. Do not remove a pressed-in bushing, bearing and/or race unless replacement is necessary.

1. Remove the gearcase as described in this chapter.

2. Drain the gearcase lubricant as described in Chapter Four.

3. Remove the water pump, pump base and shift shaft as described in this chapter.

CAUTION
The propeller shaft bearing carrier has left-hand threads. Use the spanner wrench (part No. 91-13664) to remove the carrier.

4. Unthread the propeller shaft bearing carrier from the gearcase using a spanner wrench (part No. 91-13664).

5. Remove the propeller shaft and bearing carrier assembly from the gearcase. Make sure the shift cam follower (38, **Figure 49**) comes out with the propeller shaft. If it does not, retrieve it from the gearcase cavity and temporarily reinstall it into the propeller shaft.

6. Remove the propeller shaft and reverse gear from the bearing carrier. Remove the carrier O-ring.

7. Secure the propeller shaft bearing carrier in a soft-jawed vise. Drive the seals from the carrier using a suitable punch and hammer. Drive the seals from the reverse gear side to the propeller side.

8. Lift the drive shaft out from the top of the gearcase.

9. Reach inside the gear cavity and remove the pinion gear, pinion gear thrust washer and the forward gear and roller bearing assembly.

NOTE
Do not remove the forward gear bearing race from the gearcase unless the race or bearing requires replacement.

10. If the forward gear roller bearing or bearing race requires replacement, replace them as follows:

a. Remove the race from the gearcase using a puller kit (part No. 91-83165M) in conjunction with a puller plate (part No. 91-29310) or a slide hammer puller. See **Figure 50**.

b. Remove the roller bearing from the forward gear by supporting the gear in a knife-edged bearing separator (such as part No. 91-37241). Press the gear from the bearing with a suitable mandrel. See **Figure 51**.

11. If the forward gear inner bushing requires replacement, remove the bushing by supporting the gear in a press with the roller bearing side facing upward. Position the bushing remover (part No. 91-824787 or an equivalent) against the bushing. Then press the bushing from the gear.

12. If the propeller shaft bearing carrier bushing requires replacement, remove the bushing by supporting the carrier in a press with the propeller end facing down. Position the bushing remover (part No. 91-824787 or an equivalent) against the bushing. Then press the bushing from the carrier.

13. If the propeller shaft requires disassembly, disassemble it as follows:

a. Remove the shift cam follower from the front of the shaft.

b. Place the shaft on a suitable support, such as the open, padded jaws of a vise, and carefully drive the retaining pin (40, **Figure 49**) from the sliding clutch using a suitable punch and hammer. See **Figure 52**. Drive the retaining pin from the side of the pin that is *not* grooved. Do not remove the punch from the sliding clutch at this time.

c. Position the propeller shaft upright with the cam follower end resting on a solid surface. Slowly withdraw the punch from the clutch and carefully release the spring compression. Remove the spring from the end of the shaft, then slide the clutch off the propeller shaft.

14. If the drive shaft upper bushing and sleeve, lubrication sleeve or lower needle bearing require replacement, replace them as follows:

a. Remove the drive shaft upper bushing and sleeve using a suitable bearing puller, such as a collet and expander rod, attached to a slide hammer (part No. 91-34569A-1). See **Figure 53**.

b. Lift the lubrication sleeve out the top of the gearcase. If removal proves difficult, check for the presence of burrs above the sleeve. Remove any burrs with a suitable scraper.

c. Insert the drive shaft bearing removal tool (part No. 91-824788A-1 or an equivalent) into the gearcase and into the drive shaft needle bearing. Make sure the tool's pilot is installed in place of the drive shaft upper bushing. See **Figure 48**.

d. Drive the bearing down into the gear cavity. Remove the bearing.

15. Refer to *Gearcase Cleaning and Inspection*. Clean and inspect all components as described before beginning reassembly.

Assembly (6-15 hp Models)

Lubricate all internal components with Quicksilver Premium Blend gear oil (part No. 92-13783 or an equivalent). Do not assemble components *dry*. Refer to **Figure 49** for this procedure.

1. If the drive shaft bushing and sleeve, lubrication sleeve and lower needle bearing were removed, install a new bushing and sleeve, a new lubrication sleeve and a new needle bearing as follows:

a. Lubricate a new lower needle bearing and the entire gearcase drive shaft bore.

b. Install the bearing into the gearcase from the propeller shaft bore with the numbers facing down. Pull the bearing up and into place with the bearing tool kit (part No. 91-824790A-1 or an equivalent).

c. Tighten the puller bolt until the mandrel seats in the drive shaft lower bore. The bearing is slightly recessed into its bore when it is correctly installed.

d. Install the lubrication sleeve into the drive shaft bore.

e. Lubricate a new upper drive shaft bushing and sleeve, and place them in the gearcase bore. Pull the bushing and sleeve into place with the same bearing tool kit (part No. 91-824790A-1).

f. Tighten the puller bolt until the tool's pilot seats in the drive shaft upper bore.

2. If the forward gear roller bearing and race were removed, install a new roller bearing and race as follows:

a. Lubricate a new forward gear bearing race and roller bearing.

b. Insert the race into the gearcase with the tapered side facing out.

c. Press the race into the gearcase until it seats using a mandrel (part No. 91-13658) and spanner wrench (part No. 91-13664). Place the spanner wrench over the mandrel and use the spanner as a driver rod. Make sure the race is fully seated in the gearcase bore.

d. Lubricate the forward gear hub. Place the gear in a press on a soft metal or wooden block with the gear teeth facing downward.

e. Place the roller bearing onto the gear hub with the roller bearings facing upward.

f. Press the bearing onto the gear hub until it seats using a suitable mandrel. See **Figure 54**.

3. If the forward gear bushing was removed, lubricate a new bushing and place it in the forward gear bore from the

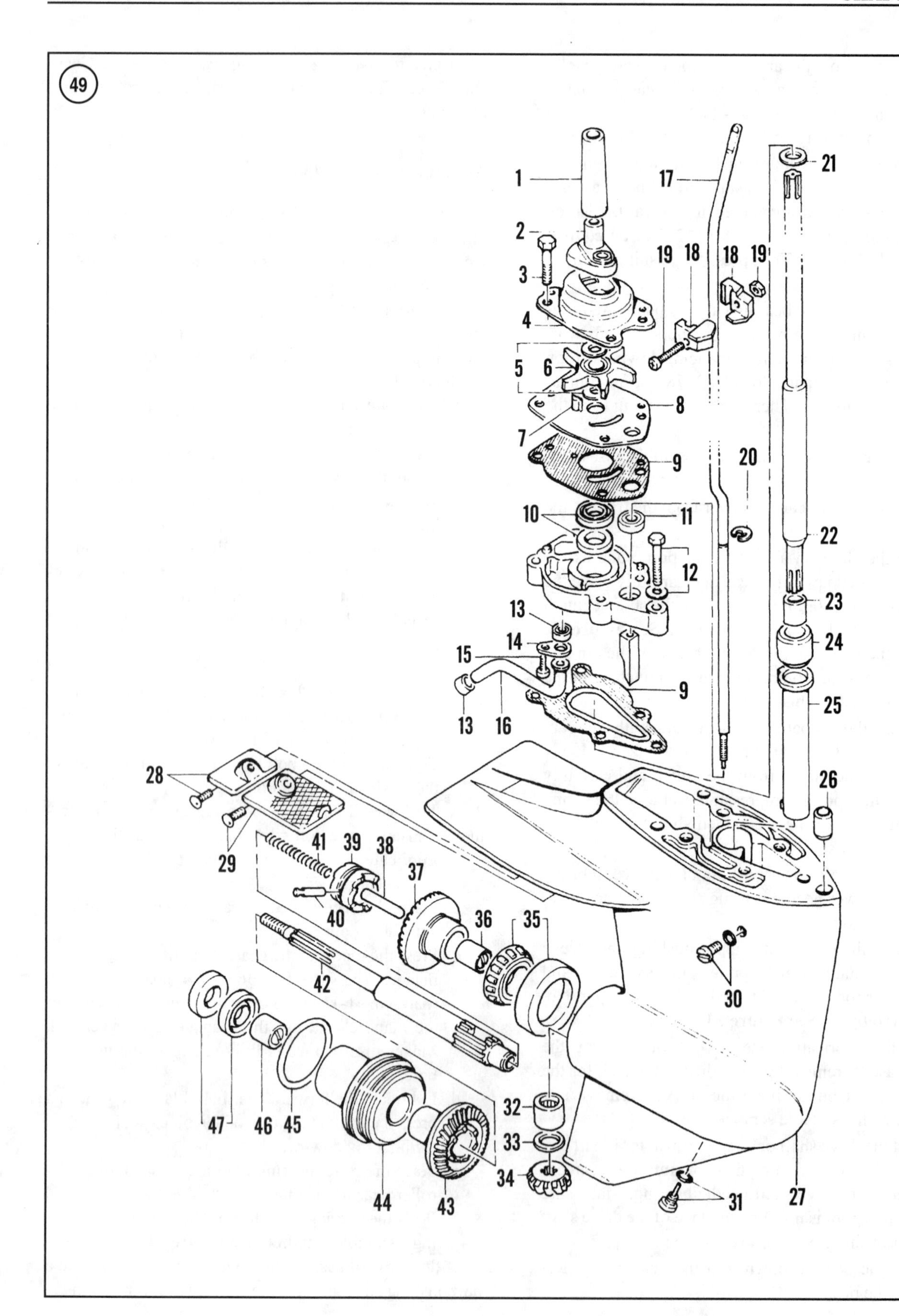
49
1
2
3
4
5
6
7
8
9
10
11
12
13
14
15
16
17
18
19
20
21
22
23
24
25
26
27
28
29
30
31
32
33
34
35
36
37
38
39
40
41
42
43
44
45
46
47

GEARCASE COMPONENTS (9-15 HP MODELS)

1. Water tube guide
2. Water tube and drive shaft seal
3. Screw
4. Pump housing
5. Nylon washers
6. Impeller
7. Impeller key
8. Impeller plate
9. Gaskets
10. Drive shaft seals
11. Shift shaft quad-ring
12. Screw and washer
13. Grommets
14. Retainer
15. Screw
16. Water inlet tube
17. Shift shaft
18. Reverse lock actuators
19. Screw and nut
20. Lock clip
21. Thrust washer
22. Drive shaft
23. Upper drive shaft bushing
24. Bushing sleeve
25. Lubrication sleeve
26. Dowel sleeve
27. Gearcase housing
28. Anode and screw
29. Water screen and screw
30. Vent plug and seal
31. Drain/fill plug and seal
32. Needle bearing
33. Thrust washer
34. Pinion gear
35. Roller bearing and race
36. Bushing
37. Forward gear
38. Shift cam follower
39. Sliding clutch
40. Retaining pin
41. Spring
42. Propeller shift
43. Reverse gear
44. Propeller shaft bearing carrier
45. O-ring
46. Bushing
47. Propeller shaft seals

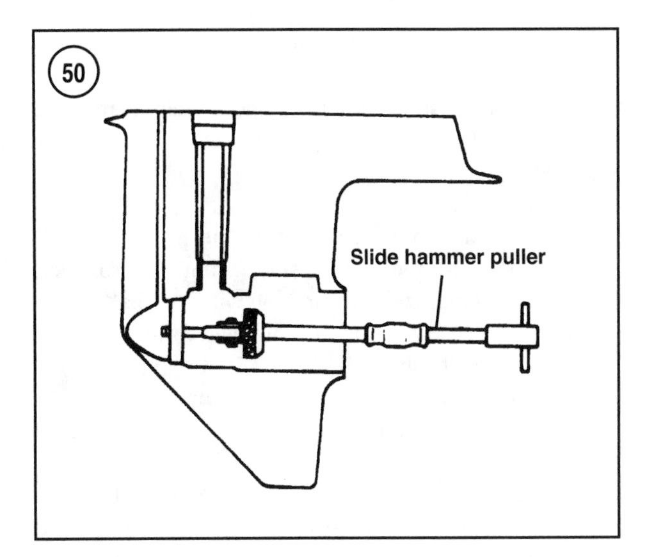

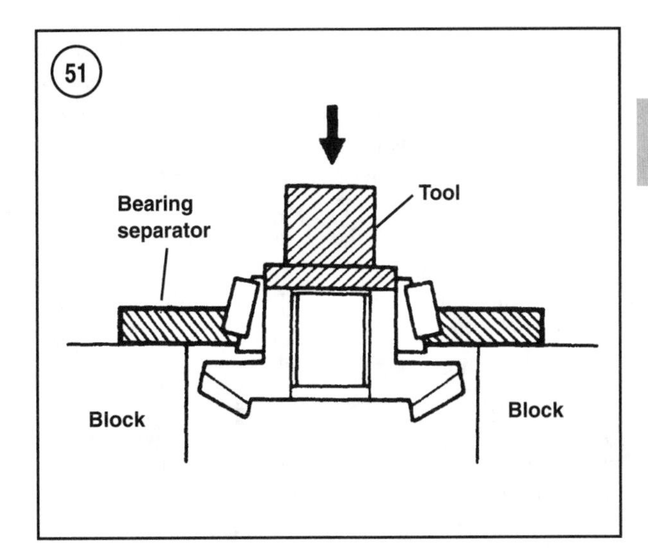

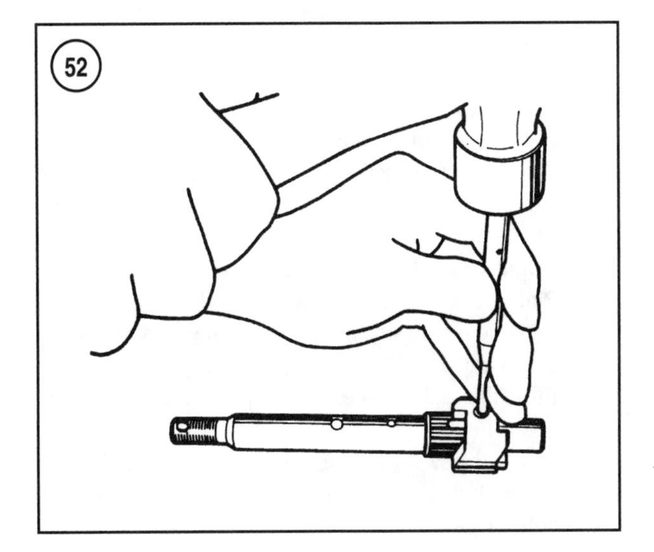

roller bearing side. Press the bushing into the gear using the small end of mandrel (part No. 91-13658 or an equivalent). Press the bushing until the mandrel seats against the gear hub. The bushing sits below flush when it is correctly installed.

4. Reassemble the propeller as follows:
   a. Install the sliding clutch over the propeller shaft splines with the longest side facing the propeller, aligning the retaining pin hole with the shaft slot.
   b. Install the spring into the shift cam follower end of the propeller shaft.
   c. Clamp a suitable pin punch horizontally in a vise. Position the shaft so the punch fits inside the cam follower bore and compresses the spring.
   d. Once the spring is compressed beyond the clutch retaining pin hole, insert another pin punch through the hole in the sliding clutch to hold the spring compressed.
   e. Install a new retaining pin into the clutch from the opposite side of the pin punch. Make sure the grooved end of the pin enters the sliding clutch *last*. Tap the retaining pin into the clutch until it contacts the pin punch, then carefully drive the retaining pin into the clutch, pushing out the pin punch. When properly installed, the cross pin is centered in the sliding clutch.
   f. Lubricate the shift cam follower with Quicksilver 2-4-C Multi-Lube grease (part No. 92-825407). Install the flat end of the cam follower into the propeller shaft bore.

5. If the propeller shaft bearing carrier bushing was removed, lubricate a new bushing and place it in the carrier bore from the gearcase side. Press the bushing into the carrier using the multistepped side of a mandrel (part No. 824785A-1) *without* the tool's removable ring. Press the bushing until the mandrel seats against the carrier.

6. Install two new propeller shaft seals into the propeller shaft bearing carrier as follows:
   a. Coat the outer diameter of both seals with Loctite 271 threadlocking adhesive (part No. 92-809819).
   b. Install the seal without the fish line cutter into the carrier using a mandrel (part No. 91-824785A-1). Make sure the removable ring is installed onto the single stepped side of the mandrel.
   c. Place the seal onto the mandrel and against the ring with the spring side of the seal facing away from the mandrel.
   d. Press the seal into the carrier until the mandrel seats against the carrier.
   e. Install the seal with the fish line cutter into the carrier with the same mandrel, but without the removable ring.

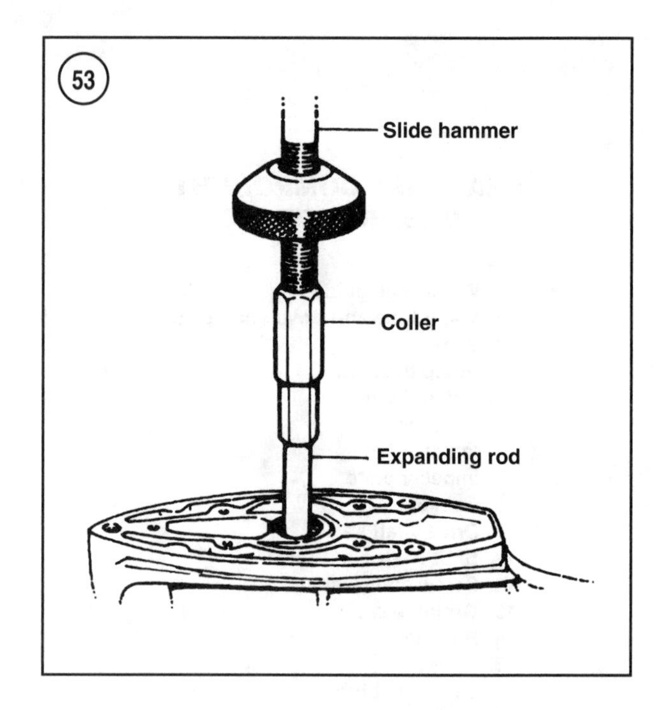

   f. Place the seal onto the single-stepped side of the mandrel with the flat metal spring side facing the mandrel.
   g. Press the seal into the carrier until the mandrel seats against the carrier.

7. Lubricate a new bearing carrier O-ring and both propeller shaft seal lips with Quicksilver 2-4-C Multi-Lube grease. Then install the O-ring into the carrier groove.

8. Install the forward gear and roller bearing assembly into the gear cavity and into the forward gear bearing race.

9. Lubricate the top of the pinion gear with Quicksilver 2-4-C Multi-Lube grease to help hold the thrust washer in place. Install the thrust washer to the pinion gear with the grooved side of the thrust washer facing the pinion gear teeth.

10. Position the pinion gear and thrust washer under the drive shaft needle bearing. While holding the pinion gear in place, install the drive shaft into the gearcase bore and through the needle bearing. Rotate the shaft as necessary to engage the pinion gear, then seat it fully in the gearcase bore.

11. Install the thrust washer on top of the upper drive shaft bushing. A small amount of grease may be used to hold the thrust washer in position.

12. Install the water pump base, shift shaft assembly and the water pump as described in this chapter. Make sure the shift cam's ramp is properly positioned toward the propeller shaft.

NOTE
The shift rod height cannot be set or verified until the propeller shaft and bearing carrier are installed.

13. Lubricate the reverse gear. Insert the gear into the bearing carrier. Insert the propeller shaft assembly into the bearing carrier.
14. Lubricate the propeller shaft bearing carrier threads with Quicksilver Special Lube 101 (part No. 92-13872A1). Then install the shaft and carrier assembly into the gearcase. Tighten the carrier counterclockwise as far as possible by hand.

NOTE
The propeller shaft bearing carrier has left-hand threads.

15. Install the spanner wrench (part No. 91-13664) over the propeller shaft and engage it to the bearing carrier. Tighten the bearing carrier counterclockwise to the specification in **Table 1**.
16. Verify the shift rod height as specified during water pump assembly procedures.
17. Pressure test the gearcase as described in *Gearcase Pressure Testing*.
18. Fill the gearcase with the recommended lubricant as described in Chapter Four.

Disassembly (20 and 25 hp Models)

The propeller shaft bearing carrier and propeller shaft can be removed without removing the gearcase from the drive shaft housing. The carrier has left-hand threads.

Refer to **Figure 55** for this procedure.

NOTE
If the forward gear or pinion gear roller bearings were replaced, replace the bearing rollers and races as assemblies. Do not remove any pressed-in bushings, bearing and/or race unless replacement is necessary.

1. Remove the gearcase as described in this chapter.
2. Drain the gearcase lubricant as described in Chapter Four.
3. Remove the water pump as described in this chapter.
4. Mark the trim tab position (**Figure 56**) and remove the trim tab retaining screw. Remove the trim tab.
5. Remove the screws securing the retaining plate to the propeller shaft bearing carrier. Remove the retainer plate and carrier O-ring.

CAUTION
The propeller shaft bearing carrier has left-hand threads. Use only the recommended spanner wrench (part No. 91-83843-1) to remove the carrier.

6. Use the spanner wrench (part No. 91-83843-1) to remove the carrier. Turn the carrier clockwise until it is free from the gearcase threads.
7. Pull the propeller shaft and bearing carrier out as an assembly. Make sure the shift cam follower (42, **Figure 55**) comes out with the propeller shaft. If it does not, retrieve it from the gearcase cavity and temporarily reinstall it into the propeller shaft.
8. Remove the reverse gear and propeller shaft bearing carrier from the propeller shaft.
9. Secure the propeller shaft bearing carrier in a vise with protective jaws. Use a suitable punch to drive both propeller shaft seals from the bearing carrier. Be careful not to damage the seal bore during the removal process.
10. Remove the screw securing the pinion nut to the drive shaft. Lift the drive shaft out of the gearcase. Remove the pinion and forward gear, and bearing assemblies from the gearcase.
11. Remove the drive shaft seals from the gearcase with a puller (part No. 91-27780 or an equivalent). See **Figure 57**.
12. Remove the shift shaft assembly from the gearcase. See **Figure 58**. Remove the outer O-ring from the shift shaft retainer. Inspect the retainer, boot, inner O-ring, detent springs and shift cam. Replace any worn or damaged parts. Do not remove the inner O-ring unless replacement is necessary. The shift retainer (8, **Figure 58**) is plastic and must be inspected carefully for cracks.
13. If propeller shaft disassembly is required, disassemble it as follows:

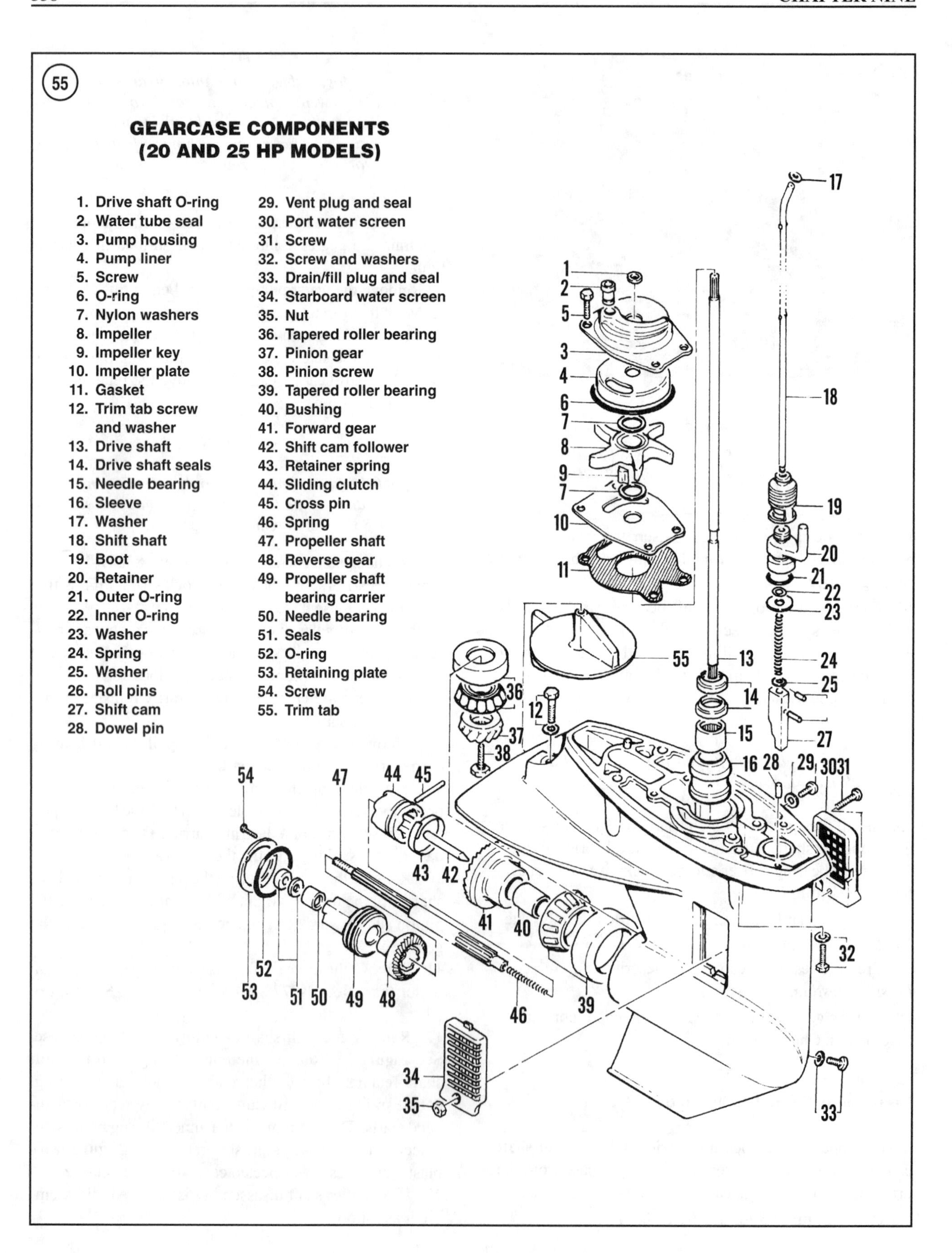
55
GEARCASE COMPONENTS
(20 AND 25 HP MODELS)
1. Drive shaft O-ring
2. Water tube seal
3. Pump housing
4. Pump liner
5. Screw
6. O-ring
7. Nylon washers
8. Impeller
9. Impeller key
10. Impeller plate
11. Gasket
12. Trim tab screw and washer
13. Drive shaft
14. Drive shaft seals
15. Needle bearing
16. Sleeve
17. Washer
18. Shift shaft
19. Boot
20. Retainer
21. Outer O-ring
22. Inner O-ring
23. Washer
24. Spring
25. Washer
26. Roll pins
27. Shift cam
28. Dowel pin
29. Vent plug and seal
30. Port water screen
31. Screw
32. Screw and washers
33. Drain/fill plug and seal
34. Starboard water screen
35. Nut
36. Tapered roller bearing
37. Pinion gear
38. Pinion screw
39. Tapered roller bearing
40. Bushing
41. Forward gear
42. Shift cam follower
43. Retainer spring
44. Sliding clutch
45. Cross pin
46. Spring
47. Propeller shaft
48. Reverse gear
49. Propeller shaft bearing carrier
50. Needle bearing
51. Seals
52. O-ring
53. Retaining plate
54. Screw
55. Trim tab

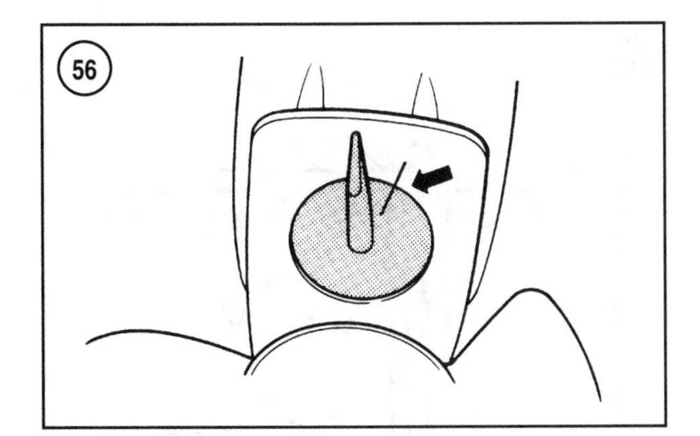

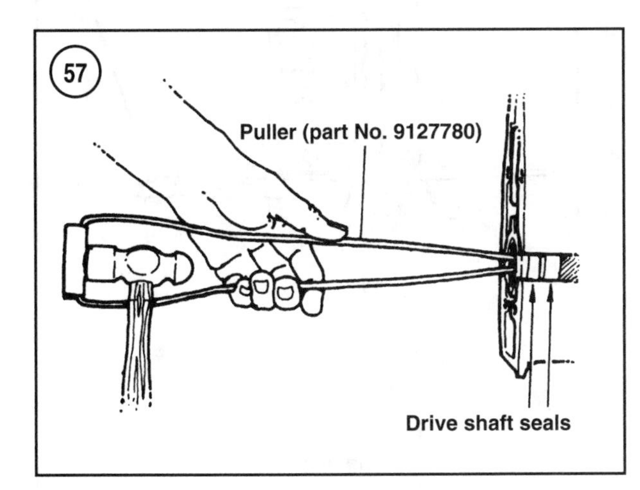

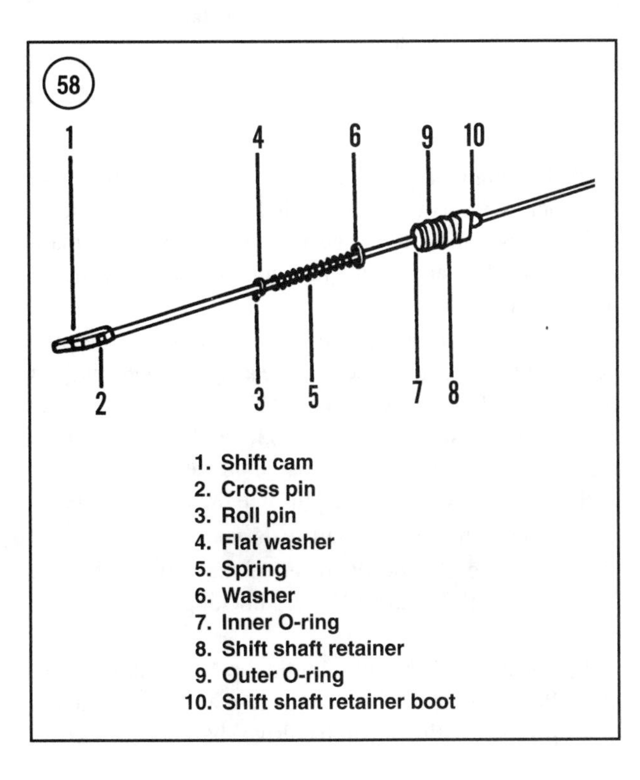

1. Shift cam
2. Cross pin
3. Roll pin
4. Flat washer
5. Spring
6. Washer
7. Inner O-ring
8. Shift shaft retainer
9. Outer O-ring
10. Shift shaft retainer boot

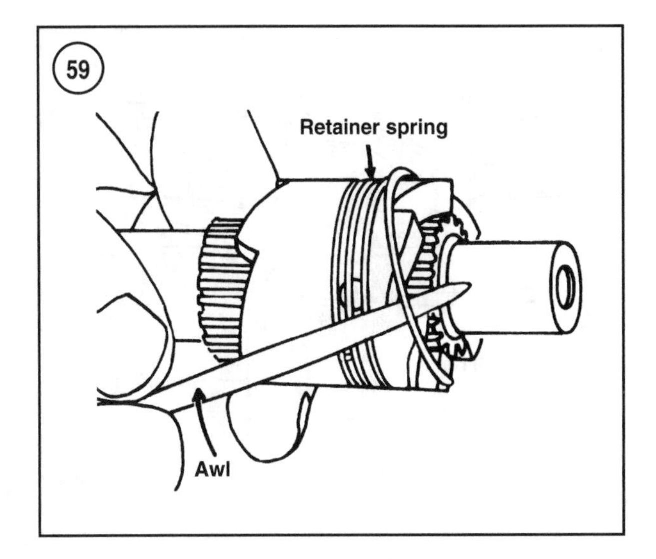

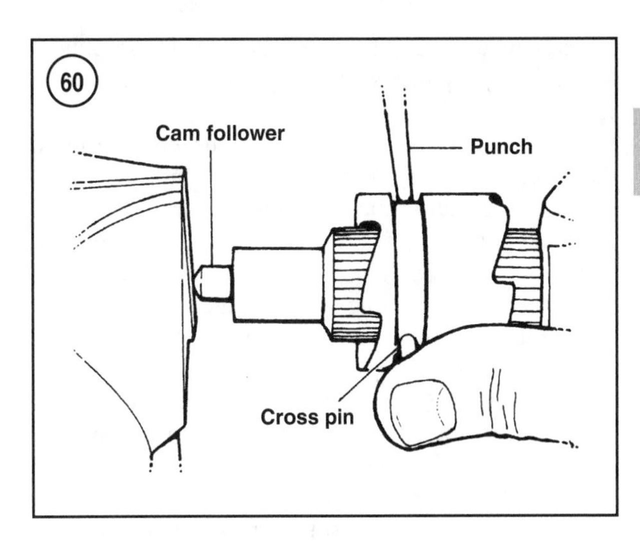

9

a. Remove the sliding clutch spring with a small screwdriver or awl. Insert the tool under one end of the spring and rotate the propeller shaft to unwind the spring. See **Figure 59**.

b. Place the shift cam follower against a solid surface and push on the propeller shaft to unload the spring pressure on the sliding clutch cross pin. Remove the pin with a suitable punch. Carefully release the pressure on the cam follower and spring. Remove the cam follower, spring and sliding clutch from the propeller shaft. See **Figure 60**.

14. If the propeller shaft bearing carrier needle bearing requires replacement, press the bearing from the carrier with a suitable mandrel, or drive the bearing out with a suitable punch.

15. If the lower drive shaft or pinion gear roller bearing must be replaced, remove the bearing race from the

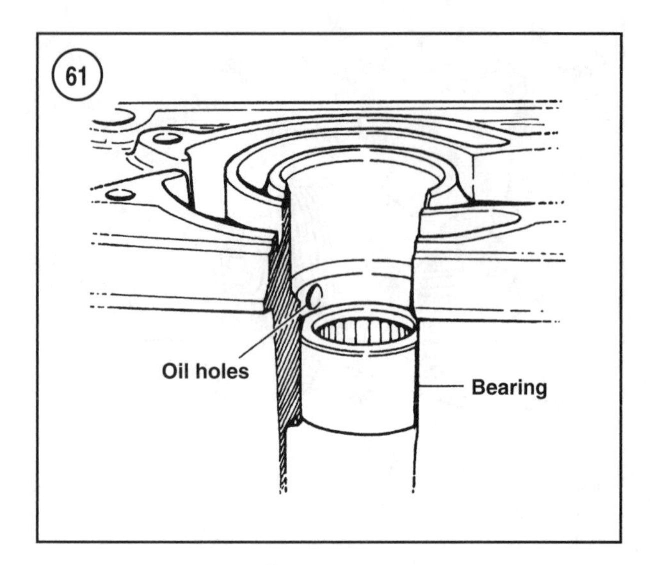

gearcase by driving it down into the propeller shaft bore with a bearing puller kit (part No. 91-31229A-7 or an equivalent).

16. If the drive shaft needle bearing at the top of the drive shaft bore must be replaced, drive it down into the propeller shaft bore with a suitable mandrel.

17. If the forward gear roller bearing must be replaced, replace it as follows:

a. Remove the bearing race from the gearcase with a suitable slide hammer, such as part No. 91-34569A-1 or an equivalent. See **Figure 50**.

b. Remove the roller bearing from the forward gear with a knife-edged bearing separator plate, such as part No. 91-37241 or an equivalent and a suitable mandrel. Support the roller bearing with the bearing plate and press against the gear hub with the mandrel. See **Figure 51**.

18. If the bushing in the forward gear bore must be replaced, remove the bushing by securing the gear in a vice with protective jaws. Drive the bushing out with a suitable punch and hammer. Do not damage the forward gear roller bearing, if it is installed during this operation.

19. Refer to *Gearcase Cleaning and Inspection*. Clean and inspect all components as directed before beginning reassembly.

Assembly (20-25 hp Models)

Lubricate all internal components with Quicksilver Premium Blend gear oil (part No. 92-13783 or an equivalent). Do not assemble components *dry*. Refer to **Figure 55** for this procedure.

1. If the drive shaft needle bearing at the top of the drive shaft bore was removed, lubricate the outside diameter of a new bearing with gear oil. Set the bearing into the drive shaft bore with its numbered side facing the power head. Drive the bearing into the gearcase using a suitable mandrel. Drive the bearing just far enough to uncover the oil hole (**Figure 61**).

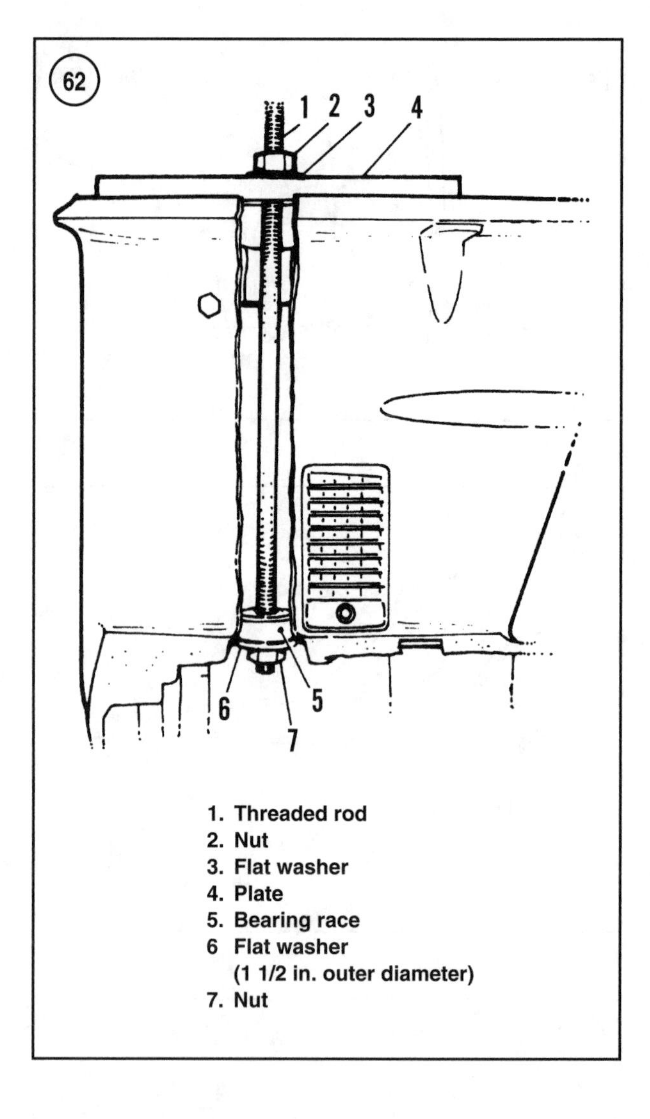

1. Threaded rod
2. Nut
3. Flat washer
4. Plate
5. Bearing race
6 Flat washer (1 1/2 in. outer diameter)
7. Nut

2. If the lower drive shaft bearing was removed, install a new race into the gearcase with bearing installer kit (part No. 91-31229A7) or with a suitable threaded rod, plate and washers as shown in **Figure 62**. Oil the outside of the race and position it with the small diameter facing the top of the gearcase. Seat the race fully in the bearing bore.

3. Install two new drive shaft seals as follows:

a. Coat the outside diameter of the new drive shaft seals with Loctite 271 threadlocking adhesive (part No. 92-809819).

b. Install the first seal with the lip facing downward. Press the seal with a suitable mandrel until it sits just below flush with the drive shaft bore.

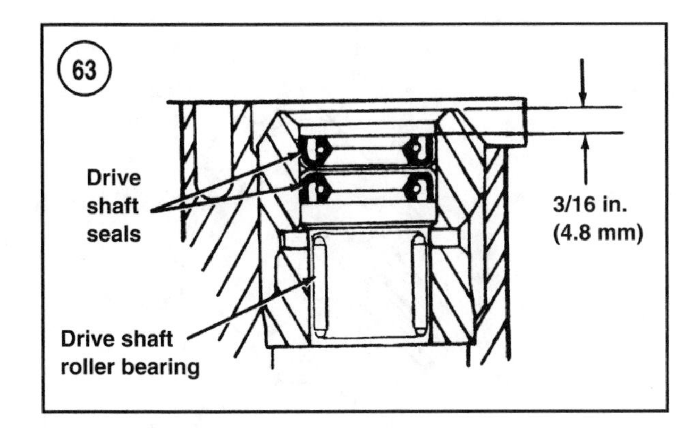

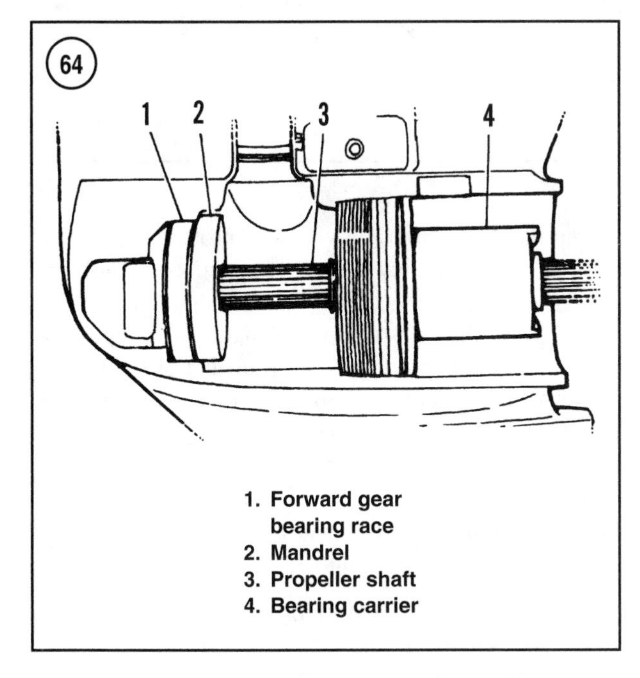

1. Forward gear bearing race
2. Mandrel
3. Propeller shaft
4. Bearing carrier

c. Install the second seal with the lip facing upward. Press both seals to a depth of 3/16 in. (4.8 mm) from the top of the drive shaft bore as shown in **Figure 63**.

d. Lubricate the seal lips with 2-4-C Multi-Lube grease (part No. 92-825407 or an equivalent).

4. If the forward gear roller bearing and race were removed, install a new bearing as follows:

a. Lubricate the new bearing race and set it into its bearing bore.

b. Place a suitable mandrel from a bearing installer kit (part No. 91-31229A7) over the race. Place the propeller shaft into the mandrel hole as shown in **Figure 64**.

c. Install the propeller shaft bearing carrier four to five full turns into the gearcase as shown in **Figure 64** to keep the propeller shaft centered.

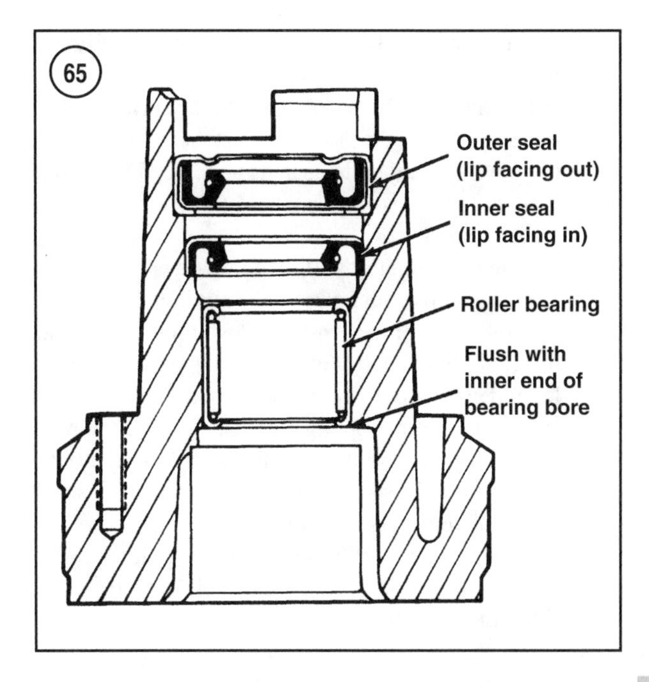

d. Thread a scrap propeller nut onto the propeller shaft. Use a mallet and drive the propeller shaft against the mandrel until the bearing race is fully seated in the gearcase bearing bore.

e. Remove the propeller nut, propeller shaft, bearing carrier and mandrel.

f. Set the forward gear in a press with the gear teeth facing downward.

g. Lubricate the new roller bearing. Then set the bearing onto the gear with the rollers facing upward.

h. Press the bearing fully onto the gear with a suitable mandrel. See **Figure 54**. Do not press on the roller cage. If a new bushing must be installed, leave the gear in the press for Step 5.

5. If the forward gear bushing was removed, lubricate a new bushing. Press the bushing into the forward gear with a suitable mandrel. When correctly installed, the bushing is flush to 0.020 in. (0.51 mm) recessed from the face of the gear hub.

6. If the propeller shaft bearing carrier's internal needle bearing was removed, lubricate a new bearing with gear oil and place it into the carrier with its lettered end facing the propeller. Use a suitable mandrel to press the bearing flush with the end of the bearing bore as shown in **Figure 65**.

7. Install two *new* propeller shaft seals as follows:

a. Coat the outer surface of the *new* propeller shaft seals with Loctite 271 threadlocking adhesive.

b. Install the small diameter seal with the spring facing the gearcase. Press the seal in with a suitable mandrel until the seal bottoms in its bore. See **Figure 65**.

c. Install the large diameter seal with the spring facing the propeller. Press the seal in with a suitable mandrel until the seal bottoms in its bore. See **Figure 65**.

d. Coat the seal lips with 2-4-C Multi-Lube grease or an equivalent.

8. Assemble the propeller shaft as follows:

a. Align the cross pin holes of the sliding clutch with the slot in the propeller shaft. Position the long end of the sliding clutch (**Figure 66**) toward the propeller and slide it onto the propeller shaft.

b. Lubricate the shift spring and install it into the propeller shaft. Install the cam follower with the beveled end out. Press the cam follower against a solid object to compress the spring.

c. Insert a small punch through the sliding clutch holes *between* the spring and the cam follower. Remove the cam follower.

d. Insert the sliding clutch cross-pin into the sliding clutch opposite of the punch and slide the pin into the sliding clutch while pushing the punch out. See **Figure 60**. When installed, the cross-pin must be in front of the spring.

NOTE

When installed correctly, the sliding clutch retaining spring must lay flat, with no overlapping coils.

9. Secure the pin to the sliding clutch with a new retainer spring. Do not open the spring any more than necessary to install it. Grease the cam follower with 2-4-C Multi-Lube grease or an equivalent and install it into the propeller shaft with the beveled end out.

10. If the shift shaft was disassembled, reassemble the shift shaft assembly with new O-rings and a *new* boot. Grease the O-rings with Quicksilver 2-4-C Multi-Lube grease. Install the shift shaft to the gearcase with the notched ramp of the shift cam facing the propeller. Seat the retainer firmly into the shift cavity.

11. Rotate the gearcase so the propeller shaft bore is pointing upward. Install the forward gear and bearing assembly into the forward bearing race.

12. Place the lower drive shaft roller bearing into the lower race. Place the pinion gear in position on top of the bearing.

13. Carefully insert the drive shaft through the drive shaft seals and into the gear housing. Hold the pinion gear and lower bearing in position and engage the drive shaft splines to the pinion gear splines.

14. Apply Loctite 271 threadlocking adhesive to the pinion screw. Install the pinion screw and hand tighten it.

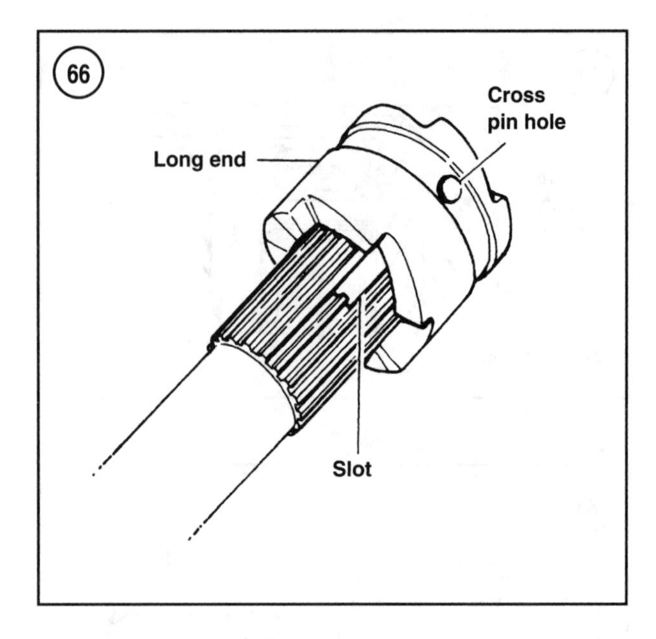

Clamp the drive shaft in a vise with protective jaws and tighten the pinion screw to the specification in **Table 1**.

15. Turn the gearcase so the propeller shaft bore is horizontal. Move the shift rod upward to the forward position (upper position) and install the propeller shaft into the center of the forward gear. Slide the reverse gear over the propeller shaft into mesh with the pinion gear.

16. Coat the threads of the propeller shaft bearing carrier with Quicksilver Special Lube 101 (part No. 92-13872A1) and thread the carrier *counterclockwise* into the gearcase as far as possible by hand.

CAUTION

The propeller shaft bearing carrier has left-hand threads. Use only the recommended spanner wrench (part No. 91-83848-1) to loosen and tighten the carrier.

17. Use the spanner wrench (part No. 91-83843-1) to tighten the carrier *counterclockwise* to the specification in **Table 1**.

18. Coat a *new* propeller shaft carrier O-ring with Quicksilver 2-4-C Multi-Lube grease and position it behind the bearing carrier in the gearcase bore. Install the retainer plate over the O-ring and align the holes with the bearing carrier. Coat the retainer screws with Quicksilver Perfect Seal (part No. 92-34227-1). Install and tighten the screws to the specification in **Table 1**.

19. Install the water pump as described previously in this chapter.

20. Pressure test the gearcase as described under *Gearcase Pressure Testing* in this chapter.

21. Fill the gearcase with the recommended lubricant as described in Chapter Four.

Disassembly (30-50 hp Models)

The propeller shaft bearing carrier and propeller shaft can be removed without removing the gearcase from the drive shaft housing. Refer to **Figure 67** for this procedure.

NOTE
If the forward gear or pinion gear roller bearings require replacement, replace the bearing rollers and races as assemblies. Do not remove a pressed in bearing and/or race unless replacement is necessary.

1. Remove the gearcase as described in this chapter.
2. Drain the gearcase lubricant as described in Chapter Four.
3. Remove the water pump as described in this chapter.
4. Inspect the exhaust deflector and seal (2, **Figure 67**). Replace the deflector and/or seal if they are damaged.
5. Remove the screws and washers (57, **Figure 67**) securing the propeller shaft bearing carrier to the gearcase. Remove the propeller shaft bearing carrier with puller (part No. 91-27780 or an equivalent).
6. Pull the propeller shaft assembly from the gearcase. If the shift cam follower falls out of the propeller shaft, retrieve it from the gearcase and reinstall it into the propeller shaft temporarily.
7. Remove the pinion nut. Use the drive shaft spline socket (part No. 91-825196) and an appropriate wrench or socket. See **Figure 68**. Then lift the drive shaft assembly from the gearcase.
8. Remove the pinion gear, pinion roller bearing, and the forward gear and bearing assembly from the propeller shaft bore.
9. Remove the shift shaft assembly from the gearcase by pulling it up and out of the gearcase bore.
10. Disassemble the shift shaft assembly. Remove the inner and outer O-rings from the shift shaft retainer. Remove the shift boot. Inspect the shift cam for wear. The shift retainer (26, **Figure 67**) is plastic and must be inspected carefully for cracks. Replace damaged or worn parts.
11. Remove the drive shaft seal carrier (19, **Figure 67**) from the drive shaft assembly. Remove the O-ring from the seal carrier. Place the large beveled end of the seal carrier down on a suitable open support and press both seals out of the carrier using a suitable mandrel.
12. Secure the propeller shaft bearing carrier in a vice with protective jaws. Remove the reverse gear and bearing assembly with a puller (part No. 91-27780 or an equivalent). See **Figure 69**. If the ball bearing (53, **Figure 67**) remains in the carrier, remove the bearing with the slide hammer puller. See **Figure 70**.

NOTE
If the bearing carrier needle bearing must be replaced, also remove the propeller shaft seals. If the seals must be replaced, but not the bearing, remove the seals using a suitable puller.

13. Place the propeller shaft bearing carrier with the seal end down on a suitable open support. Press the bearing and seals out of the carrier with driver (part No. 91-37312) and driver rod (part No. 91-37323) or an equivalent. See **Figure 71**.
14. If the propeller shaft requires disassembly, proceed as follows:
    a. Remove the cross pin retainer spring with a small screwdriver or awl. Insert the tool under one end of the spring and rotate the propeller shaft to unwind the spring. See **Figure 59**.
    b. Place the shift cam follower against a solid surface and push on the propeller shaft to unload the spring pressure on the sliding clutch cross pin. Remove the pin with a suitable punch.
    c. Carefully release the pressure on the cam follower and spring. Remove the cam follower, spring and sliding clutch from the propeller shaft. See **Figure 60**.
15. If the reverse gear ball bearing must be replaced, remove the bearing by pressing it from the gear. Support the bearing in a press with a knife-edged bearing plate, such as part No. 91-37241 or an equivalent. Press against the gear hub with a driver (part No. 91-37312 or an equivalent). See **Figure 72**.
16. If the forward gear roller bearing must be replaced, remove the race and roller bearing as follows:
    a. Remove the race from the gearcase with a puller (part No. 91-27780) or an equivalent and slide hammer. See **Figure 73**.
    b. Remove the roller bearing from the forward gear by pressing it off the gear. Support the bearing in a press with a knife-edged bearing plate, such as part No. 91-37241 or an equivalent. Press against the gear hub with a suitable mandrel. See **Figure 74**.
17. If the forward gear internal needle bearing must be replaced, remove the bearing by pressing it out of the gear with a suitable mandrel. Do not damage the forward gear roller bearing during installation.
18. If the lower drive shaft roller bearing (8, **Figure 67**) must be replaced, remove the bearing race from the gearcase by driving it down into the propeller shaft bore

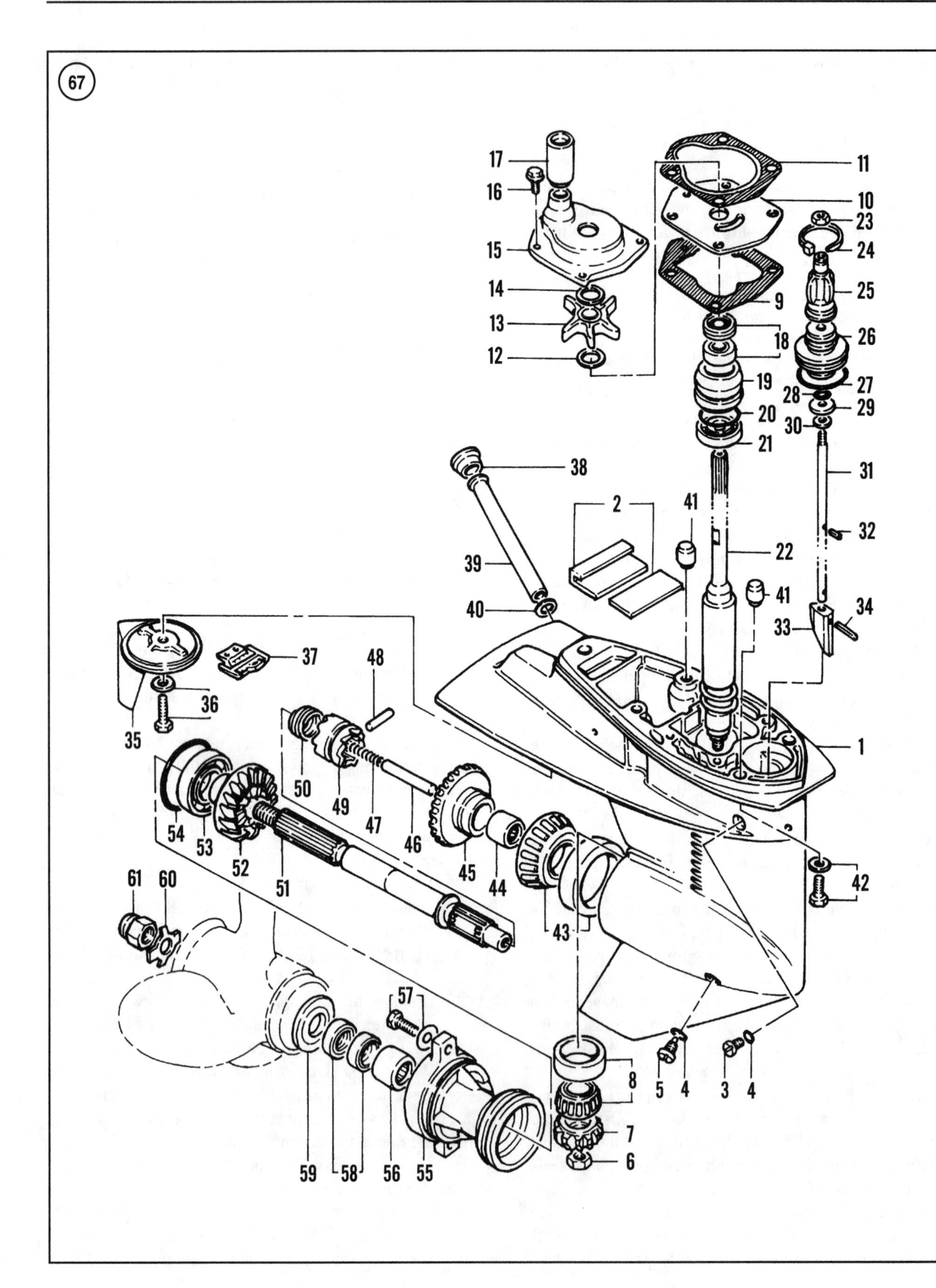
67
17
16
15
14
13
12
11
10
23
24
9
25
18
26
19
27
28
29
20
30
21
38
31
2
41
32
22
39
41
40
34
33
37
48
36
35
1
50
49
54
53
47
46
52
51
45
44
42
61
60
43
57
8
5
4
3
4
7
6
59
58
56
55

GEARCASE COMPONENTS (30-50 HP MODELS)

1. Gearcase housing
2. Exhaust deflector and seal
3. Vent/plug
4. Sealing washer
5. Drain/fill plug
6. Pinion nut
7. Pinion gear
8. Roller bearing assembly
9. Gasket
10. Impeller plate
11. Gasket
12. Washer
13. Impeller and drive key
14. Washer
15. Water pump housing
16. Screw
17. Water tube guide and seal
18. Drive shaft seals
19. Drive shaft seal carrier
20. O-ring
21. Ball bearing
22. Drive shaft
23. Nut
24. Tie-strap
25. Shift shaft boot
26. Shift shaft retainer
27. Outer O-ring
28. Inner O-ring
29. Washer
30. Washer
31. Shift shaft
32. Pin
33. Shift cam
34. Pin
35. Trim tab (anode)
36. Trim tab screw and washer
37. Water screen
38. Rubber plug
39. Water tube
40. Gasket
41. Dowel pins
42. Screws and washers
43. Roller bearing assembly
44. Needle bearing
45. Forward gear
46. Shift cam follower
47. Shift spring
48. Cross pin
49. Sliding clutch
50. Retaining spring
51. Propeller shaft
52. Reverse gear
53. Ball bearing
54. O-ring
55. Propeller shaft bearing carrier
56. Needle bearing
57. Screw and lock tab washer
58. Propeller shaft seals
59. Propeller thrust washer
60. Lock tab washer
61. Elastic locknut

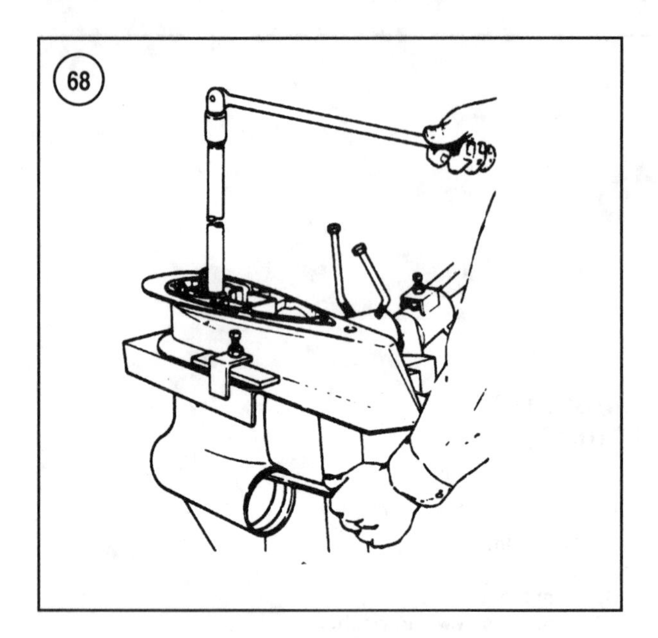
68

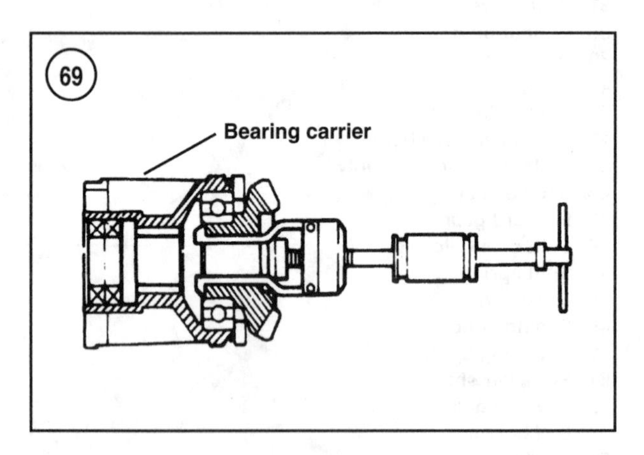

69

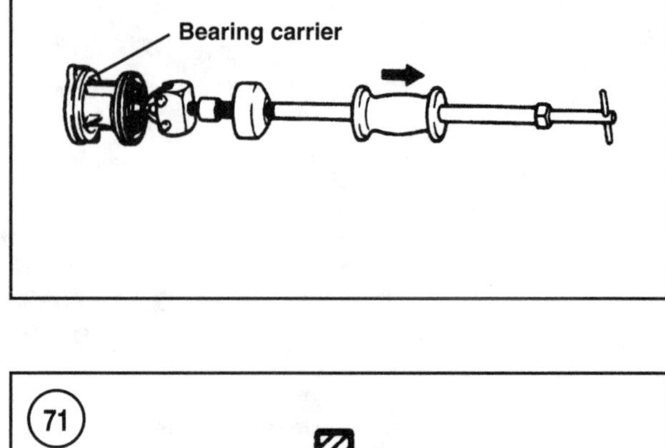

70

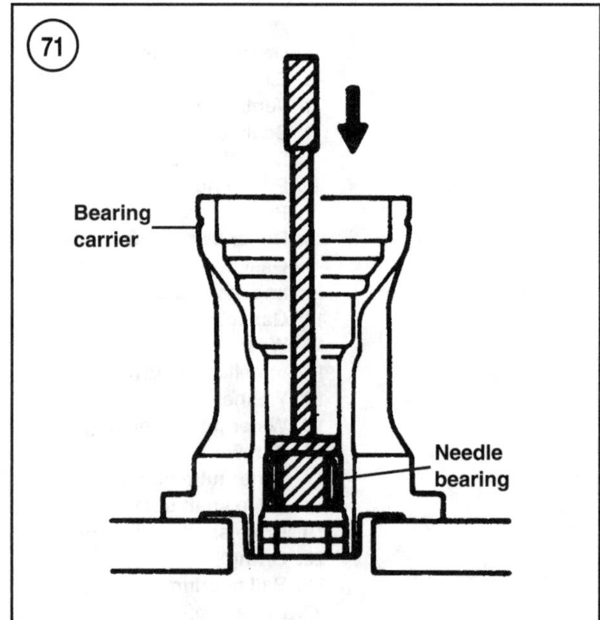

71

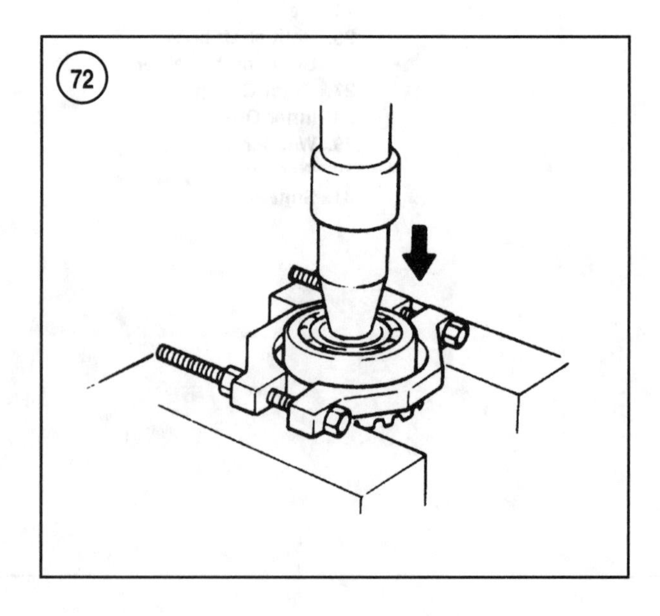
72

with bearing puller (part No. 91-825200A1) and driver (part No. 91-13779) or an equivalent. Insert the puller jaws (A, **Figure 75**) through the propeller shaft bore. Insert the driver (B, **Figure 75**) through the drive shaft bore. Place a shop cloth under the bearing puller. Use a suitable mallet to drive the bearing race out into the propeller shaft bore.

19. If the drive shaft ball bearing (21, **Figure 67**) must be replaced, place the drive shaft with the pinion end down into a vise with the jaws opened just wide enough to support the ball bearing. See **Figure 76**. Strike the power head end of the drive shaft with a lead or rawhide hammer and drive the drive shaft from the bearing. Be prepared to catch the drive shaft as it comes free.

20. Refer to *Gearcase Cleaning and Inspection* in this chapter. Clean and inspect all components as directed before beginning reassembly.

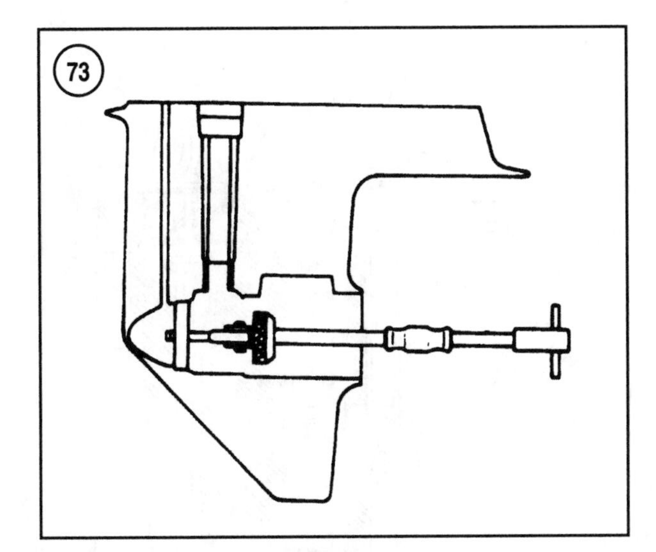

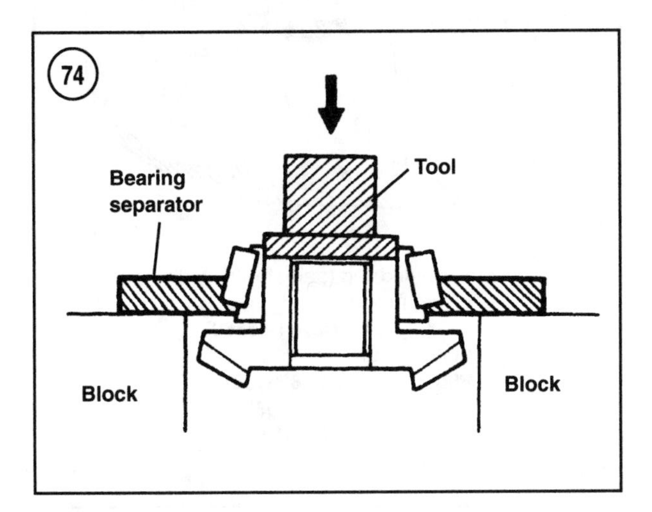

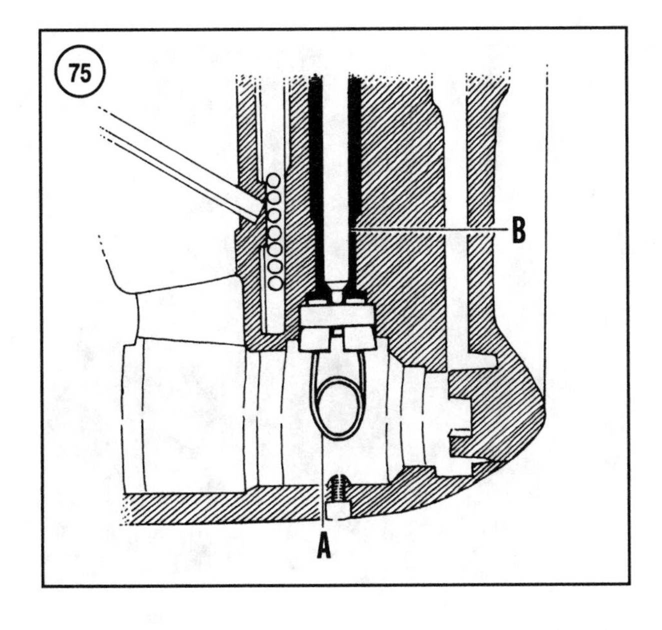

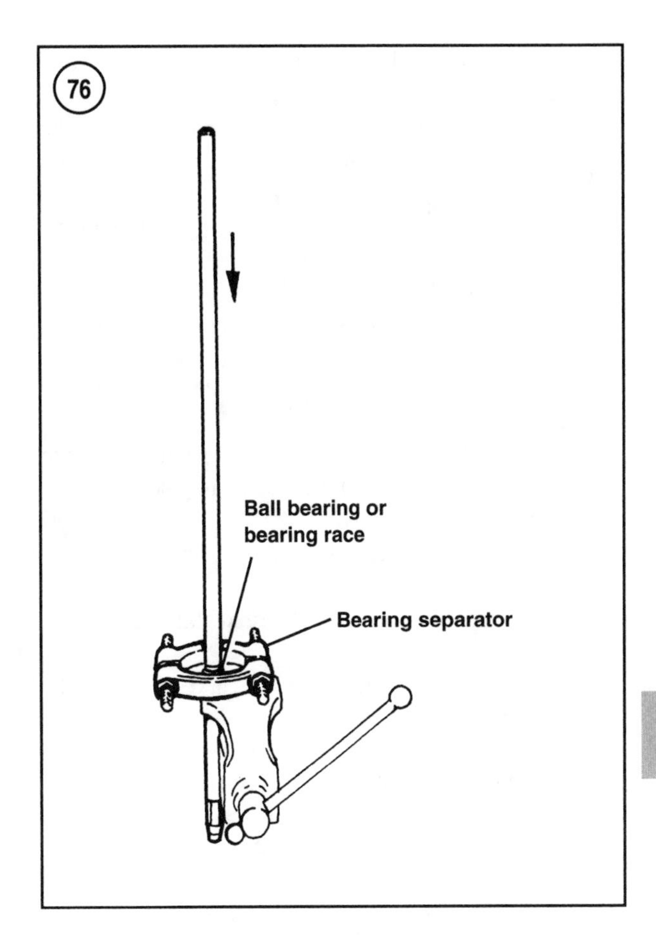

9

Assembly (30-50 hp Models)

Lubricate all internal components with Quicksilver Premium Blend gear oil (part No. 92-13783 or an equivalent). Do not assemble components *dry*. Refer to **Figure 67** for this procedure.

1. If the lower drive shaft bearing (8, **Figure 67**) was removed, install a *new* race into the gearcase as shown in **Figure 77**. Lubricate the race and install it with the small diameter facing the top of the gearcase. Tighten the nut to seat the race fully in the bearing bore.
2. If the drive shaft ball bearing was removed, install a *new* bearing as follows:
   a. Lubricate the *new* bearing and slide it over the drive shaft.
   b. Place the drive shaft and bearing in a press and support the ball bearing *inner* race with a knife-edged bearing plate, such as part No. 91-37241 or an equivalent.
   c. Thread the old pinion nut 3/4 of the way onto the drive shaft. Press against the pinion nut until the bearing seats on the drive shaft.
   d. *Do not* reuse the old pinion nut when finished.

3. If the forward gear bearing race was removed, install a *new* race as follows:
   a. Lubricate the *new* bearing race and set it into the gearcase bearing bore with the tapered side facing outward.
   b. Place a mandrel (part No. 91-36571 or an equivalent) over the race. Place the propeller shaft into the mandrel hole.
   c. Install the propeller shaft bearing carrier into the gearcase to hold the propeller shaft centered. Thread a scrap propeller nut onto the propeller shaft.
   d. Use a mallet to drive the propeller shaft against the mandrel until the bearing race is fully seated in the gearcase bearing bore.
4. If the forward gear roller bearing was removed, install a *new* bearing as follows:
   a. Set the forward gear on a press with the gear teeth facing downward.
   b. Lubricate the *new* roller bearing and set it onto the gear with the rollers facing upward.
   c. Press the bearing fully onto the gear with a suitable mandrel. See **Figure 78**. Press only on the inner race. Do not press on the roller cage.
5. If the forward gear internal needle bearing was removed, install a new bearing as follows:
   a. Set the forward gear in a press with the gear teeth facing downward.
   b. Lubricate a *new* forward gear internal needle bearing and set it into the gear with the numbered side facing upward.
   c. Press the bearing into the gear with a driver (part No. 91-826872 or an equivalent). Press until the tool seats against the gear hub.
6. If the propeller shaft bearing carrier needle bearing was removed, lubricate a new bearing and place it into the carrier with its lettered end facing the propeller. Use a mandrel (part No. 91-817011 or an equivalent) to press the bearing until the tool seats against the carrier. See **Figure 79**.
7. If the reverse gear bearing was removed, install a *new* bearing with the numbered side facing away from the gear teeth. Lubricate the bearing and set it onto the gear hub with the numbered side up. Press against the inner race with a suitable mandrel until the bearing is fully seated on the gear. See **Figure 80**.
8. Lubricate the outside diameter of the reverse gear bearing. Press the reverse gear and bearing assembly into the propeller shaft bearing carrier with a suitable mandrel. See **Figure 81**. Press until the bearing seats in the carrier.
9. Reassemble the propeller shaft as follows:

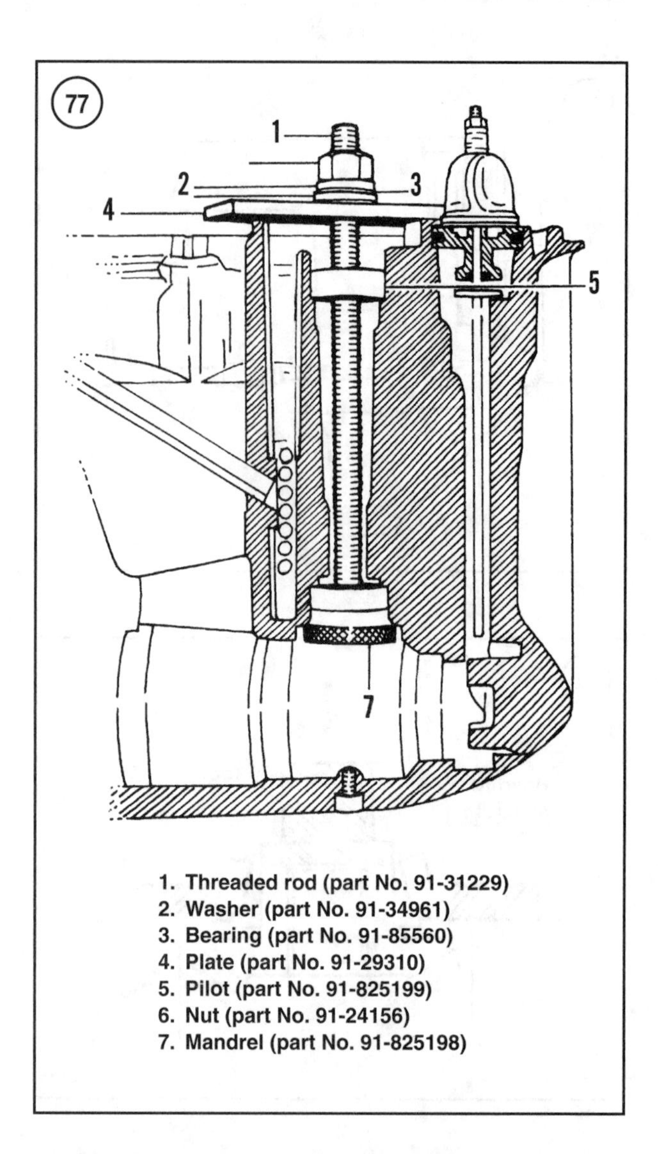

1. Threaded rod (part No. 91-31229)
2. Washer (part No. 91-34961)
3. Bearing (part No. 91-85560)
4. Plate (part No. 91-29310)
5. Pilot (part No. 91-825199)
6. Nut (part No. 91-24156)
7. Mandrel (part No. 91-825198)

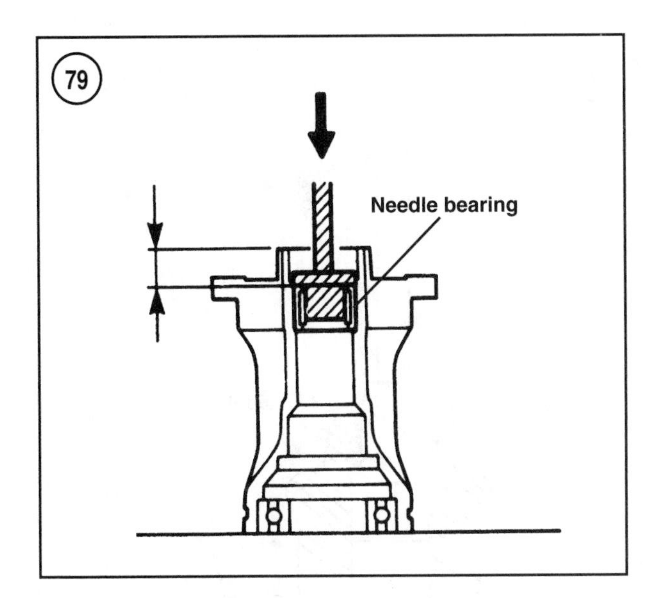

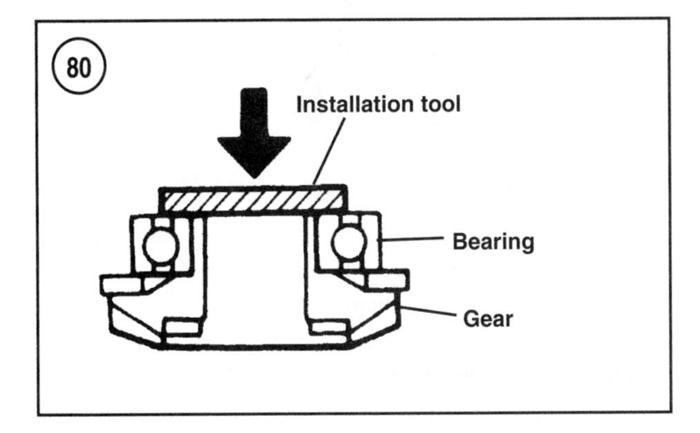

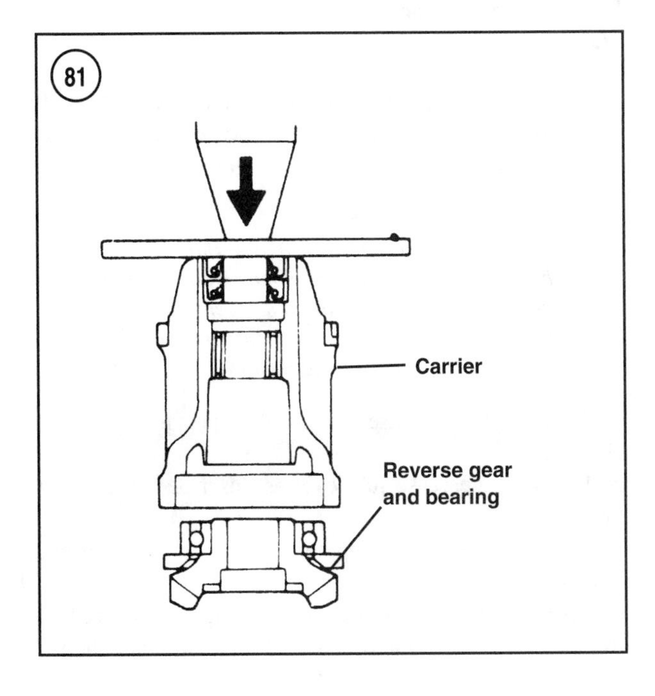

a. Align the cross pin holes of the sliding clutch with the slot in the propeller shaft. Position the long end of the sliding clutch (**Figure 66**) toward the propeller and slide it onto the propeller shaft.
b. Lubricate the shift spring and install it into the propeller shaft. Install the cam follower with the beveled end out.
c. Press the cam follower against a solid object to compress the spring. Insert a small punch through the sliding clutch holes between the spring and the cam follower. Remove the cam follower.
d. Insert the sliding clutch cross-pin into the clutch opposite of the punch and slide the pin into the sliding clutch while pushing the punch out of the clutch. See **Figure 60**. When installed, the cross-pin must be in front of the spring.

NOTE
The sliding clutch retaining spring must lay flat, with no overlapping coils.

e. Secure the pin to the sliding clutch with a *new* retainer spring. Do not open the spring any more than necessary to install it.
f. Grease the cam follower with 2-4-C Multi-Lube grease or an equivalent and install it into the propeller shaft with its beveled end facing out.

10. Install *new* propeller shaft seals as follows:
a. Coat the outer diameter of the *new* propeller shaft seals with Loctite 271 threadlocking adhesive (part No. 92-809819).
b. Install the small diameter seal with the spring facing the gearcase. Press the seal in with the large stepped end of a mandrel (part No. 91-817007 or an equivalent) until the tool bottoms against the carrier. See **Figure 82**.
c. Install the large diameter seal with the spring facing the propeller. Press the seal in with the small stepped end of a mandrel (part No. 91-817007 or an equivalent) until the tool bottoms against the carrier. See **Figure 82**.

11. Coat a new propeller shaft bearing carrier O-ring and the propeller shaft seal lips with 2-4-C Multi-Lube grease (part No. 92-825407). Then install the O-ring into the carrier groove.

12. Install *new* drive shaft seals as follows:
a. Set the drive shaft seal carrier (19, **Figure 67**) in a press with the large beveled edge facing upward.
b. Coat the outer diameter of *new* drive shaft seals with Loctite 271 threadlocking adhesive.
c. Install the metal-cased seal with the lip facing downward. Press the seal with the large stepped

side of a mandrel (part No. 91-825197 or an equivalent) until the tool bottoms on the carrier.

d. Install the rubber-cased seal with the lip facing upward. Press the seal with the small stepped side of a mandrel (part No. 91-825197 or an equivalent) until the tool bottoms on the carrier.

13. Lubricate a *new* drive shaft seal carrier O-ring and the drive shaft seal lips with 2-4-C Multi-Lube grease. Then install the O-ring into the carrier groove.

14. Rotate the gearcase so the propeller shaft bore is pointing upward. Lubricate the forward gear assembly, and place the gear and bearing assembly into the gearcase forward bearing race.

15. Place the lower drive shaft or pinion gear roller bearing into the lower race. Place the pinion gear in position on top of the bearing.

16. Spray the threads of the drive shaft with Locquic primer (part No. 92-59327-1). Insert the drive shaft into the gear housing. Hold the pinion gear and lower bearing in position and engage the drive shaft splines to the pinion gear splines.

17. Apply Loctite 271 threadlocking adhesive to a new pinion nut. Install the pinion nut with its rounded corners *against* the pinion gear and hand tighten it. Use a drive shaft spline socket (part No. 91-825196) and a suitable socket or wrench (**Figure 83**) to tighten the pinion nut to the specification in **Table 1**.

18. Reassemble the shift shaft, bushing and shift cam assembly with *new* O-rings and a *new* boot. Grease the O-rings with 2-4-C Multi-Lube grease. Install the shift rod into the gearcase with the ramp of the shift cam facing the propeller. Seat the retainer firmly into the shift cavity.

19. Turn the gearcase so the propeller shaft bore is horizontal. Move the shift rod upward to the forward position and install the propeller shaft into the center of the forward gear.

20. Install the propeller shaft bearing carrier and reverse gear assembly into the gearcase.

21. Coat the threads of the retaining screws with Loctite 271 threadlocking adhesive. Install *new* lock tab washers onto the screws and install the screw assemblies to the gearcase. Tighten the screws to the specification in **Table 1**. Then bend a lock tab over each screw.

22. Carefully slide the drive shaft seal carrier with the large beveled edge up over the drive shaft splines and seat it into the gearcase bore.

23. If the exhaust deflector and seal were removed, coat the ends of the seal and deflector with RTV sealant (part No. 92-809826) and install them into the gearcase recess.

24. Install the water pump as described previously in this chapter.

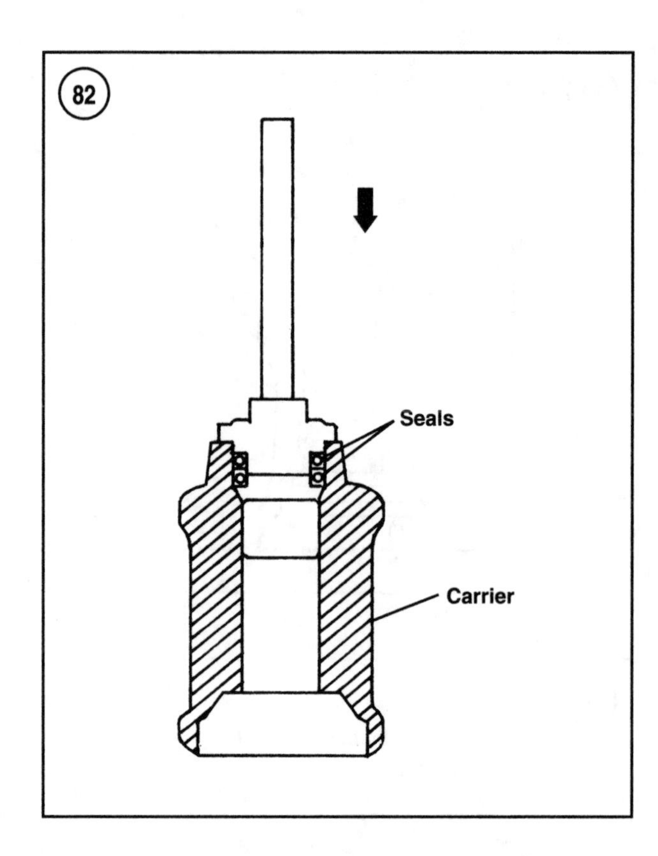

25. Pressure test the gearcase as described in *Gearcase Pressure Testing* in this chapter.

26. Fill the gearcase with the recommended lubricant as described in Chapter Four.

Disassembly (55 and 60 hp Standard Gearcase Models)

The propeller shaft bearing carrier and propeller shaft can be removed without removing the gearcase from the drive shaft housing. Refer to **Figure 84** for this procedure.

NOTE

If the forward gear or drive shaft roller bearings require replacement, replace the bearing rollers and races as assemblies. Do not remove a pressed in bearing and/or race unless replacement is necessary.

1. Remove the gearcase as described previously in this chapter.
2. Drain the gearcase lubricant as described in Chapter Four.
3. Remove the water pump and pump base as described in this chapter.
4. Inspect the exhaust deflector and seal (19, **Figure 84**). Replace the deflector and/or seal if either are damaged.

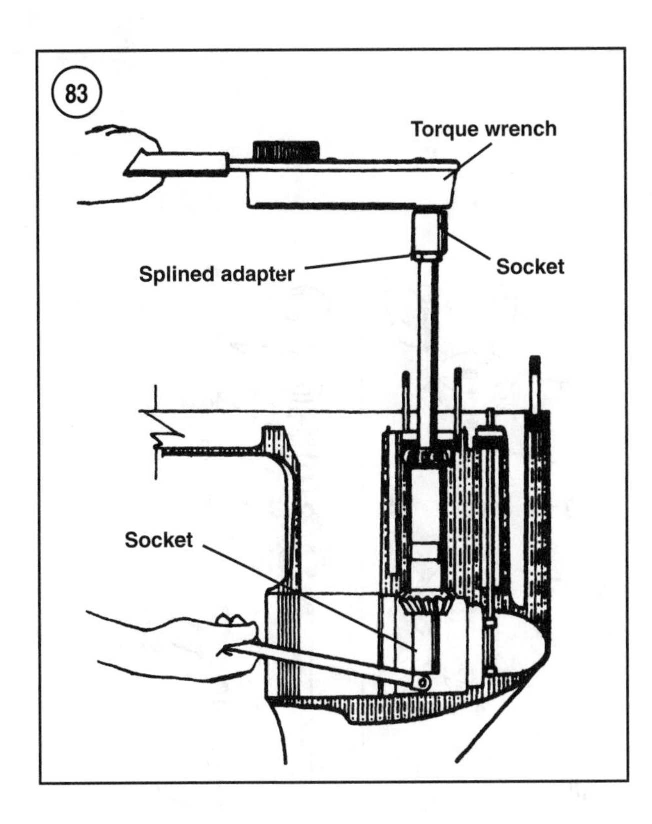

5. Remove the screws and washers securing the propeller shaft bearing carrier to the gearcase.
6. Install puller jaws (part No. 91-46086A-1) and a puller bolt (part No. 91-85716) or an equivalent and pull the bearing carrier from the gearcase. If necessary, use a propeller thrust hub (**Figure 85**, typical) to prevent the puller jaws from sliding inward. Remove the carrier O-ring.

NOTE
*If the bearing carrier's needle bearing (58, **Figure 84**) must be replaced, also remove the propeller shaft seals during the bearing removal process. If bearing replacement is not required, remove both propeller shaft seals with a suitable seal puller. Do not damage the seal bore in the process.*

7. Pull the propeller shaft assembly from the gearcase. If the shift cam follower (43, **Figure 84**) falls out of the propeller shaft, retrieve it from the gearcase and reinstall it into the propeller shaft temporarily.
8. Remove the pinion nut by holding the nut with a suitable socket or wrench, then turn the drive shaft counterclockwise with a spline socket (part No. 91-817070 or an equivalent), until the nut is free from the drive shaft. See **Figure 68**.
9. Remove the drive shaft upper bearing retainer (14, **Figure 84**) using a retainer tool (part No. 91-43506 or an equivalent). Turn the retainer until it is free from the gearcase threads.
10. Lift the drive shaft out of the top of the gearcase with the upper roller bearing and race. Remove the shim(s) from the bearing bore. Measure and record the thickness of the shims for later reference.
11. Remove the pinion gear, pinion nut and the forward gear assembly from the gear cavity.
12. Pull the shift shaft assembly from the gearcase.
13. Slide the shift shaft bushing from the shift shaft. Then remove the O-ring and seal from the bushing.
14. Reach inside the propeller shaft bore and remove the shift cam (**Figure 86**, typical).
15. If the reverse gear, reverse gear ball bearing and/or bearing carrier internal needle bearing require replacement, replace them as follows:
   a. Clamp the carrier assembly into a vise with protective jaws or between wooden blocks.
   b. Pull the reverse gear from the bearing carrier with a puller (part No. 91-27780) or a suitable slide hammer puller. See **Figure 69**.
   c. If the reverse gear ball bearing remains on the reverse gear, support the bearing with a knife-edged bearing plate, such as part No. 91-37241. Press on the gear hub with a mandrel (part No. 91-37312 or an equivalent) until the bearing is free from the gear. See **Figure 72**.
   d. If the ball bearing (54, **Figure 84**) remains in the carrier, remove the bearing with a slide hammer. See **Figure 70**.
   e. Support the bearing carrier in a press with the propeller end facing downward (**Figure 71**). Press the internal needle bearing and seals from the carrier with a driver rod (part No. 91-37323) and mandrel (part No. 91-37312) or an equivalent.
16. If the forward gear roller bearing requires removal or replacement, and/or the forward gear internal needle bearing require replacement, proceed as follows:
   a. Press the roller bearing from the forward gear. Support the bearing with a knife-edged bearing plate, such as part No. 91-37241. Press on the gear hub with a suitable mandrel. See **Figure 74**.
   b. Clamp the gear in a vise with protective jaws with the gear engagement lugs facing upward. Drive the internal needle bearing from the gear with a suitable punch and hammer.

NOTE
If the forward gear bearing race is removed only to change the shim(s) and forward gear lash, do not discard the bearing race.

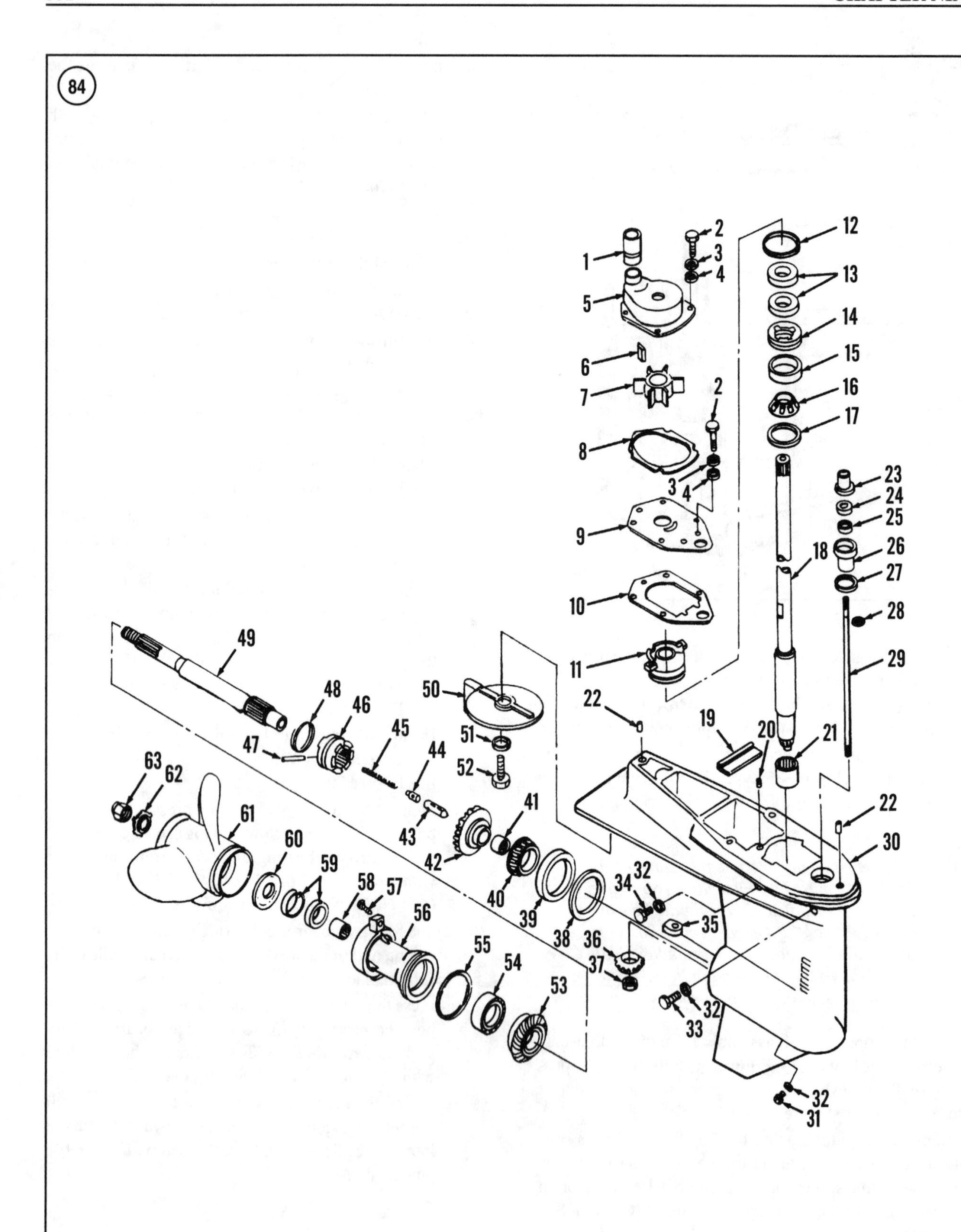
84
1
2
3
4
5
6
7
8
9
10
11
12
13
14
15
16
17
18
19
20
21
22
23
24
25
26
27
28
29
30
31
32
33
34
35
36
37
38
39
40
41
42
43
44
45
46
47
48
49
50
51
52
53
54
55
56
57
58
59
60
61
62
63

GEARCASE COMPONENTS (55 AND 60 HP MODELS [WITH A STANDARD GEARCASE])

1. Water tube seal
2. Screw
3. Washer
4. Insulator*
5. Water pump cover
6. Impeller drive key
7. Impeller
8. Gasket
9. Face plate
10. Gasket
11. Water pump base
12. O-ring
13. Drive shaft seals
14. Drive shaft upper bearing retainer
15. Bearing race
16. Tapered roller bearing
17. Shim(s)
18. Drive shaft
19. Exhaust deflector and seal
20. Plug
21. Roller bearing
22. Dowel pin
23. Shift shaft coupler
24. Spacer
25. Seal
26. Bushing
27. O-ring
28. Retaining ring
29. Shift shaft
30. Gearcase housing
31. Drain/fill plug
32. Gasket
33. Vent plug
34. Vent plug
35. Shift cam
36. Pinion gear
37. Pinion gear nut
38. Shim(s)
39. Bearing race
40. Tapered roller bearing
41. Needle bearing
42. Forward gear
43. Shift cam follower
44. Guide
45. Spring
46. Sliding clutch
47. Cross pin
48. Cross pin retainer spring
49. Propeller shaft
50. Trim tab
51. Washer
52. Screw
53. Reverse gear
54. Ball bearing
55. O-ring
56. Bearing carrier
57. Screw
58. Needle bearing
59. Seals
60. Thrust hub
61. Propeller
62. Tab washer
63. Propeller nut

*Not used on all models.

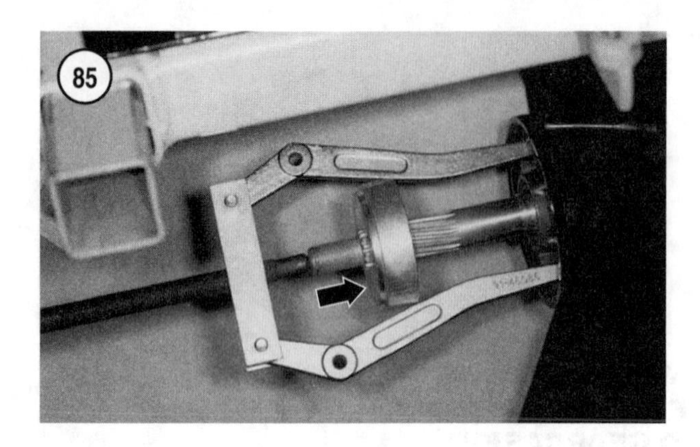

c. To remove the forward gear bearing race from the gearcase, pull the race from the front of the propeller shaft bore with a puller (part No. 91-27780) or a suitable slide hammer. See **Figure 73**. Remove the shim(s) from the bearing bore. Measure and record the thickness of the shims for later reference. Discard the race if the roller bearing assembly is being replaced.

17. If the drive shaft roller bearing requires replacement, replace it as follows:

a. Support the drive shaft roller bearing in a knife-edged bearing plate, such as part No. 91-37241. The pinion gear end of the shaft must face down. See **Figure 76**.

b. Install the drive shaft spline socket (part No. 91-817070 or an equivalent), over the drive shaft splines.

c. Press on the spline socket until the bearing is free.

18. If the needle bearing at the bottom of the drive shaft bore requires replacement, replace it as follows:

a. Install the bearing removal tool (part No. 91-817058A-1 or an equivalent) into the drive shaft bore. See **Figure 87**.

b. Make sure the tool's bushing is piloted in the drive shaft bore as shown in **Figure 87**.

c. Drive the bearing into the gear cavity with a suitable mallet.

19. If the propeller shaft requires disassembly, disassemble it as follows:

a. Remove the cross pin retainer spring with a small screwdriver or awl. Insert the tool under one end of the spring and rotate the propeller shaft to unwind the spring. See **Figure 88**.

b. Place the shift cam follower against a solid surface and push on the propeller shaft to unload the spring pressure on the sliding clutch cross pin. Remove the pin with a suitable punch. Carefully release the pressure on the cam follower and spring. Remove the cam follower, spring guide, spring and sliding clutch from the propeller shaft. See **Figure 89**.

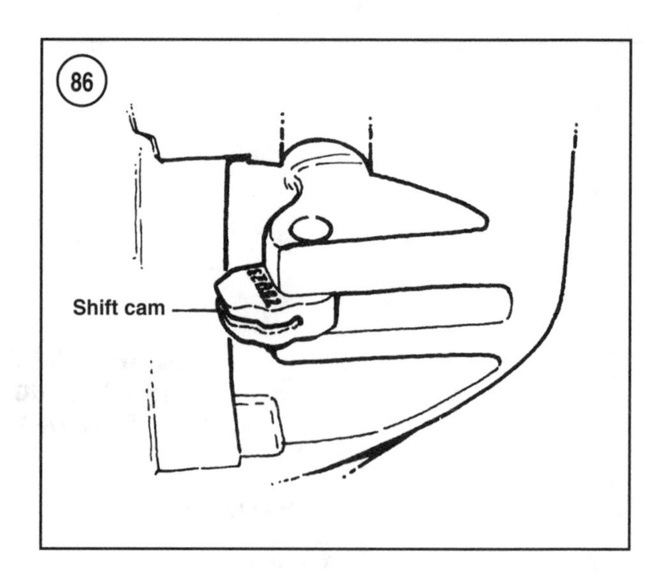

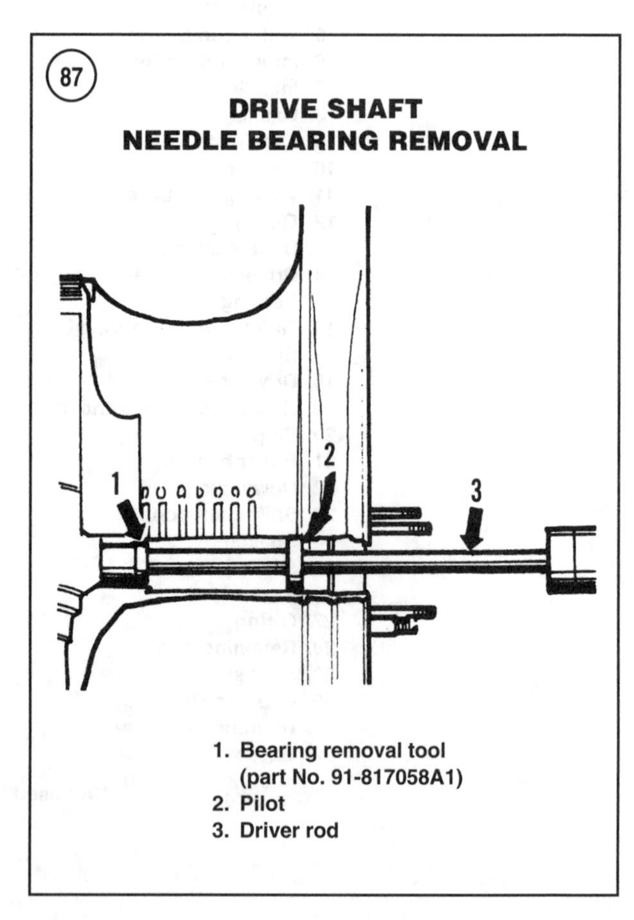

DRIVE SHAFT NEEDLE BEARING REMOVAL

1. Bearing removal tool (part No. 91-817058A1)
2. Pilot
3. Driver rod

20. Refer to *Gearcase Cleaning and Inspection* in this chapter. Clean and inspect all components as directed before beginning the reassembly procedure.

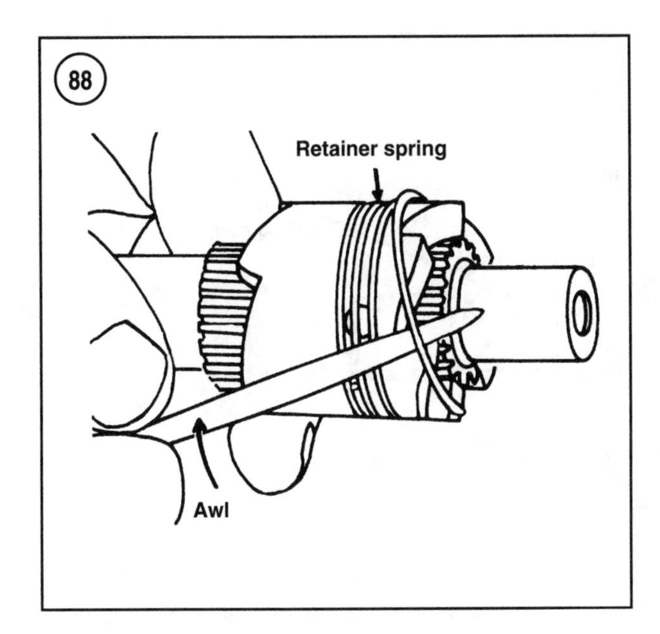

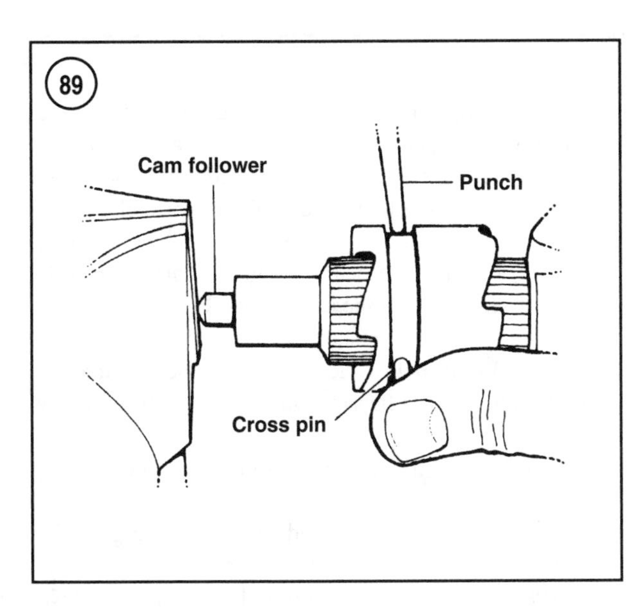

Assembly (55 and 60 hp Standard Gearcase)

Lubricate all internal components with Quicksilver Premium Blend gear oil (part No. 92-13783) or an equivalent. Do not assemble components *dry*. Refer to **Figure 84** for this procedure.

1. Refer to **Figure 90** and install a *new* drive shaft needle bearing into the gearcase bore as follows:
   a. Lubricate the bearing and set it into the drive shaft bore with its numbered side facing upward.
   b. Drive the bearing into position with an installer tool (part No. 91-817058A-1 or an equivalent).

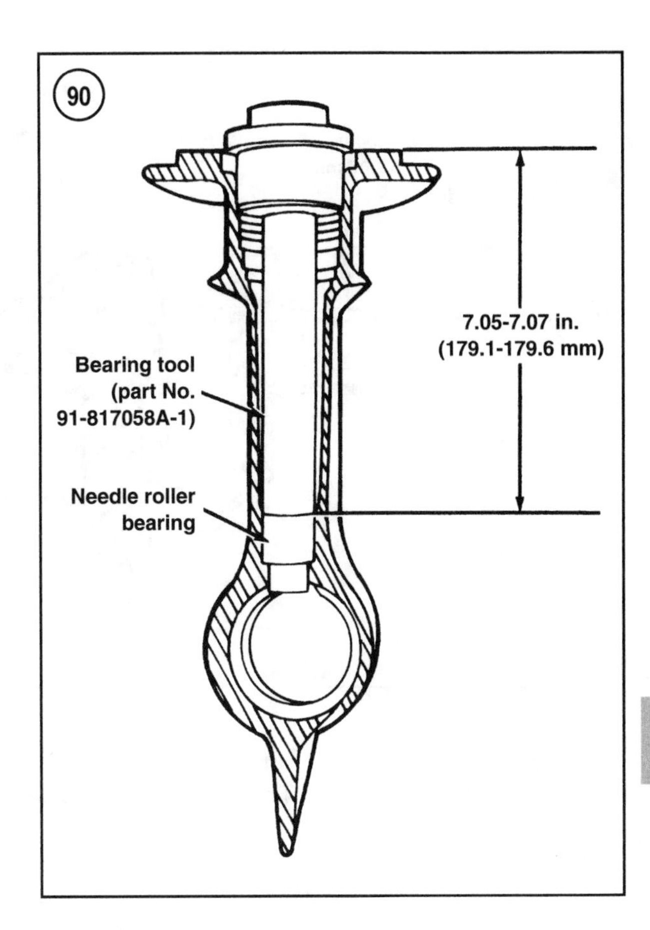

   c. When correctly installed, the top of the bearing is 7.05-7.07 in. (179.1-179.6 mm) from the top of the gearcase deck.
2. Install a new drive shaft roller bearing as follows:
   a. Lubricate the new roller bearing and slide it over the crankshaft end of the drive shaft with the rollers facing the crankshaft.
   b. Place the drive shaft and bearing in a press and support the roller bearing *inner* race with a knife-edged bearing plate, such as part No. 91-37241 or an equivalent.
   c. Install a spline socket (part No. 91-817070 or an equivalent) onto the drive shaft. Press against the spline socket until the bearing seats on the drive shaft.
3. If the forward gear bearing race was removed, install the race as follows:
   a. Install the original shim(s) into the bearing bore. If the original shims were lost or damaged (beyond measurement), install a 0.010 in. (.254 mm) shim.
   b. Lubricate the bearing race and set it into the gearcase bearing bore with its tapered side facing out.

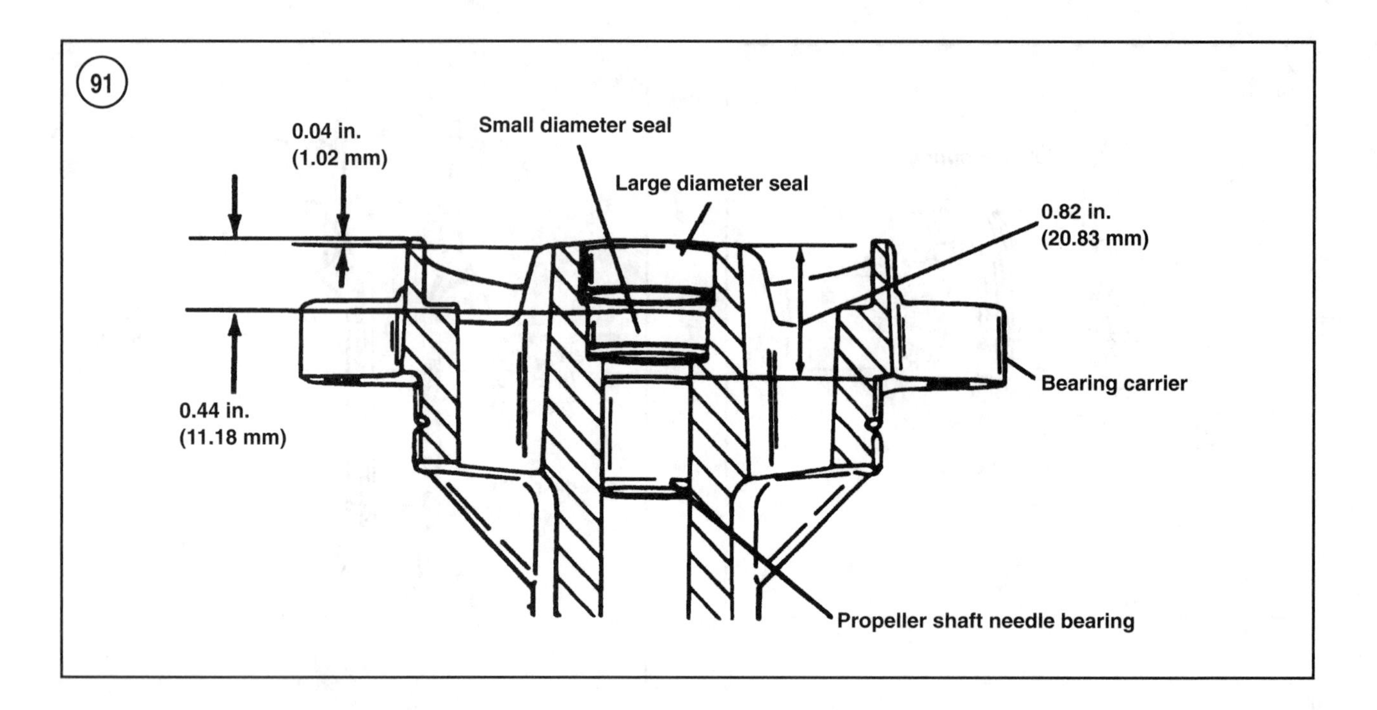

c. Place a bearing installer (part No. 91-817009 or an equivalent) over the race. Then place the propeller shaft into the mandrel hole.
d. Install the propeller shaft bearing carrier into the gearcase to hold the propeller shaft centered. Thread a scrap propeller nut onto the propeller shaft.
e. Use a mallet to drive the propeller shaft against the mandrel until the bearing race is fully seated in the gearcase bearing bore.

4. If the forward gear roller bearing was removed, install a new roller bearing as follows:
a. Set the forward gear on a press with the gear teeth facing downward.
b. Lubricate the *new* roller bearing and set it onto the gear with the rollers facing upward.
c. Press the bearing fully onto the gear with a mandrel (part No. 91-817007 or an equivalent). See **Figure 78**. Press only on the inner race.

5. If the forward gear internal needle bearing was removed, install a *new* bearing as follows:
a. Set the forward gear in a press with the gear teeth facing downward.
b. Lubricate a *new* forward gear internal needle bearing and set it into the gear with its numbered side facing upward.
c. Press the bearing into the gear with a mandrel (part No. 91-817005 or an equivalent). Press until the tool seats against the gear hub.

6. If the bearing carrier needle bearing was removed, refer to **Figure 91** and install a *new* bearing as follows:
a. Lubricate a *new* bearing and place it into the carrier bore with its lettered end facing the propeller.
b. Use a mandrel (part No. 91-817011 or an equivalent) to press the bearing into the carrier until the tool seats against the carrier. See **Figure 79**.
c. When correctly installed, the bearing is 0.82 in. (20.83 mm) below the carrier face. See **Figure 91**.

7. If the reverse gear bearing was removed, install a new bearing as follows:
a. Set the gear into a press with the gear teeth facing downward.
b. Lubricate the bearing and set it onto the gear hub with its numbered side facing upward.
c. Press against the inner race with a mandrel (part No. 91-817007 or an equivalent) until the bearing fully seats on the gear. See **Figure 80**.

8. Install the reverse gear and bearing assembly to the propeller shaft bearing carrier as follows:
a. Lubricate the outside diameter of the reverse gear bearing.
b. Set the gear in a press with the gear teeth facing downward.
c. Set the bearing carrier on top of the reverse gear ball bearing as shown in **Figure 81**.
d. Press the carrier onto the bearing until the bearing seats in the carrier bore.

9. If the propeller shaft was disassembled, reassemble it as follows:

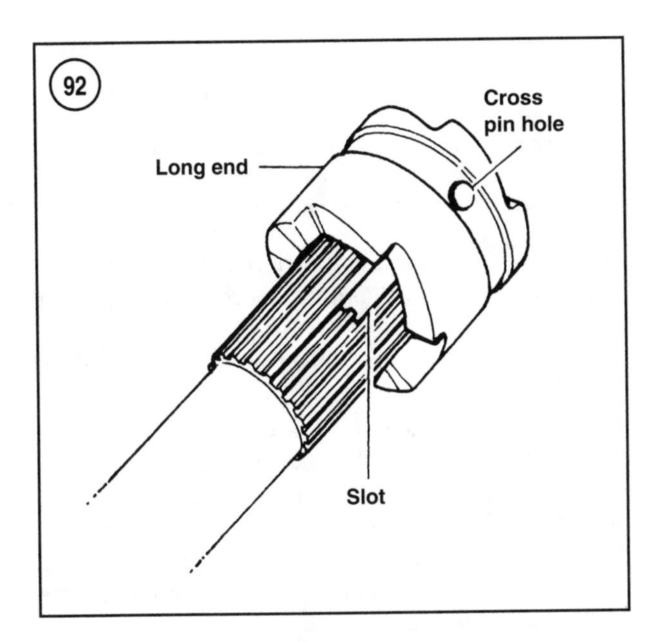

a. Lubricate all components with Quicksilver 2-4-C Multi-Lube grease (part No. 92-825407).
b. Align the cross pin holes of the sliding clutch with the slot in the propeller shaft. Position the long end of the sliding clutch toward the propeller and slide it onto the propeller shaft. See **Figure 92**.
c. Install the shift spring into the propeller shaft. Then install the shift spring guide with its narrow end facing the shift spring. Finally, install the cam follower with its beveled end facing outward.
d. Press the cam follower against a solid object to compress the spring. Align the sliding clutch holes with the spring guide's hole. Use a small punch to ease alignment. See **Figure 89**.
e. Insert the sliding clutch cross-pin into the clutch and through the spring guide's hole. The cross-pin must pass through the spring guide.

NOTE
The sliding clutch retaining spring must lay flat with no overlapping coils.

f. Secure the pin to the sliding clutch with a *new* retainer spring. Do not open the spring more than necessary to install it.

10. Install *two new* propeller shaft seals as follows:
a. Coat the outer diameter of the new propeller shaft seals with Loctite 271 threadlocking adhesive (part No. 92-809819).
b. Install the small diameter seal with its spring facing the gearcase. Press the seal in using the large stepped end of a mandrel (part No. 91-817007 or an equivalent) until the tool bottoms against the carrier (**Figure 82**).
c. When correctly installed, the small diameter seal is 0.44 in. (11.18 mm) below the carrier face. See **Figure 91**.
d. Install the large diameter seal with its spring facing the propeller. Press the seal in using the small stepped end of a mandrel (part No. 91-817007 or an equivalent) until the tool bottoms against the carrier (**Figure 82**).
e. When correctly installed, the large diameter seal is 0.04 in. (1.02 mm) below the carrier face. See **Figure 91**.

11. Coat a new bearing carrier O-ring and the propeller shaft seal lips with 2-4-C Multi-Lube grease (part No. 92-825407). Then install the O-ring into the carrier groove.
12. Press a *new* seal into the shift shaft bushing with a suitable mandrel. Lubricate the seal and a *new* bushing O-ring with 2-4-C Multi-Lube grease. Install the O-ring into the bushing groove.
13. Lubricate the shift rod with Quicksilver 2-4-C Multi-Lube grease. If the lock clip was removed, reinstall the lock clip in the shift shaft groove. Then slide the bushing over the shift shaft and position it against the lock clip.
14. Place the shift cam into the gear housing with its numbered side up as shown in **Figure 86**. Install the shift shaft assembly into the gearcase and engage the shift shaft splines to the shift cam internal splines. Seat the bushing into the gearcase bore.
15. Rotate the gearcase so the propeller shaft bore is pointing upward. Lubricate the forward gear assembly, and place the gear and bearing assembly into the gearcase forward bearing race.
16. Insert the drive shaft into the gear housing. Hold the pinion gear in position and engage the drive shaft splines to the pinion gear splines.

NOTE
Apply Locktite 271 threadlocking adhesive to a new pinion nut after the pinion gear depth and forward gear lash are verified. Install the old pinion nut without threadlocking adhesive to check the gear depth and forward gear lash.

17. Install the original pinion nut with the recessed side *toward* the pinion gear. Tighten the nut finger-tight at this time.
18. Install the original shim(s) (17, **Figure 84**) over the drive shaft and into the gearcase bore. If the original shims are lost or damaged (beyond measurement), install a 0.015 in. (0.38 mm) shim(s).

19. Install the drive shaft roller bearing race over the drive shaft and seat it against the shim(s).
20. Lubricate the drive shaft retainer, then install it into the drive shaft bore with the side marked *OFF* facing up. Use the retainer tool (part No. 91-43506) to tighten the retainer to the specification in **Table 1**.
21. Hold the pinion nut with a suitable wrench or socket. Attach a spline socket (part No. 91-817070 or an equivalent) to a suitable torque wrench. Tighten the pinion nut to the specification in **Table 1**. See **Figure 83**, typical.
22. Refer to *Gearcase Shimming* and set the *Pinion gear depth*. Do not continue until the pinion gear depth is correct.
23. Install the propeller shaft assembly into the gearcase and into the forward gear internal needle bearing. Make sure the shift cam does not fall out during assembly.
24. Liberally lubricate the front and rear flanges of the propeller shaft bearing carrier with Quicksilver Special Lube 101 (part No. 92-13872A1). Install the carrier over the propeller shaft, being careful not to damage the carrier seals on the propeller shaft splines. Make sure the screw holes are aligned with the gearcase. Rotate the drive shaft to ensure the gear teeth are meshed, then seat the carrier into the gearcase bore.
25. Secure the carrier with two screws. Tighten the screws to the specification in **Table 1**.
26. Refer to *Gearcase Shimming* in this chapter and set the *Forward gear lash*. Do not continue until the gear lash is correct.
27. After verifying the gear lash, remove the propeller shaft bearing carrier and propeller shaft. Spray the threads of the drive shaft with Locquic primer (part No. 92-59327-1). Install a *new* pinion nut with Loctite 271 threadlocking adhesive applied to the threads, as described in this procedure. Then reinstall the propeller shaft and bearing carrier as described in this procedure. Coat the two carrier screw's threads with Loctite 271 threadlocking adhesive. Install the screws and washers. Tighten the screws to the specification in **Table 1**.
28. Install the water pump assembly as described in this chapter.
29. Pressure test the gearcase as described under *Gearcase Pressure Testing* in this chapter.
30. Fill the gearcase with the recommended lubricant as described in Chapter Four.

Disassembly (60 hp Bigfoot Gearcase Models)

The propeller shaft bearing carrier and propeller shaft can be removed without removing the gearcase from the drive shaft housing. Refer to **Figure 93** for this procedure.

NOTE
If the forward gear or drive shaft roller bearings require replacement, replace the bearing rollers and races as assemblies. Do not remove a pressed in bearing and/or race unless replacement is necessary.

1. Remove the gearcase as described in this chapter.
2. Drain the gearcase lubricant as described in Chapter Four.
3. Remove the water pump and pump base as described in this chapter.
4. Remove the elastic locknuts and flat washers securing the propeller shaft bearing carrier to the gearcase.
5. Install the puller jaws (part No. 91-46086A-1) and puller bolt (part No. 91-85716) or an equivalent, and pull the bearing carrier from the gearcase. If necessary, use a propeller thrust hub (**Figure 85**, typical) to prevent the puller jaws from sliding inward.

NOTE
*If the bearing carrier's needle bearing (62, **Figure 93**) must be replaced, also remove the propeller shaft seals during the bearing removal process. If bearing replacement is not required, remove both propeller shaft seals with a suitable seal puller. Do not damage the seal bore in the process.*

6. Remove the reverse gear, thrust bearing and thrust washer from the propeller shaft bearing carrier. Then remove the O-ring.
7. Pull the propeller shaft assembly from the gearcase. If the shift cam follower (46, **Figure 93**) falls out of the propeller shaft, retrieve it from the gearcase and reinstall it into the propeller shaft.
8. Remove the pinion nut by holding the nut with a suitable socket or wrench, then turning the drive shaft counterclockwise using a spline socket (part No. 91-817070 or an equivalent), until the nut is free from the drive shaft. See **Figure 94**.
9. Pull the drive shaft assembly from gearcase.
10. Remove the pinion gear, pinion gear roller bearing and the forward gear assembly from the propeller shaft bore.
11. Remove the screws securing the shift shaft retainer/bushing (26, **Figure 93**) to the gearcase. Carefully pry the shift shaft retainer from the gearcase. Then remove the shift shaft assembly from the gearcase.
12. Remove the shift cam (40, **Figure 93**) from the front of the gearcase bore.
13. Remove the O-ring and seal from the shift shaft bushing.

14. If the bearing carrier needle bearings require replacement, remove it as follows:
   a. Clamp the carrier assembly into a vise with protective jaws or between wooden blocks.
   b. Pull the reverse gear needle bearing from the bearing carrier using a suitable slide hammer, such as part No. 91-34569A-1. See **Figure 95**.
   c. Set the carrier in a press with the propeller end facing down as shown in **Figure 96**.
   d. Assemble the mandrel (part No. 91-26569) and driver rod (part No. 91-37323) or an equivalent. Position the mandrel and driver rod in the carrier bore, on top of the needle bearing.
   e. Press the propeller shaft needle bearing and seals from the carrier.
15. To remove the forward gear roller bearing, forward gear internal needle bearing and/or forward gear bearing race, proceed as follows:
   a. Press the roller bearing from the forward gear. Support the bearing with a knife-edged bearing separator, such as part No. 91-37241. See **Figure 97**. Press on the gear hub using a suitable mandrel.
   b. Clamp the gear in a vise with protective jaws with the gear engagement lugs facing up. Drive the internal needle bearing from the gear using a suitable punch and hammer.

NOTE
If the forward gear bearing is removed only to change the shim(s) and forward gear lash, do not discard the bearing race.

   c. To remove the forward gear bearing race from the gearcase, pull the race from the front of the propeller shaft bore using a suitable slide hammer, such as part No. 91-34569A-1. See **Figure 98**. Remove the shim(s) from the bearing bore. Measure and record the thickness of the shims for later reference. Replace the race if the roller bearing assembly must be replaced.
16. If the drive shaft wear sleeve and seal (20 and 21, **Figure 93**) require replacement, replace them as follows:
   a. Support the drive shaft wear sleeve in a knife-edged bearing plate, such as part No. 91-37241. The pinion gear end of the shaft must face down.
   b. Press on the crankshaft end of the drive shaft until the sleeve is free. See **Figure 99**.
17. If the drive shaft upper bearing (17, **Figure 93**) or the lubrication sleeve (19) require replacement, replace them as follows:
   a. Remove the bearing by pulling it out of the drive shaft bore using the recommended puller (part No. 91-83165M) or an equivalent two jaw puller (**Figure 100**).
   b. Remove the bearing sleeve by also pulling it out of the drive shaft bore using the same procedure.
   c. Remove the lubrication sleeve by pulling it out of the drive shaft bore using puller part No. 91-83165M or an equivalent two jaw puller.

NOTE
If the drive shaft roller bearing race is removed only to change the shim(s) and pinion gear depth, do not discard the bearing race. The drive shaft needle bearing and lubrication sleeve do not have to be removed before removing the roller bearing race.

18. If the drive shaft roller bearing race requires removal or replacement, proceed as follows:
   a. Remove the bearing race from the gearcase by driving it down into the propeller shaft bore with a bearing remover (part No. 91-14308A-1 or an equivalent). See **Figure 101**.
   b. Insert the puller jaws (A, **Figure 102**) through the propeller shaft bore. Insert the driver (B, **Figure 102**) through the drive shaft bore.
   c. Place a shop cloth under the bearing puller. Use a suitable mallet to drive the bearing race out into the propeller shaft bore.
19. If the propeller shaft requires disassembly, disassemble it as follows:
   a. Remove the cross pin retainer spring with a small screwdriver or awl. Insert the tool under one end of the spring and rotate the propeller shaft to unwind the spring. See **Figure 88**, typical.
   b. Place the shift cam follower against a solid surface and push on the propeller shaft to unload the spring pressure on the sliding clutch cross pin. Remove the pin with a suitable punch. Carefully release the pressure from the cam follower and spring. Remove the cam follower, three steel balls, spring guide, spring and sliding clutch from the propeller shaft. See **Figure 89**, typical.
20. Refer to *Gearcase Cleaning and Inspection* in this chapter. Clean and inspect all components as described before beginning the reassembly procedure.

Assembly (60 hp Bigfoot Gearcase Models)

Lubricate all internal components with Quicksilver Premium Blend gear oil (part No. 92-13783) or an equivalent. Do not assemble components *dry*. Refer to **Figure 93** for this procedure.

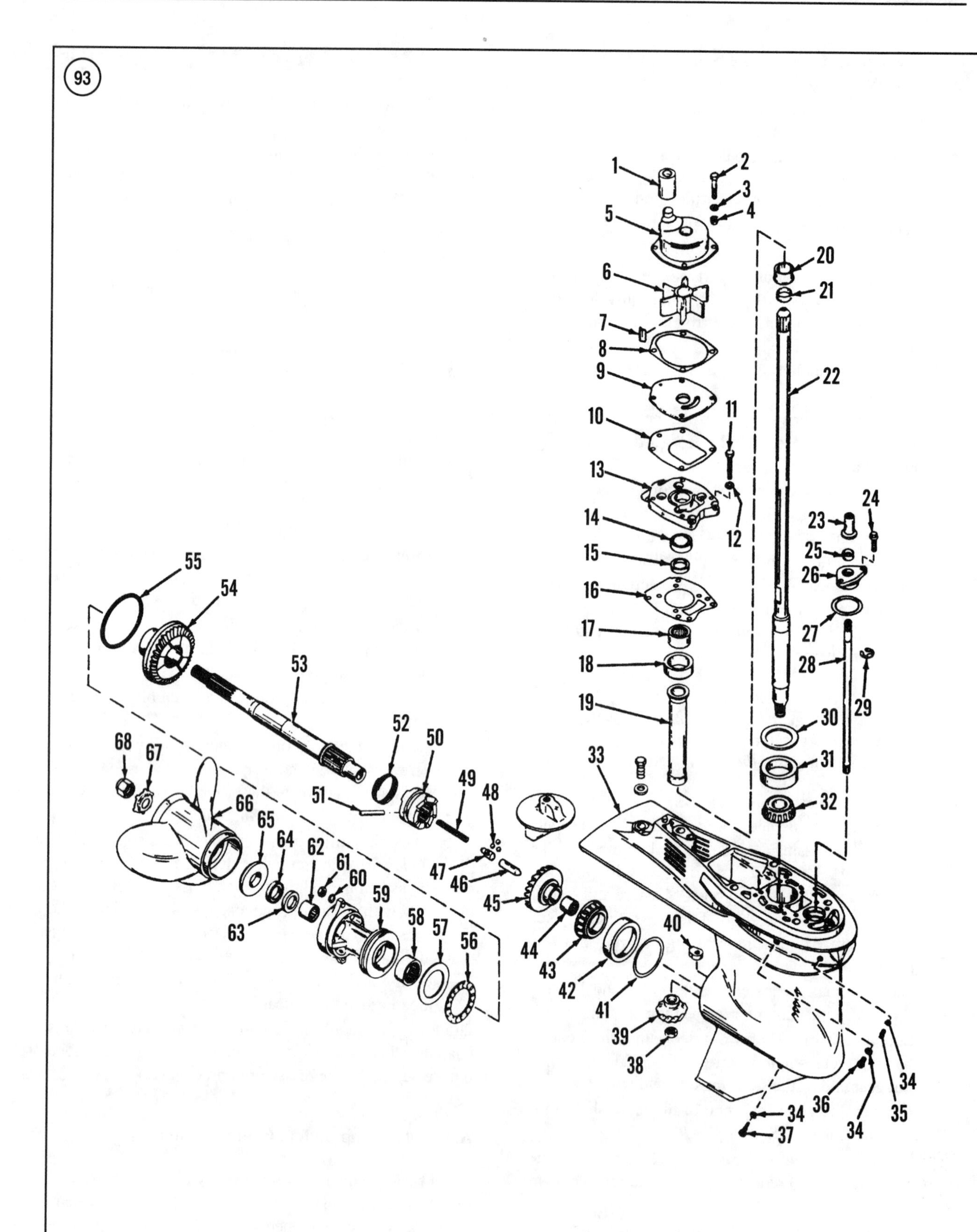
93
1
2
3
4
5
6
7
8
9
10
11
12
13
14
15
16
17
18
19
20
21
22
23
24
25
26
27
28
29
30
31
32
33
34
35
36
37
38
39
40
41
42
43
44
45
46
47
48
49
50
51
52
53
54
55
56
57
58
59
60
61
62
63
64
65
66
67
68

GEARCASE COMPONENTS (60 HP MODELS [WITH BIGFOOT GEARCASE])

1. Water tube seal
2. Screw
3. Washer*
4. Insulator*
5. Water pump cover
6. Impeller
7. Impeller drive key
8. Gasket
9. Face plate
10. Gasket
11. Screw
12. Washer
13. Water pump base
14. Seal
15. Seal
16. Gasket
17. Roller bearing
18. Carrier
19. Lubrication sleeve
20. Wear sleeve
21. Sealing ring
22. Drive shaft
23. Shift shaft coupler
24. Screw
25. Seal
26. Shift shaft bushing/retainer
27. O-ring
28. Shift shaft
29. Retaining ring
30. Shim(s)
31. Bearing race
32. Tapered roller bearing
33. Gearcase housing
34. Gasket
35. Vent plug
36. Vent plug
37. Drain/fill plug
38. Pinion gear nut
39. Pinion gear
40. Shift cam
41. Shim(s)
42. Bearing race
43. Tapered roller bearing
44. Needle bearing
45. Forward gear
46. Shift cam follower
47. Guide
48. Balls
49. Spring
50. Sliding clutch
51. Cross pin
52. Cross pin retaining spring
53. Propeller shaft
54. Reverse gear
55. O-ring
56. Thrust bearing
57. Thrust washer
58. Roller bearing
59. Bearing carrier
60. Flat washer
61. Elastic locknut
62. Needle bearing
63. Seal
64. Seal
65. Thrust hub
66. Propeller
67. Tab washer
68. Propeller nut

*Not used on all models.

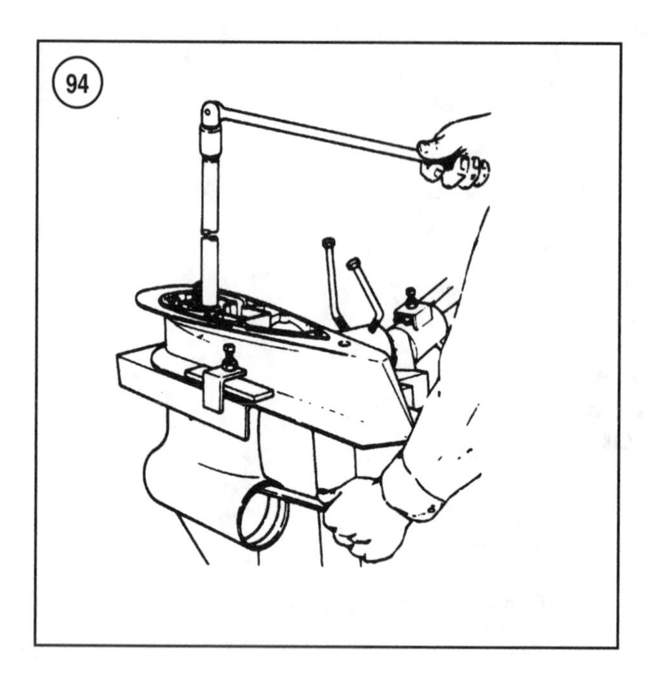

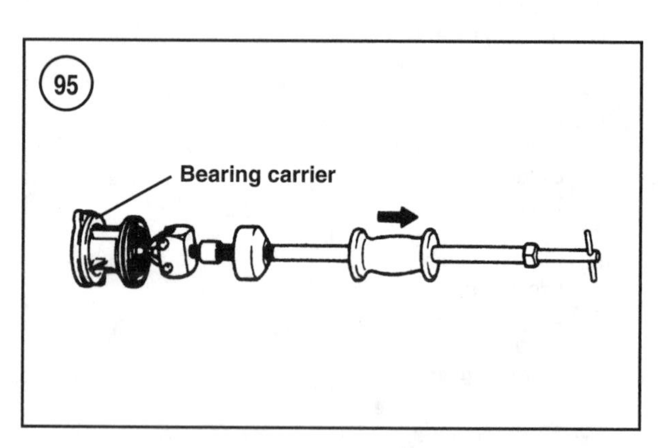

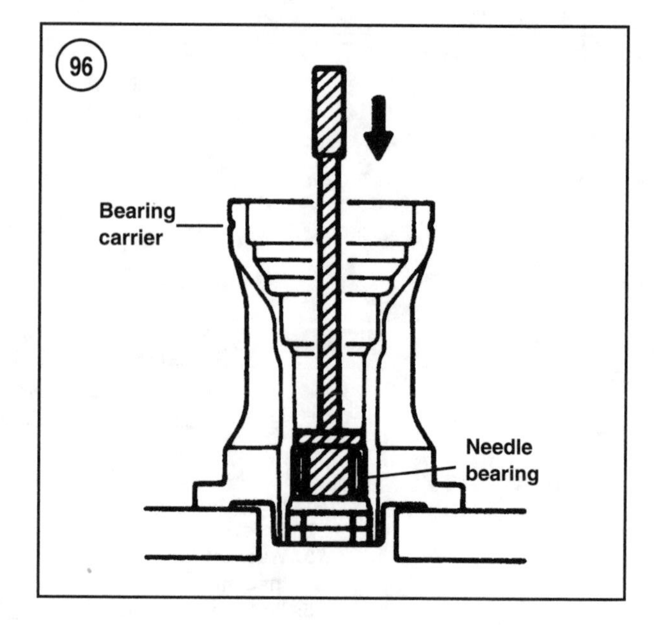

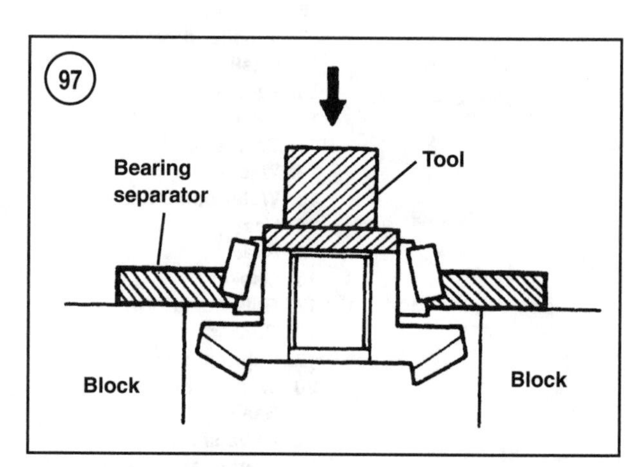

1. If the drive shaft wear sleeve and seal (20 and 21, **Figure 93**) were removed, install a *new* seal and sleeve as follows:
   a. Position a new rubber seal into the drive shaft groove. Coat the outside diameter of the seal with Loctite 271 threadlocking adhesive (part No. 92-809819).
   b. Place a *new* wear sleeve into the holder from the sleeve installation kit (part No. 91-14310A-1). Slide the drive shaft into the sleeve and holder.
   c. Place the driver from the sleeve installation kit over the pinion end of the drive shaft. Place the drive shaft and tool assembly in a press. Press the driver against the holder until they contact each other.
   d. Wipe excess Loctite from the drive shaft.
2. If the lower drive shaft bearing was removed, install the race into the gearcase as follows:

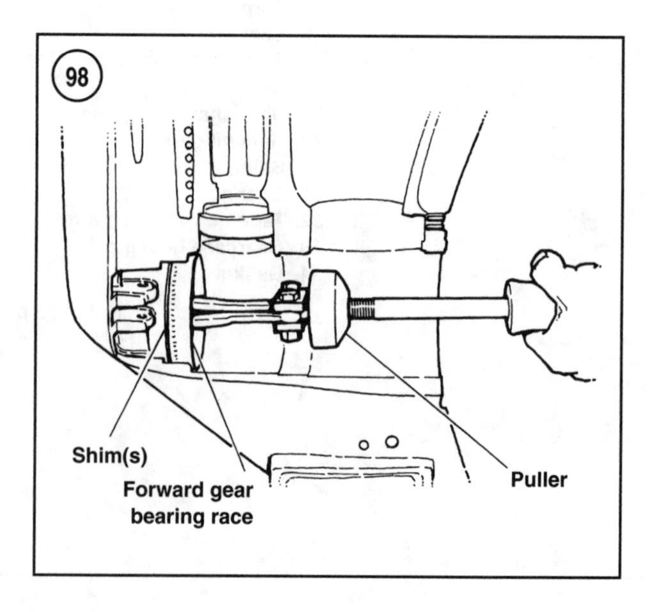

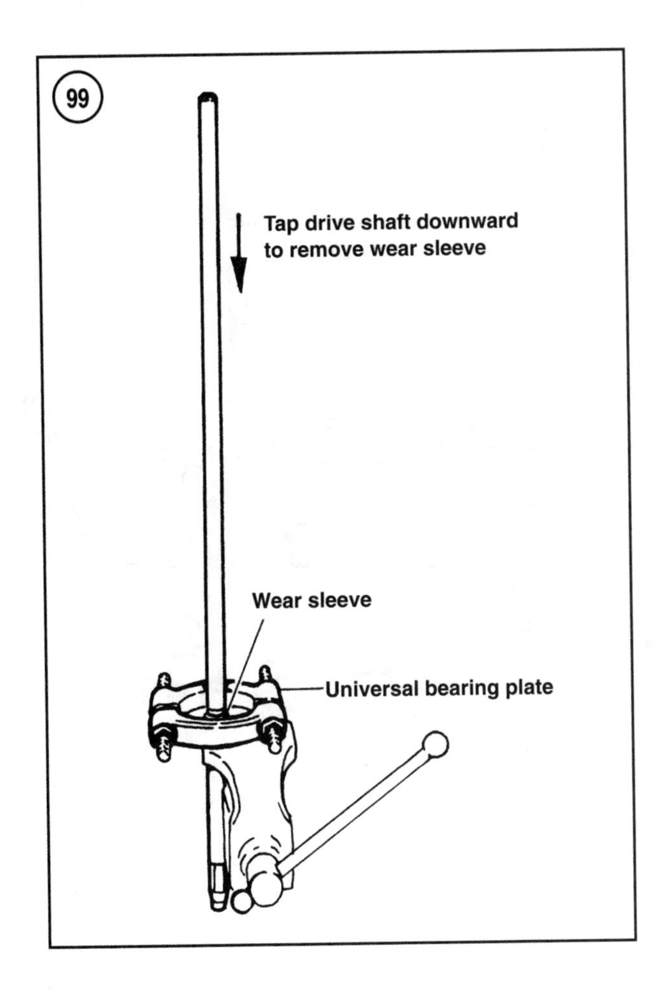

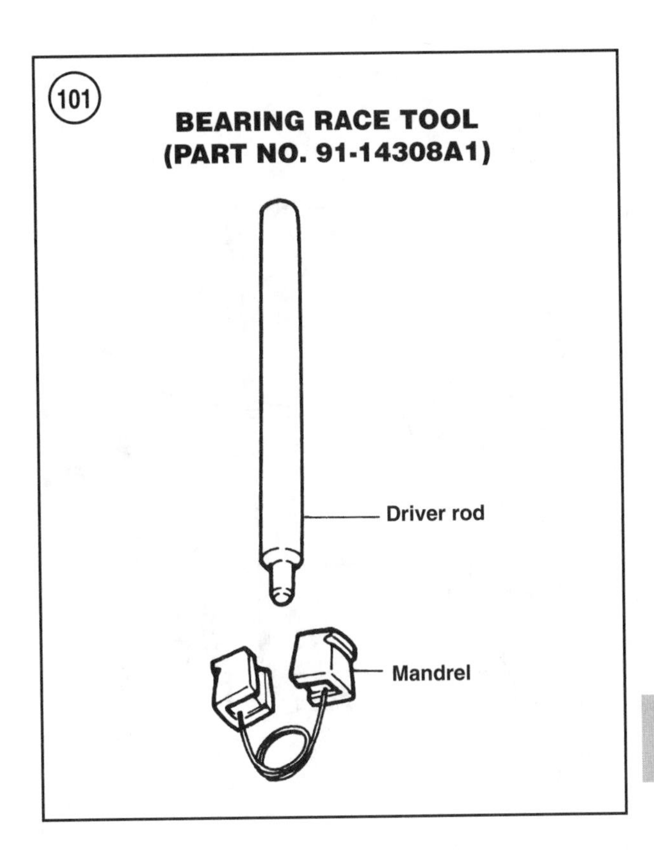

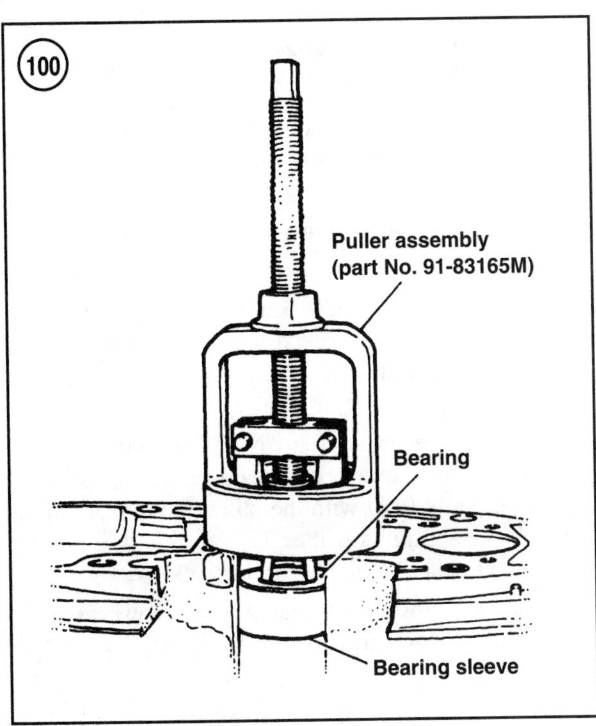

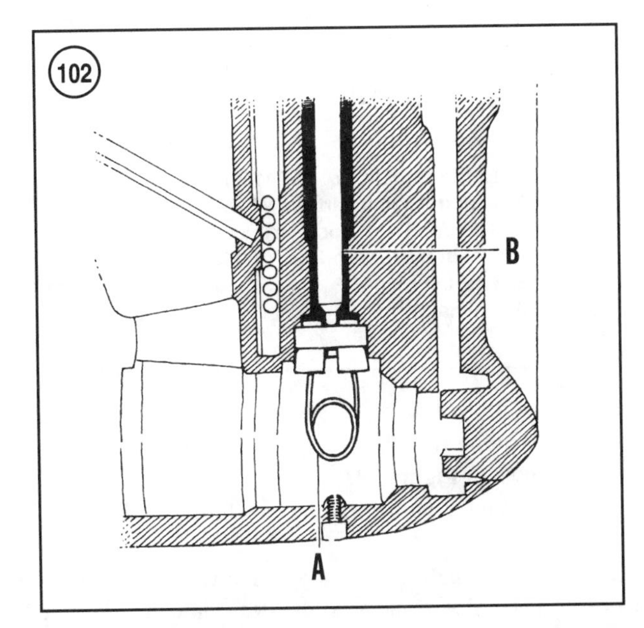

a. Lubricate the race and place the original shims on top of the race. If the original shims are lost or damaged (beyond measurement), install a 0.025 in. (0.635 mm) shim(s).

b. Position the race into its bore with the tapered side facing down.

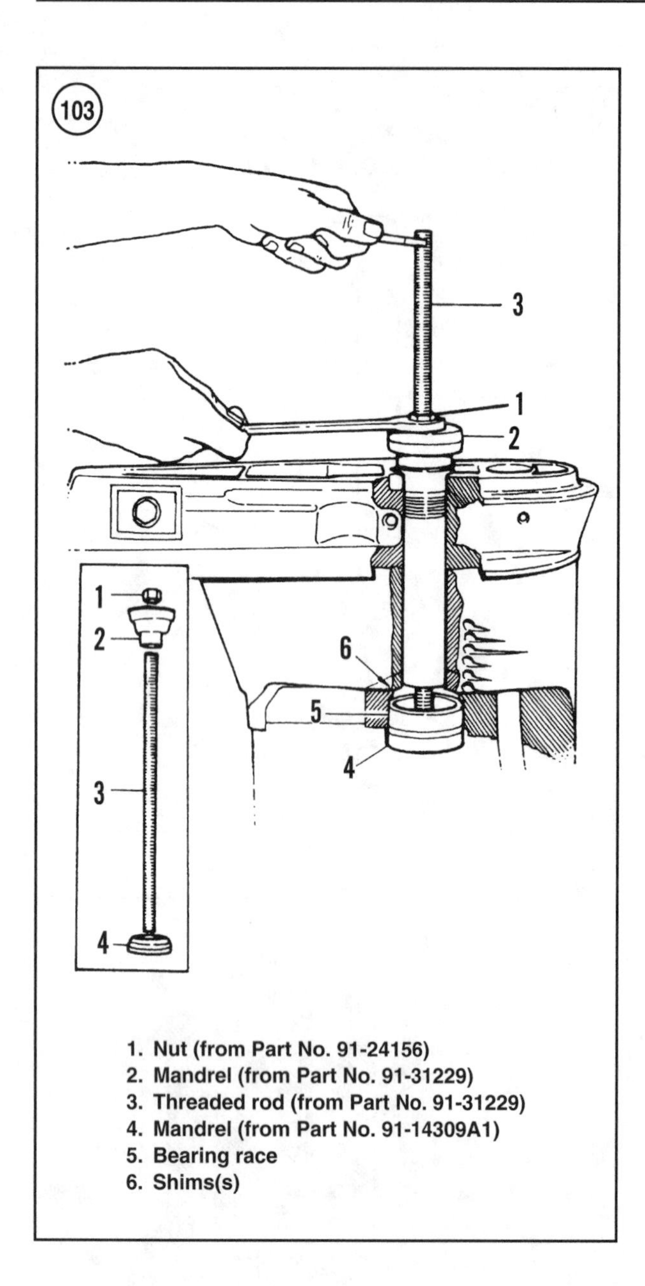

1. Nut (from Part No. 91-24156)
2. Mandrel (from Part No. 91-31229)
3. Threaded rod (from Part No. 91-31229)
4. Mandrel (from Part No. 91-14309A1)
5. Bearing race
6. Shims(s)

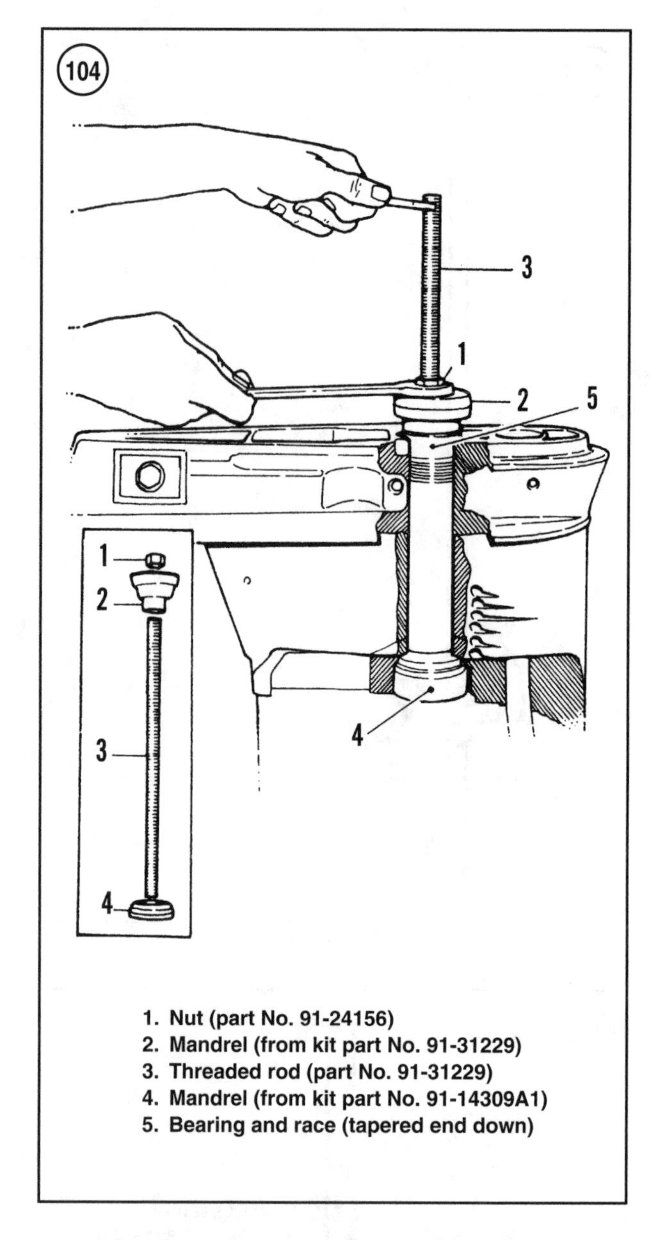

1. Nut (part No. 91-24156)
2. Mandrel (from kit part No. 91-31229)
3. Threaded rod (part No. 91-31229)
4. Mandrel (from kit part No. 91-14309A1)
5. Bearing and race (tapered end down)

c. Assemble the bearing installer components as shown in **Figure 103**.

d. Tighten the nut to seat the race fully in the bearing bore.

3. If the drive shaft lubrication sleeve was removed, lubricate a *new* sleeve and press the sleeve into the gearcase as far as possible with hand pressure. Make sure the tab at the top of the sleeve points to the rear of the gearcase. The sleeve is fully seated into the gearcase in the next step.

NOTE
The drive shaft lubrication sleeve must be installed before proceeding with the next step.

4. If the drive shaft needle bearing and sleeve were removed, install a new bearing assembly as follows:

a. If the new bearing is separate from the new bearing sleeve, lubricate both parts with Quicksilver 2-4-C Multi-Lube grease (part No. 92-825407). Set the sleeve with the tapered side down into a press. Position the bearing with the numbered side up in the sleeve and press it into the sleeve with a suitable mandrel until it is flush with the sleeve.

b. Lubricate the outside diameter of a new bearing assembly with Quicksilver 2-4-C Multi-Lube grease.

c. Position the bearing assembly in the drive shaft bore with the tapered end facing downward.

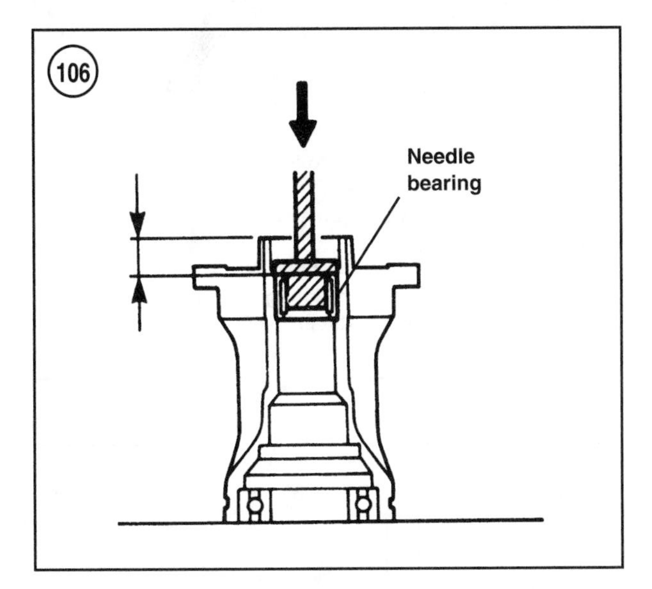

d. Assemble the installation tools as shown in **Figure 104**.

e. Tighten the nut to seat the bearing assembly (and lubrication sleeve) into the drive shaft bore.

5. If the forward gear bearing race was removed, install the race into the gearcase as follows:

a. Position the original shims in the bearing bore. If the original shims were lost, start with a 0.010 in. (0.254 mm) shim(s).

b. Lubricate the bearing race and set it into the gearcase bearing bore. Place a mandrel (part No. 91-31106 or an equivalent) over the race.

c. Place the propeller shaft into the mandrel hole. Install the propeller shaft bearing carrier into the gearcase to hold the propeller shaft centered.

d. Thread a scrap propeller nut onto the propeller shaft. Use a mallet to drive the propeller shaft against the mandrel until the bearing race is fully seated in the gearcase bearing bore.

e. Remove the propeller nut, propeller shaft, bearing carrier and mandrel.

6. If the forward gear roller bearing was removed, lubricate a new roller bearing and set it on the gear hub with the rollers facing upward. Press the bearing fully onto the gear with a mandrel (part No. 91-37350 or equivalent). See **Figure 105**. Do not press on the roller cage.

7. If the forward gear internal needle bearing was removed, install a new bearing as follows:

a. Position the forward gear in a press with the gear teeth facing down.

b. Lubricate the new needle bearing and position it in the gear bore with the numbered side facing up.

c. Press the bearing into the gear with a suitable mandrel until the bearing bottoms in the bore. Be careful not to damage the bearing by over-pressing.

8. If the bearing carrier needle bearings were removed, install new bearings as follows:

a. Set the carrier in a press with the propeller end facing upward. Lubricate a new propeller shaft needle bearing and position it into the carrier bore with the lettered end facing up.

b. Press the bearing into the carrier with a suitable mandrel (such as part No. 91-15755) until the bearing bottoms in its bore. See **Figure 106**.

c. Set the carrier in a press with the propeller end facing down. Lubricate a new reverse gear roller bearing and position it into the carrier bore with its lettered end facing up.

d. Set the carrier in a press with the propeller end facing up on a bearing installer (part No. 91-13945 or an equivalent to protect the carrier and reverse gear needle bearing.

e. Press the bearing into the carrier with the bearing installer (part No. 91-13945 or an equivalent). Press until the tool seats. See **Figure 107**.

9. If the propeller shaft was disassembled, refer to **Figure 108** and reassemble it as follows:
   a. Lubricate all components with Quicksilver 2-4-C Multi-Lube grease.
   b. Align the cross pin holes of the sliding clutch with the slot in the propeller shaft. Position the grooved end of the sliding clutch towards the propeller and slide it onto the propeller shaft.
   c. Install the shift spring into the propeller shaft. Then install the shift spring guide with the narrow end towards the shift spring. Install the three steel balls and finally install the cam follower with the beveled end facing out.
   d. Press the cam follower against a solid object to compress the spring. Align the sliding clutch holes with the spring guide's hole. A small punch may be used to ease alignment. See **Figure 89**, typical.
   e. Insert the sliding clutch cross-pin into the clutch and through the spring guide's hole. The cross-pin must pass through the spring guide.

NOTE
The sliding clutch retaining spring must lay flat with no overlapping coils.

   f. Secure the pin to the sliding clutch with a *new* retainer spring. Do not open the spring more than necessary to install it.
10. Install two *new* propeller shaft seals as follows:
   a. Coat the outer diameter of the two new propeller shaft seals with Loctite 271 threadlocking adhesive (part No. 92-809819).
   b. Set the carrier in a press with the propeller end facing up on a bearing installer (part No. 91-13945 or an equivalent) to protect the carrier and reverse gear needle bearing.
   c. Install the small diameter seal with the spring facing the gearcase. Press the seal in with the large stepped end of a mandrel (part No. 91-31108 or an equivalent) until the tool bottoms against the carrier. See **Figure 109**.
   d. Install the large diameter seal with the spring facing the propeller. Press the seal in with the small stepped end of the mandrel until the tool bottoms against the carrier. See **Figure 109**.
11. Coat a new propeller shaft bearing carrier O-ring and the propeller shaft seal lips with 2-4-C Multi-Lube grease. Then install the O-ring into the carrier groove.
12. Assemble the propeller shaft bearing carrier, reverse gear and bearings, and the propeller shaft assembly as follows:
   a. Lubricate the propeller shaft bearing carrier thrust washer with Quicksilver Needle Bearing Assembly

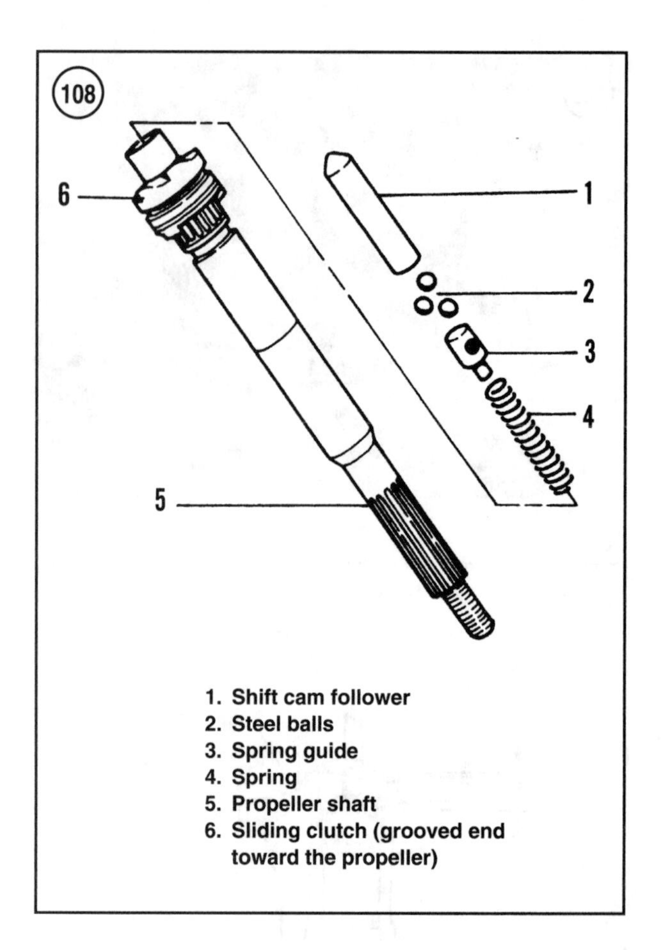

1. Shift cam follower
2. Steel balls
3. Spring guide
4. Spring
5. Propeller shaft
6. Sliding clutch (grooved end toward the propeller)

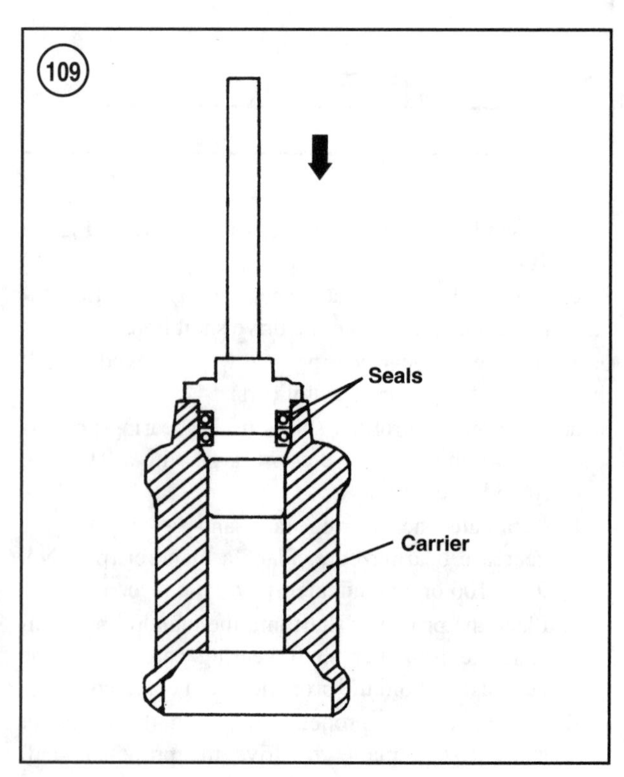

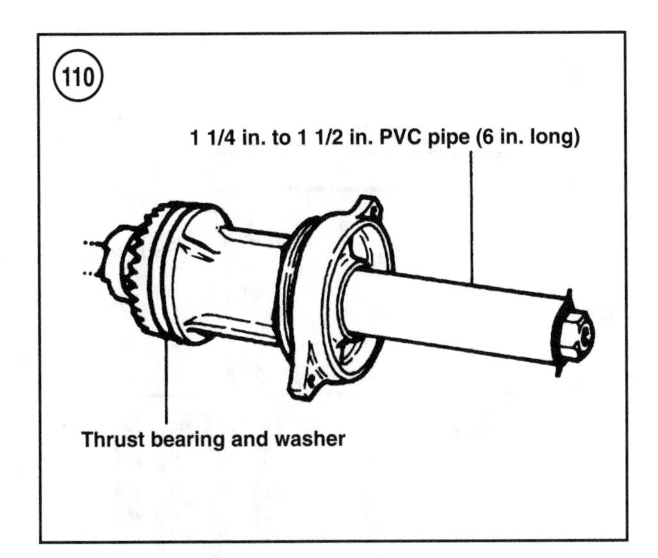

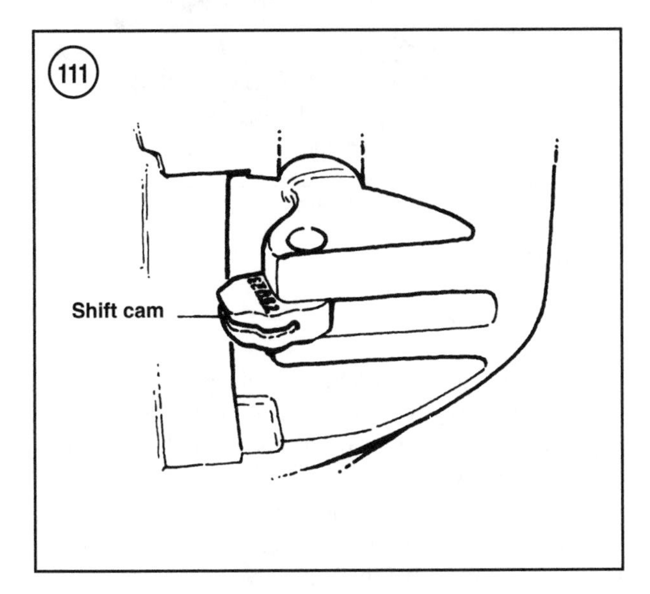

Grease (part No. 92-825265A1). Install the thrust washer onto the propeller shaft carrier.

b. Lubricate the thrust bearing with the same grease and place it on top of the thrust washer.

c. Install the reverse gear into the bearing carrier being careful not to disturb the position of the thrust washer and bearing.

d. Carefully slide the propeller shaft into the carrier assembly. Be careful not to damage the carrier seals.

e. Obtain a piece of 1 1/4 or 1 1/2 in. (31.75 or 38.10 mm) diameter, 6 in. (152.5 mm) long PVC pipe. Install the PVC pipe over the propeller shaft, then install the propeller locking tab washer and propeller nut. Hand tighten the nut to hold the propeller shaft securely into the bearing carrier. See **Figure 110**.

13. Assemble and install the shift components as follows:

a. Coat the outer diameter of a new shift shaft seal with Loctite 271 threadlocking adhesive (part No. 92-809819). Press the seal into the shift shaft retainer/bushing with a suitable mandrel until it is flush with the retainer bore.

b. Lubricate the seal lip and a *new* retainer O-ring with Quicksilver 2-4-C Multi-Lube grease (part No. 92-825407). Install the O-ring into the shift bushing groove.

c. Install the E-ring onto the shift shaft if it was removed.

d. Lubricate the shift shaft with the same grease and carefully insert the shift shaft into the shift shaft retainer and seal.

e. Place the shift cam into the gear housing with the numbered side up as shown in **Figure 111**. Install the shift shaft assembly into the gearcase and engage the shift shaft splines to the shift cam internal splines.

f. Apply Loctite 271 threadlocking adhesive to the threads of the shift shaft bushing retainer screws. Install and tighten the screws to the specification in **Table 1**.

14. Rotate the gearcase so the propeller shaft bore is pointing up. Install the forward gear assembly into the propeller shaft bore and into the forward gear bearing race.

15. Position the drive shaft roller bearing in the race at the bottom of the drive shaft bore, then place the pinion gear in position over the bearing.

16. Spray the threads of the drive shaft with Locquic primer (part No. 92-59327-1). Insert the drive shaft into the drive shaft bore while holding the pinion gear and lower bearing in position. Rotate the shaft as necessary to engage the drive shaft splines to the pinion gear splines.

NOTE

Apply Loctite threadlocking adhesive to a new pinion nut after the pinion gear depth and forward gear lash have been correctly adjusted. Install the used pinion nut without adhesive to check the gear depth and forward gear lash.

17. Install the original pinion nut with the recessed side facing *toward* the pinion gear. Tighten the nut finger-tight at this time.

18. Hold the pinion nut with a suitable wrench or socket. Attach a spline socket (part No. 91-56775 or an equiva-

lent) to a suitable torque wrench. Tighten the pinion nut to the specification in **Table 1**. See **Figure 112**, typical.

19. Refer to *Gearcase Shimming* and set the *pinion gear depth*. Do not continue until the pinion gear depth is correct.

20. Once the pinion gear depth is correct, install the propeller shaft assembly into the gearcase and into the forward gear internal needle bearing. Make sure the shift cam follower does not fall out during assembly.

21. Liberally lubricate the front and rear flanges of the propeller shaft bearing carrier with Quicksilver Special Lube 101 (part No. 92-13872A1). Install the carrier over the propeller shaft, being careful not to damage the carrier seals on the propeller shaft splines. Make sure the screw holes are aligned with the gearcase. Rotate the drive shaft to ensure the gear teeth are meshed, then seat the carrier into the gearcase bore.

22. Secure the carrier with two locknuts and washers. Tighten the locknuts to the specification in **Table 1**. Then remove the propeller nut, washer and PVC pipe.

23. Refer to *Gearcase Shimming* in this chapter and set the *forward gear lash*. Do not continue until the correct gear lash is correct.

24. Once the gear lash is correct, remove the propeller shaft and bearing carrier, and install a *new* pinion nut with Loctite 271 threadlocking adhesive as described previously in this section. Then reinstall the propeller shaft and bearing carrier as described previously in this section. Coat the bearing carrier studs with Loctite 271 threadlocking adhesive and install the locknuts and washers. Tighten the locknuts to the specification in **Table 1**.

25. Install the water pump assembly as described in this chapter.

26. Pressure test the gearcase as described in *Gearcase Pressure Testing*.

27. Fill the gearcase with the recommended lubricant as described in Chapter Four.

Gearcase Shimming

Proper pinion gear to forward/reverse gear engagement and corresponding gear lash are crucial for smooth, quiet operation and long service life. Several shimming procedures must be performed to set up the gearcase gears. The pinion gear must be shimmed to the correct height (depth) and the forward gear must be shimmed to the pinion gear for proper gear lash. Shimming operations are not required on 2.5 hp and 6-50 hp models. Proper assembly automatically positions the gears.

Refer to **Table 4** for gear lash and pinion height specifications.

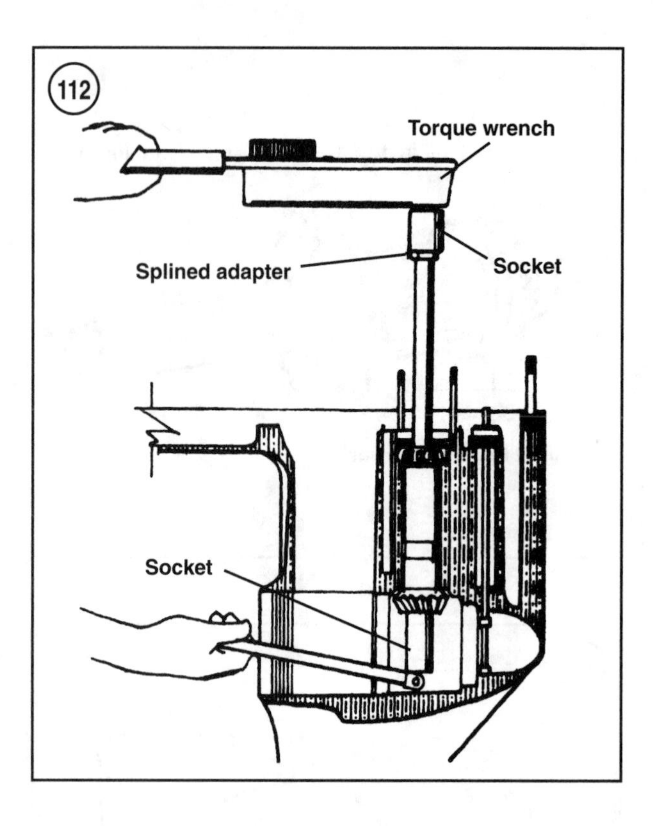

3.3 hp Models

Gear lash verification

The amount of gear lash between the pinion (drive) and forward (driven) gears is not critical as long as some lash is present. If there is no backlash between the gears, premature failure will occur.

1. To check gear lash, firmly press down on the drive shaft while pushing up on the pinion gear. The pinion gear must float on the drive shaft splines.

2. While holding the drive shaft and pinion gear in this position, have an assistant rock the propeller shaft back and forth. Approximately 0.002-0.006 in. (0.05-0.15 mm) lash must be present between the forward and pinion gears.

3. If the gear lash is excessive, over 0.012 in. (0.30 mm), recheck the pinion gear, drive gear and related components for excessive wear. Also, make sure the shim between the forward gear and the forward gear ball bearing is installed correctly.

4. If there is insufficient or no backlash, one or more of the following problems exists:

   a. The front ball bearing is not fully seated in the gearcase.

   b. The forward gear is not fully seated in the front ball bearing.

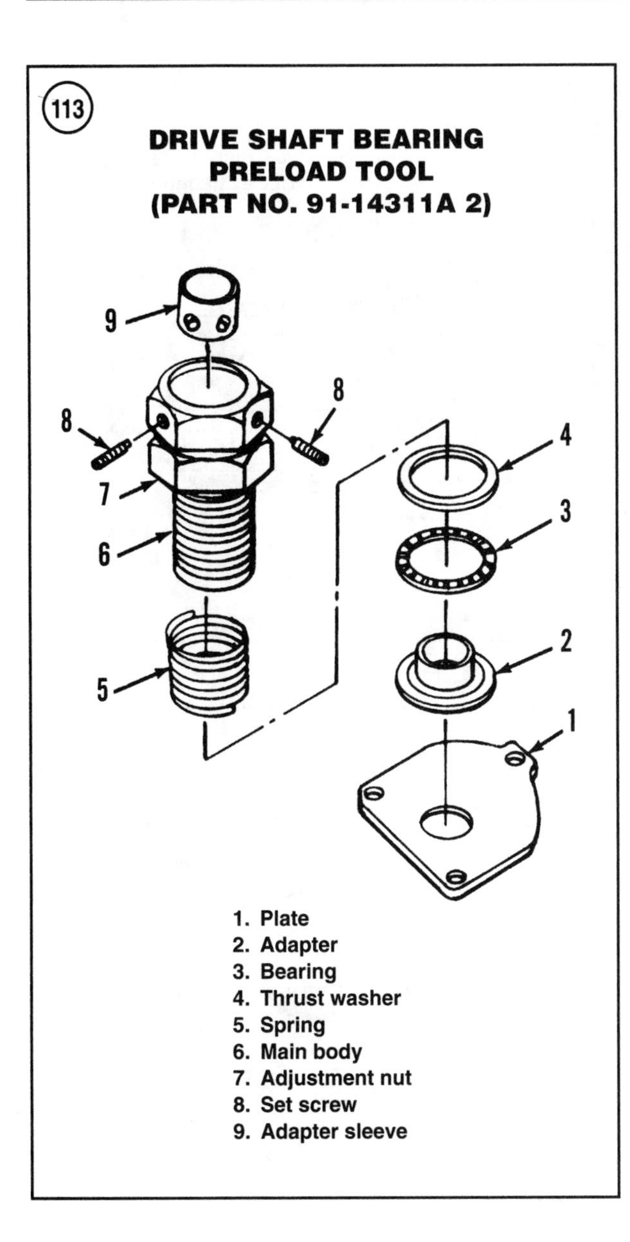

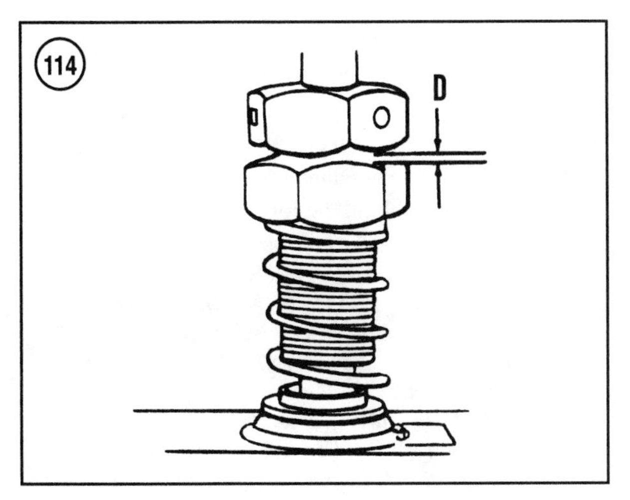

c. The lower ball bearing is too low in the gearcase bore.

d. The upper ball bearing is installed too far onto the drive shaft.

55 and 60 hp Standard Gearcase Models

Pinion gear depth

1. Use a clean shop towel to thoroughly clean the gear cavity, especially the area around the bearing carrier shoulder.
2. Position the gearcase with the drive shaft facing up.

NOTE
A drive shaft bearing preload tool (part No. 91-14311A-2) is required to check/adjust the pinion gear depth and gear lash. The tool pulls the drive shaft up into the tapered roller bearing race.

3. Install the bearing preload tool (part No. 91-14311A-2) onto the drive shaft in the order shown in **Figure 113**.
   a. Make sure the thrust bearing and washer (3 and 4, **Figure 113**) are clean and lightly oiled.
   b. Screw the nut (7, **Figure 113**) completely onto the main body (6), then securely tighten the set screws (8), making sure the holes in the sleeve (9) are aligned with the set screws.
   c. Measure the distance (D, **Figure 114**) between the top of the nut and the bottom of the bolt head. Then screw the nut downward, increasing the distance (D, **Figure 114**) by 1 in. (25.4 mm).
   d. Rotate the drive shaft 10-12 turns to seat the drive shaft bearing(s).
4. Insert a pinion locating tool (part No. 91-817008A-2) into the gearcase. Make sure the tool engages the forward gear with its access hole facing the pinion gear.

NOTE
Rotate the drive shaft and take several readings in Step 5. Then average the feeler gauge readings.

5. Insert a 0.025 in. (0.64 mm) flat feeler gauge between the gauging block and pinion gear. See **Figure 115**. The average clearance between the gear and gauging block must be 0.025 in. (0.64 mm).
6. If the average clearance is not exactly 0.025 in. (0.64 mm), proceed as follows:
   a. If clearance is over specification, remove shims as necessary from under the drive shaft roller bearing race.

b. If clearance is under specification, add shims as necessary under the drive shaft roller bearing race.

7. Reassemble the drive shaft and pinion gear as described in this chapter, then recheck pinion gear depth as described in this section.

8. Leave the drive shaft bearing preload tool installed.

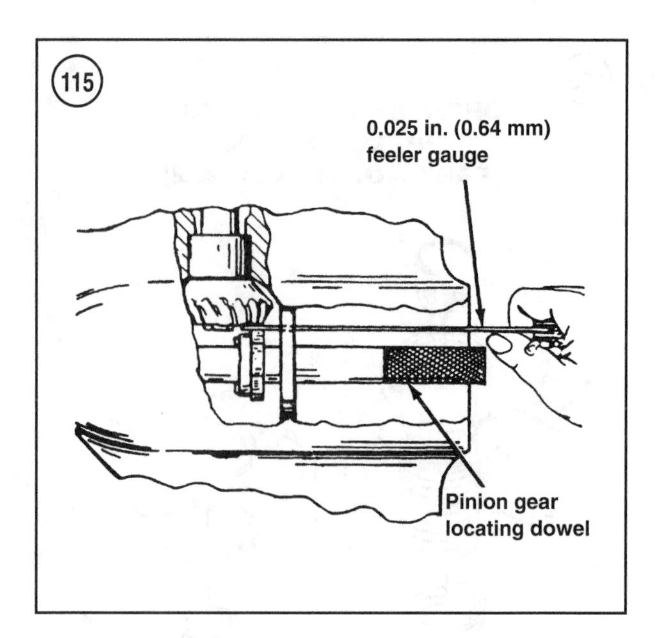

Forward gear lash

NOTE

Establish the correct pinion gear depth before attempting to adjust forward gear lash. The drive shaft bearing preload tool (part No. 91-14311A-2) must be installed for this procedure. Refer to the ***Pinion gear depth*** *for tool installation procedures.*

1. Assemble puller jaws (part No. 91-46086A1) and a threaded bolt (part No. 91-85716) or an equivalent. Install the assembly to the propeller shaft and bearing carrier as shown in **Figure 116**.

2. Tighten the puller bolt to 45 in.-lb. (5.1 N•m), then turn the drive shaft 5-10 revolutions to seat the forward gear bearing and race. This preloads the forward gear assembly into its bearing race. Recheck the torque after 5-10 revolutions.

3. Fasten a suitable threaded rod to the gearcase using flat washers and nuts. Then install a dial indicator to the threaded rod. See **Figure 117**.

4. Install a backlash indicator tool (part No. 91-19660-1) onto the drive shaft. Align the tool with the indicator plunger, then tighten the tool securely on the drive shaft. See **Figure 117**.

5. Adjust the dial indicator mounting so the plunger is aligned with line *3* on the backlash indicator tool. Then zero the indicator gauge.

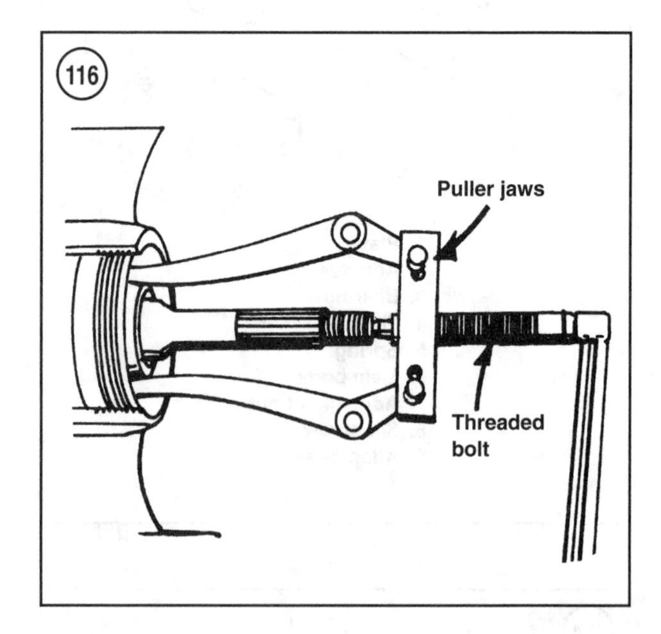

NOTE

The propeller shaft must not move during gear lash readings. Rotate the drive shaft just enough to contact a gear tooth in one direction, then rotate the drive shaft just enough in the opposite direction to contact the opposing gear tooth.

6. Lightly rotate the drive shaft back and forth while noting the dial indicator reading. The indicator should read 0.013-0.019 in. (0.33-0.48 mm).

NOTE

A 0.001 in. (0.025 mm) change in the forward gear bearing shim thickness changes the forward gear lash by approximately 0.00125 in. (0.032 mm).

7. If backlash is excessive, over 0.019 in. (0.48 mm), add shim(s) behind the forward gear bearing race as necessary. If backlash is insufficient, under 0.013 in. (0.33 mm), subtract shim(s) from behind the forward gear bearing race.

CAUTION

Once the gear lash is correct, the gearcase can be completely assembled. Install a new pinion nut, secured with Locktite 271 threadlocking adhesive (part No. 92-809819), during assembly. Also secure the propeller shaft bearing carrier's screws with Locktite 271 threadlocking adhesive.

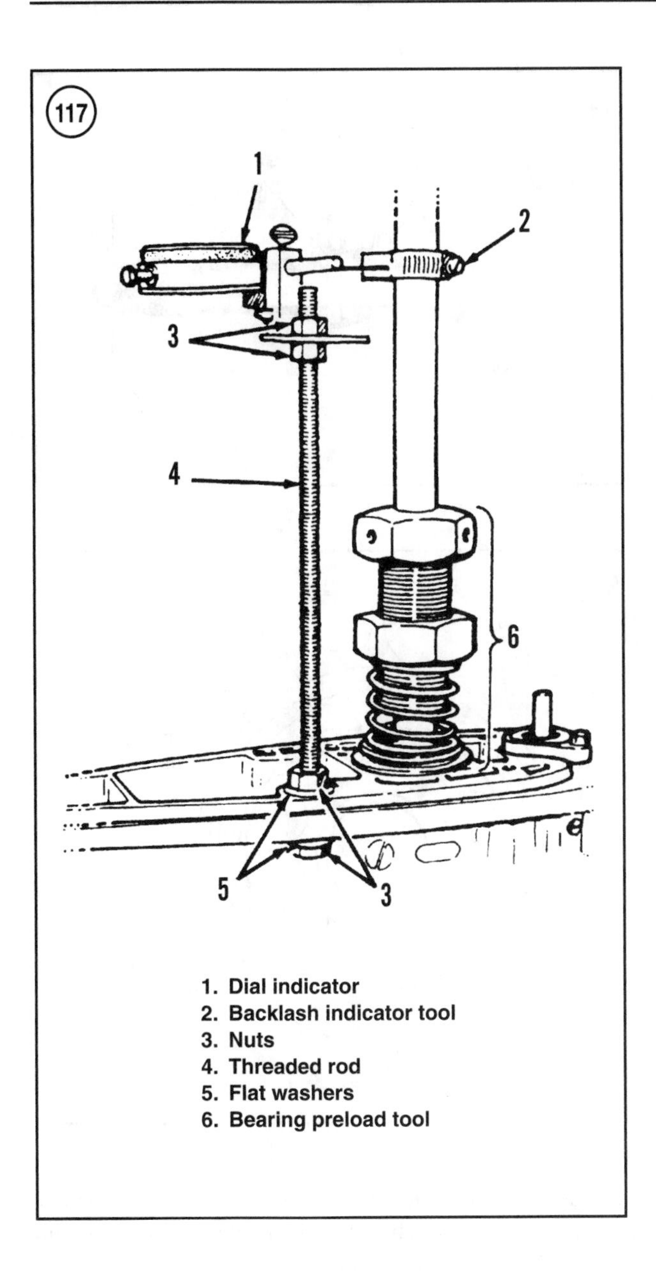

1. Dial indicator
2. Backlash indicator tool
3. Nuts
4. Threaded rod
5. Flat washers
6. Bearing preload tool

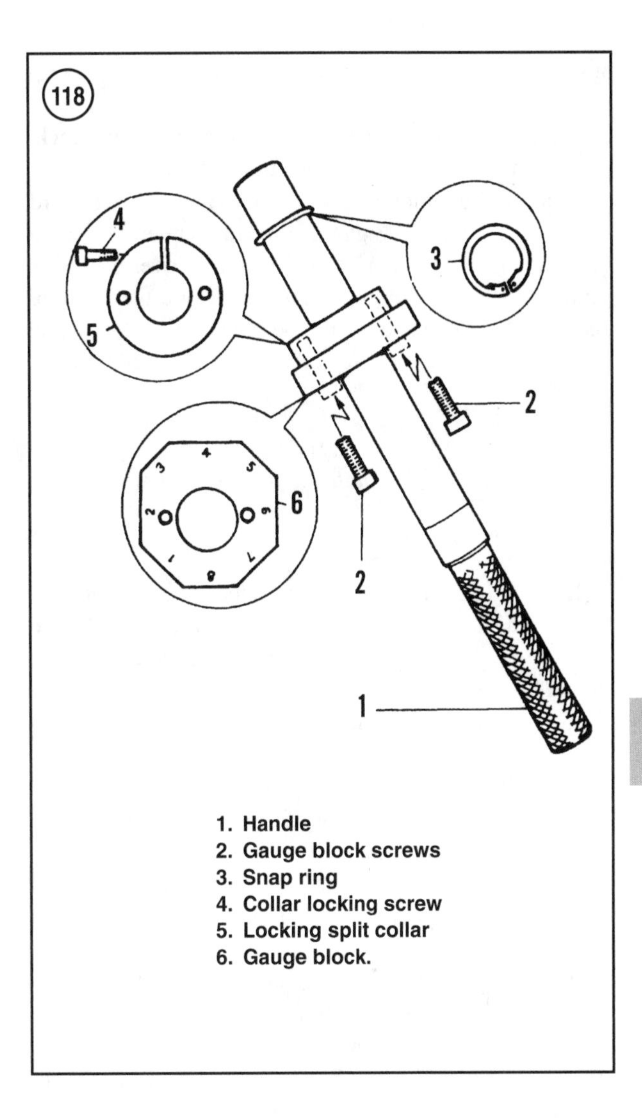

1. Handle
2. Gauge block screws
3. Snap ring
4. Collar locking screw
5. Locking split collar
6. Gauge block.

8. Complete the assembly procedure.

60 hp Bigfoot Models

Pinion gear depth

1. Position the gearcase with the drive shaft facing up.

2. Install the bearing preload tool (part No. 91-14311A-1) onto the drive shaft in the order shown in **Figure 113**. Do not install the plate (1, **Figure 113**).

   a. Make sure the thrust bearing and washer are clean and lightly oiled.

   b. Screw the nut (7, **Figure 113**) completely onto the main body (6), then securely tighten the set screws (8), making sure the holes in the sleeve (9) are aligned with the set screws.

   c. Measure the distance (D, **Figure 114**) between the top of the nut and the bottom of the bolt head. Then screw the nut down, increasing the distance (D, **Figure 114**) by 1 in. (25.4 mm).

   d. Rotate the drive shaft 10-12 turns to seat the drive shaft bearing(s).

3. Assemble the pinion gear locating tool (part No. 12349A-2) as shown in **Figure 118**. Face the numbered side of the gauge block out so the numbers can be seen as the tool is being used. Tighten the split collar retaining screw to the point where the collar can still slide back and forth on the handle with moderate hand pressure.

4. Insert the tool into the gearcase, making sure the tool pilot is in the forward gear needle bearing. Slide the gauge block back and forth as necessary to position the gauge

block directly under the pinion gear teeth as shown in **Figure 119**.

5. Without disturbing the position of the gauging block, remove the tool and tighten the collar screw.

6. Reinsert the pinion gear locating tool into the forward gear. Position the No. 8 gauge block flat under the pinion gear, then install the No. 3 alignment disc over the tool's handle as shown in **Figure 120**. Make sure the locating disc is fully seated against the bearing carrier step inside the gear cavity and the disc access hole is aligned with the pinion gear.

NOTE
Rotate the drive shaft and take several readings in Step 7. Then average the feeler gauge readings.

7. Insert a 0.025 in. (0.64 mm) flat feeler gauge between the gauging block and pinion gear. See **Figure 121**. The average clearance between the gear and gauging block should be 0.025 in. (0.64 mm).

8. If the average clearance (Step 6) is not exactly 0.025 in. (0.64 mm), proceed as follows:
   a. If clearance is less than 0.025 in. (0.64 mm), remove shims as necessary from under the drive shaft roller bearing race.
   b. If clearance is over 0.025 in. (0.64 mm), add shims as necessary under the drive shaft roller bearing race.

9. Reassemble the drive shaft and pinion gear as described previously in this chapter, then recheck pinion gear depth as described in this section.

10. Leave the drive shaft bearing preload tool installed. Continue the assembly procedure.

Forward gear lash (60 hp Bigfoot models)

NOTE
Establish the correct pinion gear depth before attempting to adjust forward gear lash. The drive shaft bearing preload tool (part No. 91-14311A-2) must be installed for this procedure. Refer to ***Pinion gear depth*** *for tool installation procedures.*

1. Assemble puller jaws (part No. 91-46086A1) and a threaded bolt (part No. 91-85716) or an equivalent. Install the assembly to the propeller shaft and bearing carrier as shown in **Figure 116**.

2. Tighten the puller bolt to 45 in.-lb. (5.1 N•m), then turn the drive shaft 5-10 revolutions to seat the forward gear bearing and race. This preloads the forward gear assembly in its bearing race. Recheck the torque after the 5-10 turns.

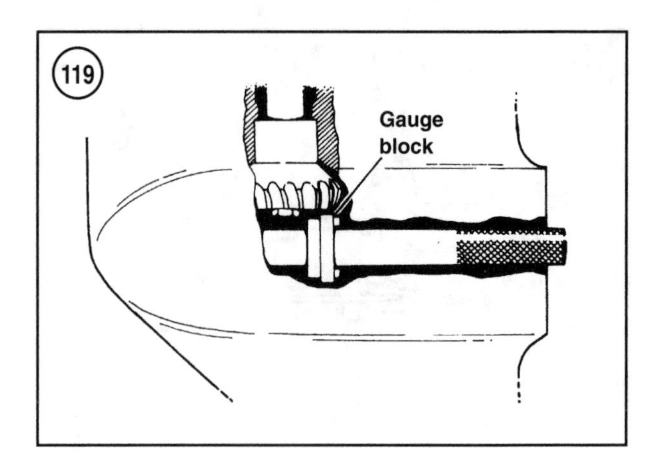

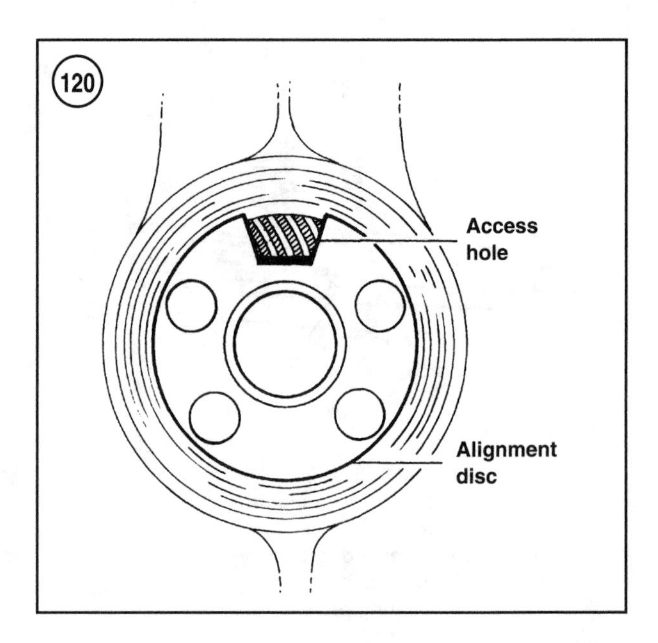

3. Fasten a suitable threaded rod to the gearcase using flat washers and nuts. Then install a dial indicator to the threaded rod. See **Figure 117**.

4. Install the backlash indicator tool (part No. 91-78473) onto the drive shaft. Align the tool with the indicator plunger, then tighten the tool securely on the drive shaft. See **Figure 117**.

5. Adjust the dial indicator mounting so the plunger is aligned with line *4* on the backlash indicator tool. Then zero the indicator dial.

NOTE
The propeller shaft must not move during gear lash measurement. Rotate the drive shaft just enough to contact a gear tooth in one direction, then rotate the drive shaft just enough in the opposite direction to contact the opposing gear tooth.

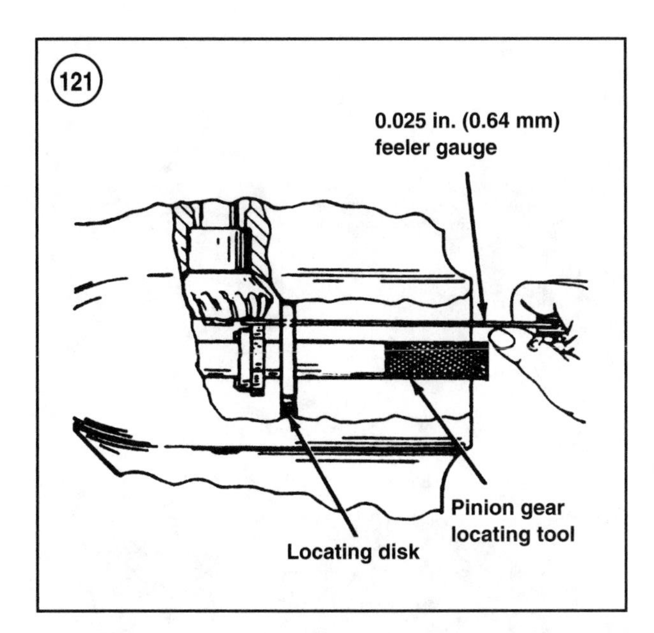

6. Lightly rotate the drive shaft back and forth while noting the gear lash measurement on the dial indicator. The gear lash should be 0.012-0.019 in. (0.30-0. 48 mm).

7. If gear lash is excessive, add shims behind the forward gear bearing race as necessary. If gear lash is insufficient, subtract shims from behind the forward gear bearing race as necessary.

CAUTION

Once the gear lash is correct, the gearcase can be completely assembled. Install a new pinion nut, secured with Locktite 271 threadlocking adhesive (part No. 92-809819) during assembly. Also secure the propeller shaft bearing carrier's screws with Locktite 271 threadlocking adhesive.

8. Complete the assembly procedure.

Gearcase Cleaning and Inspection

CAUTION

Metric and American fasteners are used on Mercury/Mariner gearcases. Always match a replacement fastener to the original. Do not run a tap or thread chaser into a hole or over a bolt without first verifying the thread size and pitch. Check all threaded holes for Heli-Coil stainless steel locking thread inserts. Never run a tap or thread chaser into a Heli-Coil equipped hole. Heli-Coil inserts are replaceable, if damaged.

NOTE

Do not remove a pressed-on roller or ball bearing, or pressed-in needle bearing, ball bearing or bushing unless replacement is necessary. A tapered roller bearing consist of the roller assembly and a bearing race. The roller and race are a matched assembly and must be replaced as an assembly.

1. Replace all seals, gaskets and O-rings removed during disassembly.

2. Clean all parts in clean solvent and dry them with compressed air. Lightly lubricate all internal components to prevent corrosion.

3. Inspect all screws, bolts, nuts and other fasteners for damaged, galled or distorted threads. Replace elastic locknuts that can be installed without the aid of a wrench. Clean all sealing compound, RTV sealant and threadlocking compound from the threaded areas. Minor thread imperfections can be corrected with an appropriate thread chaser.

4. Clean all gasket and sealant material from the gearcase housing. Make sure all water and lubricant passages are clean and unobstructed. Make sure all threaded holes are free of corrosion, gasket sealant or threadlocking adhesive. Damaged or distorted threads may be repaired with stainless steel threaded inserts.

5. Inspect the gearcase housing, propeller shaft bearing carrier and all other seal or bearing carriers for cracks, porosity, wear, distortion and mechanical damage. Replace any housing that shows evidence of having a bearing spun in its bore.

6. If the gearcase is equipped with a speedometer pickup, make sure the pickup port is not clogged with debris. Use a very small drill bit mounted in a pin vice to remove debris from the pickup port. Make sure air can flow freely from the pickup port to the speedometer hose connection.

7. Inspect all anodes as described at the beginning of this chapter. Replace any anode that has deteriorated to half its original size.

8. Inspect the water inlet screen(s) for damage or obstructions. Clean or replace the screen(s) as necessary.

9. Inspect the drive shaft and propeller shaft for worn, damaged or twisted splines. See A, **Figure 122**, typical. Excessively worn drive shaft splines are usually the result of shaft misalignment caused by a distorted drive shaft housing or lower gearcase housing due to impact with an underwater object. Replace distorted housings.

10. Inspect the drive shaft and propeller shaft threaded areas for damage. See A, **Figure 122**, typical. If equipped, check the impeller drive pin and propeller drive pin holes for wear, elongation and cracks.

11. Inspect each shaft's bearing and seal surfaces for excessive wear, grooving, metal transfer and discoloration from overheating. See B, **Figure 122**, typical.

12. Check for a bent propeller shaft by supporting the propeller shaft with V-blocks at its bearing surfaces. Mount a dial indicator to position the plunger on the area just forward of the splines. Rotate the propeller shaft while observing the dial indicator. Any noticeable wobble, or a reading of more than 0.006 in. (0.15 mm) indicates excessive runout. Replace the propeller shaft if runout is excessive.

13. Check each gear for excessive wear, corrosion and mechanical damage. Check the teeth for galling, chips, cracks, missing sections, distortion or discoloration from overheating. Check the sliding clutch and each gear's engagement lugs (**Figure 123**) for chips, cracks and excessive wear.

14. Check the pinion gear and sliding clutch splines for wear, distortion or mechanical damage.

15. Inspect all shift components and the shift linkage for excessive wear and mechanical damage. Inspect the shift cam for wear or grooving at the follower contact surfaces. Replace the shift cam and follower if they are damaged or worn. On rotary shift models, inspect the shift shaft splines for corrosion, wear, distortion or twisting. Replace the shift shaft if it is corroded, damaged or worn.

16. Inspect all roller, ball and needle bearings for water damage, pitting, discoloration from overheating, and metal transfer. Locate and inspect all internal needle bearings (**Figure 124**, typical). On models with bushings, inspect each bushing for excessive wear and mechanical damage. Replace any bushing noticeably out of round or damaged.

17. Check the propeller for nicks, cracks or damaged blades. Minor nicks can be removed with a file. Be careful to retain the original contour of the blade. Replace the propeller or have it repaired if any blades are bent or cracked. Replace the propeller if it is excessively corroded.

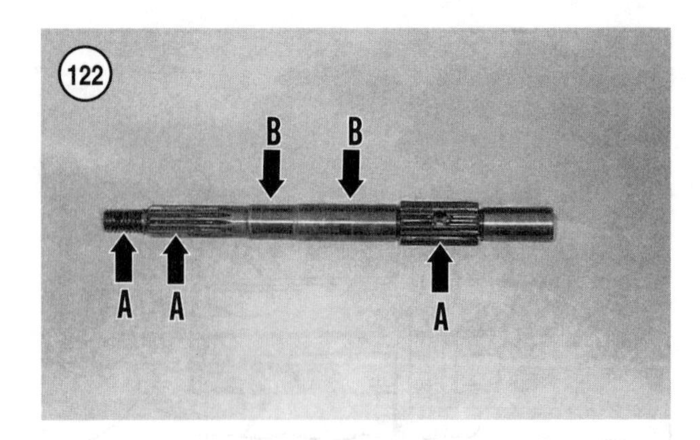

Gearcase Pressure Testing

When a gearcase is disassembled, it must be pressure tested after reassembly to ensure that no leaks are present. If the gearcase fails the pressure test, find and correct the source of leakage. Failure to correct leaks will result in major gearcase damage from water entering the gearcase or lubricant leaking out.

Do not fill the gearcase with lubricant until the pressure test has been satisfactorily completed.

To pressure test the gearcase, proceed as follows:

NOTE
Drain the gearcase lubricant before pressure testing. Refer to Chapter Four if needed.

1. Make sure the gearcase is completely drained. Then make sure the fill/drain plug is installed and properly tightened. Always use a new sealing washer or gasket on the fill plug.

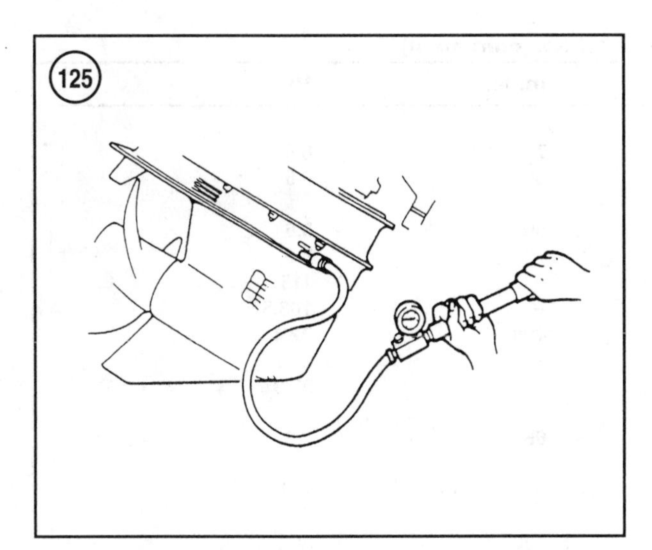

2. Remove the vent plug. Install the pressure tester (part No. FT-8950) into the vent hole. See **Figure 125**. Securely tighten the tester fitting. Always use a new sealing washer on the pressure tester fitting.
3. Pressurize the gearcase to 10 psi (69 kPa) for at least five minutes. During this time, periodically rotate the propeller and drive shafts, and move the shift linkage through its full range of travel.
4. The gearcase should hold pressure for five minutes. If it does not, pressurize the gearcase again and spray soapy water on all sealing surfaces or submerge the gearcase in water to locate the leak.
5. Replace defective seals, gaskets or repair sealing surfaces at the point of the leak. If the gearcase holds pressure as specified, fill the gearcase with lubricant as described in Chapter Four.

Table 1 TORQUE SPECIFICATIONS

Component	ft.-lb.	in.-lb.	N•m
Propeller nut			
4 and 5 hp	–	150	17
6-15 hp	–	100	11.3
20 and 25 hp	17	204	23
30-60 hp	55	–	74.6
Anode/trim tab			
2.5 hp	–	50	5.6
4 and 5 hp	–	70	8.0
6-15 hp	–	30	3.4
20 and 25 hp	–	100	11.3
30-50 hp	–	186	21.0
55 and 60 hp(standard gearcase)	22	–	29.8
60 (Bigfoot gearcase)			
Anode	–	60	6.8
Trim tab	22	–	29.8
Water intake screen			
4 and 5 hp	–	70	8.0
6-15 hp	–	30	3.4
20 and 25 hp	–	25	2.8
Gearcase mounting fasteners			
2.5 and 3.3 hp	–	50	5.6
4 and 5 hp	–	70	8.0
6-15 hp	15	180	20.3
20-60 hp	40	–	54.2
Water pump cover/housing			
2.5 hp	–	50	5.6
3.3, 4, 5 hp	–	70	8.0
9.9-15 hp	–	50	5.6
20 hp, 25 hp, 25 Jet	–	60	6.8
30-50 hp (except 45 jet)	–	60	6.8
55 hp, 60 hp, 45 Jet	–	60	6.8
Water pump base			
9.9-15 hp	–	50	5.6
60 hp (with Bigfoot gearcase)	–	60	6.8

(continued)

Table 1 TORQUE SPECIFICATIONS (continued)

Component	ft.-lb.	in.-lb.	N•m
Shift shaft bushing/retainer			
4 and 5 hp	–	70	8.0
60 hp (with Bigfoot gearcase)	–	60	6.8
Bearing carrier			
3.3 hp	–	50	5.6
4, 5 hp	–	70	8.0
9-15 hp	85	–	115
20 and 25 hp	80	–	108.5
30-50 hp	16.5	198	22.4
55 and 60 hp (standard gearcase)	19	–	25.7
60 hp (with Bigfoot gearcase)	25	–	33.9
Bearing carrier retainer plate screws			
20 and 25 hp	–	65	7.3
Drive shaft bearing retainer			
55 and 60 hp (standard gearcase)	75	–	101.7
Pinion nut			
20 and 25 hp	15	180	20.3
30-50 hp	50	–	67.8
55 and 60 hp (standard gearcase)	50	–	67.8
60 hp (with Bigfoot gearcase)	70	–	94.9
Fill/drain and vent plugs			
20 and 25 hp	–	60	6.8
30-50 hp	–	55	6.2
55 and 60 hp (standard gearcase)	–	35-80	4.0-9.0
60 hp (with Bigfoot gearcase)	–	60	6.8

Table 2 STANDARD TORQUE SPECIFICATIONS—U.S. STANDARD AND METRIC FASTENERS

Screw or nut size	ft.-lb.	in.-lb.	N•m
U.S. standard fasteners			
6-32	–	9	1
8-32	–	20	2.3
10-24	–	30	3.4
10-32	–	35	4.0
12-24	–	45	5.1
1/4-20	6	72	8.1
1/4-28	7	84	9.5
5/16-18	13	156	17.6
5/16-24	14	168	19
3/8-16	23	270	31.2
3/8-24	25	300	33.9
7/16-14	36	–	48.8
7/16-20	40	–	54
1/2-13	50	–	67.8
1/2-20	60	–	81.3
Metric fasteners			
M5	–	36	4.1
M6	6	72	8.1
M8	13	156	17.6
M10	26	312	35.3
M12	35	–	47.5
M14	60	–	81.3

Table 3 GEAR RATIO AND APPROXIMATE LUBRICANT CAPACITY

Outboard model	Gear ratio	Tooth count	Lubricant capacity
2.5 hp	1.85: 1	not available	3.0 oz. (89 ml)
3.3 hp	2.18:1	not available	2.5 oz. (74 ml)
4 and 5 hp	2.15:1	not available	6.6 oz. (195 ml)
6, 8, 9.9 15 hp	2.00:1	13:26	6.8 oz. (201 ml)
20 and 25 hp	2.25:1	12:27	7.8 oz. (231 ml)
30 and 40 hp (2-cylinder)	2.00:1	13:26	14.9 oz. (441 ml)
40 and 50 hp (3-cylinder)	1.83:1	12:22	14.9 oz. (441 ml)
55 and 60 hp			
(standard gearcase)	1.64:1	14:23	11.5 oz. (340 ml)
(bigfoot gearcase)	2.30:1	13:30	22.5 oz. (665 ml)

Table 4 GEARCASE SERVICE SPECIFICATIONS

Component	Specification
Forward gear lash	
55 and 60 hp (standard gearcase)	0.013-0.019 in.(0.33-0.48 mm)
60 hp (with Bigfoot gearcase)	0.012-0.019 in. (0.30-0.48 mm)
Propeller shaft straightness (maximum run-out)	0.006 in. (0.15 mm)
Pinion gear height	
All 55-60 hp models	0.025 in. (0.64 mm)
Shift shaft height (neutral gear)	
6-15 hp (measured from water pump base)	
Standard shaft length (15 in.)	16.5 in. (419 mm)
Long shaft length (20 in.)	22 in. (559 mm)
Extra long shaft length (25 in.)	27.5 in. (698.5 mm)

Chapter Ten

Jet Drive

Jet drive models are based on basic outboard models. The standard lower gearcase has been removed and a jet pump unit installed. An adapter plate is used on the 20 jet. The adapter plate does not normally require removal for service procedures.

1. *20 jet models*—A 25 hp power head is used.
2. *30 jet models*—A 40 hp (three-cylinder) power head is used.
3. *45 jet models*—A 60 hp power head is used.

Service to the power head, ignition, electrical, fuel, and power trim and tilt systems is the same as on propeller-driven outboard models. Refer to the appropriate chapter and service section for the engine model using a propeller gearcase. Refer to **Table 1** for specific torque specifications and **Table 2** for standard torque specifications.

Mounting Height

A jet drive outboard must be mounted higher on the transom plate than an equivalent propeller-driven outboard motor. However, if the jet drive is mounted too high, air enters the jet drive resulting in cavitation and power loss. If the jet drive is mounted too low, excessive drag, water spray and loss in speed occur.

Aftermarket water intake fin kits are available to reduce cavitation when running in rough water.

Set the initial height of the outboard motor as follows:

1. Place a straightedge against the boat bottom (not keel) and abut the end of the straightedge with the jet drive intake.
2. The front edge of the water intake housing must align with the top edge of the straightedge (**Figure 1**).
3. Secure the outboard motor at this setting, then test run the boat.
4. If cavitation occurs (over-revving and/or loss of thrust), lower the outboard in 1/4 in. (6.35 mm) increments until operation is uniform.
5. If uniform operation occurs with the initial setting, raise the outboard in 1/4 in. (6.35 mm) increments until cavitation occurs. Then lower the motor to the last uniform setting.

CAUTION

A slight amount of cavitation in rough water and during turns is normal. However, exces-

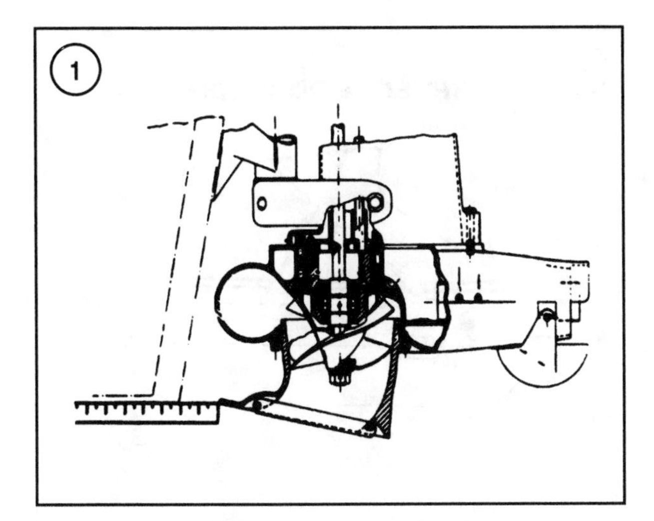

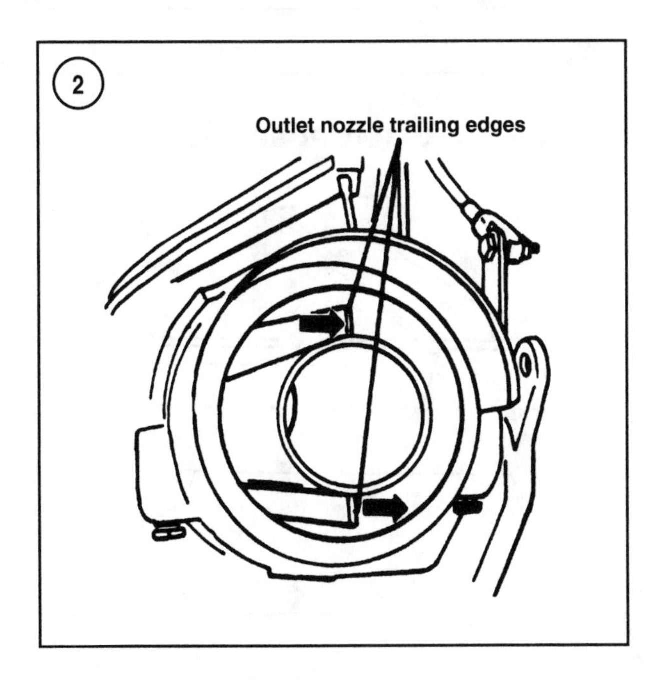

sive cavitation damages the impeller and can cause the power head to overheat.

NOTE

The outboard motor should be in a vertical position when the boat is on plane. Adjust the motor trim setting as needed. If the outboard trim setting is altered, the outboard motor height must be checked and adjusted if necessary.

Steering Torque

A minor adjustment to the trailing edge of the drive outlet nozzle may be made if the boat tends to pull in one di-

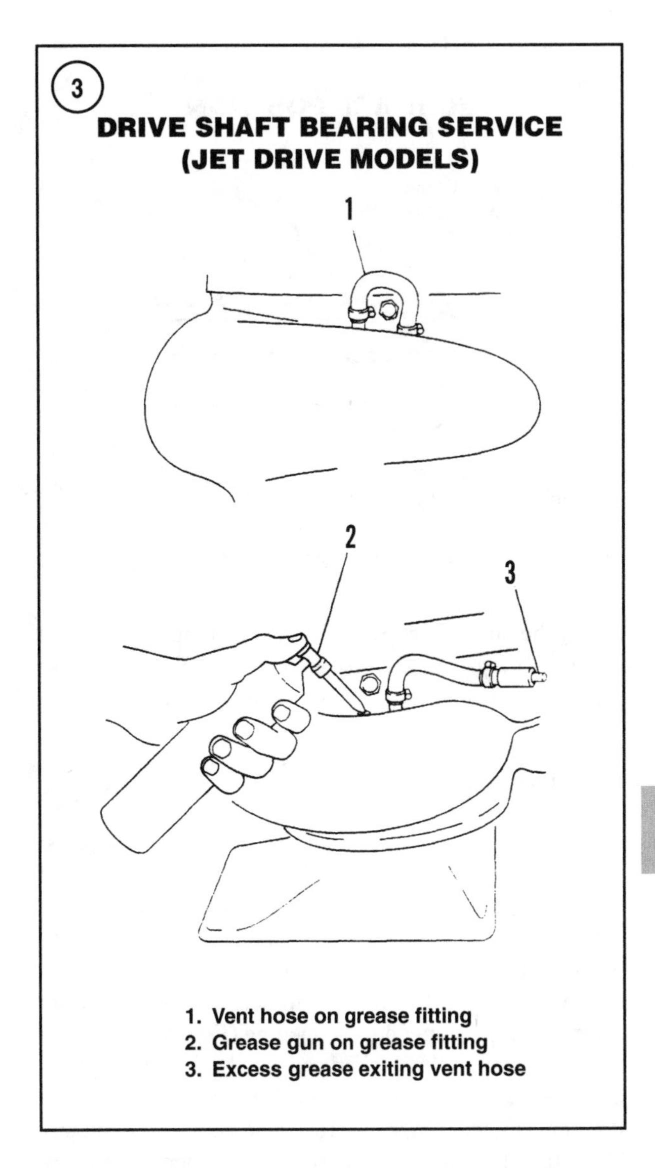

1. Vent hose on grease fitting
2. Grease gun on grease fitting
3. Excess grease exiting vent hose

rection when the boat and outboard are pointed straight ahead. If the boat tends to pull to the starboard side, bend the top and bottom trailing edge of the jet drive outlet nozzle 1/16 in. (1.6 mm) toward the starboard side of the jet drive. See **Figure 2**.

Bearing Lubrication

Lubricate the jet pump bearing(s) after *each* operating period, after every 10 hours of operation and prior to storage. In addition, after every 30 hours of operation, pump additional grease into the bearing(s) to purge moisture. Lubricate the bearing(s) by removing the vent hose on the side of the jet pump housing to expose the grease fitting. See **Figure 3**. Use a grease gun and inject good quality water-resistant grease into the fitting until grease exits the

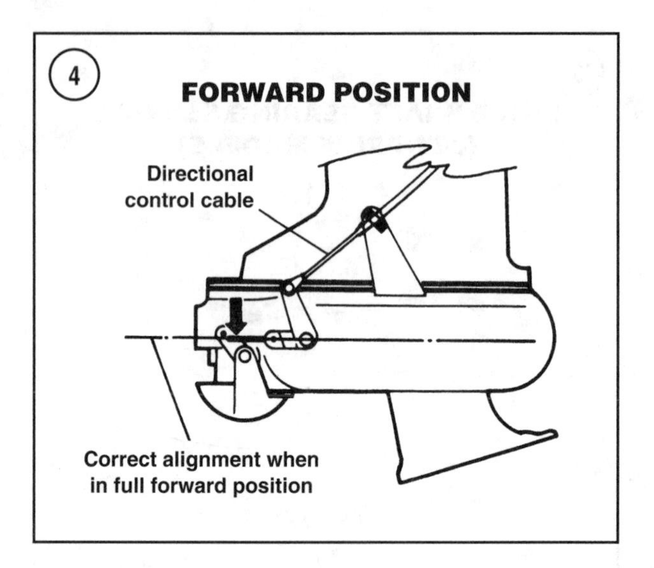

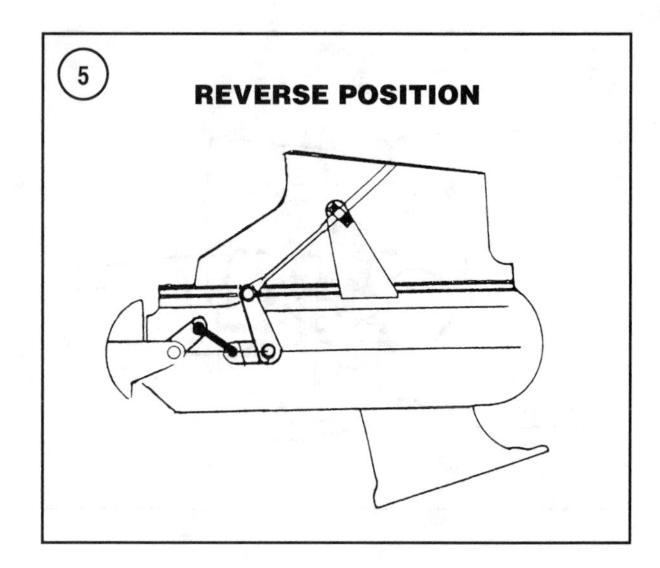

end of the hose. After every 30 hours, pump fresh grease into the fitting until all dirty grease is expelled and fresh grease exits from the end of the hose.

Directional Control

The boat's operational direction is controlled by a thrust gate. The thrust gate is controlled by an engine-mounted mechanical linkage on the 20 jet models and by a remote control shift cable on 30 and 45 Jet models. When the directional control lever is placed in the full forward position, the thrust gate should completely uncover the jet drive housing's outlet nozzle opening (**Figure 4**) and seat securely against the rubber pad on the jet drive pump housing. When the directional control lever is placed in full reverse position, the thrust gate should completely block the pump housing's outlet nozzle opening (**Figure 5**). Neutral position is midway between complete forward and complete reverse (**Figure 6**).

Shift Link Rod Adjustment (20 Jet Models)

The shift link rod is properly adjusted if, after placing the engine mounted directional control lever in the full forward position, the thrust gate *cannot* be moved into the neutral position by hand.

WARNING
Shift link rod adjustment must be correct or water pressure from the boat's forward movement can engage the thrust gate, causing reverse to engage unexpectedly.

Refer to **Figure 7** for this procedure.

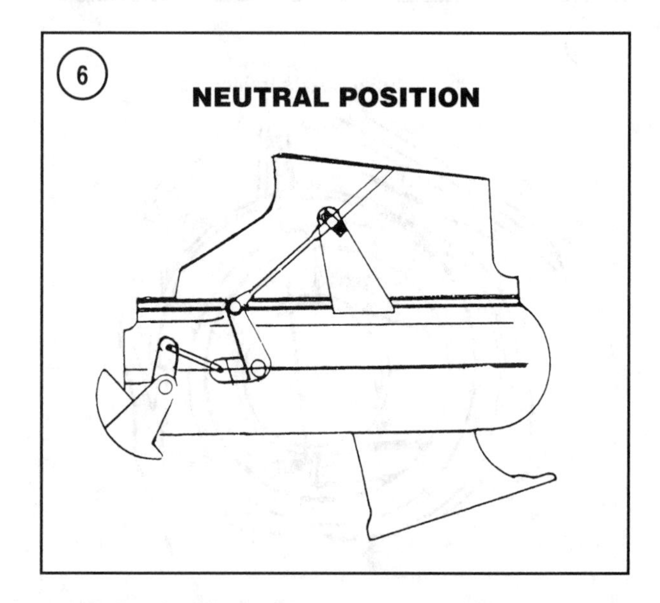

1. Place the engine-mounted directional control lever in the full forward position.
2. Remove the cotter pin and washer securing the shift link rod to the shift cam. Locate and secure the nylon bushing or tension spring.
3. Loosen the jam nut (1, **Figure 7**) and adjust the linkage by rotating the lower end of the shift link rod (2). Adjust the linkage to place the roller at the end of the shift cam slot as shown in **Figure 7**.
4. Attach the shift link rod to the shift cam. Install the washer and a new stainless steel cotter pin. Bend both prongs of the cotter pin for a secure attachment.
5. Securely tighten the link rod jam nut.
6. Attempt to move the thrust gate (5, **Figure 7**) up toward reverse. If the thrust gate can be moved up towards

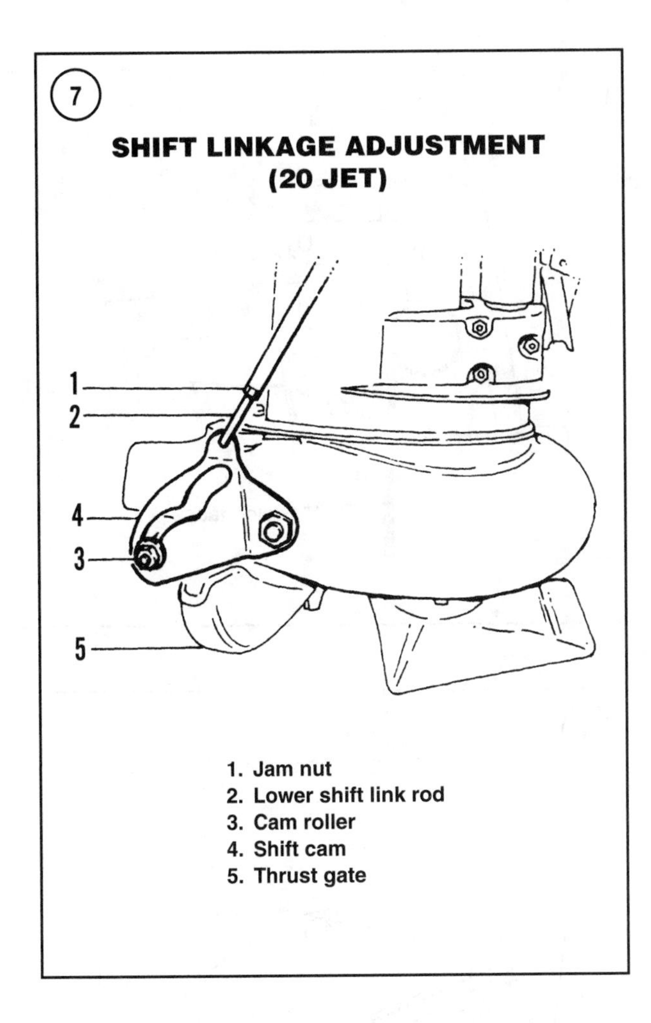

1. Jam nut
2. Lower shift link rod
3. Cam roller
4. Shift cam
5. Thrust gate

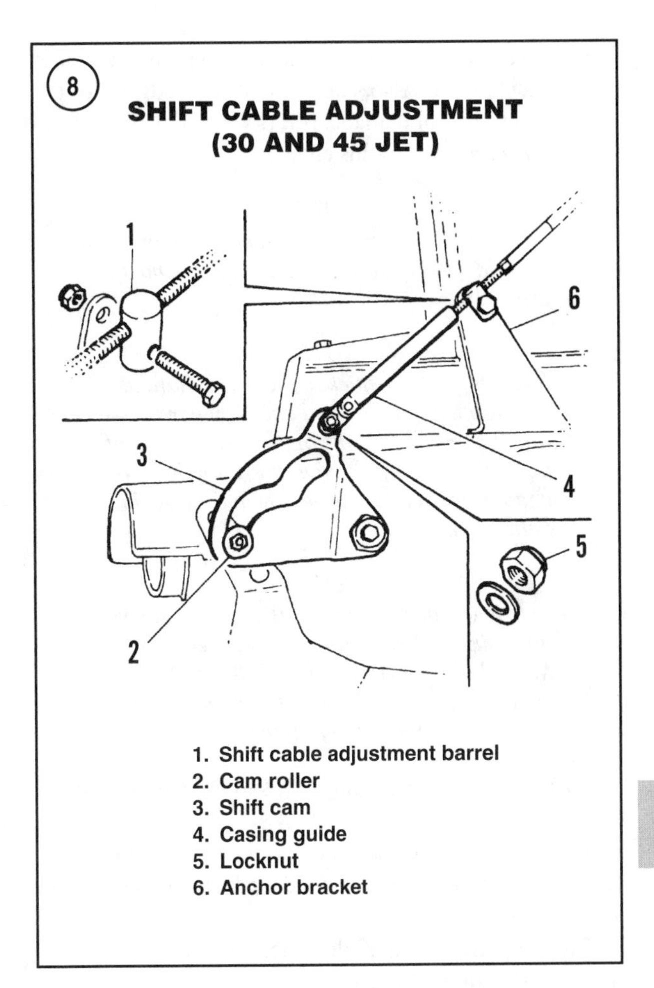

1. Shift cable adjustment barrel
2. Cam roller
3. Shift cam
4. Casing guide
5. Locknut
6. Anchor bracket

reverse, readjust the shift link rod as necessary to prevent the gate from moving toward reverse.

Shift Cable Adjustment (30 and 45 Jet Models)

The directional control cable is properly adjusted if, after placing the remote control lever in the full forward position, the thrust gate *cannot* be moved into the reverse position by hand.

WARNING
Shift cable adjustment must be correct or water pressure from the boat's forward movement can engage the thrust gate, causing reverse to engage unexpectedly.

Refer to **Figure 8** for this procedure.

1. Place the remote control shift lever into the full forward position. Remove the screw and locknut securing the shift cable barrel (1, **Figure 8**) to the shift cable anchor bracket.
2. Adjust the shift cable barrel to position the cam roller (2, **Figure 8**) at the end of the shift cam slot. Secure the cable barrel to the anchor bracket with the screw and locknut. Tighten the locknut securely.
3. Shift the remote control into neutral, then back to the full forward position.
4. Attempt to move the thrust gate up towards reverse. If the thrust gate can be moved up toward reverse, readjust the cable barrel as necessary to prevent the gate from moving toward reverse.
5. After the adjustment is correct, securely tighten the screw and locknut securing the cable barrel to the anchor bracket.
6. Tighten the cable casing guide retaining nut (5, **Figure 8**) until it bottoms, then back the nut off 1/8 to 1/4 turn.

Impeller Clearance Adjustment and Impeller Removal/Installation

If a loss of high speed performance and/or a higher than normal full throttle engine speed is evident, check the clearance between the edge of the impeller and the water intake casing liner. Also, check the leading edge(s) of the impeller for wear or damage. If it is worn or damaged, refer to *Worn impeller* in this chapter.

NOTE
Impeller wear can occur quickly when operated in water with excessive silt, sand or gravel.

NOTE
If the impeller is stuck to the drive shaft, use a suitable block of wood and hammer to rotate the impeller in the opposite direction of normal rotation. Rotate the impeller just enough to free the drive key and allow impeller removal.

NOTE
Lubricate the impeller shaft, impeller sleeve and drive key with Quicksilver 2-4-C Multi-Lube (part No. 92-825407) or Quicksilver Special Lube 101 (part No. 92-13872A1) prior to reassembly.

1. Disconnect the spark plug leads to prevent accidental starting.
2. Use a feeler gauge set to determine the clearance between the impeller blades and the intake liner. See **Figure 9**.
3. The impeller-to-liner clearance should be approximately 0.030 in. (0.8 mm).
4. If the clearance is not as specified, remove the six water intake housing mounting screws. Remove the intake housing. See **Figure 10**, typical.
5. Bend the tabs on the tab washer retaining the impeller nut to allow a suitable tool to be installed on the impeller nut. Remove the nut, tab washer, lower shims, impeller, drive key, plastic sleeve and upper shims. Note the number of lower and upper shims. See **Figure 11**.
6. If clearance is excessive, remove lower shims as necessary from below the impeller and position them above the impeller.
7. Install the impeller with the selected number of shims. Use grease to hold the upper shims to the drive shaft. Then position the plastic sleeve in the impeller and install the impeller, drive key and lower shims.
8. Install a new tab washer and impeller retaining nut on the drive shaft. Tighten the nut securely. Do not bend the tabs on the tab washer at this time.
9. Apply Quicksilver Perfect Seal (part No. 92-34227-1) to the threads of the intake housing retaining screws. Install the housing and screws. Tighten the screws finger-tight.

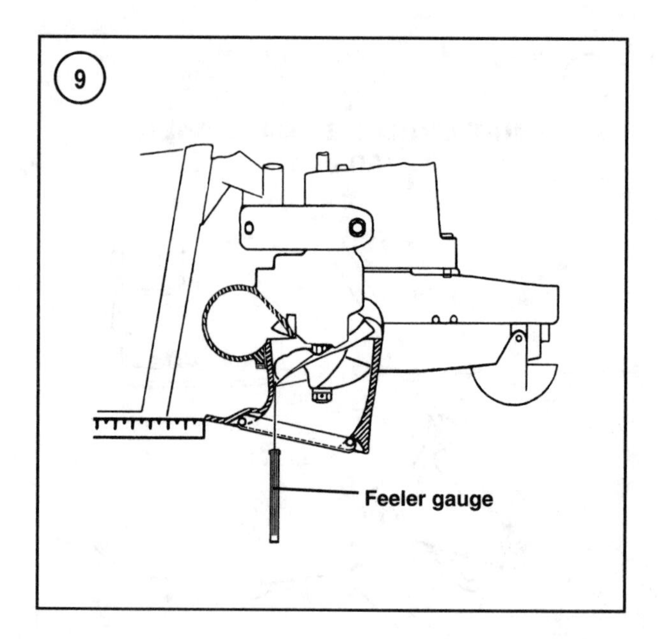

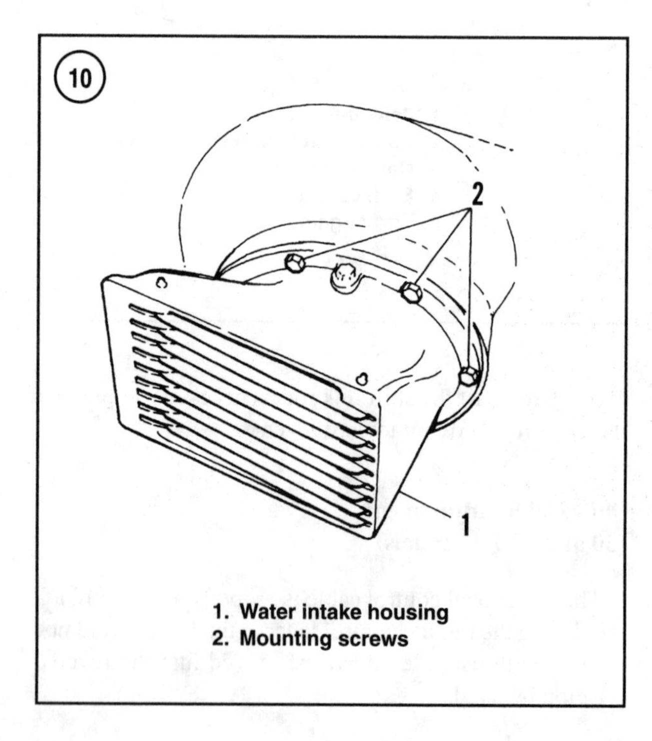

1. Water intake housing
2. Mounting screws

10. Rotate the impeller to check for rubbing or binding. Move the intake housing until it is centered over the impeller.
11. Repeat Steps 2 and 3 to recheck impeller clearance. Readjust clearance as necessary.
12. After the clearance is correct, remove the intake housing screws and housing.
13. Securely tighten the impeller nut. Bend the tabs of the tab washer against the flat surfaces of the impeller nut.

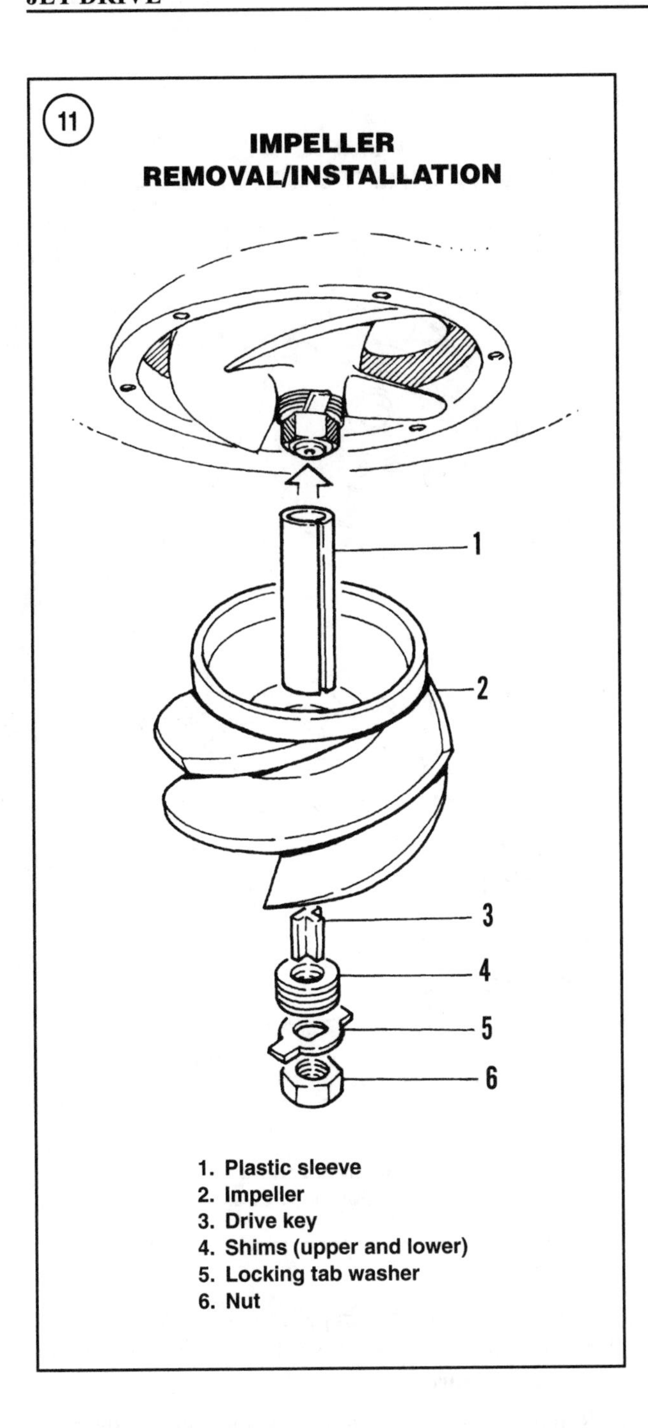

1. Plastic sleeve
2. Impeller
3. Drive key
4. Shims (upper and lower)
5. Locking tab washer
6. Nut

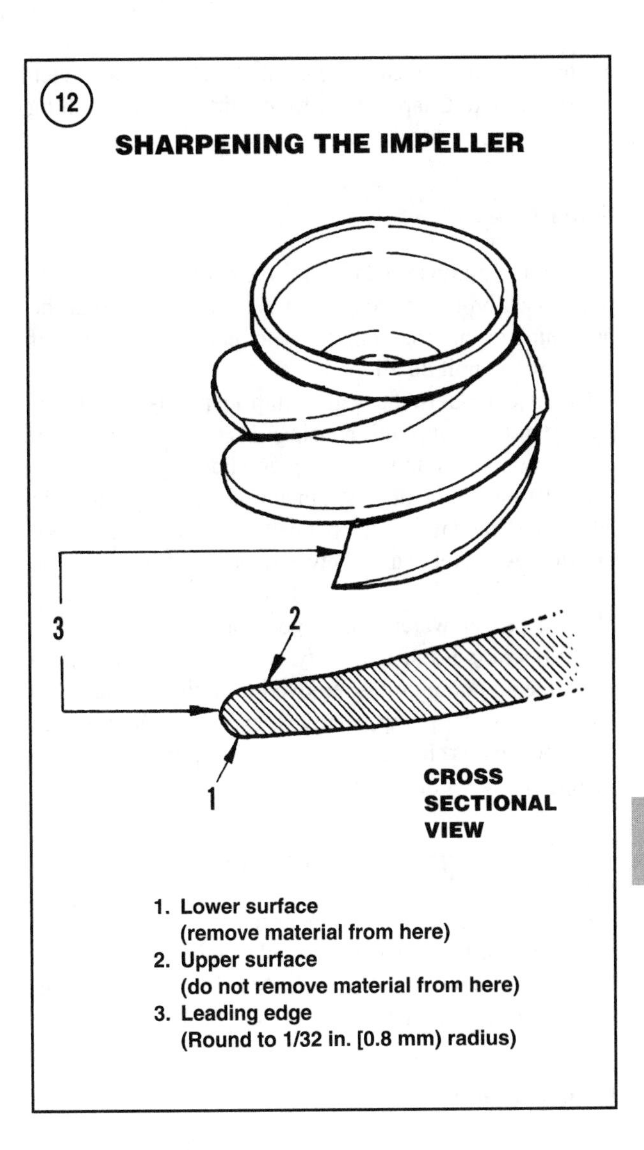

1. Lower surface (remove material from here)
2. Upper surface (do not remove material from here)
3. Leading edge (Round to 1/32 in. [0.8 mm) radius)

14. Reinstall the intake housing. Make sure the housing is centered on the impeller. Tighten the housing screws in a crossing pattern to the specification in **Table 1**.

Worn Impeller

The leading edge(s) of the impeller can become worn due to ingestion of gravel, silt and other debris. If a noticeable performance loss, increased wide-open throttle speed or difficulty in getting the boat on plane is noted, check the leading edge(s) of the impeller for wear or damage.

1. If the leading edge(s) is damaged, remove the impeller as described in the previous section.
2. Sharpen the impeller by removing material with a flat file from the lower surface of the leading edge(s) as shown in **Figure 12**. Do not remove material from the upper surface or alter the top side lifting angle of the impeller.
3. File or sand a 1/32 in. (0.8 mm) radius on the leading edge(s) as shown in **Figure 12**.
4. Reinstall the impeller and recheck the impeller clearance as described in the previous section.

Cooling System Flushing

The cooling system can become plugged by sand and salt deposits if it is not flushed occasionally. Clean the cooling system after each use in salt, brackish or silt-laden water. Refer to Chapter Four for cooling system flushing procedures.

Water Pump

All water pumps used on jet drive models are the same pump used on the related power head as described at the beginning of this chapter. The water pump is at the top of the jet pump unit on all models.

On 20 jet models, the water pump adapter is bolted separately to the pump unit housing. There is no gasket between the adapter and the pump unit housing.

On 30 and 45 jet models, a metal support ring is used to adapt the standard water pump base to the pump unit housing. A gasket is used between the pump base and the pump unit housing.

Since proper water pump operation is critical to outboard operation and durability, service the water pump when the jet pump unit is removed from the outboard. To service the water pump, remove the jet drive assembly as described in this chapter and refer to the appropriate water pump service section in this chapter.

JET PUMP UNIT SERVICE

Discard jet drive mounting fasteners if they are corroded and install new ones. Apply Quicksilver Perfect Seal (part No. 92-3427-1) to the threads of the mounting screws during installation.

Pump Unit Removal

NOTE
Note the number and location of impeller adjustment shims for reference during reassembly.

1. Disconnect and ground the spark plug leads to the power head to prevent accidental starting.
2. Tilt the outboard to the full up position and engage the tilt lock lever. Securely block the drive shaft housing or support the drive shaft housing with a suitable hoist.
3A. *20 jet models*—Remove the cotter pin and washer securing the shift link rod (2, **Figure 7**) to the shift cam (4). Disconnect the tension spring. Disconnect the shift link rod from the shift cam.
3B. *30 and 45 jet models*—Remove the shift cable adjustment barrel (1, **Figure 8**) from the anchor bracket and the casing guide from the shift cam stud.
4. Remove the water intake housing mounting screws. Remove the intake housing. See **Figure 10**, typical.

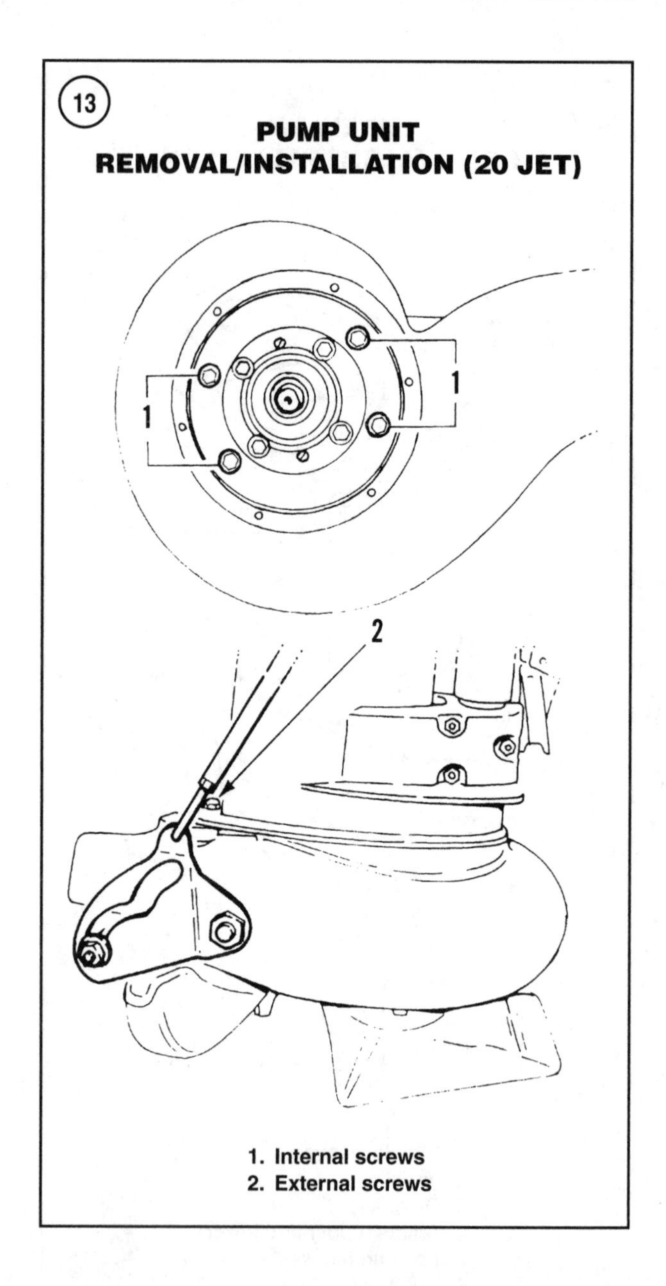

1. Internal screws
2. External screws

5. Bend the tabs on the impeller nut tab washer away from the impeller nut. Remove the impeller nut and tab washer. See **Figure 11**.

6. Remove the shims located below the impeller and note the number of shims. Remove the impeller and the shims located above the impeller and note the number of shims.

7. Slide the impeller sleeve and drive key off the drive shaft.

8A. *20 jet models*—Remove the screws (1, **Figure 13**) from the bottom of the bearing carrier in the impeller cavity. Then remove the screw (2, **Figure 13**) from the rear of the drive shaft housing. Support the pump unit as the last screw is removed.

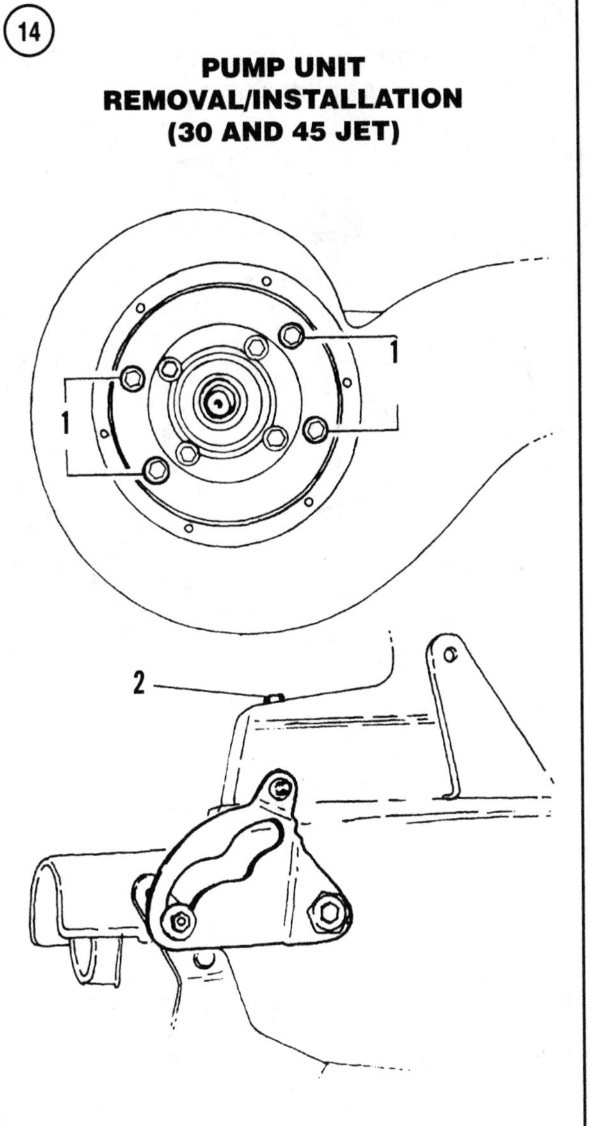

(14)

PUMP UNIT REMOVAL/INSTALLATION (30 AND 45 JET)

1. Internal screws
2. External screws

8B. *30 and 45 jet models*—Remove the screws (1, **Figure 14**) in the impeller cavity. Then remove the screw (2, **Figure 14**) from the rear of the drive shaft housing. Support the pump unit as the last screw is removed.

9. Remove the jet pump unit by pulling it straight down and away from the drive shaft housing until the drive shaft is free from the housing. Place the pump unit on a clean workbench.

10. Locate and secure the fore and aft dowel pins that align the jet pump unit to the drive shaft housing.

Pump Unit Installation

CAUTION

Do not apply lubricant to the top of the drive shaft during installation. Excess lubricant between the top of the drive shaft and the crankshaft can create a hydraulic lock, preventing the drive shaft from fully engaging the crankshaft.

NOTE

Install the original number of upper and lower impeller shims noted during disassembly if the original impeller and intake liner are used. If a new impeller or liner is installed, start with no upper shims and carefully add shims until the clearance is correct.

1. To install the pump unit, make sure the water tube guide and seal are securely attached to the water pump housing. If the guide and seal are loose, glue them to the water pump housing with Loctite 405 adhesive.

2. Clean the drive shaft splines as necessary, then coat the splines with Quicksilver 2-4-C Multi-Lube grease (part No. 92-825407). Coat the inner diameter of the water tube seal in the water pump housing with the same grease.

3. Make sure the fore and aft dowel pins are installed in either the drive shaft housing or the jet pump unit housing.

4. Position the pump unit under the drive shaft housing. Align the water tube in the water pump and the drive shaft with the crankshaft splines.

5. Push the pump unit toward the drive shaft housing. Rotate the flywheel clockwise as required to align the drive shaft and crankshaft splines.

6. Make sure the water tube is seated in the water pump seal, then push the gearcase against the drive shaft housing.

7. Coat the threads of the mounting screws with Loctite 271 threadlocking adhesive (part No. 92-32609-1).

8. Secure the pump unit to the drive shaft housing. See **Figure 13** for 20 jet models or **Figure 14** for 30 and 45 jet models. Tighten the screws securely.

9. Install the impeller and water intake housing and check the impeller clearance as described in *Impeller Clearance Adjustment and Impeller Removal/Installation.*

10. Connect and adjust the shift link rod on 20 jet models or remote control shift cable on 30-45 jet models as described previously in this section.

11. Reconnect the spark plug leads.

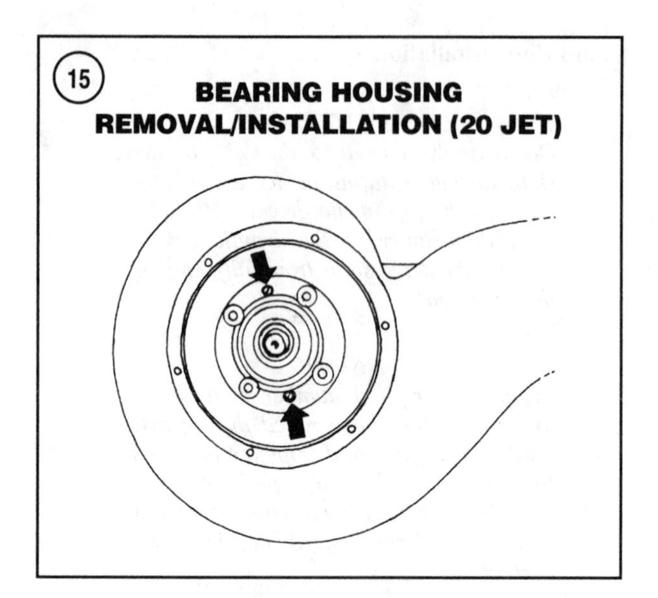

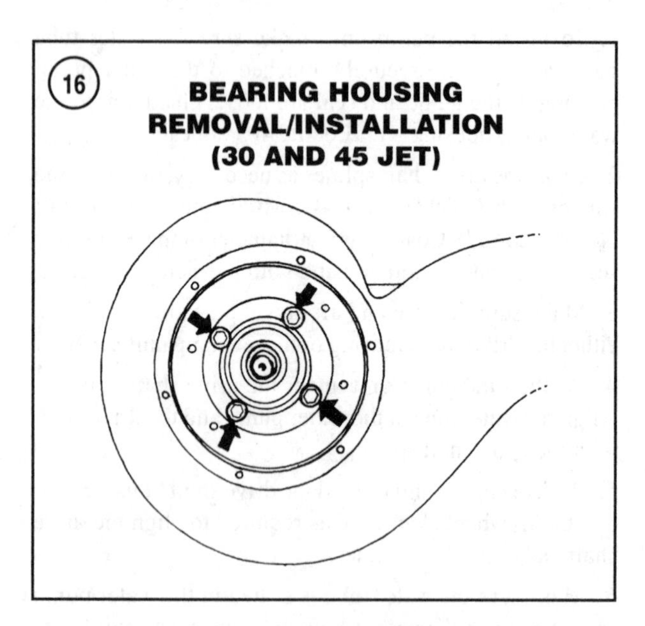

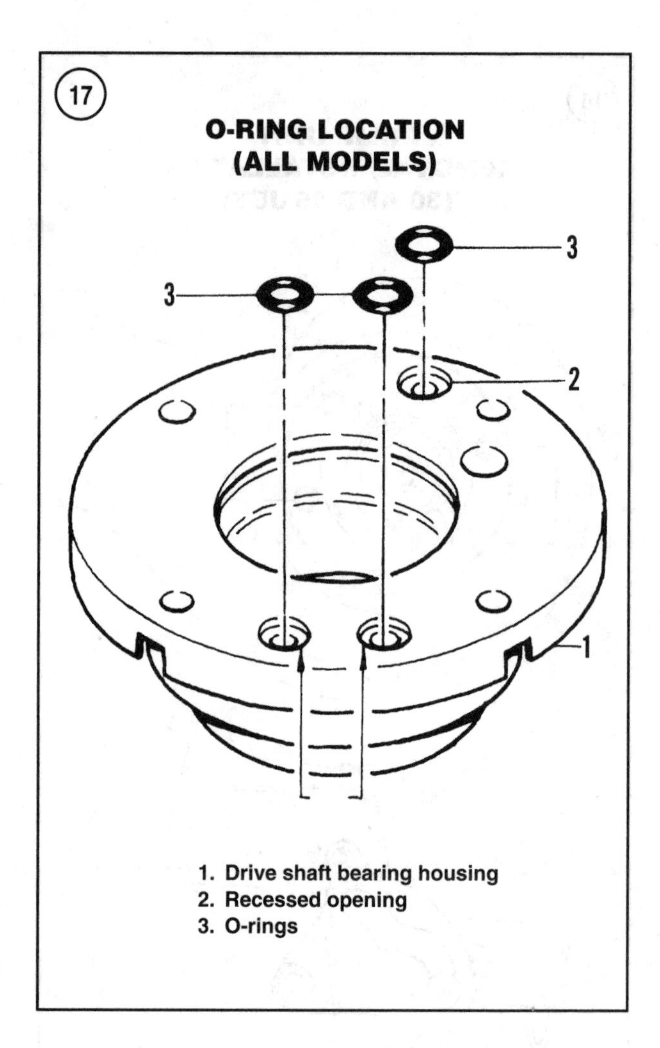

1. Drive shaft bearing housing
2. Recessed opening
3. O-rings

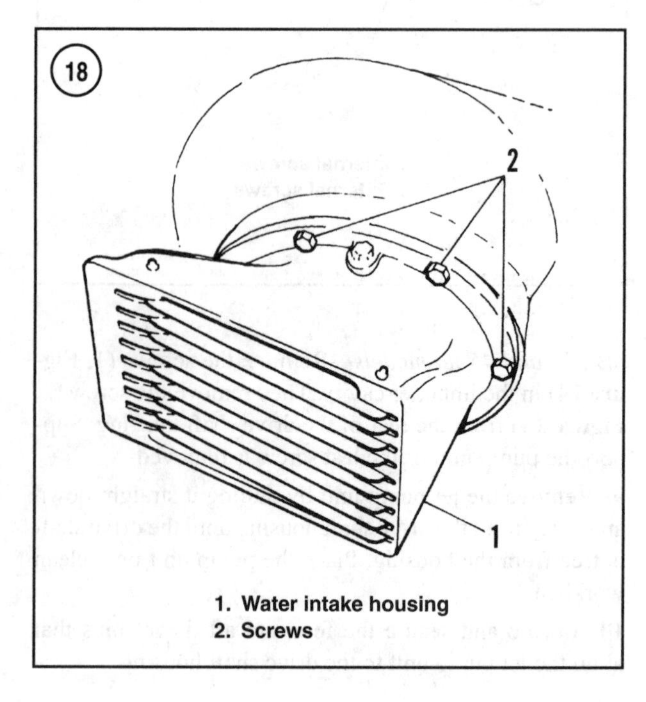

1. Water intake housing
2. Screws

Bearing Housing

Quicksilver parts and accessories only offer the bearing housing and drive shaft as an assembly. Therefore, only removal and installation of the bearing housing and drive shaft assembly is covered.

Removal

1. Remove the pump unit as described previously in this section.
2. Remove the water pump assembly. Refer to the appropriate water pump servicing section at the beginning of this chapter.

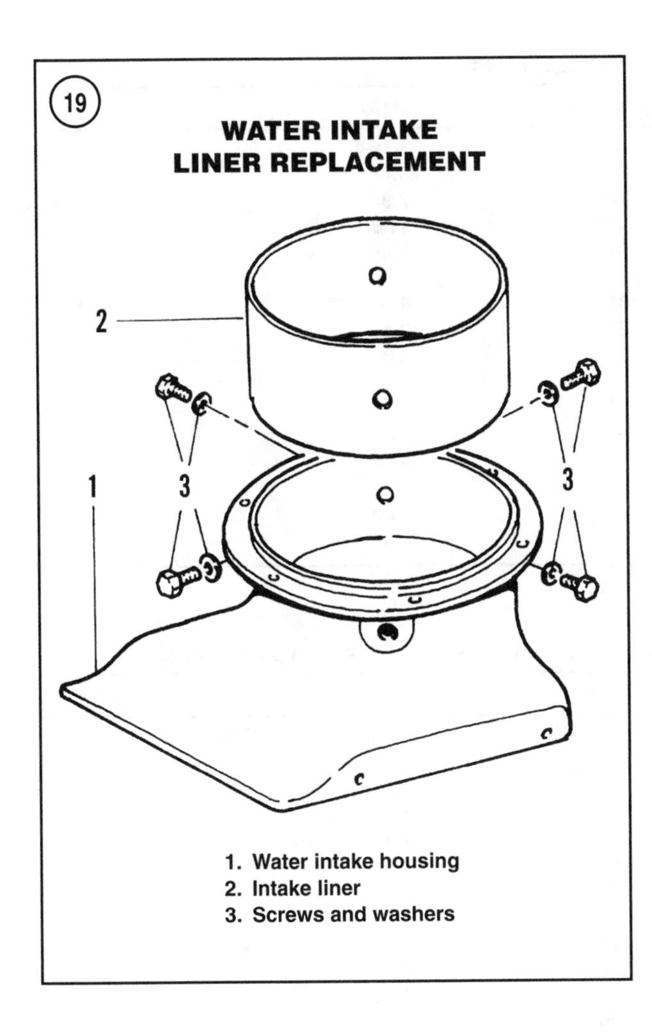

1. Water intake housing
2. Intake liner
3. Screws and washers

3. On 20 jet models, remove the screws securing the water pump base adapter to the top of the pump unit housing. Then remove the adapter.
4. On 20 jet models, remove the two screws securing the bearing housing and drive shaft assembly to the pump unit housing. See **Figure 15**. Withdraw the bearing housing and drive shaft assembly from the pump unit housing and place it on a clean work bench.
5. On 30 and 45 jet models, remove the screws securing the bearing housing and drive shaft assembly to the pump unit housing. See **Figure 16**. Withdraw the bearing housing and drive shaft assembly for the pump unit housing and place it on a clean work bench.
6. Locate and secure the three O-rings from the bearing housing-to-drive shaft housing mating surface. See **Figure 17**.

Installation

1. Lubricate the three O-rings with Quicksilver 2-4-C Multi-Lube grease (part No. 92-825407). Position the O-rings into the recessed openings on the mating surface of the bearing housing as shown in **Figure 17**.
2. Install the housing and drive shaft assembly into the pump unit housing. Make sure the retainer screw holes are aligned and the O-rings are not displaced during installation.
3. Apply Loctite 271 threadlocking adhesive (part No. 92-32609-1) to the threads of the bearing housing screws.
4A. *20 jet models*—Install the two housing retaining screws (**Figure 15**) and tighten them to the specification in **Table 1**.
4B. *30 and 45 jet models*—Install the four housing retaining screws (**Figure 16**). Evenly tighten the screws to the specification in **Table 1**.
5. On 20 jet models, install the water pump base adapter to the top of the pump unit housing. Coat the mating surfaces with Quicksilver Perfect Seal (part No. 92-34227-1) and the screws with Loctite 271 threadlocking adhesive. Tighten the screws to the specification in **Table 1**.
6. Install the water pump assembly as described in the appropriate water pump servicing section in this chapter.
7. Install the jet pump unit as described in this section. Lubricate the jet pump unit bearings as described in this chapter.

Water Intake Housing Replacement

1. Remove the water intake housing mounting screws. See **Figure 18**. Then pull the water intake housing down and away from the pump unit housing.
2. Mark or tag the liner screws for reassembly in the same location, then remove the screws and washers. See **Figure 19**.
3. Tap the liner loose by inserting a long drift punch through the intake housing grate. Place the punch on the edge of the liner and tap it with a hammer.
4. Withdraw the liner from the liner housing.
5. Install the new liner into the intake housing. See **Figure 19**.
6. Align the liner screw holes with their respective intake housing holes. Gently tap the liner into place with a soft hammer if necessary.
7. Apply Quicksilver Perfect Seal (part No. 92-34227-1) to the threads of the liner retaining screws.
8. Install the liner retaining screws and washers. Evenly tighten the screws to the specification in **Table 1**.
9. Remove burrs from the inner diameter of the liner and grind the end of the screws as necessary to ensure a flush inner surface.
10. Install the intake housing and set the impeller clearance as described under *Impeller Clearance Adjustment and Impeller Removal/Installation*.

Table 1 TORQUE SPECIFICATIONS

Component	ft.-lb.	in.-lb.	N•m
Bearing housing			
20 jet	–	30	3.4
30 jet, 45 jet	–	70	7.9
Water intake liner	–	100	11.3
Intake housing	–	96	10.8
Water pump			
20 jet	–	35	4
30 jet	–	30	3.4
45 jet	–	60	6.8
Jet pump mounting			
30 jet			
Locknuts	60	–	81
Rear bolt	23	–	31.2
45 jet			
Internal screws	25	–	33.9
External screws	23	–	31.2

Table 2 STANDARD TORQUE SPECIFICATIONS—U.S. STANDARD AND METRIC FASTENERS

Screw or nut size	ft.-lb.	in.-lb.	N•m
U.S. standard fasteners			
6-32	–	9	1
8-32	–	20	2.3
10-24	–	30	3.4
10-32	–	35	4.0
12-24	–	45	5.1
1/4-20	6	72	8.1
1/4-28	7	84	9.5
5/16-18	13	156	17.6
5/16-24	14	168	19
3/8-16	23	270	31.2
3/8-24	25	300	33.9
7/16-14	36	–	48.8
7/16-20	40	–	54
1/2-13	50	–	67.8
1/2-20	60	–	81.3
Metric fasteners			
M5	–	36	4.1
M6	6	72	8.1
M8	13	156	17.6
M10	26	312	35.3
M12	35	–	47.5
M14	60	–	81.3

Chapter Eleven

Trim and Tilt Systems

On models without a power trim and tilt, the raising and lowering of the motor is a mechanical process. The reverse lock must be released and the motor lifted manually. To change the running position of the gearcase thrust line to the boat (trim angle), the trim pin must be moved to one of the different positions available in the clamp brackets. Different operating conditions and changes to the boat load require frequent changes to the trim pin position to maximize boat performance and efficiency.

Power trim and tilt was developed to provide an easy and convenient way to change the trim angle while under way and allow hands-free tilting of the motor for trailer loading or beaching.

The term *integral* refers to components located between the clamp brackets, while the term *external* refers to components located outside of the clamp brackets or inside of the boat.

Mercury Marine uses both manual (charged accumulator) tilt systems and power (electro-hydraulic) trim and tilt systems.

This section includes maintenance, component replacement and troubleshooting procedures for manual and power trim and tilt systems. **Table 1** and **Table 2** lists torque specifications. **Table 1** and **Table 2** are located at the end of this chapter

Manual Tilt System

Manual (charged accumulator) systems use a pressurized reservoir, called an accumulator, to assist the operator in tilting the engine. The operator opens and closes the accumulator's pressure to the system by operating a mechanical linkage. The engine can be manually positioned as desired with accumulator assist and secured, returning the linkage to the closed position.

Manual tilt systems consist of a tilt cylinder, valve body or manifold, control rod, manual release lever and a nitrogen-filled accumulator. **Figure 1** shows the non-serviceable sealed system used on 30-50 hp models. **Figure 2** shows the serviceable system used on 55 and 60 hp models. Both systems are entirely contained between the outboard clamp brackets.

When the manual release lever is opened, the pressurized, nitrogen-filled accumulator eases the effort required to tilt the outboard manually. The engine can be positioned for shallow water operation, but high throttle settings cause the unit to trim in against the trim rod. A sacrificial anode is mounted at the bottom of the clamp brackets to control corrosion.

Manual tilt systems incorporate several special hydraulic functions.

1. *Impact control*—This circuit is designed to absorb and dissipate the energy of an impact with an underwater object while in forward motion. It does *not* protect the unit from impact damage in reverse. The circuit allows the hydraulic system to act as a shock absorber. High-pressure springs and check balls in the tilt cylinder piston vent hydraulic fluid to the opposite side of the piston when the pressure caused by the impact reaches a predetermined value. When fluid is vented, the engine is allowed to tilt up as necessary, dissipating the energy of the impact.
2. *Memory piston*—The memory piston works in conjunction with the impact circuit to return the engine to the trim angle it was at before the impact occurred. The memory piston stays in place during an impact as the tilt cylinder's piston pulls away during impact. After the impact, propeller thrust pushes against the tilt cylinder piston. A valve in the tilt piston allows venting of the fluid between the tilt piston and memory piston. This allows the tilt piston to move down until it seats against the memory piston, restoring the original trim angle. The memory piston contains no valves.
3. *Reverse hold down*—Reverse lock is a function of multiple circuits in the system. It holds the gearcase in the water during reverse thrust. Reverse thrust occurs during deceleration and operation in reverse gear. If the gearcase is not held in the water during deceleration and reverse gear operation, the operator will not have control of the boat.

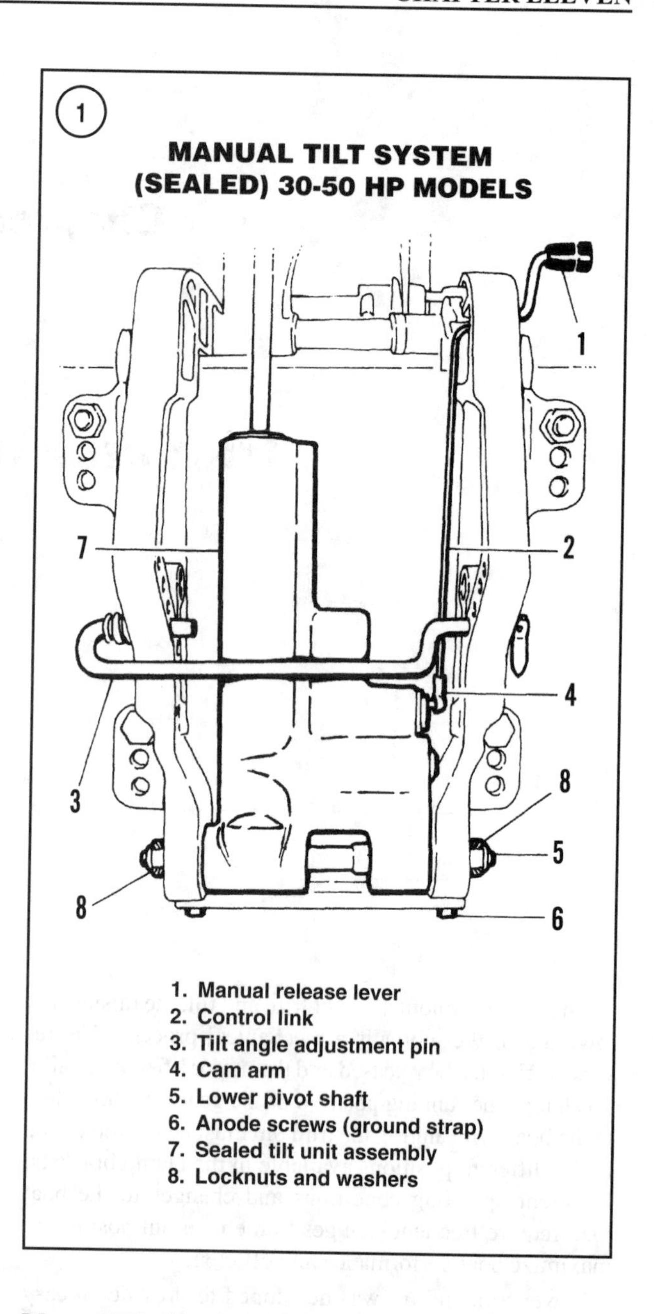

Manual Tilt System Troubleshooting

If the outboard will not stay in the tilted position, lowers during shallow water operation or trails out during reverse gear operation or during high speed deceleration, check for external fluid leaks and repair as necessary. Make sure the manual release lever and control rod open and close freely. Adjust the control rod link as necessary. If the manual release lever and control rod cannot fully close the cam lever on the valve body or manifold, the unit will leak down and not hold position.

To check for a discharged accumulator, place the manual release lever in the open position. Attach a suitable spring weight scale to the outboard as shown in **Figure 3**, then raise the outboard by pulling on the scale. If it requires more than 50 lb. (22.7 kg.) of pulling force to raise the outboard from the fully trimmed down position to the fully tilted up position, the accumulator is discharged or the system is malfunctioning.

Manual Tilt System Maintenance

Other than spraying the manual release linkage and the upper and lower pivot shafts with Quicksilver Corrosion Guard (part No. 92-815869), the system is maintenance-free.

Inspect the anode for loose mounting hardware, loose or damaged ground straps, if equipped, and excessive deterioration. Replace the anode if it is half of its original size. Make sure the anode has not been painted or coated with any substance. If paint or any other coating covers the anode, remove the anode, and strip the coating or replace the anode.

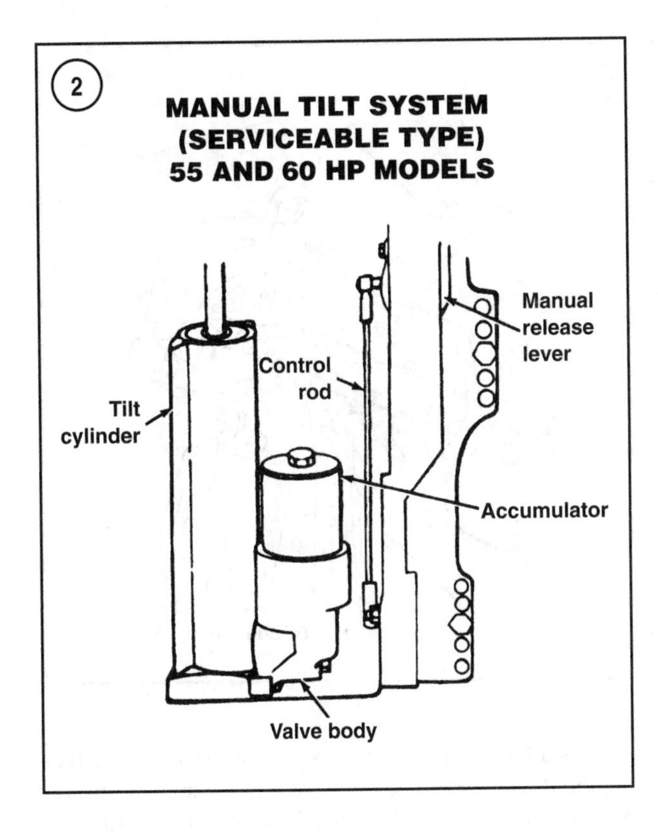

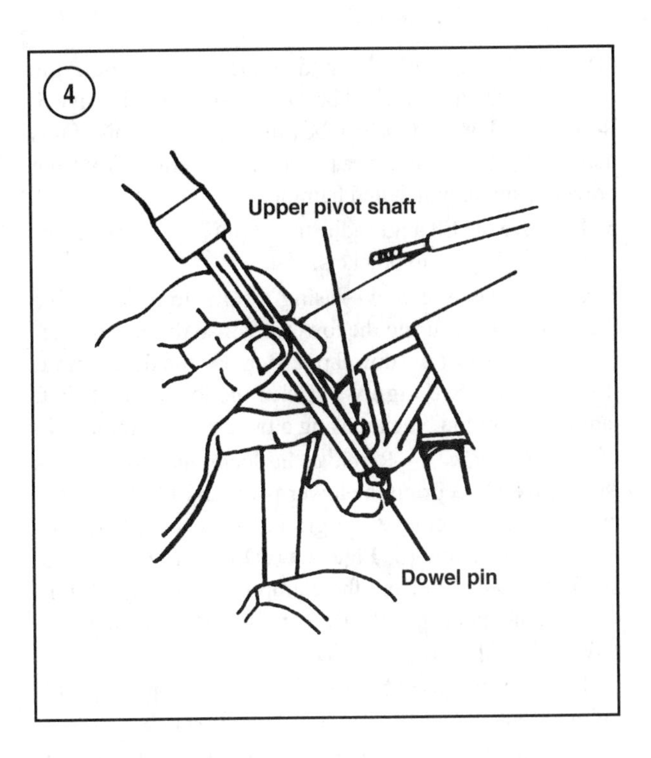

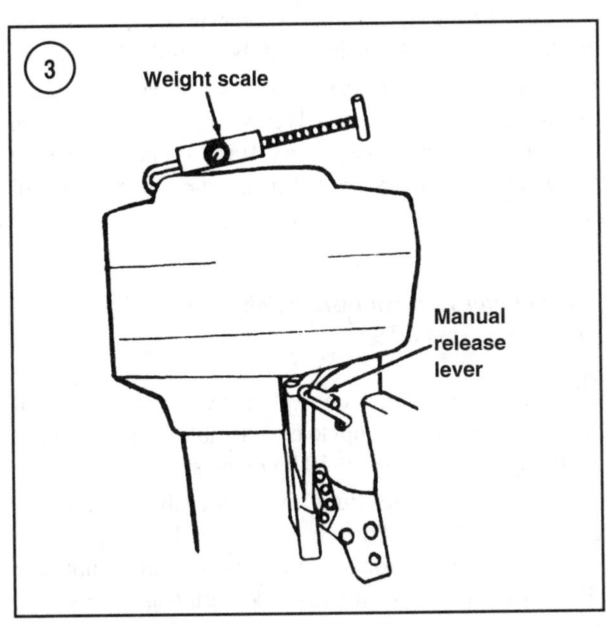

Manual Tilt System Service

WARNING
Make sure the tilt lock lever is properly engaged during all tilt system service. If the tilt lock lever is not engaged, a sudden loss of pressure in the shock cylinder can allow the outboard to fall to the full down position resulting in personal injury.

CAUTION
The system is under high pressure. The sealed system is not serviceable. Do not attempt to disassemble it. The serviceable system must have the accumulator removed as specified in this chapter to relieve internal pressure before further disassembly can be attempted.

System removal/installation

1. Disconnect and ground the spark plug leads to the power head to prevent accidental starting.

2. Raise the outboard to the fully tilted up position and engage the tilt lock lever on the starboard clamp bracket near the steering (tilt) tube. Consult the owner's manual for correct tilt lock function.

3. Disconnect the release valve control rod ball link (**Figure 1** or **Figure 2**) from the cam arm.

4A. *Serviceable models*—Use a suitable chisel or similar tool to drive the dowel pin securing the upper pivot shaft downward. Remove and inspect the dowel pin. Replace the dowel pin if it is damaged. See **Figure 4**. Drive the upper pivot shaft laterally from the bracket using a punch and hammer.

4B. *Sealed models*—Using diagonal cutters, pull the trilobe pin from the swivel bracket as shown in **Figure 5**. Remove and inspect the trilobe pin. Replace the pin if it is damaged. Drive the upper pivot shaft laterally from the bracket using a punch and hammer.

5. Remove the tilt angle adjustment pin (3, **Figure 1**, typical) from the clamp brackets.

6A. *Serviceable models*—Using a suitable punch, drive the dowel pin securing the lower pivot shaft upward. Remove and inspect the dowel pin. Replace the dowel pin if it is damaged. See **Figure 6**. Drive the lower pivot shaft laterally from the bracket using a punch and hammer.

6B. *Sealed models*—Remove the locknuts and washers (8, **Figure 1**) securing the lower pivot shaft to the clamp brackets. Then remove the ground strap from the starboard anode screw (6, **Figure 1**). Drive the lower pivot shaft (5, **Figure 1**) from the clamp brackets and tilt unit with a soft metal punch. Be careful not to damage the threaded ends of the pivot shaft.

7. Remove the assembly by pivoting the top of the tilt cylinder out and away from the clamp brackets.

8. On sealed models, remove the two flanged bushings from the tilt unit. Then remove the two bushings from the clamp bracket bores. Clean and inspect the bushings. Replace any bushings that are damaged or worn.

9. To install the manual tilt assembly, liberally apply Quicksilver 2-4-C Multi-Lube grease (part No. 92-825407) to the upper and lower pivot shafts, pivot shaft bores, dowel pin bores and dowel pins or trilobe pin.

10A. *Serviceable models*—Start the lower pivot shaft with the grooved end facing the dowel pin bore into its bore from the port side clamp bracket, then start the lower dowel pin from the top of its bore. Do not drive either pin far enough to interfere with tilt unit installation.

10B. *Sealed models*—Lubricate the four lower pivot shaft bushings with Quicksilver 2-4-C Multi-Lube grease. Make sure the grooved end of each pivot shaft is positioned on the same side of the engine as the retaining pin bore.

11. Install a straight bushing in each clamp bracket and a flanged bushing in each side of the tilt unit.

12. Install the tilt unit. Insert the bottom into the clamp brackets first, then rotate the tilt cylinder toward the transom and into position.

13A. *Serviceable models*—Drive the lower pivot shaft flush with the lower mounting bracket outer surfaces. Then, using a suitable punch, drive the dowel pin in from the top until it is fully seated.

13B. *Sealed models*—Install the lower pivot shaft through the clamp brackets and tilt unit. Install the washer and locknuts. Tighten both locknuts securely.

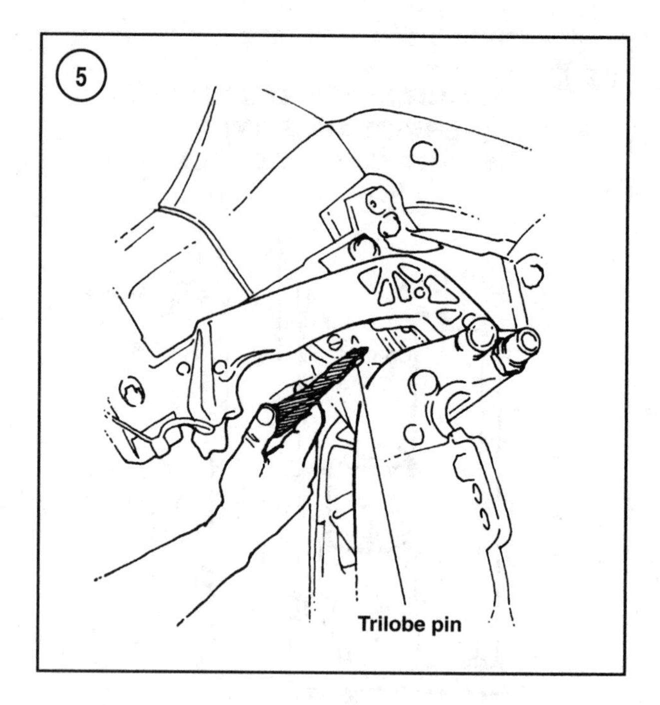

14. Install the upper pivot shaft with the grooved end facing the dowel or trilobe pin bore into the swivel bracket bore and through the tilt cylinder rod eye. The shaft must be flush with the swivel bracket outer surfaces. Then drive the dowel pin or trilobe pin into its bore until fully seated.

15. Connect the manual release link rod ball link to the cam arm. Operate the manual release lever and make sure the cam opens and closes freely. If it does not, adjust the rod ball link as necessary to change the link rod overall length.

Accumulator removal/installation (serviceable system)

Refer to **Figure 7** for an exploded view of the manual tilt system. Do not attempt to disassemble the sealed manual tilt system used on 30-50 hp models.

The recommended fluid for serviceable systems is Quicksilver Power Trim and Steering Fluid (part No. 92-90100A12) or automatic transmission fluid. Lubricate all seals and O-rings with this fluid during assembly. Apply Loctite 271 threadlocking adhesive (part No. 92-809819) to all threaded fasteners. If the cam arm shaft is removed, lubricate the shaft and O-rings with Quicksilver Special lube 101 (part No. 92-13872A1).

1. Remove the manual tilt system from the engine as described previously in this section.

2. Secure the tilt system in a vice with protective jaws or clamp the system between two suitable blocks of wood.

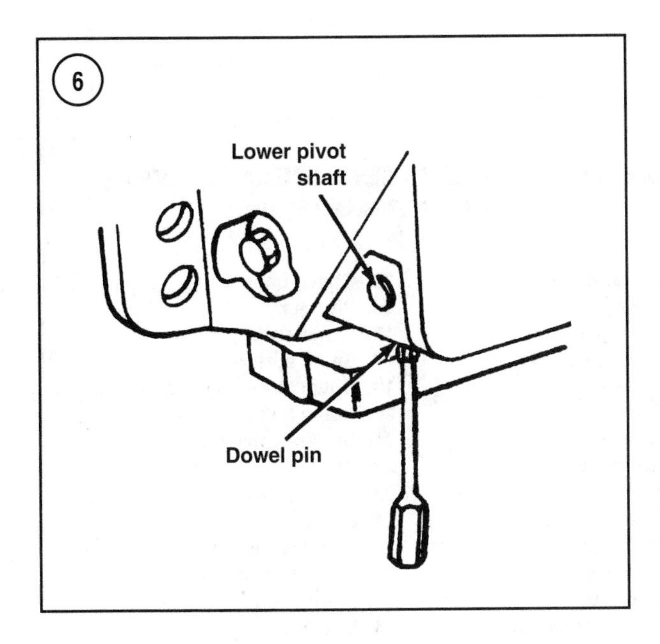

3. Open the cam shaft (manual release valve). Loosen the surge valve until it just begins to drip, then wait until the valve quits dripping.
4. Remove the accumulator with a breaker bar and the appropriate socket. Reach inside the valve body threaded bore and remove the accumulator O-ring.
5. At this point, internal pressure has been relieved and the unit can be fully disassembled.
6. To install the accumulator, install a *new* accumulator O-ring into the manifold threaded bore.
7. Tighten the surge valve plug to the specification in **Table 1**.
8. Fill the accumulator bore with Quicksilver Power Trim and Steering Fluid or automatic transmission fluid. Open and close the cam lever several times until all air bubbles are gone. Top off the bore with fluid as necessary.
9. Grease the accumulator threads with Quicksilver 2-4-C Multi-Lube grease (part No. 92-825407) and just start the accumulator into the manifold threads. Make sure the cam lever is in the down position. Then tighten the accumulator to the specification in **Table 1**.
10. Install the manual tilt system assembly as described previously in this section.

POWER TRIM AND TILT SYSTEMS

The typical system consists of the following:
1. A reversible electric motor controlled from the remote control or dash.
2. A hydraulic pump and fluid reservoir assembly.
3. A single hydraulic trim and tilt cylinder.
4. Electrical wiring, a fuse and two relays.
5. A sacrificial anode mounted at the bottom of the clamp brackets to control corrosion.

Power trim and tilt is a factory installed option on 30-60 hp models. Kits are available to add power trim and tilt to all electric start models.

Power trim and tilt systems incorporate several special hydraulic functions:
1. *Impact*—This circuit is designed to absorb and dissipate the energy of an impact with an underwater object while in forward motion. It does *not* protect the unit from impact damage while in reverse. The circuit allows the hydraulic system to act as a shock absorber. High-pressure springs and check balls in the tilt cylinder piston vent hydraulic fluid to the opposite side of the piston when the pressure caused by the impact reaches a predetermined value. When fluid is vented, the engine is allowed to tilt up as necessary, dissipating the energy of the impact.
2. *Memory piston*—The memory piston works in conjunction with the impact circuit to return the engine to the trim angle it was at before the impact occurred. The memory piston stays in place during an impact as the tilt cylinder's piston pulls away during impact. After the impact, propeller thrust pushes against the tilt cylinder piston. A valve in the tilt piston allows venting of the fluid between the tilt piston and memory piston. This allows the tilt piston to move down until it seats against the memory piston, restoring the original trim angle. Memory pistons contain no valves.
3. *Reverse lock*—Reverse lock is a function of multiple circuits in the system. It holds the gearcase in the water during reverse thrust. Reverse thrust occurs during deceleration and operation in reverse gear. If the gearcase is not held in the water during deceleration and reverse gear operation, the operator will not have control of the boat.
4. *Manual release*—The manual release valve allows the operator to raise or lower the engine if the electric motor or hydraulic system does not function. The valve can be opened and the engine positioned as desired. After positioning the engine, the manual release valve must be closed in order for the engine to hold position and for the impact and reverse lock circuits to function. Manual release valves must never be totally unscrewed except during disassembly.
5. *Hydraulic trim limit*—All trim and tilt systems use hydraulic valving to limit the maximum amount of positive trim the unit can achieve while under way. The trim range is limited to approximately 20° positive trim. The engine can be tilted higher than this when operating below planing speeds or when trailering the boat. If the unit is tilted above 20° positive trim and the operator attempts to plane out or accelerate the boat, propeller thrust overcomes the tilt relief valve(s) and the unit trims down to the maximum

(7)

MANUAL TILT SYSTEM (SERVICEABLE TYPE)
55 AND 60 HP MODELS

1. Accumulator
2. O-ring
3. Spring, check ball and push rod
4. Valve body
5. Cam arm and shaft
6. O-rings
7. Shaft seal
8. Screw
9. Retainer plate
10. Insulator
11. Springs
12. Surge valve
13. Surge valve spool
14. Push rods
15. Check balls
16. Plunger
17. Large spring
18. Small spring
19. Down slow valve cap
20. Down fast valve cap
21. Up fast valve cap
22. Alignment dowels
23. Tilt cylinder
24. Screw
25. Memory piston
26. Tilt rod assembly
27. Pivot shaft
28. Dowel pin

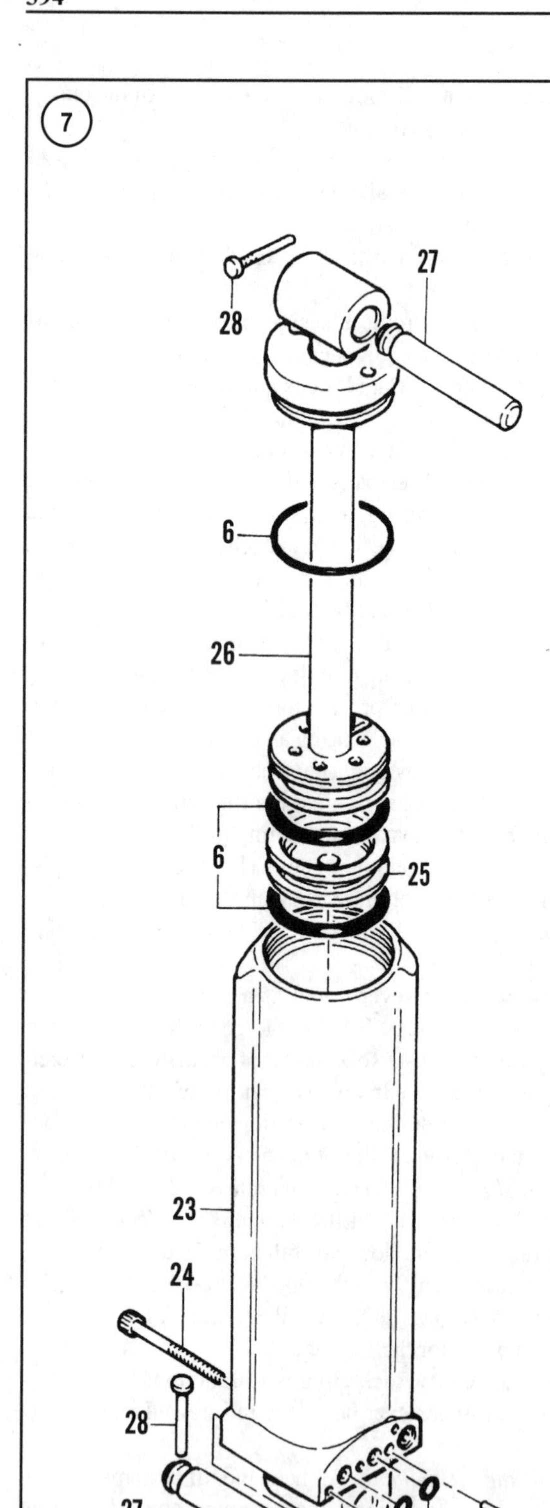

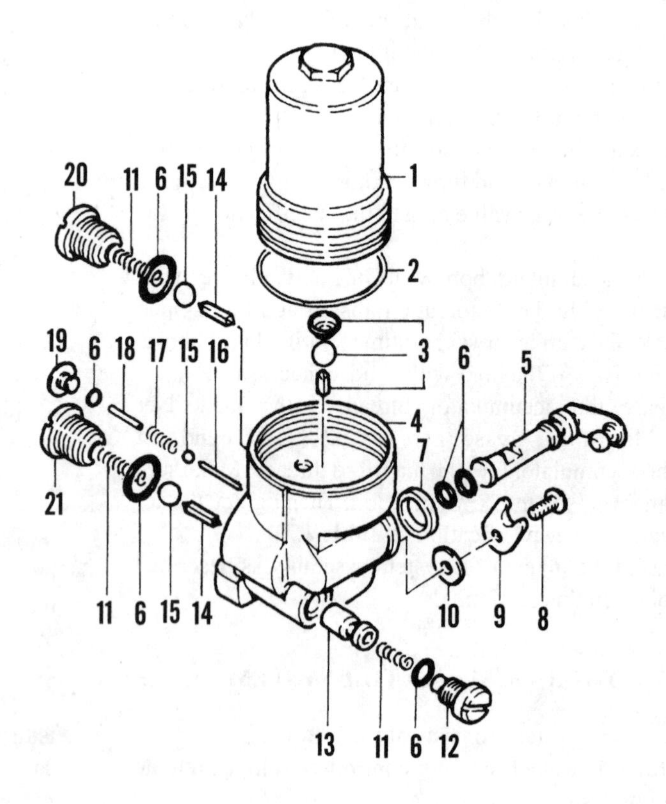

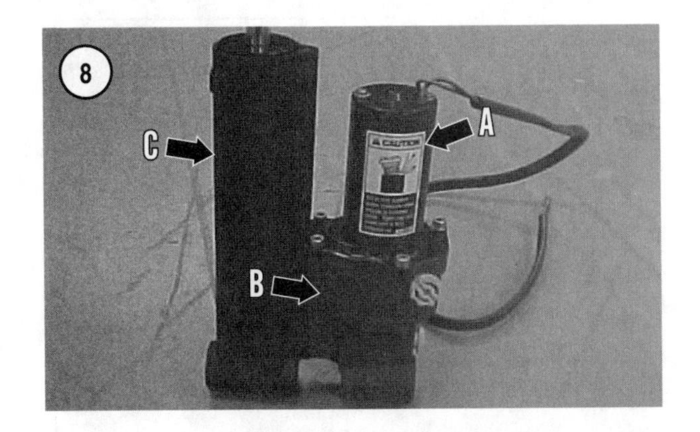

trim out position. If the operator tries to exceed the maximum trim out limit while under way, the electric motor and pump run, but the unit can trim no higher. This occurs due to internal valving that bypasses the pump's output to the hydraulic cylinder.

Power Trim and Tilt System Identification

Power trim/tilt systems can be divided into the following basic designs.

30-40 hp (two-cylinder) and 40-50 hp (except 45 jet) models—These models use an integral single ram system. The complete system is mounted between the clamp brackets. A two-wire electric motor (A, **Figure 8**) controlled by relays drives the hydraulic pump (B). The hydraulic pump and electric motor mount to the starboard side of the trim ram (C, **Figure 8**). This system is identified by the black fill plug (**Figure 9**)

55 and 60 hp, and 45 jet (three-cylinder) models—These models use an integral single ram system. The complete system is mounted between the clamp brackets. A two-wire electric motor (A, **Figure 10**) drives the hydraulic pump. The hydraulic pump and electric motor mount to the starboard side of the trim ram (B, **Figure 10**). This trim system is identified by the yellow fill plug (C, **Figure 10**). An optional trim indicator gauge sending unit is available from Quicksilver parts and accessories.

Some models use trim stop bolts (D, **Figure 10**). They limit the trim-in position to reduce undesirable handling characteristics that occur on some boats when operated with the trim in the full down position. The bolts fit into the tilt pin opening in the clamp brackets. Adjust the trim in limit as described in this chapter.

Electric Motor Operation

Both types of trim/tilt system use a permanent magnet electric motor. There are no field windings and electricity flows only through the rotating armature. Strong permanent magnets are glued to the main housing. Never strike the trim motor with a hammer as this cracks the magnets and destroys the motor. The two motor leads are blue and green. When the blue wire is connected to positive and the green wire is grounded, the motor runs in the *up* direction. When the green wire is connected to positive and the blue wire is grounded, the motor runs in the *down* direction. Two relays take care of switching the polarity of the green and blue wires to change the motor direction.

Manual Release Valve Operation

All models incorporate a manual release valve. The valve may be accessed through an opening in the starboard clamp bracket (**Figure 11**, typical). The manual release valve allows the motor to be raised or lowered to any position if the electric motor has failed.

WARNING
Do not operate the engine with the manual release valve in the open position. Otherwise, the reverse lock protection is disabled and nothing prevents the engine from tilting

out of the water while backing up in reverse gear or during deceleration. This will cause loss of directional control. Retighten the manual release valve once the engine reaches the desired position.

1. To raise or lower the motor manually, open the manual release valve (**Figure 11**) 3-4 full turns. Do not open the valve more than recommended.
2. Position the outboard at the desired tilt or trim position.
3. Retighten the manual release valve.

Maintenance

Periodically check the wiring system for corrosion and loose or damaged connections. Tighten loose connections, replace damaged components and clean corroded terminals as necessary. Coat the terminals and connections with liquid neoprene (part No. 92-25711-2). Check the reservoir fluid level as outlined in the following procedures.

Inspect the anode for loose mounting hardware, loose or damaged ground straps, if equipped, and excessive deterioration. Replace the anode if it is reduced to half of its original size. Make sure the anode is not painted or coated with any substance. If paint or any other coating covers the anode, remove the anode and strip the coating or replace the anode.

Reservoir fluid check

Use Quicksilver Power Trim and Steering Fluid (part No. 92-90100A12) or Dexron II automatic transmission fluid to fill the trim/tilt system.

The fill plug (**Figure 9** or C, **Figure 10**) is located just below the electric motor on all models. The engine must be in the full *up* position to access the plug.

1. Trim the outboard to the full *up* position. Clean the area around the fill plug. Carefully remove the fill plug while holding a shop cloth over the plug to block oil spray.
2. The fluid level must be even with the bottom of the fill plug hole (**Figure 12**). If necessary, add Quicksilver Power Trim Fluid or automatic transmission fluid to bring the level up to the bottom of the oil level hole.
3. Install the fill plug. Cycle the outboard fully down and up several times to bleed air that might be in the system.
4. Recheck the fluid level as described in Steps 1 and 2. Make sure the fill plug is tightened securely when finished.

Bleeding the hydraulic system

All trim and tilt systems are considered self-bleeding. Simply cycle the unit fully up and down a total of 3-5 times to bleed all air from the system. Make sure the fluid level in the reservoir is maintained as described in this section. If the fluid appears foamy, allow the unit to sit for a minimum of 30 minutes to allow the air to separate from the fluid.

Trim-In Limit Adjustment

On most models, adjustable stop bolts are used to limit the total amount of negative trim an outboard can obtain. Two stop bolts (one on each clamp bracket) are used (D, **Figure 10**).

The proper bolt position limits the trim-in enough to prevent undesirable handling while allowing enough trim-in to prevent excessive bow rise during acceleration.

Install the bolts into openings further from the boat transom to decrease the amount of trim-in. Install them into openings closer to the boat transom to increase the amount of trim-in.

Test run the engine with a normal load after each adjustment. Several adjustments may be required to achieve the desired results. Securely tighten the bolts after each adjustment.

Hydraulic Troubleshooting

If a problem develops in the trim/tilt system, first determine whether the problem is in the electrical system or in the hydraulic system. If the electric motor runs normally, the problem is hydraulic in nature. If the electric motor does not run or runs slowly, go to *Electrical system testing* in this chapter. It is possible for an internal hydraulic component problem to cause the pump to turn slowly or lock up completely, but not likely.

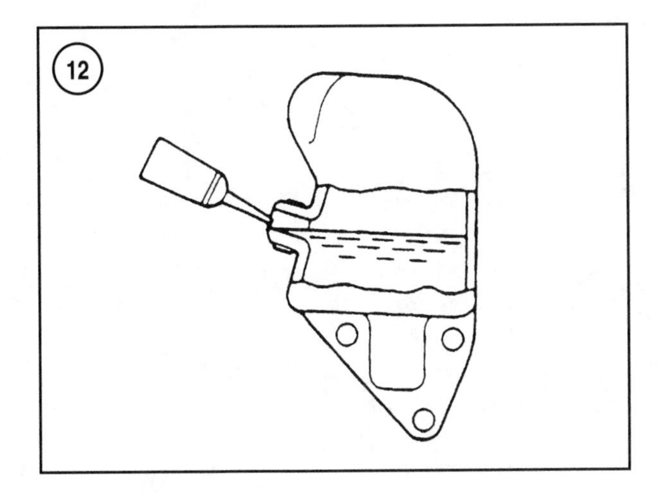

If the electric motor seems to run abnormally fast in both directions, remove the electric motor and check for a sheared pump drive shaft or coupler.

If the unit will only tilt or trim partially, or the system's movement is jerky or erratic, check the reservoir fluid level as described previously in this chapter.

If the unit will not trim out under load, but otherwise functions normally, the problem is most likely in the pump assembly. On some units, the pump can be replaced without replacing the valve body or manifold assembly. Consult a parts catalog or the appropriate illustrations in this chapter to determine the best course of action.

Lubricate all seals and O-rings with Quicksilver Power Trim and Steering Fluid (part No. 92-90100A12) or automatic transmission fluid during assembly. Apply Loctite 271 threadlocking adhesive (part No. 92-809819) to all threaded fasteners. Lubricate the manual release valve with Quicksilver Special Lube 101 (part No. 92-13872A1) if it was removed.

There is no specific troubleshooting procedure to isolate the trim/tilt cylinder from the pump and manifold. Parts are available to service most of the cylinder and pump/valve body components.

Hydraulic troubleshooting is divided into two failure modes: The unit has no reverse lock (kicks up in reverse or trails out on deceleration) or the unit leaks down (will not hold a trim position in forward gear).

NOTE

Always replace O-rings, seals and gaskets that are removed. Clean the outside of the trim system before disassembly. Always use a lint-free cloth when handling trim/tilt components. Dirt or lint can block passages causing valves to stick or prevent O-rings from sealing.

Unit has no reverse lock

Refer to **Figure 13** or **Figure 14**.

1. Remove, clean and inspect the manual release valve and O-ring(s). Replace the manual release valve if it or the O-ring(s) are damaged. Install the manual release valve and tighten the valve securely to prevent leaks. Retest the system for reverse lock function. Continue to the next step if reverse lock still does not function correctly.

2. Remove, clean and inspect the cylinder piston rod assembly for debris or damage to the impact relief (shock rod) valves. Replace the cylinder rod and piston assembly if there is any damage. Reassemble the unit with *new* O-rings and seals, and retest for reverse lock function. Continue to the next step if the reverse lock still does not function correctly.

3. Remove, clean and inspect the piston operating spool valve assemblies, if equipped. Replace the piston operating spool valves if any of the valve components are damaged. Reassemble the unit with *new* O-rings and retest for reverse lock function.

   a. *30-50 hp models (except 45 jet)*—Continue to the next step if the reverse lock still does not function correctly.

   b. *55 and 60 hp, and 45 jet models*—Replace the pump and manifold (valve body) assembly if the reverse lock still does not function correctly.

4. On 30-50 hp models (except 45 jet), remove, clean and inspect the suction seat assembly. Replace the suction seat assembly if any damage is noted. Reassemble the unit with *new* O-rings and retest the system for reverse lock function. If reverse lock still does not function properly, replace the trim/tilt unit as a complete assembly.

Leaks down

Refer to **Figure 13** or **Figure 14**.

1. Remove, clean and inspect the manual release valve and O-ring(s). Replace the manual release valve if it or the O-ring(s) are damaged. Install the manual release valve and tighten the valve securely to prevent leaks. Retest the system for leak-down. Continue to the next step if leak-down is still noted.

2. Remove, clean and inspect the tilt relief valve or pilot valve assembly components. Replace all of the tilt relief valve components if it or any of the O-rings are damaged. Reassemble the unit and retest for leak-down. Continue to the next step if leak-down is still noted.

3. Remove, clean and inspect the piston operating spool valve assembly, if equipped. Replace the piston operating spool valve if any of the components are damaged. Reas-

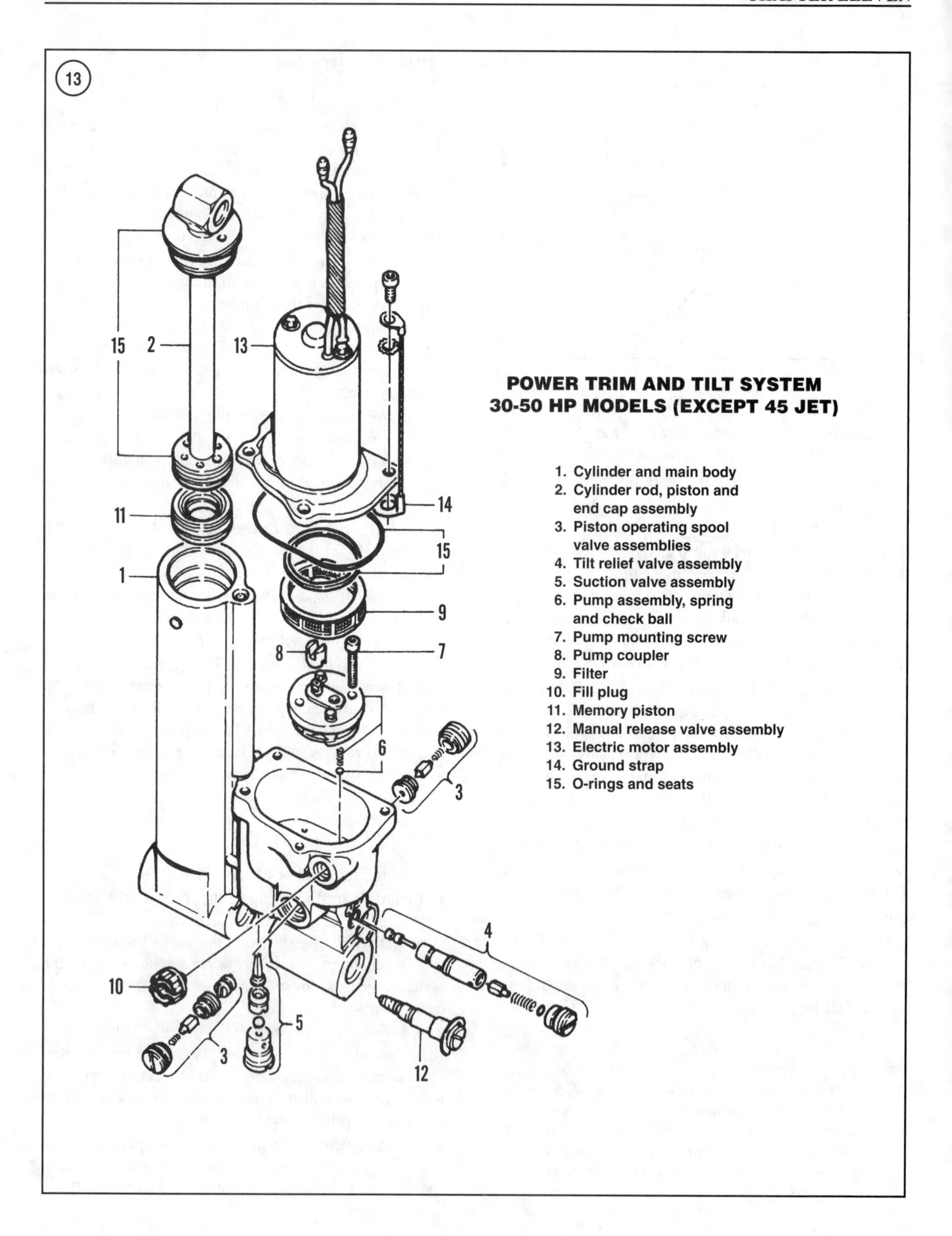
13
POWER TRIM AND TILT SYSTEM
30-50 HP MODELS (EXCEPT 45 JET)
1. Cylinder and main body
2. Cylinder rod, piston and end cap assembly
3. Piston operating spool valve assemblies
4. Tilt relief valve assembly
5. Suction valve assembly
6. Pump assembly, spring and check ball
7. Pump mounting screw
8. Pump coupler
9. Filter
10. Fill plug
11. Memory piston
12. Manual release valve assembly
13. Electric motor assembly
14. Ground strap
15. O-rings and seats
1
2
3
4
5
6
7
8
9
10
11
12
13
14
15

(14)

**POWER TRIM AND TILT SYSTEM
(55 HP, 60 HP AND 45 JET MODELS)**

1. Cylinder rod, piston and end cap assembly
2. Cylinder O-rings and seals
3. Memory piston
4. Trim/tilt cylinder
5. Screw
6. Dowel pin
7. Pump O-ring and seals
8. Valve body assembly
9. Pump assembly
10. Piston operating spool valve assemblies
11. Spring and check ball (seat not shown)
12. Manual release valve and lock clip
13. Pilot valve assemblies
14. Fill plug
15. Filter and seal
16. Screw
17. Drive shaft coupler
18. Screw and washer
19. Ground strap
20. Electric motor
21. Pivot shafts
22. Trilobe or dowel pin

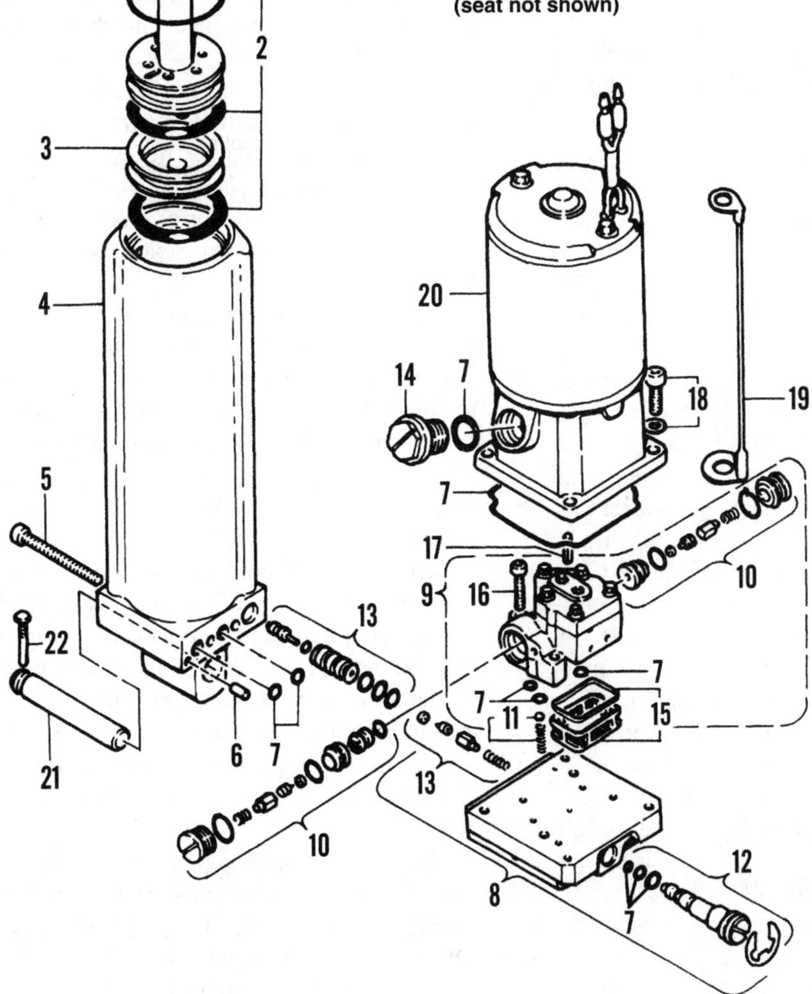

semble the unit and retest for leak-down. Continue to the next step if leak-down is still noted.

4. Remove, clean and inspect the memory piston, the cylinder rod piston and the impact relief valves. If the O-rings are undamaged and the cylinder wall is not scored, replace the memory piston and cylinder rod and piston assembly.

   a. *30-50 hp models (except 45 jet)*—If the cylinder wall is scored or damaged, replace the trim/tilt unit as an assembly.
   b. *55 and 60 hp, and 45 jet models*—If the cylinder wall is scored or damaged, replace the cylinder assembly.
   c. Reassemble the unit with *new* seals and O-rings. Retest the system for leak-down. If leak-down is still evident, continue at Step 5.

5A. *30-50 hp models (except 45 jet)*—If leak-down is still present, replace the trim system as a complete assembly.

5B. *55 and 60 hp, and 45 jet models*—If leak-down is still noted, replace the pump and valve body assembly.

Electrical System Troubleshooting

Green wires are primarily used for the down circuits. Blue wires are primarily used for the up circuits. The switching circuits for the trim/tilt system are normally protected by the main 20 amp fuse. Electrical testing is performed most accurately using a multimeter. However, a 12-volt test lamp and a self-powered continuity meter may be used if a multimeter is unavailable. Before beginning troubleshooting with a test lamp, connect the test lamp directly to the battery and observe the brightness of the bulb. Reference the rest of the readings against this test. If the bulb does not glow as brightly as when it was hooked directly to the battery, a problem is indicated. If a multimeter is used, take a battery voltage reading to reference all of the readings against. If the voltmeter reads 1 or more volts less than battery voltage, a problem is indicated. When checking continuity with an ohmmeter, a zero reading is good. The higher the ohmmeter reads above zero, the worse the condition is for that circuit.

Before attempting to troubleshoot any electrical circuit:

1. Make sure all connectors are properly engaged, and all terminals and leads are free of corrosion. Clean and tighten all connections as required.
2. Make sure the battery is fully charged. Charge or replace the battery as required.

Refer to **Figure 15** for a typical relay-controlled system wiring diagram. Refer to the *Wiring Diagrams* at the end of the manual for specific wiring diagrams for each model. Some trim/tilt system wiring diagrams are separate and others are integrated into the main wiring harness.

The two-wire motor is reversed by switching the polarity of the trim motor blue and green wires. There are two relays, one for each trim motor lead. Both relays hold their trim motor lead to ground when they are not activated.

When the up relay is activated, it takes the blue trim motor lead off of ground and connects it to positive. The down relay is inactive and holds the green lead to ground. Current can then flow from the positive terminal to the up relay to the trim motor and back to ground through the down relay causing the motor to run in the up direction.

When the down relay is activated, it takes the green trim motor lead off of ground and connects it to positive. The up relay is inactive and holds the blue lead to ground. Current can then flow from the positive terminal to the down relay to the trim motor and back to ground through the up relay causing the motor to run in the down direction. If the motor runs in one direction, but not the other, the problem cannot be the trim motor.

1. Connect the test lamp lead to the positive terminal of the battery and touch the test lamp probe to metal anywhere on the engine block. The test lamp should light. If the lamp does not light or is dim, the battery ground cable connections are loose or corroded, or there is an open circuit in the battery ground cable. Check connections on both ends of the ground cable.
2. Connect the test lamp lead to a good engine ground.
3. Connect the test lamp probe to the starter solenoid input terminal (A, **Figure 15**). The test lamp should light. If the lamp does not light or is very dim, the battery cable connections are loose or corroded, or there is an open in the cable between the battery and the solenoid. Clean and tighten connections or replace the battery cable as required.
4. Remove the 20 amp fuse and connect the test lamp probe to the input side of the 20 amp fuse (B, **Figure 15**). The test lamp should light. If it does not, repair or replace the wire between the starter solenoid and the fuse holder.
5. Reinstall the fuse and connect the test lamp probe to the output side of the fuse (C, **Figure 15**). The test lamp should light. If it does not, replace the fuse.
6. Disconnect the trim/tilt relays from their connector bodies.
7. Connect the test lamp probe to the input side of each relay (red lead terminal). The test lamp should light at each point. If it does not, repair or replace the wire from the starter solenoid to each relay.
8. Connect the test lamp lead to the positive terminal of the battery and touch the test lamp probe to each of the two black leads at each relay connector. The test lamp should light at each point of the four leads. If it does not,

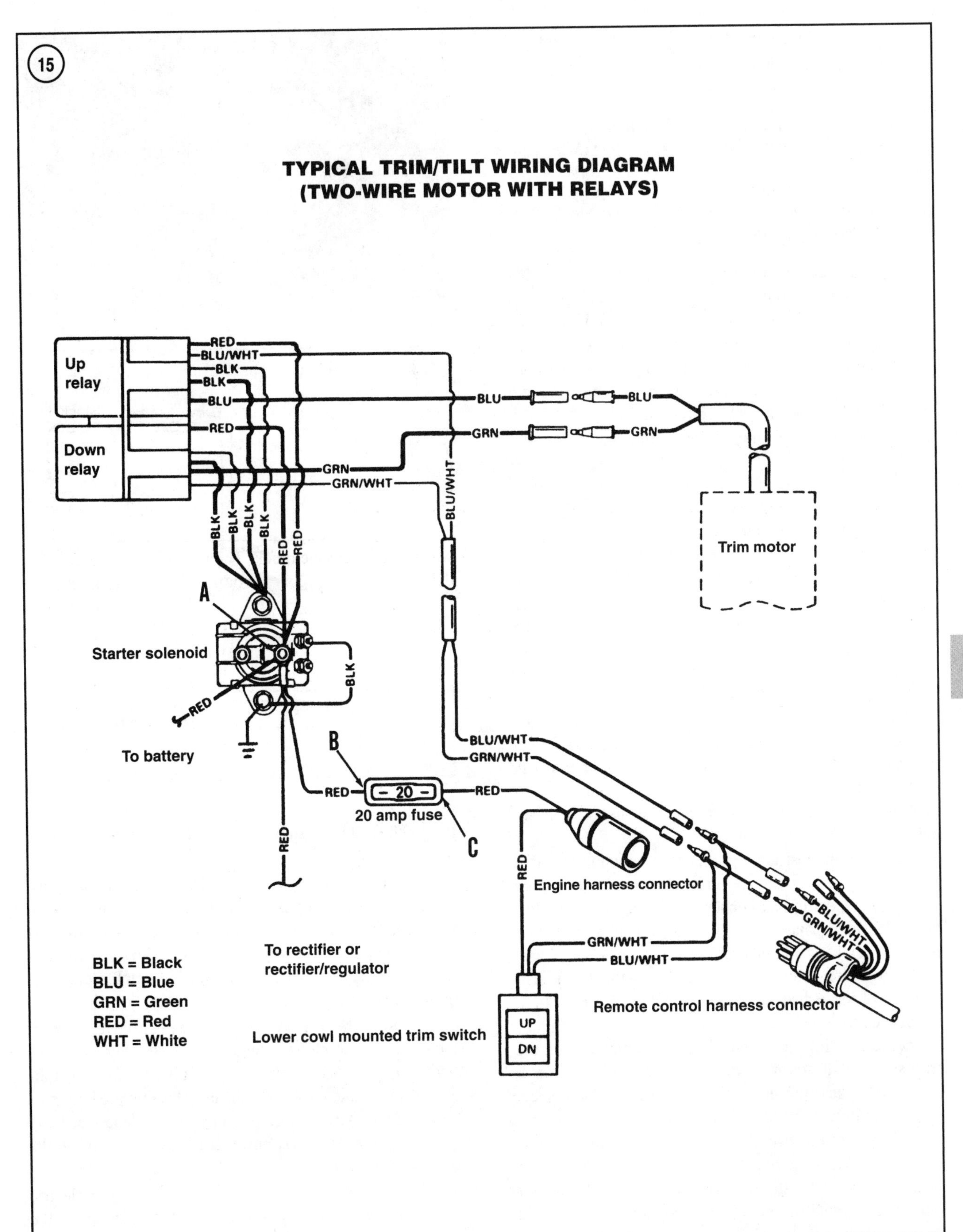
15
TYPICAL TRIM/TILT WIRING DIAGRAM
(TWO-WIRE MOTOR WITH RELAYS)
Up relay
Down relay
RED
BLU/WHT
BLK
BLK
BLU
RED
GRN
GRN/WHT
BLU
GRN
BLU/WHT
Trim motor
A
Starter solenoid
To battery
B
C
20
20 amp fuse
RED
Engine harness connector
GRN/WHT
BLU/WHT
Remote control harness connector
To rectifier or rectifier/regulator
Lower cowl mounted trim switch
UP
DN
BLK = Black
BLU = Blue
GRN = Green
RED = Red
WHT = White

repair or replace each wire from the relay connector body to the ground that failed the test.

NOTE

Refer to ***Key (Ignition) Switch Test*** *in Chapter Three for more information on testing the ignition switch and wire connections to the switch. Refer to the wiring diagrams at the end of the manual for remote control harness wiring diagrams.*

9. On remote control models, connect the test lamp lead to a clean engine ground. Connect the test lamp probe to the red or red/purple lead (B, B+ or BAT terminal) at the ignition switch (**Figure 16**). If the lamp does not light, repair or replace the wire, or the main engine harness connector terminals between the starter solenoid positive terminal and the ignition switch terminal.

10. Connect the test lamp probe to the center terminal (red or red/purple wire) of each trim/tilt switch. The test lamp should light. If it does not, repair or replace the wire from the ignition switch red or red/purple lead (B, B+ or BAT terminal) connection to the defective trim/tilt switch center terminal.

11. Connect the test lamp probe to the blue/white terminal in the up relay connector body. Hold each trim switch in the up position and observe the test lamp. The test lamp should light as each switch is activated. If it does not, connect the test lamp probe to the blue/white wire at each trim switch. Hold each trim switch in the up position and observe the test lamp. The test lamp should light when each trim switch is activated. If it does not, replace the defective trim/tilt switch. If the test lamp lights at the trim switch, but not at the up relay blue/white terminal, repair or replace the blue/white wire from the suspect trim/tilt switch to the up relay connector.

12. Connect the test lamp probe to green/white terminal in the down relay connector body. Hold each trim switch in the down position and observe the test lamp. The test lamp should light as each switch is activated. If it does not, connect the test lamp probe to the green/white wire at each trim switch. Hold each trim switch in the down position and observe the test lamp. The test lamp should light when each trim switch is activated. If it does not, replace the defective trim/tilt switch. If the test lamp lights at the trim switch, but not at down relay green/white terminal, repair or replace the green/white wire from the suspect trim/tilt switch to the relay connector.

13. Reconnect the trim/tilt relays to the wiring harness connectors. Probing from the rear of the relay connector body, connect the test lamp probe to the larger diameter blue lead terminal in the up relay connector body. Hold the trim switch in the up position and observe the test lamp. The test lamp should light. If it does not, replace the up trim relay.

14. Probing from the rear of the relay connector body, connect the test lamp probe to the larger diameter green lead terminal in the down relay connector body. Hold the trim switch in the down position and observe the test lamp. The test lamp should light. If it does not, replace the down trim relay.

15. Connect the test lamp lead to the positive terminal of the battery. Probing from the rear of each relay connector body, alternately touch the test lamp probe to the larger diameter blue lead terminal of the up relay and the larger diameter green terminal of the down relay. The test lamp should light at each test point. If it does not, replace the defective relay(s).

16. If all previous tests are satisfactory and the electric motor still does not operate correctly, repair or replace the electric motor.

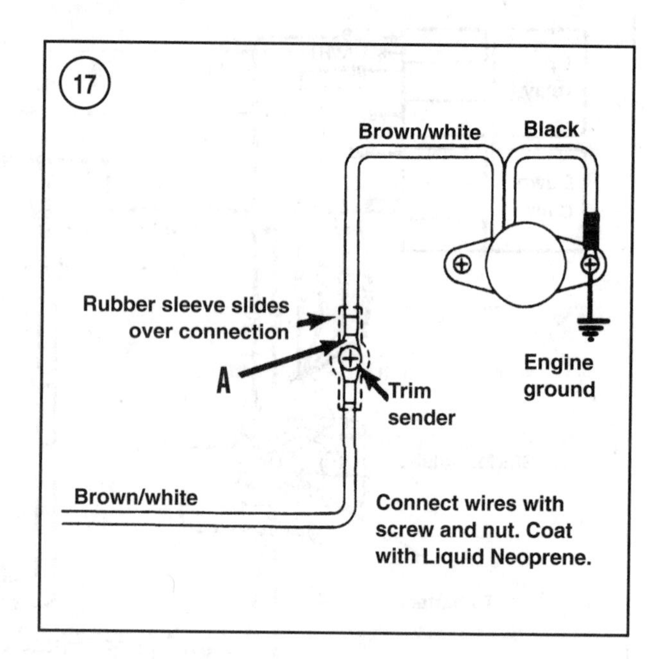

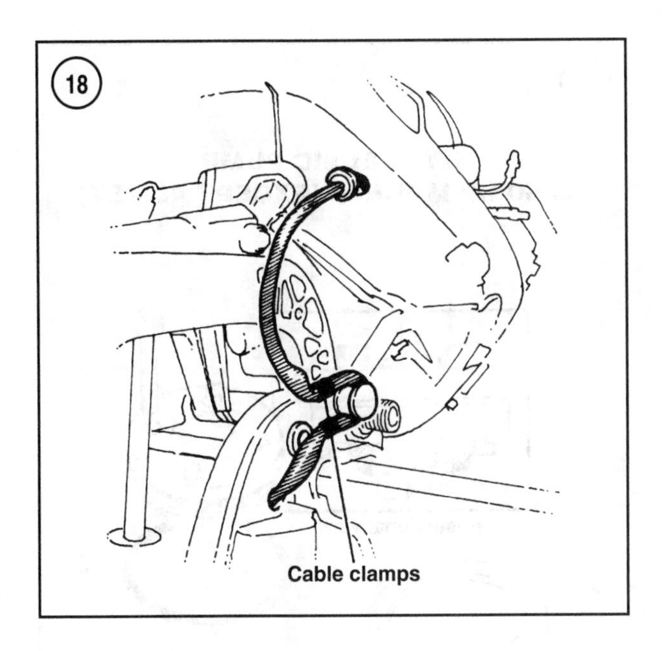

Trim Sending Unit Test

The trim sender is an optional accessory on 55 and 60 hp, and 45 jet models.

1. Make sure the black trim sender lead has a good ground connection. Clean and tighten the connection as necessary.
2. Trim the outboard to its fully down position. Make sure the ignition switch is turned off.
3. Connect an ohmmeter set to the R × 1 scale between a good engine ground and the test point (A, **Figure 17**).
4. Trim the outboard up while noting the meter reading.
5. The resistance should smoothly increase as the outboard trims up and decrease as the outboard trims down. If it does not, the trim sender is defective and must be replaced.
6. Reconnect all leads.

POWER TRIM AND TILT SYSTEM SERVICE

Metric and American fasteners are used on the clamp bracket and swivel bracket. Always match a replacement fastener to the original. Do not run a tap or thread chaser into an opening or over a bolt without first verifying the thread size and pitch.

Lubricate all seals and O-rings of the trim system with Quicksilver Power Trim and Steering Fluid (part No. 92-90100A12) or automatic transmission fluid during assembly. Apply Loctite 271 threadlocking adhesive (part No. 92-809819) to all threaded fasteners. Lubricate the manual release valve with Quicksilver Special lube 101 (part No. 92-13872) if it is removed. Refer to **Table 1** or **Table 2** for torque specifications.

NOTE
Always replace all O-ring, seals and gaskets if removed. Clean the outside of the trim/tilt system before disassembly. Always use a lint-free cloth when handling trim/tilt components. Dirt or lint can block passages, stick valves and prevent O-rings from sealing.

Relieving System Pressure

WARNING
The internal pressure must be relieved before disassembly of the trim/tilt system.

The trim/tilt cylinder ram or tilt cylinder and trim rams must be fully extended in order for internal pressure to be safely relieved. If the trim/tilt unit has already been removed from the outboard motor, simply reconnect the electric motor leads to the solenoids or relays and run the unit out until the ram fully extends. If the electric motor or pump is inoperative, slowly open the manual relief valve and manually extend the ram.

1. Trim the outboard to the full up position.
2. Engage the tilt lock, securely block the outboard or support the outboard with a suitable hoist. Refer to the owner's manual for tilt lock instructions.
3. Open the manual release valve as described previously in this chapter and allow the outboard to settle onto the tilt lock, blocks or hoist.
4. Carefully remove the fill plug while covering the plug with a shop towel.
5. Internal pressure should be relieved at this point.
6. Reinstall the fill plug and close the manual release valve to minimize fluid leaks.

System removal/installation (30-50 hp models [except 45 jet])

1. Tilt the outboard to the full tilt position. Engage the tilt lock, securely block the outboard or secure the outboard with a suitable hoist. If it is necessary to use the manual release valve to tilt the outboard, do not open the manual release valve more than 3-4 full turns.
2. Disconnect the negative battery cable.
3. Disconnect the power trim blue and green motor wires from the main engine harness bullet connectors.
4. Loosen any wire clamps or tie-straps and pull the trim motor leads through the lower support plate. Remove the trim harness wire clamp on the starboard clamp bracket (**Figure 18**). Pull the trim motor wire harness through and free from the starboard clamp bracket.

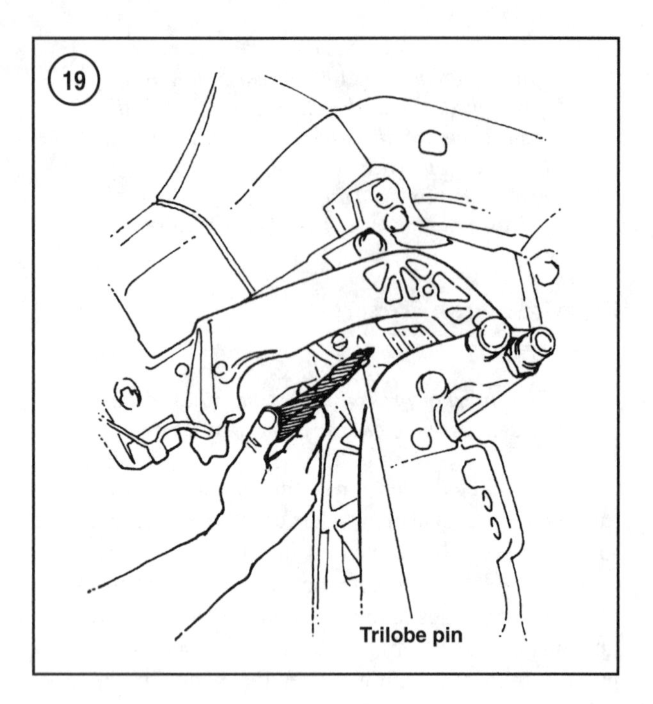

5. Use a pair of diagonal cutters to pull the trilobe pin from the swivel bracket as shown in **Figure 19**. Remove and inspect the trilobe pin, replace the pin if it is damaged. Drive the upper pivot shaft laterally from the bracket using a punch and hammer.

6. Remove the anode from the lower clamp brackets to free the trim motor ground strap.

7. Remove the nuts and washers from the lower pivot pin. Remove the pivot pin from the clamp brackets.

8. Lift the trim/tilt unit upward and out of the clamp brackets. Remove the pivot pin bushing from each clamp bracket and the two bushings from the trim/tilt unit.

9. To install the trim/tilt unit, lift the outboard motor to the full tilt position, and engage the tilt lock, securely block the outboard or secure the outboard with a suitable hoist.

10. Lubricate the pivot pin bushings with 2-4-C Multi-Lube grease (part No. 92-825407) and install a bushing into each clamp bracket bore and two bushings into the trim/tilt unit.

11. Insert the trim/tilt unit between the clamp brackets, bottom first. Lubricate the pivot pin shaft with 2-4-C Multi-Lube grease. Align the pivot pin bores and install the lower pivot pin. Install the washers and elastic stop nuts. Tighten the nuts to specification in **Table 1**.

12. Install the anode to the clamp brackets. Place the trim motor ground lead between the anode and the starboard clamp bracket. Tighten the anode screws to the specification in **Table 1**.

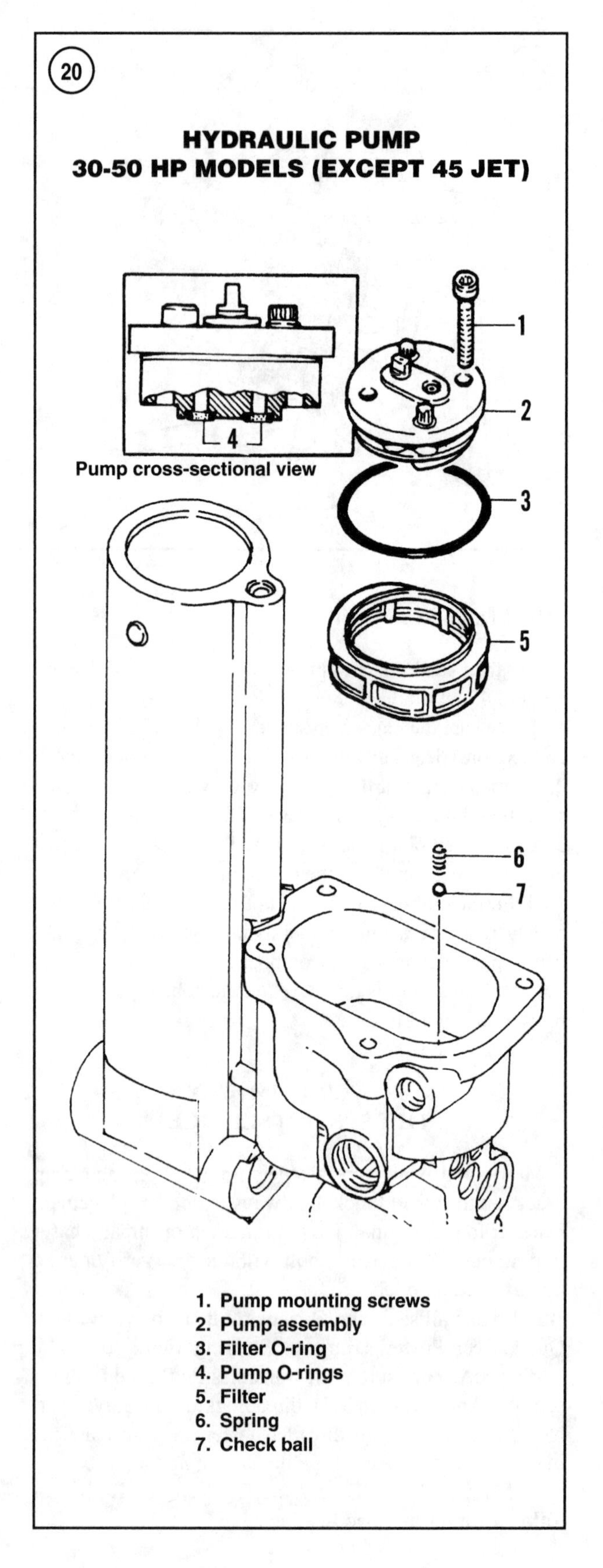

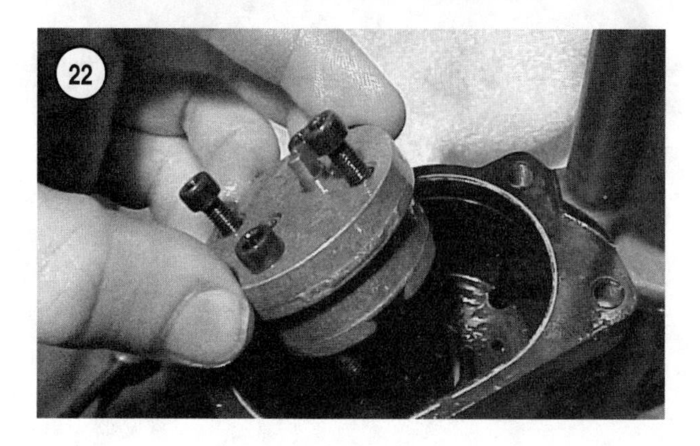

13. Route the trim motor harness through the starboard clamp bracket and into the lower support plate. Install the harness clamps to the starboard clamp brackets. The trim motor harness must loop around the steering cable as shown in **Figure 18**.
14. Lubricate the upper pivot shaft with 2-4-C Multi-Lube grease or an equivalent. Align the upper pivot shaft bore with the trim/tilt cylinder rod eye. Install the upper pivot shaft with the grooved end facing the trilobe pin bore into the swivel bracket bore and through the trim/tilt cylinder rod eye. The shaft must be flush with the swivel bracket outer surfaces.
15. Drive the dowel pin or trilobe into its bore until it is fully seated.
16. Connect the trim motor leads to the engine harness bullet connectors. Secure the harness with clamps or tie-straps.
17. Reconnect the negative battery cable.
18. Check the reservoir fluid level as described in this chapter.

Electric motor removal/installation (30-50 hp models [except 45 Jet])

Refer to **Figure 13** for this procedure.

1. Remove the trim/tilt system as described in the previous section.
2. Make sure the internal pressure has been relieved as described previously in *Relieving System Pressure*. Then drain the reservoir into a suitable container.
3. Secure the system in a vise with protective jaws or between two suitable blocks of wood.
4. Note the position of the motor ground strap, then remove the screws and washers securing the electric motor to the manifold assembly.
5. Remove the electric motor. Locate and secure the pump coupler to prevent its loss.
6. Remove the motor O-ring.
7. To install the motor, position the pump coupler onto the hydraulic pump.
8. Install a *new* O-ring on the electric motor. Carefully align the motor armature shaft with the pump coupler and install the motor to the manifold. Be careful not to pinch the O-ring.
9. Install the screws and washers, and evenly tighten them to the specification in **Table 1**. Make sure the ground strap is reinstalled under the starboard rear screw as noted on disassembly.
10. Reinstall the trim/tilt system as described in the previous section.

Hydraulic pump removal/installation (30-50 hp models [except 45 jet])

Refer to **Figure 20** for this procedure. The filter and filter O-ring are located beneath the hydraulic pump as shown in **Figure 20**.

1. Remove the electric motor as described in the previous section.
2. Remove the two socket head screws (**Figure 21**) securing the pump to the manifold. Do not remove the TORX head screws.
3. Lift the pump from the manifold (**Figure 22**). Remove the pump O-rings, filter and filter O-ring (**Figure 23**).
4. Clean and inspect the filter. Replace the filter if it cannot be cleaned or if it is damaged.
5. Remove and inspect the spring and check ball from the manifold (see **Figure 24**). Replace damaged or suspect components.
6. To reinstall the pump, place the check ball and spring into the manifold.
7. Position two new O-rings on the base of the pump. A light coat of grease may be used to hold the O-rings in place.
8. Position a new filter O-ring over the pump, then install the filter.

9. Install the pump assembly into the manifold. Be careful not to displace the pump O-rings or filter.
10. Install the pump screws. Tighten the screws evenly to the specification in **Table 1**.
11. Install the electric motor as described in the previous section.

System removal/installation (55 hp, 60 hp and 45 jet models)

1. Tilt the outboard to the full tilt position, and engage the tilt lock, securely block the outboard or secure the outboard with a suitable hoist. If it is necessary to use the manual release valve to tilt the outboard, do not open the manual release valve more than four full turns.
2. Disconnect the negative battery cable.
3. Remove the trim motor wire harness as follows:
   a. Disconnect the power trim blue and green motor wires from the main engine harness bullet connectors.
   b. Loosen any wire clamps or tie-straps and pull the trim motor wires through the lower cowl.
   c. Remove the trim harness wire clamp on the starboard clamp bracket (**Figure 25**).
   d. Pull the trim motor wire harness through and free from the starboard clamp bracket.

NOTE
If the trilobe pin proves difficult to remove in Step 4, the upper pivot shaft may be driven forcefully from its bore, shearing the trilobe pin. Remove all trilobe pin remnants, and clean and inspect the pivot shaft and trilobe pin bores for damage.

4. Use diagonal cutters to pull the trilobe pin from the swivel bracket as shown in **Figure 19**. Remove and inspect the trilobe pin. Replace the pin if it is damaged. Then drive the upper pivot shaft laterally from the bracket using a punch and hammer.
5. Use a suitable punch to drive the dowel pin securing the lower pivot shaft upward. Remove and inspect the dowel pin. Replace the dowel pin if it is damaged. See **Figure 26**.
6. Remove the lower pivot shaft from the trim/tilt unit and the clamp brackets by driving it laterally with a suitable punch and hammer.
7. Note the position and routing of the trim motor ground strap. Then remove the anode at the bottom of the clamp brackets to free the trim motor ground strap.
8. Remove the trim/tilt system by pivoting the top of the trim/tilt cylinder out and away from the clamp brackets.

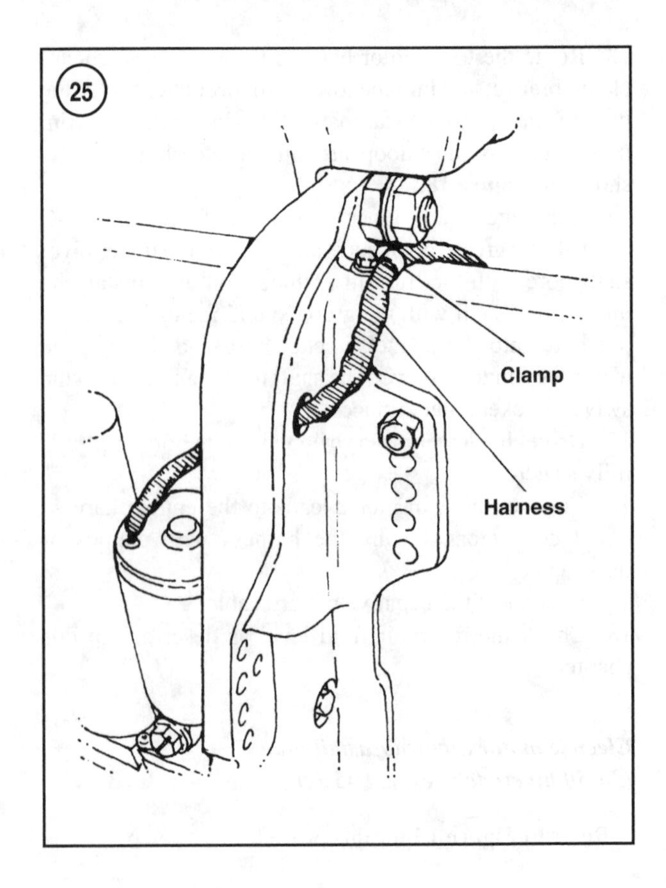

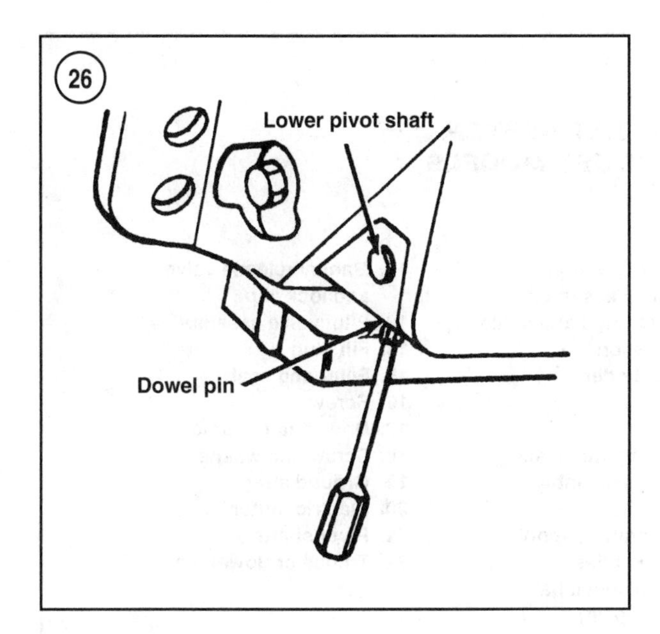

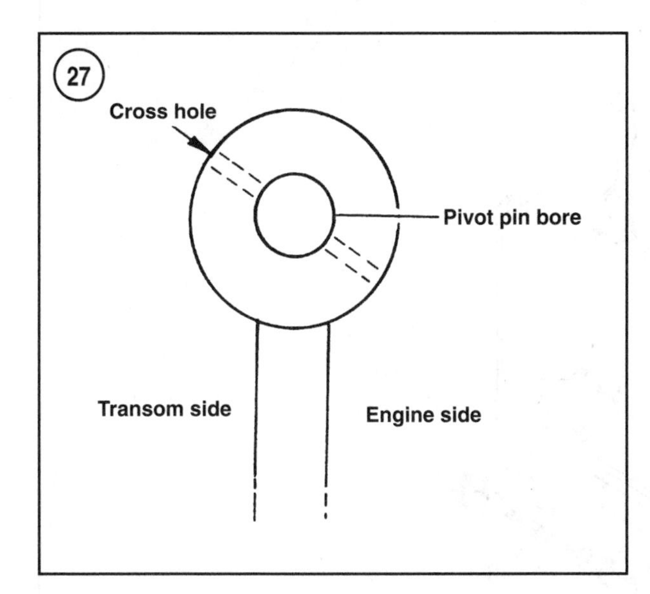

9. To install the system, liberally apply Quicksilver 2-4-C Multi-Lube grease (part No. 92-825407) to the upper and lower pivot shafts, pivot shaft bores, dowel pin and trilobe pin bores, dowel pin, and trilobe pin.

10. Start the lower pivot shaft with the grooved end facing the dowel pin bore into its bore from the port side clamp bracket, then start the lower dowel pin from the *top* of its bore. Do not drive either pin far enough to interfere with tilt unit installation.

11. Install the trim/tilt system. Insert the bottom into the clamp brackets first, then rotate the trim/tilt cylinder towards the transom and into position. Route the motor harness out through the hole in the starboard clamp bracket.

12. Drive the lower pivot shaft flush with the clamp bracket or mounting bracket outer surfaces. Then use a suitable punch to drive the dowel pin in from the top until it is fully seated.

13. Align the upper pivot shaft bore with the trim/tilt cylinder rod eye. The cylinder rod eye must be facing as shown in **Figure 27**.

14. Install the upper pivot shaft into the swivel bracket bore and through the trim/tilt cylinder rod eye. The shaft must be flush with the swivel bracket outer surfaces. Make sure the bores align, then drive the dowel pin or trilobe pin into its bore until fully seated.

15. Install the anode and connect the trim motor ground strap as noted on removal. Tighten the anode screws to the specification in **Table 1**.

16. Install the trim harness clamp to the starboard clamp bracket as shown in **Figure 25**.

17. Route the trim harness through the lower cowl and secure it with the original wire clamps or new tie-straps.

18. Connect the power trim blue and green motor leads to the main engine harness bullet connectors.

19. Reconnect the negative battery cable and disengage the tilt lock or remove the blocks or hoist. Check the reservoir oil level as described in this chapter.

Electric motor removal/installation (55 hp, 60 hp and 45 jet models)

Remove and install the electric motor and reservoir as an assembly. Replace the motor if it is defective. The electric motor is non-serviceable. Refer to **Figure 28** for this procedure.

1. Remove the trim/tilt system as described in the previous section.
2. Make sure the internal pressure has been relieved as described in this chapter. Then drain the reservoir into a suitable container.
3. Secure the system in a vise with protective jaws or between two suitable blocks of wood.
4. Note the position of the motor ground strap, then remove the screws and washers (18, **Figure 28**) securing the electric motor to the manifold assembly.
5. Remove the electric motor. Locate and secure the pump coupler to prevent its loss.
6. Remove the motor O-ring or molded seal.
7. To install the motor, position the pump coupler (drive shaft) onto the hydraulic pump.
8. Install a *new* O-ring or molded seal on the electric motor. Carefully align the motor armature shaft with the pump coupler (drive shaft) and install the motor to the manifold. Be careful not to pinch the O-ring.

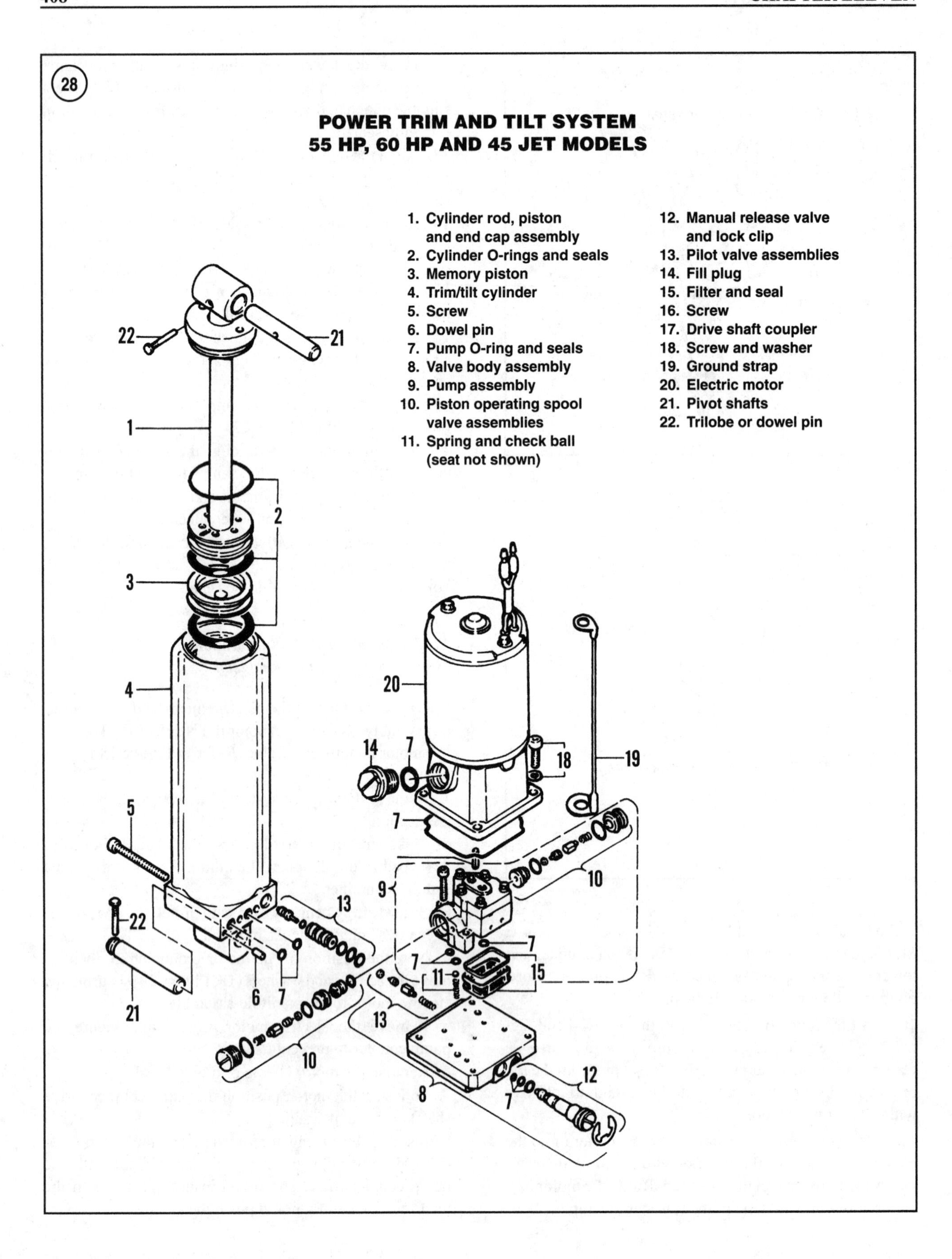
28
POWER TRIM AND TILT SYSTEM
55 HP, 60 HP AND 45 JET MODELS
1. Cylinder rod, piston and end cap assembly
2. Cylinder O-rings and seals
3. Memory piston
4. Trim/tilt cylinder
5. Screw
6. Dowel pin
7. Pump O-ring and seals
8. Valve body assembly
9. Pump assembly
10. Piston operating spool valve assemblies
11. Spring and check ball (seat not shown)
12. Manual release valve and lock clip
13. Pilot valve assemblies
14. Fill plug
15. Filter and seal
16. Screw
17. Drive shaft coupler
18. Screw and washer
19. Ground strap
20. Electric motor
21. Pivot shafts
22. Trilobe or dowel pin

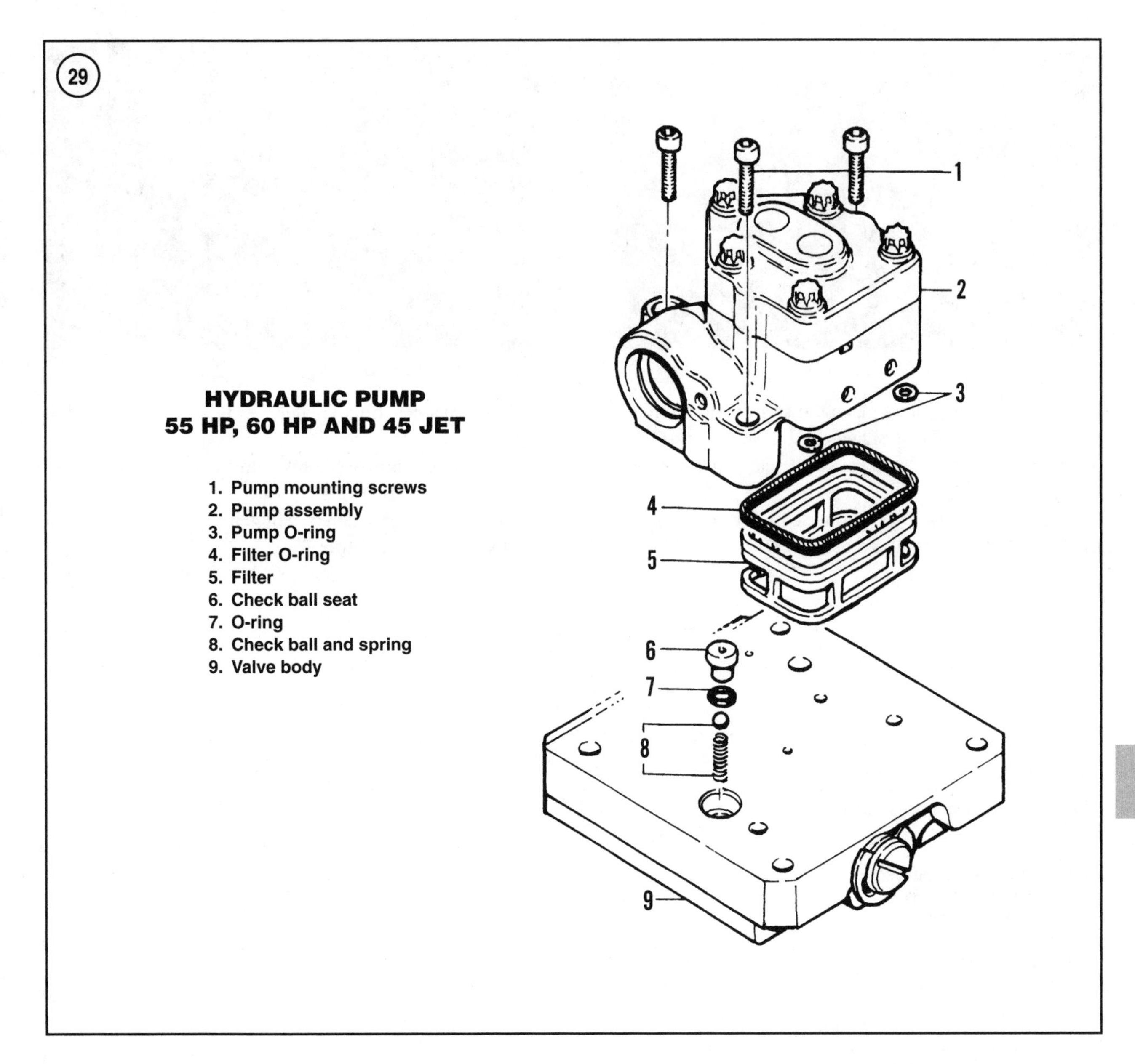

9. Install and evenly tighten the screws and washers to the specification in **Table 1**. Make sure the ground strap is reinstalled under the starboard rear screw as noted on disassembly.
10. Reinstall the trim/tilt system as described in the previous section.

Hydraulic pump removal/installation (55 hp, 60 hp and 45 jet model)

Refer to **Figure 29** for this procedure. The filter and filter O-ring are located beneath the hydraulic pump as shown in **Figure 29**.

1. Remove the electric motor as described in the previous section.
2. Remove the hex screws (1, **Figure 29**) securing the pump to the manifold. Do not remove the TORX head screws.
3. Lift the pump from the manifold (**Figure 30**). Remove the pump O-rings and filter O-ring (3 and 4, **Figure 29**).
4. Remove, clean and inspect the filter (**Figure 31**). Replace the filter if it cannot be cleaned or if it is damaged.
5. Remove the check ball and O-ring (7, **Figure 29**). Inspect the check ball seat (6). Remove the check ball and spring (8, **Figure 29**) from the manifold. Replace damaged or suspect components.

6. To reinstall the pump, place the spring into the manifold bore, followed by the check ball. Install a *new* O-ring on the check ball seat, and press the seat and O-ring into the manifold bore.
7. Position two *new* O-rings on the base of the pump. A light coat of grease may be used to hold the O-rings in place.
8. Position a *new* filter O-ring over the pump, then install the filter.
9. Install the pump assembly into the manifold. Do not displace the pump O-rings or filter.
10. Install the pump screw. Tighten the screws evenly to the specification in **Table 1**.
11. Install the electric motor as described in this chapter.

Table 1 TORQUE SPECIFICATIONS

Component	ft.-lb.	in.-lb.	N•m
Manual tilt system (serviceable)			
Accumulator	35	–	47.5
Cylinder to manifold	–	100	11.3
Cylinder piston valve screws	–	35	4.0
Tilt cylinder end cap	45	–	61
Shock piston bolt	90	–	122
Surge valve plug	–	75	8.5
Down slow valve cap	–	75	8.5
Down fast valve cap	–	75	8.5
Trim/tilt system			
30-50 hp models (except 45 Jet)			
Lower pivot pin	18	–	24.4
Anode	–	60	6.8
Cylinder end cap	45	–	61
Motor to main body	–	80	9.0
Piston operating valve plug	–	120	13.6
Pump to main body	–	70	7.9
Suction valve plug	–	120	13.6
Tilt relief valve plug	–	120	13.6
Shock piston bolt	90	–	122
55 hp, 60 hp and 45 Jet			
Anode	–	60	6.8
Cylinder end cap	45	–	61
Cylinder piston valve screws	–	35	4.0
Piston operating valve plug	–	120	13.6
Pump to valve body	–	70	7.9
Manifold to cylinder screws	–	100	11.3
Reservoir to pump body	–	80	9.0
Shock piston bolt	90	–	122

Table 2 STANDARD TORQUE SPECIFICATIONS—U.S. STANDARD AND METRIC FASTENERS

Screw or nut size	ft.-lb.	in.-lb.	N•m
U.S. standard fasteners			
6-32	–	9	1
8-32	–	20	2.3
10-24	–	30	3.4
10-32	–	35	4.0
12-24	–	45	5.1
1/4-20	6	72	8.1
1/4-28	7	84	9.5
5/16-18	13	156	17.6
5/16-24	14	168	19
3/8-16	23	270	31.2
3/8-24	25	300	33.9
7/16-14	36	–	48.8
7/16-20	40	–	54
1/2-13	50	–	67.8
1/2-20	60	–	81.3
Metric fasteners			
M5	–	36	4.1
M6	6	72	8.1
M8	13	156	17.6
M10	26	312	35.3
M12	35	–	47.5
M14	60	–	81.3

Chapter Twelve

Oil Injection Systems

All models are lubricated by mixing oil with the gasoline. On models that are not oil-injected, the oil and gasoline are premixed in the fuel tank(s) by the boat operator. The recommended fuel/oil mixture for normal operation in all models without oil-injection is 50 parts of fuel to 1 part of oil (50:1). For 6 gal. (22.7 L) of fuel, add one pint (16 fl. oz. [473 mL]) of oil.

On oil-injected models, the oil automatically mixes with the gasoline. An engine-mounted oil pump injects oil into the fuel pump inlet line. All oil-injected models are equipped with a variable-ratio oil-injection pump. The fuel/oil ratio changes with throttle position. At wide-open throttle the fuel/oil ratio is approximately 50-60:1, while at idle speeds the ratio is approximately 80-100:1.

The engine components are lubricated as the fuel and oil mixture passes through the crankcases and into the combustion chambers. The fuel/oil ratio required by the engine varies with engine load and speed. Oil demands are always highest at wide-open throttle.

The advantages of oil-injection are that the operator only has to keep the oil reservoir(s) filled. No calculations must be made as to how much oil to add when refueling. Over-oiling and under-oiling from operator miscalculations and the associated engine problems are eliminated.

Variable-ratio oil injection offers the additional benefits of reduced smoke and spark plug fouling due to the reduced oil consumption at idle and low speeds.

Refer to Chapter Four for fuel and oil requirements, engine break-in procedures and oil tank filling procedures. Refer to Chapter Five for synchronizing the oil pump and throttle linkages on variable ratio models.

Table 1 and **Table 2** lists oil injection system specifications. **Table 3** lists specific torque values. **Table 4** list standard torque values. **Tables 1-4** are at the end of the chapter.

CAUTION

If a boat-mounted electric fuel supply pump is used, fuel pressure must not exceed 2 psi (13.8 kPa) at the engine fuel line connector. If necessary, install a fuel pressure regulator between the electric fuel pump and engine fuel line connector. Adjust the fuel

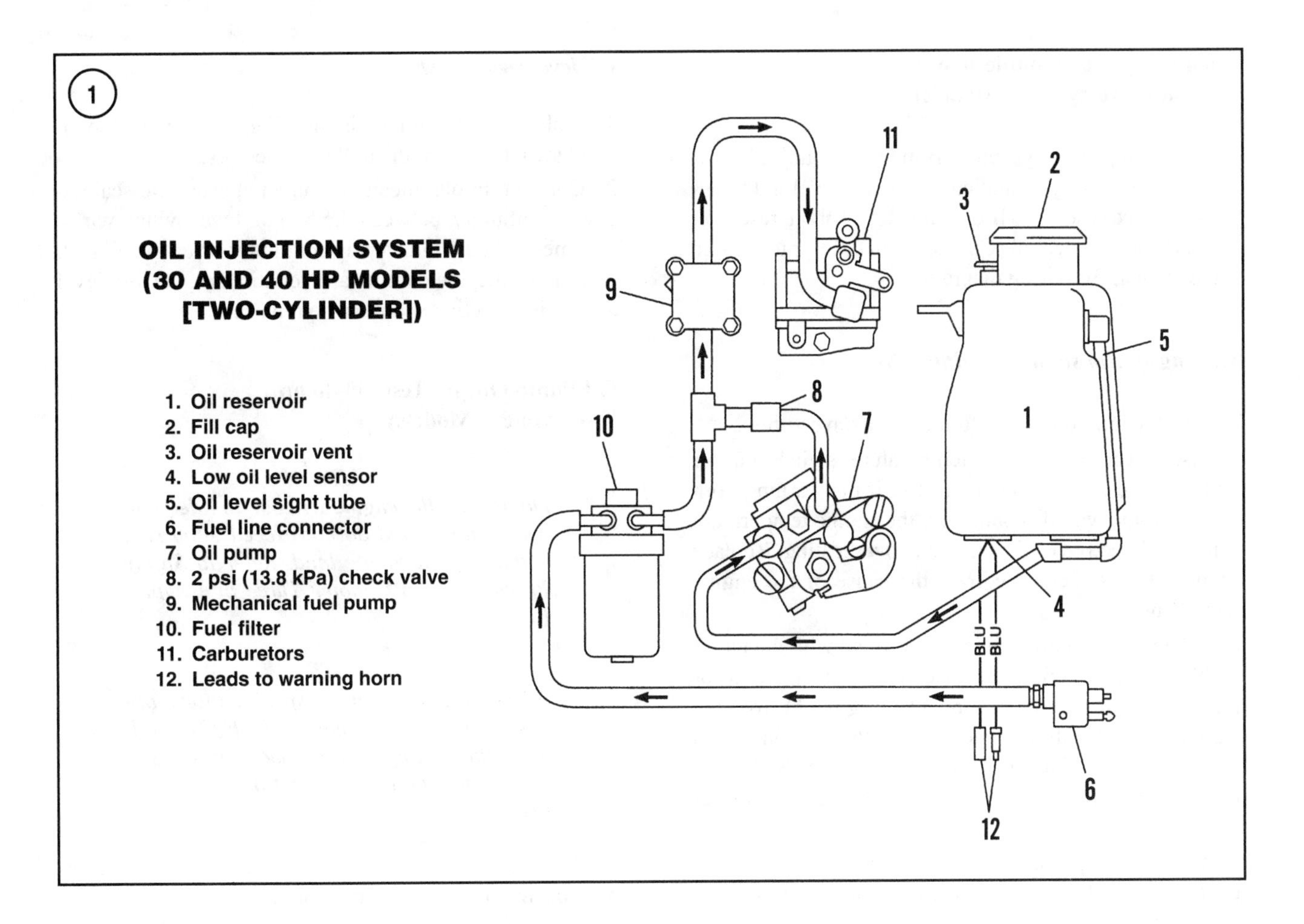

pressure regulator to a maximum of 2 psi (13.9 kPa) fuel pressure. The electric fuel pump must also conform to all applicable Coast Guard safety standard for permanently installed fuel systems.

CAUTION

If an oil-injection system malfunction is suspected, do not operate the outboard on straight gasoline. Operate the outboard on a remote fuel tank containing a 50:1 fuel/oil mixture until a oil pump output test can be performed.

Oil Injection System (30-40 hp Two-cylinder Models)

These models use a variable ratio automatic oil-injection system. The oil pump is driven by the engine crankshaft and injects oil into the fuel stream before it reaches the fuel pump as shown in **Figure 1**. The oil pump delivers oil relative to throttle lever position and engine speed. The fuel-oil ratio is approximately 100:1 at idle and approximately 60:1 at wide-open throttle.

Operation

The engine-mounted oil tank capacity is 50.5 fl. oz. (1.5 L) and provides approximately 4.7 hours of operation at wide-open throttle. The oil tank is equipped with an oil level sight gauge visible through an opening in the engine cowl. A check-valve vent at the top of the oil tank allows the reservoir to vent and prevents oil leaks when the outboard motor is tilted.

An oil level switch in the oil reservoir activates the warning horn if the oil level drops to 7 fl. oz. (207 mL). Once the low oil level warning horn sounds, the remaining oil should provide approximately 30 minutes of wide-open throttle operation.

A 2 psi (13.8 kPa) check valve is installed in the fuel line between the fuel line T-fitting and the oil pump discharge line. See **Figure 1**. The check valve prevents gasoline from entering the oil pump discharge line.

Warning System Troubleshooting (30-40 hp Two-cylinder Models)

If the oil injection system warning horn sounds, immediately stop the engine and check the oil level in the engine-mounted reservoir. If the oil is low, fill the reservoir with a recommended oil (Chapter Four). Refer to the end of the manual for wiring diagrams.

Warning system sounds continuously

1. Turn the ignition switch to the on or run position.
2. Disconnect the engine temperature switch tan or tan/blue wire at its bullet connector. If the warning horn stops with the wire disconnected, the engine temperature switch is defective or the engine is overheating. Replace the defective switch or correct the cause of the engine overheating.
3. If the warning horn continues to sound with the tan or tan/blue wire disconnected, disconnect the two blue or light blue oil level switch wires (12, **Figure 1**) from the engine harness bullet connectors. If the warning horn stops sounding with the oil level switch leads disconnected, the oil level switch is defective and must be replaced.
4. If the warning horn continues to sound after the oil level switch is disconnected, the tan/blue wire is shorted to ground.

Warning system sounds erratically

CAUTION
Do not run the engine without an adequate water supply and do not exceed 3000 rpm without an adequate load. Refer to ***Safety Precautions*** *in Chapter Three or Chapter Five.*

1. Make sure the oil tank is full. Then disconnect the oil level switch wires at the bullet connectors.
2. Run the engine and see if the warning horn still sounds erratically. If the horn no longer sounds, replace the low oil level switch in the oil tank.
3. If the horn still sounds in Step 2, disconnect the engine temperature switch tan or tan/blue wire at its bullet connector.
4. Run the engine and see if the warning horn still sounds erratically. If the horn no longer sounds, the engine temperature switch is defective or the engine is overheating. Refer to Chapter Three for troubleshooting procedures.

Oil level switch tests

1. Make sure the oil tank is full. Then disconnect the oil level switch leads at the bullet connectors.
2. Connect an ohmmeter, set on an appropriate scale to check continuity, between the two oil level switch wires. The meter should indicate no continuity when the oil tank is full. If the meter indicates continuity, the oil level switch is defective and must be replaced.

Oil Pump Output Test (30-40 hp Two-cylinder Models)

CAUTION
Do not run the engine without an adequate water supply and do not exceed 3000 rpm without an adequate load. Refer to ***Safety Precautions*** *in Chapter Three or Chapter Five.*

NOTE
Injection pump output specifications are based on tests performed at 70°F (21°C). If the ambient temperature is more or less, actual pump output may vary from that specified.

Obtain a graduated container capable of accurately measuring up to 50 cc before continuing.

1. Connect a remote fuel tank containing 50:1 fuel/oil mixture to the engine.
2. Disconnect the oil pump discharge line from the check valve fitting (8, **Figure 1**). Securely cap or plug the check valve fitting to prevent leaks.
3. Connect an accurate shop tachometer to the engine following the manufacturer's instructions.
4. Insert the disconnected end of the oil pump discharge line into the graduated container.
5. Disconnect the oil pump link rod from the oil pump control arm. Rotate the pump arm fully clockwise until the arm is against the pump casting. Hold the arm in this position.
6. Start the outboard motor and run it at 900 rpm for 10 minutes.
7. Stop the engine and check the quantity of oil in the graduated container. Oil pump output should be within the specification in **Table 2**.
8. If injection pump output is less than specified, replace the pump assembly as described in this chapter. Repeat the test. The oil injection drive gear is faulty if low output persists. Replace the drive gear as described in Chapter Eight.

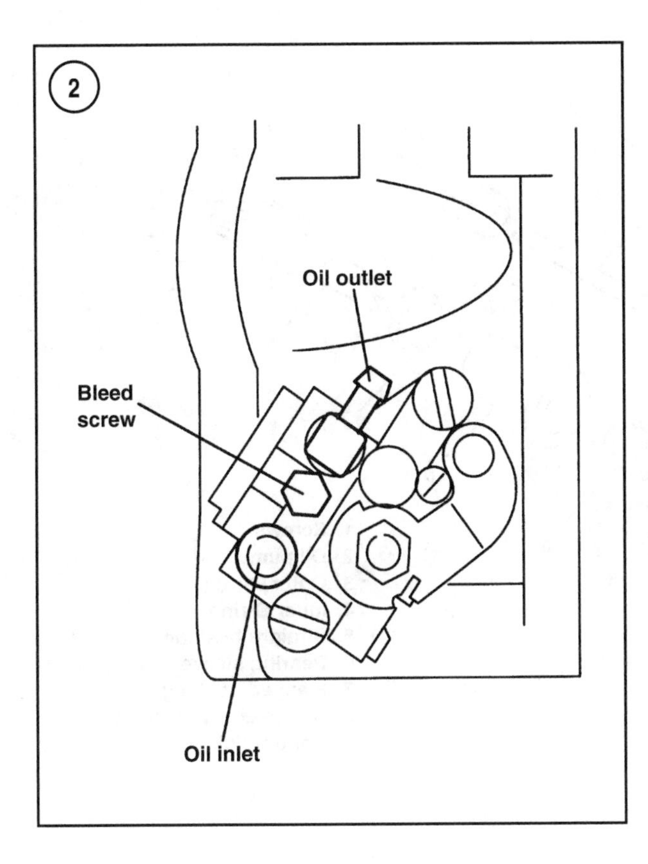

9. Reconnect the pump discharge line to the check valve fitting. Secure the connection with a new tie-strap. Then bleed the injection system as described in the next section.
10. Connect the oil pump link rod to the control arm. Adjust the link rod as specified in Chapter Five.

Oil Injection System Service (30-40 hp Two-cylinder Models)

Oil pump synchronization

Refer to Chapter Five for throttle linkage-to-oil pump linkage adjustment and synchronization procedures.

Bleeding air from the oil pump

CAUTION
Do not run the engine without an adequate water supply and do not exceed 3000 rpm without an adequate load. Refer to ***Safety Precautions*** *in Chapter Three or Chapter Five.*

If the pump has been removed, any of the lines replaced or air is present in the oil pump lines, bleed air from the oil injection pump and lines as follows:

1. Place a shop towel beneath the oil pump.
2. Loosen the oil pump bleed screw 3-4 turns (**Figure 2**).
3. Allow oil to flow from the bleed screw until no air bubbles are noted in the inlet hose.
4. Tighten the bleed screw to the specification in **Table 3**.
5. If air is present in the pump discharge line, connect a remote tank containing a 50:1 fuel/oil mixture to the engine. Start the engine and run it at idle until no air bubbles are in the discharge hose. To speed the bleeding process, disconnect the pump link rod and rotate the pump arm to the full output position.

Oil tank removal/installation

1. Disconnect the negative battery cable.
2. Disconnect and ground the spark plug leads to the power head to prevent accidental starting.
3. Remove the flywheel cover, then remove the nuts or screws securing the oil tank to the intake manifold.
4. Disconnect the oil level switch leads from the switch's bullet connectors.
5. Lift the oil tank enough to access the bottom of the tank. The tab and grommet on the bottom of the tank may stick. Do not break the tab. Cut the tie-strap securing the oil outlet hose to the oil tank, then disconnect the hose from the tank.
6. Remove the oil tank from the power head by lifting it off of its mounting studs.
7. Remove the screw securing the oil level switch to the bottom of the tank. Remove the switch.
8. To install the oil tank, install the oil level switch into the oil tank. Secure the switch with one screw. Tighten the screw securely.
9. Position the oil tank assembly on its mounting studs or brackets. Then attach the oil outlet hose to the oil tank. Secure the hose with a new tie-strap.
10. Push the oil tank all of the way down on its mounting studs or brackets, guiding the tab on the bottom of the tank into its grommet. Secure the tank with the nuts or screws. Tighten the fasteners securely.
11. Connect the oil level switch leads to the wiring harness bullet connectors.
12. Install the flywheel cover. Tighten the cover nuts securely.
13. Fill the oil tank with the recommended oil (Chapter Four), then bleed the oil pump as described in this chapter.
14. Reconnect the negative battery cable and the spark plug leads.

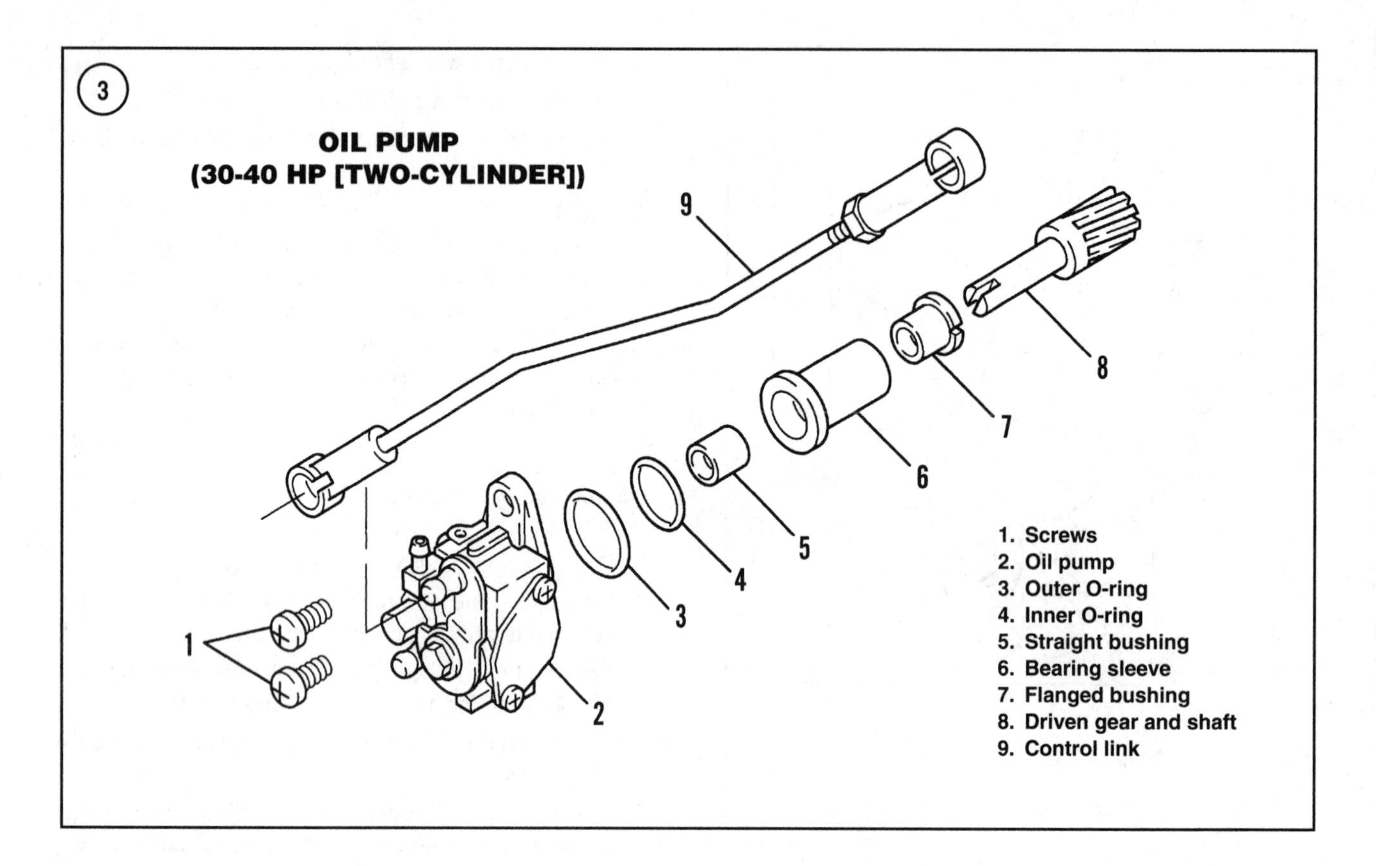

Oil pump removal/installation

If the crankshaft (drive) gear needs replacement, refer to Chapter Eight for oil pump drive gear removal/installation procedures.

Refer to **Figure 3** for this procedure.

1. Disconnect the oil inlet and discharge lines from the pump (**Figure 2**).
2. Remove the link rod from the pump control arm, then remove the pump mounting screws.
3. Remove the pump assembly from the power head. If the pump driven gear remains in the engine block, retrieve it using needlenose pliers.
4. Remove and inspect the bushings from the bearing sleeve. Replace all worn or damaged components.
5. To install the pump, thoroughly lubricate the bearing sleeve and two bushings with Quicksilver Needle Bearing Assembly Grease (part No. 92-825265A-1). Install the bushings into the sleeve. Make sure the flanged bushing faces the oil pump driven gear.
6. Thoroughly lubricate the driven gear shaft with Quicksilver Needle Bearing Assembly Grease. Insert the driven gear into the bearing assembly, making sure the gear properly engages the pump shaft.
7. Install new O-rings onto the pump. Lubricate the O-rings with Quicksilver Needle Bearing Assembly Grease.
8. Install the pump assembly onto the power head. Apply Loctite 271 threadlocking adhesive (part No. 92-809819) to the threads of the pump mounting screws. Install the screws and tighten to the specification in **Table 3**.
9. Reconnect the oil inlet and discharge hoses to the pump. Securely clamp the hoses to the pump using new tie-straps.
10. Reconnect the link rod to the pump control arm. Refer to Chapter Five for oil pump adjustment and synchronization procedures.
11. Perform the oil pump bleeding procedure as described in this chapter.

Oil Injection System (40-60 hp Three-cylinder Models)

The oil pump is driven by the engine crankshaft and injects oil into the fuel stream before it reaches the fuel pump as shown in **Figure 4**. The oil pump delivers oil relative to the carburetor throttle valve opening and engine speed. Maximum oil delivery occurs at wide-open throttle.

4

**OIL INJECTION SYSTEM
(40-60 HP MODELS [THREE-CYLINDER])**

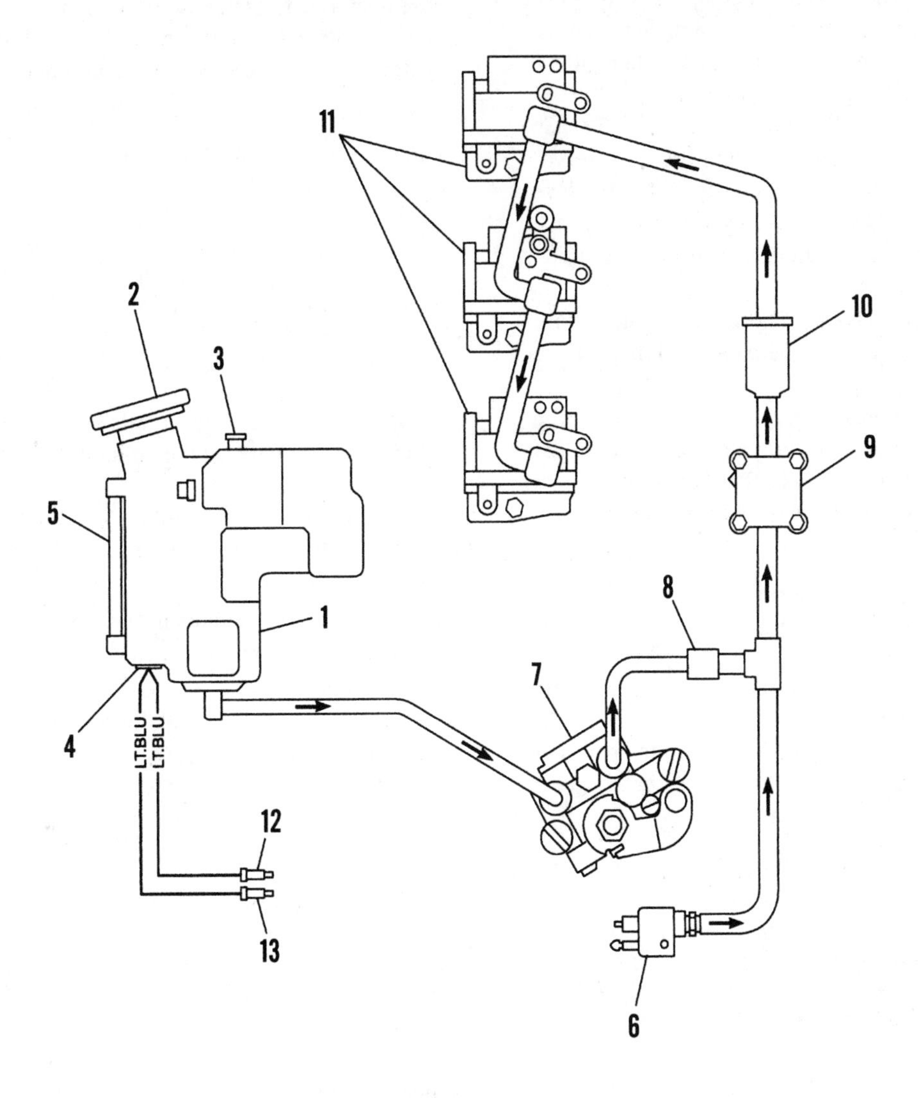

1. Oil reservoir
2. Fill cap
3. Oil reservoir vent
4. Low oil level switch
5. Oil level sight gauge
6. Fuel line connector
7. Oil pump
8. 2 psi (13.8 kPa) check valve
9. Mechanical fuel pump
10. Fuel filter
11. Carburetors
12. Connected to ground
13. Connected to warning horn

Operation

The engine-mounted oil tank capacity is 3 qt. (2.8 L) and provides approximately 7 hours operation at wide-open throttle. The oil tank is equipped with an oil level sight gauge visible through an opening in the engine cowl. A check-valve vent at the top of the oil tank allows the reservoir to vent and prevents oil leaks when the outboard motor is tilted. See **Figure 4**.

An oil level sensor contained within the oil reservoir triggers the warning horn continuously if the oil level drops to 14.5 fl. oz. (435 mL). The warning horn also sounds if the engine temperature switch in the cylinder block closes to ground.

Once the low oil level warning horn sounds, the remaining oil should provide approximately 30 minutes of wide-open throttle operation.

A 2 psi (13.8 kPa) check valve is installed in the fuel line between the fuel line T-fitting and the oil pump discharge line. See **Figure 4**. The check valve prevents gasoline from entering the oil pump discharge line.

Warning System Troubleshooting (40-60 Three-cylinder Models)

If the oil injection system warning horn sounds, immediately stop the engine and check the oil level in the engine-mounted reservoir. If the oil is low, fill the reservoir with a recommended oil (Chapter Four). Refer to the end of the manual for wiring diagrams.

CAUTION
If an oil injection system malfunction is suspected, do not operate the outboard on straight gasoline. Operate the outboard on a remote fuel tank containing a 50:1 fuel/oil mixture until an oil pump output test can be performed.

Warning system sounds continuously

1. Turn the ignition switch to the on or run position.
2. Disconnect the engine temperature switch tan or tan/blue wire at its bullet connector. If the warning horn stops with the wire disconnected, the engine temperature switch is defective or the engine is overheating. Replace the defective switch or correct the cause of the engine overheating.
3. If the warning horn continues to sound with the temperature switch wires disconnected, disconnect the two light blue wires from the oil level switch bullet connectors. If the warning horn stops with the leads disconnected, the oil level switch is defective and must be replaced.
4. If the warning horn continues to sound with the temperature and oil level switches disconnected, the tan/blue wire is shorted to ground somewhere between the warning horn and the engine temperature and oil level switch connectors. Repair or replace the lead as necessary.

Warning system sounds erratically

CAUTION
*Do not run the engine without an adequate water supply and do not exceed 3000 rpm without an adequate load. Refer to **Safety Precautions** in Chapter Three or Chapter Five.*

1. Make sure the oil tank is full. Then disconnect the two oil level switch wires at the bullet connectors (12 and 13, **Figure 4**).
2. Run the engine and see if the warning horn still sounds erratically. If the horn no longer sounds, replace the low oil level switch in the oil tank.
3. If the horn still sounds in Step 2, temporarily disconnect the engine temperature switch tan or tan/blue wire at its bullet connector.
4. Run the engine and see if the warning horn still sounds erratically. If the horn no longer sounds, the engine temperature switch is defective or the engine is overheating. Refer to Chapter Three for troubleshooting procedures.

Oil level switch tests

1. Make sure the oil tank is full. Then disconnect the oil level switch leads at the bullet connectors (12 and 13, **Figure 4**).
2. Connect an ohmmeter, set on an appropriate scale to check continuity, between the oil level switch wires. The meter should indicate no continuity when the oil tank is full. If the meter indicates continuity, the oil level switch is defective and must be replaced.

Oil Pump Output Test (40-60 hp Three-cylinder Models)

CAUTION
*Do not run the engine without an adequate water supply and do not exceed 3000 rpm without an adequate load. Refer to **Safety Precautions** in Chapter Three or Chapter Five.*

NOTE
Injection pump output specifications are based on tests performed at 70°F (21°C). If the ambient temperature is more or less, actual pump output may vary from that specified.

Obtain a graduated container capable of accurately measuring up to 50 cc before continuing.

1. Connect a remote fuel tank containing 50:1 fuel/oil mixture to the engine.
2. Disconnect the oil pump discharge line from the check valve fitting (8, **Figure 4**). Securely cap or plug the check valve fitting to prevent leaks.
3. Connect an accurate shop tachometer to the engine following the manufacturer's instructions.
4. Insert the disconnected end of the oil pump discharge line into the graduated container.
5. Disconnect the oil pump link rod from the oil pump control arm. Rotate the pump arm fully against the spring until the arm is against the pump casting. Hold the arm in this position.
6. Start the outboard motor and run it at 1500 rpm for 10 minutes.
7. Stop the engine and check the quantity of oil in the graduated container. Oil pump output should be within the specification in **Table 2**.
8. If injection pump output is less than specified, replace the pump assembly as described in this chapter. Repeat the oil pump output test. The oil injection drive gear is faulty if low output persists. Replace the drive gear as described in Chapter Eight.
9. Reconnect the pump discharge line to the check valve fitting. Secure the connection with a new tie-strap. Then bleed the injection system as described in this chapter.
10. Reconnect the oil pump link rod to the control arm. Adjust the link rod as specified in Chapter Five.

Oil Injection System Service

Oil pump synchronization

Refer to Chapter Five for throttle linkage-to-oil pump linkage adjustment and synchronization procedures.

Bleeding the oil pump

CAUTION
*Do not run the engine without an adequate water supply and do not exceed 3000 rpm without an adequate load. Refer to **Safety Precautions** in Chapter Three.*

If the pump has been removed, any of the lines replaced or air is present in the oil pump lines, bleed air from the oil injection pump and lines as follows:

1. Place a shop towel beneath the oil pump.
2. Loosen the oil pump bleed screw 3-4 turns (**Figure 2**).
3. Allow oil to flow from the bleed screw until no air bubbles are in the inlet hose.
4. Tighten the bleed screw to the specification in **Table 3**.
5. If air is present in the pump discharge line, connect a remote tank containing a 50:1 fuel/oil mixture to the engine. Start the engine and run it at idle until no air bubbles are in the discharge hose. To speed the bleeding process, disconnect the pump link rod and rotate the pump arm to the full-output position. Reconnect the link rod after air bleeding is complete.

Oil tank removal/installation

1. Disconnect the negative battery cable.
2. Disconnect and ground the spark plug leads to the power head to prevent accidental starting.

3A. *40-50 hp models*—Remove the screws securing the oil tank to the bracket just above the top carburetor.

3B. *All other models*—Remove the screws securing the oil tank bracket to the top of the electric starter motor.

4. Disconnect the oil level switch blue wires at the switch's bullet connectors.
5. Lift the oil tank enough to access the bottom of the tank. Cut the tie-strap securing the oil outlet hose to the oil tank, then disconnect the hose from the tank.
6. Remove the oil tank from the power head.
7. Remove the screw securing the oil level switch to the bottom of the tank. Remove the switch.
8. To install the oil tank, install the oil level switch into the oil tank. Secure the switch with one screw. Tighten the screw securely.
9. Position the oil tank assembly to the power head. Then attach the oil outlet hose to the oil tank. Secure the hose with a new tie-strap.
10. Secure the oil tank with screws. Tighten the screws securely.
11. Connect the oil level switch wires to the wiring harness bullet connectors.
12. Fill the oil tank with the recommended oil (Chapter Four).
13. Bleed the oil pump as described in this chapter.
14. Reconnect the negative battery cable and the spark plug leads.

Oil pump removal/installation

Refer to Chapter Eight for oil pump drive gear removal/installation procedures. Refer to **Figure 3** for this procedure.

1. Disconnect the oil inlet and discharge lines from the pump (**Figure 2**).
2. Disconnect the link rod from the pump control arm, then remove the pump mounting screws.
3. Remove the pump assembly from the power head. If the pump driven gear remains in the engine block, retrieve it using needlenose pliers.
4. Remove and inspect the bushings from the bearing sleeve. Replace all worn or damaged components.
5. To install the pump, thoroughly lubricate the bearing sleeve and two bushings with Quicksilver Needle Bearing Assembly Grease (part No. 92-825265A-1). Install the bushings into the sleeve. Make sure the flanged bushing faces the oil pump driven gear.
6. Thoroughly lubricate the driven gear shaft with Quicksilver Needle Bearing Assembly Grease. Insert the driven gear into the bearing assembly. Make sure the gear properly engages the pump shaft.
7. Install new O-rings onto the pump. Lubricate the O-rings with Quicksilver Needle Bearing Assembly Grease.
8. Install the pump assembly onto the power head. Apply Loctite 271 threadlocking adhesive (part No. 92-809819) to the threads of the pump mounting screws. Install the screws and tighten them to the specification in **Table 3**.
9. Reconnect the oil inlet and discharge hoses to the pump. Securely clamp the hoses to the pump using new tie-straps.
10. Reconnect the link rod to the pump control arm. Refer to Chapter Five for oil pump adjustment and synchronization procedures.
11. Perform the oil pump bleeding procedure as described in this chapter.

Table 1 OIL INJECTION SYSTEM CAPACITY

	fl. oz. (L)	Run time*
Oil tank capacity		
30 and 40 hp (two-cylinder)	50.5 (1.5)	4.7 hours
40-60 hp (three-cylinder)	96 (2.8)	7.0 hours
Reserve capacity		
30 and 40 hp (two-cylinder)	7 (0.21)	30 minutes
40-60 hp (three-cylinder)	14.5 (0.43)	30 minutes

*Approximate running time at wide-open throttle

Table 2 OIL PUMP OUTPUT SPECIFICATION

	Output fl. oz. (cc)	Time and engine speed
30 and 40 hp (two-cylinder)	0.25-0.32 (7.6-9.4)	10 minutes at 900 rpm
40 hp (three-cylinder)	0.41-0.61 (12-18)	10 minutes at 1500 rpm
50-60 hp and 45 Jet	0.64-0.85 (19-25)	10 minutes at 1500 rpm

Table 3 TORQUE SPECIFICATIONS

Fastener	ft.-lb.	in.-lb.	N•m
Oil pump to power head			
30 and 40 hp (two-cylinder)	–	50	5.6
40-60 hp (three-cylinder)	–	60	6.8
Oil tank			
40-60 hp (three-cylinder)	–	60	6.8
Air bleed screw	–	25	2.8

Table 4 STANDARD TORQUE SPECIFICATIONS—U.S. STANDARD AND METRIC FASTENERS

Screw or nut size	ft.-lb.	in.-lb.	N•m
U.S. standard fasteners			
6-32	–	9	1
8-32	–	20	2.3
10-24	–	30	3.4
10-32	–	35	4.0
12-24	–	45	5.1
1/4-20	6	72	8.1
1/4-28	7	84	9.5
5/16-18	13	156	17.6
5/16-24	14	168	19
3/8-16	23	270	31.2
3/8-24	25	300	33.9
7/16-14	36	–	48.8
7/16-20	40	–	54
1/2-13	50	–	67.8
1/2-20	60	–	81.3
Metric fasteners			
M5	–	36	4.1
M6	6	72	8.1
M8	13	156	17.6
M10	26	312	35.3
M12	35	–	47.5
M14	60	–	81.3

Chapter Thirteen

Manual Rewind Starters

Manual start models are equipped with a rope-operated rewind starter assembly. The starter assembly is mounted above the flywheel on all models. Pulling the rope handle causes the starter rope pulley to rotate, engage the flywheel or flywheel cup and rotate the engine.

Rewind starters are relatively trouble-free. A broken or frayed rope is the most common problem.

Table 1 lists the torque specifications for most fasteners. If the specification is not in **Table 1**, tighten the fastener to the standard torque specification listed in **Table 2**. **Table 1** and **Table 2** are at the end of the chapter.

This section covers starter removal, disassembly, cleaning and inspection, assembly and installation procedures.

WARNING
Wear suitable eye protection and gloves when servicing the manual starter. The starter spring may unexpectedly release from the housing with considerable force and result in injury. Follow all instructions carefully and wear suitable protection to minimize the risk.

2.5 and 3.3 hp Models

Removal and installation

Refer to **Figure 1** for this procedure.

1. Disconnect and ground the spark plug lead to the power head to prevent accidental starting.
2. Remove the screws securing the rewind starter assembly to the power head. Lift the starter off the power head.
3. To install the starter, place the starter on the power head. If necessary, extend the starter rope slightly to engage the ratchet with the starter cup.
4. Install the starter mounting screws. Tighten the screws securely.
5. Reconnect the spark plug lead.

Disassembly

Refer to **Figure 1** for this procedure.

1. Remove the starter as described in this chapter. Untie the knot securing the rope handle to the rope. Keep the rope pulley from turning and remove the handle from the rope.
2. Allow the rope pulley to slowly unwind, releasing the tension on the rewind spring.
3. Invert the starter and remove the lock clip (10, **Figure 1**), thrust washer (9) and friction plate (8).
4. Remove the return spring (7, **Figure 1**), cover (6) and friction spring (5).
5. Carefully lift the rope pulley approximately 1/2 in. (13 mm) out of the housing, then turn the pulley back and forth to disengage the rewind spring from the pulley.

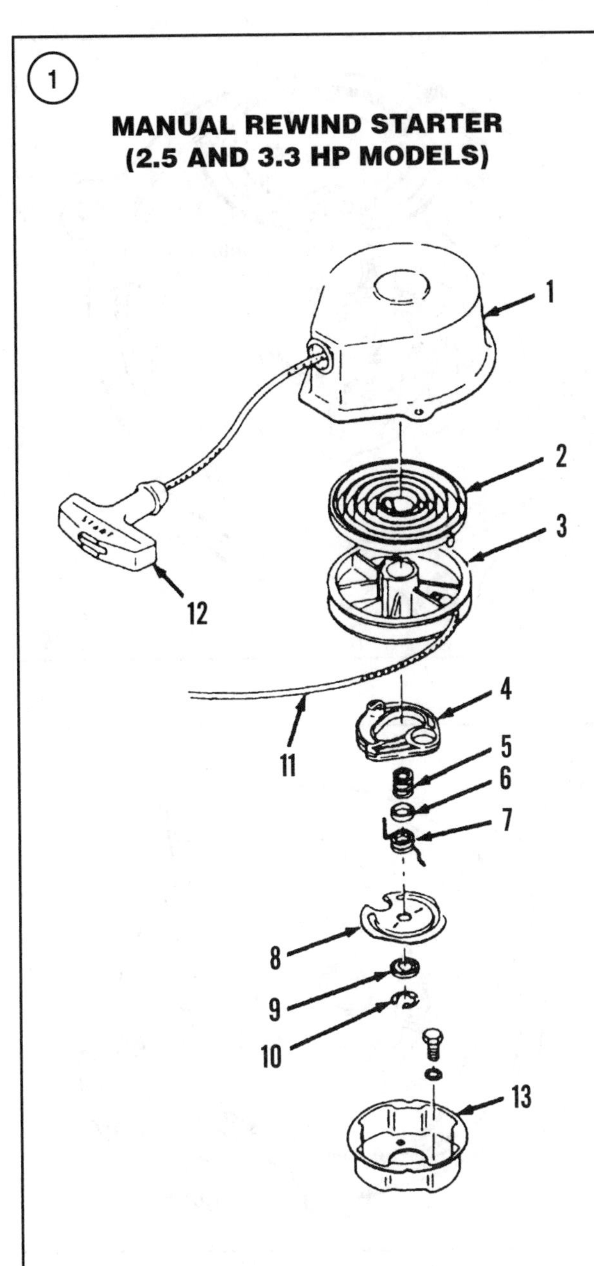

1. Starter housing
2. Rewind spring
3. Rope pulley
4. Ratchet
5. Friction spring
6. Friction spring cover
7. Return spring
8. Friction plate
9. Thrust washer
10. Lock clip
11. Starter rope
12. Rope handle
13. Starter cup (on flywheel)

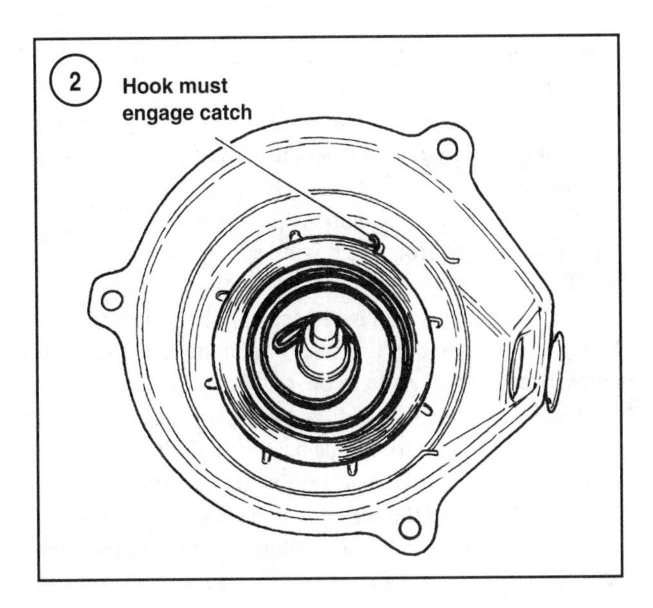

6. Remove the pulley from the housing. Remove the rope from the pulley.
7. If rewind spring replacement is necessary, place the starter housing upright with the rewind spring facing down on a suitable bench. Tap the top of the housing until the rewind spring falls out and unwinds inside the housing.

Cleaning and inspection

1. Clean all components in solvent, then dry them with compressed air.
2. Inspect the rewind spring for kinks, burrs, corrosion, cracks or other damage.
3. Inspect the starter pulley and housing for nicks, cracks, excessive wear or other damage.
4. Inspect the neutral start interlock components for excessive wear or other damage.
5. Inspect the starter rope for excessive wear, cuts, fraying or other damage.

Assembly

NOTE

Quicksilver replacement starter rope is precut to the correct length. When using an aftermarket starter rope, cut the rope to the same length as the original.

1. Lubricate the rewind spring area of the starter housing with Quicksilver 2-4-C Multi-Lube grease (part No. 92-825407).
2. Carefully wind the spring (2, **Figure 1**) into the starter housing in a counterclockwise direction, starting from the outer coil as shown in **Figure 2**. Make sure the hook in the

outer coil of the spring is properly engaged with the catch in the housing as shown.

3. Insert the rope through the hole in the pulley. Tie a single knot in the rope, then push the knot into the recess of the pulley.
4. Wind the starter rope onto the pulley in a clockwise direction as viewed from the rewind spring side of the pulley.
5. Install the pulley and rope assembly into the housing. Make sure the hook on the inner coil of the rewind spring properly engages the slot in the pulley.
6. Install the ratchet (4, **Figure 1**), friction spring (5) and cover (6) onto the starter housing center shaft.
7. Hook one end of the return spring into the friction plate and the other end of the spring into the ratchet. See **Figure 3**. Install the friction plate and return spring onto the center shaft. Install the thrust washer (9, **Figure 1**) and the lock clip (10).
8. Insert the rope into the notch (**Figure 4**) in the pulley. While holding the rope in the notch, rotate the rope pulley three turns counterclockwise to apply tension to the rewind spring.
9. Hold the rope securely, then pass the rope through the rope guide in the housing and into the rope handle. Fix the handle to the rope with a knot as shown in **Figure 5**.
10. Install the starter assembly onto the power head as described previously in this section.

4 and 5 hp Models

This rewind starter has a starter interlock system to prevent the starter from operating if the engine shift linkage is in forward or reverse gear.

The starter rope may be replaced without completely disassembling the starter. Refer to *Starter rope replacement only (4 and 5 hp models)*.

Starter removal and installation

Refer to **Figure 6** for this procedure.

1. Disconnect and ground the spark plug lead to the power head to prevent accidental starting.
2. Disconnect the interlock rod from the lever (**Figure 7**).
3. Remove the screws holding the starter assembly to the power head. Lift the starter off the power head.
4. Position the starter on the power head and secure it with screws and lockwashers. Tighten the starter mounting screws to the specification in **Table 1**.
5. Reconnect the interlock rod to the lever.
6. Reconnect the spark plug lead.

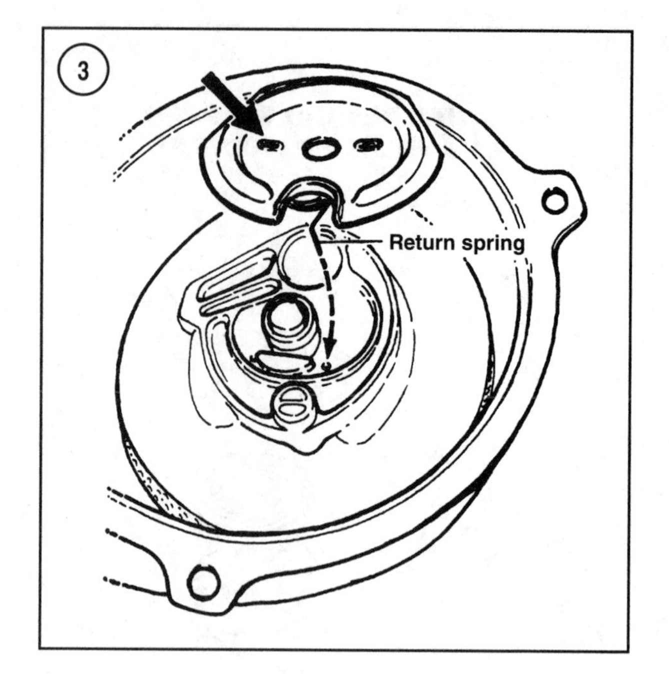

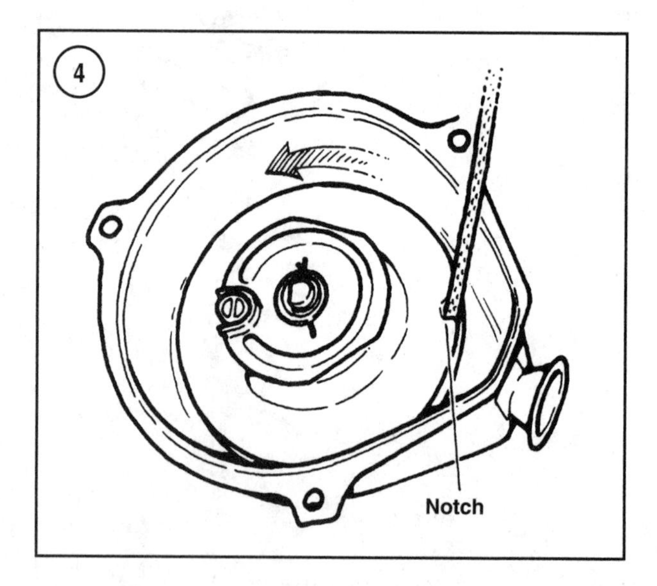

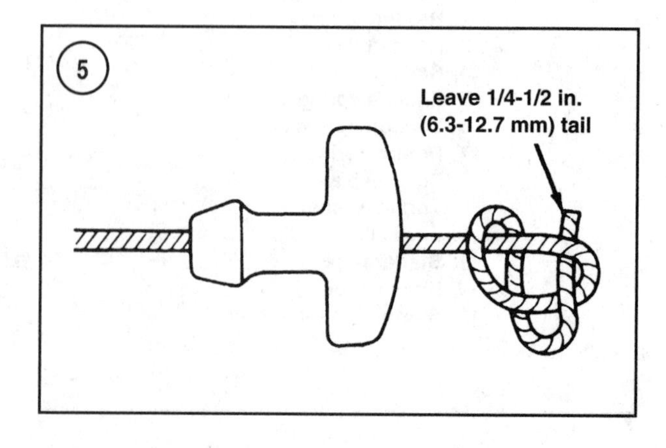

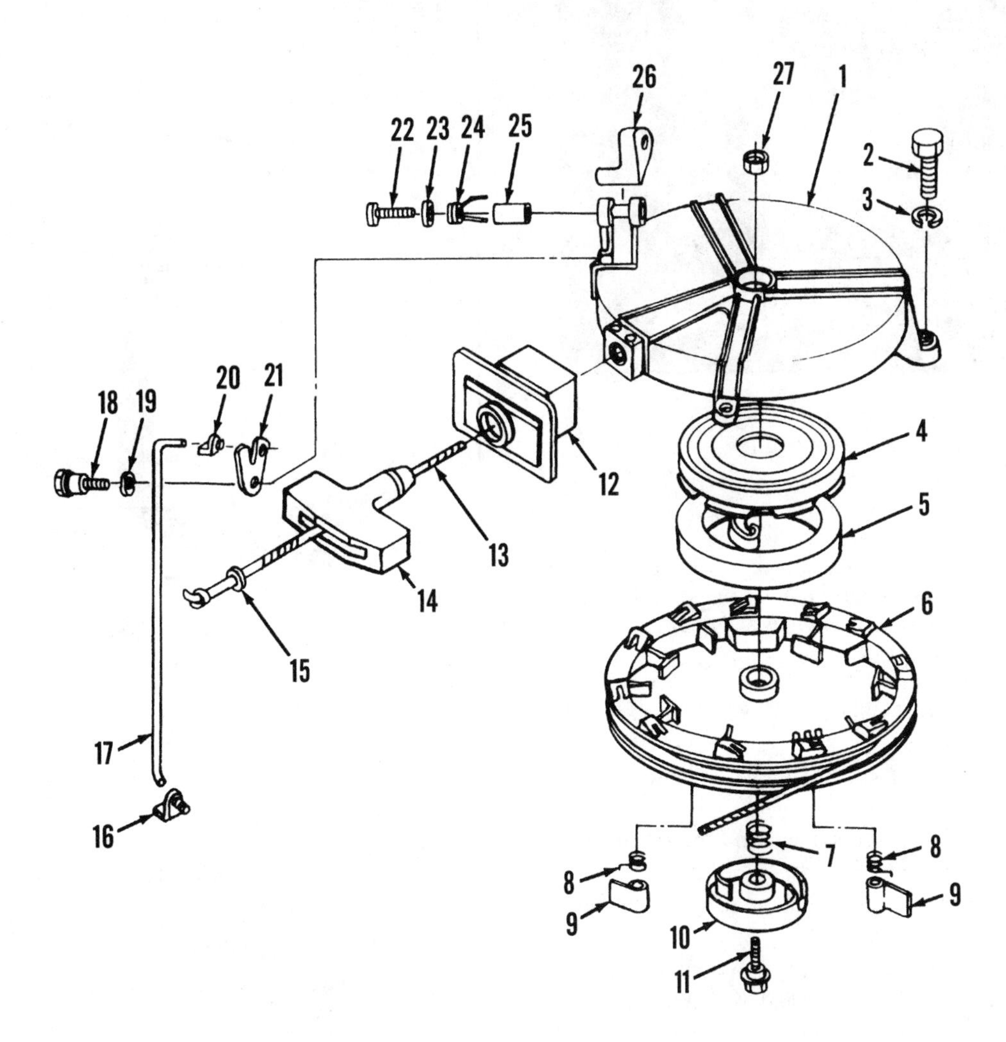

1. Starter housing
2. Screw
3. Lockwasher
4. Rewind spring case
5. Rewind spring
6. Rope pulley
7. Friction spring
8. Pawl return springs
9. Starter pawls
10. Friction plate
11. Shoulder bolt
12. Rope guide
13. Starter rope
14. Starter handle
15. Washer
16. Locking clip
17. Interlock link
18. Shoulder screw
19. Wave washer
20. Locking clip
21. Lever
22. Screw
23. Washer
24. Spring
25. Collar
26. Interlock arm
27. Nut

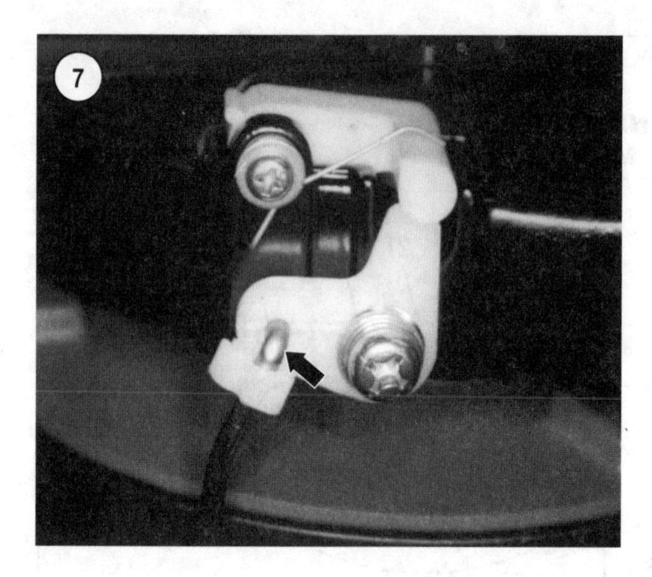

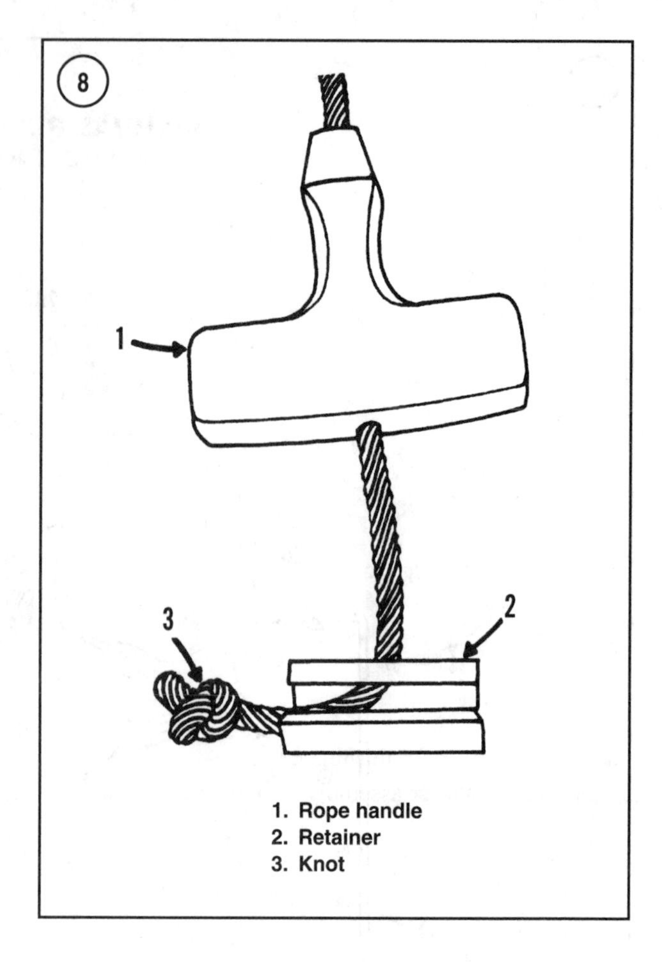

Starter rope replacement only (4 and 5 hp models)

NOTE
Quicksilver replacement starter rope is precut to the correct length. When using an aftermarket starter rope, cut the rope to the same length as the original.

1. Remove the starter assembly as described previously in this section.

2A. If the rope is not broken, pull the starter handle to extend the rope fully.

a. Have an assistant or a suitable tool keep the rope pulley from turning.
b. Lift the knot in the rope from the recess in the pulley.
c. Untie or cut the knot, then untie or cut the knot securing the rope to the handle and washer or retainer.

2B. If the rope is broken, remove the remaining rope from the starter pulley by untying or cutting the knot. Then untie or cut the knot securing the rope to the handle and washer or retainer.

3. Attach the replacement rope to the rope handle and retainer as shown in **Figure 8**. Secure it with a knot as shown in **Figure 5**. Insert the rope into the rope guide and through the hole in the rope pulley as shown in **Figure 9**. Tie a knot as shown in **Figure 5**, then push the knot into the recess in the pulley.

4A. If the original rope was not broken, slowly release the pulley, allowing it to rewind the remaining rope slowly. Make sure the rope rewinds completely.

4B. If the original rope was broken, turn the rope pulley clockwise as viewed from the bottom or pulley side to wind the rope onto the pulley, then hold the rope into the slot in the outer diameter of the pulley. Turn the pulley counterclockwise to tension the spring. The spring must have at least a half turn of tension, but less than 1 1/2 turns of tension. When it is properly tensioned, release the rope from the slot and allow the rope to rewind completely.

5. Reinstall the starter as described previously in this section.

Disassembly

WARNING
Wear suitable eye protection and gloves when servicing the manual starter. The starter spring may unexpectedly release from the housing with considerable force and result in injury. Follow all instructions carefully and wear suitable protection to minimize the risk.

CAUTION
Do not remove the rewind spring from the case. The spring and case are replaced as an assembly.

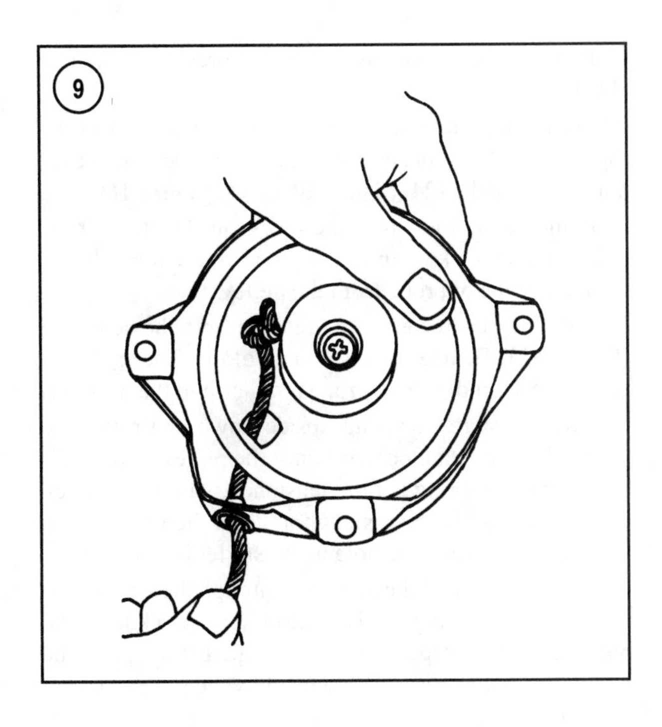

Refer to **Figure 6** for this procedure.

1. Remove the starter assembly as described previously in this section.
2. Fully extend the starter rope, then keep the rope pulley from turning.
3. Lift the knot in the rope from the recess in the pulley. Untie or cut the knot, then allow the rope pulley to unwind slowly, relieving the tension on the rewind spring.
4. If necessary, untie or cut the knot securing the rope to the handle and retainer or washer. See **Figure 8**.
5. While holding the nut (27, **Figure 6**), remove the shoulder bolt (11). Remove the friction plate (10, **Figure 6**) and friction spring (7). Remove the pawls and pawl return springs from the rope pulley. Carefully remove the rope pulley from the starter housing. Do not dislodge the rewind spring.
6. Lift the rewind spring and spring case out of the rope pulley if they are corroded, broken or otherwise defective.
7. Further disassembly is not necessary to replace the rewind spring or rope. If necessary, remove the rope guide (12, **Figure 6**) and interlock components (21-26) from the housing.

Cleaning and inspection

1. Clean all components in solvent, then dry them with compressed air.
2. Inspect the rewind spring for kinks, burrs, corrosion, cracks or other damage.
3. Inspect the starter pulley and housing for nicks, cracks, excessive wear or other damage.
4. Inspect the neutral start interlock components for excessive wear or other damage.
5. Inspect the starter rope for excessive wear, cuts, fraying or other damage.

Assembly

Refer to **Figure 6** for this procedure.

1. If the spring was inadvertently removed from the spring case:
   a. Lubricate the spring cavity of the rewind spring case with a thin coat of Quicksilver 2-4-C Multi-Lube grease (part No. 92-825407).
   b. Insert the hook in the outer coil of the rewind spring into the slot in the outer rim of the spring case.
   c. Wind the rewind spring into the case in a clockwise direction.
2. Reinstall the rope guide and neutral start interlock components if they were removed. Apply Loctite 242 threadlocking adhesive (part No. 92-809821) to the threads of the screw (22, **Figure 6**) and tighten it securely.
3. Lubricate the inner hub of the starter housing with a small amount of Quicksilver 2-4-C Multi-Lube grease.
4. Install the starter pawls and pawl return springs into the rope pulley.
5. Install the rewind spring and case assembly into the rope pulley. Install the rope pulley into the starter housing.
6. Lubricate the friction plate with a small amount of Quicksilver 2-4-C Multi-Lube, then install the friction spring and plate. Apply Loctite 242 threadlocking adhesive to the threads of the shoulder bolt (11, **Figure 6**). Install the bolt and nut (27), and tighten them to the specification in **Table 1**.
7. Attach the rope to the rope handle and retainer as shown in **Figure 8**. Secure it with a knot as shown in **Figure 5**. Insert the rope into the rope guide and through the hole in the rope pulley as shown in **Figure 9**. Tie a knot as shown in **Figure 5**, then push the knot into the recess in the pulley.
8. Wind the rope onto the pulley by turning the pulley 2 1/2 turns clockwise as viewed from the bottom of the starter. Then, place the rope into the notch in the outer rim of the pulley and turn the pulley three turns counterclockwise. Slowly release the pulley, allowing it to rewind the remaining rope. Make sure the rope rewinds completely.
9. Install the starter assembly onto the power head as described previously in this section.

6-25 hp and 20 Jet Models

These rewind starters uses a starter interlock system to prevent the starter from operating if the engine shift linkage is in forward or reverse gear. There are slight differences between the 6-15 hp, 20-25 hp and 20 jet models, but the basic service procedures are the same.

The starter rope may be replaced without disassembling the starter. Refer to *Starter rope replacement only (6-25 hp models)*.

Removal and installation

Refer to **Figure 10** for 6-15 hp models or **Figure 11** for 20 and 25 hp, and 20 jet models for the following procedures.

1. Disconnect and ground the spark plug leads to the power head to prevent accidental starting.
2. On models so equipped, pry and/or pull the fuel filter assembly straight down from the rope guide. Do not disconnect the fuel hoses.
3. Disconnect the interlock rod from the lower lock lever (**Figure 12**, typical).
4. Remove the starter mounting screws, then lift the starter assembly from the power head.
5. To reinstall, place the starter in position on the power head. Install the mounting fasteners and tighten them to the specification in **Table 1**.
6. Reconnect the interlock rod to the lower lock lever.
7. On models so equipped, reinstall the fuel filter assembly by pushing the filter straight up into its socket on the rope handle guide until it snaps into place. Do not break the fuel line fittings on the filter housing.
8. Reconnect the spark plug leads.

Starter rope replacement only (6-25 hp models)

CAUTION
Hold the rope pulley continuously in the following procedure to prevent sudden unwinding of the spring. Enlist the aid of an assistant or hold the rope pulley with a suitable clamping tool during rope replacement.

NOTE
Quicksilver replacement starter rope is precut to the correct length. When using an aftermarket starter rope, cut the rope to the same length as the original.

1. Remove the starter as described previously in this chapter.
2. Pry the rope retainer from the starter handle. Cut the rope or untie the knot and remove the old rope from the retainer and handle (24, **Figure 10**, or 14, **Figure 11**).
3. Remove the knot from the recess in the starter rope pulley. Untie the knot or cut the rope and remove the remainder of the old rope from the pulley.
4. Install the handle and rope retainer onto the new rope (**Figure 13**). Tie a knot into the end of the rope as shown in **Figure 14** and push the rope retainer into the handle.
5. Turn the starter housing upside down and rotate the rope pulley counterclockwise until the pulley stops, indicating that the spring is fully tensioned. Allow the pulley to rotate clockwise just enough to align the rope hole in the pulley with the rope hole in the starter handle rest.
6. Thread the end of the new rope through the starter handle rest, rope bushing and into the pulley. Tie a knot in the pulley end of the rope as shown in **Figure 9** and place the knot in the pulley recess. Slowly allow the pulley to wind the remaining rope.
7. Reinstall the starter as described previously in this section.

Disassembly

WARNING
Wear suitable eye protection and gloves when servicing the manual starter. The starter spring may unexpectedly release from the housing with considerable force and result in injury. Follow all instructions carefully and wear suitable protection to minimize the risk.

Refer to **Figure 10** or **Figure 11** for this procedure.

1. Remove the starter assembly from the power head as described in this chapter.
2. Invert the starter. Fully extend the starter rope and keep the rope pulley from turning. Pull the knot in the rope from the recess in the pulley, then untie or cut the knot from the rope. Allow the pulley to unwind slowly, relieving the rewind spring tension.
3. Remove the shoulder bolt (12, **Figure 10** or 20, **Figure 11**). Remove the cam plate and friction spring.
4. Lift the rope pulley off the housing center shaft.
5. Lift the rewind spring and case assembly from the housing, then remove the felt pad.
6. If necessary, remove the lock clips securing the starter pawls (8, **Figure 10** or 16, **Figure 11**). Remove the pawls and pawl return springs from the rope pulley.

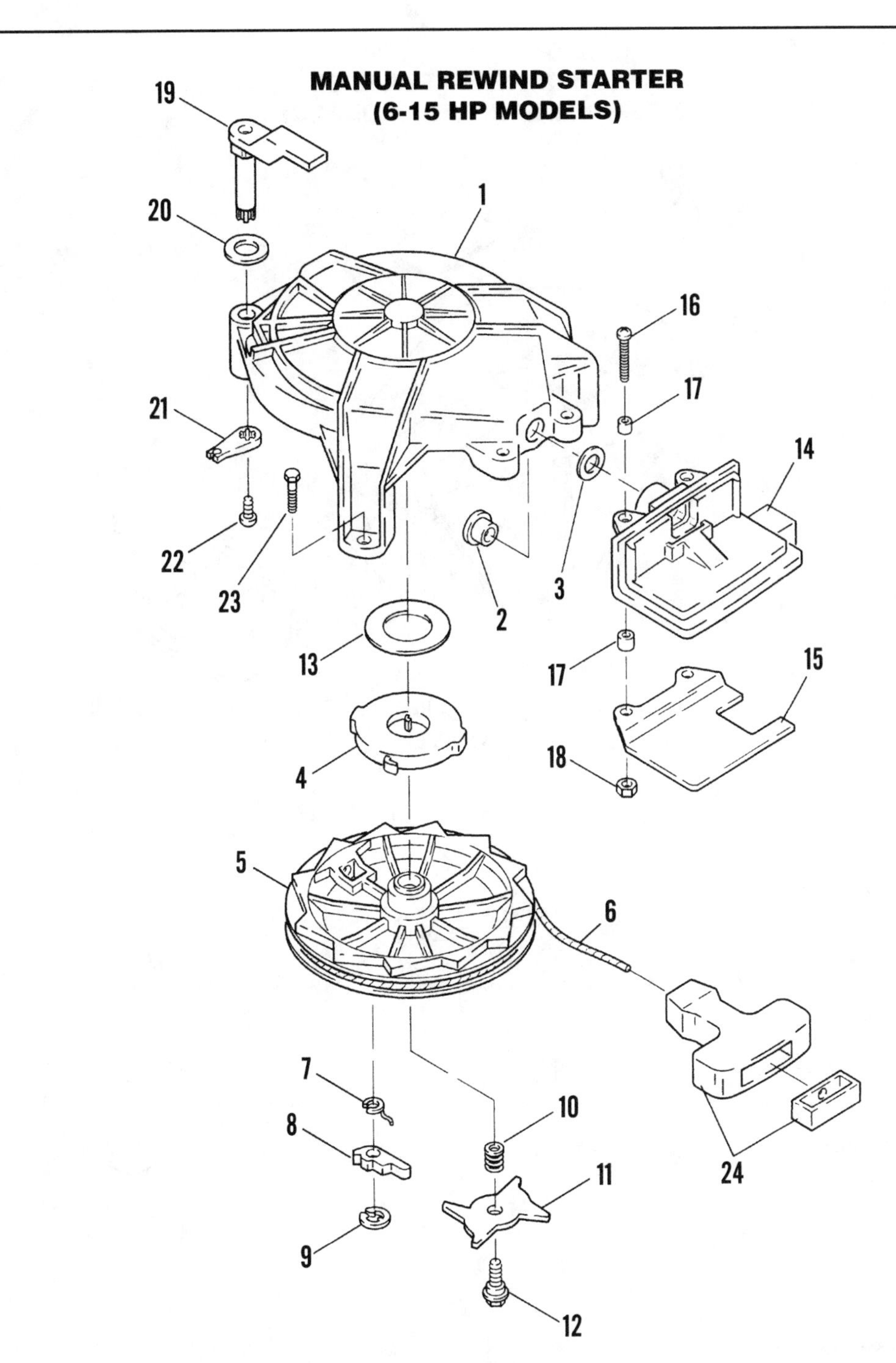

1. Starter housing
2. Rope bushing
3. Bushing retaining ring
4. Starter spring assembly
5. Rope pulley
6. Starter rope
7. Pawl return spring
8. Starter pawl
9. Lock clip
10. Friction spring
11. Cam plate
12. Shoulder bolt
13. Felt washer
14. Starter handle rest
15. Support plate
16. Screw
17. Spacer
18. Locknut
19. Upper interlock lever
20. Washer
21. Lower interlock lever
22. Screw
23. Screw
24. Handle and rope retainer

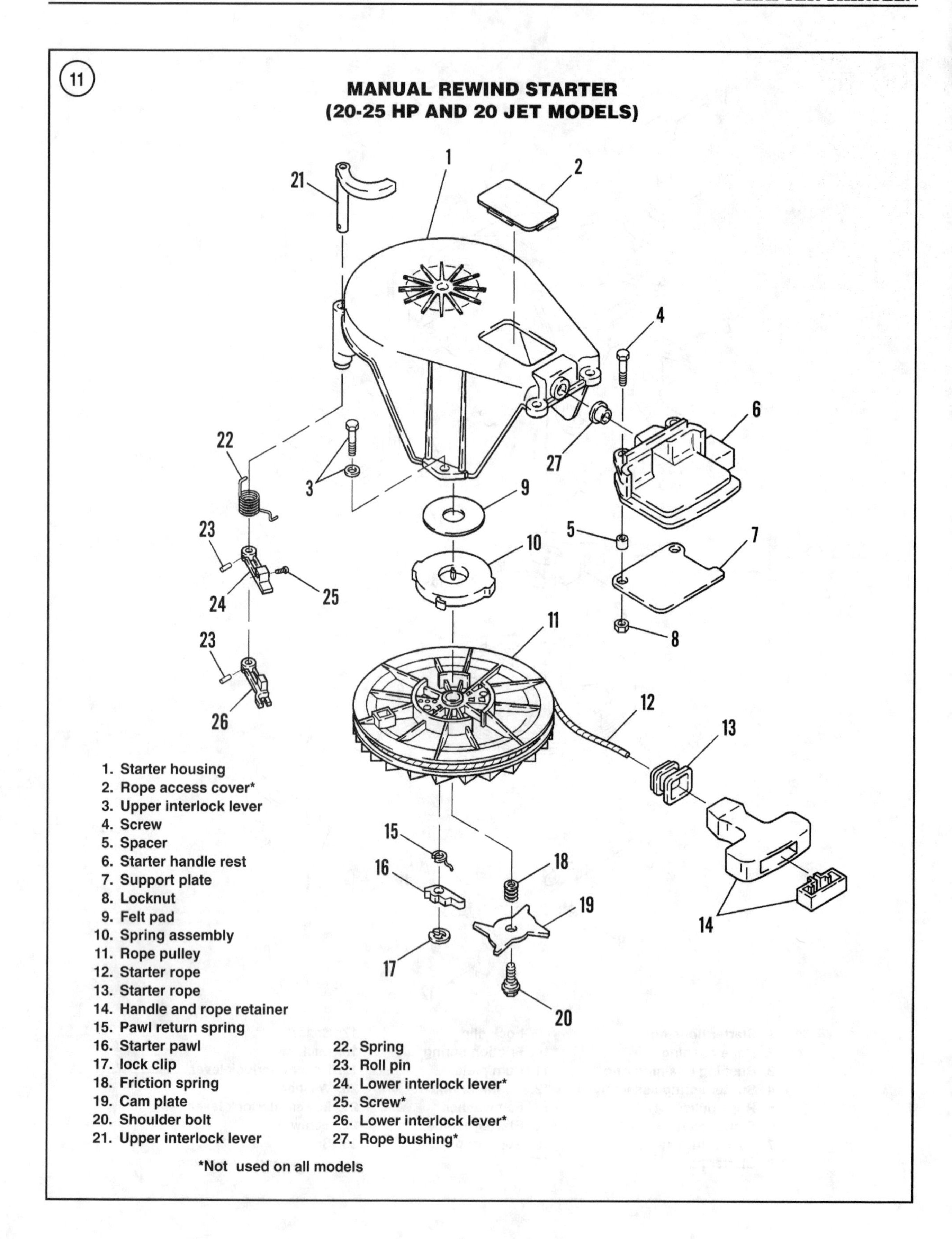
11
MANUAL REWIND STARTER
(20-25 HP AND 20 JET MODELS)
1
2
21
4
6
22
27
3
9
23
5
10
7
24
25
11
23
8
12
26
13
15
18
16
19
14
17
20
1. Starter housing
2. Rope access cover*
3. Upper interlock lever
4. Screw
5. Spacer
6. Starter handle rest
7. Support plate
8. Locknut
9. Felt pad
10. Spring assembly
11. Rope pulley
12. Starter rope
13. Starter rope
14. Handle and rope retainer
15. Pawl return spring
16. Starter pawl
17. lock clip
18. Friction spring
19. Cam plate
20. Shoulder bolt
21. Upper interlock lever
22. Spring
23. Roll pin
24. Lower interlock lever*
25. Screw*
26. Lower interlock lever*
27. Rope bushing*
*Not used on all models

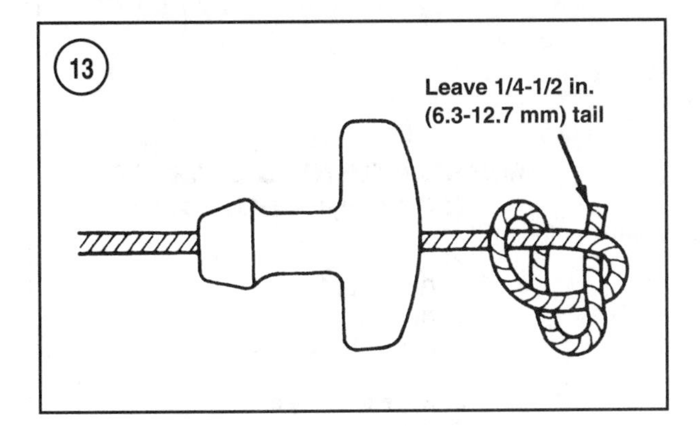

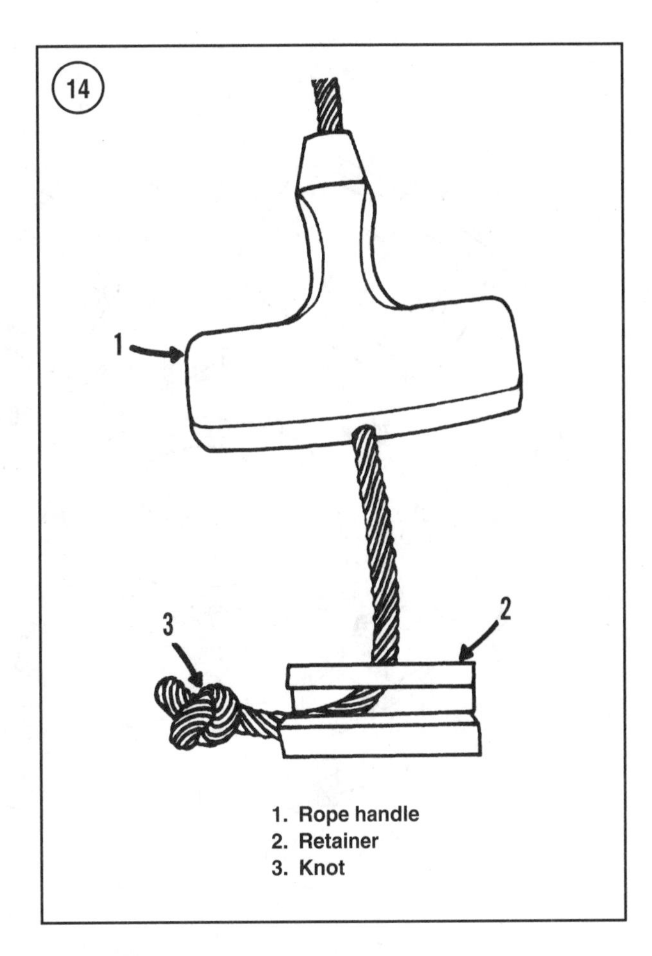

7. If necessary, remove the interlock components by removing the lower lever screw on 6-15 hp models or driving the roll pin from the lower lever on 20 and 25 hp, and 20 jet models.
8. If necessary, remove the starter handle rest and support plate by removing the two screws and locknuts, and the four spacers. The rope bushing in the starter housing on 6-15 hp models and in the starter handle rest on 20 and 25 hp, and 20 jet models is replaceable.

NOTE

It is not necessary to remove the rewind spring from its case for inspection. If the spring requires replacement, replace the spring and case as an assembly.

Cleaning and inspection

1. Clean all components in solvent and dry them with compressed air.
2. Inspect the rewind spring in its case for cracks, burrs, breakage or other damage. Replace the spring and case as an assembly if necessary.
3. Inspect the rope pulley and starter housing for cracks, breaks, nicks, grooves, distortion or other damage. Inspect the pawl pivot pins on the pulley for excessive wear. Replace the rope pulley as necessary.
4. Check the starter pawls and pawl springs for excessive wear or breaks and replace them as necessary. Check the starter cam plate for excessive wear and replace it if necessary.
5. Inspect the starter rope for excessive wear. Replace the rope if necessary.

Assembly

CAUTION

Hold the rope pulley continuously during rope installation to prevent sudden unwinding of the spring. Enlist the aid of an assistant or hold the rope pulley with a suitable clamping tool during rope installation.

NOTE

Quicksilver replacement starter rope is precut to the correct length. When using an af-

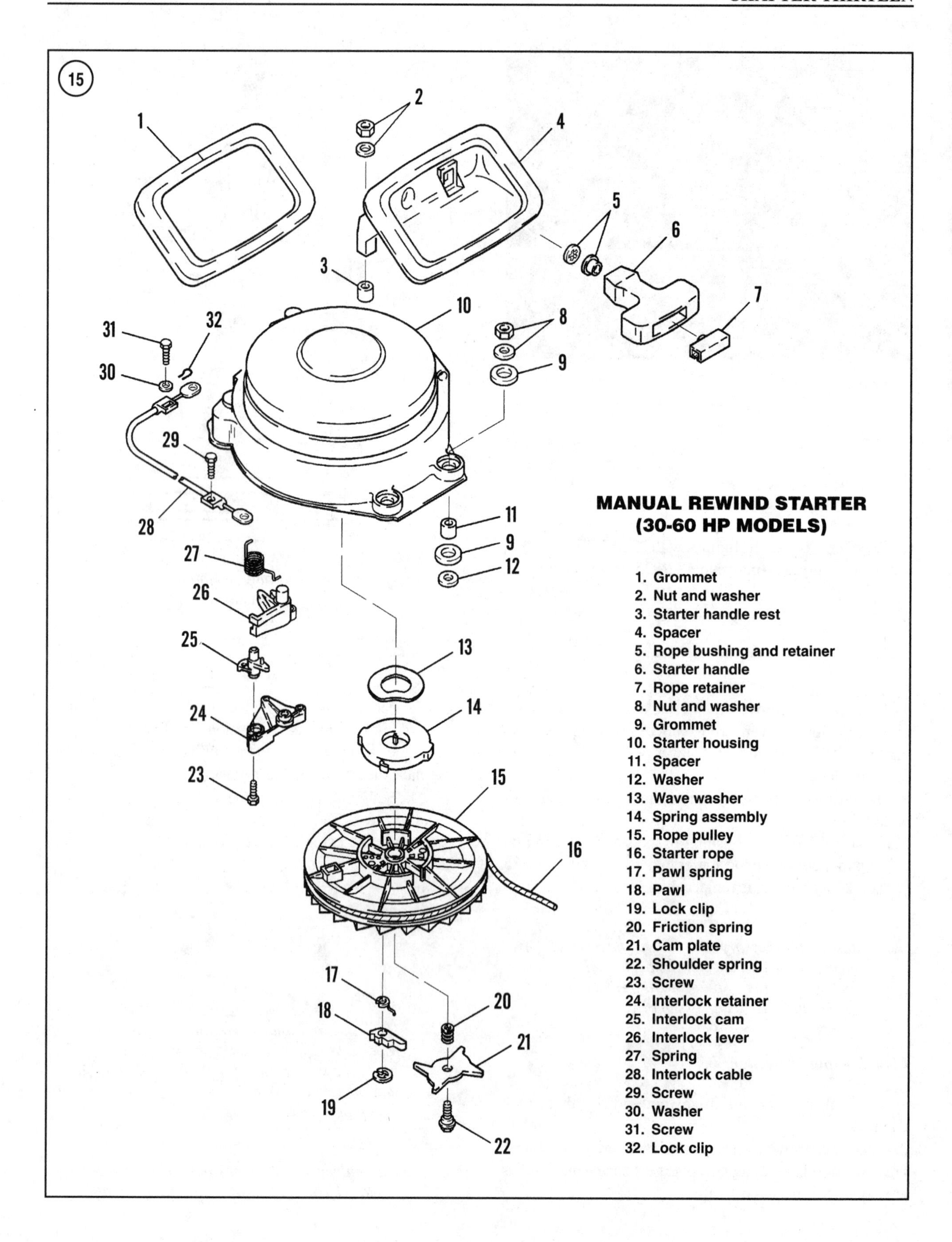
15
1
2
3
4
5
6
7
8
9
10
11
9
12
13
14
15
16
17
18
19
20
21
22
23
24
25
26
27
28
29
30
31
32
MANUAL REWIND STARTER
(30-60 HP MODELS)
1. Grommet
2. Nut and washer
3. Starter handle rest
4. Spacer
5. Rope bushing and retainer
6. Starter handle
7. Rope retainer
8. Nut and washer
9. Grommet
10. Starter housing
11. Spacer
12. Washer
13. Wave washer
14. Spring assembly
15. Rope pulley
16. Starter rope
17. Pawl spring
18. Pawl
19. Lock clip
20. Friction spring
21. Cam plate
22. Shoulder spring
23. Screw
24. Interlock retainer
25. Interlock cam
26. Interlock lever
27. Spring
28. Interlock cable
29. Screw
30. Washer
31. Screw
32. Lock clip

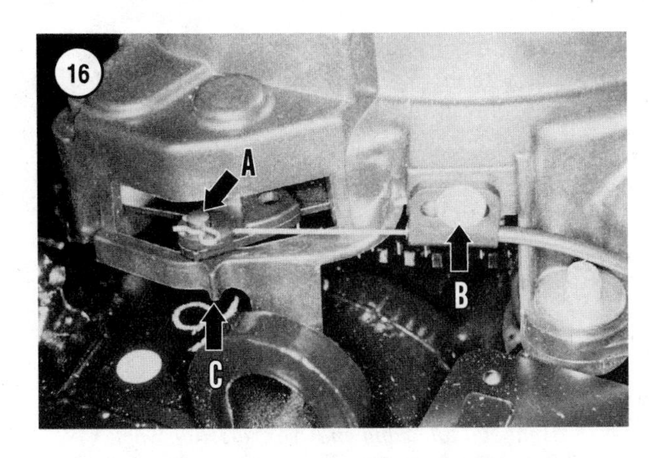

termarket starter rope, cut the rope to the same length as the original.

Refer to **Figure 10** or **Figure 11** for this procedure.

1. Install the starter handle rest and support plate if they were removed. Tighten the screws and locknuts securely.

2A. *6-15 hp models*—Install the interlock components to the starter housing if they were removed. Lubricate the upper lever and shaft lightly with Quicksilver 2-4-C Multi-Lube grease (part No. 92-825407). Secure the lower lever to the shaft with a screw. Securely tighten the screw.

2B. *20-25 and 20 jet models*—Install the interlock components to the starter housing if they were removed. Lubricate the upper lever and shaft lightly with Quicksilver 2-4-C Multi-Lube grease (part No. 92-825407). Install the spring, then the lower lever arm. Secure the arm in position with the roll pin. The spring must hold the interlock lever inward, toward the starter housing.

3. Install the starter pawls and springs onto the rope pulley pivot pins if they were removed. Lubricate the pivot pins with Quicksilver 2-4-C Multi-Lube grease and secure the pawls with the lock clips.

4. Place the felt pad in place over the housing center shaft. Hold the felt pad in place with a small amount of 2-4-C Multi-Lube grease.

5. Lubricate the rewind spring with 2-4-C Multi-Lube grease. Install the spring and case assembly into the rope pulley. The spring side should face the pulley. Carefully align the tabs with rope pulley recesses and seat the spring case into the rope pulley.

6. Install the rope pulley and rewind spring assembly onto the housing center shaft. Turn the pulley to engage and align the rewind spring to the housing shaft. If the rope pulley and rewind spring are properly seated, the housing center shaft and the pawl side of the rope pulley will be flush. If the rope pulley does not seat, hold the upper interlock lever up to prevent interference with the pulley.

7. Install the friction spring and cam plate. Install the shoulder bolt and tighten it to the specification in **Table 1**.

8. Install the handle and rope retainer onto the rope (**Figure 13**). Tie a knot in the end of the rope as shown in **Figure 14** and push the rope retainer into the handle.

9. Turn the starter housing upside down and rotate the rope pulley counterclockwise until the pulley stops, indicating that the spring is fully tensioned. Allow the pulley to rotate clockwise just enough to align the rope hole in the pulley with the rope hole in the starter handle rest.

10. Thread the end of the new rope through the starter handle rest, rope bushing and into the pulley. Tie a knot in the pulley end of the rope as shown in **Figure 9** and place the knot in the pulley recess. Allow the pulley to slowly wind up the remaining rope.

11. Reinstall the starter as described in this chapter.

30-60 hp Models

These rewind starters use a starter interlock system to prevent the starter from operating if the engine shift linkage is in forward or reverse gear. The starter rope may be replaced without disassembling the starter. Refer to *Starter rope replacement only (30-60 hp models)*.

Refer to **Figure 15** for the following procedures.

Starter removal/installation

1. Disconnect and ground the spark plug leads to the power head to prevent accidental starting.

2. Disconnect the interlock cable from the interlock cam by removing the lock clip and washer (A, **Figure 16**). Then remove the screw (B, **Figure 16**) securing the cable to the starter housing.

3. Pull the starter rope out approximately 12 in. (304 mm) between the starter handle rest and the starter housing. Tie a slip knot in the starter rope and let the rewind spring pull the slip knot against the starter housing.

4. Pry the rope retainer from the starter handle. Cut the rope or untie the knot, and remove the old rope from the retainer and handle.

5. Remove the starter mounting nuts, washers and upper grommets, then lift the starter assembly from the power head. Do not lose the lower washers and grommets.

6. To reinstall, make sure the lower washer and grommets are in place on the four mounting studs. Then set the starter into position on the studs. Install the four upper grommets and washers, then the four nuts. Tighten the nuts to the specification in **Table 1**.

7. Reconnect the interlock cable to the interlock cam and secure it with a washer and lock clip. Then secure the ca-

ble to the starter housing with a screw. Tighten the screw hand-tight at this time.

8. Move the shift linkage fully into reverse gear while rotating the propeller. Then shift into neutral without going past the neutral detent.

9. Align the raised rib on the interlock cam with the raised rib on the starter housing (C, **Figure 16**) by moving the cable back and forth under the screw (B, **Figure 16**). Tighten the screw when the ribs are aligned. In **Figure 16** the ribs are not yet aligned.

10. Route the starter rope through the starter handle rest's rope hole.

11. Install the handle and rope retainer onto the rope (**Figure 13**). Tie a knot in the end of the rope as shown in **Figure 14** and push the rope retainer into the handle.

12. Untie the slip knot and allow the starter rope to retract fully.

13. Reconnect the spark plug leads.

Starter rope replacement only (30-60 hp models)

1. Remove the starter as described in this chapter.

2. With the starter inverted, fully extend the starter rope and keep the rope pulley from turning. Pull the knot from the recess in the pulley, then untie or cut the knot from the rope. Remove the rope.

3. Allow the pulley to unwind slowly, relieving the rewind spring tension.

CAUTION

Hold the rope pulley continuously in the following steps to prevent it from suddenly unwinding. Enlist the aid of an assistant or hold the rope pulley with a suitable tool to prevent it from unwinding while installing the rope.

NOTE

Quicksilver replacement starter rope is precut to the correct length. When using an aftermarket starter rope, cut the rope to the same length as the original.

4. Turn the starter housing upside down and rotate the rope pulley counterclockwise until the pulley stops, indicating that the spring is fully tensioned. Allow the pulley to rotate clockwise one full turn, then just enough to align the rope hole in the pulley with the rope hole in the starter handle rest.

5. Thread the end of the new rope through the starter housing rope hole and into the rope pulley recess. Tie a knot in the pulley end of the rope as shown in **Figure 14** and place the knot in the pulley recess.

6. Tie a slip knot in the starter handle end of the rope, approximately 12 in. (304 mm) from the end of the rope.

7. Slowly allow the pulley to wind the rope until the slip knot is against the starter housing.

8. Reinstall the starter as described previously in this section.

Disassembly

Refer to **Figure 15** for this procedure.

WARNING

Wear suitable hand and eye protection when servicing the starter. The starter spring may unexpectedly release from the housing with considerable force and result in injury. Follow all instructions carefully and wear suitable protection to minimize the risk.

NOTE

It is not necessary to remove the rewind spring from its case for inspection. If the spring requires replacement, replace the spring and case as an assembly.

1. Remove the starter assembly from the power head as described in this chapter.

2. Invert the starter and remove the screws securing the interlock retainer (24, **Figure 15**) from the starter housing. Then remove the cam retainer, the interlock cam and spring, and finally the interlock lever from the starter housing.

3. With the starter still inverted, fully extend the starter rope and keep the rope pulley from turning. Pull the knot from the recess in the pulley, then untie or cut the knot from the rope. Remove the rope. Allow the pulley to unwind slowly, relieving the rewind spring tension.

4. Remove the shoulder bolt (22, **Figure 15**). Remove the cam plate and friction spring.

5. Lift the rope pulley from the housing center shaft.

NOTE

The rewind spring and case are serviced as an assembly. Do not remove the spring from the case.

6. Lift the rewind spring and case assembly from the housing, then remove the wave washer.

7. If necessary, remove the lock clips (19, **Figure 15**) securing the starter pawls. Remove the pawls and pawl return springs from the rope pulley.

8. If necessary, remove the starter handle rest from the engine by removing the two locknuts, washers and spacers. The rope bushing in the starter hand rest is replaceable.

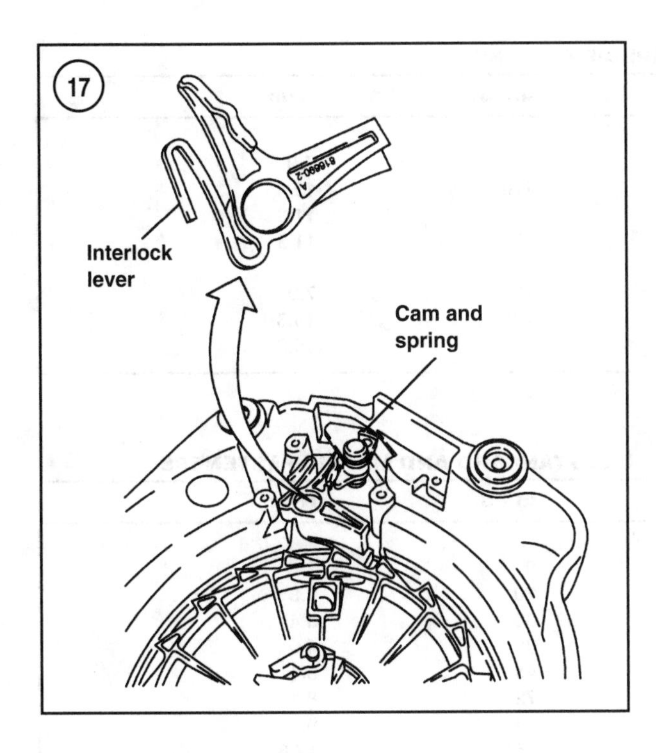

Cleaning and inspection

1. Clean all components in solvent and dry them with compressed air.
2. Inspect the rewind spring in its case for cracks, burrs, breaks or other damage. Replace the spring and case as an assembly if necessary.
3. Inspect the rope pulley and starter housing for cracks, breaks, nicks, grooves, distortion or other damage. Inspect the pawl pivot pins on the pulley for excessive wear. Replace the rope pulley if necessary.
4. Check the starter pawls and pawl springs for excessive wear or breaks and replace them as necessary. Check the starter cam plate for excessive wear and replace it if necessary.
5. Inspect the starter rope for excessive wear. Replace the rope as necessary.

Assembly

Refer to **Figure 15** for this procedure.

1. Install the starter handle rest if it was removed. Tighten the locknuts securely.
2. Install the starter pawls and springs onto the rope pulley pivot pins if they were removed. Lubricate the pivot pins with Quicksilver 2-4-C Multi-Lube grease (part No. 92-825407) and secure the pawls with the lock clips.
3. Place the wave washer in place over the housing center shaft. Hold the wave washer in place with a small amount of 2-4-C Multi-Lube grease.
4. Lubricate the rewind spring with 2-4-C Multi-Lube grease. Install the spring and case assembly into the rope pulley. The spring side must face the pulley. Carefully align the tabs with rope pulley recesses and seat the spring case into the rope pulley.
5. Install the rope pulley and rewind spring assembly onto the housing center shaft. Turn the pulley to engage and align the rewind spring to the housing shaft. If the rope pulley and rewind spring are properly seated, the housing center shaft and the pawl side of the rope pulley will be flush.
6. Install the friction spring and cam plate. Install the shoulder bolt and tighten it to the specification in **Table 1**.

CAUTION

Hold the rope pulley continuously in the following steps to prevent it from suddenly unwinding. Enlist the aid of an assistant or hold the rope with a suitable tool while installing the rope.

NOTE

Quicksilver replacement starter rope is precut to the correct length. When using an aftermarket starter rope, cut the rope to the same length as the original.

7. Turn the starter housing upside down and rotate the rope pulley counterclockwise until the pulley stops, indicating that the spring is fully tensioned. Allow the pulley to rotate clockwise one full turn, then just enough to align the rope hole in the pulley with the rope hole in the starter handle rest.
8. Thread the end of the new rope through the starter housing rope hole and into the rope pulley recess. Tie a knot in the pulley end of the rope as shown in **Figure 14** and place the knot in the pulley recess.
9. Tie a slip knot in the starter handle end of the rope, approximately 12 in. (304 mm) from the end of the rope.
10. Slowly allow the pulley to wind the rope until the slip knot is against the starter housing.
11. Lubricate all of the bearing surfaces of the interlock components with Quicksilver 2-4-C Multi-Lube grease. Install the interlock lever with the part number facing up and the spring arm contacting the starter housing as shown in **Figure 17**.
12. Install the interlock cam and spring into the housing. The spring and cam should push the interlock lever into the starter pulley when correctly installed. See **Figure 17**.
13. Install the interlock retainer over the cam and secure it with two screws. Tighten the screws securely.
14. Reinstall the starter as described previously in this section.

Table 1 TORQUE SPECIFICATIONS

Component	ft.-lb.	in.-lb.	N•m
Starter housing to motor			
4-15 hp	–	70	7.9
20-25 hp and 20 Jet	–	110	12.4
30-40 (two-cylinder)	–	100	11.3
40-60 (three-cylinder, 30 Jet and 45 Jet	–	100	11.3
Starter shoulder (center bolt)			
4-15 hp	–	70	7.9
20-25 hp and 20 Jet	–	135	15.3
40-60 (three-cylinder, 30 Jet and 45 Jet	–	135	15.3

Table 2 STANDARD TORQUE SPECIFICATIONS—U.S. STANDARD AND METRIC FASTENERS

Screw or nut size	ft.-lb.	in.-lb.	N•m
U.S. standard fasteners			
6-32	–	9	1
8-32	–	20	2.3
10-24	–	30	3.4
10-32	–	35	4.0
12-24	–	45	5.1
1/4-20	6	72	8.1
1/4-28	7	84	9.5
5/16-18	13	156	17.6
5/16-24	14	168	19
3/8-16	23	270	31.2
3/8-24	25	300	33.9
7/16-14	36	–	48.8
7/16-20	40	–	54
1/2-13	50	–	67.8
1/2-20	60	–	81.3
Metric fasteners			
M5	–	36	4.1
M6	6	72	8.1
M8	13	156	17.6
M10	26	312	35.3
M12	35	–	47.5
M14	60	–	81.3

Chapter Fourteen

Remote Control

The 1998-2001 model year Mercury/Marine outboard motors primarily use the Quicksilver Commander 2000 or 3000 series remote control. These controls use standard Quicksilver Mercury Marine style control cables. Mercury Marine control cables do not require any adjustments in the control box, only at the engine.

The Commander 2000 remote control (**Figure 1**) is side-mounted onto the boat structure and the entire unit is exposed. This control incorporates the ignition switch, neutral safety switch, warning horn, primer switch, emergency lanyard switch, trim/tilt switch, on models so equipped, and related wiring within the control housing. The main harness connector on these prewired control boxes is the standard Mercury Marine eight-pin connector.

The Commander 3000 series remote control (**Figure 2**) is flush mounted. The main body of the control fits behind the mounting panel or boat structure. Only the control handle, emergency lanyard switch and tilt/trim switch, on models so equipped, are exposed. This control incorporates the neutral safety switch, lanyard switch and trim/tilt switch, on models so equipped. The ignition switch, primer switch and warning horn are dash mounted. A separate instrument harness connects the dash and control mounted components to the engine using the standard Mercury Marine eight-pin connector.

There are usually additional connectors at both ends of the main engine harness for accessory gauges and warning systems. Control box wiring diagrams are at the end of the manual.

This section primarily covers throttle and shift cable removal, installation and adjustment at the engine.

Control Box Service

When servicing Mercury control boxes, lubricate all internal friction points with Quicksilver 2-4-C Multi-Lube grease (part No. 92-825407 or an equivalent). Apply Loctite 242 threadlocking adhesive (part No. 92-809821) to all internal threaded fasteners. Refer to **Table 1** for torque specifications on Commander 3000 series control boxes. On Commander 2000 series control boxes, securely tighten all fasteners to the standard torque specifications in **Table 2**.

Refer to **Figure 3** for an exploded view of the Commander 2000 series control box internal components and **Figure 4** for an exploded view of the Commander 3000 series control box internal components.

Control Cable Removal/Installation/Adjustment

Lubricate the control cable moveable casing guides, moveable barrel threads and attachment points with Quicksilver 2-4-C Multi-Lube grease before installation.

NOTE
The control cables must be installed into the control box before installation and adjustment at the engine.

CAUTION
Always install and adjust the shift cable first and the throttle cable last.

Control cable removal (6-25 hp and 20 Jet models)

The control cable's casing guide on 6-15 hp models is retained by the locking clip shown in **Figure 5**. Simply slide the locking clip away from the control arm attachment pin to remove the cable. Install a tie strap (3, **Figure 5**) to prevent the locking clip from dislodging from the casing guide when the cables are removed.

The control cable's casing guide on 20-25 hp models and all 20 Jet models are retained by a swivel locking arm. Simply rotate the swivel lock up and away from the control arm attachment pin to remove the cable.

The anchor barrel end of the control cable is attached with one of three styles of simple swivel locks.

1. Disconnect the negative battery cable. Disconnect and ground the spark plug leads to the power head to prevent accidental starting.

2A. *6-15 hp models*—Slide the cable retainers away from the control arm attachment pins, then pull the cable casing guides from the control arm pins. See **Figure 5**.

2B. *20-25 hp and 20 jet models*—Rotate the swivel locks up from the control arm attachment pins, then pull the cable casing guides from the control arm pins.

3A. *6-15 hp models*—Unlock both cable's anchor barrels by first removing the lock clip (1, **Figure 6**) from the latch (2). Then lift the latch and rotate the retainer (3, **Figure 6**) straight away and down from the cable barrels. Pull both control cables free from the bracket.

3B. *20-25 hp and 20 jet models*—Unlock both cable's anchor barrels by rotating the cable retainer plate away from both barrels as shown in **Figure 7**. Then lift both cables straight up and out of the anchor pocket. Remove the cable barrel holder from the cable barrels. Remove both cables from the lower cowl grommet, then remove the cables from the engine.

4. To install the cables, route both cables through the lower cowl grommet, if applicable, and position each cable near the appropriate control arm pin and cable anchor barrel retainer bracket. Then, refer to the appropriate cable installation/adjustment procedure in this chapter.

Control cable removal (30-60 hp models)

1. Disconnect the negative battery cable. Disconnect and ground the spark plug leads to the power head to prevent accidental starting.

2A. *Jet models*—Disconnect the shift cable from the jet drive as described in Chapter Ten under *Pump Unit Removal*. Rotate the swivel lock up then pull the throttle cable casing guide from the control arm pin.

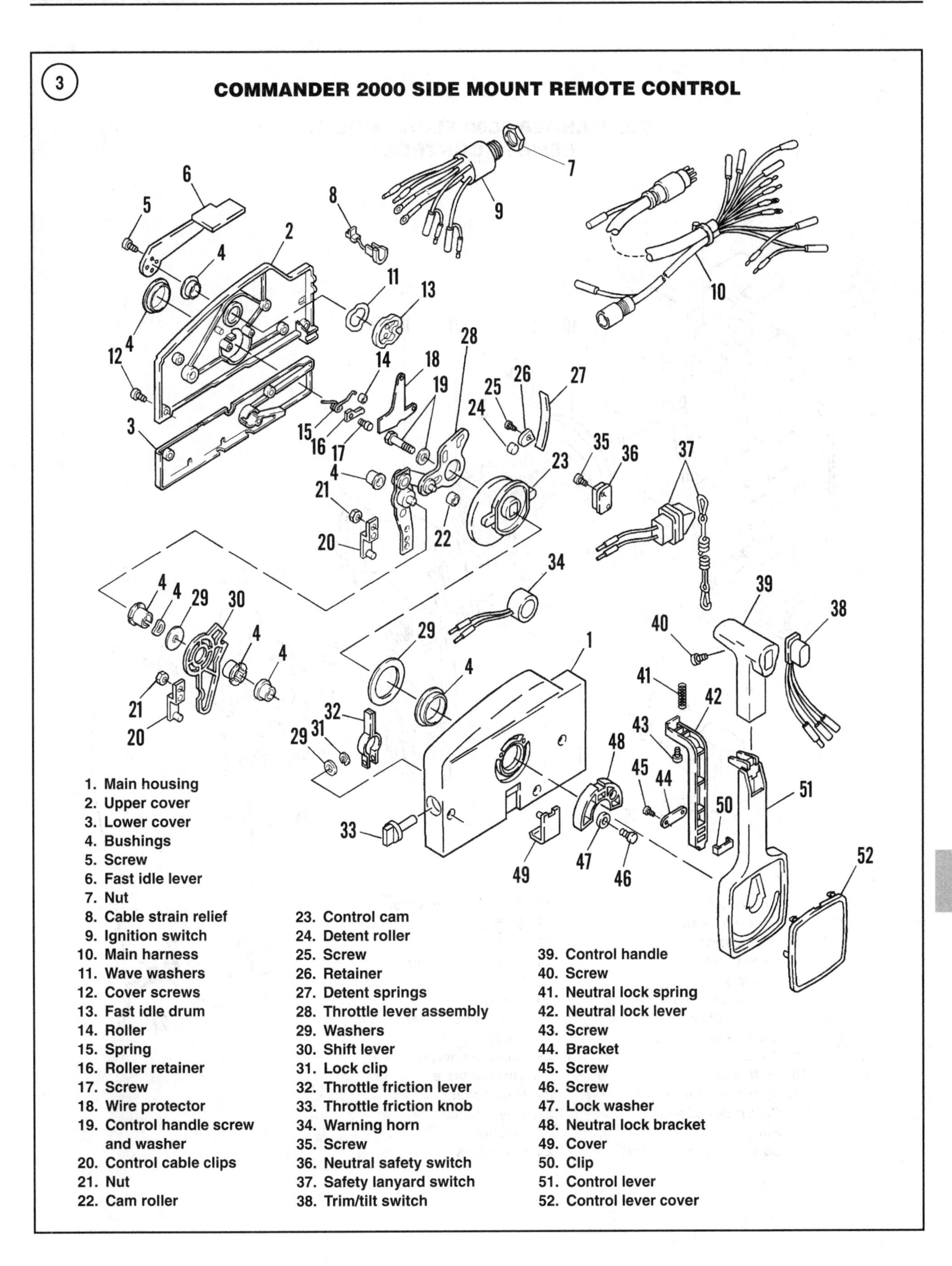
3
COMMANDER 2000 SIDE MOUNT REMOTE CONTROL
1. Main housing
2. Upper cover
3. Lower cover
4. Bushings
5. Screw
6. Fast idle lever
7. Nut
8. Cable strain relief
9. Ignition switch
10. Main harness
11. Wave washers
12. Cover screws
13. Fast idle drum
14. Roller
15. Spring
16. Roller retainer
17. Screw
18. Wire protector
19. Control handle screw and washer
20. Control cable clips
21. Nut
22. Cam roller
23. Control cam
24. Detent roller
25. Screw
26. Retainer
27. Detent springs
28. Throttle lever assembly
29. Washers
30. Shift lever
31. Lock clip
32. Throttle friction lever
33. Throttle friction knob
34. Warning horn
35. Screw
36. Neutral safety switch
37. Safety lanyard switch
38. Trim/tilt switch
39. Control handle
40. Screw
41. Neutral lock spring
42. Neutral lock lever
43. Screw
44. Bracket
45. Screw
46. Screw
47. Lock washer
48. Neutral lock bracket
49. Cover
50. Clip
51. Control lever
52. Control lever cover

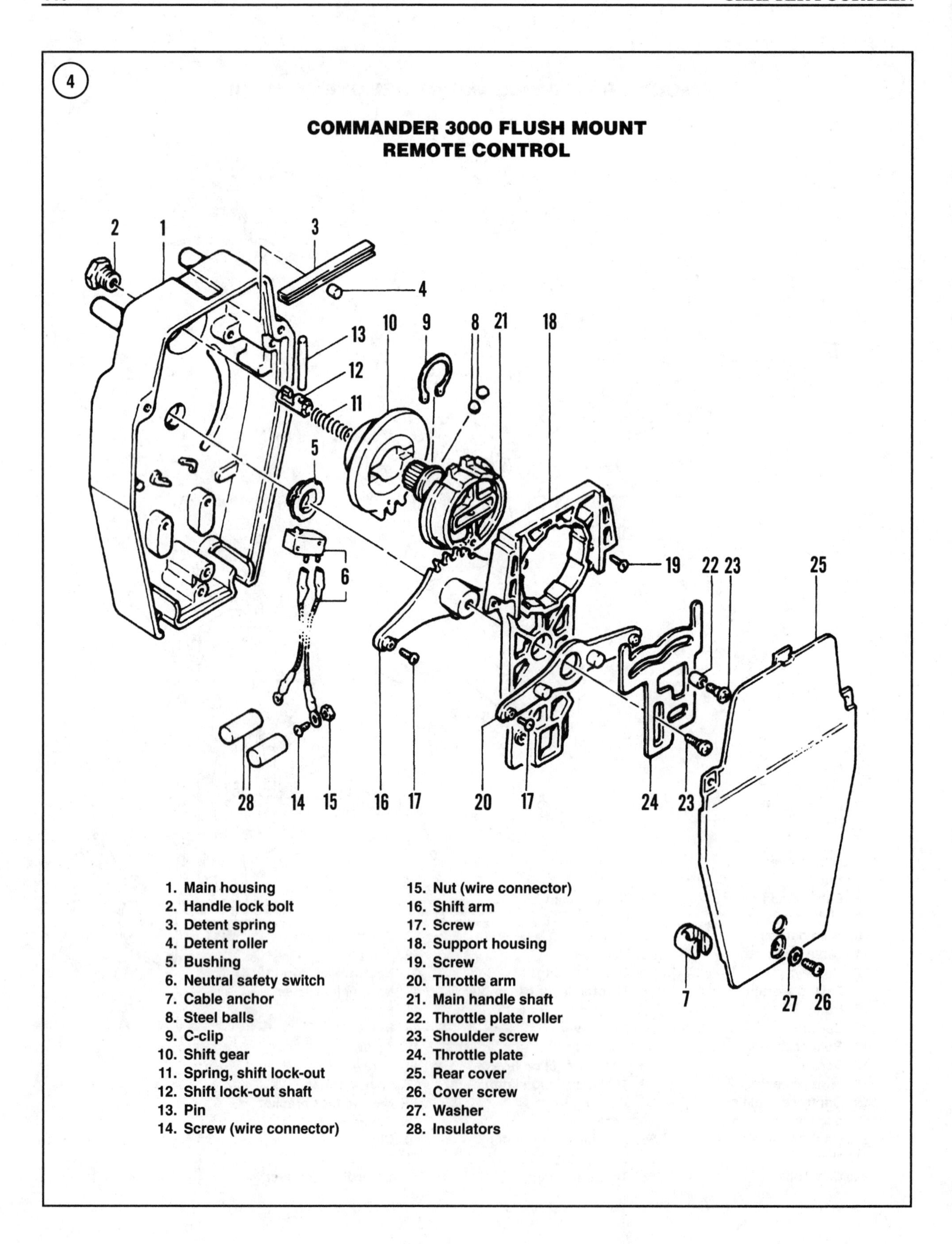

1. Main housing
2. Handle lock bolt
3. Detent spring
4. Detent roller
5. Bushing
6. Neutral safety switch
7. Cable anchor
8. Steel balls
9. C-clip
10. Shift gear
11. Spring, shift lock-out
12. Shift lock-out shaft
13. Pin
14. Screw (wire connector)
15. Nut (wire connector)
16. Shift arm
17. Screw
18. Support housing
19. Screw
20. Throttle arm
21. Main handle shaft
22. Throttle plate roller
23. Shoulder screw
24. Throttle plate
25. Rear cover
26. Cover screw
27. Washer
28. Insulators

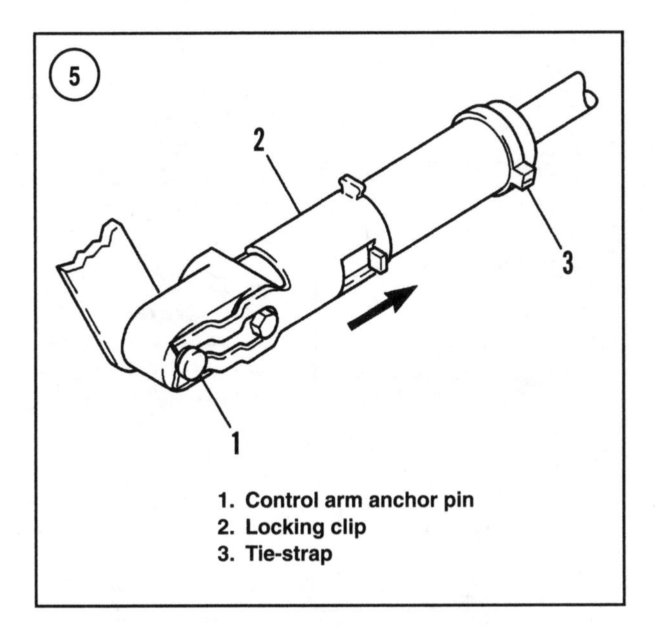

1. Control arm anchor pin
2. Locking clip
3. Tie-strap

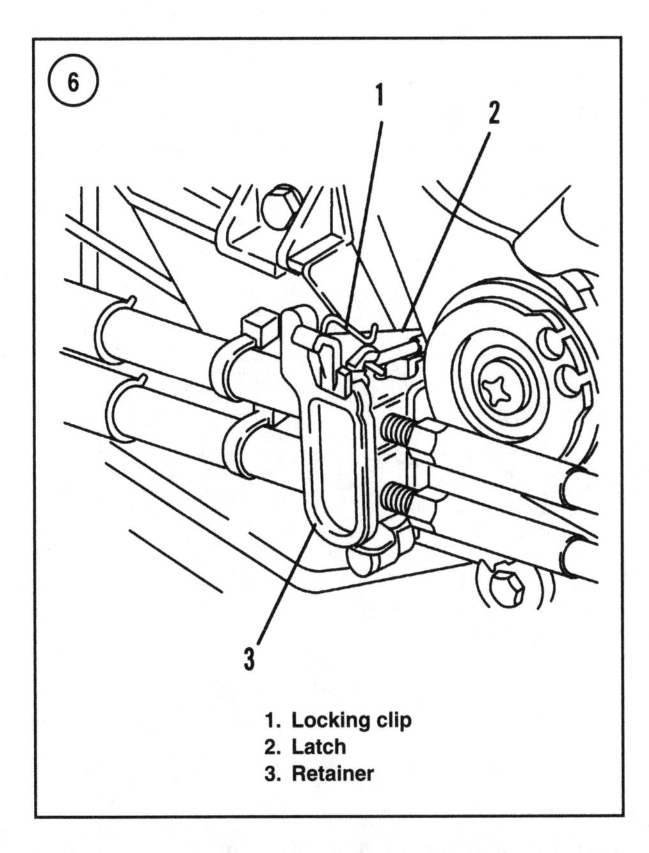

1. Locking clip
2. Latch
3. Retainer

2B. *Non-jet models*—Rotate the swivel locks up from the control arm attachment pins, then pull the throttle cable casing guides from the control arm pins. Remove the elastic locknut and nylon washer from the shift control arm, then pull the shift cable casing guide from the shift arm pin.

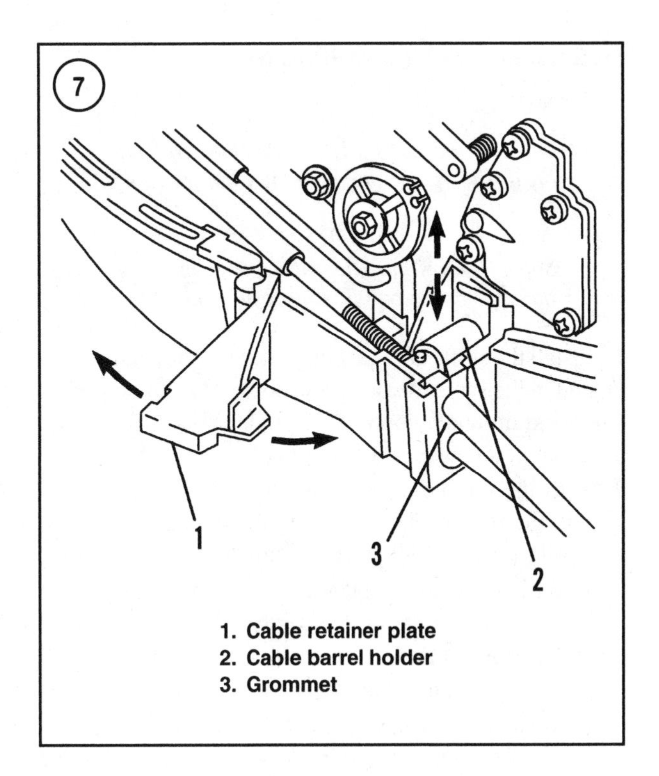

1. Cable retainer plate
2. Cable barrel holder
3. Grommet

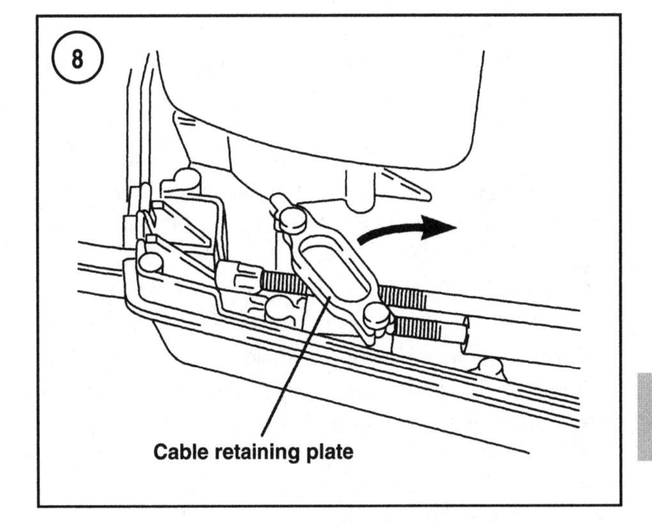

Cable retaining plate

3. Unlock the cable barrel(s) by rotating the cable retainer plate to the full open position. See **Figure 8**, typical. Then lift the cable(s) straight up and out of the anchor pocket. Remove the cable barrel holder from the cable barrel(s). Remove the cable(s) from the lower cowl grommet, then remove the cables from the engine.

4. To install the cables, route both cables through the lower cowl grommet and position the cable(s) near the appropriate control arm pin and cable anchor barrel retainer bracket. Then, refer to the appropriate cable installation/adjustment procedure in this chapter.

Shift Cable Installation/Adjustment

NOTE
On jet models, install and adjust the shift cable or link rod as described in Chapter Ten.

NOTE
Some models may have a case-in mark on the shift linkage indicating true neutral.

1. Install the shift cable casing guide onto the actuator pin or stud in the same manner as it was removed. On models with locknuts and nylon washers, tighten the nut securely, then loosen the nut 1/4 turn to prevent cable binding.
2. Shift the remote control into neutral. Center the free play in the shift cable by moving the casing guide back and forth, and positioning it in the middle of its free play.
   a. Push in on the casing guide as shown in A, **Figure 9** and make a mark on the cable sleeve as shown.
   b. Pull out on the casing guide as shown in B, **Figure 9** and make a mark on the cable sleeve as shown.
   c. Make a third mark on the cable sleeve in the exact middle of the *A* and *B* marks as shown in C, **Figure 9**. The third mark is true neutral of the remote control system.
3. Manually move the engine shift linkage to the neutral position (the exact center of total shift linkage travel). The propeller must spin freely in both directions. Adjust the shift cable barrel to fit into its anchor when the center mark (C, **Figure 9**) is aligned with the casing guide. Rotate the barrel two to three turns to slightly preload toward the reverse gear.
4. Install the shift cable barrel into its anchor in the same manner as it was removed.
5. Shift the remote control into forward gear while rotating the propeller and check gear engagement. Return the remote control to neutral and make sure the propeller spins freely in each direction. Shift the control box into reverse gear while rotating the propeller and make sure the gear engages. Adjust the shift cable barrel as necessary if gear engagements are not satisfactory.

NOTE
Shift verification is best done with the boat in the water. Both forward and reverse gear engagement must require the same amount of control handle movement from the neutral detent. Both gears must fully engage before throttle cable movement occurs. If one gear requires more control handle movement than the other gear, adjust the shift cable barrel to transfer some movement to the gear that requires move movement.

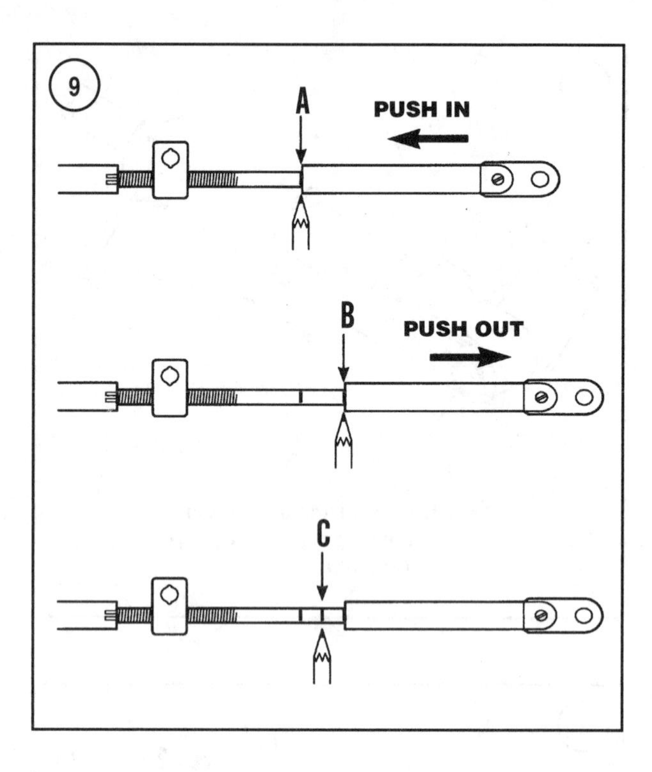

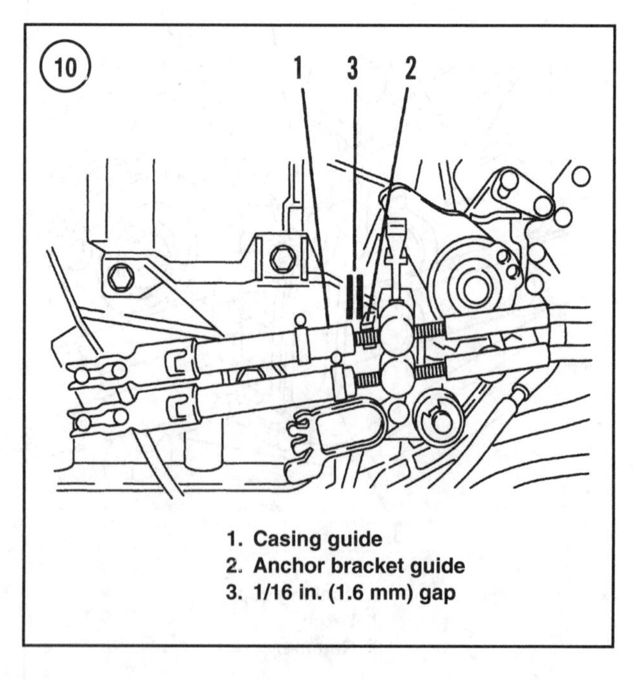

1. Casing guide
2. Anchor bracket guide
3. 1/16 in. (1.6 mm) gap

Throttle Cable Installation/Adjustment (6-25 hp and 20 Jet Models)

Refer to Chapter Five for identification of the wide-open throttle stop. The shift cable must be installed and correctly adjusted before attempting throttle cable installation and adjustments.

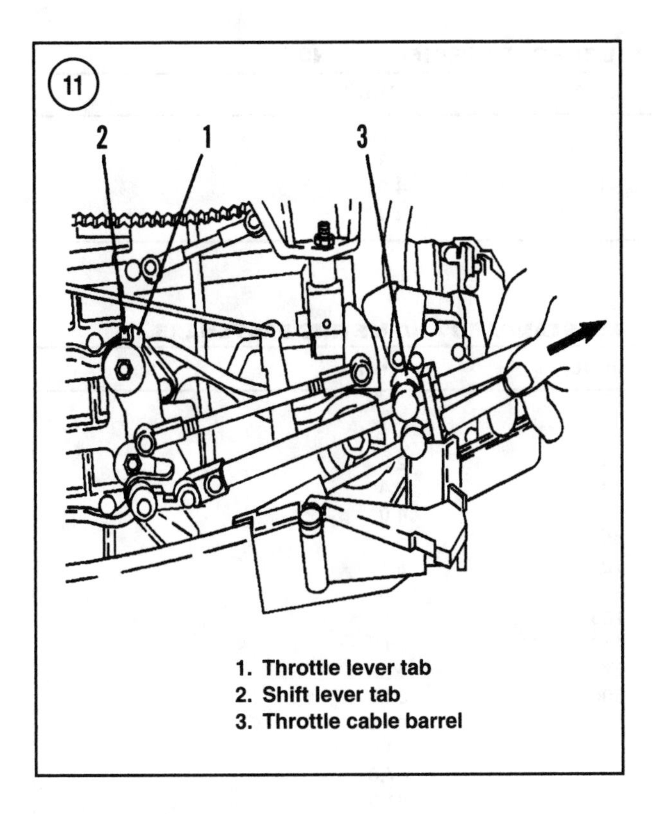

1. Throttle lever tab
2. Shift lever tab
3. Throttle cable barrel

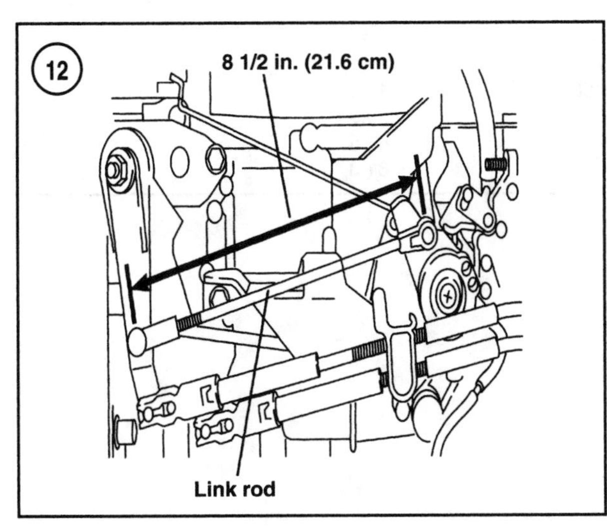

1. Install the throttle cable casing guide onto the actuator pin in the same manner as it was removed.
2. Shift the remote control into forward gear and wide-open throttle.
3A. *6-15 hp models*—Adjust the throttle cable barrel to provide a 1/16 in. (1.6 mm) gap between the end of the casing guide and the anchor bracket guide as shown in **Figure 10**.
3B. *20-25 hp and 20 jet models*—Pull the throttle cable toward the front of the engine until the throttle lever stop tab contacts the shift lever stop tab as shown in **Figure 11**. Adjust the throttle cable barrel to allow the barrel to fit easily into its cable barrel holder and anchor pocket without preload on the cable.
4. Install the throttle cable barrel into its anchor in the same manner as it was removed.
5. On 6-15 hp models, adjust the throttle link rod to a dimension of 8.5 in. (21.6 cm) as shown in **Figure 12**.
6. Make sure the throttle cable and shift cable barrels are both secured in the appropriate manner.
7. Reconnect the negative battery cable and the spark plug leads.

Throttle Cable Installation/Adjustment (30-60 hp, 30 jet and 45 jet models)

Refer to Chapter Five for identification of the idle stop screw. The shift cable must be installed and properly adjusted before attempting throttle cable installation and adjustment.

1. Install the throttle cable casing guide onto the actuator pin or stud in the same manner as it was removed. On models with locknuts and nylon washers, tighten the nut securely, then loosen the nut 1/4 turn to prevent cable binding.
2. Shift the remote control into the neutral position. Hold the throttle linkage on the engine against the idle stop screw. Adjust the throttle cable barrel to provide a slight preload of the throttle lever against the idle stop screw. This slight preload ensures that the throttle returns to the idle stop without binding the control system.
3. Install the throttle cable barrel in the same manner as it was removed.
4. Shift the remote control into forward gear and full throttle while rotating the propeller to ensure full gear engagement. Return the remote control into neutral gear.

NOTE
The throttle cable must positively return the throttle linkage and position the idle stop screw against the idle stop on the power head.

5. Insert a thin piece of paper between the power head and the idle stop screw. Throttle cable preload is correct when the paper can be removed without tearing, but a noticeable drag can be felt. Adjust the throttle cable barrel and repeat Steps 4 and 5 as necessary.
6. Make sure the throttle cable and shift cable barrels are both secured in the appropriate manner.
7. Reconnect the negative battery cable and the spark plug leads if they were disconnected.

Table 1 COMMANDER 3000 CONTROL TORQUE SPECIFICATIONS

	ft.-lb.	in.-lb.	N•m
Control handle lock bolt	–	150	16.9
Control arm screw	–	25	2.8
Support assembly screw	–	35	4.0
Shoulder screw	–	35	4.0

Table 2 STANDARD TORQUE SPECIFICATIONS—U.S. STANDARD AND METRIC FASTENERS

Screw or nut size	ft.-lb.	in.-lb.	N•m
U.S. standard fasteners			
6-32	–	9	1
8-32	–	20	2.3
10-24	–	30	3.4
10-32	–	35	4.0
12-24	–	45	5.1
1/4-20	6	72	8.1
1/4-28	7	84	9.5
5/16-18	13	156	17.6
5/16-24	14	168	19
3/8-16	23	270	31.2
3/8-24	25	300	33.9
7/16-14	36	–	48.8
7/16-20	40	–	54
1/2-13	50	–	67.8
1/2-20	60	–	81.3
Metric fasteners			
M5	–	36	4.1
M6	6	72	8.1
M8	13	156	17.6
M10	26	312	35.3
M12	35	–	47.5
M14	60	–	81.3

Index

A

Adhesives, lubricants
 and sealants 218-219
Alternator driven capacitor
 discharge ignition system. 202-208
Alternator driven ignition
 troubleshooting 58-67
Anticorrosion maintenance 107-108
Antisiphon devices 163-164

B

Battery 178-185
 cable recommendations 214
 capacity hours 214
 state of charge 215
 troubleshooting
  cable recommendations 87
  charging system 47-51
Bleed (recirculation) system
 fuel. 172-174
Break-in, power head 217

C

Capacitor discharge
 ignition 196-202
  troubleshooting,
   2.5, 3.3, 4 and 5 hp 53-58
 module (CDM) ignition. 208-212
  troubleshooting 67-76
   system specifications
    (30-60 hp models) 91
Carburetors 146-162
 specifications 175-177
CDI ignition system
 troubleshooting specifications. . . . 90-91
Charging system 185-188
 troubleshooting. 88
  specifications. 89
Cleaning and inspection
 gearcase. 373-374
 power head 247-257
Connecting rod
 service specifications 290
Cooling system, flushing 109-110
Corrosion
 control. 295-296
 galvanic. 8-10
Crankshaft
 service specifications 290
Cylinder block
 service specifications 290-291

E

Electrical and ignition systems
 accessories, troubleshooting. 51
 alternator driven capacitor
  discharge ignition system . . . 202-208
 battery 178-185
  cable recommendations 214
  capacity, hours 214
  state of charge 215
 boat wiring harness
  standard wire colors 215
 capacitor discharge
  ignition. 196-202
  module (CDM) ignition 208-212
 charging system. 185-188
 fuses 188
 starter, troubleshooting,
  specifications 89
 starting systems. 188-196
 torque specifications 213-214
 wiring diagrams 448-465
Engine
 mounted primer bulb,
  30-55 hp manual start models 163
 operation. 3
 submersion 108-109
 troubleshooting 83-87
  temperature and overheating . . . 80-83

F

Fasteners 3-6
 and torque, power head 220-221
Filters, fuel 164
Fuel system
 and lubrication 94-104
 antisiphon devices. 163-164
 bleed, recirculation, system. . . 172-174
 carburetors 146-162
 filters 164
 hose and primer bulb 164-166
 primer
  and enrichment systems,
   troubleshooting 78-80
  engine mounted bulb,
   30-55 hp manual start models. . . 163
  solenoid, 6-25 hp and 20 jet
   remote control models 162-163
  valve, 30-60 hp
   remote control models 163
 pump 142-146
 reed valve service 167-172
 specifications
  carburetor 175-177
  reed valve. 177
  torque 174-175
 tanks 164
 troubleshooting 76-78, 88-89
Fuses 188

G

Galvanic corrosion. 8-10
Gasket sealant 7-8
Gear ratio and approximate
  lubricant capacity 377
Gearcase
  3.3 hp models 368-369
  55 and 60 hp standard models. . . 369-371
  60 hp Bigfoot models 371-373
  assembly 321-375
    2.5 hp models 323-324
    3.3 hp models 325-328
    4 and 5 hp models 329-332
    6-15 hp models 333-337
    20-25 hp models 340-343
    30-50 hp models 347-350
    55 and 60 hp standard 355-358
    60 hp Bigfoot models 359-368
  cleaning and inspection 373-374
  corrosion control 295-296
  disassembly. 321-375
    2.5 hp models 321-323
    3.3 hp models 324-325
    4 and 5 hp models 328-329
    6-15 hp models 332-333
    20 and 25 hp models 337-340
    30-50 hp models 343-346
    55 and 60 hp
      standard models 350-354
    60 hp Bigfoot models 358-359
  gear
    housing. 300-309
    ratio 293-294
      and approximate
        lubricant capacity 377
  high-altitude operation 294
  lubrication 296
  operation 292-293
  pressure testing 374-375
  propeller 296-299
  service
    precautions 294-295
    specifications. 377
  shimming 368
  torque specifications 375-376
  trim tab adjustment 299-300
  water pump 309-321
General information
  engine operation 3
  fasteners 3-6
  galvanic corrosion. 8-10
  gasket sealant 7-8
  lubricants. 6-7
  propellers 10-16
  torque specifications. 3

H

High-altitude operation 294
Hose and primer bulb, fuel 164-166
Hour meter 93

I

Ignition and electrical systems
  alternator driven capacitor
    discharge ignition system . . . 202-208
  battery 178-185
    cable recommendations 214
    capacity, hours 214
    state of charge 215
  boat wiring harness
    standard wire colors 215
  capacitor discharge
    ignition 196-202
    module (CDM) ignition 208-212
  charging system. 185-188
  fuses 188
  starting systems. 188-196
  torque specifications 213-214
  troubleshooting
    ignition system 51-53, 88
      identification. 92
      stator resistance specifications . . . 91
Inspection and cleaning
  gearcase. 373-374
  power head 247-257

J

Jet drive 378-388
  pump unit service 384-387
  torque specifications. 388

L

Lighting coil, troubleshooting 46
Lubrication 6-7
  and fuels 94-104
  and sealants and adhesives . . . 218-219
  capacity 119
  gearcase 296
  gear ration and
    approximate capacity 377

M

Maintenance
  anticorrosion 107-108
  cooling system flushing. 109-110
  engine submersion 108-109
  hour meter 93
  off-season storage 104-107
  schedule 118
  torque specifications. 118
Manual rewind starters 422-435
  torque specifications. 436

O

Oil injection systems 412-420
  capacity 420
  pump output specification 420
  torque specifications. 420
Operating requirements
  troubleshooting. 35-36

P

Piston
  ring end gap, all models. 258-259
  service specifications 291
Power head
  assembly 257-288
  break-in 217
  cleaning and inspection 247-257
  connecting rod
    service specifications 290
  crankshaft service specifications. . . . 290
  cylinder block
    service specifications 290-291
  disassembly. 229-246
  fasteners and torque. 220-221
  lubricants, sealants
    and adhesives 218-219
  model
    identification 217
    number codes 291
  piston
    ring end gap, all models 258-259
    service specifications. 291
  removal/installation 221-229
  sealing surfaces 219-220
  service
    considerations. 216-217
    recommendations 217-218
  torque specifications 288-289
Pressure testing, gearcase. 374-375
Primer
  bulb, and hose, fuel 164-166
    engine mounted,
      30-55 hp manual start models. . . 163
  solenoid, fuel, 6-25 hp and 20 jet
    remote control models 162-163
  valve, fuel, 30-60 hp
    remote control models 163
Propeller 10-16, 296-299

Pump
fuel 142-146
unit, jet, service 384-387

R

Reed valve
specifications 177
service 167-172
Remote control 437-443
Commander 3000
control torque specifications 444
Rod service,
connecting, specifications 290

S

Safety 17-19
precautions, troubleshooting 31-32
Sealants, lubricants
and adhesives 218-219
Service
hints 25-27
precautions, gearcase 294-295
recommendations,
power head 217-218
specifications, gearcase 377
Shimming, gearcase 368
Solenoid, fuel primer,
6-25 hp and 20 jet
remote control models 162-163
Spark plug, recommendations 119
Specifications
general, synchronization
and linkage adjustments 138-141
standard torque, U.S. standard and
metric fasteners . . 175, 214, 289, 376, 388, 411, 436, 444
torque 118
Commander 3000 control 444
fuel system 174-175
gearcase 375-376
general 3
ignition and
electrical systems 213-214
jet drive 388
manual rewind starters 436
oil injection systems 420
power head 288-289
Starting system 188-196
troubleshooting 36-46, 87-88
Stator resistance specifications,
troubleshooting,
battery charging/lighting circuit . . . 90
Storage, off-season 104-107
Synchronization and
linkage adjustments
general
information 121-122
specifications 138-141
procedures 123-138
safety precautions 120-121

T

Tanks, fuel 164
Test equipment 22-25
Tools, basic hand 19-22
Torque
and fasteners, power head 220-221
specifications 118
Commander 3000 control 444
fuel system 174-175
gearcase 375-376
general 3
ignition and
electrical systems 213-214
jet drive 388
manual rewind starters 436
oil injection systems 420
power head 288-289
standard, U.S. standard and
metric fasteners . . 175, 214, 289, 376, 388, 411, 436, 444
trim and tilt systems 410
Trim and tilt systems
manual 389-393
power 393-403
service 403-410
torque specifications 410
Trim tab, adjustment 299-300
Troubleshooting
alternator driven ignition 58-67
battery
cable recommendations 87
battery charging system 47-51
boat wiring harness
standard wire colors 87
capacitor discharge
ignition, 2.5, 3.3, 4 and 5 hp . . . 53-58
module (CDM) ignition 67-76
CDI ignition system
specifications 90-91
charging system
specifications 89
troubleshooting 88
electric
accessories 51
starter specifications 89
engine 83-87
temperature and
overheating 80-83
fuel system 76-78
and primer and
enrichment systems 78-80
troubleshooting 88-89
ignition system 51-53
identification 92
stator resistance specifications 91
troubleshooting 88
lighting coil 46
operating requirements 35-36
safety precautions 31-32
starting system 36-46
troubleshooting 87-88
stator resistance specifications,
battery charging/lighting circuit . . . 90
terminology and test equipment . . . 32-35
Tune-up 110-117
spark plug recommendations 119

V

Valve
fuel primer, 30-60 hp
remote control models 163
reed
specifications 177
service 167-172

W

Water pump 309-321
Wiring diagrams 448-465
Wiring harness
standard wire colors 215
troubleshooting 87

IGNITION SYSTEM (2.5 AND 3.3 HP)

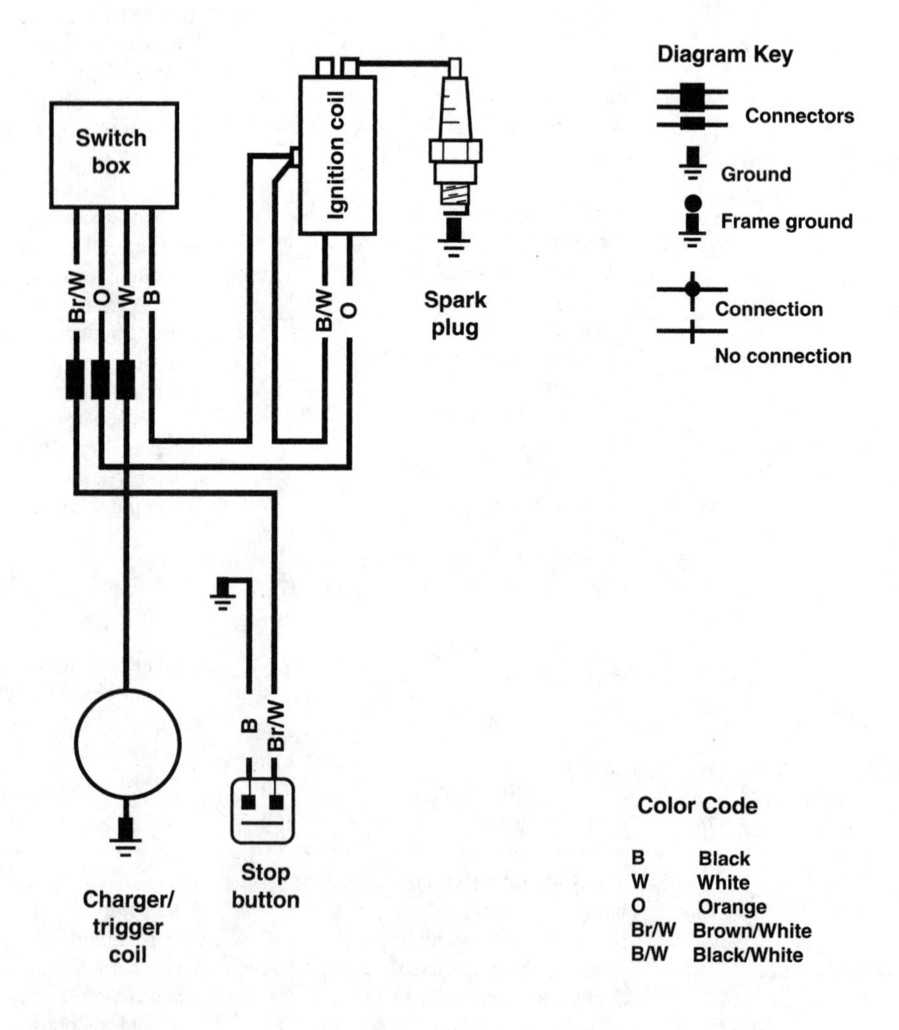

IGNITION SYSTEM (4 AND 5 HP)

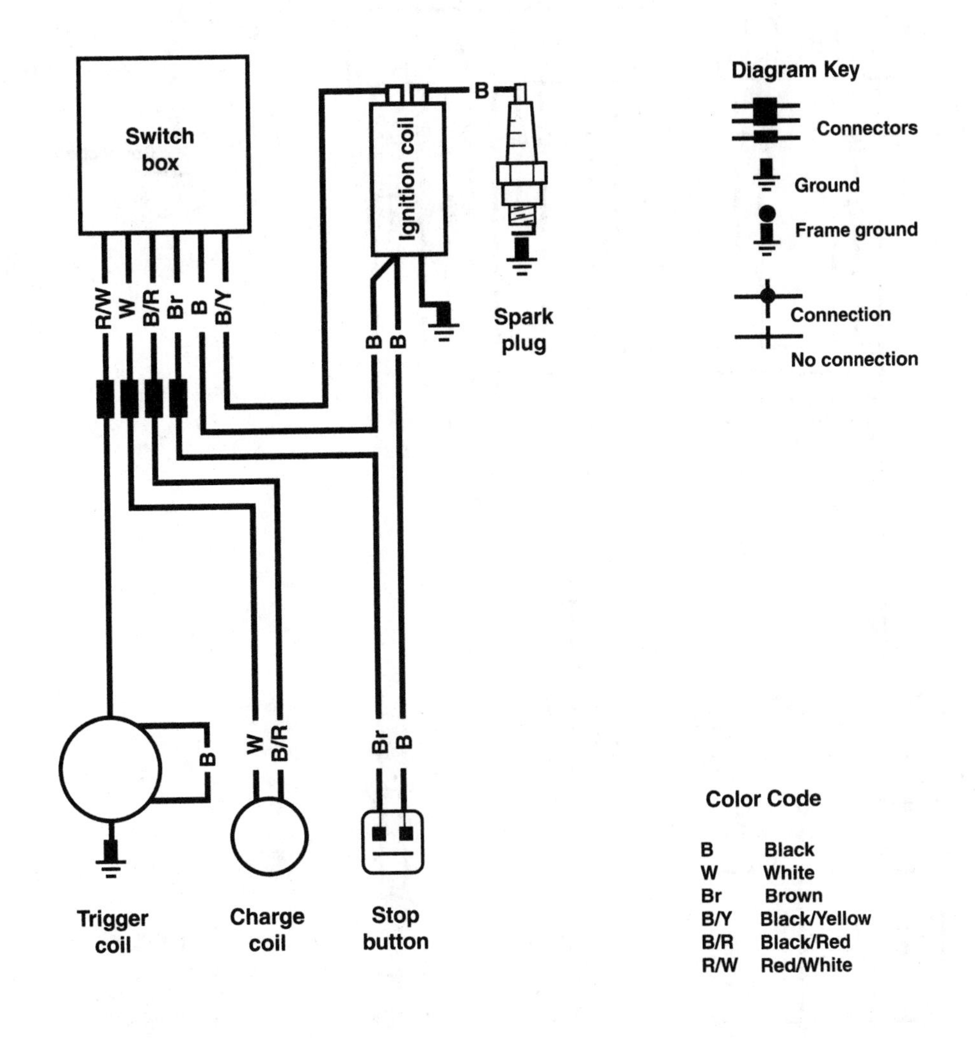

6-25 HP (1998 ONLY) MANUAL START (EXCEPT 20 JET) BLACK IGNITION STATOR

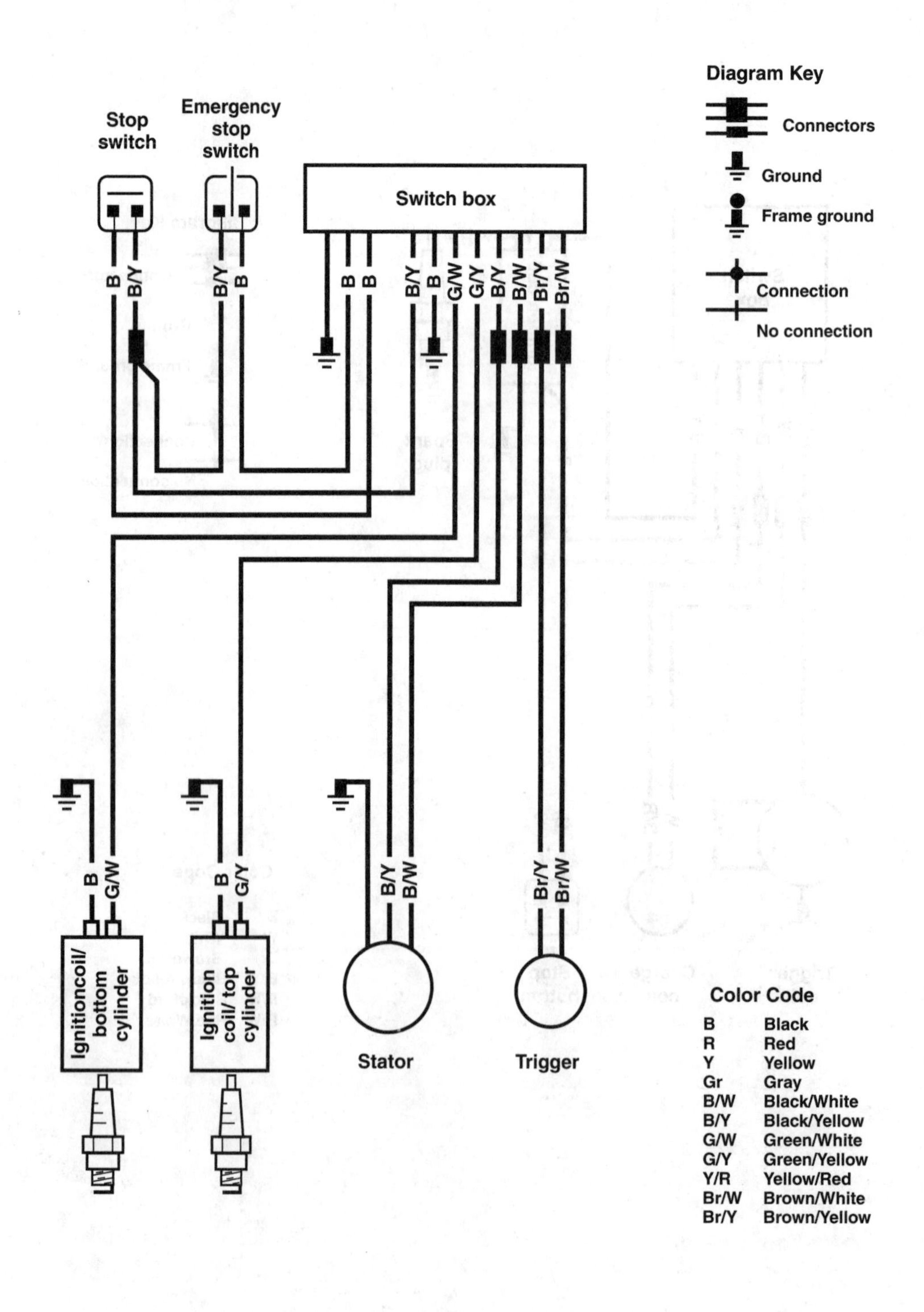

6-25 HP (1998 ONLY) TILLER CONTROL ELECTRIC START (EXCEPT 20 JET) BLACK IGNITION STATOR

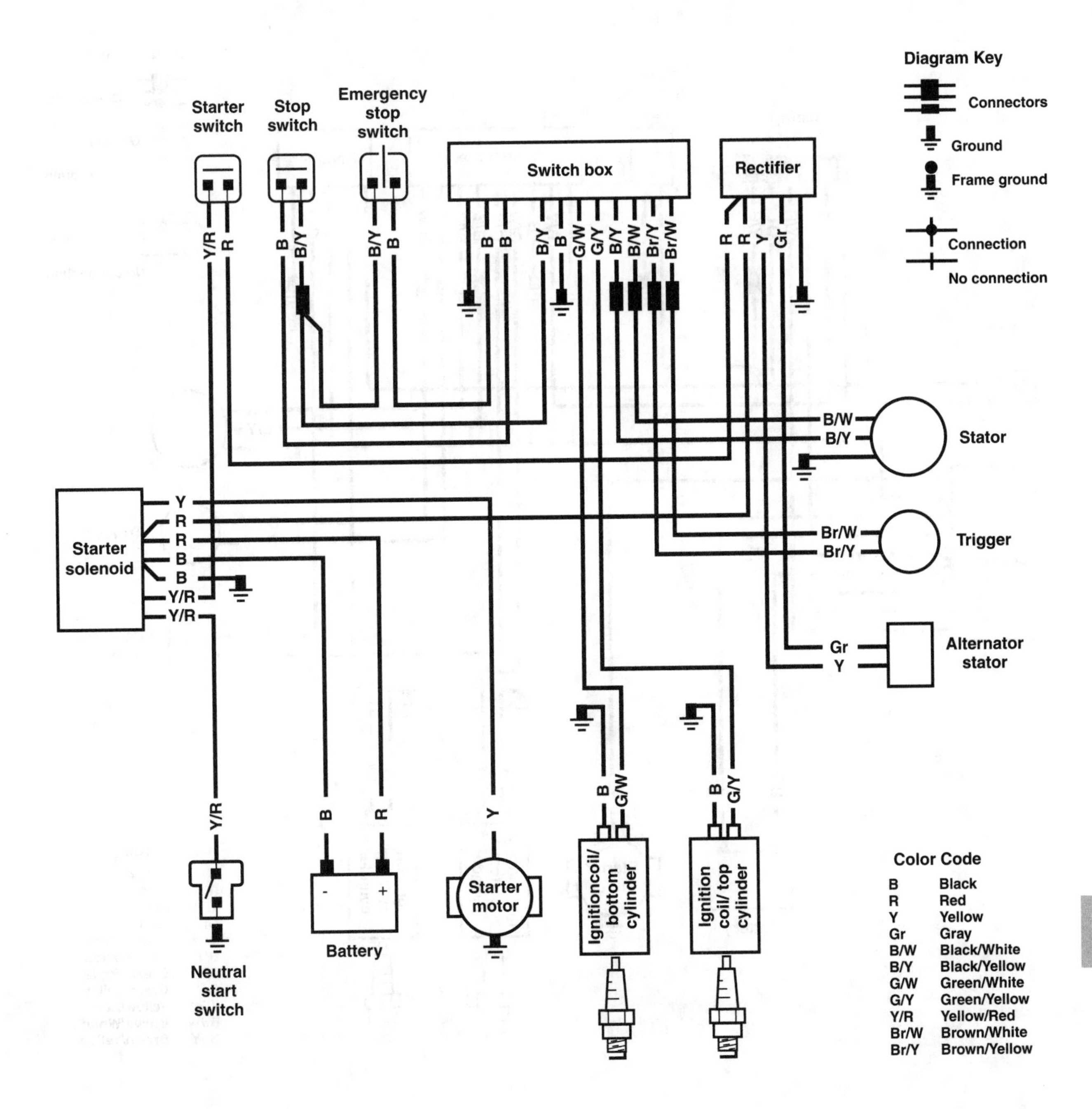

6-25 HP (1998 ONLY) REMOTE CONTROL ELECTRIC START (EXCEPT 20 JET) BLACK IGNITION STATOR

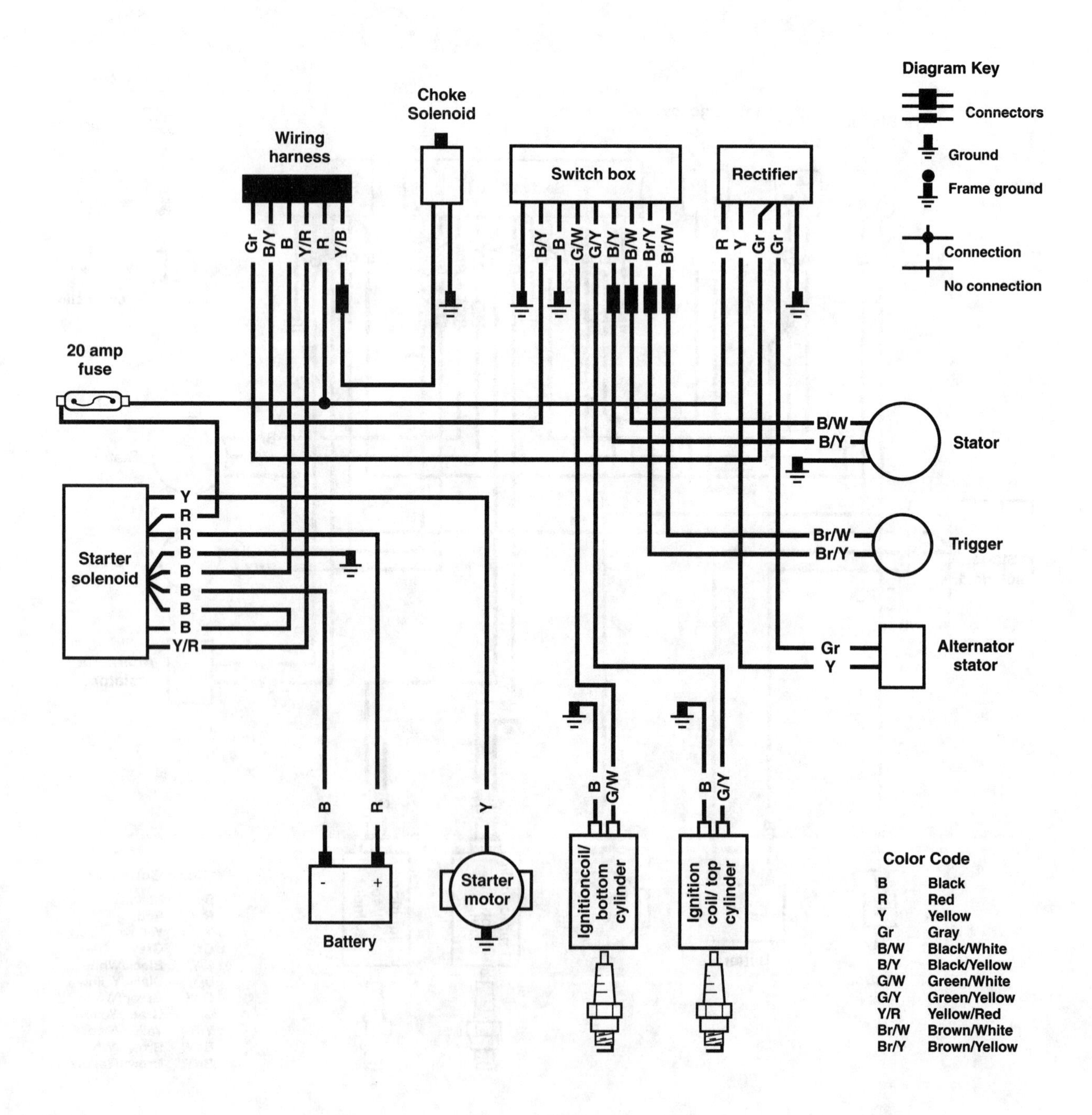

6-25 HP (1999-ON) TILLER CONTROL MANUAL START (RED IGNITION STATOR)

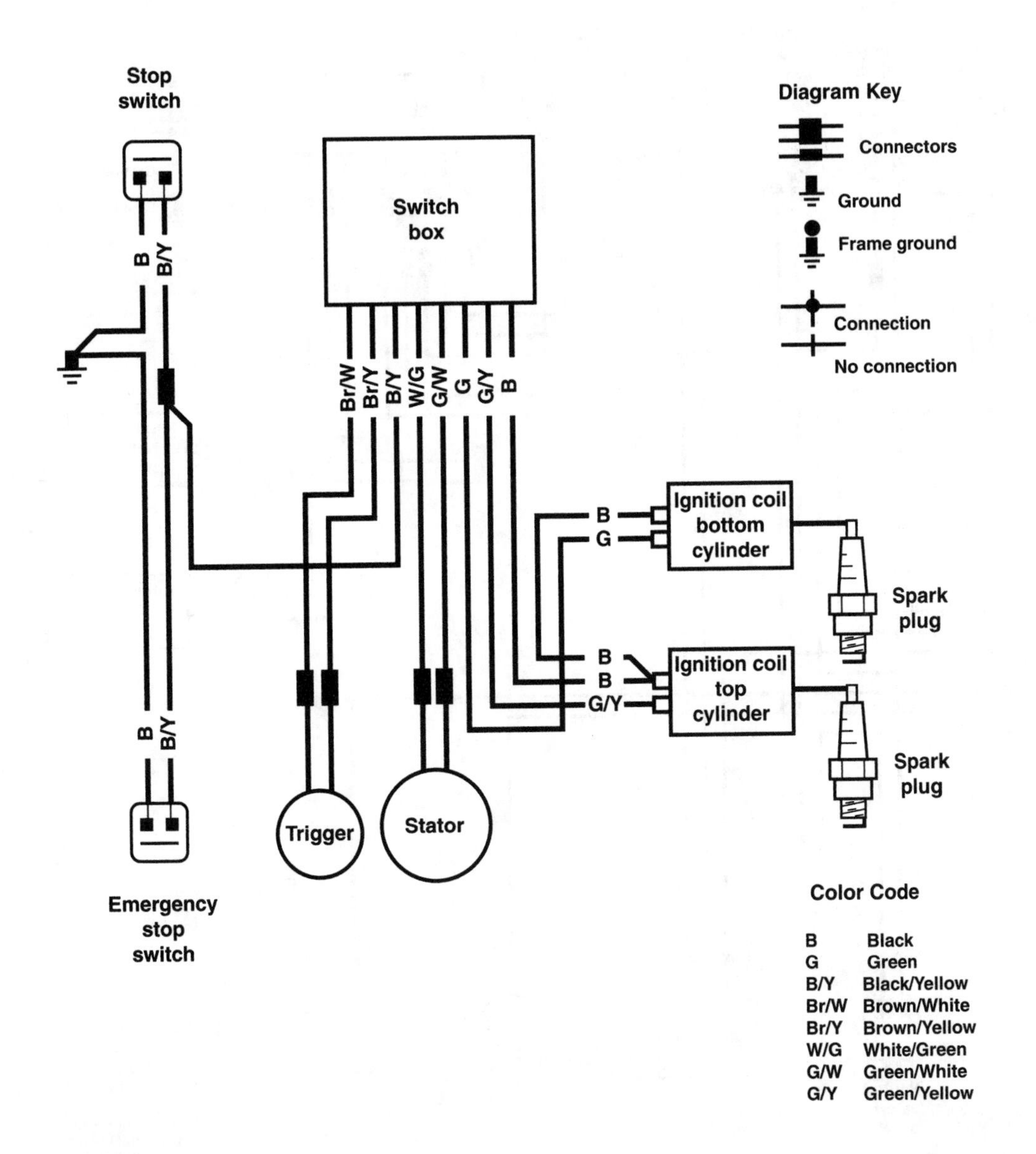

Color Code

B	Black
G	Green
B/Y	Black/Yellow
Br/W	Brown/White
Br/Y	Brown/Yellow
W/G	White/Green
G/W	Green/White
G/Y	Green/Yellow

6-25 HP (1999-ON) TILLER CONTROL ELECTRIC START (RED IGNITION STATOR)

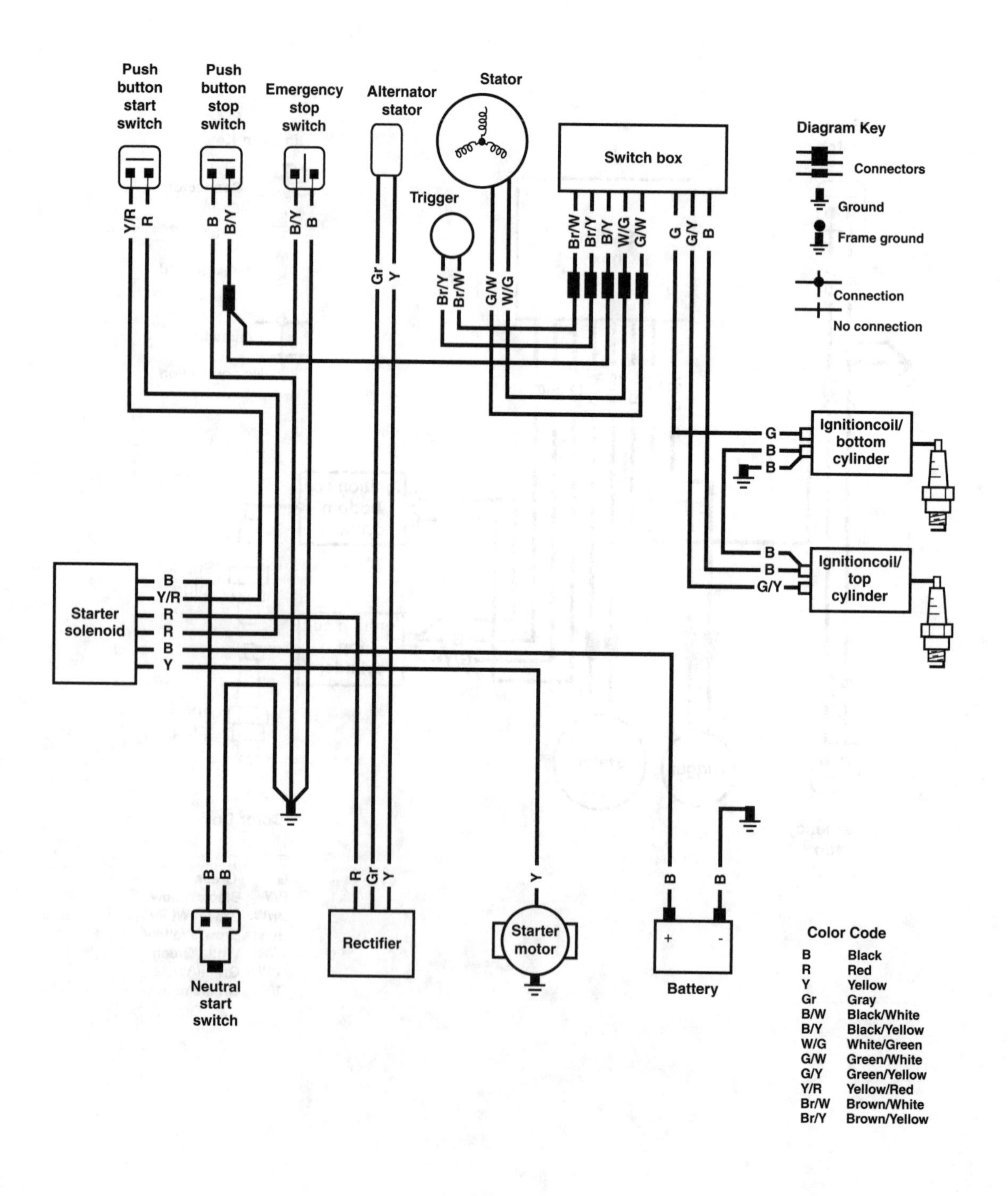

6-25 HP (1999-ON) REMOTE CONTROL ELECTRIC START (RED IGNITION STATOR)

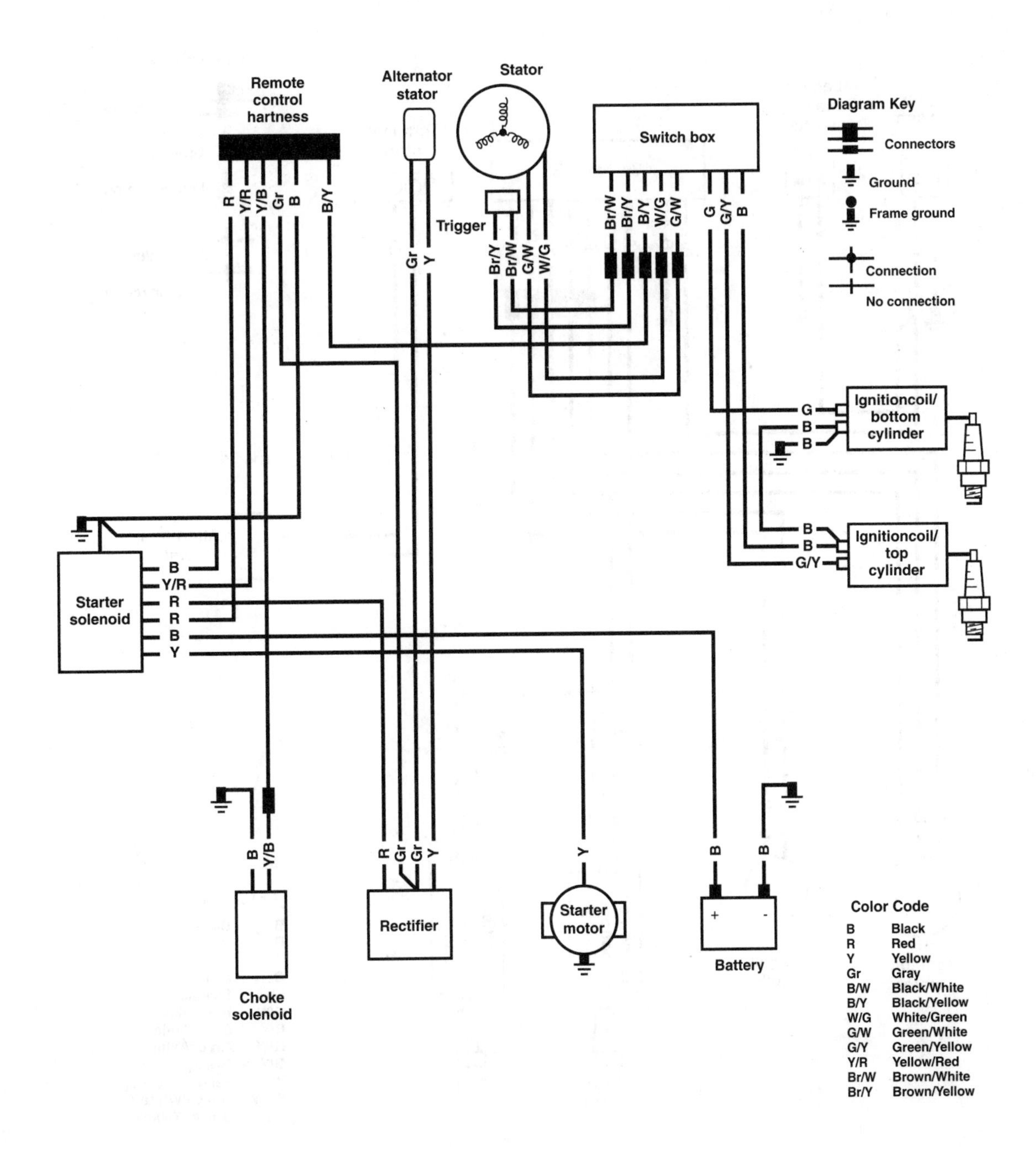

20 JET (1998 ONLY) BLACK IGNITION STATOR

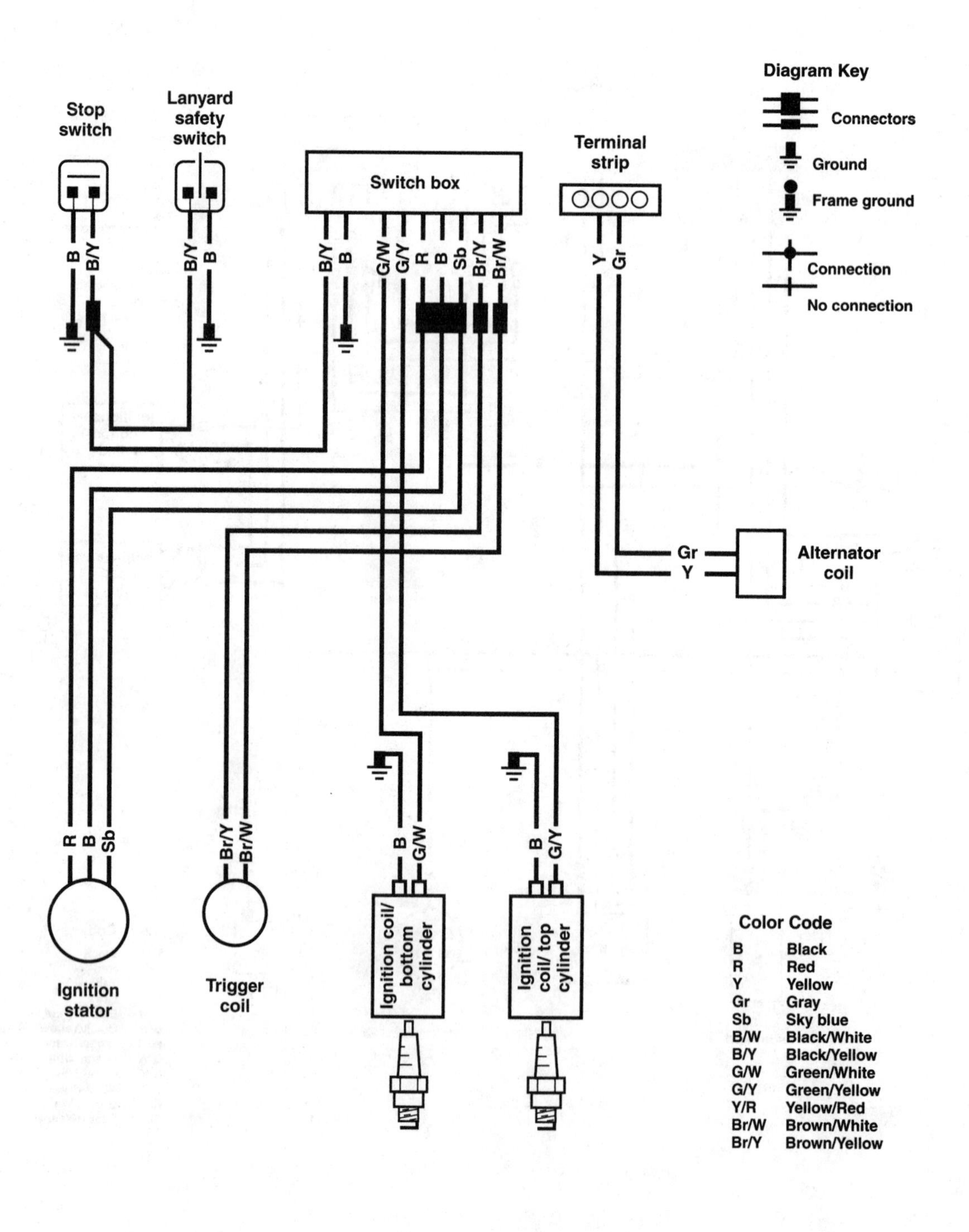

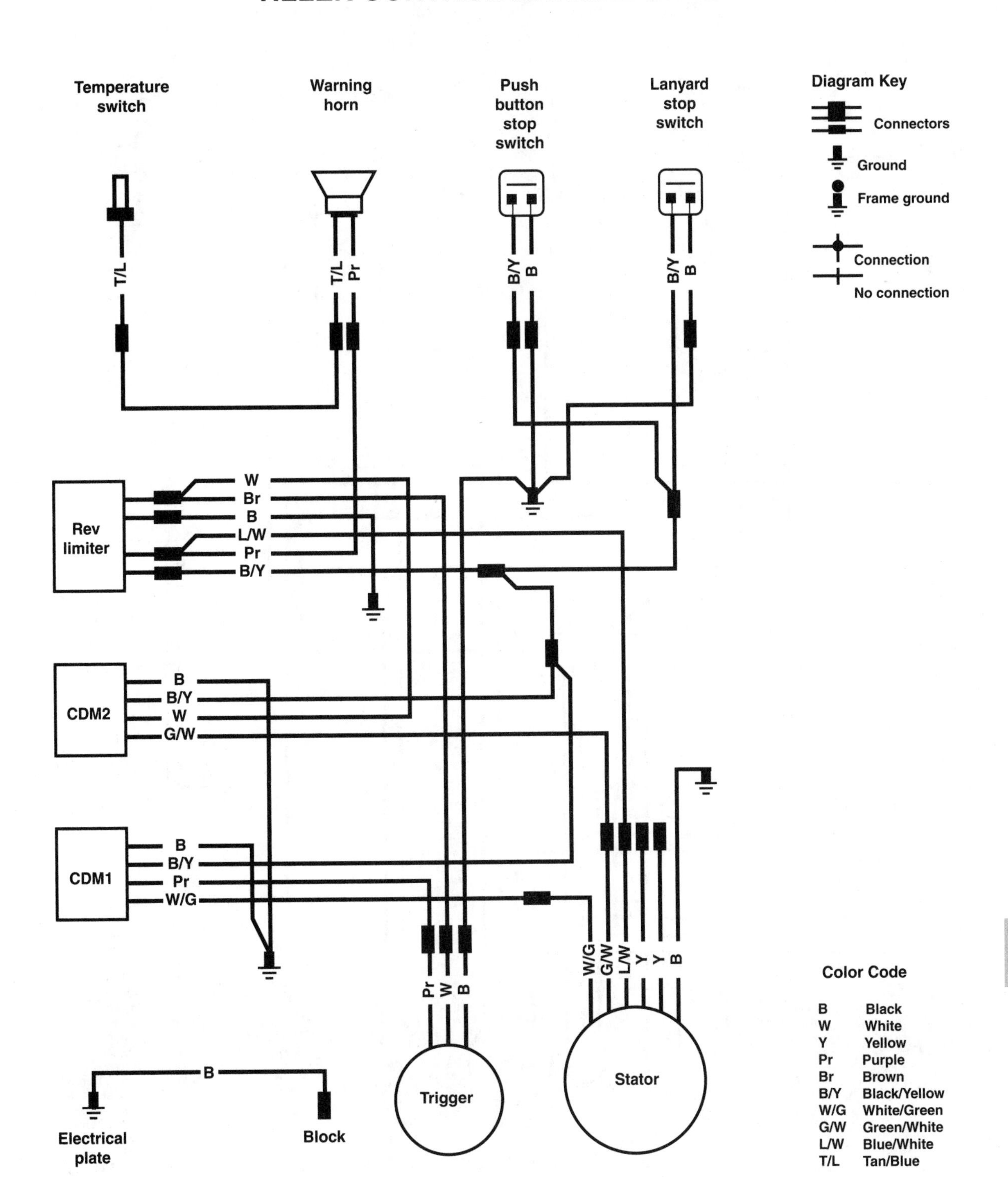

Color Code

B	Black
W	White
Y	Yellow
Pr	Purple
Br	Brown
B/Y	Black/Yellow
W/G	White/Green
G/W	Green/White
L/W	Blue/White
T/L	Tan/Blue

30 HP (TWO-CYLINDER) TILLER CONTROL ELECTRIC START

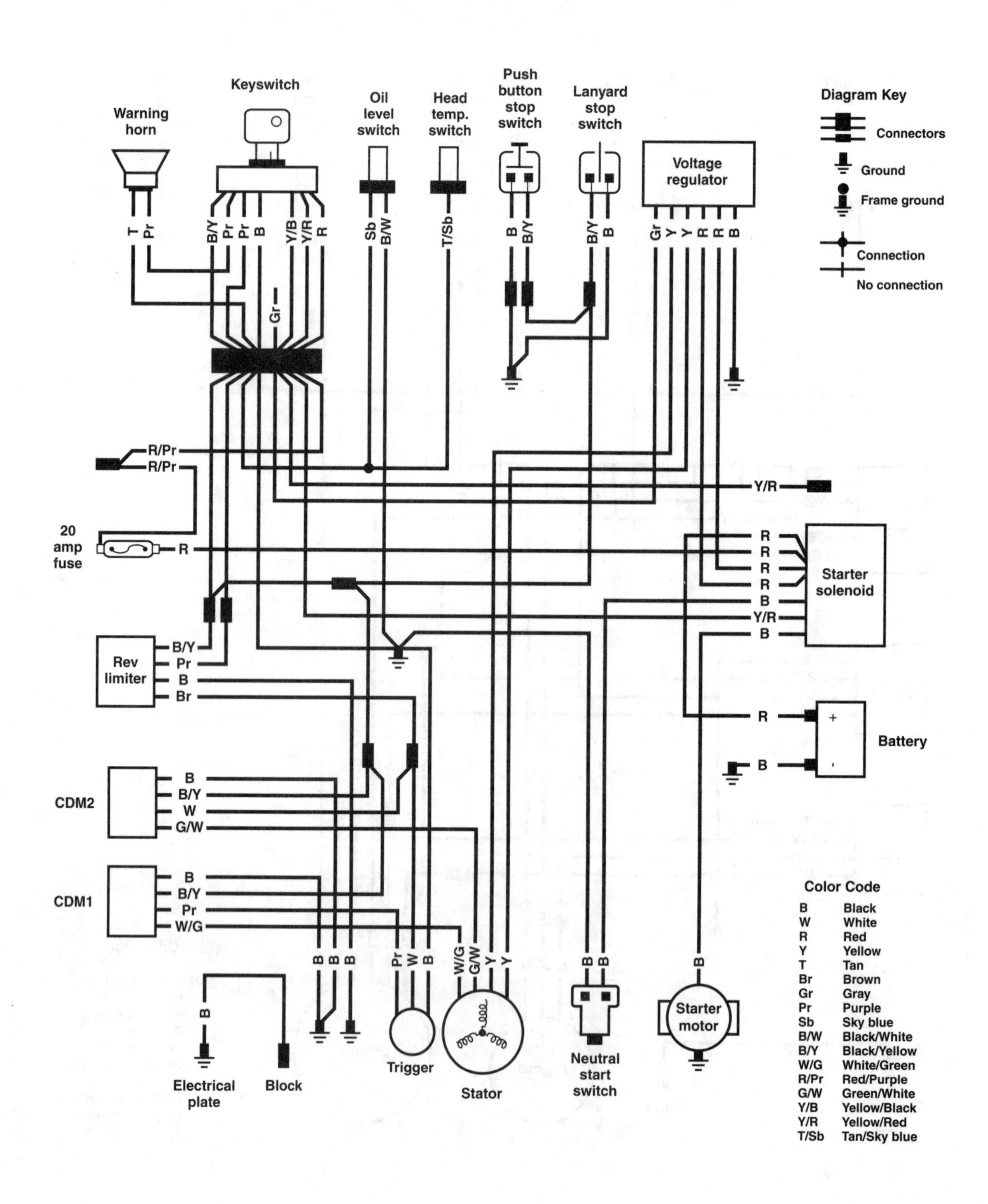

30 AND 40 HP (TWO-CYLINDER) REMOTE CONTROL (WITHOUT POWER TRIM)

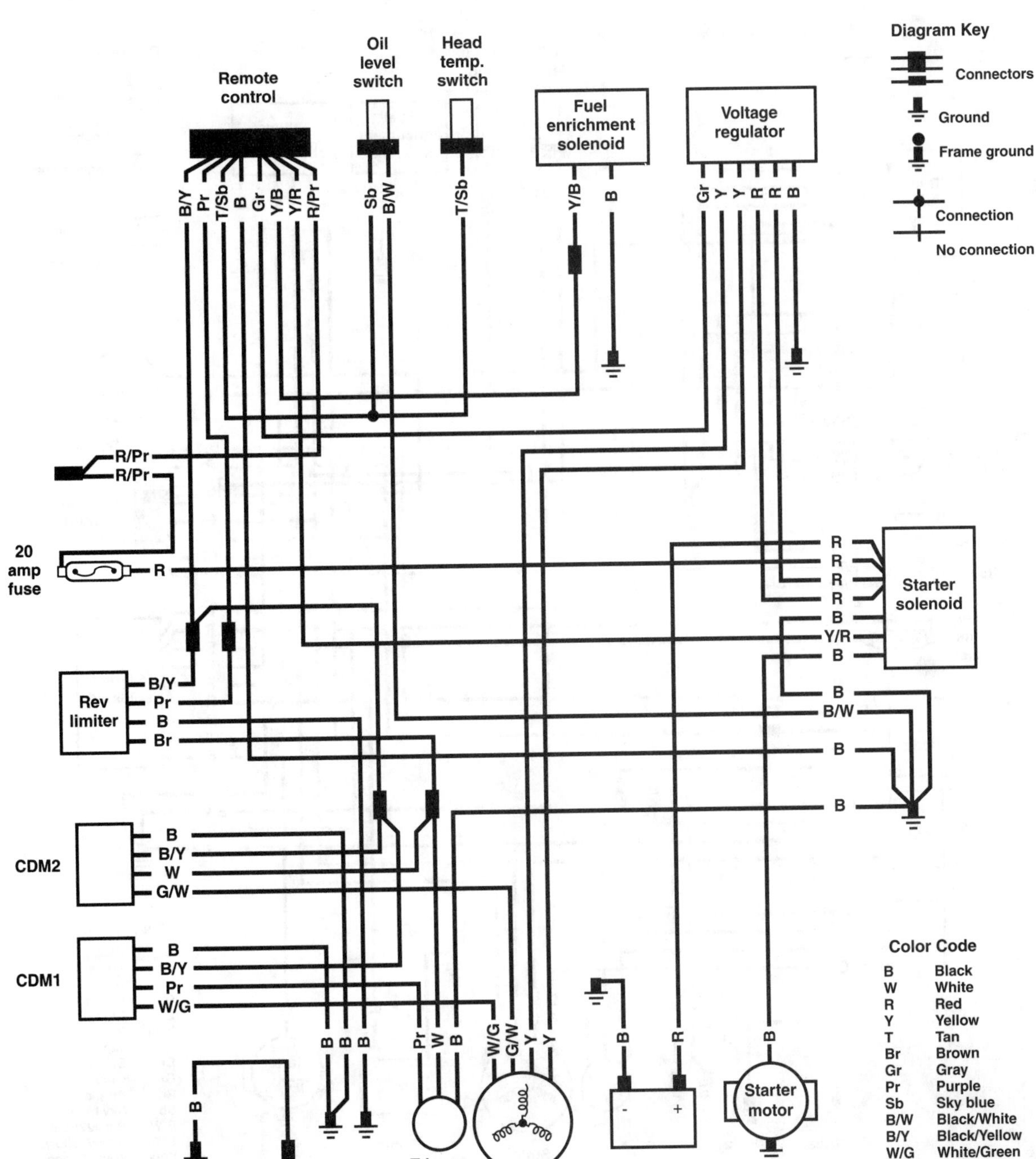

30 AND 40 HP (TWO-CYLINDER) REMOTE CONTROL (WITH POWER TRIM AND TILT)

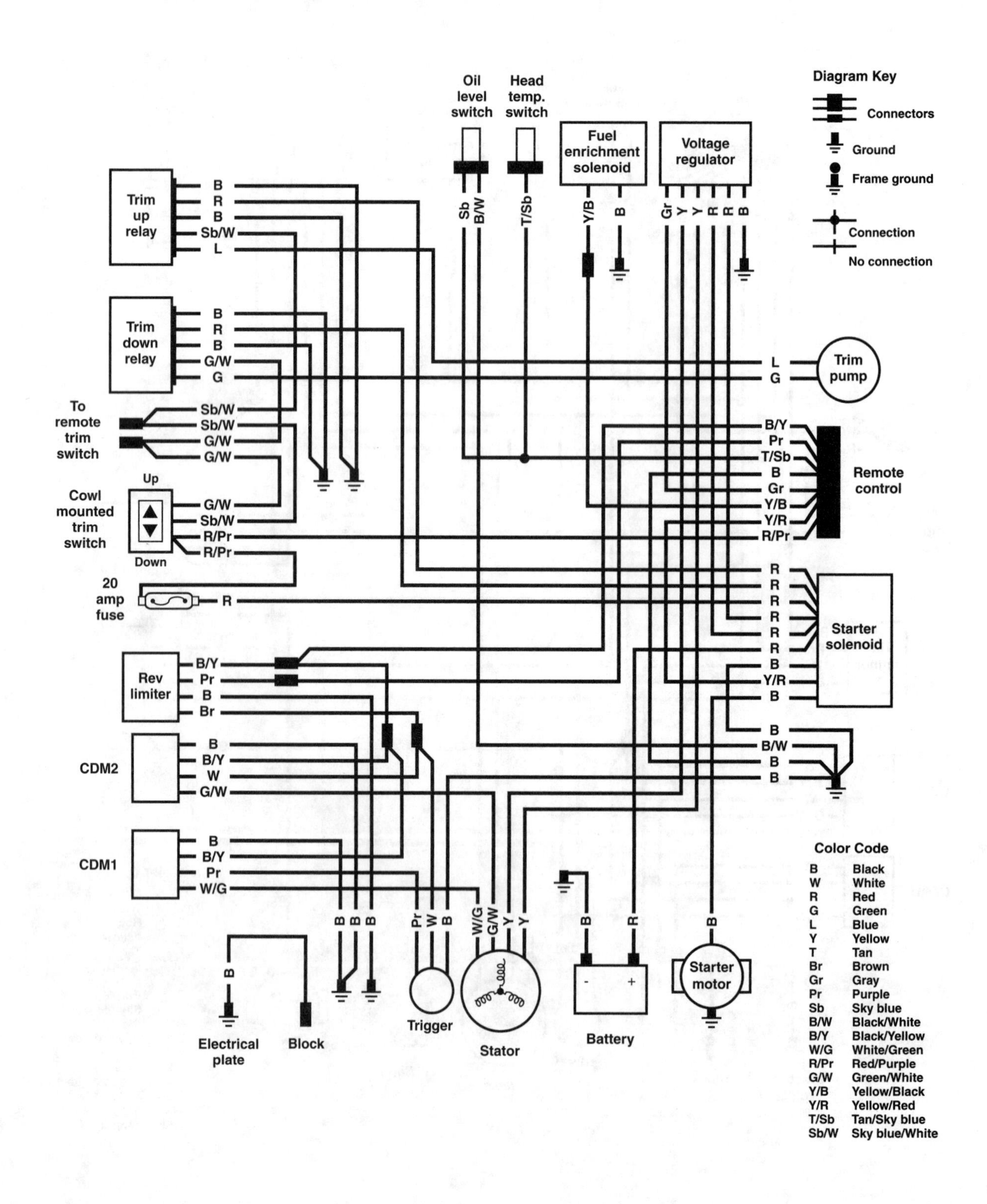

40 AND 55 HP (THREE-CYLINDER) TILLER CONTROL MANUAL START

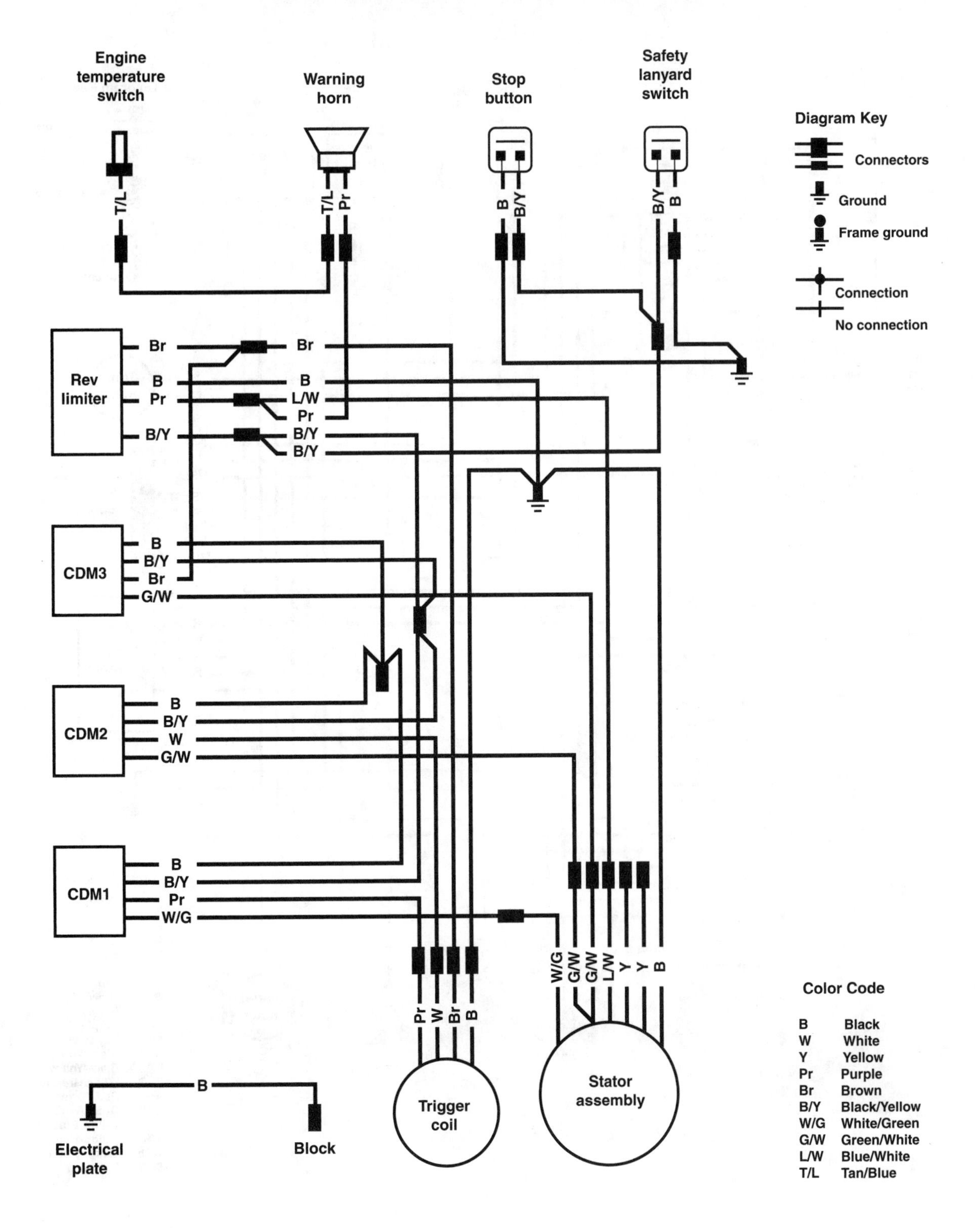

40-60 HP AND 30-45 JET (THREE-CYLINDER) REMOTE CONTROL ELECTRIC START

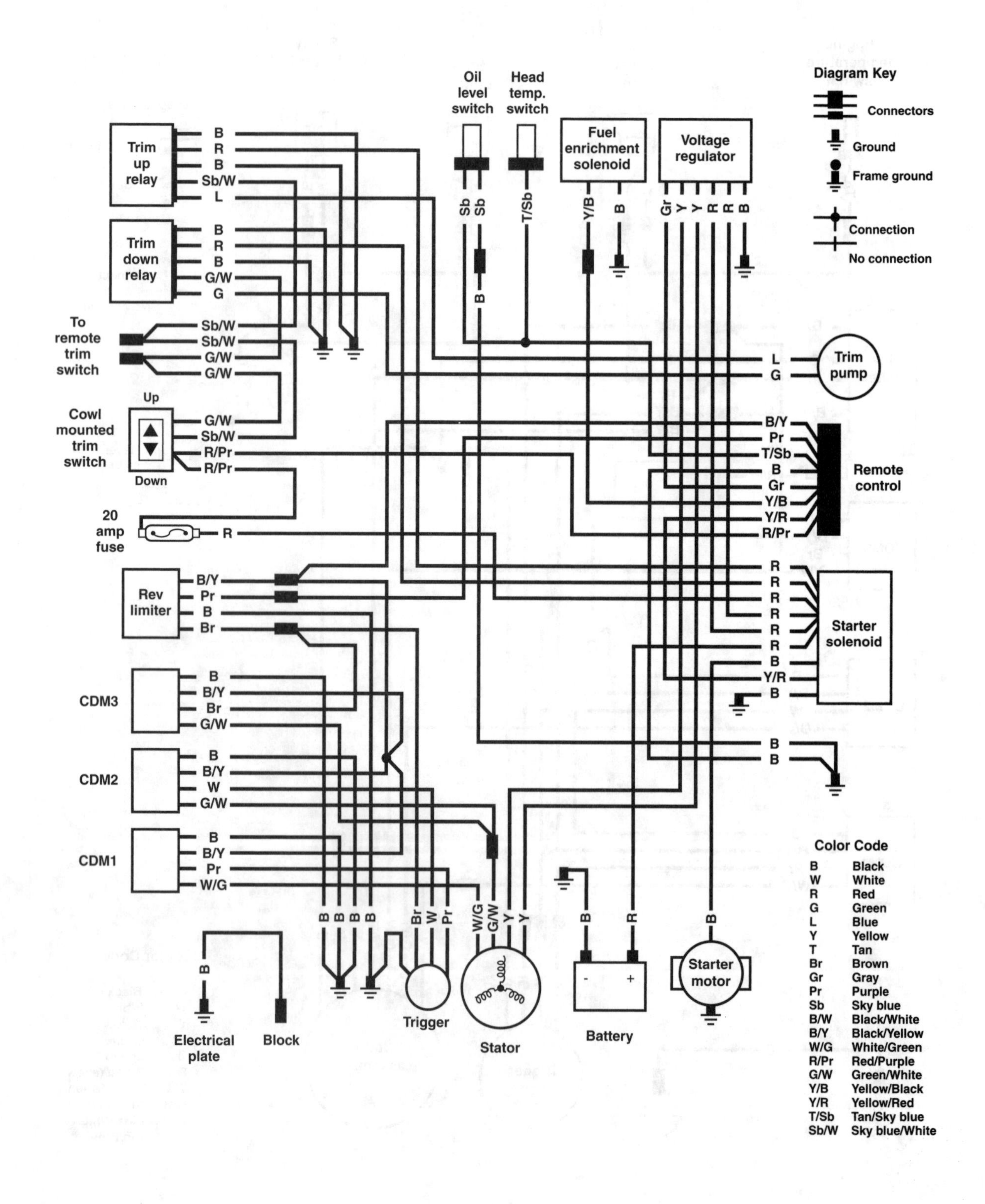

40 AND 50 HP (THREE-CYLINDER) TILLER CONTROL ELECTRIC START

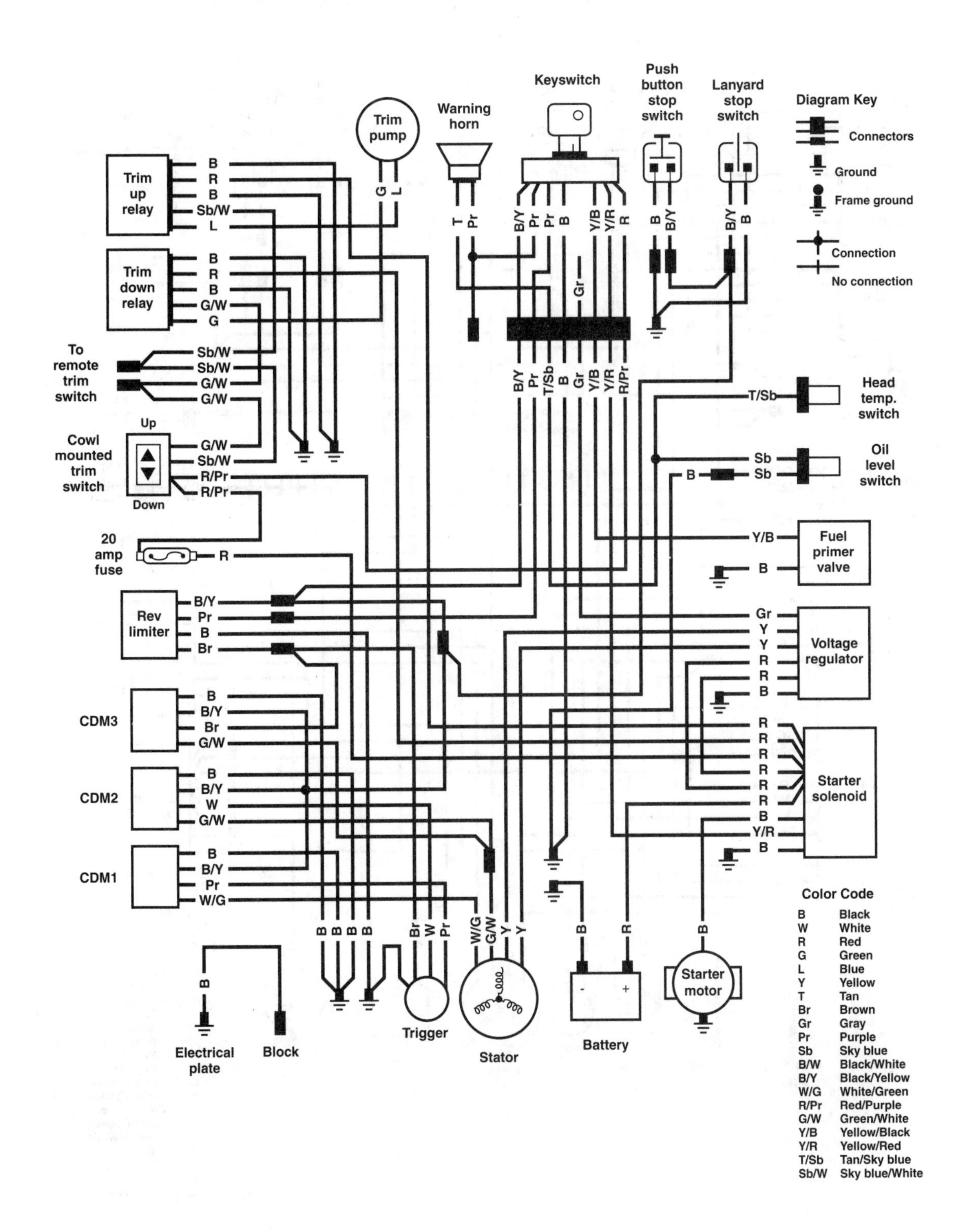

60 HP TILLER CONTROL

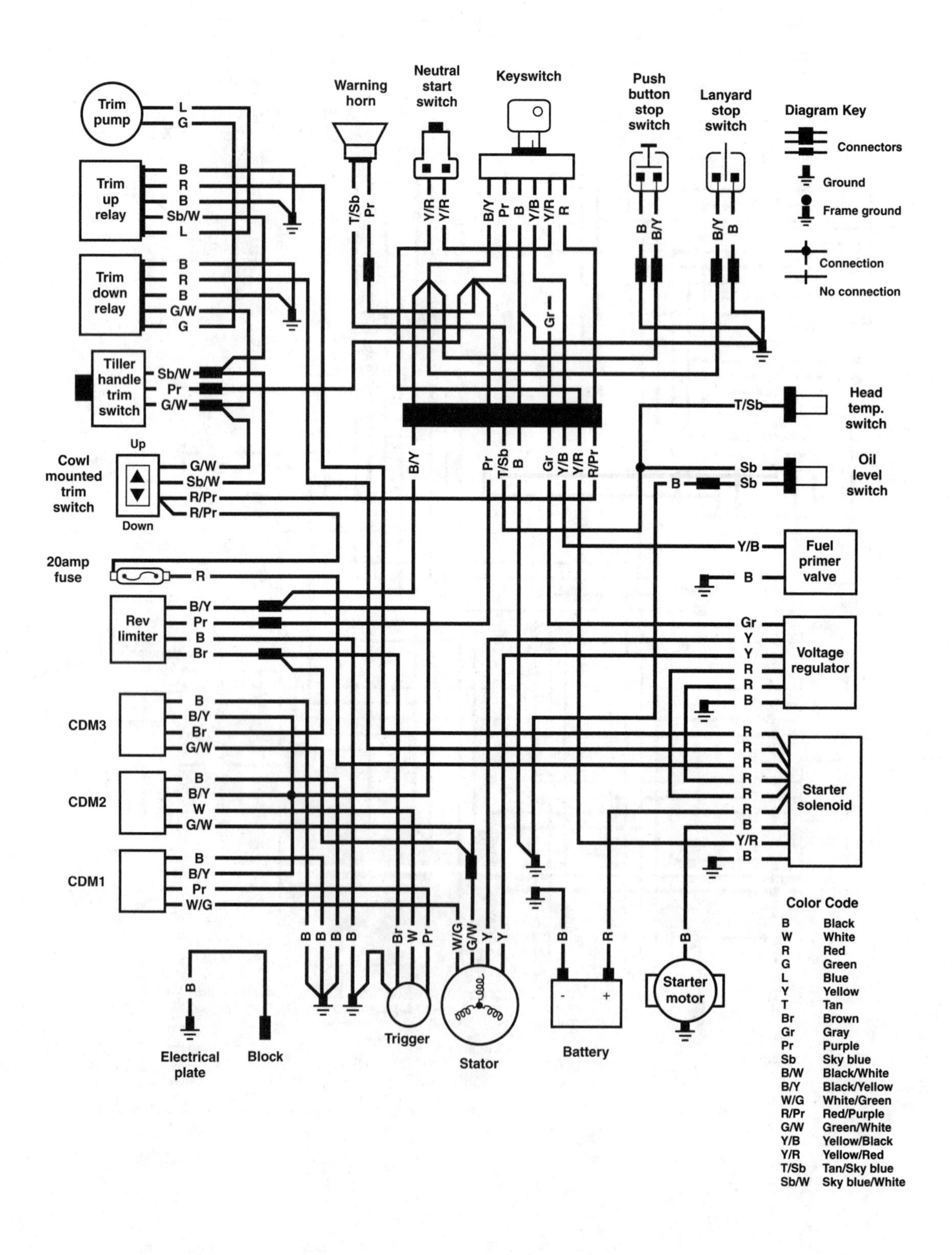

QUICKSILVER REMOTE CONTROL

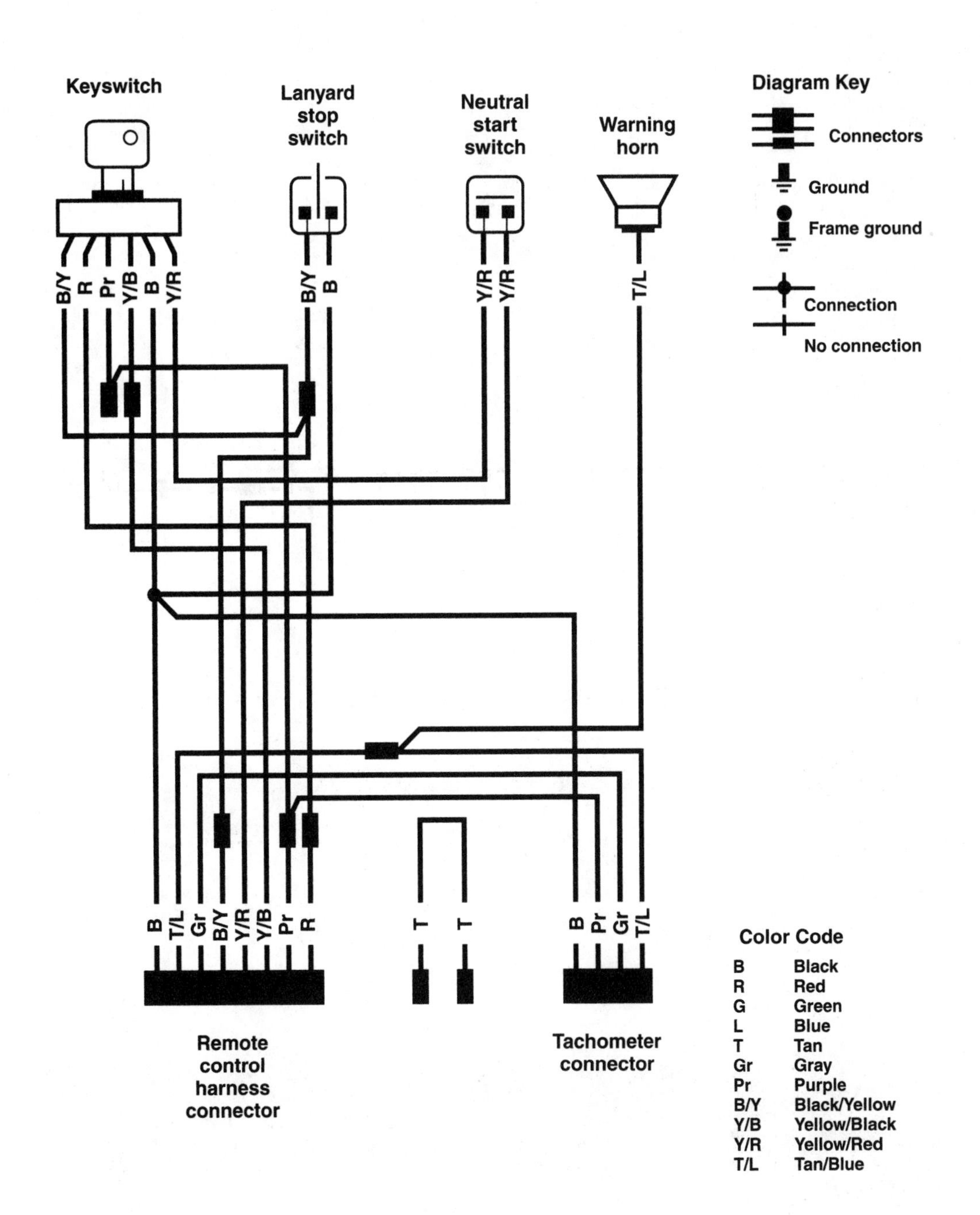

NOTES

NOTES

MAINTENANCE LOG

Date	Maintenance performed	Engine hours